STUDIES OF

Mascarene Island birds

T0206121

EDITED BY A.W.DIAMOND

*for the British Ornithologists' Union
with the assistance of
A. S. Cheke and Sir H. F. I. Elliott*

*The right of the
University of Cambridge
to print and sell
all manner of books
was granted by
Henry VIII in 1534.
The University has printed
and published continuously
since 1584.*

CAMBRIDGE UNIVERSITY PRESS

Cambridge

London   New York   New Rochelle

Melbourne   Sydney

CAMBRIDGE UNIVERSITY PRESS
Cambridge, New York, Melbourne, Madrid, Cape Town, Singapore, São Paulo, Delhi

Cambridge University Press
The Edinburgh Building, Cambridge CB2 8RU, UK

Published in the United States of America by Cambridge University Press, New York

www.cambridge.org
Information on this title: www.cambridge.org/9780521113311

First published 1987
This digitally printed version 2009

A catalogue record for this publication is available from the British Library

Library of Congress Cataloguing in Publication data
Studies of Mascarene Island birds.
  Bibliography
  Includes index.
  1. Birds – Mascarene Islands.   I. Diamond, A.W.
(Antony William), 1944–   .   II. Cheke, A.S.
(Anthony S.)   III. Elliott, Hugh F.I.   IV. British
Ornithologists' Union.
QL695.5.S78 1987   598.2969'8   86-6860

ISBN 978-0-521-25808-1 hardback
ISBN 978-0-521-11331-1 paperback

CONTENTS

Contents

CONTRIBUTORS

FOREWORD

Anthony S. Cheke
139 Hurst Street, Oxford OX4 1HE, UK

Graham S. Cowles
Sub-department of Ornithology, British Museum (Natural History), Tring, Hertfordshire, UK

Jennifer F. M. Horne
National Museums of Kenya, PO Box 40658, Nairobi, Kenya

Christian Jouanin
Muséum National d'Histoire Naturelle, Zoologie, 55 Rue de Buffon, 75-Paris (V^e), France

Carl G. Jones
ICBP, c/o Forestry Quarters, Black River, Mauritius

I have been invited to introduce the book arising out of the BOU Mascarene Island Expedition because this project took place during my term of office as President of the BOU. I am pleased to take advantage of this gesture.

The BOU has a long tradition of ornithological exploration. During the latter part of the last century many distinguished members made important collections in remote parts of the world, and *The Ibis* in consequence was full of long faunistic lists. Such ventures were privately financed and leisurely in tempo. From time to time, the BOU organized its own larger-scale expeditions, financed at least in part from its own funds, notable examples being the Jubilee Expedition to Dutch New Guinea (1909–11) and the Centenary Expeditions to Ascension Island (1957–9) and the Comoro Islands (1958). The Jubilee Expedition was over-ambitious and failed to reach its final destination, though doing useful work in an alternative location. Both the Centenary Expeditions of 1958/9, however, were highly successful, thanks to the care with which they were organized and to the skilled leadership of Dr Bernard Stonehouse in Ascension and of C. W. Benson in the Comoros. Previous expeditions had been chiefly concerned with collecting, at a time when museum collections were still far from complete. This was, however, a diminishing requirement; concern for the conservation of rare species was gaining momentum and studies of birds in the field had emerged as the primary interest among ornithologists. Although limited collecting was necessarily involved in the Comoros in order to fill important gaps in existing material, both Centenary Expeditions concentrated on ecological and behavioural studies. The published results amply illustrate the value of a broadly based approach to ornithological exploration in island communities.

In 1970 the policy of concentrating the very limited resources for ornithological exploration on areas containing endangered species was advocated (*Ibis* 112: 445–7). It was pointed out that many of the 315 rare species or races listed in the IUCN *Red Data Book (Aves)* might otherwise disappear completely before anything was known of their ecology or behaviour. This proposal led to the decision to organize a BOU expedition to study the gravely endangered and biologically little known land-birds of the Mascarene Islands. Four endemic species were listed by the IUCN as nearing extinction and seven others as suffering very severely from the increasing human exploitation of the islands. After a preliminary reconnaissance, the Mascarenes Expedition took the field on 1 January 1974, under the leadership of Anthony Cheke. The leader had already gained experience on expeditions to Ethiopia, the Ivory Coast and Corsica, and had taught ecology at the University of Chiangmai in Thailand. As in the case of the Centenary Expeditions, this new venture by the BOU was graced by the patronage of HRH the Duke of Edinburgh, who also contributed to the special fund which was created. Field studies were made over a period of about seventeen months, a total of nine official or voluntary members taking part for varying lengths of time. Two others, who were working on their own projects, contributed to the work of the expedition and used its facilities, while several residents of Mauritius, Réunion or Rodrigues also made valuable contributions. The names of all who participated are mentioned and their roles acknowledged in the chapters that follow. The association with Dr Stanley Temple, working for the ICBP and WWF on the Mauritius Kestrel, proved to be particularly valuable. Two attached specialists who deserve special mention were Jennifer Horne of the National Museum of Kenya, who recorded the voices of all the endemic birds, and Graham Cowles, an osteologist from the British Museum (Natural History), who worked on subfossil remains of extinct species. The work of the expedition was greatly facilitated by the generous co-operation of the various local government officials and agencies in each of the islands. Their help is gratefully acknowledged.

The Mascarenes Expedition was by no means a simple matter either to organize or to execute. It was planned at a time of financial crisis in Britain and would not have been possible but for the generosity of BOU members and of the supporting individuals and institutions. The fact that several participants were able to find their own funds was a great help. Tribute must also be paid to the painstaking work of the BOU Expedition Committee and its Chairman, Dr David Snow, who, with the Hon. Treasurer, Archie Bryson, the Hon. Secretary, Peter Olney, and the Administrator of the expedition, Peter Hogg, handled the problems of planning and administration with great skill.

It is curious that the Mascarene Islands, which are renowned as the home of the extinct Dodo *Raphus cucullatus*, should have remained relatively little studied for so long. Such ornithological literature as existed was fragmentary and long out of date. Mauritius, Réunion and Rodrigues were once richly forested and supported a very varied fauna, much of which was unique. The present very high density of the human population and the disastrous exploitation of forests now gravely threaten the survival of what little is left of the original fauna. At least thirty endemic species of birds are already extinct and the populations of several others which survive in the forest remnants may now be so small as to be no longer viable. The importance and urgency of the work of the BOU expedition in the light of this situation was obvious. It is gratifying to know that as a result of the intensive study made of the endangered species, both the local authorities and the international conservation organizations such as the IUCN, the World Wildlife Fund and the International Council for Bird Preservation, have now been provided with up-to-date information on which conservation planning can be based.

That the expedition was opportune in its timing and also highly successful can be seen from the following pages. Its success is a tribute to the planners, to the leader and to all who took part in the fieldwork, much of which was carried out under far from comfortable conditions. The BOU can take pride in having promoted this expedition and I hope it will be able to make further such contributions to both science and conservation in the future.

Guy Mountfort

Introduction

A.W.DIAMOND and A.S.CHEKE

Background to the expedition (AWD)

In 1970 the Council of the British Ornithologists' Union decided on a policy of promoting expeditions to study ecologically little-known endangered species (Mountfort 1970). The first proposed survey was to be in Cuba, but for various reasons this plan had to be abandoned. However, a timely visit to Dr D. W. Snow, then Chairman of the Research Committee, by Dr France Staub, then Chairman of the Mauritian National Section of the International Council for Bird Preservation, drew the Union's attention to the Mascarenes and the proposed expedition was formally announced in January 1972 (*Ibis* 114: 138). The *IUCN Red Data Book* vol. 2 (*Aves*) (Vincent 1966, partly revised 1971) listed 11 species or forms from the Mascarenes, four of which were given 'three-star' ratings, their rarity giving rise to 'very grave anxiety' as to their prospects for survival. None of these birds was well known biologically and there was thus a strong case for further investigation. The urgency of action and a plan for research had already been outlined by F. R. G. Rountree as early as 1950 (Rountree 1951), but it is doubtful whether his report was seen by many outside Mauritius, and the matter thus lay fallow for another 20 years, there being no one available locally to do the necessary work. Since 1950 the natural habitat has been substantially reduced, and the birds thus made rarer.

Coincidentally, and independently, Dr Stanley Temple then of Cornell University had put forward a project to study the endangered birds of prey in the Indian Ocean, with particular attention to the Mauritius Kestrel *Falco punctatus*, the rarest of them.

Funding proved a problem for the BOU; shortage of money and the exigencies of the leader's other work delayed the start of the expedition and made it impossible to send out the three people initially envisaged. However the leader, Anthony S. Cheke, was able to fly out to Mauritius at the end of September 1973 to confirm the feasibility of the project, and spent two months in Mauritius and Réunion. By this time Dr Temple was well established in Mauritius and it was clear that a useful if not complete survey could be done with the manpower and funds then available.

The expedition began officially on 1 January 1974, when Dr Temple also became formally associated with the BOU project. Both Dr Temple and Anthony Cheke left Mauritius in February 1975, their work in the meantime being supplemented by Jennifer F. M. Horne (sound recordist) and Jennifer Shopland (field assistant), whose visits were made possible by additional funds donated after the start of the expedition. Several unofficial helpers also visited the islands and helped wth the fieldwork. Barbara Temple helped her husband in the field and by giving talks on birds and conservation to schools before leaving in June 1974. During the course of his stay, Dr Temple's project developed from a study only of the Kestrel into a programme to breed in captivity the three endangered non-passerines (Kestrel, Pink Pigeon *Nesoenas mayeri* and Echo Parakeet *Psittacula echo*). This programme has since been continued and, in collaboration with the Jersey Wildlife Preservation Trust, expanded to include the Fody *Foudia flavicans* and the flying fox *Pteropus rodricensis* of Rodrigues, and endangered snakes and lizards from Round Is.

This book is long delayed, and in the course of its development has become rather different from the publication originally envisaged. The work done by the members of the BOU Expedition *sensu stricto*, and that of the independent collaborators, has for the most part not been followed by more detailed work and is not diminished by the decade that has passed since the fieldwork was carried out; in most cases it has been brought up to date where necessary. But Dr Temple's work initiated a long-term programme by ICBP on the most desperately threatened Mauritian birds (the Kestrel, Pink Pigeon and Echo Parakeet); thus, in addition to difficulties experienced in publishing the results of Dr Temple's own work (which he has since decided to publish separately), it was rapidly outdated by more recent studies. The BOU is indebted to Carl G. Jones, the latest to work on these species, for the immense amount of trouble he has taken to bring the accounts of these species up to date. Without his contributions the book could not have appeared.

The objective of the BOU Expedition was primarily scientific; yet it took place in the Mascarenes because of their unique importance to bird conservation. The scientific results presented here will enable the appropriate government departments and conservation bodies to proceed on the basis of a sound knowledge of the needs of the threatened birds. Some members of the expedition also played active roles in local conservation and Anthony Cheke produced a number of important reports and articles on conservation in addition to his scientific work reported here.

Although birds are the focus of both the expedition and this book, they cannot be considered in isolation from their ecological and historical background. Anthony Cheke was also active in scientific work on groups other than birds – notably lizards and fruit bats – and has made a major contribution to knowledge of the Mascarene biota as a whole. The BOU was fortunate to find someone of his breadth of interest and expertise to lead this expedition; his special qualities are reflected in the first chapter reviewing the ecological history of the islands. This contains much material on groups other than birds; no apology is surely necessary for this, rather it is offered as a model of a comprehensive review of the history of an avifauna against a suitably detailed background of the ecology of entire island communities.

Conventions followed (ASC)
Bird names
Scientific nomenclature and sequence
Nomenclature, classification and sequence of endemic Mascarene birds follow Morony *et al.* (1975), unless there has been a more recent revision of a particular group. Forshaw & Cooper (1978) is thus followed for parrots and Jouanin & Mougin (1979) for petrels. In addition three forms treated by Morony *et al.* as subspecies are accorded full specific status: *Circus maillardi*, *Saxicola tectes* and *Hypsipetes olivaceus*. The genera *Nesoenas*, *Hypsipetes* and *Collocalia* are retained on grounds of established usage, although Morony *et al.* (following Goodwin 1967) merged *Nesoenas* in *Col-*

umba, Mees (1969: 302n) revived *Ixos* for the bulbuls (*contra* Deignan 1942), and Brooke (1972) *Aerodramus* for echo-locating swiftlets; Goodwin (1983) has restored *Nesoenas*. *Acrocephalus* is used for the Rodrigues Warbler following Diamond (1980). Reasons for these decisions are discussed in the species accounts. The names *Zosterops* and *Hypsipetes* are considered masculine following amendments to the code of zoological nomenclature (International Committee for Zoological Nomenclature 1974).

Recent comprehensive regional lists are used for introduced species, migrants and seabirds. Thus White (1963, 1965) is followed for African species, Ali & Ripley (1968–74) for Asian, Milon *et al.* (1973) for Malagasy, Voous (1973) for Palaearctic migrants and Serventy *et al.* (1971) for seabirds.

English names
As far as possible the English names used follow standard reference works, although various changes have been made to take taxonomic changes into account and to conform with the current usage in giving English names to full species rather than to subspecies.

For both native and introduced Mascarene birds Staub (1976) has been followed with a few exceptions. Seabird names follow Tuck & Heinzel (1978) except for the tern *Gygis alba*, which is called Fairy Tern in accordance with established practice. Palaearctic migrants follow Heinzel *et al.* (1972).

A special exception to the 'official' English name is used for three species: the Madagascar Fody *Foudia madagascariensis* and the two *Hypsipetes* bulbuls. They are called by their Créole names 'Cardinal' and 'Merle' throughout to avoid any possible confusion with, respectively, the endemic fodies (called 'Fody') and the introduced Red-whiskered Bulbul *Pycnonotus jocosus* (called 'Bulbul').

Unless there is risk of confusion, geographical prefixes of names are dropped after the first mention in any discussion.

Local names
Créole names are those in current use as found by the authors. Only the most widespread and well understood name or names are cited, given in the familiar 'bastard-French' orthography, followed in brackets in species headings by the phonetic form used by Cheke (1982*b*). Certain names used by local French-speakers

and well established in the ornithological literature are in fact obsolete among speakers of Créole (Cheke 1982*b*); these are indicated by double quotation marks. One Mauritian bird that has lost its local name has been given a new one by Staub (1973*a*, 1976); this is given in square brackets. In the past both *oiseau* and *zozo* have been given as 'Créole' prefixes to bird names. Most Créole speakers now use *zoiseau* (*zwazo*).

Examples of the above are as follows:

Terpsiphone b. bourbonnensis = *çakoit* (*sakwat*), "*oiseau de la Vièrge*"

Zosterops chloronothos = ['*Zoiseau linette*'], "*oiseau pit-pit*"

French speakers on Mauritius and Réunion generally use a Frenchified version of the Créole name, but occasionally employ the metropolitan French version: *puffin* for *fouquet* (shearwater) or *foudi* for *cardinal*, for instance. Attempts to standardise French names (Joly 1977, Devillers 1976–) are not widely known on the islands. Migrant waders and seabirds generally do not have very specific Créole names, and the French one is perforce used when known.

Other taxa

Mammal, reptile and amphibian nomenclature normally follows that used by the authors of the most recent local study as cited in the text (mostly in Chapter 1).

The nomenclature and classification of the nineteenth-century floras are now hopelessly out of date, and the new but incomplete *Flore des Mascareignes* (Bosser *et al.* 1978–) is followed as far as possible, but a fuller recent list is *Endemic Species of Mascarenes (partial list), Seychelles and Socotra* produced by the IUCN Threatened Plants Committee Botanic Gardens Conservation Co-ordinating Body at Kew (*c.* 1982). Cadet (1977, 1981) is generally followed for introduced species and native ones not covered by the IUCN list or the published parts of the *Flore*.

Maps and place names

The basic working maps used of the islands were as follows (see special section below for explanation of abbreviations):

Mauritius

1. DOS 1:25 000 (series DCS 29/Y881: 12 sheets) (1957–8) [now replaced by series Y881 (DOS 329), ed. 4-DOS, 1981: 14 sheets]

2. DMS 1:100 000 series Y682 'Mauritius' 2-GSGS edition (1971) [now replaced by Y682 (DOS 529) ed. 4-DOS, 1983]

3. IGN 1:100 000 Carte Touristique 515 'Ile Maurice' (1972, revised 1977)

4. Admiralty Chart No. 711 (1879, frequent minor revision)

For field work the 1:10 000 blueprint maps produced by the Forestry Services of the Ministry of Agriculture were also used.

Réunion

1. IGN 1:50 000 'La Réunion' (4 sheets) (1968)

2. IGN 1:100 000 Carte Touristique 512 'La Réunion' (1971, revised 1978)

Rodrigues

1. 1:24 000 map (blueprint) produced by Mauritius Survey Department (No. K.D./2ª)

2. IGN 1:10 000 'Ile Rodrigues' (6 sheets) (1978)

3. Admiralty Chart No. 711 (1876, occasional minor revision)

Rodriguan place names are rather fluid and not well known to map-makers; the ones used are clarified as far as possible on the maps published in the chapter on Rodrigues (Chapter 8).

General

As published figures vary so widely, all inter-island distances have been re-measured using Admiralty Chart No. 4702 'Chagos Archipelago to Madagascar' (1974 edition), correcting for scale changes with latitude.

Wind and current data are from Admiralty Routeing charts for the Indian Ocean (No. 5126(1–12), a separate chart for each month).

Units

Units in general use on the island in question are used, with the metric equivalent in brackets where applicable.

The standard measurement of area in Mauritius and Rodrigues is the arpent; this is the pre-metric French 'acre', and was somewhat variable in extent. In Mauritius the arpent is fixed at 1.04 acres.

Abbreviations

The following abbreviations are frequently used without further explanation in the chapters that follow:

DOS Directorate of Overseas Survey (UK)
DMS Directorate of Military Survey (UK)
ICBP International Council for Bird Preservation
IGN Institut de Géographie Nationale (France)
IUCN International Union for Conservation of Nature and Natural Resources
ONF Office National des Forêts (France, Réunion)
WWF World Wildlife Fund

For abbreviations of museum names see Chapter 9, Table 3.

References

Text references use the 'Harvard' system of author followed by date in round brackets. Where a name is followed by a date in square brackets this is not a reference but a date pertaining to the person's activities: e.g. 'other visitors were Rountree [1943], Jauffret [1947] and Courtois [1955]'.

General acknowledgements

The expedition would not have been possible without the generous funding of the following organisations: Overseas Development Administration of the British Foreign and Commonwealth Office, the World Wildlife Fund (US and UK Appeals), National Geographic Society (US), the Fauna Preservation Society, the British Museum (Natural History), the Gilchrist Trust. The BOU contributed from its Research and Publications Fund and its then President, Mr G. R. Mountfort, contributed personally, as did Sir Landsborough Thomson, D. A. Bannermann and C. W. Mackworth-Praed. The British Institute of Recorded Sound gave tapes for use by Mrs J. F. M. Horne. His Royal Highness the Duke of Edinburgh graciously consented to be patron of the expedition, and also contributed financially. Dr Temple's work was financed by the ICBP, WWF (US Appeal) and the New York Zoological Society.

Mr Peter Hogg was home administrator for the expedition, and also did much of the organising before the project was in the field; the fieldworkers are most grateful to him for this.

We would like to thank the Director of the Edward Grey Institute, Dr C. M. Perrins, for giving us space in Oxford to work on the results of the expedition. During the protracted preparation of this book, Sir Hugh Elliott bore the brunt of correspondence and preliminary editorial work; ill-health prevented him from completing the task but his contribution made the final editing proceed much more smoothly than would otherwise have been possible.

The editor settled in Canada in July 1983 and would like to acknowledge the considerable amount of work undertaken by Anthony Cheke, particularly on the material contributed since then, which would normally have fallen to the editor's lot. We both thank Martin Walters and Jane Farrell of Cambridge University Press for their forbearance and industry with more than usually difficult material.

1

An ecological history of the Mascarene Islands, with particular reference to extinctions and introductions of land vertebrates

A.S.CHEKE

Part I

THE NATIVE AVIFAUNA OF THE MASCARENE ISLANDS

Although a great deal has been written about the history, both human and biological, of the Mascarenes, there has been no synthesis of the available material from an ecological point of view; this chapter is an attempt to fill this gap. I make no apologies for covering ground familiar to Mascarene specialists, since there is at present no suitable introductory work for biologists new to the area; also much of the original material is rare, and even many of the secondary sources are not easily accessible outside the islands themselves.

Already known to Arab navigators, the Mascarenes were 'discovered' by Europeans in the early sixteenth century. The exact dates at which each island was first sighted by the Portuguese have long been a matter of debate, but are of little concern here as no useful accounts of Portuguese visits apparently survive. North-Coombes (1980b) reviewed the history of Portuguese movements in the Indian Ocean in the sixteenth century and concluded that they very rarely landed in the Mascarenes, and then probably only on Réunion. The most they appear to have left us by way of description is a note in 1728 that "Santa Apelonia" (= Réunion) had "plenty of fresh water, trees, birds and fish" (ibid.).

The next description of the islands dates from the Dutch visit to Mauritius in 1598 (see e.g. Pitot 1905, Barnwell 1948). Réunion was described briefly by Verhuff in 1611 but the earliest adequate account is Tatton's in 1613 (Tatton 1625, Lougnon 1970). Only the scantiest accounts of Rodrigues exist prior to Leguat's

2-year stay on the island in 1691–3 (North-Coombes 1971).

Although the course of events in each island has been broadly similar I have treated the three islands separately, cross-referring where appropriate. While this chapter covers the whole period from the islands' discovery to the present day, more detailed information on recent events can be found in the chapters which follow.

Sources
The early literature (1598–1750)

In the nineteenth century a great deal of work was done in searching out old manuscripts relevant to the Mascarenes, and interesting material is still turning up. Many of these records have been published in the compilations listed below. Also during the past century many of the published accounts of early voyagers, now very rare and available only in a few libraries, have been reprinted; sometimes in full, but often only selectively in anthologies of Mascarene literature. In general I have used this secondary literature, including several works missed by previous students of Mascarene natural history.

For descriptions of the islands in the seventeenth and early eighteenth centuries the most extensive source is Grandidier *et al.*'s *Collection des ouvrages anciens concernant Madagascar* (1903–20). For Mauritius Pitot's *T'Eylandt Mauritius* (1905) gives the best coverage of the Dutch period; other important compilations are Barnwell (1948), Grant (1801), Brouard (1963) and Strickland's part of the classic book on the Dodo (Strickland & Melville 1848). Almost all the anthologies of old material about Réunion are due to Albert Lougnon, his *Sous la signe de la tortue* (1970), the most useful, covering nearly all recorded visits up to 1725. Barré & Barau (1982) have recently reprinted many important early texts referring to Réunion birds. Rodrigues is poorly served, though Dupon (1969) and North-Coombes (1971, 1980a) are useful. Both Mauritius and Réunion have had journals wholly or partly devoted to reprinting old or manuscript material, although the journals themselves are now very rare. The *Revue historique et littéraire de l'Ile Maurice* was published from 1887 to 1894, the *Nouvelle revue* from 1897 to 1908, and later (1950–5) Noel Regnard produced a similar magazine the *Revue retrospective de l'Ile Maurice*. In Réunion Lougnon published the *Receuil trimestriel*

from 1932 to 1953; publication continues as the *Receuil de documents et travaux inédites*.

An invaluable source-book is Toussaint & Adolphe's *Bibliography of Mauritius* (1956), which also covers Rodrigues and, to a lesser extent, Réunion. Defos du Rau (1960) is the best general source of published material on Réunion, as is North-Coombes (1971, 1980a) for Rodrigues; North-Coombes (1971) provides in addition an extensive bibliography of manuscripts. Hachisuka (1953) also has an extensive bibliography, and Ronsil (1948–9) provided a number of useful leads.

Manuscripts concerning Réunion have undoubtedly not yet yielded all their natural history content, and interesting discoveries may yet be made in manuscripts cited, but not quoted, by Lougnon (1957) and Kaeppelin (1908). The Mauritian material has been more thoroughly worked, but again would repay research; where for instance is the material on natural history sent by Le Juge to the Duc d'Orleans in 1751, referred to in Le Juge's papers (MS 2619, Museum d'Histoire Naturelle library, Paris)? Lougnon (1953, 1956) and Toussaint (1956, 1965) cited a large number of unpublished manuscripts pertaining to both islands, whose contents they only hinted at. There is probably little more to be found concerning Rodrigues during this period.

Individual books are usually cited on their own, but I have in only one case consulted the Dutch literature so most material of Dutch origin is cited in translation (French or English) from secondary or tertiary sources, though Strickland (1848) also gave some passages in the original Dutch. In the case of difficult animal names the translators have in most cases indicated the Dutch originals, which is often essential to the correct interpretation of a passage. In the References I have tried to refer to reprints as well as original editions of books, but they are all cited by the original date in the text. In some cases the bibliographical details given here differ from those in other published lists; this is because I have given the information as it appears in the copy of the book I consulted, usually in the Bodleian Library and its branches (Oxford) or the British Library (London).

The middle period (1750–1850)

The first scientific contributions to Mascarene zoology were made in 1676 (Perrault 1676) and 1737 (Petit 1741), but it was not until Brisson (1756, 1760) and Buffon (1770–83, 1776, 1789) published their books that the fauna became fairly well known. No reasonably complete surveys are available before the mid-nineteenth century (Clark 1859: Mauritius; Maillard 1862: Réunion; Transit of Venus reports, 1879: Rodrigues), so information for this period on the status and ecology of the fauna is still confined to travellers' accounts, most of which were published as books, many (e.g. La Caille 1763, Bernardin 1773, Sonnerat 1782, Bory 1804, Milbert 1812 well known, but others (e.g. Cossigny 1764) almost forgotten. Some important manuscript material has appeared in the anthologies and journals mentioned in the previous section, and there is some material in French literary and scientific periodicals (e.g. d'Héguerty 1754, Morel 1778). The increasing familiarity of the islands led travellers to give less detailed descriptions; many of the books have only snippets of information and there is particularly little on the natural history of either Mauritius or Réunion in the last quarter of the eighteenth century, and virtually nothing for Réunion between 1801 and 1862.

In Mauritius a natural history society was founded in 1829 (Ly-Tio-Fane 1972) and circulated reports more or less annually thereafter. One of the founder members, Julien Desjardins, intended to write a zoologically inclined natural history of the island, but died in 1840 before it was finished; his collections formed the basis of a natural history museum which still thrives. His manuscripts, of which the bird parts were written up by Oustalet (1897), are now apparently lost (Cheke & Dahl 1981: 210).

The manuscripts of Philibert Commerson from the early 1770s, preserved in Paris and recently catalogued by Laissus (1978), would repay further study. I have examined a microfilm of MS 2127 (*ibid.*: 156, no. 57) which contains, according to the catalogue, descriptions of mammals, birds and plants of (by implication) Mauritius, but the animals covered are specimens or captives originating elsewhere. The *pièce de resistance* is MS 282 (*ibid.*: 161, no. 73), the substantial collection of drawings (259 vertebrates) by Jossigny and Sonnerat annotated by Commerson. The birds were studied by Oustalet (1897) and the tortoises by Vaillant (1898), but no one seems to have been through them for Mascarene mammals or other reptiles. Also, reading between the lines, Oustalet may not have picked up all the birds of Mascarene interest (Cheke 1983*b*).

During the first half of the nineteenth century various French scientific expeditions touched at Mauritius, providing specimens written up rather fragmentarily later (e.g. Quoy & Gaimard 1824, Dumeril & Bibron 1834–54). The only good general accounts of the island arose from the 1801 expedition, which also covered Réunion (Bory 1804, Milbert 1812, Chapotin 1812).

For Rodrigues most of the information for this period is available only in manuscripts, some of which have been partially published (e.g. Dupon 1969); North-Coombes (1971) gave a useful guide to these materials. Perhaps the most significant are the diary and accounts of Abbé Pingré, a French astronomer who visited Rodrigues to observe the transit of Venus in 1761 and was forcibly detained there for several months by the British navy. His manuscripts (Pingré 1760–2, 1763) are in the Bibliothèque Sainte-Geneviève in Paris; although much quoted (Milne-Edwards 1873, A. & E. Newton 1876, Vinson 1964*b*, Bourne 1968, Nelson 1974, etc.) only an abridged version of the description of Rodrigues has appeared (Dupon 1969). The Mauritian naturalist Paul Carié prepared the manuscripts for publication before he died (Crépin 1931), but they never appeared in print.

The recent period (1850 on)

The literature of the last 130 years consists largely of books and papers written as scientific contributions to the islands' fauna, though some more general works, such as Bertuchi's (1923) account of Rodrigues, are useful. This literature is very scattered, and here I bring together all the land vertebrate material in one bibliography. There are valuable though more restricted bibliographies in Rountree *et al.* (1952) and Vinson & Vinson (1969). Alfred Newton's voluminous papers, covering half a century to 1907 and including considerable correspondence on the Mascarenes from his brother Edward and others, remain uncatalogued and inaccessible to researchers at the Zoology Department, Cambridge University.

Carié (1916) reviewed the history and status of birds and mammals introduced to Mauritius, but much more material has come to light since then; Moutou (1983*b*) has summarised extinctions and introductions in Réunion. Hachisuka (1953) reviewed the extinctions of birds on all the islands, but his work requires considerable further comment; the birds are also covered in various tertiary compilations (e.g. Halliday 1978, Day 1981). There are also useful reviews of the fauna of Rodrigues (Vinson 1964*b*, Staub 1973*b*) and the northern islets of Mauritius (Vinson 1950, 1964*a*). Biogeographical introductions to the native Mascarene fauna in general, and that of Réunion in greater detail, were given by Moutou (1983*a*, 1982*a* respectively). Owadally (1981*a*) summarised the history of ecological studies in Mauritius, and Dupon (1978) provided a very general (and somewhat outdated) survey of the ecological history of all three islands.

After half a century of neglect natural history in Réunion was revived in 1855 by the establishment of a natural history museum, and in the 1860s Auguste Vinson, Charles Coquerel and François Pollen were active. In 1861 Edward Newton arrived in Mauritius and stimulated the activities of the local society, by then the Royal Society of Arts and Sciences. This active phase was followed in both islands by quiet periods, lasting in Mauritius from about 1875 to the turn of the century. In Réunion it lasted from 1870 till 1948, relieved only by the collections of Lantz, the museum curator from 1862 to 1893 (Jouanin 1970*a*), and Carié in about 1906 (Koenig 1932). Lantz wrote remarkably little, but the manuscript accessions book and associated papers of the museum in St.-Denis are a useful source of information (Jouanin 1970*a*, pers. obs.). Carié did not write up his Réunion trip, but his bird specimens are in the natural history museum in Paris. In Mauritius the only work in the last quarter of the nineteenth century was excavation of subfossil bones, which added substantially to our knowledge of the extinct fauna (e.g. Newton & Gadow 1893); some of these bone collections have remained inadequately studied until very recently (e.g. Arnold 1980). After Paul Carié's departure from Mauritius in *c*. 1914 (Crépin 1931) there was little work on land vertebrates until after the 1939–45 war, since when many biologists have been active. Papers giving details of status are cited here, but fuller references to ecological conservation and other studies will be found in Procter & Salm (1975), Cheke (1978*a*), Jones (1980*a*), Jones & Owadally (1982*a*), Jones & Pasquier (1982), Cheke & Dahl (1981), Temple (1981), Diamond (1984) and Chapters 4, 5, 6 and 8. For seabirds Staub (1973*b*), Vinson (1976*a*) and Jouanin (Chapter 7) are useful. The recent studies on the fauna of Round Is. (Mauritius) have yet to be brought together, though Vinson (1975), Durrell (1977*a*), Bullock (1977, 1982, *et al.* 1983) and Bullock & North (1977, 1984) should be consulted.

Long's (1981) comprehensive review of bird introductions world-wide was compiled from easily accessible sources, and new material in this chapter alters the picture for the Mascarenes, as it does for extinctions, where King (1980) listed the cause as 'unknown' in 65% of taxa (mostly Mascarene) on Indian Ocean islands. Recent research (Arnold 1979, Bour 1980, 1981) has rendered useless the section on Mascarene tortoises in Honegger's review (1981) of reptile extinctions. Most currently endangered taxa are treated in the *Red Data Books* (King 1978–9, Honegger 1975, Thornback 1978, Collar & Stuart 1985). Collar & Stuart (*ibid.*) also list recent small-circulation conservation reports not all of which are cited in this book.

AN ECOLOGICAL HISTORY OF THE ISLANDS

More has probably been written on the biological history of the Mascarenes than of any other oceanic islands apart perhaps from the Galapagos, New Zealand and Hawaii. Much of this material has either been very incomplete or devoted solely to a particular aspect of the islands' biota. Although I concentrate on the land vertebrates, my objective is to place their history in the context of man's activities.

Man's principal documented impacts on the existing biota have been threefold: hunting, habitat alteration and destruction, and introduction of alien species. The effects of these are interlinked and synergistic, but I will consider them separately for ease of discussion. It is appropriate first to consider the animals known to have become extinct before the first historical accounts.

Prehistoric extinctions
Mauritius
Extensive subfossil bone deposits in Mauritius have confirmed most of the species of birds, mammals and reptiles described by early travellers, but they also include a number of forms that are not mentioned in the old accounts. Hachisuka (1953) and Greenway (1967) summarised the bird material, which Cowles has brought up to date in Chapter 2. Hoffstetter (1946*a*, *b*, *c*, 1949) and Arnold (1980) have considered the lizards and snakes, and Arnold (1979) and Bour (1980) the tortoises. There are also bat (*Pteropus* spp.) bones,

but the only publications refer to material from Round Is. (Cheke & Dahl 1981), and Rodrigues (Andersen 1913).

Two species of land-bird found as subfossil remains are nowhere mentioned in travellers' accounts: a cormorant *Phalacrocorax* cf. *africanus*, originally described as an anhinga (*Anhinga nana*) but re-assessed by Olson (1975), and a coot *Fulica* (= *Paludiphilus* = *Palaeolimnas*) *newtoni*. Early references to "moorhens", however, probably refer to this coot rather than to the still-extant *Gallinula chloropus*; this was certainly the case in Réunion (see below). In addition there are no unequivocal references in the early reports to the heron *Nycticorax mauritianus*, the owl *'Tyto' sauzieri* or the harrier *Circus alphonsi*. There were apparently two owl species on the island: the bones found at the Mare aux Songes represent an undescribed endemic genus near *Ninox* (Chapter 2), but the drawing and description of a form surviving to the early nineteenth century ('*Scops commersoni*') clearly represent an eared owl near *Asio* (see discussion in Oustalet 1897, and Hachisuka 1953). Abbott's Booby *Sula abbotti* is known from subfossil bones (Bourne 1976), but there is no unequivocal reference to this large and conspicuous tree-nester in travellers' accounts. A bone originally described as from an extinct endemic grebe ('*Podiceps gadowi*') was actually from a Whimbrel *Numenius phaeopus* (Chapter 2).

Of the other animals it is possible that the bat *Pteropus rodricensis* (Mason 1907, Cheke & Dahl 1981) and the blind-snake *Typhops cariei* (Hoffstetter 1946c) could have been overlooked, but it seems that the giant skink *Didosaurus* (= *Leiolopisma*) *mauritiana* (Günther 1877, Hoffstetter 1949, Arnold 1980) must have already become extinct before 1598. Although they survive to this day on Round Is., there are also no reports in the literature of snakes, the large skink *Leiolopisma telfairii*, nor the large gecko *Phelsuma guentheri* on the mainland, although remains have been found at the Mare aux Songes and elsewhere (Hoffstetter 1946a, b, Arnold 1980). These reptiles were probably exterminated by rats from shipwrecks, which were reported as very numerous in 1606 (Matelief; see Bonaparte 1890, Pitot 1905). Rats were not mentioned in 1598, but the reports of this voyage failed also to refer to the abundant tortoises (although they did appear in an illustration) (*ibid.*, also Strickland 1848).

The Dutch visitors in 1598 found evidence of a recent shipwreck (Toussaint *in* Visdelou-Guimbeau

1948, Pitot 1905, Grandidier 1903–20, vol. 1) but no sign of any attempt at settlement. There is no hint among the early reports that man had altered the island at all before the Dutch arrived.

Réunion

Little subfossil material has come to light in Réunion, and the only animals so represented and not reported by travellers were a stork (see below under flamingos), an owl *'Tyto' sauzieri* (Chapter 2), and the large skink *Leiolopisma telfairii* (Arnold 1980, Bour & Moutou 1982). It is not known whether the lizards became extinct before the island was settled, or were simply overlooked in early accounts and fell victim to rats in the late seventeenth century.

Rodrigues

Two passerines, a *Hypsipetes* bulbul and a form of uncertain affinities, probably a babbler, were discovered amongst bones collected in 1974 (Chapter 2), but not reported by visitors to the island; they perhaps succumbed to the rats that colonised the island well before Leguat (1708) gave us the first account. There is no evidence that *Pteropus niger*, found as subfossil material (Andersen 1913, Cheke & Dahl 1981, Moutou 1982b), survived into historical times.

Introduction of alien species
Predators

For several decades on both Réunion and Mauritius, the number of human settlers was very small: there were never more than 300 people in the Dutch colony on Mauritius (1638–1710) and there were still under 1000 in Réunion in 1708 (Toussaint 1972). Large tracts of Réunion remained unexplored well into the middle 1700s or later (Defos du Rau 1960), and most of the Mauritian forests held their secrets until well after the French colonisation (1721 onwards). The direct human impact at first was quite small, yet many species disappeared astonishingly quickly. One can fairly attribute the early extinctions to the effects of introduced animals which multiplied rapidly and spread throughout the islands. The animals most likely to be destructive to land vertebrates were rats, cats and pigs, and, in Mauritius, monkeys. Cattle and goats and, in Mauritius, deer, were also released on the islands (see below); these herbivores probably had only an indirect effect on the fauna through gradual degradation of the vegetation and are unlikely to have contributed to early extinctions. Goats were reportedly very numerous and may have competed for food with the native giant tortoises, as has happened with devastating results in the Galapagos Islands (MacFarland *et al*. 1974, Lewin 1978), though not on Aldabra, where the goat population has remained low (Gould & Swingland 1980). Tortoises themselves, particularly at the high densities observed on Aldabra (Coe *et al*. 1979) and by early visitors to the Mascarenes, have a major effect on the vegetation (Merton *et al*. 1976, Hnatiuk *et al*. 1976, Garnett 1977) and their extinction could have affected the habitat significantly for other native species.

Early inhabitants of both Mauritius and Réunion blamed pigs for preventing tortoises and sea turtles from breeding (Pitot 1905, Lougnon 1957, Dellon *in* Lougnon 1970). Around 1670 tortoises were plentiful on the pig-free lagoon islets off the east coast of Mauritius, but pigs ate the eggs of the mainland ones (Pitot 1905). In 1673 Governor Hugo reported that pigs came out onto the shore to eat turtle eggs as they were being laid, and that they also raided tortoise eggs laid in "hollow trees exposed to the sun" (*ibid.*). It appears that some tortoises did not even bury their eggs!

Pigs are the worst predators on giant tortoises in the Galapagos: "wherever present [they] destroy the vast majority of nests and also kill large numbers of young tortoises up to at least 35–40 cm in curved carapace length" (MacFarland *et al*. 1974). The incubation period is so long (3–8 months) that even at low density pigs rarely miss a nest (*ibid.*). Unlike Aldabra tortoises, which nest in soil (Swingland & Coe 1978), Galapagos tortoises require soft sand for laying in (Perry 1964), migrating to coastal sandy areas to breed, but retreating inland out of season as the dry coastal vegetation is too sparse to support many animals (Brosset 1963, MacFarland *et al*. 1974). Although some Mascarene tortoises evidently did not bury their eggs (see above), the reports indicate that the majority laid in coastal sands and so may well have had a similar type of seasonal migration. If so, the eggs, hatchlings and young up to 10–15 years old (MacFarland *et al*. 1974) would have been concentrated in such a way as to increase enormously their vulnerability to predation. As the pigs were so numerous (80 men killed 1500 in a day in 1709: La Merveille *in* La Roque 1716), they must have severely curtailed tortoise and turtle production and probably that of ground-nesting birds such as the Dodo *Raphus cucullatus*, the Red Rail

Aphanapteryx bonasia, and the native ducks and geese. Tortoises, lizards and ground-nesting birds also had to contend with feral cats, introduced in the hope of controlling rats. No direct observations were made, but Après de Mannevillette (1775, Dupon 1969), Marragon (1795, Dupon 1969) and Cossigny (1799) blamed the near disappearance of tortoises on Rodrigues to cats eating their eggs and young, Marragon adding that he had seen cats fishing and presumed they would take young turtle also. Giant tortoise young would certainly be vulnerable to cats for several weeks or months after hatching. However, giant tortoises on Aldabra have bred and thrived in the presence of feral cats (Bourn 1976, Bourn & Coe 1979) since the 1890s (Stoddart & Wright 1967), so it seems unlikely that cats would be responsible for the extinction of a population. In the Galapagos, cat predation on young is the worst threat to the Marine Iguana (Laurie 1983; also Gibbons (1984) for Fijian iguanas), a lizard comparable in size to the extinct Mascarene skinks, and cats also take tortoises up to 2 years old but do not destroy nests (MacFarland *et al*. 1974). Rodrigues also supported feral pigs in the late 1700s, and was swept by fires (North-Coombes 1971), from which tortoises would have been unable to escape. Cossigny (1732–55) wrote in 1755 of being told that feral cats were responsible for the near extinction of Rodrigues Solitaires *Pezophaps solitaria*, though he himself blamed over-hunting. Pigs were apparently not feral in Rodrigues at that date. Pingré (1763) attributed the disappearance of Rodrigues rails and herons to cats; the owl '*Athene*' *murivora* was possibly another victim. The sudden extinction of the once-abundant native doves in Réunion was blamed by Borghesi (in 1703; Lougnon 1970) and Feuilley (1705) on the cats recently introduced for rat control. The Réunion Solitaire '*Ornithaptera*' *solitaria* and the *oiseau bleu* (a rail) apparently survived feral pigs successfully, but did not last long after cats were introduced. Cats were introduced too late to have been responsible for the extinction of the Mauritian Dodo, but the Red Rail disappeared around the time feral cats were first reported.

The most devastating introduced predator in the Mascarenes is surely the rat. Long blamed by biologists for extinctions on islands, rats were partly exonerated in a review by Norman (1975) of their interactions with birds, but the evidence of Whitaker (1973), Atkinson (1977), Diamond & Feare (1980), Bourne (1981) and especially Dingwall *et al*. (1978) is

clear enough: rats are disastrous to many island populations of birds and reptiles. In the Seychelles (Cheke
1984) and New Zealand (Whitaker 1973, 1978) several
native reptiles are now confined to the rat-free vegetated offshore islets; in Mauritius the only such refuge
is Round Is. (Vinson 1964*a*, 1975). Their former occurrence on islets now infested with rats is well attested
(see below), as also on the mainland prior to the
advent of man (see above). Rats are the only consistent
explanation for these extinctions and, as the only
non-native predators antedating man's colonisation
(of Mauritius and Rodrigues), are the most probable
cause of pre-settlement reptile extinctions. In Rodrigues the large *Phelsuma gigas* seems to have become
confined to offshore islets before 1761. As early as 1691
Leguat (1708, Hachisuka 1953) observed that the
island's doves (*'Alectroenas' rodericana*) nested only on
offshore islets "to avoid the persecution of the rats".
Tafforet (1726) confirmed that doves were confined to
islets, and said the same of the larger parrot (*Necropsittacus rodericanus*) and the starling (*Necropsar leguati*).
This was before the introduction of other predators, so
rats were surely responsible, though we do not know
when rats reached the southern islets or whether they
were responsible for the final extinction of these
species, which occurred soon after Tafforet's report
(see below). Rats could also have exterminated the two
passerines recently discovered in subfossil deposits
(Cowles, Chapter 2). In Réunion there is no mention of
the formerly abundant *moineau* (*Foudia* sp.) after the
introduction of rats; the Seychelles Fody *F. seychellarum* likewise does not survive in the presence of rats
(Diamond & Feare 1980) and *F. eminentissima* on
Aldabra is very vulnerable to rat predation on its nests
(Frith 1976). The only mention of the native kestrel and
rail on Réunion (Dubois 1674) dates from shortly
before rats were introduced, though cats, brought in
shortly afterwards, could also have been to blame.

The Ship Rat *R. rattus* is notorious for its devastating effect on arboreal birds, especially passerines
(Atkinson 1973, 1977, 1978, 1979, Bell 1978, Mills &
Williams 1978, Moors 1983). This species (see below)
was the first to reach the islands. There is no sign that
the arrival of Common Rats *R. norvegicus* in the 1730s
further affected endemic animal populations, potentially vulnerable ground-nesting birds having already
vanished. However, the Common Rat is a serious
predator of nesting seabirds (Imber 1975, 1978) and
may have contributed to the disappearance of Audu-

bon's Shearwaters *Puffinus lherminieri* from Mauritius;
the rarefaction of Wedge-tailed Shearwaters *P.
pacificus* on its offshore islets (Gunner's Quoin, Coin
de Mire, Flat Is. and the Grand Port group) and those
of Rodrigues (Frégate) is probably due to Ship Rats.
Predation by man and, on the Mauritian mainland, the
mongoose *Herpestes edwardsii*, may also have contributed. Two petrels (*P. lherminieri, Pterodroma baraui*)
still nest commonly on Réunion in rugged areas
unfavourable to Common Rats (Moutou 1979), but
within the range of Ship Rats. Ship Rats are the main
predators in Hawaii of Dark-rumped Petrels *Pterodroma phaeogpygia* which breed in conditions similar to
those on Réunion (Berger 1981), but their predation is
not sufficient to eliminate the population.

In the Galapagos *R. rattus*, not a serious menace
on most islands, appears to have been responsible, by
eating hatchlings, for the total reproductive standstill
of giant tortoises on one island, Pinzón (= Duncan),
where the rats were introduced in the 1920s and still
seem to be in the phase of superabundance (MacFarland *et al.* 1974, Lewin 1978). There is nothing to
suggest the Mascarene populations were vulnerable
to this species, and tortoises co-exist with Ship Rats on
Aldabra (Bourn 1976). However, *R. norvegicus* is likely
to have been a more formidable predator of hatchlings;
it is possible that it reached Rodrigues before the
tortoises became extinct and that Marragon (1795,
Dupon 1969) and Cossigny (1799) were right to blame
cats and rats jointly for their demise. In Mauritius it is
difficult to distinguish the effects of the two arboreal
predators, the Ship Rat and the monkey *Macaca fascicularis*, both introduced at an early date. In 1741 the
elder Grant (*in* Grant 1801, Pope-Hennessy 1886)
reported diminishing bird numbers "as the monkies
[*sic*] which are in great numbers, devour their eggs".

Recent studies (McKelvey 1976, Temple 1977*a*)
have suggested that monkeys are persistent nest predators on the native kestrel *Falco punctatus*, pigeon
Nesoenas mayeri and parakeet *Psittacula echo*, though
there is little direct evidence (Chapter 5). Grant (1801)
attributed the continuing abundance of parrots to
their habit of making "their nests in the holes of the
rocks which the monkies cannot ascend". The parakeet is purely a tree nester today (Chapter 5), but we
do not know where the grey parrot *'Lophopsittacus'
bensoni* nested. The lower density of all forest birds in
Mauritius as compared with Réunion (pers. obs.)
suggests that monkeys are an important additional

burden on all the native birds. Although the surviving species have co-existed with monkeys for 350 years or so, monkeys could have been responsible for the very rapid disappearance of the large parrot *Lophopsittacus mauritianus*. Slow attrition may have contributed to the disappearance of such birds as the grey parrot '*L.' bensoni*, the pigeon *Alectroenas nitidissima* and the owl '*Scops' commersoni* (see below).

The only other important introduced predator in the islands is the mongoose *Herpestes edwardsii*, introduced to Mauritius around 1900 (see below). It has not affected the endemic fauna, potentially susceptible forms being already long extinct, but rapidly eliminated several species of introduced game birds (Antelme 1914, Carié 1916; see below). It is likely also to have given the *coup de grâce* to the small population of Audubon's Shearwaters surviving in the late nineteenth century, as its relative *H. auropunctatus* is the worst predator on petrels breeding at under about 5000 ft (1540 m) on the main islands of the Hawaiian chain (Berger 1981). Its introduction to Réunion would be disastrous to the large Audubon's Shearwater populations there.

The House Shrew *Suncus murinus* eats chicks, mice and baby rats in Guam (Barbehenn 1962), so could have had an adverse effect on small terrestrial animals (e.g. skinks *Scelotes bojerii*), as could the Tenrec *Tenrec ecaudatus* (Eisenberg & Gould 1970); Tenrecs are probably important predators on skinks in the Seychelles (A. S. Gardner *in litt.*). Mynahs *Acridotheres tristis* are serious predators on Wedge-tailed Shearwaters in Hawaii (Byrd & Moriarty 1980).

Herbivores and habitat degradation

A more insidious effect of alien species is slow alteration of habitat. In vegetation that was originally browsed only by tortoises, large mammals may have drastic effects. Introduced plants compete with native ones, especially if means of seed dispersal are also introduced. These factors have all been at work in the Mascarenes.

Early navigators who put in at isolated islands for revictualling and repairs, often released enough livestock to establish wild breeding herds as food for future visits. Réunion and Mauritius were thus for a long time overrun with goats, cattle and pigs, Mauritius also with deer. These have been virtually hunted out in Réunion, but pigs and deer are still abundant (and protected as game) in Mauritius. In

Rodrigues feral livestock no longer exists, but overstocked domestic animals continue to affect vegetation (North-Coombes 1971, Cheke 1974a, 1978b; see also below for histories and references for all three islands).

Although no critical work has been done in the Mascarenes, studies elsewhere have established the disastrous effects on vegetation of goats (Coblentz 1978), deer (Allen *et al.* 1984, Howard 1964, 1965, 1967, Mitchell *et al.* 1977), cattle (in Hawaii: Berger 1981) and pigs (Howard 1964, 1965, 1967, Griffin 1978).

Monkeys too can prevent many native Mauritian trees from seeding (Carié 1916, King 1946, Owadally 1973, 1980, Proctor & Salm 1975), and rats are also important selective predators of seeds and seedlings in forests (Edgerley 1961, Campbell 1978). Even without taking into account damage by introduced insects, predation by introduced snails *Achatina* spp. (Thompson 1880, Brouard 1963, Owadally 1980), competition from exotic plants and direct deforestation by man, it is no wonder that Mascarene forests are degraded. Gradual but inexorable habitat modification has been going on ever since man and his animals reached the islands, to the extent that only in the upper vegetation zones of Réunion do any apparently unchanged areas remain anywhere in the Mascarenes (Cadet 1977). If any of the islands' native animals were closely adapted to the original vegetation, the gradual changes may have affected their survival. For instance the disappearance on both islands of the flying-fox *Pteropus subniger* and the parrot '*Lophopsittacus' bensoni*, on Réunion the parakeet *Psittacula eques*, and on Mauritius the owl '*Scops' commersoni* and the pigeon *Alectroenas nitidissima*, cannot be attributed directly to any one cause.

On Réunion the initial livestock introductions were all but hunted out by the mid-nineteenth century, leaving only rats as habitat modifiers. However, deer have been re-introduced at high density (30/km^2) in certain places (see below), allowing a direct comparison of areas with and without these herbivores. Superficial comparisons showed striking depressive effects of deer on regeneration and understorey growth, most conspicuous in the endemic bamboo *Nastus borbonicus* (Cheke 1975f, 1976); further dramatic deterioration of the *Nastus* stands was obvious 4 years later (Oct. 1978: pers. obs.).

Mauritius by contrast is still overrun with deer *Cervus timorensis*, pigs, and monkeys, as well as rats, and native forest regeneration is at a standstill in most areas (Owadally 1973, Chapters 4 and 5). Deer are a major game industry (Owadally & Bützler 1972, Douglas 1982), stocked at very high densities (70–200/km^2: Noël 1974) in forests operated as hunting estates (up to 600/km^2 on overstocked estates with supplementary feeding: M. J. W. Douglas *in litt.*), and even averaging 10–40/km^2 in the areas where most of the native forest survives (Douglas 1982, Noël 1974). These densities far exceed those regarded as tolerable in production forests in Europe, where 2.5/km^2 is normally considered maximal (Müller 1965, Holloway 1968, Mitchell *et al.* 1977, M. J. W. Douglas *in litt.*; up to 3.75/km^2 with supplementary feeding: du Boisrouvray 1975). In New Zealand, which, like the Mascarenes, is a forested island without native herbivores, a density of 5–7/km^2 is considered compatible with regeneration of the (modified) native forest; some regeneration may occur, especially with 'habitat manipulation', under densities up to 15/km^2, but over 20/km^2 is definitely unacceptable (Challies 1974; M. J. W. Douglas *in litt.*).

In south-east Asian forests, which are adapted to browsers, various species of deer occur at combined densities of 4.7–6.9/km^2 in semi-deciduous forest equivalent to the Mauritian lowlands, but only 1.4/km^2 in moist evergreen forest; the extreme upper limit under unusually fertile natural conditions is 21.4/km^2. In its native Java *Cervus timorensis* reaches only 0.9/km^2 (all Asian data from Eisenberg & Seidensticker 1976). Despite this evidence, there was general agreement in Mauritius until recently that deer were not harmful to native forests (e.g. Owadally & Bützler 1972, Procter & Salm 1975), although there was acknowledgement of some damage to plantations (Edgerley 1961, Owadally 1971a, anon. 1974a, Owadally 1980) and, now also, to native saplings (Owadally 1980). Long forgotten were the observations of Thompson (1880) who commented that "unfortunately the deer seem to prefer eating such species as *Makak*, *Natte* and *Ebony*" (all native species) and complained that "whole acres of young transplants of certain [native] species of which they are fond are found cut back". In 1882 the Woods and Forests Board considered destroying deer completely in regenerating forests because damage to young trees was so bad (Brouard 1963). King (1946) reported severe damage by deer to native seedlings transplanted out in plantations, though he stated that "in natural forest attack is inconspicuous". It is precisely this inconspicuousness of the damage that led to the view that deer are not injurious: when Koenig (1895) wrote that

"the rangers are unanimous in saying that the damage done by deer in our forests is practically nil" he was not quoting the result of experiments!

Although deer are only one of several factors involved in forest degradation, they are clearly one of the most serious; nevertheless, it was recently suggested that deer density in upland forests be increased to 95–120/km^2 for meat production (Noël 1974). However, a draft work-plan for the proposed (but now shelved) Black River Gorges National Park did suggest a reduction of deer within the park to 1/km^2 (M. J. W. Douglas *in litt.*) which, if implemented, would be an important step forward. Some hard evidence on deer impact is still needed, such as the exclosure experiment which Dr W. Bützler has offered to pay for (*in litt.*, 1976, to me and the Conservator of Forests; Cheke 1978a). The need to limit herbivore densities in forest to a level in equilibrium with the native vegetation is now recognised (Owadally 1980, Jones & Owadally 1982). An earlier exclosure study, set up by Vaughan & Wiehé (1941), was unfortunately not followed through. Jones (1980a) reported increasing deer density in the Black River Gorges area and Pasquier (1980) considered the deer a major threat to forest stability.

Pigs and monkeys are also very important in the degradation of Mauritian forests, both as predators and dispersal agents of invasive exotics (Thompson 1880, Gleadow 1904, Vaughan & Wiehé 1941, Owadally 1973, 1980). There is no information on pig density in Mauritian forests, but it must be very much higher (pers. obs.) than the 0.3–1.1/km^2 found in the natural state in south-east Asia (Eisenberg & Seidensticker 1976). The monkey's ecology has been studied only in exotic secondary forest (Sussmann & Tattersall 1980).

In Rodrigues any regeneration of native forest at low elevations has long been prevented by the large numbers of goats and cattle pastured in the 'cattle-walk' (North-Coombes 1971, Cheke 1974a, 1978b, Chapter 8). The habitat has been so altered, from forest to open woodland with no understorey (where any trees remain at all), that of the three surviving native vertebrates, only the bat makes even irregular use of it (Cheke 1974a, 1979a, Cheke & Dahl 1981, Carroll 1982b). The forests on higher ground, which had been free for a time from herbivores, were opened to grazing during the drought of the 1970s with unfortunate results (Cheke 1979a). Natural vegetation severely damaged by goats can recover rapidly if goat numbers are greatly reduced (e.g. Pinta Is., Galapagos: Hamann 1979).

Alien plants and habitat invasion

In New Zealand, forests altered by herbivores pass through three principal phases (Howard 1965): (1) herbivores increase rapidly to a level of overpopulation (2) which increases browsing and eliminates all ground vegetation, promoting erosion; starvation follows and the population falls, eventually recovering (3) to around the carrying capacity of an altered forest "still composed of native species but more stable in the face of animal browsing and grazing". Selective browsing eliminates the more palatable species and the carrying capacity of the new 'climax' is consequently far below that of the original forest (*ibid.*).

In the Mascarenes this pattern seems to have been followed at first, modified by hunting pressure, but the progression towards a 'browse-climax' was soon rendered unstable by invading exotic species of plant. Invasion of native forests by alien plants is therefore inextricably linked with the spread of introduced herbivores; to understand the spread of exotic plants it is necessary first to examine the history of the herbivores in more detail. This is possible for Mauritius (Fig. 1), but the data for the other islands are inadequate. The evidence suggests a massive build-up of all species to a peak about 1690–1710, contributing to the Dutch decision to leave the island (Bernardin 1773, Grant 1801, Pitot 1905; this interpretation is not, however, taken seriously by recent historians, e.g. Barnwell & Toussaint 1949, Toussaint 1972). During the last years of Dutch occupation there were devastating epidemics affecting ungulates (in 1697 and 1704: Pitot 1905), suggesting overpopulation and low resistance. Numbers then crashed, aided by a severe drought in 1706 (Pitot 1905, Barnwell 1948), apparently fluctuating (La Merveille reported large numbers of some species in 1709: La Roque 1716), before reaching a new low. By the time the French started settling 10 years later ungulate populations may have stabilised somewhat, though goats were numerous enough to be troublesome at Port Louis (Nyon & Hauville 1722–23). Rats and monkeys were still at very high levels in some parts of the island, though not everywhere; rats were devastating at Grand Port, but monkeys abounded at Port Louis (*ibid.*), which explains the contradictions between the accounts of Ducros (1725, Barnwell 1948) and Jonchée and St.-Martin (Brouard 1963).

The third phase may then have begun, though the longevity of existing canopy trees would cause the final climax to take centuries to stabilise. However, the French brought in many kinds of exotic plants, some of which, notably the Strawberry or 'Chinese' Guava *Psidium cattleyanum* (introduced *c*. 1750: Brouard 1963), adapted well to the island. The guava is distributed by monkeys, pigs and deer (Clark 1859, Vaughan & Wiehé 1937, Owadally & Bützler 1972, Owadally 1980) and has penetrated all the upland forests; in many places it forms thickets so dense that no regeneration of other species occurs (Owadally 1973). On Réunion, with only man and birds to spread it, guava is much more restricted, and is aggressive only in man-modified forest and on fresh lava flows (Cadet 1977, Lavergne 1978). Rodrigues has neither mammals nor suitable birds to distribute it, and the guava is widespread but only weakly invasive (Wiehé 1949, Cadet 1975, pers. obs.).

At lower elevations in Mauritius the wind-dispersed creeper *Hiptage benghalensis* and the Mauritius Hemp *Furcraea gigantea* are as troublesome as guava is higher up. Other important invasive exotics are the bird-dispersed *Litsea* spp., the privet *Ligustrum robustum* var. *walkeri*, the bramble *Rubus mollucanus* and the undershrubs *Ardisia crenata* and *Wikstroemia indica*. Many other trees which become dominant on cleared part-felled forest (e.g. *Ravenala madagascariensis*) have been introduced but most of these do not invade undamaged native forest (Vaughan & Wiehé 1937, Brouard 1963, Vaughan 1968, Owadally 1973, Procter & Salm 1975).

The afflictions of the native forests are synergistic. Invasive plants need dispersal agents and reduced competition by selective browsing to help them spread. In turn it seems likely that the extraordinarily high fruit production of the Strawberry Guava has helped sustain high populations of pigs (Owadally 1980) and monkeys, forcing them to continue damaging the native flora outside the guava's fruiting season. I have seen areas in native forest where pigs have grubbed up every seedling over thousands of square metres. Some indication of the way the introduced species interact can be seen by comparing the different islands. The introduced Rose-Apple *Eugenia* (= *Syzygium*) *jambos* is common on all three islands, but due to an introduced fruit fly *Ceratitis rosa* (Etienne 1972) rarely sets good fruit on Mauritius or Réunion (Balfour 1879, Lavergne 1978, pers. obs.). On Rodrigues where fruit set is prolific, it is spread by the

Fig. 1. Evolution of population density of monkeys, rats and feral ungulates in Mauritius over the first 200 years after introduction.

Abundance categories: 1, present, low density; 2, common, density in line with long-term carrying capacity; 3, abundant enough for the animal to be a pest or easy hunting; 4, over-abundant, severe damage to crops etc.; 5, plague numbers, extreme damage.

Symbols: D, drought; E, epidemic; N, *Rattus norvegicus*; ▷, introduction.

Notes:

1. This diagram is somewhat speculative as it is derived from the subjective remarks of early visitors and residents of the island; however I believe it correctly interprets their perceptions. See text for references.

2. The unshaded parts for 1720–30 for monkeys, rats and goats indicate differences in density at Grand Port (G) and Port Louis (P).

3. The persistent high density of rats throughout most of the eighteenth century was due to the introduction of a second species, *Rattus norvegicus*, probably in the 1730s.

4. Ungulates were badly affected by severe drought in 1706. Deer and cattle were decimated by epidemics in 1697 and 1704; goats were probably also affected but

this was not reported. There was also a deer epidemic in 1880 that nearly eliminated herds in some areas (Antelme 1914); this also followed a 20-year period of high population density (*ibid.*).

5. The increase in deer to current densities in the mid-nineteenth century was due to deliberate management for hunting (see also note 6).

6. The increase in monkey and pig density and subsequent agricultural damage between 1810 and 1859 may have been due to a concentration of these animals into too small an area of forest by the massive deforestation of that period (see text and Fig. 2).

7. I do not know why ungulates reached peak densities more quickly than monkeys and rats. Goats on Pinta Is., Galapagos, increased from 3 to over 20 000 in 12 years (1959–71; Hamann 1979).

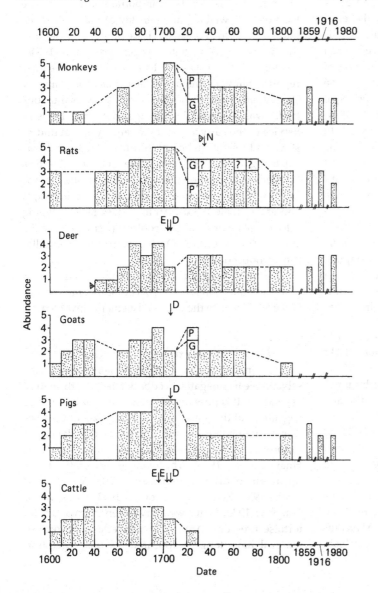

endemic flying-fox *Pteropus rodricensis* (Cheke 1974a, Cheke & Dahl 1981, Carroll 1982b). In Mauritius Vaughan & Wiehé (1937) noted that, formerly invasive, it was retreating before the Strawberry Guava. In Réunion it is, like the guava, largely confined to man-altered areas (Cadet 1977, Lavergne 1978), perhaps dispersed by rats. Another species helped by endemic fruit-bats is the Travellers' Palm *Ravenala madagascariensis*, which is very widespread in Mauritius (Vaughan & Wiehé 1937) and spreading in Rodrigues (pers. obs.). In Réunion, where the flying-foxes are now extinct and were already rare 200 years ago, *Ravenala* is not invasive (pers. obs.). Native as well as introduced birds are responsible for spreading *Litsea* spp., *Lantana camara*, *Rubus mollucanus*, privet etc. (Chapters 4 and 5). A disastrous example of the introduction of a dispersal agent for a plant already present but quiescent is provided by the explosive spread of the weed *Cordia interrupta* in Mauritius following the introduction of the Red-whiskered Bulbul *Pycnonotus jocosus* (Wiehé 1946). This bulbul is also considered to be the principal dispersal agent of the privet (Vaughan & Wiehé 1939) and was blamed by d'Emmerez (1914) and Carié (1910) for the disappearance of the large spiders of the genus *Neophilia* (which are still common in Réunion and Rodrigues, where the bird is absent: pers. obs.), and for dramatic reductions in populations of native White-eyes *Zosterops* spp. Introduced birds do not in general seem to have had a serious direct effect, although some niche restriction has occurred (*Foudia* spp., *Hypsipetes*/*Pycnonotus*, *Psittacula* spp.: Chapters 4, 5 and 8). In New Zealand success of exotic birds in forest follows habitat modification by browsers and is not due to direct competition with native birds (Diamond & Veitch 1981).

Many of these alien plants are so perniciously successful only because their new environment lacks their natural diseases and predators (Lavergne 1978). *Cordia interrupta* was eventually controlled by introducing the predatory insects *Schematiza cordiae* and *Eurytoma attiva* (Greathead 1971), as the Prickly Pear *Opuntia tuna* had been much earlier (c. 1797: de Guignes 1808) after threatening to overrun the drier parts of the island in the mid- to late 1700s (Brouard 1963). Prickly Pears were controlled by a cochineal *Dactylopius* sp.; later eruptions were also controlled by cochineal species and *Cactoblastis* sp. (Greathead 1971). When I suggested unofficially to Mauritius Forestry Service staff that the Commonwealth Insti-

tute of Biological Control might be able to provide a control insect for the Strawberry Guava, the answer was that this might be so, but what about the economic importance of the plant? As Clark (1859) wrote, referring to pigs and guava: "Their fondness for this fruit has been of great advantage to the colony. Many thousands of this most useful tree have been planted by the wild pigs, nor are they found in any part of the forests except where they have been so planted. This tree supplies wood of unrivalled excellence for the shafts and poles of carriages." The wood is nowadays coppiced for charcoal as well as being used widely for implement handles, joinery, poles etc. (Sale 1935a, pers. obs.), and the collection of the fruit (rich in vitamin C) in season is an important addition to the local diet, as well as the mainstay of the 'fruit-wine' industry. Yet the Forestry Service now recognises it as the 'worst weed' in native vegetation (anon. 1974a, Owadally 1980; it was not listed as a weed in earlier reports). Similarly the privet *Ligustrum robustum*, now classed amongst the 'worst weeds' in plantations (*ibid.*) was deliberately propagated by the Forestry Service in the early years of the century as "an understory and useful fuel crop" (Brouard 1963), and was still being used in 1935 as a "nurse for conifers" and as "a check against the spread of *Rubus roridus* [= *mollucanus*], *Lantana aculeata* [= *camara*], *Cordia interrupta* and grass" (Sale 1935a). An important factor keeping *Rubus mollucanus* under control (Réunion: Cheke 1975f, 1976; Mauritius: Owadally 1980) is, ironically, the introduced deer.

This account of the history of the effects of alien plants in Mauritian forests is of course incomplete, but serves to illustrate the principal elements involved.

Effect of faunal extinctions on vegetation

Extinctions of the indigenous fauna may themselves have had negative feedback effects on the native vegetation. It is possible, for instance, that the coastal vegetation of the Mascarenes was a graze- or browse-climax dominated by the enormous population of tortoises, as it is in parts of Aldabra (Merton *et al.* 1976, Hnatiuk *et al.* 1976). The open aspect of the palm savannah in drier areas (Leguat 1708, Vaughan & Wiehé 1937, Rivals 1952, Brouard 1963, Houssaye *in* Lougnon 1970) could well have been maintained by tortoise pressure. Any observable effects of tortoise removal have long disappeared, as this vegetation zone has been completely obliterated on all the islands by the actions of man, goats and fire.

A subtler but equally important effect of extinctions could have been the removal of pollinating, dispersal or germination-facilitating agents from the ecosystem. Of the nectarivorous birds likely to be pollinators (*Zosterops* spp., *Foudia* spp., *Hypsipetes* spp.: Chapters 5, 6 and 8) only the Réunion Fody and the Rodrigues Bulbul are extinct, although the numbers of the Mauritian species (except *Z. borbonicus*) are low enough to be a danger to any plant dependent on them, and the Fody *F. rubra* and the Olive White-eye *Z. chloronothos* are quite absent from some extensive areas of native forest (Chapter 4). In Rodrigues the tiny surviving areas of native vegetation (Cadet 1975, Cheke 1974a, 1978b) are outside the present range of the Fody *F. flavicans*. Some Mascarene trees, notably *Tambourissa* spp. and several of the *Eugenias*, have flowers and fruit arising directly out of the trunk near the ground (Baker 1877, pers. obs.); this is characteristic of plants pollinated and dispersed by reptiles (Pijl 1972), in this case presumably the extinct large lizards *Leiolopisma* spp. and tortoises. Insects and geckos seem now to effect sufficient pollination, as these plants do set fruit (pers. obs., and discussion with various botanists).

One cannot be so optimistic about dispersal. Many Mascarene trees produce fleshy fruits (Baker 1877) clearly adapted for animal consumption. In Réunion all native frugivores are extinct except the small bulbul *Hypsipetes borbonicus* which can tackle only small fruit up to about 1 cm across (Cadet 1977, Chapter 6); no frugivorous aliens have been introduced apart from the Mynah *Acridotheres tristis*, which is little larger than the bulbul and absent from native forests (Barré 1983, Chapter 6). Cadet (1977: 78) has blamed the loss of frugivores for the failure of many native species, though fruiting well nearby, to colonise recent (less than 100-year-old) lava flows where they would formerly have formed part of the succession. Mauritius and Rodrigues retain only their flying-foxes *Pteropus* spp. whose arboreal foraging methods prevent them from tackling many kinds of fruit, though in Mauritius monkeys and pigs may offset the losses but probably destroy more native seeds than they disperse. Other animals important in dispersal would have been the fruit-pigeons *Alectroenas* spp., parrots (Psittacidae), dodos (Raphidae), skinks *Leiolopisma* spp. and tortoises *Geochelonia* spp. In contrast to fruit-

pigeons, Columbine pigeons, even if frugivorous like the Mauritian *Nesoenas mayeri*, normally destroy seeds in their gizzards (Goodwin 1983). Parrots are normally seed predators, but may carry fruits and drop seeds as well as crush them, and can swallow and pass small seeds (e.g. *Coracopsis nigra* in Seychelles: Evans 1979). Cossigny (1732–55) found intact seeds of "tacamacha" (*Calophyllum*) and "bois de natte à petite feuilles" (? *Labourdonnaisia revoluta*) in the oesophagus and gizzard of *Alectroenas nitidissima* in 1755. It may be relevant that Malayan species of *Calophyllum* and *Canarium* need to have the flesh of the fruit removed by animals before they will germinate (Ng 1983). Tortoises on Aldabra pass many species of seeds intact through their gut, including several tree species (Hnatiuk 1978). On Rodrigues tortoises used to eat *Latania* palm fruits (Pingré 1763), but it is not clear whether they digested or passed the seeds. On Aldabra and in the Galapagos some seeds germinate readily only if they have passed through the gut of a tortoise (Hnatiuk 1978). Any Mascarene fruit formerly eaten by tortoises but which is unattractive to pigs or monkeys would now be deprived of a means of dispersal, and possibly also of germination potential. The same may also apply to fruits formerly favoured by *Alectroenas* spp. and the dodo family: for instance the lowland *Pandanus* thought by Pitot (1914) to be the Dodo's principal food (though no doubt also taken by tortoises).

Vaughan & Wiehé (1941) suggested that the germination and distribution of the remarkable woody seeds of the endemic *tambalacoque* tree *Sideroxylon grandiflorum* (= *Calvaria major*) were probably assisted by their passage through the alimentary canal of the Dodo, and Temple (1977b) further claimed that the extinction of the Dodo was responsible for the failure of this tree to reproduce. Temple argued that the extremely hard seeds of this tree "fail to germinate because the thick endocarp mechanically resists the expansion of the embryo within", and invoked the abrasive power of a Dodo's gut to weaken the nut's shell and facilitate germination. He successfully germinated three seeds out of ten passed or regurgitated by force-fed turkeys *Meleagris gallopavo*, and claimed thereby to have produced the first seedlings of this tree for 300 years.

Although this attractive story rapidly became folklore in the semi-popular press (e.g. *Sunday Times* 16.10.1977; *Oryx* 14: 292–3, 1978), it contains several flaws which, while not disproving mutualism, put

Temple's claims in perspective (Cheke *et al.* 1984). First, while it is true that *tambalacoque* and some other species (e.g. the *colophane (Canarium paniculatum)*, two more *Sideroxylon* spp. and *Antidesma madagascariensis*) do not at present regenerate naturally (Thompson 1880, Koenig 1895, Vaughan & Wiehé 1941, King 1946, Coode 1979, Friedmann 1981), this may be due to problems of seed production. Koenig (1914*a*) included the *tambalacoque* amongst native trees producing few or no fertile seeds, King (1946) and Friedmann (1981) blaming predation by monkeys (and bats) for the absence of seeds of this tree as well as *Canarium* and *Antidesma* (see also Vaughan & Wiehé 1941). King, like Temple (Staub 1977*a*), was able to obtain seed only from trees in private gardens in Curepipe. Owadally (1977) also blamed cyclones for the poor seed set, but this is one (Friedmann 1981) of many Mascarene trees that are stimulated to flower abundantly by cyclones (Vaughan & Wiehé 1937, Rivals 1952), and a rare good harvest from forest trees followed the 1975 cyclone Gervaise (T. A. M. Gardner pers. comm.). Flowers or fruit on the tree at the time of the storm are lost.

Secondly, *tambalacoque* seeds do not need the aid of a bird's gut to germinate. There is no evidence, notwithstanding Temple's speculations (1984), that King (1946) or Hill (1941) had special problems with germination or that they prepared their seeds in any way. Hill germinated a seed in England in 1941, and King reported established seedlings in Mauritius in 1940; Temple's claim to the distinction of first germination must give way to King's. King (1946) quoted a germination time of "3–6 months", confirmed more recently (4–6 months) by Gardner (*in* Staub 1977*a*, Horn 1978). The Forestry Service has recently been propagating *tambalacoques* again (anon. 1972, Staub 1977*a*, Owadally 1972, 1977, 1979). Although Vaughan & Wiehé (1941) reported the germination capacity of *Calvaria* to be similar to that of *Canarium* (about 20%; "germinates well": King 1946), others give much lower figures ("very low": King 1946; 2.5% in two samples: Owadally 1977). Against this Temple's three in ten (1977*b*, 1984) at first seem a considerable improvement. However, a further seven seeds were digested by his turkey, and the Dodo had a larger and more powerful gizzard containing a large grinding stone (Hachisuka 1953) which would surely have led to the digestion of such seeds rather than their dispersal! Owadally (1977, 1979) considered his 2.5% germination rate sufficient to maintain the tree at its naturally

low stocking rate (2%: Brouard 1963, Vaughan & Wiehé 1941) in the aboriginal Mauritian forests, and that this rate was equivalent to other local forest trees (though this is true of only some species: King 1946). Under present conditions any rare seed germinating would be subject to the adverse effects of competitive alien plants and pigs (Owadally 1977, 1979, Staub 1977*a*), also deer, which ate seedlings recently planted in Macabé (W. Strahm pers. comm.). Temple (1979) also claimed, citing T. A. M. Gardner as his source, that the Forestry Service artificially abraded the seeds before planting to aid germination, strengthening the case for mutualism. None of the germinated seeds Gardner showed me in 1978 showed any signs of abrasion; the practice (suggested by Temple) had been abandoned when it was found that it made no difference to germination. *Tambalacoque* seeds have a natural zone of weakness in the woody seed coat (Hill 1941) where they always split on germination (T. A. M. Gardner pers. comm.; pers. obs. Friedman 1981: pl. 5).

Thirdly Owadally (1977, 1979) considered that the small trees (around 10 cm diameter at breast height) in Vaughan & Wiehé's (1941) sample plots were 75–100 years old (30–50 years according to Friedmann 1981), i.e. germinated long after the Dodo died out; Temple (1979) preferred not to believe him. While Vaughan & Wiehé may have underestimated the age of their trees a little (Thompson (1880) was already noting an absence of seedlings of *tambalacoque* and *colophane* 60 years earlier), it does seem improbable that such small trees should be 300 years old, as Temple claimed. Finally Temple (1977*b*) made a number of statements in support of his case that had no basis in fact. There are many more than 13 trees surviving (total in the low hundreds; W. Strahm per CGJ 1985), and many are far from old and overmature, though young specimens are easy to confuse with related species (pers. obs., Owadally, Gardner and Strahum pers. comm.). Mauritian scientific folklore (e.g. Koenig 1932) suggests that *tambalacoque* seeds were found in association with Dodo remains (and were its food). Temple (1977*b*) re-stated this association without a supporting reference, though he later (1979) cited a personal communication from F. Friedmann as his source; yet Friedmann was cited by Staub (1977*a*), Owadally (1979) and R. E. Vaughan (*in litt.* 1977) as denying any such presence! The seeds found with Dodo remains were called *Calvaria globosa* and *C.*

hexangularis (Hemsley 1899, Hill 1941). The form-species C. *hexangularis* has been identified as belonging to another nearly extinct tree *Sideroxylon sessiliflorum* (Staub 1977a, Owadally 1979, Friedmann 1981, Vaughan 1984). It is a moot point as to whether the *tambalacoque* itself occurred at low elevations; the general view of botanists is that it did not (Vaughan & Wiehé 1937, Vaughan *in litt.*, Owadally 1977, Friedmann 1981), though Temple (1979) claimed that an extant tree stands at only 150 m altitude, and cited Brouard (1963) as referring to its former occurrence in the lowlands. Temple did not give the location of his tree, and Brouard (1963) mentioned *Calvaria major* three times, once taxonomically and twice in connection with upland forests. Likewise it has been claimed that the Dodo rarely visited the plateau forests (Pitot 1914, Staub 1977a, Owadally 1977, 1979), but this is inference from the early reports which do not give adequate details (Strickland 1848, Pitot 1905, Hachisuka 1953).

Clearly it requires more than a turkey and a few seeds to prove "obligatory mutualism" (Temple 1977b) between an extinct bird and a rare tree. The dynamics of Mauritian vegetation do deserve proper study, and indeed the forests will certainly not survive unless this is done soon and the results acted upon. Vaughan's work over many years at the tiny (1.5 ha) Perrier nature reserve, mentioned by Vaughan (1968) and Owadally (1973), demonstrates that the native vegetation can recover if exotic plants are removed and herbivores fenced out. Life-cycle studies of many species will be needed to discover whether disappearance of frugivores is a significant factor in addition to the other pressures to which the forest is subject. Future research must be devoted to conservation rather than to creating scientific folklore of a kind that, like Strawberry Guava, is extraordinarily hard to root out once established (Gould 1980: 288). If any animal was necessary to the *tambalacoque*, was it not the tortoise?

Hunting

Early visitors to all three islands found the birds tame and easy to kill, and had no trouble rounding up the numerous tortoises; the accounts show that human predation was unrestrained. However, it is very difficult to say what effect this had on the hunted species. On both Mauritius and Réunion *merles* (native bulbuls *Hypsipetes* spp.) parrots, pigeons, rails, ducks,

geese, dodos and flamingos are listed among the favourite prey of hunters; the last five became extinct within a few decades of settlement, most of the pigeons and parrots soon after. As the numbers of men were initially small (only a few hundred in Mauritius and Réunion in the 1600s (Toussaint 1972), and the forests thick and difficult to penetrate, it seems unlikely that man would have been responsible for the total extinction of any animals which were widespread and could reproduce successfully. Those limited to certain habitats or traditional sites (ducks, geese, flamingos and colonial tree-nesting seabirds) may well have been eliminated by hunting, especially in Réunion where there are very few wetlands, but in general reduced breeding output due to introduced animals (see above) is a more probable ultimate reason. Rodrigues is a much smaller island and easier to cover, but even there most of the birds died out (see below) when only a handful of men lived on the island (1725–65: North-Coombes 1971).

The tortoises, which had a special value as a tonic food for sick mariners ashore, and an ability to stay alive for months without sustenance on ships and so provide fresh meat (see e.g. North-Coombes 1971) are a special case. They almost certainly were hunted to near destruction by man, despite attempts at conservation (Pitot 1905, Lougnon 1957). However there is evidence that they too were failing to breed (see above) and so again hunting was not necessarily the ultimate factor. Some indication of how difficult it would have been for the early settlers to have wiped out any successfully breeding species can be seen in the attempts by the Dutch to eliminate wild pigs. The pigs were eating the eggs of both tortoises and turtle, and thought also to devour any ambergris (a very valuable commodity) that was washed up, so attempts were made in 1673 (Pitot 1905) and 1709 (La Roque 1716) to exterminate them – with no success at all. Dubois (1674) despaired of destroying the countless *moineaux* (*Foudia* sp.) which raided the crops in Réunion; within a few years they had vanished, apparently wiped out by rats (see species account). No attempt at destruction on any island made any impression on rat numbers, nor in Mauritius on monkeys.

After the initial massacres of native animals the hunters, their population dramatically increasing in the early 1700s (Toussaint 1972), turned to introduced game, and succeeded over time in eliminating wild cattle from all the islands, wild pigs from Réunion, wild goats from Mauritius and both from Rodrigues (see below). Deer introduced to Réunion were wiped out (Bory 1804), re-introduced, eliminated again, and again re-introduced (see below). Between about 1760 and 1840 there was a second wave of extinctions on Mauritius and Réunion, largely of edible, popularly hunted species. La Nux (1772) blamed hunting for the decline in Réunion of *Pteropus subniger*; this flying-fox roosted gregariously in hollow trees in which it was easily trapped (Cheke & Dahl 1981). La Nux was also concerned about the commoner *P. niger*, which in the event disappeared first (*ibid.*; see below). This species roosted gregariously in trees in the open and could easily be shot or netted out by determined hunters. It seems that this did happen in Réunion, although such tactics are considered unsporting in Mauritius (pers. obs.) where the species still survives, and do not seem to be practised in the Seychelles, according to recent studies of bat-hunting there (Maisels 1979, Racey 1979).

In 1755 Cossigny (1732–55) blamed predation by *marrons* (fugitive slaves), as well as deforestation, for the rarefaction over the previous 23 years of the pigeon *Alectroenas nitidissima*. *Marrons* were numerous in the forests in the mid-eighteenth century, despite being hunted like game by their former masters, and while they inhibited forest penetration by settlers they had themselves to live off the wildlife (Pingré 1763, Bernardin 1773, Grant 1801, Bulpin 1958). Cossigny (1764) considered green and grey parrots still common in the forest, suggesting they were no longer hunted as they had once been – yet the grey, '*Lophopsittacus*' *bensoni*, was not reported again. In Réunion similar parrots and the larger *Mascarinus mascarinus* disappeared over the same time period. These animals had all co-existed with rats and other predators for 100 years or more, and were acknowledged game. However, the Mauritian owl '*Scops*' *commersoni*, hardly a game animal, also disappeared during this period, and even the tasty *Alectroenas* survived to the end of it (1830s), so hunting was again probably only a contributory factor. Although we cannot now know what the other factors were, deforestation of the lowlands may well have been more significant (see next section). Over the same period Rodrigues lost the Solitaire and Abbott's Booby, in both cases probably partly due to hunting, though cats (Cossigny 1732–55) and fires (North-Coombes 1971) are also blamed for the loss of the Solitaire.

When a population becomes very small and restricted in distribution it is very vulnerable to persecution. In 1973 only seven Mauritius Kestrel *Falco punctatus* survived (Jones, Chapter 5), and when a pair of these were shot (Temple 1977a) the population was at once reduced by 29%. Shooting for food is now a major threat to the few remaining Rodrigues flying-foxes *Pteropus rodricensis* (Cheke 1974a, 1979a, Cheke & Dahl 1981, Carroll 1982b), although prior to massive deforestation in the period 1955–68 the population, then much larger, could sustain the hunting pressure (Cheke & Dahl 1981). The two extant skins of the Rodrigues Parakeet *Psittacula exsul* were shot from amongst the last few surviving birds (A. & E. Newton 1876, Hachisuka 1953), but the rarefaction was probably due to other causes (see below).

The marine animals I have considered – Dugongs *Dugong dugong* and turtles *Chelonia mydas* and *Eretmochelys imbricata* – were subject from the start to intensive human predation, but persisted nonetheless well past the mid-eighteenth century in Mauritius and to the turn of the century in Rodrigues. Although pigs may have been serious predators on turtles nesting on the main islands (see above), the offshore islets of Mauritius and Rodrigues would have provided safe nesting until human population pressure drove fishermen to these refuges. I believe hunting was the reason for the disappearance of turtles and Dugongs, but that their long resistance is further evidence that it required not only intensive hunting by individuals but a large number of hunters to eliminate a thriving population.

I conclude that hunting was not the ultimate factor in the extinction of any Mascarene land vertebrate except *Pteropus niger* in Réunion, although it was certainly a contributory element in many of the bird species, and undoubtedly the proximate factor for the tortoises. Turtles and Dugongs were probably direct victims of over-hunting.

Destruction of habitat
Mauritius
The seventeenth-century Dutch settlers cut a lot of trees, but were principally interested in marketable wood (especially Ebony *Diospyros tesselaria*), so their fellings were selective and confined to areas from which they could easily extract the timber (Vaughan & Wiehé 1937, Brouard 1963). They made no large clearances and it is doubtful whether they had any major direct impact, except very locally, on the vegetation

structure. They imported most of their food (apart from local meat) and practised little agriculture (Pitot 1905, Barnwell 1948, Barnwell & Toussaint 1949). The island's habitats thus remained more or less intact until the French granted large-scale concessions for agriculture in the 1730s, which resulted in widespread clearances by slash and burn in the lowlands (Crépin 1922, Brouard 1963, Toussaint 1972). Much timber was also extracted over large areas which remained wooded though degraded (Brouard 1963). By 1766 45% of the island had been conceded, and a quarter of this (i.e. *c*. 11% of the whole) was being cultivated, the remainder "still uncultivated and mostly wooded" (*ibid.*; Bernardin (1773) gave a higher figure (a quarter) for land under cultivation in 1768, but it appears too large in relation to other evidence).

The central, southern and eastern parts of the island remained fairly intact, but the north, the west southwards to the Corps de Garde, and the uplands behind Port Louis (Moka) were about half cleared with most of the timber gone in the wooded remainder (Brouard 1963; see also Bellin's map of 1763 *in* Staub 1976). Thus the palm savannah of the dry north-western coastal belt (Vaughan & Wiehé 1937) had probably already gone, and much of the drier lowland forest was badly degraded. Dry season fires were widespread in cleared areas and degraded open woodland (Bernardin 1773, Brouard 1963). No native vertebrate dependent on lowland forest could have survived this period. We do not know the habitat of the parrot '*Lophopsittacus*' *bensoni* which died out at this time, but it may well have been drier forests, as the bird was clearly common in coastal areas according to the early Dutch accounts (Strickland 1848, Pitot 1905, etc.).

Some attempt was made at forest conservation when the island was ceded by the Compagnie des Indes to the French Crown in 1767, but Poivre's admirable regulations were not seriously enforced (Brouard 1963) and forest destruction actually increased during the last 30 years of the century, reflecting the steadily increasing population pressure (Fig. 2). There followed a slight respite during the economic stagnation of the Napoleonic war and the early period of British rule, so that by the mid-1830s some 65% of the island (Gleadow 1904, Vaughan & Wiehé 1937), essentially the uplands over 200 m altitude with the exception of the Moka area, was still under forest (compare Fraser's 1835 map (*ibid.*) with the 200 m contour in Venkatasamy 1971). Around this time a number of events

occurred which proved disastrous for the forests. In 1825 a discriminatory tax on sugar on the London market was removed (Toussaint 1966, 1972) resulting in a rapid increase in the area under cane. At first this was at the expense of other crops (Toussaint 1966, 1972, Roy 1960), but by the mid-1830s the spare agricultural land had been used up and forest clearance began in earnest. At the same time there was increasing demand for wood to fuel the new steam-powered sugar mills (Toussaint 1972), and a road was opened in 1830–2 from Port Louis to Souillac and Mahébourg (Barnwell & Toussaint 1949, Toussaint 1966), giving access to previously inaccessible forests.

In 1835 slavery was abolished, and after 4 years' 'apprenticeship' the slaves were fully freed in 1839, leaving their former masters and setting up as shifting cultivators on unclaimed land (Barnwell & Toussaint 1949). Enormous numbers of Indians were then imported to replace the slaves; the population doubled between 1830 and 1850, and tripled by 1860 (Fig. 2), hugely increasing the demand for wood-fuel (Mann 1860). Between 1835 and 1846 more than half the forest on the island was felled (Fig. 2), and it may be no coincidence that the extinction of the Pigeon Hollandais *Alectroenas nitidissima* and the owl '*Scops*' *commersoni* occurred during this period. The main clearances for sugar stopped after about 1858 (Brouard 1963), but the assaults on the forests were kept up by the opening in 1864 and 1865 of railways powered by wood-burning steam engines (Koenig 1914*a*). The malaria epidemics of 1865–8 also sent much of the population of the lowlands fleeing to build on higher ground (Toussaint 1966, 1972), and resulted in an extension of sugar at the expense of forests in the uplands (Walter 1914*b*).

The period 1864–75 was particularly severe, with a peak period of devastation in 1871–2 (Koenig 1914*a*; also 1872 map in Gleadow 1904). The flying-fox *Pteropus subniger* disappeared during this phase (Cheke & Dahl 1981; see also below). Only after 1880 was some restraint introduced, but by this time the untouched forests were down to some 16 000 acres (6500 ha), only 3.6% of the island's area, although a good deal more land was still covered in degraded forest suitable for at least some of the endemic birds (Fig. 2). Since that time the area under native forest has continued to decline.

The last good tall forests on the island, Kanaka–Grand Bassin, evocatively described in almost poetic terms by Thompson (1880: 34–5), were felled over the period 1895–1924 (Koenig 1895, 1914a, 1926, Edgerley 1962), as was the last tall timber in Midlands (Koenig 1924) on the site of a reservoir that was never built (pers. obs.). Apart from the reservoir site, felling of native trees was stopped in 1923 (*ibid.*), and all but ceased after 1928 (Sale 1935b). Failure of timber imports during the 1939–45 war resulted in extensive use of native hardwoods again, "mainly from poor relic

forests" due for replacement with pine (King 1947); this timber included 200 trees of *tambalacoque* and *colophane* (*ibid.*), now seriously endangered. The construction of Mare Longue reservoir resulted in a further loss of native forest (Carver 1948).

After 1950 a resurgence of forestry activity resulted in a policy of "replacing useless scrub areas by timber plantations" (Brouard 1963), which on higher ground meant the clearance of degraded forest (long

Fig. 2. Deforestation in Mauritius, compared with the growth of sugar cane cultivation and human population.
Notes:

1. Forest area figures are compiled from many sources (Meldrum 1868, Thompson 1880, Gleadow 1904, Koenig 1914a, Vaughan & Wiehé 1937, Brouard 1963, anon. 1961, 1967, 1974a, Procter & Salm 1975, Pasquier 1980). Information for the period 1840–1880 is conflicting, and I have taken conservative figures (those indicating less destruction) for the total native forest area (including degraded forest). The standstill in deforestation between 1852 and 1872 indicated by Procter & Salm (1975) is an artefact arising from their not checking original sources: Gleadow (1904) included a map showing forest in 1852 and 1872, only about half remaining at the latter date. Thompson's (1880) figure for total forest area in 1880 of 35 000 acres

(14 000ha) does not square with the 1874 survey quoted by Gleadow (1904), nor with the area still under native forest in 1912 (Koenig 1914a); I believe Thompson exaggerated the amount of clearance between 1872 and 1880.

2. Areas under cane taken from Walter (1914b), Lamusse (1958), Ramdin (1969), Venkatasamy (1971), Toussaint (1972) and Mannick (1979). The curve is largely based on 10-year averages, but there appear to have been peaks in 1830–1 and 1858 (Toussaint 1966, Brouard 1963) which indicate that land was being cleared faster than the averages suggest.

3. The human population figures are from Toussaint (1972).

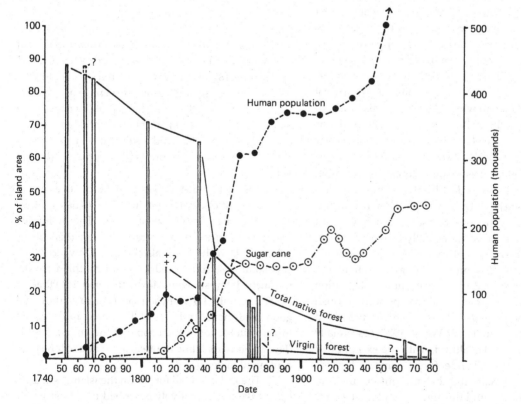

since cut over for timber) which was still acceptable habitat for many endemic birds. Large areas south of Curepipe disappeared to pine (Edgerley 1962), and in the Midlands region also to tea (anon. 1961, Tilbrook 1968, Roy 1969). In the 1970s a job-creation programme (Travail Pour Tous) operated by the Development Works Corporation (anon. 1971c) and funded by the World Bank (Temple 1976b, Procter & Salm 1975) extended and accelerated clearances for exotic plantations into key habitat areas such as the remnants of the Kanaka and Grand Bassin forests and Les Mares. The latter low marshy forest was partly cut over in the 1890s (Koenig 1895), but was still of largely native composition and an important habitat for the Pink Pigeon *Nesoenas mayeri* – the 'Pigeon des Mares', named after this area (Temple 1976b; Chapter 5). Fortunately international interest in the by now precarious position of the surviving endemic birds has helped stabilise the situation (Temple 1976b, Jones 1980). More clearances will surely occur, but for the fauna to survive must stop well before the limit of the 2–2.5% of the island currently in nature reserves (4018 ha/9930 acres, Owadally 1981b). However, no more species have so far become extinct, though the parakeet *Psittacula echo* appears doomed (Jones 1980c, Chapter 5).

Most of the surviving endemic vertebrates have failed to adjust to the secondary habitats on Mauritius. Only the *Phelsuma* geckos (Vinson 1976b), the Grey White-eye *Zosterops borbonicus* (Staub 1976) and the two aerial feeders (Chapter 5) are widespread in or over man-made or secondary woodland, though the flycatcher *Terpsiphone bourbonnensis* has accommodated well in limited areas (Staub 1976, Chapter 4). Several other native passerines, namely the Merle *Hypsipetes olivaceus*, the Cuckoo-shrike *Coracina typica*, the Fody *Foudia rubra* and the Olive White-eye *Z. chloronothos*, would probably adapt to man-modified vegetation if the right mix of evergreen broad-leaved fruit and nectar-bearing trees were provided (Cheke 1978a, Chapter 4), as also no doubt would the hill forest race, *rosagularis*, of the gecko *Phelsuma guimbeaui*. The Pink Pigeon and native parakeet appear to require native vegetation for feeding (Chapter 5), though the Pigeon has catholic feeding habits in captivity and may prove capable of adapting to the right kind of replacement vegetation (Jones 1980a). The endemic kestrel, feeding as it does largely on day-geckos (Chapter 5), would appear capable of living in most areas provided it was not persecuted, but is

nevertheless at present confined to areas of native forest. Temple (1974a, c, 1976b, 1977a) and Newton (1958a, b) assumed that kestrels need native vegetation, but it has long seemed to me more probable (Pasquier 1980) that the catastrophic post-war decline to near extinction is attributable to the use of DDT and BHC during the successful campaign to eliminate human malaria on the island. A very intensive spraying programme started in 1948 had largely eradicated malaria and the main vector *Anopheles funestus* by 1952 (Toussaint 1972, Bruce-Chwatt 1974) but DDT and BHC formulations were used on a large scale until 1965 to control *A. gambiae* and thereafter at a much reduced intensity until the island was declared malaria-free in 1973 (Bruce-Chwatt 1974, Mamet 1979), and DDT was also widely used in agriculture for a time (C. M. Courtois *in litt.*) until restricted by the Pesticide Control Act of 1970 (Ricaud 1975). Since kestrels occurred on the Moka range near Port Louis in the 1950s (Jones 1980a, Chapter 5), I believe their restriction to the unpopulated forests around the Black River Gorges was due to this being the one area of the island not sprayed with DDT. However, even there it seems that fertility must have been reduced sufficiently for predation by monkeys to become a serious threat to survival; high levels of pesticides have been found in recent eggs in captivity (Chapter 5). In the Seychelles *Falco araea*, another small forest-adapted kestrel that eats lizards and birds, remains fairly common (Watson 1984) despite habitat changes, and even nests on buildings (Feare *et al.* 1974). The Seychelles have always been free of malaria and its vector (Lionnet 1972, High 1976), and insecticides have not been widely used.

Réunion

The history of man's destruction of the native vegetation on Réunion roughly parallels that on Mauritius, but was somewhat constrained by the very rugged topography which made access to many parts of the interior difficult. As a result a proportionately much larger area of native forest survives uncut, but nearly all of this is at an altitude of over 500 m, and most over 1000 m (Cadet 1977).

Defos du Rau (1960) mapped and discussed the progressive occupation of the land in Réunion. As forest was not cleared independently of settlement, I have taken his stages of settlement to be equivalent to those of forest destruction.

Unlike Mauritius where the original Dutch

colonisation was expressly to exploit ebony, Réunion was occupied for the purely political reason that King Louis XIV wanted a French presence in the Indian Ocean (Toussaint 1972). The main political object was Madagascar, and Réunion was originally little more than a bolt-hole from there. The embryo colony was persistently forgotten by France, and the few hundred inhabitants had to fend for themselves, often not seeing a visiting ship for 5 years at a time (Toussaint 1972). Hence the settlement pattern was an agricultural one from the start, needing land clearance on a fairly large scale (Defos du Rau 1960).

Settlement began at St.-Paul with a second focus at St.-Denis. By 1715 (*ibid.*) the northern and north-western coastal zone from St.-Gilles round to Bras Panon was occupied, up to 500 m in altitude above St.-Gilles–St.-Paul but mostly under 200 m elsewhere. The dry west coast palm savannah would have been about half-destroyed, together with an important section of the western lowland dry forest and the north-eastern part of the lowland wet forest (interpretation from maps in Cadet 1977).

The next phase, 1715–89, was critical for the wildlife. With the introduction of coffee in 1715 the colony suddenly became potentially important, though it was not until about 1725 that cultivation gave economic returns and the population began to rise dramatically (Toussaint 1972). By 1788 there were 45 800 people, spread round the entire coast except by the volcano in the south-east; the palm savannah and the dry forest ("forêt mégatherme semi-xérophile" of Cadet 1977) and the lowland wet forest (south and east) were all but gone. Bory (1804) described the sorry state of these zones, and noted that wood-fired lime-kilns were a particular feature of the west coast. Gruchet (1977) also stressed the effects of fires in this dry zone and the long period of overgrazing by the numerous feral goats. The centre of the island was populated only by small roving bands of fugitive slaves, fewer than in Mauritius (Pingré 1763, Bory 1804, Bulpin 1958) though enough for La Nux (1772) to consider them a danger to flying-fox populations. The beginning of this period saw the extinction of the Solitaire and *oiseau bleu*, and later, of the three native parrots. The first two, as indicated above, were probably victims of feral cats, but the parrots may have been dependent on the lowland habitats. Evidence is conflicting but the flying-fox *Pteropus niger* and most of the tortoises probably also disappeared during this period (Cheke & Dahl 1981, and below).

The next period, 1789–1848, was largely one of consolidation in already settled areas, though in the latter part of the period the cirques, three huge caldera-like valleys surrounded by cliff walls up to 2000 m high, were colonised for the first time (Defos du Rau 1960). By 1850 Salazie, the first cirque to be settled, was devastated (Scherer 1965). As in Mauritius sugar cane was developed as a crop in the first half of the nineteenth century, but here it appears to have spread until about 1851 on land already cleared for other purposes (Defos du Rau 1960). The limited forest damage is reflected in the lack of extinctions during this period, though the last captive *Mascarinus* parrots, and with them the species, died out (see species account).

In Réunion, as in Mauritius, the emancipation of slaves (1848 here) brought a new wave of forest destruction. By 1852 over 30 000 ex-slaves had established smallholdings in montane forest areas, and a large number of poor whites, whose holdings became uneconomical after they lost their slaves, left the lowlands to settle higher up, usually in much more remote areas than the slaves (gorge bottoms, cirques, the Plaine des Palmistes and part of the Plaine des Cafres: Defos du Rau 1960, Toussaint 1972). These refugees from the coast largely practised shifting cultivation, thus destroying far more forest, and causing more erosion, than their population might suggest (*ibid.*). By 1880 little remained of the original vegetation below 1000 m in the west and 500 m in the east, though forest survived to below 200 m between Le Baril and Bois Blanc around the base of Le Volcan. Although in the mid-1860s the islands' naturalists were still optimistically claiming that *huppes Fregilupus varius*, black parrots *Coracopsis* sp. (introduced) and the last of the flying-foxes *Pteropus subniger* were still to be found in the forests of the cirques (Maillard 1862, Schlegel & Pollen 1868), it appears the naturalists had not visited the cirques themselves. Salazie had been deforested by 1850 and the other two laid waste by 1868 (Defos du Rau 1960), Mafatte in just 6 years (1862–8: Miguet 1973); all three species were probably extinct before 1860 (see below).

Although there was some attempt to limit *déboisement* in 1874, the authorities were unable or unwilling to prevent the next major wave of forest clearance that swept the western parts of the island between 1900 and 1925. This was due to the development of the Scented Geranium *Pelargonium graveolens* as a major cash crop amongst the shifting cultivators in the highlands (Defos du Rau 1960, Lauret 1978*b*). This crop exhausts the soil very fast, and in addition the distillation of the geranium oil requires large amounts of fuel wood (*ibid.*). The rest of the Plaine des Cafres was cleared by 1913, and the culture spread along the west coast massif, the foresters managing to hold the clearances at the 1500 m contour above St.-Leu (*ibid.*), north to Trois Bassins, and rather lower further north (pers. obs.; anon. 1962 and other recent maps). This wave of destruction swept to the Dos d'Ane, then jumped to the Moka area east of St.-Denis, where it stopped around 1924–5 but restarted with a massive rise in the price of geranium oil after 1945 (Defos du Rau 1960, Lauret 1978*b*); Rivals (1952) published photographs of virgin forest freshly cleared for geranium at Moka in the 1940s. Further areas were cut for timber during the 1939–45 war (Miguet 1973).

Since the war a policy of replacing 'uneconomic' native forest by productive plantations (Miguet 1980) has resulted in some of the finest remaining areas of mid-altitude forest being lost (e.g. the Bélouve plateau above Salazie) and also the destruction of the last remnants of mature lowland forest, apart from one ludicrously small (*c.* 20 ha) nature reserve near St.-Philippe in the south-east (Gruchet 1973, Cadet 1977).

The Office National des Forêts has recently cut back on some of its plans for extending the replacement of native forest (ed. note in *Info-Nature* 15: 38 (1977); Cadet 1977) and further nature reserves are planned (Cadet 1977), though implementation is still awaited (Barré & Barau 1982). This has to a large extent removed the immediate threat to the habitat of Réunion's only remaining endangered bird species, the cuckoo-shrike *Coracina newtoni* (Cheke 1976, 1978*a*, Chapter 6).

The depauperate vertebrate fauna that survives on Réunion has adapted rather better to changed conditions than has its counterpart on Mauritius. The Grey White-eye, Paradise Flycatcher and stonechat *Saxicola tectes* are widespread in exotic vegetation (Barré & Barau 1982, Barré 1983, Chapter 6), as is the Harrier *Circus maillardi* where not persecuted (Clouet 1978, Chapter 6). The Merle *Hypsipetes borbonicus* and the Olive White-eye *Zosterops olivaceus* are also commonly seen in secondary forest, but may require native forest intact at higher elevations at certain seasons (Barré 1983, Chapter 6). The cuckoo-shrike is extremely limited in its distribution for reasons as yet not understood (Cheke 1976, Chapter 6). As in Mauritius the *Phelsuma* geckos have adapted to exotic

vegetation, but *P. ornata inexpectata* is restricted to a very small area near Manapany-Les-Bains in the south-west (Vinson & Vinson 1969, Cheke 1978*a*, 1979*a*, Bour & Moutou 1982, Moutou 1983*c*, 1984), "the zone where the dry forest originally reached the coast" (Cadet 1977). Reintroduction of species that still survive on Mauritius (kestrel, Pink Pigeon, Echo Parakeet, flying-fox) has been proposed (Cheke 1974*b*, 1978*a*, Temple 1981, Moutou 1983*d*) and a trial release of Mauritius Fodies was made in 1975 (Cheke 1975*a*, 1979*a*, Chapter 4).

Rodrigues

For some time after its discovery Rodrigues remained well wooded (Leguat 1708, Tafforet 1726, Pingré 1763). Pingré (1763) did, however, report a large area deforested by fire west of Baie aux Huitres and also remarked that he had seen no regenerating *Latanias* or *palmistes* (*Dictyosperma album*, *Mascarena verschaffeltii*), though *Pandanus* seedlings were everywhere. This suggests that the abundant rats were eating the seeds and preventing reproduction, as happens today on Gunner's Quoin off Mauritius (Bullock *et al.* 1983) (pers. obs. of chewed seeds); no doubt the goats ate any that did sprout. As late as the end of the eighteenth century the forests were still "thick and difficult to penetrate" over much of the island (Marragon 1795, Dupon 1969). By this time all but one of the birds which Rodrigues was destined to lose had vanished, the tortoises were reduced to a handful and the big gecko *Phelsuma gigas* confined to offshore islets – all victims of predators combined with hunting, rather than habitat loss.

By 1825 the cumulative effects of feral livestock and shifting cultivation since the beginning of permanent settlement in the early 1790s had reduced the vegetation in much of the island to a savannah with scattered trees (Hoart 1825, North-Coombes 1971). Corby's description (1845; North-Coombes 1971 in part) is of an island largely deforested with pockets of woodland here and there, the only well-wooded part he saw being around what is now La Ferme. Traces of fire were widespread and he encountered large numbers of feral cattle and pigs. It is clear that by then the feral animals and fires had prevented regeneration for a long time. Thorough exploration in 1874 (Balfour 1879*a*, *b*) did reveal forested valleys missed by Corby, but these were a tiny proportion of the area. Balfour wrote that the island was "now . . . a bare parched volcanic pile with deep stream courses for the most part dry, in place of the verdant well-watered island of 200 years ago". He also noted the invasive spread of the exotics *Leucena glauca* and *Eugenia* (= *Syzygium*) *jambos*, apparently introduced since Corby's visit. Around this time the parakeet *Psittacula exsul* and the last mainland geckos *Phelsuma edwardnewtonii* disappeared, the habitat, and with it their numbers, too far reduced to withstand predation and cyclones (see below). The remnants of native vegetation have progressively shrunk (Koenig 1914*b*, Wiehé 1949, Cadet 1975, Cheke 1974*a*, 1978*b*, Tirvengadum 1980, Strahm 1983), so that today only a few hectares survive, and these are mostly heavily overgrazed with no regeneration. Many of the endemic plant species are extinct or gravely threatened (Friedmann & Guého 1977, Tirvengadum 1980, Strahm 1983). The two surviving native birds have adapted to exotic vegetation (Cheke 1974*a*, 1978*a*, 1979*a*, Chapter 8). Plantations of the mixed evergreen species they favour were initiated following Koenig's recommendations in 1914 (North-Coombes 1971), but nearly destroyed again by the prevailing land-use policy during 1955–68 (Cheke & Dahl 1981: 219, Chapter 8).

Disease and pollution

Both introduced micro-organisms and contamination of the environment by man-made chemicals may have had their part to play in the extinction and rarefaction of endemic Mascarene vertebrates. We have, however, no direct evidence of either, though there are certain indications.

The remarkably rapid disappearance of the Réunion Starling *Fregilupus varius*, from abundance to extinction in 10 years (see below), led to the suggestion that a disease may have been responsible (Brasil 1912). The introduction of avian malaria and avian pox and their vectors to the Hawaiian islands has had disastrous effects on endemic birds there (Warner 1968), though Atkinson (1977) has claimed for the Ship Rat *Rattus rattus* a greater responsibility for extinctions. A number of avian parasites have been introduced to the Mascarenes (two spp. of *Leucocytozoon*, *Haemoproteus columbae*, *Plasmodium* spp. incl. *relictum*: Peirce *et al.* 1977, Peirce 1979), but others appear to be native (*Leucocytozoon zosteropis*, a *Plasmodium* and two species of *Trypanosoma* at least: *ibid.*); we know nothing of their pathogenicity. There is some suggestion in the distribution of avian malaria in native birds in Mauritius that birds in little-frequented upland forests are not

infected, unlike those in disturbed forests and in the lowlands (Chapter 4). This distribution corresponds with the habits of the introduced vector *Culex pipiens quinquefasciatus* (= *fatigans*) (Peirce *et al.* 1977). Some native birds which are less widespread than one might expect from their habits (Olive White-eye, Fody) occur where the vector is least abundant; malaria may be a restricting factor in their distribution, though clearly not as dramatically so as in Hawaii (Warner 1968).

The usual gamut of pathogens and endoparasites exists in domestic poultry and pigeons, at least in Réunion (Pourquier 1963); some were no doubt introduced at an early date and may well have infected native species, though helminths are rare in wild birds (Barré 1982).

Mammal diseases are better known because of their implications for man and livestock. However, little work has been done on wild species apart from Moutou's (1980) study on rats and leptospirosis, and a survey of helminth parasites (Barré & Moutou 1982). The only native mammals potentially at risk would have been bats, whose contact with introduced animals must always have been minimal, apart from interactions between rats and *Tadarida acetabulosus*, which often roosts in roofs (Moutou 1979, 1982*b*, Cheke & Dahl 1981). No work has been done on parasites or diseases of native bats.

Reptilian pathogens may have been brought in with introduced species. The mysterious nineteenth-century disappearance of *Scelotes bojerii* on both Réunion and Mauritius could have had such a cause, likewise the decline of *Phelsuma borbonica* in Réunion (see below), though this last was attributed by Vinson (1868) to the introduction of snakes. Native reptiles were not covered in the only parasite survey so far conducted (Barré 1982).

Temple (1976*a*) reported 14.7 ppm dry weight of DDE in seven addled eggs of the Trindade Petrel *Pterodroma arminjoniana* collected on Round Is. in 1974. He tentatively related this to pollution of the waters round Mauritius by DDT used in agriculture and mosquito control. It is difficult to know the significance of this observation as Bourne *et al.* (1977) found only relatively minute quantities of DDE in two other oceanic feeders (*Puffinus pacificus*, *Sterna fuscata*) from on or near Round Is. in 1975. The possible role of DDT in the decline of the Mauritius Kestrel has already been discussed under 'Destruction of habitat'. On Réunion, heavily treated with DDT against malarial mosquitos during 1949–67 (Defos du Rau 1960, Gerard & Picot 1975), there was a severe post-war decline of the endemic harrier *Circus maillardi*, which led to special protection in 1966 (Gruchet 1975). The harrier has since increased considerably (Chapter 6), though whether this is due to reduced persecution or the cessation of DDT spraying is not known. Other possible victims of DDT are insectivorous bats, which during 1973–5 were noticeably commoner in Réunion than in Mauritius where DDT was widely used until 1973 (see above). Since then, observations by Jones (1980*a*) suggest an increase at least in one species, *Tadarida acetabulosus*. As the islands are free of heavy industry there appear to be no other widespread serious pollutants.

Wildlife protection in the islands
Mauritius

Although there had been sporadic attempts by the Dutch to protect tortoises and game (Pitot 1905), the first comprehensive game law was proclaimed in October 1767. This restricted hunting of ungulates, protected game birds and their eggs during the breeding season and gave the Common Mynah *Acridotheres tristis* special protection as it was expected to save the islands from a grasshopper plague (Rouillard 1866–9, Ly-Tio-Fane 1968). Other recently introduced granivorous birds had become such agricultural pests that in 1770 it became compulsory for landowners to produce each year a minimum of ten birds' heads per slave owned, the largest contributor getting a prize (*ibid.*); this law was still in effect in 1825 (Oustalet 1897). Mynahs (again!), Merles, doves and 'mesanges' (?) were protected in 1792 (d'Unienville 1982). The eighteenth century game laws were apparently not revised until the Game Ordinance of 1869, subsequently variously amended (Vinson 1956), including adding the protection of tenrecs in 1900 (Lane 1946 and later eds.). There was a provision for the governor to prohibit killing of any wild bird or animal by proclamation (Vinson 1946, Lane 1946) and it was under this regulation that Edward Newton first gave complete protection to kestrels and parakeets, and protection during the breeding season to a further three endemic species (E. Newton 1878*a, b*). By 1910 all the endemic birds were fully protected (Meinertzhagen 1912). Vinson (1956*a*) discussed the development of legislation to that date, and proposed revisions, which were largely followed by the 1957 proclamation (given in

full by Newton 1958*b*), renewed in 1967 (Staub 1976). These rules allowed game birds and various introduced pests to be freely killed (though the game birds were covered elsewhere in the Game Ordinance: Lane 1946), and certain common introduced species to be kept as cage birds. Up to this time the proclamations all lapsed after a period and had to be renewed, but a new law in 1977 has given permanent protection to the native birds (Staub 1980). The endemic reptiles were protected in 1973 (Procter & Salm 1975), but the flying-fox remained outside conservation laws until the 1983 Wildlife Act (Act 33 of 1983), which consolidated and extended previous game and wildlife legislation.

An Ancient Monuments and National Reserves Ordnance was promulgated in 1944 (Carver 1945, Brouard 1963, Vaughan 1968) and the first nature reserves established in 1951 (anon. 1961). These were chosen on purely botanical considerations (Vaughan 1968), and included little of the best bird habitat (Procter & Salm 1975), and almost no bats. Round Is. was added in 1957 (Vaughan 1968), but it was only in 1974 that Bel Ombre and Macabé were linked by the intervening land to provide a reserve of sufficient size and diversity for birds (Procter & Salm 1975, Owadally 1976*b*, Staub 1980). A list of the reserves and their sizes was given by Owadally (1981*b*).

Réunion

Lougnon (1957) mentioned many of the early attempts at tortoise and game protection through legislation in Réunion. Killing of bird pests was made compulsory around 1760 (Pingré 1763). In 1767 the twin colonies of Mauritius and Réunion were ceded to the French Crown, and ruled jointly from Mauritius under the same laws until 1810 (Toussaint 1972; see above). In 1820 new stricter bird-pest legislation banned the import of any granivorous bird, and even required seed-eating cage-birds to be killed (various finches, ploceids and estrildines were specified); at the same time mynahs, doves and stonechats *Saxicola tectes* were specifically excluded from the provisions (Decary 1962). In 1946 the island became an overseas *département* of France and thus subject to metropolitan French laws (Defos du Rau 1960, Toussaint 1972). The endemic harrier *Circus maillardi* was listed as a pest species until 1966 (Gruchet 1975)! All birds are fully protected under current legislation (Servat 1974, Staub 1976), except three introduced pests, and the endemic bulbul *Hypsipetes borbonicus* which is considered as game, and may be shot during a 3-month season (Barré & Barau 1982). The only declared nature reserves are tiny botanical ones (Cadet 1977), but a network of larger reserves has been proposed (Bosser 1982), including one specifically for the endangered cuckoo-shrike *Coracina newtoni* (Cheke 1976).

Rodrigues

The first wildlife legislation in Rodrigues was the Rodrigues Game Regulations of 1883, under which hunters required a licence and the Magistrate (the island's chief executive) was empowered to impose close seasons for any species he defined as game (Lane 1946). Around the same time (Reg. no. 6 of 1882) Mountain and River Reserves were established to protect watersheds (Koenig 1914b), so incidentally preserving some suitable habitat for the endemic birds and bats. Subsidiary regulations in 1923, protecting a number of non-game birds (Lane 1946), were most memorable for their failure to cover the two endemic species (Cheke 1974a). This anomaly was not removed until the 1983 Wildlife Act, which, *inter alia*, protected the two endemic birds and the flying-fox and repealed the 1923 regulations. Two nature reserves for breeding seabirds were declared in 1981 (Owadally 1984), but the best remaining areas of native vegetation and the bird and bat habitat are covered only by ordinary forest legislation, although many proposals have been made for reserves (Staub 1973b, Cheke 1974a, 1978b, Lesouëf 1975, Strahm 1983 etc.).

EXTINCTIONS AND INTRODUCTIONS: SPECIES ACCOUNTS

In this section the history of mammals, birds and reptiles in the islands is given in detail. Extinctions and introductions are documented where possible, and the subsequent history of introduced animals followed. Least emphasis is given to the surviving native fauna, which is considered in detail elsewhere (Staub 1976, Jones 1980a, Cheke & Dahl 1981, Chapters 4–8). The accounts are by taxonomic group, subdivided into separate island histories where appropriate. Although they are marine, I have included sea-cows and turtles in the accounts as they form a coherent part of the pattern, and like seabirds require land or its close proximity for feeding and for breeding.

The native fauna
MAMMALS
Bats: Megachiroptera

Flying-foxes are reported by the first Dutch visitors to Mauritius in 1598 (van Neck *in* Strickland 1848 and Grandidier *et al.* 1903–20, vol. 1) and are regularly mentioned by travellers thereafter, though the two species, *Pteropus niger* and *P. subniger*, while well known on Réunion, were not distinguished by Mauritius by any writer before Geoffroy (1806). *P. niger* survives in much reduced numbers, but *P. subniger*, a much smaller animal with the strange habit of roosting in hollow trees and caves, was last recorded in 1864 by H. Whiteley, and is extinct (Cheke & Dahl 1981). A third species, *P. rodricensis*, is known on Mauritius only from subfossil bones found on Round Is. (Mason 1907, Andersen 1912, Cheke & Dahl 1981); no visitor to Round Is. ever reported seeing flying-foxes, though "large bats" were reported on nearby Flat Is. in 1703 (log of the *Scarborough*).

On Réunion Carpeau du Saussay in 1666 was the first to mention flying-foxes (Lougnon 1970) but most subsequent travellers also did so. Although the existence of two kinds was known to the inhabitants at least as early as 1722 (La Nux 1772), this fact never comes out in travellers' writings, but only in scientific works. Brisson (1756), using specimens sent by La Nux to Réaumur, was the first to describe the two species. By the end of the century *P. niger* was extinct, though *P. subniger* persisted in small numbers until rather later, possibly into the 1860s (Cheke & Dahl 1981, Moutou 1982b).

The Rodrigues flying-fox, *P. rodricensis*, referred to as abundant in 1691 by Leguat (1708), was mentioned by most subsequent visitors. It is now reduced to very small numbers (Carroll 1982b); a full history of its decline was given in Cheke & Dahl (1981). Subfossil remains of *P. niger* have been found (Andersen 1913, Moutou 1982b), but there is no evidence that the species still survived when man first reached the island (Cheke & Dahl 1981).

Bats: Microchiroptera

Small bats are not mentioned by travellers to Mauritius before La Caille's account (1763), presumably by oversight. For Réunion the first mention is much earlier: Borghesi in 1703 (Lougnon 1970) mentioned seeing bats which were clearly *Taphozous mauritianus*, as, a century later, did Bory (1804), the next writer to report them.

The species involved (*T. mauritianus*, *Tadarida acetabulosus* and *Scotophilus 'borbonicus'* (= *S. leucogaster*)) do not differ from the same species on Madagascar and/or the African mainland (Cheke & Dahl 1981, Moutou 1982*b*) and could perhaps be recent colonists. Apart from *S. 'borbonicus'* (Réunion only), which has not been reported since 1867, these bats are still widespread today on both islands (Moutou 1979, 1982*b*; Jones 1980*a*, Cheke & Dahl 1981). No microchiropterans have reached Rodrigues.

Sea-cows (Sirenia)
Mauritius

Records of sea-cows in the Mascarenes were reviewed by Stoddart (1972). While casting doubt on Leguat's (1708) and Herbert's (1634 etc.) descriptions, he came to the conclusion that the animals usually described by French writers as *lamentins*, were indeed probably Dugongs *Dugong dugong* and not seals. Having had access to more accounts than Stoddart, my view is that the animals were certainly, rather than probably, sea-cows. Stoddart gave the earliest record for Mauritius as Matelief's in 1606 (Pitot 1905), but they had already been noted by van West Zanen in 1602 (Strickland 1848). I have found four additional seventeenth-century mentions of sea-cows unknown to Stoddart (accounts of the survivors from the *Arnhem* (in 1662; Grandidier *et al.* 1903–20, vol. 3), Granaet in 1666 and Bonnell in 1690 (both in Barnwell 1948), and Hoffman's (1680) long and detailed account around 1674). Hoffman very clearly distinguished the Mauritian sea-cows from the seals he had seen at the Cape. Governor Momber, in 1706, complained of only having seen two sea-cows in 2½ years (Barnwell 1948). In letters written to Réaumur in 1732 and 1734 Cossigny (1732–55) mentioned *lamentins*, and in the second wrote "In certain places and at certain particular times [or weather?; the original is "en certains temps marqués"] *lamentins* are taken. This is an advantage we lack altogether at Mascarin or Bourbon [= Réunion], where the shore is almost everywhere too steep" (my translation). *Lamentins* are also referred to in the log of the *Courrier de Bourbon* in 1721, by Jonchée (1729), by Abbé Gandon (1732), and by La Motte (1754–7). Cossigny (1764) wrote: "There are still taken, though much more rarely than formerly, sea turtles weighing 2 to 300 [French] pounds, and *lamentins* much heavier still which have very firm blubber three finger-widths thick" (my translation); he goes on to praise *lamentin* as a stew or in soup. Although, as Stoddart (1972) notes, Bernardin's

(1773) record of a *lamentin* at Black River in or about 1768 is the last positive one in Mauritius, Morel (1778) mentioned them, and the younger Cossigny (1799) wrote that "sea turtles and *lamentins*" were "nowadays very rare", suggesting that they were occasionally seen up to the turn of the century.

Réunion

No doubt for the reason stated by Cossigny (1732–55), mentioned above, sea-cows appear never to have frequented Réunion.

Rodrigues

For Rodrigues, Stoddart (1972) cited the accounts or mentions of *lamantins* (Leguat's spelling) by Leguat, Tafforet (the 'anonymous *Relation . . .*'), Nichelsen, Pingré and Marragon, but was wary of Leguat's and Tafforet's evidence in the light of Atkinson's (1922) dismissal of Leguat's book as fiction, and stigmatisation of Tafforet as derivative. North-Coombes (1971) and Holthuis *et al.* (1971) have reviewed the literature of this controversy, and more recently North-Coombes (1980*a*) has written a powerful apology for Leguat. On the basis of these studies there seems to be no longer the slightest reason for doubting Leguat's testimony. In addition to the eighteenth-century records mentioned by Stoddart, d'Héguerty (1754) said of Rodrigues in the 1730s that "the *lamentin* fishery is considerable and a major resource for the Isle de France, whither many are transported salted together with turtles" (my translation), and North-Coombes (1980*a*) noted La Bourdonnais's instructions on *lamentin* fishing there. This no doubt explains the diminution from the hundreds seen by Leguat to their scarcity in 1761; Pingré (1763) saw some but could not get near them, although later in the same year the crews of Tiddeman's squadron were more successful (Nichelson 1780, Dupon 1969). Marragon (1795, Dupon 1969) saw them occasionally; his is the last record I have found of Dugongs at Rodrigues, though it could be rewarding to search for items of natural history interest amongst the papers of the British fleet which assembled there in 1810 prior to the assault on Mauritius (North-Coombes 1971).

BIRDS
Seabirds
Mauritius

Apart from Audubon's Shearwaters *Puffinus lherminieri*, frigate-birds *Fregata* spp., and Abbott's Booby

Sula abbotti, which have gone, there is no evidence that the seabird fauna of Mauritius when first discovered differed from what survives today, except that sea-birds (terns and shearwaters) once nested in good numbers on the islets off Mahébourg (Leguat 1708), and in much larger quantities on the northern islets (Pike 1873). The Trindade Petrel *Pterodroma arminjoniana* may have colonised Round Is. during the early part of this century (Vinson 1976*a*).

Early visitors to Mauritius mentioned the *rabos forcados* (frigate-birds) and one is depicted in the famous engraving accompanying the text of van Neck's voyage of 1598 (reproduced by Strickland 1848, Bonaparte 1890, Hachisuka 1953, etc.). At this time they presumably bred on Mauritius or the offshore islets. After 1640 (Cauche's visit; Cauche 1651, Oustalet 1897) there are no further records until 1753 (La Caille 1763), when they were said to be common. After that there is a gap until they were seen in 1844 by Lloyd (1846) and again in 1860 by E. Newton (1861*a*) on Round Is.; however, as Desjardins (*c.* 1830s; Oustalet 1897) and Clark (1859) stated that they were seen only in bad weather or after strong gales, it is likely that they had long ceased to breed.

Bourne (1976) reported that the subfossil bones of an unspecified 'gannet' previously reported from the Mare aux Songes were actually those of Abbott's Booby, a species now confined to Christmas Is. (Indian Ocean). There is also good evidence that it occurred in Rodrigues (see below). Only one early account might refer to Abbott's Booby: in 1668 John Marshall wrote (Khan 1927): "I see upon the Island 2 birds by a nest upon a very high tree. They were much bigger than geese as seemed to mee, had long beakes and nests [*sic*; = necks], and were of a whitish colour." Perhaps referring to Red-footed Boobies *Sula sula*, he also described from Mauritian waters "birds which they call Boobos which sometimes light upon ships, are as big as a kite, have a long bill and are of a reddish greene and some part white colour." Under '*Sula piscatrix*' (= *S. sula*) Oustalet (1897) cited Desjardins (MSS) as saying that "*fous*" were common on islets off Mauritius and that they nested in holes in rocks. *Fou* is the normal French name for boobies, but it was also used in the Mascarenes for shearwaters *Puffinus* spp. and even noddies *Anous* spp. (Cheke 1982*b*), and the nest-site suggests they were shearwaters (or possibly tropic-birds *Phaethon* spp.). Meinertzhagen (1912) misquoted Oustalet (1897) as stating that Red-footed

Boobies once bred on Flat Is.: Oustalet was referring to Ile Platte in the Seychelles. A specimen listed by Hartlaub (1877) as from Mauritius actually came from Agalega (Cheke & Lawley 1984). These errors were repeated by Newton (1958*a*) and Feare (1978); there is no evidence that Red-foots ever bred in Mauritius. The Masked Booby *S. dactylatra* still breeds in small numbers on Serpent Is. (Staub 1976, Temple 1976*a*).

The eighteenth-century Mascarene name for Abbott's Booby was *boeuf* from its call (Cheke 1982*b*), but the nestling that Cossigny (1732–55) sent to Réaumur under that name was apparently a tropic-bird. It may be that there had been tree-nesting Abbott's Boobies within living memory, and that the name has transferred to another white, tree-nesting seabird, the White-tailed Tropic-bird *P. lepturus*. *Boeuf* also transferred, more durably, to the Masked Booby (Cheke 1982*b*).

The only other early description of nesting sea-birds in Mauritius is that of Leguat (1708; misinterpreted by Oustalet 1897), who gave details of the breeding of "ferrets" (= ?Sooty Terns *Sterna fuscata*) and "*plutons*" (= ?Wedge-tailed Shearwaters *Puffinus pacificus*) on the islets of Mahébourg in 1693. Subsequent details are all nineteenth-century or later and are considered elsewhere (e.g. Newton 1956, Vinson 1964*a*, Gill *et al.* 1970, Staub 1976, Temple 1976*a*, Vinson 1976*a*, Rowlands 1982). Audubon's Shearwaters, not recorded by earlier visitors, were breeding in small numbers in the nineteenth century (Benson 1970–1), but have not been nesting recently.

Réunion

Amongst early visitors to Réunion seabirds are mentioned only by Feuilley (1705), who reported tropic-birds *Phaethon* sp. Both d'Héguerty (1754) and Lebel (1740) referred to nocturnal "*fouquets*" (petrels), and Bory (1804) gave a now famous (Jouanin & Gill 1967) account of the killing of Barau's Petrels *Pterodroma baraui* by attracting them to fires at night. Bory also wrote of "*maquois*" (Brown Noddies *Anous stolidus*) nesting at Le Baril.

The White-tailed Tropic-bird *P. lepturus*, four petrels (*Puffinus lherminieri*, *P. pacificus*, *Pterodroma aterrima* and *P. baraui*) and the noddy still nest on the island (Milon 1951, Jadin & Billiet 1980, Barré & Barau 1982, Chapter 7). Barau's Petrel probably became very rare early in the nineteenth century, as none were reported until Milon (1951) found some specimens in the St.-Denis museum. The species was not identified until 1963 (Jouanin 1964a, Jouanin & Gill 1967), since when, to judge by increasing numbers seen around the coast and sightings on neighbouring islands (Chapter 7), the population has been expanding.

Rodrigues

Staub (1973b) reviewed the history of nesting seabirds on Rodrigues, but several items have since come to light. Staub listed the petrel *Pterodroma aterrima*, the Red-footed Booby, two frigate-birds and the Sooty Tern *Sterna fuscata* as seabirds formerly breeding but now extinct; the Wedge-tailed Shearwater, two tropic-birds, two noddies and the Fairy and Roseate Terns *Gygis alba* and *Sterna dougallii* still survive (Staub 1973b, 1976, Cheke 1974a, 1979a).

Bourne (1968) referred a petrel skull from Rodrigues to the Réunion Petrel *Pterodroma aterrima*, and suggested that other *Pterodroma* species may also have formerly frequented the island; Staub (1973b) pointed out that Tafforet's (1726) rare nocturnal *"fouquets de montagne"* were probably gadfly petrels, presumably *P. aterrima*. Apart from Tafforet's observation there is nothing else in the literature suggesting the presence of any petrel apart from the Wedge-tailed Shearwater. Bourne's prediction has proved correct. Barau's Petrel *P. baraui* was found nesting in the hills in 1974 (Cheke 1974a), though there have been no observations (despite searches) since December that year (pers. obs.) and there is no local lore to suggest a breeding population of long standing. The other surprise was Cowles' discovery (Chapter 2) of a new extinct gadfly petrel amongst subfossil bones collected in 1974.

Bourne (1968) considered Tafforet's (1726) and Pingré's (1763) two boobies, the *tra-tra* and the *boeuf*, to refer respectively to *Sula sula* and *S. dactylatra*; Staub (1973b) thought them more likely to represent brown and white phases respectively of the former, since *boeufs*, as well as *tra-tras*, were said to nest up trees (Tafforet 1726). However, since then Nelson (1974), subsequently supported by Bourne (1976), has argued that the *boeuf* was actually Abbott's Booby *S. abbotti*, also an arboreal nester. Nelson based his view on Tafforet's and Pingré's descriptions of the *boeuf*'s bill and feet (as quoted by Bourne 1968), and on the cry, like that of a bull (Pingré, Tafforet), which accords well with Abbott's Booby.

Bourne (1976) confirmed the occurrence of Abbott's Booby in the Mascarenes from a subfossil

bone from Mauritius (see above); another bone from Rodrigues was less certainly attributable. Further convincing details from the original descriptions, not available to Bourne or Nelson, support these conclusions.

Tafforet (1726) described the inside of the beak as like a saw and gave its length as "about 5 inches": the tooth-like serrations of the mandibles are bigger and coarser in *abbotti* than other boobies and its bill reaches 120 mm (4.7 in.), whereas the biggest Masked Booby bill (apart from Galapagos populations) is only 107 mm (4.2 in) (Nelson 1978).

Pingré (1763) described its eyes as "handsome, black and large", which fits *abbotti* perfectly (Nelson 1978; see also colour photographs in Nelson 1971), whereas *dactylatra* has "piercing yellow" irises (Nelson 1978; photos in Staub & Guého 1968, Staub 1976). The white phase of the Red-footed Booby is ruled out as both Tafforet and Pingré clearly mentioned black in the tail as well as the wing, a detail omitted by Bourne (1968); in addition *boeufs* were said by Tafforet to be somewhat larger than *tra-tras*, which were definitely dark-phase Red-foots.

Pingré's description is the last certain reference to Abbott's Boobies in Rodrigues, although the name *boeuf* survived until 1832 at least (Dawkins *in* Telfair 1833), but the bird of that name produced by Rodriguans turned out to be a Red-foot (ed. note following Telfair 1833). Marragon mentioned only *fous* (= boobies spp.) in 1795, and apart from the name there are no indications in the nineteenth century of two species of booby at Rodrigues. As the name *boeuf* is used nowadays in Mauritius for *S. dactylatra* (Staub 1976, Cheke 1982b) it may well have transferred in Rodrigues to *S. sula*, but the possibility that Abbott's Booby survived until the 1830s cannot be ruled out; the locals who caught the Red-foot may not have been able to distinguish the two, especially as it appears to have been a white-phase bird.

The Red-footed Booby itself survived at least until 1874, when there were still large numbers nesting on Ile Frégate (Slater 1875, Balfour 1879b: 365). Many young were slaughtered for their down by the crew of HMS *Shearwater* (Slater 1875), though Slater himself took only one adult (Sharpe 1879). They were extinct by 1916, as Bertuchi (1923), who explored the seabird islets, made no mention of them.

Frigates, almost certainly *Fregata minor* and possibly also *F. ariel* (Bourne 1968, Staub 1973b), bred on

Ile Frégate in the eighteenth century. Tafforet (1726) definitively stated they bred only on Frégate; Pingré (1763) saw frigate-birds on other islets, but from his description of the birds he evidently saw no breeding males, and it may be that out of season the birds would visit any other island with boobies on them. The subsequent presence of frigates, without any supporting details, was reported by Marragon (1795), Higgin (1849) and E. Newton (1865a) who saw only one on a visit in 1864. Slater (1875) saw only non-breeders in the booby colony in 1874, and failed to collect any (Sharpe 1879). By 1916 Bertuchi (1923) described them as rare visitors. Vagrants still occur sporadically (Staub 1976).

Of the original seabird fauna, the only other species no longer breeding regularly is the Sooty Tern *Sterna fuscata*. Tafforet (1726) and Pingré (1763) testify to the huge colony once covering one or both of the two large southern islets: Tafforet reported it on "Ile aux Fols et Fouquets" (= Ile Pierrot) but not on "Ile au Mat" (= Ile Combrani), whereas Pingré, who visited only the latter ("Mombrani") found the birds there and also saw large numbers on Ile Coco. Duncan (1857) reported that an important source of food was "the eggs of a species of sea gull which are to be had in abundance", and Newton (1865a) referred to the collection of "innumerable birds' eggs chiefly of terns". By 1874 Bouton (1875, citing C. E. Bewsher) commented on the threat to the "clouds of *golettes*" from excessive egg-robbing; Slater collected two birds in the same year (Sharpe 1879), but made no comment on their abundance (Slater 1875).

By 1916 Bertuchi (1923) reported that "large colonies of Sooty Terns used at one time to visit the island yearly. Their favourite spots were Sandy and Cocos Islands, and also Gombranis island where they laid their eggs. The eggs were collected in boat-loads and eaten by the natives. Their visits are not so frequent now, owing undoubtedly to the bad treatment they received at the hands of the natives." Although common at sea around the island (pers. obs.) they breed only sporadically and in very small numbers (Staub 1973b).

There are two nineteenth-century collections of Crested Terns *Sterna bergii* from Rodrigues, one dated '1845 or 1846' in Cambridge (labelled Round Is., Mauritius, with a query, but in the same lot as other birds from Rodrigues only; C. W. Benson *in litt.*) and two collected by Slater in 1874 (Sharpe 1879). However, all these are in non-breeding plumage, and there

is no evidence that the species has ever bred in Rodrigues, although Slater (1875) described them as "pretty common".

Herons and egrets (Ardeidae)

Early travellers to the Mascarenes often mentioned ardeids, and their presence has been confirmed by bones found subfossil in Mauritius and Rodrigues, but most travellers' descriptions cannot be identified to species.

Mauritius

The subfossil remains consist of an endemic night-heron *Nycticorax* (formerly 'Butorides') *mauritianus* (Chapter 2), the egret "*Egretta garzetta gularis*" (= *E. (gularis) dimorpha*; see Hancock & Elliott 1978) and two other species mentioned by E. Newton & Gadow (1893) but not described or identified. No remains of the extant Little Green Heron *Butorides striatus* are recorded in the literature; the only other heron recorded regularly is the Cattle Egret *Bubulcus ibis* (Oustalet 1897, Temple 1976a, Barré & Barau 1982).

'Egrets' and 'herons' and 'bitterns' were reported in the seventeenth century by various writers, but not thereafter until Milbert (1812) reported a "*martin-pêcheur*", which appears to have been the Little Green Heron. In 1602 (Strickland 1848) Van West Zanen mentioned "white and black herons" which were presumably Dimorphic Egrets *E. (g.) dimorpha*; Herbert (1634 etc.) in 1627 reported "white and beautiful" herons, and van Hagen in 1607 "herons of various colours" (Oustalet 1897), which are not identifiable. Other authors, including Leguat (1708) who saw "great flights of bitterns" in 1693, left no descriptions of the birds they saw.

As the Little Green Heron in the Mascarenes is of the Javan race (Salomonsen 1934, White 1951, Ripley 1969) it may be a recent immigrant and genuinely not have been present until the late eighteenth century; it was certainly well established by the 1830s (Desjardins *in* Oustalet 1897), but at that time apparently unknown along some parts of the coast (Savanne) where it can certainly be seen today (pers. obs.).

Réunion

A few early travellers to Réunion reported herons or egrets. Ruelle was the first (in 1667, Lougnon 1970), but only Dubois' (1674) descriptions in 1671–2 are good enough to allow possible identifications. His white and grey egrets, confirmed by Feuilley (1705), must be Dimorphic Egrets as Berlioz (1946) and Milon (1951) have already noted, but the "butors ou grands gosiers" as big as capons, with grey plumage spotted white, and green feet, have led to considerable speculation. Rothschild (1907b), followed by Hachisuka (1953), believed them to be an endemic heron now extinct, and named it *Megaphoyx duboisi*. Berlioz (1946) suggested the extinct '*Butorides*' (= *Nycticorax*) *mauritianus*, Milon (1951) a young night-heron and Cowles (Chapter 2) has recently confirmed the former presence of an endemic *Nycticorax* from subfossil bones. There are no records after Feuilley's of ardeids of any sort, until Maillard (1862) and Coquerel (1864) listed *B. striatus* as very rare. At present the species is a sporadic breeder in the island (Jouanin 1964b, Barré & Barau 1982); Moutou's (1984) reference to *Ardeola idae* was an error for *B. striatus*.

Rodrigues

I have little to add to Hachisuka's (1953) and Staub's (1973b) accounts of the extinct endemic *Nycticorax* ('*Megaphoyx*') *megacephalus*, except to note that Pingré (1763) specifically stated that there were no longer any *butors* in 1761. Hachisuka made much of this heron being flightless, dismissing Tafforet's statement (1726) that it did fly a little, and quoting but ignoring Günther & Newton's remark (1879) that the heron and the owl "without losing the power of flight, . . . became brevipennate". The species was described from subfossil bones found on the island (details in Hachisuka and Chapter 2).

Tafforet's "*sentinelle*", as Staub (1973b) pointed out, was probably the Little Green Heron *Butorides striatus*, as no doubt was Marragon's (1795) "*crabier*". It is still common in the island (Staub 1973b).

Flamingos (Phoenicopteridae) and storks (Ciconiidae)

Flamingos are mentioned by early travellers to both Mauritius and Réunion, and *Phoenicopterus* bones have been found in the Mare aux Songes deposits (Hachisuka 1953).

In Mauritius the birds were reported as common into the 1730s (*Courrier de Bourbon* 1721; Jonchée 1729; Cossigny (1732–55) in 1732), and even as late as 1753 La Caille (1763) considered *flamants* to be amongst the most "ordinary" birds of the island. However by 1768 flamingos were reported to have been reduced to three

individuals (Bernardin 1773). Subsequent reports of vagrants are frequent (Oustalet 1897, Guérin 1940–53) but it appears that the resident population was wiped out in the middle of the eighteenth century.

From Carpeau du Saussay in 1666 onwards (Lougnon 1970) visitors to Réunion reported "*flamants*", often described as of the height of a man, so presumably *Phoenicopterus ruber*. Unlike the other local birds they were hard to approach, and Feuilley (1705) described hunting them at night by crawling stomach to ground up to the edge of the mere; he reported up to 3000–4000 on the Etang du Gol. They were unanimously praised as excellent eating, and disappeared early in the eighteenth century. De Villers (quoted by La Merveille; La Roque 1716) still reported them in 1708, but Boucher (1710) considered them extinct. Cossigny (1732–55) recorded them in 1732, but it is not clear whether he was including Réunion (as well as Mauritius) in his reference to flamingos as common, although the letter in question was written from Réunion. It is likely that the population moved between the islands (and Madagascar?), and that it was disturbance rather than annihilation that shifted them from the scarce meres of Réunion to better pastures at Mauritius, until they declined there too towards the middle of the century.

Flamingos have never been other than vagrants in Rodrigues (Carié 1930, Guérin 1940–53).

A word is necessary at this point on Leguat's "*géan*" (giant), an entity which has spawned a considerable literature. The name *géan* or *géant*, and a description, first appeared in a memoir on Réunion by du Quesne published in 1689 (Carié 1930, Lougnon 1970), copied, somewhat abridged, from Dubois (1674; see comments in Lougnon 1970 and North-Coombes 1983), including the substitution of *géant* for Dubois's name *flamant*. Leguat (1708, Carié 1930) cited du Quesne's description (further abbreviated) with reference to Réunion, and gave a new one of his own for birds he saw in Mauritius, adding several details not mentioned by Dubois/du Quesne. Dubois's *flamants* were goose-sized with very long neck and legs, white with black wing-tips. Leguat added a red patch under the wing (diagnostic for flamingos; see Cramp & Simmons 1977), but also a goose-like bill (but more pointed) and very long fully separate toes. He illustrated the bird by borrowing (without acknowledgement) a picture of Collaert's originally published in 1598, and apparently representing a large undescribed probably Asian rail

(Holthuis *et al.* 1971, Cheke 1983c). Schlegel (1858) managed to unite the description and the picture in a single form, which he described as a giant extinct Mascarene rail, *'Leguatia gigantea'*. There has been much subsequent dispute as to the existence of *'Leguatia'* (e.g. Rothschild 1907b, Hachisuka 1953, North-Coombes 1983) as against flamingos (e.g. Buffon 1770–83, Strickland 1848, A. Newton 1907a, Carié 1930, Mortensen 1934, Holthuis *et al.* 1971), the origin of Collaert's bird (Holthuis *et al.* 1971, Cheke 1983c, North-Coombes 1983), and indeed the whole question of Leguat's veracity (see North-Coombes 1980a). Leguat's description of the bill and feet has always been the mainstay of the argument in favour of *Leguatia*'s existence, but I suspect he added it to what he remembered of the size and plumage colour to conform with the picture he (or his editor) had found as illustration. This, but for one fact, leaves the Mascarene birds as flamingos, as described by all other visitors, and Collaert's bird as a mystery Asian rail. The awkward fact is Cowles's recent discovery of a stork bone in material from a Réunion cave (Chapter 2). The cave is known to have been used by early settlers, and so the bone seems likely to have been from a bird eaten by colonists rather than one that died there naturally. This gives some credence to North-Coombes's view (1983) that both flamingos and *géants* (?storks) inhabited the islands, and that the *colons* used the same name, *flamant*, for both. However, the facts that only Leguat out of numerous visitors saw and described *géants*, that he failed to see any flamingos, that none of the other travellers mentioned two types of *flamant* nor described a *flamant* or *vlamink* different from European ones, nor ever used the name *cigogne*, suggests to me that there were in fact no storks or *géants* wild on the island. Cowles's bone appears therefore to represent either a prehistoric extinction, or, more probably, the remains of a bird killed and salted elsewhere and eventually eaten on Réunion.

Ducks and geese (Anatidae)

Early travellers to both Mauritius and Réunion repeatedly mentioned the presence of evidently very edible ducks and geese.

Mauritius
Two endemic species, *Anas theodori* and *Alopochen* (formerly *'Sarcidiornis'*) *mauritianus*, have been

described from the Mare aux Songes deposits (E. Newton & Gadow 1893, Oustalet 1897, Hachisuka 1953, Chapter 2), but little is known about them in life.

The only account to give any indication of the appearance of the Mauritian duck, was the log of the *President* (1681) which referred to "gray teal". At that date there were "great numbers" on lakes and ponds in the woods, but by 1693 Leguat (1708) placed them amongst the birds formerly abundant but by then rare. The last mention is by Governor Deodati in 1696 (Barnwell 1948).

Only Marshall (in 1668; Khan 1927, Oustalet 1897) described the Mauritian sheldgoose: "Here are many geese, the half of their wings towards the end, are black, and the other halfe white. They are not large, but fat and good." The log of the *President* (1681) gave their habitat as "most[ly] in the woods or dry ponds". Like the ducks they were plentiful in 1681 but declined rapidly thereafter, Leguat (1708) listing them as rare in 1693, and Deodati stating categorically in 1698 that they were extinct (Barnwell 1948).

Réunion
As in Mauritius, early visitors to Réunion reported ducks and geese, especially on the Etang de St.-Paul. Tatton (1625) in 1613 called the ducks "mallards"; the log of the *Breton* in 1671 (Lougnon 1970) called them *"sarcelles"* (= teal) and added that the geese were smaller than European ones. However, the only good descriptions are from Dubois (1674) in 1671–2. His geese (plumage as European ones, bill and feet red) can be related to a new genus of sheldgoose found subfossil in 1974 (Chapter 2); his ducks, small and 'teal'-plumaged, may well have been *Anas theodori*, though Garganey *A. querquedula* are regular visitors in small numbers (Barré & Barau 1982). The wildfowl were rapidly wiped out: as early as 1667 Martin (Lougnon 1970) was complaining of massive destruction of geese on the Etang de St.-Paul. La Merveille in 1709 (La Roque 1716) stated that "geese and ducks in quantity" could be seen on the same mere, but I suspect outdated hearsay, as wildfowl are strikingly absent from Feuilley's catalogue of wildlife (1705) and Boucher (1710) specifically declared them extinct. The last positive mention of either ducks or geese, if La Merveille is to be discounted, is that of Père Bernardin (1687). Both Boucher and Bernardin (*locs. cit.*) mention *canards* and *sarcelles* as well as geese, raising the possi-

bility of there having been more than one species of duck.

Rodrigues
No ducks or geese are recorded as ever being resident at Rodrigues, which lacks lakes or marshes.

Other water-birds (including aquatic Rallidae)
Mauritius
The Mare aux Songes deposits have yielded remains of a cormorant *Phalacrocorax africanus* and a coot *Fulica* ('*Paludiphilus*') *newtoni* (Hachisuka 1953; Olson (1975) assigned the bones formerly described as '*Anhinga nana*' to the Reed Cormorant). Nothing that corresponds to cormorants was ever described by early travellers to Mauritius; Oustalet (1897) assigned Leguat's "*plutons*" on the islets off Mahébourg to an unidentified cormorant, but the description accords much better with Wedge-tailed Shearwaters. Three travellers did mention aquatic rails, describing them as "moorehennes" (in 1628; Mundy 1608–67), "waterhens" (1666; Granaet *in* Barnwell 1948) and "*poules d'eau*" (1693; Leguat 1708). The obvious identification would be Moorhens *Gallinula chloropus*, but there are two important reasons to question this. First, in Réunion (see below) *poule d'eau* rather than the usual French term *foulque* was used to describe a coot; second, there was a long gap from Leguat's report in 1693 to the end of the eighteenth century before *poules d'eau* are mentioned again (Cossigny 1799). There is thus a strong possibility that the seventeenth-century birds were coots and that the Malagasy race of the Moorhen, *G. c. pyrrhorrhoa* did not colonise the island until the late eighteenth century. There is no record of its deliberate introduction, so it probably arrived, if in fact not already present, on its own. There are some bones from the Mare aux Songes deposits which E. Newton & Gadow (1893) assigned to 'moorhen' and which are labelled '*G. c. pyrrhorrhoa*' in the Cambridge museum, but doubt was cast on this identification by Carié (1916), and Cowles (Chapter 2) has assigned them to *Dryolimnas cuvieri* (see below).

Réunion
Among the early accounts of Réunion birds are references to *cormorans* and *poules d'eau*. The cormorants were mentioned only by Dubois (1674) and Feuilley (1705), the latter stating that they were "the size of a

duck and [of] the same appearance, except that their neck is a little longer. They inhabit the meres like the flamingos and live in the same way. They are not eaten unless very young. They smell strongly of marshes and of the wild" (my translation). There is still, near St.-Gilles, a pool called Bassin du Cormoran. These birds were probably *Phalacrocorax africanus* (Milon 1951; and see above).

Poules d'eau are mentioned by several travellers (Martin, Carpeau du Saussay, Dubois, log of the *Breton*; all *in* Lougnon 1970), and were it not for Dubois' description one might well expect them to have been Moorhens. Dubois, however, described them as "entirely black with a big white crest on the head": clearly coots, and recently confirmed as *Fulica newtoni* (Chapter 2). Martin complained in 1667 (Lougnon 1970) that they had been wiped out on the Etang de St.-Paul, and there are no reports after 1672, despite the fact that the log of the *Breton* described them as "not good to eat".

Hawks and falcons (Falconiformes)
Mauritius
Van West-Zanen and Matelief (Strickland 1848, Grandidier *et al.* 1903–20 vol. 1, Pitot 1905) reported "hawks" and "falcons" as did Herbert (1634 etc.) in 1627, who added "hobbies". However Pitot (1905) suggested, as seems likely, that Herbert borrowed his list in part from Matelief. Apart from Granaet, who mentioned "falcons" in 1666 (Barnwell 1948), there is no further mention of diurnal raptors until La Caille (1763) wrote of the "*mangeur de poule*" (= the kestrel *Falco punctatus*) in 1753. These reports are not sufficiently detailed to indicate whether or not the harrier found in the Mare aux Songes deposits, *Circus* ('*Astur*') *alphonsi* (Hachisuka 1953; ? = *C. maillardi*, Chapter 2), was ever seen in life. The long disappearance from travellers' reports of the endemic Kestrel, a tame and conspicuous bird, suggests that it quickly became rare (?from monkeys raiding its nests), perhaps not recovering for a century.

Réunion
The only raptor present in Réunion today is the endemic harrier *Circus m. maillardi* (Clouet 1976, 1978), though migratory falcons appear occasionally (Barré & Barau 1982). The harrier is mentioned by a few early visitors, the first being Père Vachet in 1669 (Lougnon 1970); Dubois (1674) in 1671–2 was the first to use its

current local name *papangue*. Dubois also talks of *"pieds jaunes"* (apparently migratory falcons: Cheke 1982*b* *contra* Berlioz 1946) and of *"emerillons"*, which were no doubt, as Berlioz (1946) surmised, Kestrels; the endemic species was not confirmed until bones were found in 1974 (Chapter 2).

No other author mentioned small hawks, and Dubois was the only one to distinguish clearly between harriers and falcons; Feuilley (1705) used the names *Papangue* and *pied jaune* for the same bird, the harrier. It is astonishing that no writer mentioned the harrier between Feuilley's report and 1834 when Desjardins (1835) reported on it. As the bird is conspicuous and had a reputation for chicken stealing, its absence from lists of the island's *fléaux* (plagues) suggests it had a long period of rarity.

Rodrigues
No diurnal bird of prey has ever been reported from Rodrigues.

Button-quails (Turnicidae)
Many early travellers to Réunion reported *perdrix*, described by Carpeau du Saussay as "much smaller than ours, but with much prettier plumage", by Dubois as "little grey partridges, the size of quails", and by Bellanger as "very small" (all *in* Lougnon 1970). Unfortunately none of them left a detailed plumage description, and after 1671 there are no more reports of *perdrix* or *cailles* until some were introduced in 1714 (Foucherolle *et al.* 1714). The most likely 'quail' to be indigenous would be *Turnix nigricollis*, native to Madagascar and still present on Réunion (Berlioz 1946; *contra* Milon (1951) who postulated an extinct endemic form). Perhaps cats introduced in the latter part of the seventeenth century wiped them out in accessible regions of the island, or indeed altogether? *Turnix* was not listed amongst the island's birds until 1916 (Carié 1916), when it was said to be abundant.

Terrestrial rails (Rallidae)
1. Aphanapteryx spp.
Much has been written about the extinct flightless Red Rail *Aphanapteryx bonasia* of Mauritius; Hachisuka (1953) provided a very full bibliography. He also split the travellers' reports into three species, but there seems no reason (Greenway 1967, Olson 1977) to maintain the extra two ('*Kuina mundyi*' and '*Pezocrex*

herberti') as they fall well within the expected variation in descriptions by non-ornithological voyagers, and there is no osteological support for their existence.

There are several more references to Red Rails in the old reports than were known to Hachisuka. The *"feldhüner"* of Cornelisz and Verhuffen (in 1602 and 1611; Strickland 1848) were presumably this species (the English referred to them as hens: Herbert 1634 etc., Mundy 1608–67, Sclater 1915). Gooyer, the first Dutch commandant, referred in 1638 to what Bonaparte (1890) translated as *"perdrix"* (= *feldhüner* in Dutch), as did the survivors of the wrecked *Aarnhem* in 1662 (Grandidier *et al.* 1903–20, vol. 3). Marshall, in 1668 (Khan 1927), gave a very good description: "Here are also great plenty of Dodos or red hens which are larger a little than our English henns, have long beakes and no, or very little, Tayles. Their fethers are like down, and their wings so little that it is not able to support their bodies; but they have long leggs and will runn very fast, that a man shall not take them, they will turn so about the trees. They are good meate when roasted, tasting something like pig, and their skin like pig skin when roosted [*sic*], being hard."

Marshall was the first to use 'Dodo' for the Red Rail, indicating a transfer of names some time before Hoffman's (1680) well-known use of *Todaersen* for this species in 1675. For reasons discussed below I believe the real Dodo *Raphus cucullatus* became extinct rather earlier than generally thought, hence Benjamin Harry's famous remark that amongst the island's "Producks" in 1681 were "Dodos whose fflesh is very hard" (Strickland 1848, Hachisuka 1953) must refer to Red Rails, the description of the cooked bird recalling Marshall's final comment. Leguat (1708) listed *"gelinottes"* amongst the birds formerly common but rare by 1693; he also used this term for *A. leguati* on Rodrigues, so it is reasonable to assume he was referring to Red Rails in Mauritius. There were no subsequent reports.

Ripley (1977) raised a historical conundrum when he referred to a painting in the Prado, Madrid, depicting, *inter alia*, a Red Rail. This is Bassano's *Noah's Ark*. According to Ripley this painting antedates 1584, when it was "left to the church of Santa Maria Maggiore in Venice". Giacoma da Ponte, known as Bassano, lived and worked in Bassano, his home town, and Venice (anon. 1862), so one might wonder how he came by a Red Rail when only the Portuguese had visited Mauritius. No records of Mauritian birds have been found in any Portuguese archive or painting. Bassano died in 1592, six years before the first Dutch visit to Mauritius. However, he had four sons, also artists who used the name Bassano, so it is possible that there has been some historical error in the dating or attribution of the painting of *Noah's Ark*.

Dubois (1674) was the only visitor to Réunion to report rails, but gave no description of his *"râles des bois"*. They may well have been an *Aphanapteryx*-like form, but until subfossil remains are found their identity will remain unknown.

I have nothing to add to Hachisuka's (1953) and Staub's (1973*b*) accounts of *Aphanapteryx (Erythromachus) leguati* of Rodrigues, seen by Leguat (1708) and· Tafforet (1726), except to point out that Pingré (1763) stated that it no longer existed in 1761. I have followed Cowles (Chapter 2) in putting *leguati* in the genus *Aphanapteryx*, but Olson (1977), while doing the same, considered the birds probably distinct enough to be kept in separate genera.

2. *Oiseau bleu of Réunion*

To the four well-known mentions of the *oiseau bleu* in the Réunion travel literature (Dubois in 1671–2, de Villers in *c.* 1708 and Le Gentil in 1717: all *in* Lougnon 1970; Brown 1773: see Hachisuka 1953) I can add two more: those of Feuilley (1705) and Hébert (1708). Neither adds much to what is already known. Hébert qualified the colour as "dark blue". Feuilley said the bird was the size of a "large capon", i.e. considerably smaller than a Solitaire ("size of an average turkey cock"); this conflicts with Dubois' statement that they were as big as Solitaires. Despite the unlikely montane habitat of the Plaine des Cafres, attested to by all writers except Dubois, Berlioz (1946), from Dubois' description, considered them to be Purple Gallinules *Porphyrio porphyrio*. Other authors (Rothschild 1907*b*, Milon 1951, Hachisuka 1953, etc.) used Dubois' remarks on size to argue for something more akin to a Takahe *P. mantelli*, giving it the name '*Cyanornis coerulescens*' (= '*P. coerulescens*', Olson 1977). Feuilley's account casts doubt on Dubois' size assessment, and as no other author suggested they were unusually large, I conclude that something close to *P. porphyrio* is a reasonable assessment in the absence of subfossil material. Milon (1951) also argued against Purple Gallinule on the grounds that Dubois said the *oiseau bleu*

was good eating, which Purple Gallinules are not. The other authors (except Hébert) were more reserved, Feuilley distinguishing between tough inedible adults and excellent young birds. Hachisuka (1953), following many previous writers, considered the bird flightless, despite quoting Père Brown's (1773) testimony that it could and did fly. This is another important point in which Brown's account adds to Le Gentil's (see below, under Lizards). The assumption of flightlessness was based on Dubois' statement "they do not fly"; none of the other writers mentioned the subject, though Feuilley (1705) described hunting them as "not difficult as they can be killed with batons or with stones"; this of course applied equally to many Mascarene birds fully capable of flight.

There is no positive reference to the birds after about 1730 (Brown 1773), and it is unlikely that this information was up to date. However Grant (1801, Strickland 1848, Hachisuka 1953) quoted an anonymous officer of the British Navy as telling in 1763, of a "curious bird" from the Plaine des Cafres which "suffer themselves to be killed by the stroke of a walking-stick". These are generally assumed to have been *oiseaux bleus* (Strickland and Hachisuka *locs. cit.*), but whether or not the officer saw them himself is impossible to say. They are not mentioned by other mid-eighteenth century writers on Réunion.

3. *White-throated rail Canirallus (Dryolimnas) cuvieri*

Subfossil bones found in Mauritius and ascribed by E. Newton & Gadow (1893) to *Gallinula chloropus* are in fact *Dryolimnas c. cuvieri* (Cowles, Chapter 2). The only historical record of this species on the island is a single bird (the type!) caught in 1809 (Schlegel & Pollen 1868, Hartlaub 1877), from which has arisen the belief that it was a "one-time indigenous resident" (Rountree *et al.* 1952). The 1809 specimen, which does not differ from Malagasy birds (Schlegel & Pollen 1868, Ripley 1977) may, despite an enlarged ovary (Hartlaub 1877), have been a vagrant, but the existence of subfossil remains suggests that Mauritius did once support a population of this species.

Whimbrels

I have included mentions of "curlewes", "corbijeaux" etc. (= Whimbrel *Numenius phaeopus*) in the diagrams (Figs. 1:3–5) for comparison with endemic herons, rails etc. of similar size and gustatory potential. The lack of early seventeenth-century references may be due to their being ignored when easier (i.e. flightless and tame) game was available. Later writers often commented that the Whimbrels were excellent eating but hard to shoot (Tafforet 1726, Bernardin 1773). They appear always to have been, as they still are, common visitors to the islands (Staub 1976, Temple 1976a, Barré 1984); the old reports no doubt include the rarer Curlew *N. arquata*.

Dodos and Solitaires (Raphidae)
Mauritius

More has been written on the Dodo *Raphus cucullatus* than any other extinct bird (see Hachisuka 1953 for a bibliography), and it has become the symbol of extinction. I have re-examined the evidence for the survival of Dodos into the second half of the seventeenth century and find that the accepted date between 1681 and 1693 for their extinction is no longer tenable.

Dodos were reported by almost every traveller to Mauritius up to and including 1640, but only once thereafter. Hachisuka (1953) gave a long list of accounts (to which should be added Almeida's in 1616: Grandidier *et al.* 1903–20, vol. 2), but his later records need re-examination.

Volquard Iversen, or Evertsz in Dutch (the version I consulted is Olearius 1670), was shipwrecked on Mauritius in 1662 (not 1669 as Hachisuka stated) after his vessel the *Aarnhem* was destroyed in a storm. His party stayed 5 days on the island before being rescued by an English ship. No Dodos were reported on the mainland, but a good description is given of some they found on an islet (off the east coast?) accessible on foot at low tide. Although this account has been alluded to in the literature (A. Newton 1868b, Hachisuka 1953) the full description justifying the identification as Dodos has not been given in English. In view of the transfer of Dodo names to the Red Rail, I give it here in full, as it is the last eye-witness account of living Dodos. "Amongst other birds were those which men in the Indies call *doddaerssen*; they were larger than geese but not able to fly. Instead of wings they had small flaps [vlimmen]; but they could run very fast" (translation by M. Visser). He goes on to describe catching them and others running up when they screamed. Iversen also saw *velthoenders* (not described) which were presumably Red Rails.

Cauche (1651), who visited the island in 1640

(see Lougnon (1970) for reasons for disbelieving his own date of 1638), is the last visitor to claim to have seen a Dodo on the Mauritian mainland. In 1638 Mundy (1608–67, Sclater 1915) expressly stated that he saw none there, though he had seen two in captivity at Surat (India) – of which an Indian miniature painting survives (Ivanov 1958, Stresemann 1958, Das 1973). Apart from Iversen's account and an uncertain record from the Dutch living on the island (see below) all post-1640 records of live Dodos appear to stem from captive animals seen abroad. We have no evidence of when these animals left the island, nor how long Dodos survived in captivity – possibly many years. In 1638 Sir Hamon Lestrange saw a live "Dodo" exhibited in London (Strickland 1848); Hachisuka considered this to have been a Réunion Solitaire *'Victorionis imperialis'*, but the description does not seem to me to rule out a Mauritian Dodo, as accepted by most authors. A living Dodo was sent from Batavia (= Jakarta) to Japan in 1647 (Millies 1868, A. Newton 1868b) and Nieuhoff (1682) gave a detailed description of a bird which he appears to have seen alive in Batavia between 1653 and 1658. *Contra* Hachisuka (1953: 52) I can find no evidence from his voyages that Nieuhoff ever touched Mauritius, or indeed any other Indian Ocean island apart from the Comores, which he passed without landing on his final return in 1671. Hoffman's *"todaersen"* seen during 1673–5 have long been recognised as *Aphanapteryx bonasia* (A. Newton 1868b, Milne-Edwards 1868), but earlier, in 1668, John Marshall also used the term 'Dodo' for Red Rails (see above). Finally it appears that Benjamin Harry's 'Dodos', usually accepted as the last record, were also Red Rails (see above).

Hachisuka cited Henry Neville's *Island of Pines* (published in 1668) as evidence of Dodos at that date, but apart from this book being quite clearly a romantic novel, the supposed Dodos were found when the character George Pine first landed on the island in 1559, not when his progeny left.

This leaves Governor Hugo's account (Pitot 1905, 1914), unknown to Hachisuka, of the interrogation in 1674 of a slave recaptured after 11 years in the forests. He reported having seen a Dodo only twice during this period. Although the name Dodo had transferred to *Aphanapteryx* as early as 1668, the rails were still common around 1674 (Hoffman 1680), and it does seem more probable that Hugo and the slave were talking about *Raphus*. Although Dodos survived on offshore islets until the 1660s (Iversen's account) they must have been very rare on the mainland after 1640, as they are never mentioned by Dutch settlers, nor by the visitors who gave good faunal lists: Granaet in 1666 (Barnwell 1948), Marshall in 1668 (Khan 1927), Hoffman (1680) in 1673–5 or the sailors of the *President* in 1681. Pitot (1905, 1914) considered the Hugo report as evidence of the bird's disappearance well before Harry's visit of 1681, and I agree.

There remains the question of how long captive Dodos survived outside the island. Hachisuka considered that paintings executed as late as 1651, 1655, 1658 and even 1666 were drawn from life; the last but one is Nieuhoff's bird in Batavia, which he could have sketched as early as 1653. The 1651 bird, by Jan Savery, now in the University Museum, Oxford, with the only surviving head (and a foot), is more likely to have been based on his uncle's paintings (Strickland 1848) done in the late 1620s (Hachisuka 1953, Friedmann 1956). The basis of the *c.* 1655 painting by Kessel, about which Hachisuka was doubtful, is supported by the existence of a better one by the same artist in Madrid, dated *c.* 1660 (Friedmann 1956); Hachisuka considered there was a live bird in Amsterdam around this time. Finally Hachisuka considered it possible that there was a live bird in Berlin in 1666 to serve as a model for Ruthardt, though he admitted this painting may be based on Savery's work. Piso's 1658 commentary on Bontius (accompanied by an engraving), quoted at length by Strickland (1848), is clearly a compilation and does not indicate that either author was familiar with Dodos in life, though Piso, at least, must have seen a good picture (? a Savery).

To sum up, Dodos became very scarce on the Mauritian mainland after 1640, a few apparently surviving in the interior until the 1660s. Others persisted until at least 1662 on islets in the lagoon. In the 1620s, 1630s and possibly the early 1640s a number of Dodos were shipped abroad, and one probably survived until 1655 or a little later; this bird, in Holland, could have been brought from Batavia where one was certainly present in 1653. References to 'Dodos' in accounts of Mauritius from the mid-1660s onwards refer to flightless Red Rails *Aphanapteryx bonasia*.

Réunion

Discrepancies between the four well-known accounts of Solitaires in Réunion led Rothschild (1907*b*) and Hachisuka (1937, 1953) to describe two species of Dodo from the island. Berlioz (1946) and Greenway (1967) rejected this interpretation on the grounds that travellers untrained as naturalists can hardly be expected to produce fully consistent reports. None of the existing paintings ascribed to Réunion birds (Hachisuka 1953) has supporting documentation (see Greenway 1967 *re* Withroos's aquarelles), and those Hachisuka used for his 'Réunion Solitaire' (*Ornithaptera solitaria sensu* Hachisuka) are so poor as to defy identification. The Holsteyn and Withroos paintings of the 'White Dodo' ('*Victoriornis imperialis*') are clearly cognate: the bird has the same position and stance in both works. If there was a white dodo that Holsteyn painted, was it from Réunion, where the Dutch scarcely went, or was it an albinistic Mauritian bird?

Hachisuka's main reason for separating two forms was that Tatton (1625) and Bontekoe (Lougnon 1970) described very fat birds, whereas Carré and Dubois (*ibid.*) described tall athletic birds. A remarkable fat cycle was postulated by Oudemans (1917) for the Mauritian Dodo and is suggested for the Rodrigues Solitaire by Leguat's evidence (1708), so the Réunion bird is likely to have had a similar physiology (as does the large flightless Kakapo parrot in New Zealand (Merton *et al.* 1984: 278)). Tatton visited Réunion in March, Carré in October; Bontekoe gave no dates, and Dubois was on the island for 16 months but did not describe the Solitaire's bulk clearly.

Six further references to Réunion Dodos can be added to the four familiar accounts. Two called the birds "*lourdes*" (= heavy, stupid; Ruelle in 1667, log of the *Navarre* 1671; both *in* Lougnon 1970), but gave no descriptions. The other four are more explicit, and use the name *solitaire*. Bellanger (Lougnon 1970; North-Coombes 1980*a*) referred in 1671 (Apr.–Jan.) to "another sort of bird which is excellent and fat, which is always found alone in the mountains. One takes them by hand. They are called *solitaires*." The log of the *Breton* (in Sept.–Oct. 1671; *ibid.*) reported "solitaires which are found in the mountains, which are very tasty, and we were given one that had the flavour of a turkey cock". Feuilley's (1705) account is the most complete: "The *solitaires* are the size of an average turkey cock, grey and white in colour. [They] inhabit the tops of mountains. Their food is but worms and filth [*saleté*] taken on or in the soil. [These] birds are not eaten, being of a very unpleasant flavour and very tough. They are so called because their retreat is on the summit of the mountains. Although there are large numbers of them, one sees them but little as these areas are little frequented." Hébert (1708), the last to report the bird, simply said "There are also [birds] called *solitaires*: it is a kind of little ostrich" (my translations).

The accounts agree, except for Feuilley's low opinion of the bird's edibility, and Feuilley confirmed the general appearance of whiteness described in the earlier accounts, but they do not clarify the number of species involved. Bellanger, visiting in April–June, called the birds fat, but the other accounts did not mention the fat condition.

Later writers did not mention the species, even though d'Héguerty (1754) and Cossigny (1732–55) were both resident in Réunion in the 1730s for many years and discussed the Rodrigues Solitaire at some length. Table 1 shows that Réunion Solitaires retreated progressively from inhabited areas; I suggest that by the early 1700s the few remaining birds were prevented from breeding by introduced predators, and died out around 1710–15. Billiard (1829; discussed by Strickland 1848, Hachisuka 1953) claimed that La Bourdonnais (governor of Mauritius and Réunion 1735–42) sent a Solitaire to the directors of the French East India Company as a curiosity. This has not been verified, but any such bird is much more likely to have been from Rodrigues. D'Héguerty (1754), who was commandant on Réunion during La Bourdonnais' governorship (Thomas 1828, Azéma 1859), described seeing Rodrigues Solitaires in captivity. A regular tortoise-run to Rodrigues was in operation at that time (North-Coombes 1971), and the boats would no doubt have brought back such curiosities as Solitaires.

As Storer (1970) pointed out, there is no evidence to assign the Réunion Solitaire to either of the well-known genera of Rhaphidae, so it is best to use Bonaparte's genus *Ornithaptera* for the species. '*Victoriornis imperialis*' Hachisuka 1937 becomes a synonym of *Ornithaptera solitaria* (Selys 1848).

Rodrigues

I can add two important accounts of the Rodrigues Solitaire *Pezophaps solitaria* to the well-known reports of Leguat (1708), Tafforet (1726) and d'Héguerty (1754). Dupon (1969) and North-Coombes (1971) added a brief mention by Captain Valleau, who left Leguat on the island in 1691. Leguat's long account, which has been reproduced many times (e.g. Strickland 1848, Hachisuka 1953), gave a remarkably full

picture of the habits of the Solitaire, and included the earliest description of territoriality in birds (Armstrong 1953, Halliday 1978).

Gennes (1735) left a fourth independent eye-witness description of the Rodrigues Solitaire in March 1735. Men from his ship went ashore and returned with game:

> Our men told of having seen goats and a large quantity of birds of different kinds; they brought, amongst others, two which were bigger by a third than the largest turkey; they appeared, nevertheless to be still very young, still having down on the neck and head; their wingtips were but sparsely feathered, without any proper tail. Three sailors told me of having seen two others, of the same species, as big as the biggest ostrich. The young ones that were brought had the head made more or less like the latter animal, but their feet were similar to those of turkeys, instead of that of the ostrich which is forked and cloven in the shape of a hind's foot. These two birds, when skinned, had an inch of fat on the body. One was made into a pie, which turned out to be so tough that it was uneatable. (my translation)

The amount of fat confirms Leguat's mention of a fat period from March to September, but Leguat found Solitaires "of excellent taste" (Hachisuka 1953).

As explained under the Réunion Solitaire I believe d'Héguerty (1754) saw his captive Rodrigues Solitaires on Réunion, and it would have been a Rodrigues bird that La Bourdonnais sent to France (Billiard 1829). Jonchée (1729) listed the *Solitaire* among the birds of Mauritius; presumably he too saw captive birds from Rodrigues.

By the 1750s the birds had become very rare, and Cossigny (1732–55) wrote in 1755 that:

> for 18 months I have been trying without success to procure a *Solitaire* from Rodrigues island . . . I have promised all one could want, in spirits or *piastres*, to whoever brings me at least one alive. It is claimed that cats, which have gone wild on this little island, have destroyed this species of bird that only has stumps for wings, but I am strongly inclined to believe that these cats are the men of the post who have eaten all those they have found, as they are very good. At last,

Table 1. *Summary of observations on the Réunion Solitaire* Ornithaptera solitaria

Observer (Reference)	Date	Season	Whether birds fat	Whether good eating	Colour	Size	Legs	Habitat
Castleton (Tatton 1625)	1613	Late summer	Yes	—	White	Turkey	—	(Not stated, but must have been near coast)
Bontekoe (Lougnon 1970)	1619	?	Yes	—	—	(So fat can scarcely walk)	—	(Not stated but must have been near coast)
Carré (Lougnon 1970)	1667	Late winter	—	Yes	Yellowish	Turkey	Like turkey, longer	Remote areas
Dubois (1674)	1671–2	[All year]	—	Yes	White; black wing-tips and tail	Large goose	Like turkey	—
Bellanger (Lougnon 1970)	1671	Late summer–early winter	Yes	Yes	—	—	—	Mountains
the Breton (Lougnon 1970)	1671	Late winter	—	Yes	—	—	—	Mountains
Feuilley (1705)	1704–5	[All year]	—	No	Grey and white	Turkey	—	Mountain tops

I have been given hope of obtaining one, which, so it is said, has been spotted. (my translation)

Pingré was on the island for 3 months in 1761, but saw no Solitaires. He was told they still survived in out of the way areas (Pingré 1763, Dupon 1969), and attributed their disappearance, along with that of the rails and herons, to the introduction of cats. Morel (1778) was very clear that Solitaires no longer existed when he was writing, and Marragon (1795, Dupon 1969) did not mention them. Gorry, who arrived in 1793 (North-Coombes 1971), stated in 1832, when shown bones found in caves, that he had never seen a bird of that size on the island (Telfair 1833, Strickland 1848, North-Coombes 1971). The Solitaire was presumably extinct in 1786 ('1789', Strickland 1848, Hachisuka 1953) when the first bones were found in caves (Desjardins 1831*b*, North-Coombes 1971, 1980*a*) as the finders did not relate them to anything then living. The birds had presumably died out in the 1760s; North-Coombes (1971) added slaves (cf. Cossigny's account) and pigs to cats as the destructive agents, concluding that the last birds were driven by fires to take refuge in the famous caves where so many skeletons have subsequently been found (Chapter 2).

Pigeons (Columbidae)
Mauritius

Early travellers frequently reported pigeons' and 'turtle-doves' in their various languages, but it is rarely possible to identify them. Considerable attention was given to the ill-effects from eating the local pigeons; although latterly this reputation became firmly attached to the Pink Pigeon *Nesoenas mayeri*, the early reports suggest that the Pigeon Hollandais *Alectroenas nitidissima* (and even Echo Parakeets *Psittacula echo* (La Motte 1754–7)) may also have been involved. Cornelisz in 1602 implicated 'doves' with red tails (Strickland 1848) which could have been either species, but van Hagen in 1607 referred to red bodies as well as tails (Oustalet 1897). D'Héguerty (1754), who visited the Mascarenes in the 1730s, wrote:

The *ramiers* [wood pigeons] which are very common there [Mauritius] are very tasty; but it is dangerous to eat them all the year round; there is a season when they eat a seed that intoxicates them. If one eats them during this critical time, one is left with distressing effects, of which the least are contractions in the nerves and muscles,

and often convulsions from which one is only cured by resource to remedies and care. (my translation)

In 1732 Cossigny (1732–55) wrote:

The *pigeons-ramiers* in Mauritius are of great beauty, but there is a season of the year in which one must respect them. Many having eaten them at this time have found themselves very sick. They provoke in you a kind of paralysis with contortions of all the limbs and intolerable pains, which lasts for the 24 hours following; I don't however know of anyone [afflicted] who actually died. These pigeons must no doubt feed themselves on various seeds that produce these effects. They are never so fat as at this time, and it is true they are very tempting as they are so appetising. As for me, I don't want to taste them at any season. (my translation)

Later (*ibid.*) he distinguished between the "*pigeon hollandois*" and the "*pigeon ramier*", sending Réaumur stomach contents of both (fruit and seeds), adding:

I do not know which of these two fruits which the *ramiers* swallow makes them so pernicious that one thinks twice about eating them. The least which happens is to experience awful convulsions: many people have been caught, especially in the past [when they thought] all was good [to eat]. It is claimed that it is only during a certain time of the year, when they find in the woods seeds of *morelle* ['nightshade'] which they eat, that these *ramiers* are pernicious, and that one can eat them at all other times without danger and that they are very tasty. At least, I know well that they are often very fat. The officer that gave me this one assured me he was very incommoded from having eaten of it. A fortnight ago he sent me 4 or 5 by his black [slave]. I gave them him back; the black was delighted, as he ate them without his master knowing, and was in no way indisposed. However, at this season (March) all is in seed and in flower in the woods, and the nightshade, if it is that which gives them this dangerous quality, should not be lacking. (my translation)

La Caille (1763) and Bernardin (1773) mentioned Mauritian pigeons with toxic flesh; Bernardin, like Cossigny, ascribed the effects to the species (not described) that was not the *pigeon hollandais*. Milbert (1812) gave the *pigeon hollandais* the bad reputation, but said it was false as he had eaten plenty without ill-effects; he did not mention the Pink Pigeon. Oustalet (1897) quoted Desjardins's manuscripts of the 1830s as drawing attention to the local natural history society's intention in 1806 to investigate "the *pigeon ramier* said to be poisonous" (the society in question must have been the Société Libre d'Emulation de l'Ile de France: Ly-Tio-Fane 1972). Desjardins himself (Oustalet 1897) was assured that the Pink Pigeon was poisonous unless its head was cut off immediately after death. He also commented that the bird was rare, and indeed thereafter it was all but forgotten for some time; Clark (1859), whose natural history notes were so complete, omits any mention of it. Edward Newton (1861*b*, 1865*b*, 1875, 1878*b*) persistently remarked on the Pink Pigeon's rarity, and the position has remained much the same with some fluctuations ever since, the bird being restricted to the remaining tracts of native forest (Hartley 1978, Chapter 5). Temple (1978*c*) has investigated the cause of intoxication and poisoning but his conclusions are open to alternative explanations (Jones, Chapter 5).

Much less attention has been paid to the Pigeon Hollandais *Alectroenas nitidissima*. Cauche (1651) in 1640 describing Malagasy and Mauritian pigeons together, referred to "white, black, red pigeons, *ramiers* and *bisets*"; assuming the three colours belong together to the 'pigeon', then it was a good description of the Pigeon Hollandais. Thereafter there is no clear reference to it until Cossigny (1732–55) wrote in 1755, sending Réamur a specimen and its gut contents, nuts he understood to be of "tacamacha" (= *Calophyllum tacamahaca*) or "bois de natte à petite feuilles" (= *Labourdonnaisia calophylloides*); he found three such nuts in the crop and another filled the gizzard. The only other mention of its food is Milbert (1812), who said it lived solitarily in river gorges and lived on fruit and freshwater molluscs. The latter seemed somewhat unlikely to Greenway (1967), but Goodwin (1967, 1983) accepted it as other fruit-eating pigeons are known to eat molluscs.

Cossigny (1732–55) noted in 1755 a marked decline in Pigeon Hollandais numbers since 1732, which he ascribed to deforestation and consumption by fugitive slaves in the uncut forests. The numbers were clearly sufficient for Milbert (1812) to have eaten a good many in 1801, but by the mid 1830s it was extinct. The last specimen known was collected by E. Geoffrey in 1826 and described by Desjardins (Milne-Edwards & Oustalet 1893, Oustalet 1897). Desjardins (1832) reported the birds as "still found towards the centre of the island in the middle of those fine forests which by their remoteness, have escaped the devastating axe" (my translation), but there were no later records. Doyen (1870) claimed to have seen one caught 20 years previously in the woods of Curepipe, but his description is clearly of a broad-billed roller *Eurystomus glaucurus*. E. Newton (1865*b*) considered it certainly extinct.

Réunion

Early visitors to Réunion regularly reported the presence of *ramiers, pigeons* and *tourterelles*, but we have only two surviving plumage descriptions. Bontekoe in 1619 (Lougnon 1970) mentioned an abundance of "*ramiers* of the species with blue wings", and Dubois (1674) described slate-coloured and rusty-red pigeons in addition to undescribed *ramiers* and *tourterelles*. The red birds have often been considered as akin to *Nesoenas mayeri* of Mauritius (Rothschild 1907*b*, Berlioz 1946, Hachisuka 1953) and the slaty or blue-winged birds are likely to have been an *Alectroenas*. The identity of the others remains a mystery: Berlioz (1946) and Milon (1951) suggested *Streptopelia picturata*, still called *ramier* on the island. However, as there was a very long period in which no pigeons but introduced *Columba livia* were recorded, it does not seem probable that such a continuity really exists.

Most authors simply reported 'pigeons' without mentioning different species. Boyer in 1671 (Lougnon 1670) reported a vertical migration, birds spending November to March in the lowlands; Dubois (1674) mentioned a fat cycle. Bernardin (1687) and Houssaye (in 1689; Lougnon 1970) were the last eye-witnesses to report native pigeons; Borghesi's account in 1703 (*ibid.*), although referring to pigeons "*de tour*" (ostensibly feral *Columba livia*), appears in fact to relate to the native slate-coloured *Alectroenas*. As it stands, his account is contradictory, but this hangs on one word possibly either misprinted in the original Italian or mistranslated into French. "In ["dans", but ? = 'avant': before] the days we passed in the island, there were also found those wild pigeons we call *de tour* in such large numbers that the women, at the very moment when they were preparing the meal, killed them by dozens with a stick, even right inside their kitchen." He then referred to the introduction of rats, followed by cats, which latter spread into the forests and "entirely destroyed the above-mentioned pigeons". Feuilley (1705) stated "*Ramiers* have not been seen for some time; either they have left the island or the [feral] cats have destroyed them." He also mentioned the existence of but a few domestic pigeons. Boucher (1710) asserted that both *ramiers* and *tourterelles* were extinct, but Cossigny (1732–55) mentioned *tourterelles* in the woods in 1732 (though it is not entirely clear whether he was referring only to Réunion). Apart from Bory's (1804) observation of feral pigeons in 1801, it is not until the 1860s that pigeons of several species are reported again (Maillard 1862). In summary it appears likely that of perhaps three original species, two became extinct by the early 1700s, and the third (if not *Streptopelia picturata*) persisted perhaps to the 1730s.

Rodrigues
The only two accounts of native pigeons on Rodrigues are by Leguat (1708) and Tafforet (1726). Hachisuka (1953) considered the former in some detail but ignored the latter, which, as Staub (1973b) pointed out, is nevertheless of interest as it confirmed Leguat's belief that the pigeons nested only on offshore islets; this may, as Leguat supposed, have been due to the already abundant rats. On the basis of subfossil bones, this pigeon is generally considered to be a small *Alectroenas* (*A. rodericana*), but the generic attribution is doubtful (Chapter 2).

Parrots (Psittacidae)
Mauritius
Past interpretations of early travellers' reports have left a rather confused picture of the former Mauritian parrot fauna. In clarifying the situation I will dispose first of a straightforward mistake, and then of a common misapprehension. The mistake was by Rothschild (1907a, b), who based a new species '*Necropsittacus francicus*' on supposed old reports of parrots with "head and tail fiery red, rest of body and wings green"; Hachisuka (1953) tentatively followed this treatment. In fact, as in several other cases in his paper (1907a), Rothschild confused reports from different islands, the reports in this case being from Réunion. There are no accounts from Mauritius of parrots with the coloration mentioned (Greenway

1967, pers. obs.). The misapprehension is the common assignment to *Agapornis cana* of a small parrot described from Mauritius by Cauche (1651). I believe it to be unidentifiable (see below under introduced parrots).

This leaves three kinds of parrots described by travellers. The first, referred to as cockatoos or 'Indian ravens' by early visitors, have been identified with sub-fossil remains of a large form known as *Lophopsittacus mauritianus* (Hachisuka 1953). The second, 'grey parrots', have usually been either ignored or assimilated with *L. mauritianus*. The third, 'green parrots' with a dark neck ring, are the still extant *Psittacula echo*.

The identity of the pictures of parrot-like birds with crests, the birds called 'Indian ravens' and *Lophopsittacus* bones is well established (A. & E. Newton 1876, A. Newton & Gadow 1896, Hachisuka 1953), but two errors need correcting. Oustalet (1897), followed by Rothschild (1907b), Guérin (1940–53) and Hachisuka (1953) asserted that the sketches from life taken during Harmanzoon's voyage included a note to the effect that the birds were wholly of a blue-grey colour. None of the descriptions of the sketches by Alfred Newton (1875a, b, A. & E. Newton 1876, A. Newton & Gadow 1896) mentioned colour at all, and I believe that the colour assignation is an error. Few of the early travellers described the birds' plumage but some that did certainly gave the impression that the bird was largely grey. However Mundy (1608–67) said he saw "russett parratts" in 1638, and Hoffman (1680) described them in 1673–5 as "red crows with recurved beaks and with blue heads, which fly with difficulty and have received from the Dutch the name of 'Indian ravens'". The caption to the famous plate from van Neck's voyage in 1598 (Strickland 1848, Bonaparte 1890, Hachisuka 1953) named a crested bird up a tree as a "bird we called the Indian Crow, being in size as big again as the parakeets, of two and three colours", which strongly suggests it wasn't just grey! Perhaps there was sexual dimorphism in plumage, as there was in size (Holyoak 1971).

The other common misconception is that it was flightless (Hachisuka 1953, Holyoak 1971, etc.). While early visitors to Mauritius invariably pointed out that Dodos and Red Rails were flightless, they never said the same of any parrot, the nearest being Hoffman's comment quoted above. As mentioned, it was pictured in 1598 at the top of a tree. The supposition of flightlessness seems to have arisen from conclusions drawn from anatomical evidence (e.g. A. & E. Newton 1876).

Hoffman's account, cited above, is the last to refer unequivocally to *L. mauritianus*. Leguat (1708) referred to the presence of "parrots of all sorts" in 1693, and Gennes (1735) reported three species, but neither give any indication that any of the birds they mentioned was large and spectacular as was *L. mauritianus*. Ducros (1725) and most importantly Cossigny (1732–55, 1764) mentioned only two native species, neither of which in either case could be *L. mauritianus*. I conclude that the 'Indian ravens' died out around 1680.

La Caille (1763) referred to large and colourful "*Amazones*" amongst the parrots of Mauritius, but as no other contemporary mentioned any such bird, I suspect he was referring to captive macaws or other large parrots from elsewhere.

The very first Dutch accounts refer to grey parrots, sometimes in large flocks, clearly in addition to the 'Indian ravens' (accounts of van Neck, van Warwyck and Cornelisz; Strickland 1848, Pitot 1905, Barnwell 1948). Subsequent accounts did not always distinguish between different kinds of parrots, and seem to have led recent commentators into believing that the grey parrots were the same as the supposedly grey *L. mauritianus*. Where 'parrots' were distinguished from 'ravens' the former were often not described (e.g. accounts of van West-Zanen and Granaet; Strickland 1848, Barnwell 1948), although Hoffman (in 1673–5; Grandidier *et al.* 1903–20, vol. 2) clearly mentioned two species, "spotted and green parrots", in addition to the 'ravens'. The mystery is clarified by early eighteenth-century documents not consulted by earlier workers. Ducros' account (1725) of two kinds of parrot, grey and green, has long been known, but little significance has been attached to it, though Grandidier *et al.* (1903–20, vol. 2: 374n) ascribed the grey birds tentatively to *Coracopsis*; La Caille (1763) also mentioned grey parrots. Of the 'new' sources Gennes (1735) remarked that of the three species of parrot the grey ones were good to eat. Cossigny (1732–55) described them in 1732 as "gris-fauve" (fawn-grey), and later (1764) made the following remarks: "The woods are full of parrots, either completely grey or completely green. One used to eat them a lot formerly, the grey especially; but the one

and the other are always thin and very tough whatever sauce one puts on them" (my translation). Clearly there was nothing special about these grey parrots, and no indication that they differed markedly in size from the green ones (*Psittacula echo*). By contrast *Lophopsittacus mauritianus* was very large, had a conspicuous crest and a huge bill, and it is impossible that Cossigny, one of the best eighteenth-century writers on the Mauritian fauna should not have made some comment had these been the birds he described.

What then was this grey parrot? Until recently there was nothing reported from subfossil remains that would fit this bird, but Holyoak (1973) re-examined the *Lophopsittacus* material and described a new much smaller species, *L. bensoni*, though Cowles (Chapter 2) has cast doubt on the generic attribution. I believe this was the grey parrot, and that it survived until at least the middle 1760s, Cossigny's account in 1764 being the last reference. Bernardin (1773) referred to several species of parrot, and Milbert (1812) in a rather unclear passage, mentioned "several" species of "*perruche*", as well as some "*perroquets*" which were rare, including "a black species which comes from Ile Bonaparte" (= Réunion: see below). I suspect that imported cage-birds were confused with native wild ones.

There is no reason to suppose that the frequently mentioned green parakeets were anything other than *Psittacula echo*. Once popular as food (a hunter could kill 3–4 dozen a day: La Motte 1754–7), they are now very rare (Jones 1980c, Chapter 5).

Réunion

Two parrots are definitely known from Réunion: *Mascarinus mascarinus*, of which two skins exist (Sassi 1940, Jouanin 1962, Greenway 1967), and *Psittacula eques*, named from Martinet's plate (Buffon 1770–83, *Planches enluminées* edition) drawn from a specimen (now lost) in the Cabinet Aubry (later part of the Cabinet du Roi) in Paris (Brisson 1760, Buffon and Greenway *locs. cit.*). Stresemann (1952) considered the Paris parakeet to have come from Mauritius on the grounds that Aubry's collection was often mislabelled; however, he neglected the eye-witness reports of a green parakeet in Réunion. Travellers described both the above forms and also others, notably grey and black parrots.

The history of the Mascarin is well known, and I have little to add to the account in Milne-Edwards &

Oustalet (1893). The birds were well described by Dubois (1674) in 1671–2 and mentioned by Borghesi in 1703 (Lougnon 1970) and Feuilley (1705). Later in the eighteenth century several specimens reached France alive and were described by Brisson (1760), Mauduyt (1784) and, once stuffed, by Buffon (1770–83). De Querhoënt (1770s; *ibid.*) was the last witness to report living birds in Réunion, though there were "several" alive in Paris around 1784 (Mauduyt). There is no suggestion in Bory's account (1804) that he heard of any such bird in 1801, and I suspect that by then it was already extinct in the wild. The captive birds in Paris had died by the turn of the century, but three stuffed specimens remained (Levaillant 1801–5, Milne-Edwards & Oustalet 1893). Very remarkably, one bird, no doubt very old, survived until at least 1834 in the King of Bavaria's menagerie (*ibid.*).

The Réunion Ring-necked Parakeet fared no better. It was well described by Dubois (1674) and mentioned by Borghesi (in 1703; Lougnon 1970), Feuilley (1705), and Cossigny (1732–55) in 1732, and at least three specimens reached France to be described by Brisson (1760), Buffon (1770–83) and Levaillant (1801–5). Cossigny's report is the last reasonably definite indication of the bird alive in Réunion (and even he may have been including Mauritius in his list), though d'Héguerty (1754), who was on the island around 1740, mentioned the existence of several species of parrot. This *Psittacula*, apparently conspecific with *P. echo* of Mauritius (Chapter 5), probably preceded *Mascarinus* into extinction.

Dubois (1674) also described a plain grey parrot, and a green one the size of *Psittacula eques* with red head, wings and tail. In addition there was a plain green one, no doubt a young *P. eques*. His mysterious (Berlioz 1946) "parrots of three sorts as above no bigger than *merles*" are surely another hunter's description of the 'above' pigeon-sized three: i.e. adult *P. eques*, the red and green bird, and young *P. eques*. As the size of *P. eques*, assuming it was more or less identical to *P. echo*, was between that of a *Hypsipetes bulbul* (*merle*) and a pigeon, it is understandable that two informants might give different size comparisons.

The red and green bird was called '*Necropsittacus borbonicus*' by Rothschild (1907*a*, *b*), and Hachisuka (1953) followed this. At first sight one might consider this bird a plumage phase of the *Psittacula* (some species have reddish heads and most have red wing patches), but as no other writer referring to green parrots mentioned birds with these striking colours, they must have been a separate species. Greenway (1967) suggested an escaped lory (perhaps *Domicella garrula*). If there was a wild parrot of this description in Réunion it must have become extinct by about 1680.

The grey parrot is much better attested. In addition to Dubois' account it is mentioned by Borghesi (in 1703; Lougnon 1970), Feuilley (1705), and Cossigny (1732–55) in 1732. Dubois singled out the grey birds as the only parrot worth eating (cf. Cossigny 1764 for the Mauritian species), though Feuilley asserted all species were good to eat. Dubois did not give the size of the grey parrot, but Feuilley clearly stated they were the size of a *ramier* (pigeon), larger than the green ones (*P. eques*), but smaller than those "the size of a chicken, grey in colour, with a red beak" (*M. mascarinus*). He also reported a fat cycle, birds being fat during June to September "because at this period the trees produce a certain wild seed on which the birds feed". As with the Ring-necked Parakeet, the last clear mention of grey parrots is that of Cossigny (1732–55) in 1732, though again d'Héguerty's (1754) mention of several species presumably included them. No specimens are known.

Nineteenth-century authors (Vinson 1861, 1868, Coquerel 1864) confused this native grey parrot with Bory's (1804) "*perroquet noir*" seen, shot and eaten at the Plaine des Chicots in 1801. This species, always referred to in the nineteenth-century as a 'black parrot', is widely assumed (Coquerel and Vinson *locs. cit.*, Maillard 1862, Berlioz 1946) to have been *Coracopsis vasa*, or (Hartlaub 1877, Milon 1951) *C. nigra*, although I have not been able to ascertain that specimens were ever collected and the identification, even to genus, confirmed. It was no doubt introduced (Berlioz and Milton *locs. cit.*). I believe the grey parrot to have been different, and most probably closely akin to or identical with the Mauritian species I have attributed to '*Lophopsittacus' bensoni*. Cossigny (1732–5, 1764) wrote about grey and green parrots on both islands, and apparently noticed no difference between them.

Rodrigues

Two native parrots, both extinct, are known from Rodrigues. The larger one, *Necropsittacus rodericanus*, apparently plain green in colour, is known only from the accounts of Leguat (1708) and Tafforet (1726) and from sub-fossil bones (Günther & Newton 1879,

Hachisuka 1953, Chapter 2). According to Pingré (1763, A. Newton 1872) a few still existed in 1761, but that was the last report.

The smaller *Psittacula exsul*, with two colour morphs (Jones, Chapter 5), on the other hand, survived into the second half of the nineteenth century, although it was already scarce in 1761, Pingré bemoaning this as it was such good eating (A. Newton 1872, Staub 1973b). Commerson described a specimen around 1770 but his manuscript was subsequently lost for 120 years (Oustalet 1897). Several nineteenth-century visitors mentioned the bird, but it was not until the 1870s that specimens reached Europe and the species was described (A. Newton 1872, A. & E. Newton 1876). However they were by then so rare that Slater saw only one in his 3 months on the island in 1874 (Slater 1875, A. Newton 1875c, Sharpe 1879). Caldwell (1875) saw several but was unable to obtain a specimen himself, though he did get one from a resident (A. & E. Newton 1876). This was the last record, and it seems probable that the last few were wiped out within the year when the island suffered "the worst cyclonic season of the nineteenth century, with four cyclones in two months, the last – on 27th February 1876 – being one of the most severe of all time" (North-Coombes 1971). Colin, in a letter to Admiral Kennedy (1893), blamed the extinction of "large green parrots" on "the large fire which destroyed the forests on the western side of the island", but since he referred to their occurrence "in former times" he may have been alluding to *Necropsittacus rodericanus*; he gave no date for the fire. Despite recent claims that the parakeet might survive on outlying islets (Watson *et al.* 1963, Greenway 1967) it is quite certainly long extinct (Oustalet 1897, Vinson 1964b, Gill 1967, Staub 1973b).

Tafforet (1726) claimed three species of parrot, the last being like *P. exsul* (of which he described the green and red morph, still unknown to science) but small, plain green, with a black bill. A. Newton (1875c), Oustalet (1897) and Staub (1973b) considered this to be evidence for an early introduction of *Agapornis cana*, but I believe Tafforet was simply describing a female *P. exsul* of the blue-green morph; these authors based their belief on the size (though Tafforet did not say it was smaller than *P. exsul*) and ignored the fact that *A. cana* has a pale-grey bill whereas female *P. exsul* had a black bill (Hachisuka 1953).

A. Newton (1872) stated that Pingré did not describe the parrots he saw. While this is true for the *perroquet*, he did describe the *perruche* in a second draft of the chapter on Rodrigues, entitled 'Essai sur l'histoire naturelle de Rodrigue' bound in at the end of MS 1804 (Pingré 1763), which was presumably not the part transcribed for Newton. He wrote: "the *perruches* I saw at Rodrigues were entirely green, without any [word illegible] of red or other colour" (my translation). He presumably saw, at close range, only the blue-green morph.

Owls (Strigidae)
Mauritius
Owls were mentioned by two of the early Dutch visitors to Mauritius, van West-Zanen (in 1602) and Matelief (in 1606) (Strickland 1848, Pitot 1905), and by Thomas Herbert (1634 etc.) in 1627. There were no further references to owls until around 1770 when Jossigny's famous sketch, later to be named *Scops commersoni* by Oustalet (1897), was drawn. The younger Cossigny is reported to have mentioned them (Desjardins 1837, Oustalet 1897) in a tract, *lettre à M. Sonnerat*, dated 1774, of which I have been unable to find a copy. Desjardins left a very full description of a specimen of the same eared owl killed in the Bamboo Mountains in October 1836, and added that in 1837 various inhabitants of the Savanne area had told him they had seen them, and that a Dr Dobson had killed one near Curepipe (Oustalet 1897). The birds were evidently widespread in the forests in the 1830s, but had disappeared by 1859, when Clark wrote: "a species of horned owl existed here as lately as the beginning of this century, and was tolerably plentiful in the woods, but I believe there are no more remaining". He quoted a Mr Dalais as telling him he had shot several in his youth.

The eighteenth- and nineteenth-century reports all refer to 'Scops' (? = *Otus*) *commersoni*, described from Jossigny's drawing and Desjardins' description, but the only subfossil owl remains are of a quite different genus (Chapter 2). This is '*Tyto*' *sauzieri* (A. Newton & Gadow 1893); Rothschild's (1907b) '*T.*' *newtoni*, based on a pair of somewhat smaller metatarsi was probably a small individual of the same (Greenway 1967, Chapter 2). The seventeenth-century reports of owls give no description of plumage, so there is no indication whether or not '*T.*' *sauzieri* survived into historical times.

Réunion

No owls were ever reported by travellers but Cowles (Chapter 2) has identified bones found in 1974 with '*Tyto*' *sauzieri* of Mauritius.

Rodrigues

There is nothing to add to what has been written (e.g. Hachisuka 1953, Staub 1973*b*) on the extinct Rodrigues Owl, '*Athene*' *murivora*, mentioned by Leguat (1708), last reported in 1726 by Tafforet (1726), and confirmed by the finding of bones. Cowles (Chapter 2) points out that it should be placed in the same endemic genus (to be described) as '*Tyto*' *sauzieri* of Mauritius, which is closer to *Ninox* than to *Athene*. An alleged second species of owl was named *Bubo leguati* by Rothschild (1907*b*) on the basis of a single *Bubo*-like tibiotarsus described by Milne-Edwards (1873). The meagre details are treated fully by Hachisuka (1953).

Starlings (Sturnidae)
Réunion

The *huppe*, as the Réunion Starling *Fregilupus varius* was always called, figures in travellers' and residents' descriptions of Réunion from 1669 onwards. During the eighteenth and nineteenth centuries there was considerable confusion amongst ornithologists as to the true country of origin of the specimens that had reached Europe (see discussion in Hachisuka 1953), but it was finally established that Réunion was its only home. It was apparently a common bird well into the 1830s, though Bory's (1804) failure to mention it is odd and Bouton (1878) was told in 1817 that it was getting rare, though birds were on sale in St.-Denis market. In 1835 Desjardins (Milne-Edwards & Oustalet 1893) received four live birds from Réunion, of which two (or all four; Desjardins 1837) escaped into the Mauritian forests, where one was subsequently shot in 1837 (*ibid.*). The most recent specimens are probably those in Italy collected by Lombardi and given to Savi in Pisa in 1844 (Hartlaub 1877, Legendre 1929). The extinction of this species was reviewed by Brasil (1912) and Legendre (1929), but a few points need clarification.

The key witness to the extinction was 'Dr Jacob de Cortimoy', who wrote a letter in 1910 to Manders (1911). He described hunting the birds as a child, but then remarked that "after ten years spent in Paris I did not find a single one in the forests where formerly they flew about in flocks". This was clearly a very rapid

extinction, and it is important to discover the dates during which the fatal decline occurred. Legendre (1929) suggested the bird died out in the period 1848–58, which, in context, looks like a reference to Jacob's ten years in Paris. However these dates, and indeed the whole sentence almost verbatim, are taken from Milne-Edwards & Oustalet (1893), except that Legendre changed the dates from '1838–1858' to '1848–1858', probably a typographical error. In order to determine the date of extinction, I have tried to discover when Jacob was in Paris.

Firstly, who was this "Dr Jacob de Cortimoy" said by Manders (1911) to have been "verging on his ninetieth year"? The Jacob de Cordemoy (correct spelling) family was prominent in the sciences in Réunion in the nineteenth century, so the individual in question requires identification before his history can be checked. Fortunately there is a manuscript letter in the British Museum (Natural History), kept with a specimen of the extinct flying-fox *Pteropus subniger*, that was written to George Mason in 1908 from Hellbourg and signed "Dr Jacob de Cordemoy" (Cheke & Dahl 1981). The handwriting is that of Eugène Jacob, author of the island's flora (Jacob 1895), and it is reasonable to assume that the letter to Manders was written by the same man (*ibid.*).

Eugène Jacob was actually 75 in 1910; he was born in 1835 and died in Hellbourg in 1911 (Lincoln, n.d.). I have not been able to establish exactly when he was in Paris, but Lincoln (*loc. cit.*) gave the date of his degree as 1859, and he was certainly back in Réunion in 1861 (M. Chabin, Archives de la Réunion, *in litt.*). He is unlikely to have gone to Paris much before 1850 so the dates of the starling's disappearance can be taken as 1850–60. Reports of its continued existence into the 1860s are but hearsay: Vinson (1861), Maillard (1862) and Coquerel (1864) listed it as very rare, and Pollen (Schlegel & Pollen 1868) was unable to find any in 1865 though he was told that some still existed in the interior near St.-Joseph. Pollen (*ibid.*) also quoted Legras (1861), though without giving the reference, as saying that in much wandering around after birds he had only seen a dozen *huppes*; no dates are mentioned. Legendre (1929) quoted another similar remark from the *Album de la Réunion*. Edward Newton, speaking in Mauritius in 1874 (Bouton 1875), remarked that "at Réunion also, one species, 'The Huppe' (*Fregilupus capensis*), which, twenty-five years ago was said to be tolerably common [i.e. *c.* 1849], has now been destroyed". His brother Alfred, in 1896, put the date of extinction as "some 40 years ago" (i.e., in principle, *c.* 1856; A. Newton & Gadow 1896), but this figure seems to have been taken from an earlier article (A. Newton 1875*b*) which quoted Murie (1874), who had only partial information, as its authority. The accessions book of the natural history museum in St.-Denis starts in 1855, but no *huppes* are recorded therein (pers. obs.), and there is no mention of the species in a paper on the island's fauna by Lantz (1887), the museum's director from 1862 to 1893 (Jouanin 1970*a*), although he had written to Milne-Edwards in the 1870s (Hartlaub 1877) claiming the bird was not then extinct.

For a bird to go from abundance to extinction in a decade is rather surprising, and, as Brasil (1912) pointed out, neither man, rats nor Mynahs *Acridotheres tristis* are likely to have been to blame, as the *huppe* had co-existed with all of them for many generations. Brasil suggested an introduced disease or parasite might have been responsible. I think this is the most likely cause, but there was also a major human population movement into remote areas which may have been important. To the last (Jacob de Cordemoy *in* Manders 1911), the *huppe* was unafraid of man and easily slaughtered. The invasion of previously uninhabited areas following the emancipation of slaves in 1848 (Defos de Rau 1960, and above) may well have contributed to the disaster, as Jacob (*in* Manders 1911) stated that even in his youth the birds lived only well away from habitation. Vinson (1877) claimed that forest fires and drought, in addition to deforestation, had contributed to the extinction.

Legendre (1929) and Hachisuka (1953) listed 24–25 specimens of *Fregilupus varius* in museums around the world. They recorded two in the Port Louis and "2 or 3" in the St.-Denis museums. Both of these now preserve only one each (Barré & Barau 1982). The famous Caen specimen did not survive the last war in Europe (Jouanin 1962), nor did those in Turin and Leghorn (Violani *in litt.*). This leaves still extant 16 dry skins, one skeleton and two alcohol specimens. *Fregilupus* has generally been considered a somewhat aberrant starling (Sturnidae), but a detailed anatomical study by Berger (1957) cast doubt on this assessment, though without assigning it to any existing family.

Rodrigues

There has been considerable confusion in the literature about the relationship between starling bones found on Rodrigues (*Necropsar rodericanus*, Chapter 2), a unique and enigmatic skin in the Liverpool museum (*N.* ('*Orphanopsar*') *leguati*) and Tafforet's description (1726) of lively egg-eating birds on the Ile du Mât (= Ile Combrani; Staub 1973*b*).

Hachisuka (1953) sought to establish in Tafforet's birds a sort of 'chough' ('*Testudophaga bicolor*') instead of equating them, as previous writers had done, with the starling bones found in the caves. His claim was based solely on the birds' carnivorous and scavenging habits, which he thought impossible for a starling; I follow subsequent writers (Vinson 1964*b*, Greenway 1967, Bourne 1968, Staub 1973*b*) in rejecting '*Testudophaga bicolor*' and regarding Tafforet's birds as identical with the starling of the subfossil remains.

More problematical is the solitary skin described as *Necropsar leguati* by Forbes (1898; see Wagstaffe 1978 for a recent photograph). It has no known history before being bought off the French dealer Verreaux by Lord Derby in 1850, and bears the label 'Madagascar' in his hand. Forbes allied *leguati* closely to the known subfossil *N. rodericanus* and assumed it represented the Ile du Mât bird, while the other, *rodericanus*, lived on the mainland. Rothschild (1907*b*) and Hachisuka (1953), disliking the idea of two starlings in Rodrigues, blithely banished it to Mauritius where there were none! Others have been more cautious (e.g. Greenway 1967), and recent examination has shown that the bird may not be a starling at all (P. J. Morgan pers. comm.; it remains *incertae sedis*), and is in any case an albino, the isabelline wing and tail feathers being (*contra* Forbes and Hachisuka) due to discoloration (Morgan pers. comm.; pers. obs.). Forbes supposed the bird to have been collected by Verreaux himself, but there is no evidence that he visited Rodrigues, nor anything to suggest such a bird existed there in the nineteenth century (see below). The only bird known to have been collected in Rodrigues in the eighteenth century was the parakeet *Psittacula exsul*, collected for Commerson around 1770 (Oustalet 1897). Commerson's bird specimens were subsequently lost (*ibid.*), and it seems most unlikely that the only survivor should have been an unidentifiable albino! *Necropsar leguati* is best left in limbo until more documentation on its history and more positive evidence of its taxonomic status can be found.

What then became of the starlings on Rodrigues? Nothing is heard of them after Tafforet's account; Staub (1973*b*) pointed out that Pingré visited the Ile du Mât in 1761, and although finding huge numbers of nesting terns and other seabirds, he did not report any starlings; no such bird was known to Marragon (1795; Dupon 1969).

The Réunion native fody (Ploceidae)

Two travellers only tell of *moineaux* in Réunion, Dubois (1674) describing them in 1671–2 as having "plumage like those of Europe, except that the males in the breeding season have the throat, the head and the upper part of the wings the colour of fire". This description is clearly of a *Foudia*, and Dellon (Lougnon 1970) described in 1668 their inquisitive familiarity inside houses, recalling the behaviour of *F. seychellarum* on Cousin Is., Seychelles (Staub 1976, pers. obs.): "While the *moineaux* ('sparrows') are not any larger in Réunion than in other countries, the quantity makes them an inconvenience. They ravage sown fields, and the houses are full of them, as ours are of flies. One often sees them falling into pots and platters and burning their wings in fires, lit outside" (my translation). Dubois said that only one harvest was possible per year, instead of three, because of the ravages of these birds – the sole harvest took place when the birds were up in the hills nesting.

There is a long silence in the literature before the publication of Buffon's encyclopaedia (1770–83) in which appeared a bird he called the *mordoré*, depicted by Martinet in the *planches enluminées* (No. 321) as the *bruant* (= bunting) *de l'île de Bourbon*. This bird is a dull red without any black pigment, and I concur with Berlioz (1946) that it is an anomalous Cardinal, *Foudia madagascariensis*. Plumage anomalies of this kind are common in this species (Delacour 1933, Berlioz 1946, Forbes-Watson 1969, pers. obs. in Mauritius). There is no corroboration of the alleged origin of this specimen, and no further records of fodies in Réunion until the mid-nineteenth century.

Moreau (1960) accepted the existence of two endemic fodies in Réunion, arguing that the Martinet plate was not a *madagascariensis* as its bill was too slender, it had no streaks on the back, and that Réunion was large and diverse enough to support two

forms. Unstreaked cardinals, *contra* Moreau, are not unusual (see above), and to my eye the *mordoré's* beak is almost as big as that of the same artist's cardinal (*Planches enluminées* No. 134). Other writers (e.g. Rothschild 1907*b*, Hachisuka 1953) assimilated both Dubois' and Buffon's birds as '*Foudia bruante*'. Greenway (1967) rejected any native fody on Réunion. I follow Berlioz (1946) in allowing the Dellon/Dubois birds, but not accepting the *mordoré*. This fody must have become extinct shortly after Dubois' visit, as a pest of its severity would have been mentioned later had it survived. Its disappearance coincides with the introduction of rats (see below). The Réunion Fody has no scientific name, *F. bruante* P. L. S. Muller belonging to the enigmatic *mordoré*. Like Berlioz (1946) and Moreau (1960), I believe Dubois' "*moyneaux*" were a form akin to *F. omissa*, analogous to *F. eminentissima* in the Comores and Aldabra, and *F. rubra* in Mauritius. The native fodies on Mauritius and Rodrigues are mentioned below (Chapters 4 and 8, respectively).

Other passerines and swifts
Mauritius

As far as is known no Mauritian passerines have become extinct. However, no small birds, except one 'finch' (A. Newton & Gadow 1893) have been recovered in subfossil deposits, and early travellers almost entirely ignored passerines other than the highly edible Merle *Hypsipetes olivaceus*. I have included the mentions of this species, 'swallows' and 'sparrows' etc. in Fig. 3 for comparison with the records of larger birds. The long silence, 1628 to *c*. 1770, noted for owls recurs with 'swallows' (presumably *Phedina borbonica* and *Collocalia francica*); one can be certain both were present throughout. I have taken early mentions of '*moineaux*', 'sparrows', 'lynnetts' etc. to refer to the fody *Foudia rubra*. None of the other passerines receives any mention until described by Buffon (1770–86) from stuffed specimens, supplemented by field notes from de Querhoënt. Their current status is discussed elsewhere (Staub 1976, Chapter 4).

Réunion

Apart from the fody already discussed, few of the small birds of Réunion are mentioned by travellers. The Merle *Hypsipetes borbonicus*, a favourite food, was regularly recorded, but apart from Borghesi (Lougnon

1970), who recorded Grey White-eyes *Zosterops bor-bonicus* and two species of 'swallow' in 1703, no one reported on the other species until Brisson (1760) described several from specimens sent by La Nux (Stresemann 1952). None of the known smaller Réunion birds has become extinct; their present status is considered elsewhere (Barré & Barau 1982, Barré 1983, Chapter 6).

Rodrigues

The references to small birds in Rodrigues in the old literature are somewhat enigmatic, but can be related reasonably well to the two endemic passerines surviving today: a warbler *Acrocephalus rodericanus* and a fody *Foudia flavicans* (Staub 1973b, 1976, Cheke 1979a, Chapter 8). Leguat (1708) claimed to have seen a few 'swallows', which Staub (1973b) suggested may have been *Phedina borbonica*; there are no other such observations. Bourne (1968) argued that the birds described by Leguat and Pingré respectively as like a 'canary' and a 'tit' (Vinson 1964b) were more likely to have been a white-eye *Zosterops* than either of the surviving passerines. The fody fits both descriptions: it is yellow and behaves like a tit *Parus* sp., but Bourne also commented that the old authors denied it a song, which is odd as *F. flavicans* is very vocal (Vinson 1964b, Chapters 3 and 8). Tafforet's descriptions of both birds are much less ambiguous (*ibid.*) and Marragon's *moineaux* must have been fodies, though Staub (1973b) thought they were introduced house sparrows.

Corby (1845, North-Coombes 1971, 1980a) aptly described the fody as "a small bird half canary and half cardinal", and it must have been he who collected the first specimen, transmitted shortly afterwards to Alfred Newton in Cambridge but not appreciated as a new species until 1864 (A. Newton 1865a, E. Newton 1865a). The Newton brothers, however, gave the sender of the skin as 'Col. Lloyd', whom one might reasonably presume was Lt.-Col. J. A. Lloyd, Mauritius Surveyor-General of the time (Lloyd 1846, Vinson 1964a). He was the man to whom Corby addressed his report, and to whom, to judge by a memoir on plants attached to the manuscript (Corby 1845), he gave all his specimens (though no birds are mentioned). However, in Cambridge there are several bird specimens from Rodrigues (two Red-footed Boobies, a Crested Tern, a Sooty Tern, and the fody; C. W. Benson *in litt.*) dated '1845 or 1846' and labelled

as collected by 'Col. M. B. S. Lloyd' or 'Col. M. Lloyd'. There was no Lloyd, Colonel or otherwise, on HMS *Conway*, the ship on which Corby travelled to Rodrigues (I have checked the log in the Public Record Office), and it is most unlikely that there should have been two Col. Lloyds in Mauritius at the same time, both involved with the natural history of Rodrigues. I can only suggest that in writing to Alfred Newton the Colonel signed himself 'Lloyd, M. B. S.', the 'initials' denoting 'Mauritius B(..?..) Surveyor', or that the Lloyd in Mauritius had a military relative in Britain who forwarded the birds to Newton.

We now know that at one time there were also other passerines on Rodrigues. Cowles (Chapter 2) found subfossil remains of a *Hypsipetes* bulbul and a possible babbler (? Timaliinae) in the caves of Plaine Corail. No traveller mentioned *merles* and it seems likely that both these were extinct before Leguat's arrival on the island.

REPTILES
Tortoises (Testudinidae)

Arnold (1979, 1980) and Bour (1980) have reviewed the species of Mascarene tortoises, rationalising the number of species accepted to two each on Mauritius and Rodrigues and one on Réunion, where there was probably also a second (Bour *in litt.*1982). Bour (1978, 1980) revived the genus *Cylindraspis* for the Mascarene tortoises, but I follow Arnold (1979, 1980) in treating it as a subgenus of *Geochelone*. The complicated nomenclature has been clarified by these authors for Rodrigues (*G. peltastes, G. vosmaeri*) and Réunion (*G. borbonica*), but differing interpretations have left the Mauritian nomenclature confused. I follow Arnold (1979, 1980) in using *G. inepta* and *G. triserrata*, names founded on subfossil material, for the two species there, as new doubts (Bour *in litt.*) have arisen about the allocation (Bour 1980) of specimens derived from living animals in the seventeenth and eighteenth centuries. The name *gigantea*, usually applied to Aldabra tortoises, turns out to have been based on a Mascarene specimen (Bour 1984a). Of the travellers who saw tortoises in the Mascarenes, only three visitors to Rodrigues distinguished different species. All other visitors simply referred to 'land turtle' or '*tortue du terre*', as though there were a single form on all islands. The history of tortoise populations on Indian Ocean islands was reviewed by Stoddart &

Peake (1979) using only well-known sources, but Bour (1981) has covered Réunion more thoroughly. The following account brings together a number of sources not known to these authors.

The first visitors on all three islands reported tortoises in enormous abundance. Unfortunately for these reptiles, they were much sought after by ships' crews as fresh meat, for they could be kept alive without food or water for 3 or 4 months aboard ship. They were also renowned as a quick cure for scurvy, then a terrible affliction for travellers; the sick, landed on a tortoise island, were given a pure tortoise diet until they recovered. Tortoises (and turtles) also yielded an oil of high quality, and the liver was a delicacy (details from accounts in Gennes 1735, Pitot 1905, Lougnon 1970, North-Coombes 1971).

Mauritius

As early as 1639, only a year after Mauritius was settled, the commander, Van der Stel, was ordered to husband the tortoises (Pitot 1905). By 1652 they were reported almost gone (*ibid.*), meaning presumably near the settlement. In the late 1660s and early 1700s although common on east-coast islets and Flat Is., they were scarce on the mainland, their eggs preyed on by feral pigs (*ibid.*). Governor Hugo, in 1673 (*ibid.*, Bour 1981), reported a massacre of tortoises and observed that they were killed only for their grease, the rest wasted; 400–500 were needed for a barrel of grease, 30–40 for half a pint, up to 50 being killed before a fat one was found! By contrast, when fat, according to Dubois (Lougnon 1970), each (Réunion) tortoise could yield "deux pots" (quantity?) of oil. Leguat (1708) reported both turtles and tortoises as very rare in 1693. By 1698, Governor Deodati was complaining that he had eaten tortoise only twice in 5 years and that none could be found to supply the ship on which he sent his despatch (Barnwell 1948). La Merveille in 1709 (La Roque 1716) is silent about tortoises, and the last mention of them on the Mauritian mainland comes from the log of the *Courrier de Bourbon* (1721), when they were reported scarce; good numbers were, however, found on offshore islets (see below). Gennes (1735) stated categorically that "the tortoises are entirely destroyed", La Motte (1754–7) mentioned only domesticated ones and Cossigny (1732–55, 1764) made no reference to them at all; Pingré (1763) wrote that "since the island has been populated by man, it has been depopulated in land tortoises, but a few sea turtles are fished". Finally the collection of Mauritian laws from 1722 onwards compiled by Rouillard (1866–9) contains nothing on tortoises, whereas on Réunion in the 1720s there was a rash of legislation aimed at tortoise conservation (Lougnon 1957, Bour 1981); I infer from this that tortoises were already past saving in Mauritius by 1722.

In 1741 Grant (Grant 1801, Pope-Hennessy 1886) wrote "we posses [*sic*] a great abundance of fowl, as well as both land and sea turtle, which are not only a great resource for our daily wants, and serve to barter with the crews of ships who put in here for refreshment in their voyage to India". On the basis of this statement, supported by an ambiguous remark in a similar vein by Gandon (1732; "tortoises begin to become rare although they are still pretty numerous"), the date of extinction is generally put at well after 1741 (Stoddart & Peake 1979), extending indeed to 1778, when Morel (Sauzier 1893) reported tortoises "rare". However, Morel worked for the hospital in Port Louis, and he would certainly have been referring to supplies from Rodrigues or the Seychelles rather than local abundance. In fact it is clear from other passages in his paper (Morel 1778) which Sauzier omitted to transcribe, that he was lumping all the islands together in his assessment of scarcity, although he did in addition make specific reference to Rodrigues (q.v. below). Subsequent writers appear not to have referred back to Morel's original writings.

I believe that these post-1721 statements all refer to the availability of tortoises, and not to their presence on Mauritius. After the settlement of Mauritius by the French a tremendous trade in tortoises from Rodrigues developed (see North-Coombes 1971) and these were kept at Port Louis for local purposes and for revictualling ships; Dalrymple (1755) described the pounds constructed near the port for Rodrigues turtles and "a particular spot of ground inclosed here for keeping and breeding [*sic*] the land tortoise". It is to these animals, not native Mauritian ones, that Gandon, Grant and Morel referred; in the 1730s there was also a residual supply on the northern islets (see below).

Réunion

In Réunion early reports extolled the abundance of tortoises (Froidevaux 1899, Bour 1981), but by 1689 they had gone from around the settlement at St.-Paul and had to be fetched from St.-Gilles and St.-Leu (Houssaye *in* Lougnon 1970). In 1703–5 they had to be fetched from up in the mountains (Luillier and Durot *in* Lougnon 1970), and by 1708 Hébert (*ibid.*) reported that they could only be had 7 leagues (39 km assuming a *lieue marine*; most travel around the island was by boat) from St.-Paul and 14 leagues (78 km, i.e. half-way round the island) from St.-Denis. He recommended a ban on taking tortoises, which was enacted in 1710 (Bour 1981), and in 1715 the local assembly reiterated the prohibition, on pain of fines and lashing, on taking the animals for "domestic uses" (as opposed to supplying the hospital) (Lougnon 1957, Bour 1981). In 1717 tortoises were reported as rare, found only in the east of the island (Le Gentil 1727, Lougnon 1970), though the crew of the *Athalante* was able to find some at Grand Bois (south-west) in 1722. There were no longer any in the lowlands by 1725–6 (Lougnon 1957) and by 1732 they were considered totally extinct (Gandon 1732, Cossigny 1732–55, Gennes 1735); Caulier (1764) stated they had been "extinct down to the last hatchling for more than 30 years". However, Petit (1741) received and described two living animals from Réunion in 1736 (Bour 1979), and recent discoveries of bones from the island (*ibid.*) confirm that Petit's animals were native, and not from Flat Is. or Rodrigues. This was the last unequivocal record of Réunion tortoises for a century. Travellers continued to make either vague statements not certainly referring to wild or even native tortoises (e.g., allegedly, La Motte in 1754 (Vaillant 1899, Bour 1981) though he was in fact writing only of Mauritius (La Motte 1754–7)), or declarations that the animals were extinct. Bory (1804), who was very keen to find tortoises and explored a large part of the island in 1801, considered the tortoises totally extinct, discovering nothing better than a small carapace, found in 1777 near St.-Philippe, in use as a lamp. He illustrated a 'similar' one he saw in a French provincial museum, which has been identified as the Asian terrapin *Geoemyda spengleri* (Wermuth & Mertens 1977, Bour 1981), but Bour (1981) argued that the lamp was the tooth-edged carapace of a very young native tortoise, noting that the discovery site was at that time in a dense forest. Stoddart & Peake (1979) quoted Bory (1804) as having seen "animals up to about 230 kg, perhaps domesticated", suggesting they were surviving native ones, but the passage in question (vol. 1: 248) is in quotation marks: Bory was using

Du Quesne's words from the mid-1600s to convey an impression of the island's past. Billiard (1829, Sauzier 1893), who was in the island 1818–20, reported that no tortoises had been seen for a long time.

In the latter part of the nineteenth century accounts were published referring to tortoises surviving around the turn of the century in the heights between St.-Denis and St.-Paul (Vinson 1868) and until the mid-1800s in the fastnesses of the Cirque de Cilaos (*ibid.*, Hermann *in* Bour 1981), an area hardly penetrated by man before 1832 (Bour 1981). Hermann (*ibid.*) cited an old man who had seen tortoises scrambling up the vegetated cliffs in (*fide* Bour) about 1840. He also claimed to have found bones at the Ilet à Cordes in Cilaos, but none was preserved. Huge numbers of Malagasy tortoises *Geochelone* (*Asterochelys*) *radiata* and some also from Aldabra (*G.* (*Dipsochelys*) *elephantina*) were being imported in the mid-1800s (*ibid.*, Vinson 1868), and one might suppose reports of feral tortoises to refer to escapes, especially as Vinson (1861), and so no doubt others, apparently thought the native animals were identical to *G. radiata*. However, the fact that the animals were seen in Cilaos, an area with almost impregnable natural defences against penetration from outside (gorges, torrents, cliffs), and that they disappeared during precisely the period of human exploration of the area (1832–45, Bour 1981), suggests that it genuinely was a refuge in which native tortoises survived for many decades after they had vanished elsewhere on the island.

Rodrigues

All early visitors to Rodrigues mentioned the abundance of tortoises. By the 1730s, when the animals had all but disappeared on Mauritius and Réunion, a major trade had developed bringing tortoises from Rodrigues to the other islands (North-Coombes 1971, Bour 1981). Such were the quantities available that as late as 1761 Pingré (Dupon 1969, Stoddart & Peake 1979) complained of having almost nothing else to eat for over 3 months. North-Coombes (1971) has reviewed some details of the trade; at its peak (1740s) some 10 000 a year were being exported, dropping to 7000–8000 in the early 1750s (La Caille 1763), and further to 4000–5000 up to 1770 (Stoddart & Peake 1979), although we know at least 12 000 were taken in 1761 (North-Coombes 1971). A cyclone killed 4000–5000 in 1739 (Mahé 1740) in the keeping pounds for Rodrigues turtles and tortoises in Port Louis harbour described in the 1750s by Dalrymple (1755). By 1769 the tortoise station was almost closed down, and in 1771 orders given from France to husband the few remaining animals (North-Coombes 1971). Around the same time Jossigny drew a fresh specimen for Commerson in Mauritius (Vaillant 1898). After 1768, when a shipment of 1215 left the island (North-Coombes 1971), there is only second-hand information until the finding of a single animal on Plaine Corail in 1786 (*ibid.*: 269; not 1804 as erroneously re-cited by North-Coombes 1980*a* and Bour 1981).

Après de Mannevillette (1775, Dupon 1969) blamed the decline of tortoises on cats and rats, while Cossigny (1799) put the blame on exploitation and only secondarily on introduced predators; Morel (1778), given to sweeping statements, considered the animals "destroyed". Philibert Marragon took over as civil agent in 1794 (North-Coombes 1971), and the following year wrote that he had found only two tortoises in over a year (Marragon 1795, Dupon 1969). This is the last eye-witness report of tortoises from the island, although 8 years later Marragon (1803) was writing as though a few animals still survived, blaming their continued destruction on random fires lit in the scrub by slaves wandering the islands on their days off. Well into the nineteenth century a myth persisted that tortoises still survived, apparently through endless copying of Après de Mannevillette's statement of 1775; thus in 1830 d'Unienville (Dupon 1969) referred to the "present scarcity" of tortoises and repeated the blame on rats and cats, and even Froberville (1848) appeared to think a few still survived. Corby (1845), who explored the island thoroughly, finally pronounced the tortoises extinct.

On Rodrigues both Leguat (1708) and Tafforet (1726) claimed to distinguish three species of tortoise, and Pingré (1760–2, 1763) two. Neither Leguat nor Tafforet described the differences they noted, but Pingré clearly distinguished larger rare *carosses* from the commoner smaller kind (no special name) which measured a foot and a half (46 cm) long by a foot (30 cm) wide. The *carosses* were formerly commoner, but had been selectively removed for export. These distinctions fit well with remains of the two osteologically distinguishable species, *G. vosmaeri*, a large high-backed form up to 85 cm in length, and *G. peltastes*, a smaller domed tortoise reaching only 42 cm (Arnold 1979).

Marine turtles (Cheloniidae)
Mauritius
Records of breeding turtles, presumably primarily *Chelonia mydas* since they were always eaten without ill-effects, follow a similar pattern to those of sea-cows (q.v.). While landings are often mentioned by the Dutch in the seventeenth century (Pitot 1905), and there are repeated references to turtle eggs being devoured by wild pigs, there are no positive records of turtles breeding in Mauritius in the eighteenth century. However, they evidently came ashore regularly throughout with decreasing frequency, and are often mentioned with sea-cows (q.v. for quotations) as becoming increasingly rare. Clark (1859) referred to both Hawksbill (i.e. *Eretmochelys imbricata*) and Green Turtle as being "rarely met with at present, though once abundant"; the same is true today (D. Ardill, pers. comm.). The same informant told me of a successful nesting of a Green Turtle, the first for many years, near Cannoniers Point in 1977 – though many of the emergent young were caught for captive rearing.

Réunion
As in Mauritius most of the early visitors mentioned the presence of both land tortoises and sea turtles. Réunion has relatively few suitable sandy beaches for turtles to use for laying. In the 1660s and 1670s, according to Martin and Dubois (Lougnon 1970), the majority came up at St. Paul, which was also the site of the biggest settlement. Dubois (1674; *ibid.*) stated that they came to land throughout the year, but Feuilley (1705) noted a peak in March and April, and said they were given some protection (two per person per week allowed!). Boucher (1710) was the last writer to mention turtles onshore in Réunion, but Lougnon (1957) noted that female turtles were protected under a law of 1720, which suggests landings continued into the 1720s. No doubt turtles continued to make sporadic appearances, as in Mauritius: Maillard (1862) listed *Chelonia mydas* as rare. In recent years a project for a turtle hatchery and farm (using imported eggs) has been set up (Lebeau & Lebrun, 1974, Bour & Moutou 1982). Bellanger in 1691 (Lougnon 1970) was the only early author to mention "*carrets*" Hawksbills), but there are a few recent records (Bour & Moutou 1982).

Rodrigues
Turtles are mentioned in all early accounts of Rodrigues, and suffered heavy exploitation during the

eighteenth century (North-Coombes 1971; 500–600 a year in the 1750s according to La Caille 1763). When the tortoise trade fell off, turtles were given a respite, though some trading continued into the 1790s (North-Coombes 1971) and Marragon (1795; Dupon 1969) stressed their value as an article for export from the island. Although they are mentioned thereafter, there is no indication of numbers, which must have been fairly low throughout the nineteenth century or they would have attracted attention; as it was, Mauritius was supplied with turtle from the Seychelles and St. Brandon (Sauzier 1893, Barnwell 1948, Staub & Guého 1968), and indeed 300–500 were still being imported annually from the latter in the 1950s and 1960s (Scott 1961, Staub & Guého 1968). Both Green Turtles and Hawksbills still come ashore at Rodrigues, juveniles of the latter in sufficient numbers that each time the *Mauritius* leaves the island there are five or six stuffed ones brought on board as souvenirs (pers. obs. 1974). *Carets* (i.e. Hawksbills) were mentioned in the eighteenth century by Pingré (1760–2) and Marragon (1795), but not at all in the nineteenth-century literature.

Lizards (Sauria)

The native lizard fauna of the islands consists of skinks and geckos, only geckos having reached Rodrigues.

Mauritius: mainland

Early travellers to Mauritius rarely mentioned lizards, the first to do so being Peter Mundy (1608–67) in 1638, Hoffman (1680) seeing "lizards of all colours". Not until the mid-eighteenth century are there any descriptions that can be identified to species: Cossigny's (1732–55) clear description of *Phelsuma ornata* in 1736 was the first. Subfossil remains have however been found of a very large skink *Leiolopisma (Didosaurus) mauritiana* (Hoffstetter 1949), which could hardly have avoided notice by early travellers, especially as such skinks are usually bold and inquisitive. Arnold (1980) has examined further bones and found a second large skink, referable to the extant *Leiolopisma telfairii*, now confined to Round Is. (Vinson 1964a). Both Hoffstetter and Arnold (*locs. cit.*) also identified remains of *Scelotes (Gongylomorphus) bojerii* amongst the subfossil bones. This endemic species was common and widespread in the mid-nineteenth century (Desjardins 1830, Clark 1859) but was thought to have become extinct on the

mainland until recently rediscovered (Vinson 1973) in very small numbers in Macabé Forest. It survives also on Ile Fouquets off Mahébourg and on all the northern islets (Vinson & Vinson 1969). Remains have also been found (Hoffstetter 1946b, Arnold 1980) of the large gecko *Phelsuma guentheri*, also now found only on Round Is. (Vinson 1973). Hoffstetter (1946b) referred remains of smaller geckos to *Phelsuma cepediana* and *Hemidactylus frenatus*. The former has since been split into three endemic species (*P. cepediana, P. ornata* and *P. guimbeaui*; see Vinson & Vinson 1969); these may not be fully separable osteologically, but Arnold's (1980) material appears to be referable to *P. cepediana* and *P. guimbeaui*. Arnold (1980) also found jaws of *Nactus serpensinsula* (formerly *Cyrtodactylus*, see Kluge 1983), now confined to the northern islets, amongst mainland deposits, and considered that the material Hoffstetter (1949) referred to *Hemidactylus frenatus* was in fact *N. serpensinsula*, a species discovered subsequently (Loveridge 1951).

The extinction of *Leiolopisma mauritiana, L. telfairii, Phelsuma guentheri* and *Nactus serpensinsula* before the advent of man can probably be attributed to rats accidentally introduced in the sixteenth century (see above). The near-destruction of *Scelotes bojerii*, and its total disappearance from its former haunts (open areas, especially near the coast; Desjardins 1831) was tentatively attributed by Vinson & Vinson (1969) to mongooses or the agamid lizard *Calotes versicolor*, both introduced around 1900 (see below). Pike (1870b, 1873), however, clearly stated that the only skink he ever saw on the Mauritian mainland was *Cryptoblepharus boutonii*, so it seems that *S. bojerii*'s decline must have occurred much earlier (on p. 161 of *Subtropical Rambles* (1873), Pike confused the names *S. bojerii* and *C. boutonii*, although the descriptions make it clear which species he was really discussing). Of the other native lizards, the three endemic *Phelsumas* and *Hemidactylus frenatus* (native at least on islands to the north; Cheke 1984) remain widely distributed and generally common (Vinson & Vinson 1969, Vinson 1976b). *C. boutonii* is rather scarce, being now found only on the north and north-east coasts (Vinson & Vinson 1969).

Réunion

In Réunion, Dubois (Lougnon 1970) mentioned the existence of lizards in 1670, but there are no further allusions to them before Brown's report (1773).

Père Brown's account of the island has been widely quoted in the past, but d'Aglosse (1891; see also Lougnon 1970) has discovered that no such Jesuit missionary existed, and considered the passage to be simply a copy of Le Gentil's visit in 1717 (Le Gentil 1727, Lougnon 1970). I consider d'Aglosse's and Lougnon's assessment to be too simple, and this warrants a brief digression. Despite various assertions that Brown's letter was written in 1667 (Vinson 1868) and/or published in 1724 (Coquerel 1865a, Hachisuka 1953, etc.), it did not in fact appear before 1773 (d'Aglosse 1891, Grandidier *et al.* 1903–20, vol. 3; J. Griffin of the British Library *in litt.*). The text is clearly cognate with Le Gentil's, but Brown's text is fuller on the *oiseau bleu* (see above), and only he mentions large colourful lizards (see below). His visit was clearly later than Le Gentil's as the human population was given as larger, Parat was no longer governor and the date of discovery of native coffee was given as "22 years ago". This last would put the date as late as 1733 (discovery in 1711; Toussaint 1972), but other figures in both texts are wrong (e.g. discovery date of island; population details incompatible with table in Toussaint (1972)), so a date some years after 1717 is the best guess. I suggest that, rather than one copying the other, both authors used the same unidentified source for their reports, selecting different parts and paraphrasing differently.

The lizard story is rather fanciful, and Brown admitted not having seen the beast, but it suggests that the inhabitants knew of a large colourful arboreal gecko, possibly not, from the size given, any existing *Phelsuma*, but something closer to *P. edwardnewtonii* of Rodrigues. Bones would be worth looking for.

> It resembles our European lizards, except that it is fatter and longer, and that the colour of its body is infinitely brighter and more varied. I was assured that its head was flat and pierced through the middle, in such a way that one could pass a wire through without upsetting it [the animal]. This animal is most common towards the southern part [of the island], where it is also fatter and longer; it is claimed that some can be found there up to a foot and a half long.

Brown was also told the lizard had wings and flew from tree to tree – geckos often jump from branch to branch. No traveller reported anything resembling *Leiolopisma telfairii*, recently found subfossil (Arnold 1980).

Bory (1804) reported three species of lizards – a house gecko, a green gecko (*Phelsuma*), and a skink (presumably *Scelotes bojerii*) living on tracks and rocks – but he did not clearly state whether these were in Réunion or Mauritius, or both. Two species of *Phelsuma* are known today, endemic races of *P. borbonica* and *P. ornata* (Vinson & Vinson 1969, Cheke 1982a, Bour & Moutou 1982), but the skink has not been recorded for a long time (Vinson & Vinson 1969, Bour & Moutou 1982). There are specimens in the Paris museum collected in the early nineteenth century (*ibid.*), the newest, from Rosseau, dating from 1839 (F. Roux-Estève pers. comm. from museum MSS); Desjardins reported it common in 1837 (Vinson & Vinson 1969). By mid-century Maillard (1862) considered it rare; there are no more recent reports (Bour & Moutou 1982). Maillard (1862) was the only author to mention *Cryptoblepharus boutonii* (Bour & Moutou 1982).

As the *Phelsumas* were thought extinct for nearly a century until 'rediscovered' in 1964 (Mertens 1966), and *S. bojerii* has only recently been rediscovered on the Mauritian mainland (Vinson 1973), there is perhaps still hope of the skink re-appearing on Réunion. Before their decline to 'extinction' the *Phelsumas* had been common, Vinson (1868) attributing the decrease to the depredations of a (then) recently introduced Malagasy snake (see below).

Rodrigues

Vinson & Vinson (1969) have given details of the history and extinction of the two large geckos, *Phelsuma gigas* and *P. edwardnewtonii*, which once inhabited Rodrigues; they also reprinted the observations of early travellers and Liénard's (1843) detailed descriptions. The very large (total length up to 540 mm) nocturnal greyish *P. gigas*, probably extinct on the mainland by 1761, survived on Ile Frégate until at least 1841, but was not reported thereafter. Rats were abundant on this islet in 1916 (Bertuchi 1923) and it was no doubt their introduction that eliminated the lizards (Vinson & Vinson 1969). *Phelsuma edwardnewtonii*, brightly coloured and diurnal, was reported by all writers up to and including Marragon (1795) as abundant on the main island. The specimen discussed by Liénard (1843) was of unknown provenance, but Corby (1845), who did not mention visiting any offshore islets, reported seeing 'beautiful . . . lizards', and the specimen collected by Desmarais in July 1876 was collected in the "jungle of Rodrigues" (Günther

1879). However, the lizards were by then extremely rare on the mainland as none was obtained by the Transit of Venus expedition in 1874, and Caldwell (1875) got none in 3 months on the island in 1875 despite offering a good reward for "a specimen of the long lizard called *coulevec*, supposed to be extinct but which I imagine still exists".

Vinson & Vinson (1969) supposed that the specimens collected in 1884 and 1917 were found on offshore islets, which seems likely. Bertuchi (1923) maintained that by his time (1916) the "large and beautifully coloured lizards" were extinct, although in fact two were collected by Thirioux the following year (Vinson & Vinson 1969). They were the last to be recorded.

The only other lizards apparently native to Rodrigues are the small geckos *Lepidodactylus lugubris* (*ibid.*), and possibly *Hemidactylus frenatus* (Cheke 1984 *contra* Vinson (1964b) and Vinson & Vinson (1969), who considered it introduced). Either or both may have been the small "*lizard de murs*" mentioned by Pingré (1763, Vinson & Vinson 1969), but not commented on by any other writer before Bertuchi (1923); the Transit of Venus expedition collected only the introduced *Gehyra mutilata* (Günther 1879).

Snakes (Ophidia: Boidae)

Early travellers to the Mascarenes were unanimous in rejoicing in the absence of snakes, and Cossigny (1764) related a legend that a Portuguese monk on discovery of Mauritius exorcised all snakes. D'Héguerty (1754) opined that, in Réunion, snakes that people tried to introduce "died at the sight of land, and as soon as they began to breathe its air". In fact snakes were discovered on the offshore islets of Mauritius in the 1670s (Pitot 1905; see below), but the fact remains that none was ever reported alive on the mainland until they were introduced in the nineteenth century.

Subfossil remains of two snakes have been found in the Mare aux Songes, Mauritius; Hoffstetter (1946a) referred one of these to the endemic subfamily Bolyerinae, and Arnold (1980) confirmed from further material that it is *Casarea dussumieri*, which still occurs on Round Is. The other Round Is. snake, *Bolyeria multicarinata*, the only other surviving bolyerine, has not been found amongst the subfossil remains, although it also presumably once inhabited the main island. The history of snakes in offshore islets is dis-

cussed in the next section. From the other snake bones, Hoffstetter (1946c) described a blind-snake *Typhlops cariei*, now extinct. It was larger than *Typhlina bramina*, which occurs today on all three islands (Bour & Moutou 1982), and was probably introduced.

The early introduction of rats may have been responsible for the elimination of the native snakes on Mauritius, as of the lizards. Barkly (1870) reported that a snake "not unlike" one of the Round Is. species had been found (no date) at Pointe aux Piments, and was at that time in the Desjardins Museum, but it has since vanished (Vinson 1950). It is possible that it was an individual sea-drifted from Round Is., but more probably an introduction from Madagascar as it had been labelled "*Herpetodryas Bernierii*" (i.e. presumably, *Dromicodryas bernieri*: see Guibé 1958).

There are no indications either from travellers' observations or from subfossil remains that snakes ever occurred on Réunion or Rodrigues.

The reptile fauna of the Mauritius offshore islets

While on the Mauritian mainland the endemic snakes, and the large lizards *Leiolopisma mauritiana*, *L. telfairii* and *Phelsuma guentheri*, became extinct before man's arrival, and the tortoise disappeared soon after 1721 (see above), these reptiles persisted a great deal longer on offshore islets that were free of rats and, in some cases, difficult to land on.

Snakes were first reported, on Flat Is., in 1672 and 1674 (Pitot 1905); around the turn of that century this islet was generally known as Snake Is. (or Long Is.; Barnwell 1955). In 1703 men from the *Scarborough* (1703) reported that on "Snake Is." there were "large bats and doves and a prodigious number of large snakes that we were fearful of sleeping at night".

The same 1672 and 1674 reports on Flat Is. recorded good numbers of tortoises there, whereas they were by then getting scarce on the mainland. Around 1700 Flat Is. was a regular collecting place for tortoises; there were "great plenty of large land turtle" there in 1702 (log of the *Rising Sun*), the *Scarborough* (see above) did well in 1703, and the crew of *Westmoreland* collected 68 in 1706 (Barnwell 1948). This continued into the 1720s; they were abundant in 1714 (Foucherolle *et al.* 1714), in "grande quantité" in 1721 (Garnier 1722, *Courrier de Bourbon* 1721), and the *Lyon* took 120 in 1722. The *Courrier de Bourbon* also reported an abundance of tortoises on Gunner's Quoin and on "a small islet two leagues from Grand Port covered in nothing but coconuts". This cannot have been the Ile de la Passe group, as Leguat (1708) would surely have mentioned (and eaten) any tortoises; it may have been Ile aux Aigrettes or Ile des Deux Cocos; Ile Cerf seems too far off to have been described as "two leagues" away.

As late as 1732 tortoises were still to be had in reasonable numbers on Round and Flat Is., as the Compagnie des Indes ordered that local needs should be met from there rather than from Rodrigues (reserved for shipping), although they admitted the local tortoises were smaller and inferior (Lougnon 1933–7).

The ban on collecting from Rodrigues was lifted in 1734, the trade from there developing enormously (North-Coombes 1971, 1980a); the offshore islets were apparently forgotten, the tortoise numbers presumably being too low to warrant attention. The next mention of these islands was in 1753, when La Caille (1763) reported from hearsay the presence of snakes on Round Is., Flat Is. and Gunner's Quoin. He added, however, that on Gunner's Quoin he had seen lizards a foot (30 cm) long and an inch (2.5 cm) thick (i.e. *Leiolopisma telfairii*), whereas he had only seen small ones on the mainland. Dalrymple (1755) reported that sailors from the *Sumatra*, wrecked "a few years ago" on Round Is., found it "full of large serpents and snakes".

The native snakes have generally been thought to have been confined in historical times to the northern islets, but the following passage from Cossigny (1764) shows that they also occurred on the islets off Grand Port: "I should say also that there are snakes, quite long and big, marked with brown and green, on the islets marked on my map. I have seen several shot on the Gunner's Quoin. There were also some on the islet by the Grand Port pass [= Ile de la Passe]. The workmen working at the battery killed several" (my translation).

The northern islets are not mentioned again until 1790, when Flat Is. was visited by Lislet Geoffroy (1790; Vinson 1950), who reported snakes much diminished as a result of fires; Cossigny (1799) and Chapotin (1812) mentioned large lizards on Flat Is. Bory (1804) and Quoy & Gaimard (1824) recorded hearsay reports of snakes on Ile aux Serpents. It is not clear how this name became attached to the one islet on which snakes almost certainly never occurred; there is virtually no vegetation, landing is extremely hard and the whole dome-shaped rock is covered in

hundreds of thousands of breeding terns (Vinson 1950; pers. obs.). The Dutch aptly called it Meeuwe Klip (Gull Rock); in the mid-eighteenth century it was called Ile Parasol, and only acquired the misnomer Serpent Is. about 1800 (Barnwell 1955).

Prior (1820) wrote that on Flat Is. in 1810 "myriads of lizards occupy the ground, the rocks, the shrubs, the grass and the sand", describing them as up to 10 or 12 inches long and an inch in diameter – evidently *Leiolopissma telfairii*.

The endemic snakes were first scientifically collected on Round Is. by Peron & Lesueur and Dussumier (Dumeril & Bibron 1834–54, Guibé 1858) in the early 1800s; in 1830 Desjardins (1830, 1831a) formally described the endemic lizards *L. telfairii*, *Scelotes* (*Gongylomorphus*) *bojerii* and the widespread *Cryptoblepharus boutonii*, adding that the first was common on, but confined to, Gunner's Quoin, Flat Is. and Round Is. Magon (1839) confirmed the presence of snakes on Gunner's Quoin and Flat Is. in the late 1830s, and Thomas Corby, probably around 1844, collected a snake on the former which he presented to the Desjardins Museum (Barkly 1870), but is now lost (Vinson 1950). Magon's was the last report of snakes on Flat Is.; they presumably disappeared, together with the larger lizards, when a quarantine station was established in 1856 (Bulpin 1958), no doubt bringing with it rats. Bouton, in 1872 (footnote to the printing in that year of Geoffroy 1790), was the first to record that snakes had gone from there. The date rats (*Rattus rattus*; Bullock 1982) reached Gunner's Quoin is unknown, but by 1879 the snakes (and presumably lizards) were confined to Round Is. (Daruty 1883b). Lloyd (1846) gave the first full account of Round Is., and it has since been the subject of numerous reports (refs. in Vinson 1964a, also Temple 1974b, Vinson 1974, 1975, Bullock 1977, 1982, Bullock & North 1977, 1984, Alexander *et al.* 1978). The large crepuscular gecko *Phelsuma guentheri* was not collected until 1869 (Vinson & Vinson 1969), and it is not known if it ever occurred on Flat Is. or Gunner's Quoin. *Nactus serpensinsula*, first collected on Serpent Is. in 1948 (Vinson 1950, Loveridge 1951), was found also on Round Is. in 1952 (Vinson 1953b). A related but much smaller form, now considered a separate species (Bullock *et al.* 1985), was discovered on Gunner's Quoin in 1982 (Bloxam 1983, Bullock & North 1984).

Over the same period the last tortoises also

vanished. They are not mentioned after 1732 until 1844, when Lloyd (1846) noted that on Round Is. "there are besides [rabbits and geckos] a few very large species [*sic*] of the land tortoise of the Angola description [*sic*]". Corby, who was on the island with Lloyd, brought one of these tortoises back to Mauritius, where it laid eggs and produced young (Barkly 1870). Unfortunately all trace of these specimens had already disappeared by 1870 (*ibid.*) and we cannot now be certain, however probable it seems, that they were endemic Mauritian tortoises. Mason (1907) reported tortoise bones amongst a collection of subfossil material from Round Is., but attempts to trace the present whereabouts of these specimens have failed (E. N. Arnold *in litt.*). By 1865 when Pike (1870*a*, 1873) visited the island, there were no longer any tortoises; they presumably succumbed over the years to grazing competition from rabbits (introduced before 1810; see below) and goats (introduced between 1844 and 1865). On the other islets, easier to land on, it seems likely that the tortoises were destroyed much earlier by human predation.

A number of Round Is. reptiles are now maintained at the Jersey Wildlife Preservation Trust for captive breeding and to preserve the species should disaster strike the island (Durrell 1977*a*, 1978, Bloxam 1983). *Leiolopisma telfairii* and *Phelsuma guentheri* have been bred successfully (Bloxam 1977, 1982, Bloxam & Vokins 1979), but hatchlings have yet to be reared from captive-laid eggs of the snake *Casarea dussumieri* (anon. 1982).

Note added in proof

1. The gecko *Nactus coindemirensis* (Bullock *et al.* 1985) was described too late to be included in the appendix table.

2. A subfossil egg-cluster of *Phelsuma guentheri* has recently been found by CGJ on a rock face in the Black River Gorges (Mauritius mainland).

3. New editions of Bernardin (1773) and Leguat (1708) have been published (1983, 1984) by Editions la Decouverte, Paris. The Leguat has an important historical introduction by J.-M. Racault, and also includes Duquesne's memoir for colonising Réunion.

4. Jersey Wildlife Preservation Trust are now successfully rearing young *Casarea dussumieri* snakes from Round Is. (*On the Edge* 51, 1986; see p. 59).

5. I overlooked an important paper by Tuijn (1969) which includes notes on the behaviour of the extinct pigeon *Alectroenas nitidissima* in captivity from Vosmaer's MSS, and good drawings from Harmansz's visit to Mauritius in 1602, the earliest positive identification (cf. Fig. 3 and text). Tuijn also gave details of major papers left unpublished by Oudemans. Reference: Tuijn, P. (1969). Notes on the extinct pigeon from Mauritius, *Alectroenas nitidissima* (Scopoli 1786). *Beaufortia* 16: 163–70.

Key to symbols used in Figs. 3–7:

⊙ Definite record of species/genus.

× Definite record of absence of species ('no-record').

(⊙) Positive record rendered doubtful by later writers.

(X) No-record rendered doubtful by later writers.

?⊙? Uncertain record.

(?) Very uncertain record.

● Where more than one species is represented on one line, a filled circle indicates that the author identified all the species.

⊙• One or another of two or more species identified.

⊡ Species in captivity.

⊠ Captives recorded as having died.

△ Date of recorded introductions.

△△ Repeated introductions.

-?-?- Presence uncertain.

-?- Before a first record, indicates that the species appeared on that date to be already well established and had thus been present for some time. After a last record, indicates that the species probably survived somewhat beyond that occasion.

--- Sporadic occurrence of native species after it has ceased to breed or be resident.

+ Positive record of a generic nature embracing several species not distinguished, e.g. 'parrots', 'pigeons'.

// Separates two taxa on the same line.

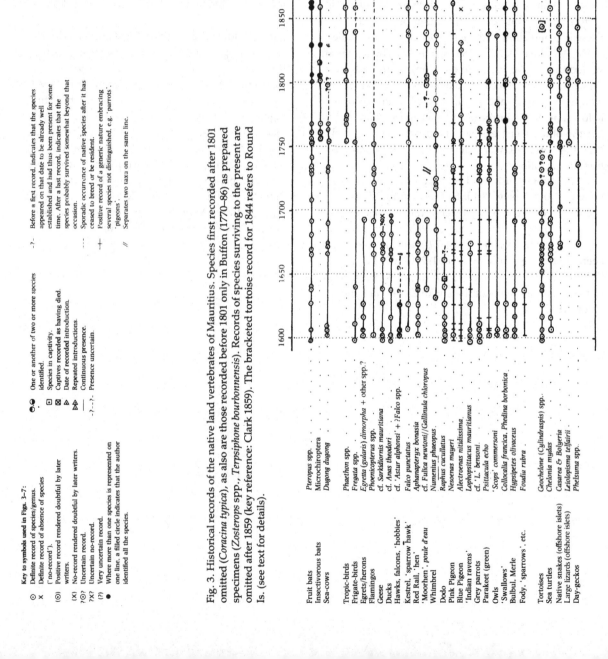

Fig. 3. Historical records of the native land vertebrates of Mauritius. Species first recorded after 1801 omitted (*Coracina typica*), as also are those recorded before 1801 only in Buffon (1770–86) as prepared specimens (*Zosterops* spp., *Terpsiphone bourbonnensis*). Records of species surviving to the present are omitted after 1859 (key reference: Clark 1859). The bracketed tortoise record for 1844 refers to Round Is. (see text for details).

Fruit bats	*Pteropus* spp.
Insectivorous bats	Microchiroptera
Sea-cows	*Dugong dugon*
Tropic-birds	*Phaethon* spp.
Frigate-birds	*Fregata* spp.
Egrets/herons	*Egretta* (*gularis*) *dimorpha* + other spp.?
Flamingos	*Phoenicopterus* spp.
Geese	cf. *Sarkidiornis mauritiana*
Ducks	cf. *Anas theodori*
Hawks, falcons, 'hobbies'	cf. '*Astur alphonsi*' + ?*Falco* spp.
Kestrel, 'sparrow hawk'	*Falco punctatus*
Red Rail, 'hen'	*Aphanapteryx bonasia*
'Moorhen', *poule d'eau*	cf. *Fulica newtoni*//*Gallinula chloropus*
Whimbrel	*Numenius phaeopus*
Dodo	*Raphus cucullatus*
Pink Pigeon	*Nesoenas mayeri*
Blue Pigeon	*Alectroenas nitidissima*
'Indian ravens'	*Lophopsittacus mauritianus*
Grey parrots	cf. '*L.*' *bensoni*
Parakeet (green)	*Psittacula echo*
Owls	'*Scops*' *commersoni*
'Swallows'	*Collocalia francica, Phedina borbonica*
Bulbul, Merle	*Hypsipetes olivaceus*
Fody, 'sparrows', etc.	*Foudia rubra*
Tortoises	*Geochelone* (*Cylindraspis*) spp.
Sea turtles	*Chelonia mydas*
Native snakes (offshore islets)	*Casarea & Bolyeria*
Large lizards (offshore islets)	*Leiolopisma telfairii*
Day-geckos	*Phelsuma* spp.

1600　1650　1700　1750　1800　1850　1900　1950　1980

Fig. 4. Historical records of the native land vertebrates of Réunion. Species first recorded after 1801 omitted (*Coracina newtoni*, *Gallinula chloropus*), as also those recorded before that date only in Brisson (1760) and Buffon (1770–86) even if noted by Bory (1804) (*Zosterops olivaceus*, *Terpsiphone bourbonnensis*, *Saxicola tectes*). Records of species surviving to the present are omitted after 1862 (key reference: Maillard 1862). Observations by Fréri (1751) are included in addition to citations in the text.

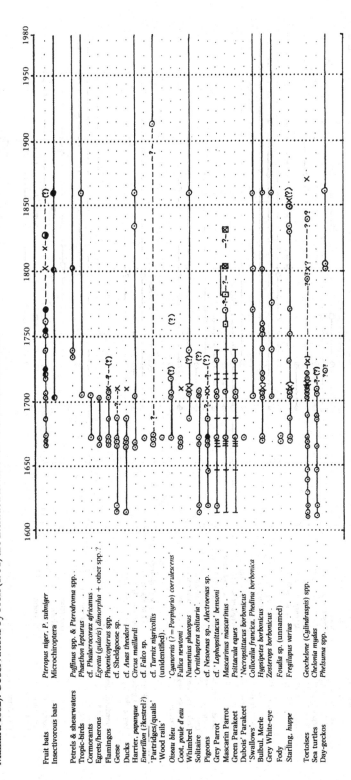

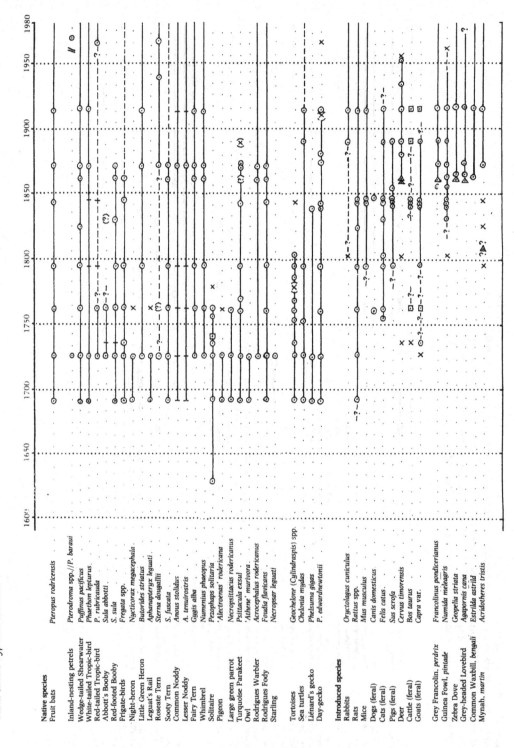

Fig. 5. Historical records of land vertebrates on Rodrigues. Records of species surviving to the present are omitted after 1916 unless Bertuchi (1923) failed to record them. Animals introduced after 1874 omitted (*Serinus mozambicus*, *Passer domesticus*, *Foudia madagascariensis*), as are house geckos and the blind-snake *Typhlina bramina*.

Fig. 6. Historical records of land vertebrates introduced to Mauritius. Records of species surviving to the present are largely omitted after 1859 (key reference: Clark 1859). Animals introduced after 1850 (several birds, *Calotes versicolor*, terrapins) are omitted apart from the mongoose, included for comparison with game bird extinctions; *Typhlina bramina* is also omitted. The square bracket 1810 rabbit record, and the post-1844 rabbit and goat records refer to the northern offshore islets only (see text for details).

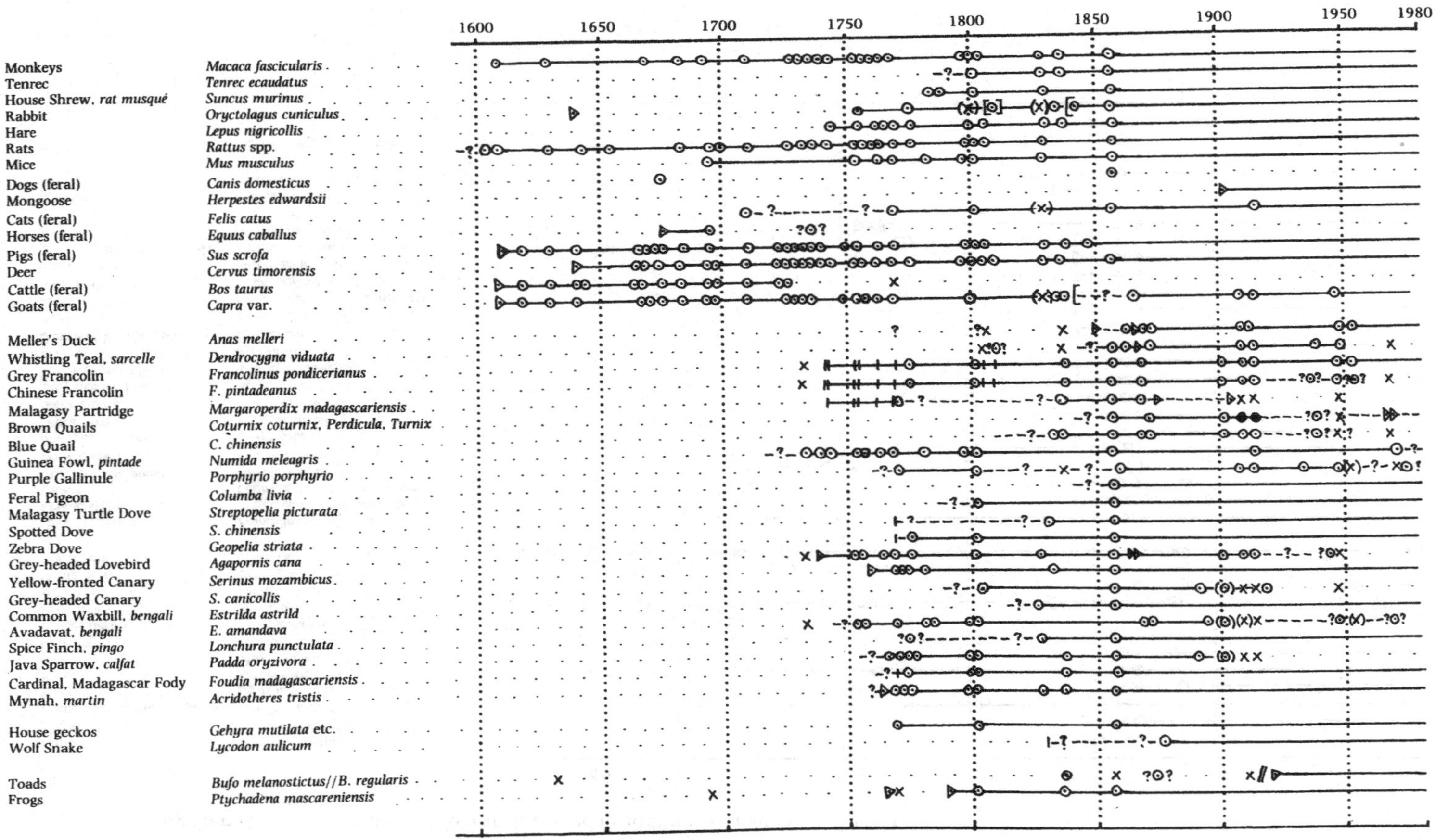

Monkeys	*Macaca fascicularis*
Tenrec	*Tenrec ecaudatus*
House Shrew, *rat musqué*	*Suncus murinus*
Rabbit	*Oryctolagus cuniculus*
Hare	*Lepus nigricollis*
Rats	*Rattus* spp.
Mice	*Mus musculus*
Dogs (feral)	*Canis domesticus*
Mongoose	*Herpestes edwardsii*
Cats (feral)	*Felis catus*
Horses (feral)	*Equus caballus*
Pigs (feral)	*Sus scrofa*
Deer	*Cervus timorensis*
Cattle (feral)	*Bos taurus*
Goats (feral)	*Capra var.*
Meller's Duck	*Anas melleri*
Whistling Teal, *sarcelle*	*Dendrocygna viduata*
Grey Francolin	*Francolinus pondicerianus*
Chinese Francolin	*F. pintadeanus*
Malagasy Partridge	*Margaroperdix madagascariensis*
Brown Quails	*Coturnix coturnix, Perdicula, Turnix*
Blue Quail	*C. chinensis*
Guinea Fowl, *pintade*	*Numida meleagris*
Purple Gallinule	*Porphyrio porphyrio*
Feral Pigeon	*Columba livia*
Malagasy Turtle Dove	*Streptopelia picturata*
Spotted Dove	*S. chinensis*
Zebra Dove	*Geopelia striata*
Grey-headed Lovebird	*Agapornis cana*
Yellow-fronted Canary	*Serinus mozambicus*
Grey-headed Canary	*S. canicollis*
Common Waxbill, *bengali*	*Estrilda astrild*
Avadavat, *bengali*	*E. amandava*
Spice Finch, *pingo*	*Lonchura punctulata*
Java Sparrow. *calfat*	*Padda oryzivora*
Cardinal, Madagascar Fody	*Foudia madagascariensis*
Mynah. *martin*	*Acridotheres tristis*
House geckos	*Gehyra mutilata* etc.
Wolf Snake	*Lycodon aulicum*
Toads	*Bufo melanostictus* // *B. regularis*
Frogs	*Ptychadena mascareniensis*

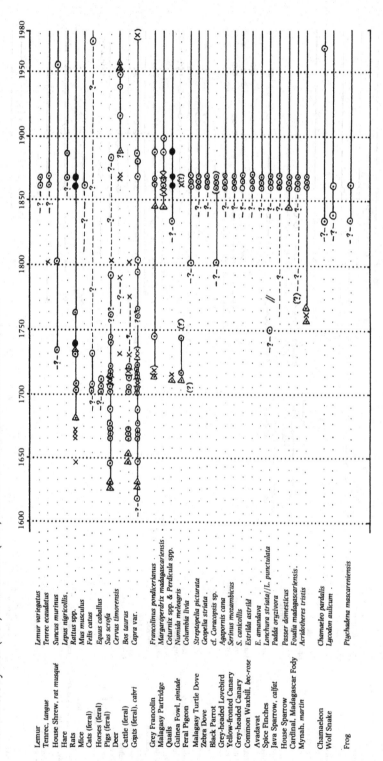

Fig. 7. Historical records of land vertebrates introduced to Réunion. Records of species surviving to the present are largely omitted after 1862 (key reference: Maillard 1862). Animals introduced after 1850 are omitted (four birds, *Calotes versicolor*, *Bufo regularis*), as are house geckos and *Typhlina bramina*. Observations by Fréri (1751) and Crémont (1768) are included in addition to citations in the text.

The introduced fauna
MAMMALS
Primates: lemurs (Lemuridae)

Maillard (1862) and Vinson (1868) reported feral lemurs in Réunion, the former mentioning the Plaine des Makes ('Make' = *maki*, Malagasy for lemur) and the Gorge of the Rivière des Marsouins, the latter only the forests near St.-Benoît, presumably in the same river gorge. The authors differed on the species involved, but Vinson's identification, *maki vari* (= Lemur variegatus) appears from his description to be correct (Moutou 1981). A list of wildlife sent by Lantz to Maillard in 1864, preserved in the Museum in St.-Denis, also called them *maki vari*. The name 'Plaine des Makes' suggests that *makes* had been present for a considerable time prior to the 1860s, but nothing is known of the previous history of these animals, nor of their subsequent fate. They no longer exist in Réunion (Moutou 1979).

Carié (1916) reported the presence of recently escaped but wild-breeding *L. catta* at the Réduit, Mauritius. There is no evidence of how long they survived. In the 1850s pet lemurs were evidently commonplace on the island (Clark 1859).

Primates: monkeys (Cercopithidae)

The Crab-eating Macaque *Macaca fascicularis*, still abundant on the island today (Sussman & Tattersall 1980), is generally supposed to have been introduced to Mauritius by the Portuguese in the sixteenth century (La Caille 1763, Pitot 1905). The grounds for this belief are the early mention of monkeys by the Dutch (Matelief in 1606; Pitot 1905), and the alleged fact that the Portuguese, but not the Dutch, were fond of monkey-meat (Carié 1916). The first Dutch visitors did not mention monkeys, but this (Pitot 1905) is not evidence of their absence, as they hardly ever mentioned these animals in their reports throughout their period of colonisation (1638–1710), although other visitors referred to them regularly (Herbert (1634 etc.) in 1627; Marshall in 1666 (Khan 1927), Harry in 1687 (Barnwell 1948), Leguat (1708) in 1793, etc.). North-Coombes (1980b) argued that introduction by the Portuguese was most unlikely as: (a) the Portuguese frequented Réunion, where there are no monkeys, rather than Mauritius; (b) there is no evidence of their having introduced monkeys to any other island; and (c) they did not introduce bananas, something they normally did at stopping places, so if it wasn't worth doing that,

why leave monkeys? The animals have recently been identified as of Javan origin (Sussman & Tattersall 1980), so early Dutch visitors now seem almost certain to have been the original source.

No primates have been introduced to Rodrigues.

Tenrecs (Tenrecidae)
Mauritius

Although the Tenrec *Tenrec ecaudatus* was not mentioned by late-eighteenth century writers, Bory (1804) and Milbert (1812) encountered it commonly over wide areas of the island in 1801; Clark (1859) gave the introduction date as "late 18th-century", and named a Mr Mayeur as responsible. Carié (1916) observed that the younger Cossigny was in 1803 belatedly advocating its introduction, evidently unaware that it was already well established. The Tenrec remains abundant today (pers. obs.).

Three Tenrec specimens in Vienna of *Hemicentetes semispinosus*, presented by Bojer in his bequest in 1860, are said to come from Mauritius (accessions register *per* K. Bauer). Bojer usually labelled his material with its place of origin, so the immediate locality is probably correct, though whether the animals were wild or captive is unknown. It is possible that other species of Tenrecs were introduced to Mauritius in the eighteenth and nineteenth centuries; Bojer himself often visited Madagascar in the 1820s and 1830s (Vaughan 1958).

Réunion

Tenrecs were introduced into Réunion later than Mauritius. Bory (1804), who covered the island very thoroughly, never heard of any in Bourbon in 1801. The date of introduction is unknown, but the animals were so well established by the 1860s that Vinson (1868) considered them native. Tenrecs were listed by Maillard (1862), and Pollen (*in* Schlegel & Pollen 1868) gave a classic account of hunting hibernating tenrecs. They remain abundant today, and are still hunted as food (Moutou 1979, pers. obs.).

Shrews (Soricidae)
Mauritius

The Indian House Shrew *Suncus murinus*, known in the Mascarenes as *rat musquée*, is first mentioned for Mauritius by Buffon (1776). He was quoting de Querhoënt, writing probably in 1773, who reported that "Since a while back a rat from India has begun to establish itself; it has the strongest odour of musk . . . and it is believed that when it passes through a place where there is wine it sours it" (my translation). The only animal with this reputation (cf. Milbert 1812, Carié 1916 and below) is the House Shrew, although Buffon himself did not describe the species until Sonnerat brought one back from India some time later (Buffon 1789). Carié (1916) erroneously assigned this specimen to Mauritius, from which the earliest specimens were actually brought by Péron & Le Sueur (Geoffroy 1827), who were in the island during 1800–4 (Toussaint 1972). Another early reference is Céré (1781; cited as 1783 by Carié 1916). The shrew must have been introduced in the 1760s and has remained common ever since (Milbert 1812, Desjardins 1830, Clark 1859, Carié 1916, Michel 1972, pers. obs.).

Réunion

D'Héguerty (1754), describing the various kinds of "rat" present in Réunion in the 1730s, referred to one that lives in cellars and "whose smell leaves traces so strong, that if they pass over a stack of bottles, all the wine in them becomes musky to the point that it is no longer possible to drink it" (my translation). This must have been the House Shrew, which was evidently well established in Réunion by the middle 1730s. There may have been a subsequent decline, as, although the species was mentioned by Bory (1804), neither Maillard (1862) nor Schlegel & Pollen (1868) listed it. It is common throughout the island today (Heim de Balsac & Heim de Balsac 1956, Moutou 1979, 1980, pers. obs.).

Rodrigues

No insectivores have been introduced to Rodrigues, although Corby (1845) suggested introducing the House Shrew to control insect pests.

Hares and rabbits (Lagomorpha)
Mauritius

Six pairs of tame rabbits *Oryctolagus cuniculus* were brought in by the Dutch in 1639 (Pitot 1905, Carié 1916), but these did not become feral. According to Cossigny (1732–55) there were neither hares nor rabbits on the island in 1732, but by 1741 Grant (Grant 1801, Pope-Hennessy 1886) was referring to a species of "white hare" formerly common but becoming rare. Carié (1916) considered that references to "white [-fleshed] hares" described rabbits, but if Grant did

mean rabbits, there is no further mention of them until 1754 ("hares and rabbits", La Motte 1754–7), and around 1775 when Sonnerat (1782) reported the presence of two species of "*lièvre*" (= hare) one of which was "as much like a rabbit [as a hare]". In the intervals several writers (La Caille 1763, Pingré 1763, Cossigny 1764, Bernardin 1773) reported the presence of introduced 'hares' without describing them. The use of the word '*lièvre*' suggests they were true hares, which is confirmed by de Querhoënt (in Buffon 1776) who gave an excellent description of *Lepus nigricollis*, the hare present today (Owadally 1980). Milbert (1812) also gave a description that fits *L. nigricollis*, and mentioned the existence of a smaller species (rabbit?) which he did not see, though another visitor in 1801, Chapotin (1812), stated that the only rabbits were domesticated. Desjardins (1830) mentioned only hares *L. nigricollis* as wild, though Backhouse (1844) referred also to wild rabbits on the east coast in 1838, which Clark (1859) and Pike (1873) reported only from Ile d'Ambre. Desjardins (1829) referred, presumably wrongly, to the Ile d'Ambre animals as "*Lepus nigricollis*". It appears that rabbits never prospered on the mainland, perhaps because of predation by rats on their young (Carié 1916). By contrast on rat-free Round Is., where they were already abundant in 1844 (Lloyd 1846), they have thrived and still do so (Vinson 1964a, Temple 1974b, Procter & Salm 1975, Bullock 1977, 1982, Bullock & North 1984). The Round Is. rabbits are a mixture of breeds with a high component of Argenté de Champagne stock (J. A. G. Flux *in litt.*).

Gunner's Quoin and Flat Is. were populated with "small but not ill-flavoured hares" (i.e. rabbits) in 1810 (Prior 1820, Barnwell 1948) and the Round Is. ones presumably derive from the same introduction. The oft-repeated belief that Thomas Corby introduced rabbits to Round Is. in 1844 (Temple 1974b, Procter & Salm 1975, Staub 1973a, Durrell 1977a) derives from Barkly's mistaken statement to that effect (1870). As mentioned above, and as noted by Vinson (1964a, 1974), Lloyd (1846) who was on the island with Corby in 1844, recorded rabbits as already present. Horne (1887) implied there were still rabbits on Flat Is. in 1885, and they still survive on adjacent Ilot Gabriel (Vinson 1953a, Procter & Salm 1975: 'hares'). The animals seen recently on Gunner's Quoin are hares (Bullock *et al.* 1983, Bloxam 1983), though a skull I picked up in October 1978 was identified at the British Museum as from a rabbit (P. D. Jenkins *in litt.*, Cheke

1979*a*), so both species may now be present, the hares representing a more recent introduction. Rats are abundant and lagomorphs are at very low density (Bullock *et al.* 1983, Bloxam 1983).

Réunion

The first to mention lagomorphs in Réunion was Le Gentil (1727) who visited the island in 1714 and reported that rabbits had been introduced but had not established. The next report was not until Vinson (1868) mentioned hares, imported from India a long time previously, as "very abundant". There was no mention of a hare by either Maillard (1862) or Schlegel & Pollen (1868), but Lantz (1887) identifying it as *Lepus nigricollis*, confirmed that it was common in coastal areas, occurring occasionally to 1500 m. The hares are rather local today (Moutou 1979). A second species, *L. capensis*, introduced around 1960, apparently failed to establish (Moutou 1981).

Rodrigues

In 1803 Marragon (1803) wrote to the administration in Mauritius detailing *inter alia* his plans to introduce game, including hares and rabbits, to Rodrigues, and his decision to put the animals on offshore islets to prevent damage to crops and predation by feral cats and off-duty slaves. Whether he did so is unrecorded. North-Coombes (1971: 58) believed he brought in rabbits, although they were not mentioned by any visitor before Kennedy (1893). The Transit of Venus expedition found subfossil rabbit bones in the caves in 1874 (Dobson 1879), showing that the original introduction was not confined to the islets to which Kennedy (1893) said they were restricted in 1892. Bertuchi (1923) said they were only found on Iles Sable, Coco and Frégate in 1916. Presumably Marragon's prediction of predation by cats and people on the mainland was borne out. Rabbits still survived until recently on at least Ile Frégate (Vinson 1964*b*, pers. obs. 1974), though they were extremely scarce in 1978 (pers. obs.), and had apparently died out by 1983 (W. Strahm, pers. comm.). There is no recent evidence of them on Ile Coco or Ile Sable. Hares appear never to have been introduced.

Squirrels (Sciuridae)

Maillard (1862, Moutou 1979) mentioned the introduction to Réunion of an Indian squirrel "*Sciurus tristriatus*" (= *Funambulus palmarum*; Prater 1965) "four

or five years ago" (i.e. *c*. 1857. A specimen was accessioned by the natural history museum in St.-Denis in 1863 (manuscript accessions book), but they presumably died out soon afterwards, as there were no further reports.

Rats (Muridae)
Mauritius

Rats are generally assumed to have reached Mauritius from Portuguese ships or shipwrecks in the sixteenth century, and indeed the first Dutch visitors in 1598 reported finding the wreckage of a ship (Pitot 1905). They did not report the presence of rats, despite giving good lists of other fauna, and it was not until 1606 that they were first mentioned, by which time they were already abundant (Matelief *in* Pitot 1905, Bonaparte 1890). However, it seems probable that the "*Katten*" (= cats) reported by van West-Zanen in 1602 as the only quadrupeds present on the island (Strickland 1848) was a transcription error for '*Ratten*' (= rats). The best evidence of the presence of rats before the arrival of the Dutch is the extinction of snakes and large endemic lizards, none of which were ever reported alive on the Mauritian mainland.

There are no seventeenth-century descriptions, but Mauritian rats were described by Grant in 1741 as "very large" (Grant 1801, Pope-Hennessy 1886) and by de Reine (eighteenth century, date unknown) as being the size of rabbits, with white bellies (Pitot 1899), echoing early European descriptions of *Rattus norvegicus* (Twigg 1975). The first arrival of *R. norvegicus* in Réunion was about 1735 (see below), so they probably also reached Mauritius at the same time; the sixteenth-century arrivals were doubtless Ship Rats *R. rattus*. Bernardin (1773) reported rats as climbing trees after young birds, behaviour typical of Ship Rats. Carié (1916) reported the Ship Rat to be very rare, but today they are quite common, living arboreally in the forest (Owadally 1980). Desjardins (1830) wrongly assigned all Mauritian rats to *R. rattus*.

Rats were astonishingly abundant in the seventeenth and eighteenth centuries, and though Cossigny (1764) suggested there were many fewer than previously, his son (Cosssigny 1799) described their numbers as "prodigious" 35 years later. The abandonment of cereal cultivation in the nineteenth century was probably responsible for a general decline, although Clark (1859) and Carié (1916) still reported large numbers of *R. norvegicus*, and they are very common today, though not regarded as a serious pest (Williams 1953, Owadally 1980).

Réunion

Rats were introduced to Réunion some considerable time after the island was settled. As late as 1671–2 Dubois and the log of the *Breton* testify to their absence (Lougnon 1970). François Martin, who had visited Réunion in 1665 and 1667 when it was rat-free (*ibid.*), wrote in about 1684–5 (Kaeppelin 1908) of the rats coming ashore from a wrecked longboat "some years ago". By 1678 the rats had reached sufficient numbers to cause serious agricultural losses and famine (Bulpin 1958, Reydellet 1978, Moutou 1980, 1981). An introduction date around 1675 seems likely, though Moutou (1981) put it earlier, being unaware of the date Martin actually wrote his memoirs. Feuilley (1705) reported that rats had been overwhelmingly abundant before the introduction of cats, which had then exerted some control. Schlegel & Pollen (1868) and Vinson (1868) erroneously referred to early colonists in Réunion (1648–64) being driven off by rats; this story belongs to Mauritius (Toussaint 1972).

There are no descriptions of the rat originally introduced, but Lebel (1740) reported the introduction 5 years before of a new larger species from a shipwreck. He described them as "as big as little cats . . . They have a white belly and a very short tail". They were reported to take chickens and even attack suckling pigs and goats; cats could not take them, only good dogs. D'Héguerty (1754) described them as attacking adult goats and pigs and sucking their brains out! They lived in burrows. The Compagnie des Indes acknowledged the islanders' complaints about these "*gros rats*" in 1741, when the rats were still confined to the eastern part of the island around St.-Denis and Ste.-Suzanne (Lougnon 1933–7). This species must have been the Common Rat *R. norvegicus*, which first reached European ports from Russia in 1716 (Twigg 1975, Moutou 1980). The earlier introduction was therefore of Ship Rats *R. rattus*. Both species are common today, even in forests far from human habitation, though the Ship Rat is more widespread (Moutou 1979, 1980, 1981).

Rodrigues

Leguat (1708), Tafforet (1726) and Pingré (1763) were plagued by enormous numbers of rats, and they were still mentioned as serious pests throughout the nineteenth century and as late as 1916 (Bertuchi 1923).

Leguat (1708) described the rats as like those in Europe, which at the time would have meant *R. rattus*. No other visitor gave any description, though Marragon (1795, Dupon 1969) mentioned their 'nesting' in hollow tree trunks and in burrows, suggesting the presence by then of both *R. rattus* and *R. norvegicus*. The first definite identification came in the 1870s when Milne-Edwards (1873) described subfossil bones from caves on the island as "not the surmulot [= *R. norvegicus*] but the *Mus alexandrinus* [= *R. rattus*]". Confusingly a few years later Dobson (1879) referred a new collection of rat bones from the caves to '*R. decumanus*' (= *R. norvegicus*), but P. D. Jenkins (*in litt.*) has re-examined these and wrote that "although labelled as *R. norvegicus* they agree closely with old *R. rattus*".

The common rat on the Rodriguan mainland today is *R. norvegicus* (Cheke 1979b). Carroll (1982b) claimed that, from their climbing behaviour, the rats he saw were Ship Rats, but the local Common Rats climb trees (pers. obs.), and the presence of Ship Rats was not confirmed by specimens until 1983 (C. G. Jones pers. comm.). Ship Rats may have re-invaded after the new wharf was built in Port Mathurin in 1980 (Rivière 1982a), as I had feared (Cheke 1979b; cf. Atkinson 1977). Forests on Rodrigues had become so depauperate that previously Ship Rats may have been unable to survive in the presence of *R. norvegicus*. The high nesting success of the two endemic passerines in the mid-1970s (Chapter 8) compared well with rat-free islands in the Seychelles (Prŷs-Jones & Diamond 1984), suggesting an absence of effective arboreal predators. There are unidentified rats on islets in the lagoon (Vinson 1964b, Cheke 1979b).

Mice (Muridae)

Mauritius

The only mouse reliably recorded in the Mascarenes is the House Mouse *Mus musculus*. Mice were first reported in Mauritius by Leguat (1708) in 1693, but not again until 1753 (La Caille 1763), after which they were mentioned frequently (e.g. Carié 1916); Leguat may possibly have seen only rats, but added "and mice" assuming the two were likely to go together. The House Mouse remains common today, in the forest (pers. obs.) as well as in inhabited areas.

Réunion

Apart from a few early travellers who noted an absence of both rats and mice in Réunion, mice were not mentioned until Maillard (1862) included them in a list of the island's animals. He reported two species, *Mus musculus* and '*M. sylvaticus*' (= *Apodemus sylvaticus*), but it seems likely that only the former has ever occurred (Moutou 1979). The date of introduction is unknown; the species remains common today (Moutou 1979, 1980, 1981).

Rodrigues

There is no mention of mice on Rodrigues before Marragon's report (1795, Dupon 1969), and only scant references thereafter; Vinson (1964*b*) claimed that the only surviving introduced mammals were rats and rabbits. Mice are in fact common on the island (pers. obs.) and one I caught at Solitude in 1978 was confirmed as *Mus musculus* (P. D. Jenkins *in litt.*), as were subfossil bones collected in 1874 (Dobson 1879).

Cats and dogs (Felidae, Canidae)
Mauritius

Cats were reported as the only quadruped on Mauritius by Van West Zanen in 1602 (Strickland 1848, Grandidier *et al.* 1903–20, vol. 1), but I believe this was an error for rats (see above). There is no further mention of feral cats until La Merveille's account in 1709 (La Roque 1716). Numbers may have been low for a long time, as they are mentioned in the literature only sporadically, most writers referring only to tame cats. Milbert (1812) mentioned feral cats, though Desjardins (1830) did not include them in his list of wild mammals. By the mid-nineteenth century they may have increased (possibly with increased land clearance and introductions of quails), as Clark (1859) and Carié (1916) described them as abundant. Cats are still common in the forest (Temple 1974*c*, C. G. Jones pers. comm.) but appear to have declined over the island as a whole during the twentieth century.

Dogs escaped and were released at various times during the Dutch occupation (Pitot 1905, Barnwell 1948) but apparently died out, perhaps due to disease (Barnwell 1948). However, Clark (1859) reported that in the mid-nineteenth century packs of wild dogs were numerous around Mahébourg. They have again disappeared; Carié (1916) made no mention of feral dogs.

Réunion

The first mention of the introduction of cats to Réunion, to control rats, was by Borghesi who visited the island in 1703 (Lougnon 1970); this is confirmed by Feuilley (1705). Feral cats are mentioned again by Cossigny (1732–55) in 1732, but not thereafter. This is surprising, as such cats still exist (Moutou 1979, 1981), although in small numbers, possibly supplemented from domestic stock. Feral dogs are not reported at all in the literature, but are not uncommon today in some areas (e.g. the Plaine des Chicots); dogs used by poachers for digging up hibernating tenrecs frequently escape into the forest and apparently survive quite well and breed (T. Bègue pers. comm.).

Rodrigues

Cossigny (1732–55) reported feral cats in 1755 and Pingré (1763) confirmed this in 1761. Marragon (1795) used trained dogs to hunt them, so destructive were they to poultry. They were still present in 1916, but being reduced by persecution (Bertuchi 1923), and appear to have died out (or become very scarce) since. Dogs were released by British ships in 1761 (North-Coombes 1980*a*) but presumably died out. Feral dogs were again rife in 1848 (North-Coombes 1971), derived from animals left by people returning to Mauritius. There is no other report of wild dogs, although semi-feral strays remain common (*ibid.*, pers. obs.).

Mongooses (Herpestidae)

Mongooses *Herpestes* spp. were originally brought to Mauritius in the mid-nineteenth century as rat-catchers (Clark 1859); they were apparently kept like cats, rather than let loose. Clark commented that they were "more destructive to game and poultry than to vermin; and some persons who have had them have been glad to get an opportunity of destroying them".

With this experience evidently forgotten, and despite the disastrous experiences in Jamaica [1880s] and Hawaii (1883 onwards; Berger 1981) and a vigorous press campaign against the idea (Carié 1916), mongooses *H.* '*griseus*' (= *edwardsii*, Prater 1965) were released in 1900. The government was under pressure to control rats because of the outbreak of plague in the previous year (Carié 1916, Toussaint 1972) and the instructions were officially for a trial release of animals of one sex only. However sixteen males and three females were released (*ibid.*) and they rapidly populated the island. After complaints from poultry-keepers and hunters that their chickens and game were being wiped out, the government put a price on each mongoose killed in 1905, but it was too late; the pest remains (Temple 1974*a*, pers. obs.), its victims do not (see section on game birds, pp. 72–5).

Mongooses have not been introduced to the other Mascarene islands, though there was a near miss in Rodrigues (North-Coombes 1971) in 1887 when two pairs sent to control rats were destroyed by the civil commissioner after he was warned of the danger to poultry and game.

Horses and donkeys (Equidae)
Mauritius

Horses *Equus caballus* were released into Mauritian forests in 1672 (Pitot 1905), and reported as still feral in 1693 by Leguat (1708). There were no subsequent reports of feral horses, though Gennes (1735) reported that the French colonists had found "wild donkeys the size of mules" on the island, which they caught and tamed; some were still at large in 1754 (La Motte 1754–7).

Réunion

Feral horses were reported as present in small numbers in the early 1700s (Luillier *in* Lougnon 1970, Feuilley 1705, Lougnon 1957 (report of 1713)) but Luillier reported that they had subsequently been domesticated.

Rodrigues

Duncan (1857) reported donkeys as feral in Rodrigues, but it seems more likely that they were owned animals roaming the 'cattle-walk', as seen by Corby (1845).

Deer (Cervus timorensis)
Mauritius

Deer *Cervus timorensis* were brought to Mauritius by van der Stel in 1639 (Pitot 1905, Carié 1916), and have remained an important component of the wild fauna ever since. They have at various times become rare and had to be protected by law (e.g. La Motte 1754–7, Sonnerat 1782), but nowadays thrive under active game management (Owadally & Bützler 1972).

Réunion

There were no deer in Réunion in 1732 (Cossigny 1732–55), but by 1761 they were being kept by the governor and some were probably released at the same time, for Bory (1804) reported the killing of the last wild one in 1793 at the Ravine Blanche, adding that

they had formerly been common. Vinson (1868) mentioned that introduced deer failed to survive and Bouton (1869) also affirmed their absence. There have been subsequent introductions. Carié (1916) mentioned that a herd of *Cervus timorensis* "recently introduced" from Mauritius was thriving in the mountains of Salazie; a few, perhaps of this origin, survived on the Plaine des Chicots in the 1940s (P. Rivals *in litt.*) and early 1950s, when they were said to be all too old to reproduce (Cazal 1974). Five new releases of Mauritian deer (*C. timorensis*), and one of *C. elaphus* from France, were made from 1954 onwards (*ibid.*). The *C. timorensis* thrive today due to gamekeeping against poachers. Further deer were released recently at Bélouve and the Rivière des Roches (Moutou 1979, 1981), the former, in an *Acacia heterophylla* plantation, having rapidly to be culled to contain damage to the trees (*ibid.*).

Rodrigues

No deer were introduced into Rodrigues before single pairs in both 1862 and 1863 (Colin *in* Kennedy 1893); these had increased to 1500–2000 by 1892. After a decline due to over-hunting (Bower 1903, Bertuchi 1923) they had recovered with special protection by 1916 (Bertuchi 1923). The recovery was temporary, as only a handful were left by 1937 (A. North-Coombes *in litt.*), and these, still present in 1954, had died (or been shot) by 1956 (C. M. Courtois *in litt.*). Colin (*loc. cit.*) claimed that the first pair introduced came from Borneo, and only the second from the obvious source, Mauritius.

Feral livestock: cattle, goats and pigs
Mauritius

Twenty-four goats and nine pigs were released on Mauritius from Matelief's ships in 1606 (Pitot 1905, Bonaparte 1890), and a further 20 goats and 10–12 pigs the following year (Barnwell 1948: 19). Also in 1606, van Wybrandt released the first cattle (Pitot 1905, Carié 1916). There is no evidence to support Carié's assertion (1916) that goats and pigs were introduced by the Portuguese; they were not reported by the Dutch in 1598 and 1602 (Pitot 1905, Barnwell 1948), and it was surely their absence that led the next batch of ships to carry livestock for release.

Wild cattle *Bos taurus* (zebus; Carié 1916) remained at large until the early years of the French occupation, being last reported in 1726 (Brouard 1963); they were at various times caught and tamed (Pitot 1905, Carié 1916), as well as hunted.

Goats *Capra 'hircus'* maintained a feral existence much longer. Although not listed among wild mammals by Desjardins (1830), they were by Desnoyers (1837), and Backhouse (1844) saw some on the Pouce mountain in 1838. There were no subsequent reports, and Clark (1859) mentioned goats only as domestic animals. Although they died out on the mainland, a population of feral goats survived until very recently on Round Is. off the north coast. There were none on this islet in 1844 (Lloyd 1846), though Barkly (1870) wrongly attributed their introduction to Thomas Corby, one of the party who had landed in 1844. The next visitor, E. Newton (1861*a*), mentioned only birds, but goats were abundant in 1868 (Pike 1973). Although at one time around a hundred in number (Vinson 1950), by 1978 they had been reduced by shooting to two females, not seen since (Bullock & North 1984). Goats, probably feral, were present also on Flat Is. in 1885 (Horne 1887).

Pigs *Sus scrofa* thrived, reaching plague proportions in the early 1700s (La Merveille *in* La Roque 1716), and have remained common in the forests despite continuous hunting (Carié 1916, Owadally 1980).

Réunion

Bontekoe reported goats already present on Réunion in 1619 (Lougnon 1970); we have no record of when they were released, although Vinson (1868), citing no evidence, claimed it was by the Portuguese. British ships added further goats and also pigs in 1627 (Herbert 1634 etc.) and 1629 (log of the *Hart*). These had multiplied considerably by the time Flacourt sent four cows and a bull for release in 1649 (Lougnon 1970). In 1654 he sent a further consignment, by which time the previous introductions had increased to 25–30 (*ibid*.). These animals were heavily hunted after the island was permanently settled, and by 1705 there was a ban on hunting them to persuade settlers to develop their own herds (Feuilley 1705). By 1712 (Hardancourt 1712) the wild animals were further depleted. The only subsequent eye-witness report of wild cattle was in 1722 (log of the *Athalante*), but Bory (1804) was told they survived until about 1775 near St.-Paul before being wiped out by pitfall traps. It is probable that

cattle, goats and pigs repeatedly went feral. Wild goats must also have been scarce in the early eighteenth century, as both Cossigny (1732–55) and d'Héguerty (1754) denied their existence; wild pigs continued to be common during this period, but later in the century Caulier (1764) reported both pigs and goats scarce. Although wild pigs were recorded by Petit-Radel (1801) in 1794, Bory (1804) was told they were extinct; he saw and shot some of the rare goats. Vinson (1868) also reported wild pigs extinct, but Leal (1878) saw some in the Cirque de Salazie. Vinson and Leal (*locs. cit.*) and Lantz (1887) reported goats in mountainous areas. They survive to this day in small numbers around the Volcan (A. de Villèle pers. comm., *contra* Moutou 1979). No wild pigs have been seen since Leal's report, but a new release of wild boar projected for the Plaine des Chicots was recently averted (Moutou 1981). Wild boar are kept domestically at the Tampon (Moutou 1979).

Rodrigues

The first report of feral livestock on Rodrigues was of goats seen in 1735 (Gennes 1735), but they were few, presumably recently left by tortoise-collecting expeditions. Pingré (1763) saw only domestic goats in 1761. By 1795 Marragon reported feral goats again; they persisted through the nineteenth century, and were last reported by Kennedy (1893). By 1916 the only goats mentioned by Bertuchi (1923) were domestic, though they were and still are given free range in the 'cattle-walk'.

Puvigny had a small herd of cattle in 1761 (Pingré 1763). It is not clear whether they were turned loose deliberately, but feral cattle were mentioned in several mid-nineteenth-century reports (Self 1841, Corby 1845, Higgin 1849). By the end of the century apparently only domestic animals remained (Kennedy 1893). From an early stage (North-Coombes 1971) domestic cattle were allowed free range in the zone known as the 'cattle-walk', and this is still so today.

There is no mention of feral pigs before 1795 (Marragon 1795). In 1761 there were apparently not even domestic ones (Pingré 1763, who listed Puvigny's livestock). They were clearly common throughout the nineteenth century, but Kennedy (1893) was the last to report them, and Bertuchi (1923) implied that the only pigs in 1916 were domestic, though free-ranging, as they often still are today (pers. obs.).

BIRDS
Ducks (Anatidae)
Mauritius

In the 1830s Desjardins formally contradicted Milbert's statement (1812) that the ducks in Mauritius were "wild, numerous and varied", adding that the only ducks were domestic (Oustalet 1897). Bernardin (1773) referred to ducks and geese released on certain meres, but no other eighteenth-century authors mentioned wildfowl in their lists of game animals, and it seems probable that whatever species were introduced did not establish. Carié (1916) doubted Desjardins' assessment on the grounds that he had not visited the remote interior of the island, but he would have heard of any wildfowl from deer hunters and from his colleague W. Bojer, who certainly did explore remote areas (Vaughan 1958). Flinders (1814) also explored these areas in his 6½ years in Mauritius (1803–10), and found that "neither wild geese nor ducks are known in the island", and Desnoyers (1837), listing game birds, made no mention of ducks. It is possible that occasional migrant groups appeared, as Prior (1820) reported that in 1810 "teal and a few wild geese give employment to a keen sportsman".

By the mid-nineteenth century, however, ducks were imported from Madagascar and released on a large scale. The Whistling Teal *Dendrocygna viduata* was already established on coastal meres in 1859 (Clark 1859); Edward Newton (1863, anon. 1873) referred to large numbers of this species and *D. bicolor* being imported on "nearly every bullock ship", adding that despite this he had only seen the former wild. Newton collected eggs (now in Cambridge) from wild-breeding Meller's Duck *Anas melleri* at Mare Longue in 1863; this was another species being widely imported (anon. 1873) and which Meinertzhagen (1912) suggested was first released around 1850. H. H. Slater found Meller's Ducks common at Mare aux Vacoas in 1875 (Hartlaub 1877), but they were always confined to the remote uplands (Carié 1916) and are now very rare (Staub 1976, McKelvey 1977b). The Whistling Teal, once common (Carié 1916), was rare by 1952 (Rountree et al. 1952), and had disappeared through over-hunting by 1973 (Staub 1973a) or possibly earlier, as Newton (1958a) did not record it. It is possible that both of these species occasionally arrive naturally from Madagascar.

Domestic ducks and geese were introduced by the Dutch in 1639 (Pitot 1905, Carié 1916), and were already doing very well by 1641 (Bonaparte 1890).

The French also brought in Muscovy Ducks *Cairina moschata* at an early date (Bernardin 1733), and the younger Cossigny apparently introduced "mountain ducks from the Cape of Good Hope" (Carié 1916). None of these became feral.

Réunion

Vinson (1868) bemoaned the lack of attention given to protecting ducks released on Réunion waters. He referred to four species (not separately identified) of *sarcelles* frequently imported from Madagascar, all of which apparently flew off as soon as their clipped wings were regrown; Egyptian Geese *Alopochen aegyptiacus*, which were also bred in captivity, did the same. No ducks established feral populations on the island. The species imported most frequently was probably *Dendrocygna viduata*, as it was recorded the most often in the manuscript accessions book of the St.-Denis museum in the late nineteenth century.

Rodrigues

Only domestic ducks have been introduced to Rodrigues; they were present in 1761 (Pingré 1763) and have remained domestic.

Partridges: Francolinus spp. and Margaroperdix
Mauritius

'Partridges' were present about 1740 (d'Héguerty 1754), and Baron Grant wrote of three species in February 1741 (Grant 1801, Pope-Hennessy 1886). They had been introduced only recently, as Cossigny in 1732 expressly stated that there were no "*perdrix*". Although several more mid-eighteenth-century writers mention them, they cannot be identified until Sonnerat (1782) described the Chinese Francolin *Francolinus pintadeanus* in detail, and contrasted it with the "*perdrix commune*" (presumably Grey Francolin *F. pondicerianus*). Grant's third species may have been the Malagasy Partridge *Margaroperdix madagascariensis*, but it is not recorded for certain until around 1770, from a drawing by Sonnerat amongst the Commerson manuscripts (Oustalet 1897). Chinese Francolins and Malagasy Partridges are extinct in Mauritius; the former was a victim of the mongoose (Carié 1916, Guérin 1940–53), possibly surviving in small numbers until 1956 (Newton 1958a), while the latter appears never to have been common. It was described as rare by Clark (1859), and a re-introduction in 1876 (Daruty 1878) apparently failed, as the birds were reported "as extinct for many years" in 1910 (Meinertzhagen 1912). Carié's attempt to re-introduce them again in 1906 also failed (Carié 1916).

Réunion

Le Gentil (1727), who was in Réunion in 1717, mentioned that *perdrix* had been introduced, but had disappeared. The birds may have been brought with Guinea-fowl (q.v.) a couple of years earlier. Either Le Gentil was wrong about the birds having disappeared, or more were introduced, as d'Héguerty (1754) described *perdrix* as "common" in the late 1730s. There is no further report until Maillard (1862) included '*Francolinus perlatus*' (= *F. pintadeanus*, but see below) and *Margaroperdix madagascariensis* in his list of Réunion birds, Coquerel (1864) adding *F. pondicerianus*. Legras (1863) reported the introduction of one pair of the Malagasy Partridge "nearly twenty years ago" (i.e. *c.* 1845) at St.-Pierre, with many being killed on a shoot 2 years after. However, Vinson (1861, 1868) and Coquerel (1864) both considered the species extinct despite frequent introductions. Pollen recorded it in 1865 (Schlegel & Pollen 1868), and Lantz (1887) and Crépin (1887) pointed out that it lived in the heights of the interior of the island, avoiding coastal areas, which may be why other writers believed it extinct. Vinson (1868) put the date of introduction of the Grey Francolin as some 20 years previously (*c.* 1850); an editorial note in Maillard (1862) referred to "recent" releases of this species (Barré & Barau 1982).

In the mid-nineteenth century there was considerable confusion between the Chinese Francolin and the Malagasy Partridge (Carié 1916); this was due to similarities in plumage, and the fact that Sonnerat (1782), in his original description, gave the origin of the Chinese bird as Madagascar. I suggest that although they listed both species, Maillard (1862) and Coquerel (1864) were really referring only to the Malagasy Partridge, especially as Vinson (1876) wrote to Mauritius in 1876 requesting a pair of "perdrix pintadées which Réunion does not yet possess". This was presumably not done, as Lantz (1887) did not list the bird. Crépin (1887) wrongly used the name '*perlatus*' as a synonym of *M. striata* (= *madagascariensis*).

To summarise, it seems most probable that the original eighteenth-century introduction was of Grey Francolins (cf. Mauritius), and that these died out. In the 1840s and 1850s this species was re-introduced and the Madagascar Partridge added. The Chinese Francolin, introduced at an unknown date after 1887, was fairly common in 1948 (Milon 1951). However F. B. Gill (*in litt.*) saw none in 1967, and though still listed by Staub (1976), it appears to have died out (Barré & Barau 1982). The Grey Francolin, now rare, and Malagasy Partridge remain (*ibid.*).

Rodrigues

Marragon (1803) intended to introduce *perdrix* to Rodrigues, but if he did so they died out. According to Colin (*in* Kennedy 1893) 36 brought on the *Teemayma* from Tranquebar (S. India) were released at Baie aux Huitres during Mr Jenner's magistracy (1862–71; North-Coombes 1971). Bertuchi (1923) gave the date as 1862 and the ship's name as *Gemima*. They were well established by 1874 (Slater 1875; "imported from Mauritius", Sharpe 1879) when they were identified as *Francolinus pondicerianus*. Today they remain widespread but reduced in numbers (Staub 1976); like Guinea-fowl, they are hunted to prevent depredations in maize fields (Staub 1973*b*).

Quails: *Coturnix, Perdicula* and *Turnix*
Mauritius

The first clear evidence of quails in Mauritius is Desjardins' notes of a specimen of the Painted Quail *Coturnix sinensis* killed in 1834 (Oustalet 1897). Desnoyers (1837) also refers to 'quails', one being Painted Quail ("*Caille de chine*"), the other Malagasy Partridge. The word "*caille*" was often used for the large Malagasy Partridge, but there is no converse reason to suppose that eighteenth-century references to *perdrix* should have applied to true quails. It thus seems likely that quails were not introduced until after 1800, otherwise their presence would have been noted by earlier writers. Bush Quails *Perdicula argoondah* were first recorded by A. Newton (1861*a*), there are eggs attributed to Common Quail *Coturnix coturnix* in Cambridge collected in 1866, and the Malagasy Button-Quail *Turnix nigricollis* was first mentioned by Meinertzhagen (1912) as "introduced from Madagascar at a fairly recent date".

Three of these species have died out, and Common Quails are maintained only by release from quail farms (Straub 1973*a*). Bush Quails were common at the turn of the century (Carié 1904) but already rare by 1910 (Meinertzhagen 1912). Carié (1916) introduced more without success, but these were probably Jungle Bush-Quail *P. asiatica* (Guérin 1940–53, Rountree *et al.* 1952). Guérin (1940–53) considered *P. argoondah* extinct in 1940. The Painted Quail, from being the commonest of the quails in 1910 (Meinertzhagen 1912), had almost vanished by 1916 (Carié 1916). It was said still to survive in 1942 (Guérin 1940–53), but had probably disappeared by 1952 (Rountree *et al.* 1952), and has certainly not been recorded since (Staub 1973a). The Malagasy Button-Quail was uncommon in 1910 (Meinertzhagen 1912), rare by 1916 (Carié 1916) and extinct by 1942 (Guérin 1940–53). These rapid extinctions of successfully established species were attributed by d'Emmerez (1914) and Carié (1916) to the recent introduction of the mongoose *Herpestes edwardsii* (see above).

Réunion

The first quails were released on Réunion in January 1714 (Foucherolle *et al. c.* 1714), but according to Le Gentil (1727) they had disappeared by 1717. There is no further mention of quails until Desjardins (1834) commented that Painted Quail occurred there as well as in Mauritius. Vinson (1868) put their introduction as "in the earliest days", but we have no information on the identity or origin of the 1714 birds. By 1864 two more species, *C. 'textilis'* (error for *C. coturnix*: Carié 1916, Guérin 1940–53) and *'C. cambayensis'* (= *Perdicula asiatica* or *argoondah*), had been added (Coquerel 1864); Vinson (1868) stated that the Bush Quail (*'Perdix gularis'*!) had been introduced some 20 years earlier (i.e. *c.* 1850), and implied that Common Quails had been established longer. Lantz (1887) mentioned only two species, omitting the Common Quail, whereas Milon (1951), also admitting two, omitted the Bush Quail. All three are present today (Barré & Barau 1982), the Bush Quail confirmed as *P. asiatica*.

For a discussion of the Malagasy Button-Quail in Réunion see the section on native birds.

Rodrigues

Colin (*in* Kennedy 1893) reported that 12 "quails" (species unknown) had been released in Rodrigues in the 1860s at the same time as partridges were successfully introduced, but were rapidly "destroyed by wild cats".

Feral chickens and Jungle Fowl
Mauritius

Chickens *Gallus gallus* were left by the Dutch in 1598 (Pitot 1905) but apparently failed to establish a feral population. La Merveille reported feral chickens in 1709 (La Roque 1716). There were no other reports.

Réunion

The existence of a wild jungle fowl *Gallus gallus* was suspected by Milon (1951) and confirmed by Jouanin (1964b), who was told that it had been first released at Bras Panon at the beginning of the century. There were certainly birds present earlier, as a *'poule de Cochinchine'* was accessioned by the St.-Denis Museum in 1892 (manuscript accessions book), but this may have been from captive stock. Feral domestic chickens are not recorded in the literature, but I have heard them calling in formerly inhabited areas, now abandoned, of the Rivière des Remparts.

Rodrigues

Feral chickens have never been reported here.

Guinea-fowl
Mauritius

The first mention of wild Guinea-fowl *Numida meleagris* dates from 1732 (Cossigny 1732–55) and d'Héguerty (1754) also implied their presence in the 1730s. They were presumably brought from Réunion by the first French colonists in the early 1720s. Formerly widespread (Carié 1916), the species seems to have suffered, like other game birds, from mongoose predation, though a small population survived until at least 1970 (Staub 1976).

The form on all the islands was the Malagasy race *N. m. mitrata* (E. Newton 1865a, Rountree *et al.* 1952).

Réunion

The first report of Guinea-fowl was in 1717 by Le Gentil (1727, Lougnon 1970), who remarked that they were the only introduced game bird to have successfully established. In 1714 or 1715 an order had been given to ships returning through Bourbon from elsewhere in the east to pick up Guinea-fowl and other game birds to release in Bourbon (Foucherolle *et al. c.* 1714), and this was presumably done soon afterwards. D'Héguerty (1754) confirmed their presence in the 1730s, but the only subsequent report is second-hand (Buffon 1770–83), citing Aublet, a resident of Mauritius in the 1750s (Ly-Tio-Fane 1958). Guinea-fowl presumably died out in the second half of the eighteenth century; although Vinson (1868) implied their continued feral breeding, Coquerel (1864) clearly stated that, although formerly abundant in the wild, they "are now only found in a state of domestication".

Rodrigues

Marragon (1803) reported that he intended to introduce, amongst other game animals, Guinea-fowl, and presumably did so (North-Coombes 1971); by 1832 they were well established (Dawkins *in* Telfair 1833, Strickland 1848), and they were reported as common throughout the rest of the century. By 1916, however, they were "becoming rare owing to the destruction of their nests by pigs, which in the past have been allowed to wander free all over the island" (Bertuchi 1923). they became extinct shortly before 1964 (Gill 1967); Gill's informants blamed cyclones, but mine concurred with Staub (1973b, 1976) that the birds were perceived as a threat to sown maize and were systematically destroyed during the 'agricultural revolution' in Rodrigues (1955–68; under P. L. Hotchin, North-Coombes 1971).

Miscellaneous game birds

Partridges (47 "*perdrix rouges*": partridges, ? *Alectoris* sp.) were introduced to Mauritius from Spain by Poivre in 1768 (Bouton 1871, E. Newton 1871, Carié 1916) but evidently failed to establish. Carié (1916) reported various unsuccessful attempts to introduce pheasants of several species.

Vinson (1863, 1868) apparently brought Silver Pheasants *Lophura nycthemera* and California Quail *Lophortyx californicus* into Réunion, but it is not clear whether any were released. Chukars *Alectoris chukar* (or perhaps *A. graeca*) are widely kept today on both Mauritius and Réunion (pers. obs.) and I saw a feral *Alectoris* partridge on Réunion in 1974. There have been recent attempts to establish Ring-necked Pheasants *Phasianus colchicus* in Réunion (Barré & Barau 1982).

Rails (Rallidae)

Purple Gallinule Porphyrio (p.) madagascariensis
Although the Madagascar Purple Gallinule has been regarded as native (A. Newton 1861a, E. Newton 1888), the first writers to mention the species in Mauritius clearly considered it introduced: de Querhoënt (in 1773; Buffon 1770–83) said the Purple Gallinule "had produced young" on the island, implying captive breeding, and Milbert (1812) explicitly stated that it had been introduced from Madagascar, and was kept for its appearance. In his manuscript notes (1830s) Desjardins commented that the Purple

Gallinule did not exist in Mauritius (Oustalet 1897), so the feral birds well known later in the nineteenth century (A. Newton 1861a, etc.) may have come from subsequent introductions. Carié (1916) commented that the mongoose had drastically reduced their numbers, but they persisted in several places until the late 1930s (Guérin 1940–53), and in very small numbers at Wolmar/Flic-en-flac until very recently (seen in 1951 (F. R. G. Rountree pers. comm.) and 1976 (Michel 1981)). Staub (1976) considered it extinct through overhunting. Drainage was perhaps more important, but certainly it was a popular game bird in its time (Meinertzhagen 1910–11).

This species seems never to have established itself in Réunion. Coquerel (1864) mentioned birds occasionally seen on the Etang de St.-Paul, but Vinson (1868) referred only to captive individuals. The manuscript accessions book of the St.-Denis museum records numerous specimens in the late nineteenth century.

Moorhen Gallinula chloropus

I have treated the Moorhen amongst the birds native to the islands, but there must remain some doubt about this. Carié (1916) commented on the absence of Moorhen bones in the Mare aux Songes deposits on Mauritius, in what must surely have been suitable habitat. The seventeenth-century references to 'moorehennes', 'water-hens' and '*poules-d'eau*' probably referred to the extinct coot *Fulica newtoni*, and after these there are no mentions of *poules-d'eau* until the end of the eighteenth century (Cossigny 1799). Likewise on Réunion the first definite mentions of Moorhens date from the mid-nineteenth century (Maillard 1862, Coquerel 1864). It seems possible that *G. chloropus* was not present on Mauritius until the late eighteenth century, reaching Réunion in the first half of the nineteenth. Desjardins (1830s; Oustalet 1897) remarked that Moorhens were absent from Réunion. As there is no mention whatever in the literature of them being introduced, they may have invaded by themselves; the race present on both islands is the Malagasy *G. c. pyrrhorrhoa* (Benson 1970–1).

Pigeons (Columbidae)
Mauritius
Very few eighteenth-century writers described the various columbids, '*tourterelles*', '*ramiers*' and '*pigeons*' they saw in Mauritius, so it is impossible to be sure of

dates of introduction. Indeed the information is too poor even to establish whether the Malagasy Turtle Dove *Streptopelia picturata* is introduced or native. The fact that La Caille (1763) in 1753 and Cossigny (1732–55) in 1755 referred to only two species of pigeon suggests that the Malagasy Turtle Dove had not at that time been introduced; Bernardin (1773), writing of his visit during 1768–70, referred in addition to two introduced "*tourterelles*", one of which was no doubt the Zebra Dove *Geopelia striata* noted by Sonnerat (1782) shortly afterwards. Since Sonnerat also described Spotted Doves *Streptopelia chinensis* without referring to their occurrence in Mauritius, I suspect that species was a later arrival, and indeed Desjardins (Oustalet 1897) reported that the younger Cossigny had brought *tourterelles* from Bengal in 1781. These cannot have been *G. striata*, which does not occur there (Goodwin 1983), so Oustalet (1897) suggested they were Spotted Doves, which seems likely. The first definite record of this species is in 1834 (*ibid.*). Bernardin's second species may therefore have been *S. picturata*, which is the bird to which Carié (1916) assigned Milbert's (1812) "*Columba triangularis*". There is a Mauritian specimen of this species in the Dufresne collection in the Edinburgh Museum (pers obs.); assuming it reached Dufresne in the same manner as the *Alectroenas nitidissima* also now in Edinburgh (Milne-Edwards & Oustalet 1893), it would have been collected between 1802 and 1811. All three species are common today (Staub 1976).

Although 'pigeons' were brought to Mauritius in 1839 by Van der Stel (Pitot 1905, Carié 1916) and were thriving in 1641 (Bonaparte 1890), there is no evidence that these birds, presumably *Columba livia*, reverted to the wild. Domestic pigeons are referred to occasionally in the literature (e.g. Bernardin 1773, Milbert 1812), but feral birds were not mentioned before Clark's account (1859). They were then and still are (Staub 1976) common in Port Louis, and had also established on cliffs elsewhere (e.g. Corps de Garde).

A free-living group of the Diamond Dove *Geopelia cuneata* existed briefly at Rivière Noire after over 100 were released from aviaries at La Balise in 1976 (S. D. McKelvey pers. comm.); they had disappeared by 1978 (F. Steele pers. comm.; pers. obs.).

Réunion

Assuming that the '*tourterelles*' seen by early visitors were a native species that became extinct, as Borghesi's (Lougnon 1970) and Feuilley's (1705) accounts

suggest, one must consider Malagasy Turtle Doves as introduced at a later date (*contra* Vinson 1868, 1887). There are no certain eighteenth-century records of any kind of pigeon in Réunion, though Bory (1804) saw birds that answer to feral *Columba livia* in 1801. Feuilley (1705) mentioned the existence of a few domestic pigeons, but there is no information on when they became feral; Cossigny (1732–55) mentioned *tourterelles* in 1732, but it is impossible from the context to say whether he was referring to both islands or only to Mauritius. By the 1860s Malagasy Turtle Doves, Zebra Doves and feral pigeons were clearly all well established (Vinson 1861, 1868, Maillard 1862, Coquerel 1864, Schlegel & Pollen 1868) and remain so today (Staub 1976, Barré & Barau 1982), though the feral pigeon seems to have nearly disappeared 30 years ago (Milon 1951).

Vinson (1887) mentioned several species of pigeon regularly imported into Réunion and bred in aviaries; none established feral populations.

Rodrigues

The only dove introduced to Rodrigues is the Zebra Dove, which was said by Bertuchi (1923) to have been released along with other birds by the crew of the "*Gemima*" (= *Teemayma*) in 1862. E. Newton (1865a) was told of a 'dove' in 1864, though Slater (1875) saw none in 1874. The species is very common today (Staub 1976).

Parrots (Psittacidae)
Mauritius

Two species of parrot have at various times established wild populations. It has been commonly accepted (Oustalet 1897, Guérin 1940–53, Staub 1973a, 1976) that the Grey-headed Lovebird *Agapornis cana* was introduced to Mauritius early in the seventeenth century, or even (Carié 1916) that it is native to the Mascarenes. These views are based on Cauche's description (1651) of "little parakeets in Mauritius which have a yellow neck and the rest green. It [sic] does not exceed the size of a lark [*alouette*]"; he saw these in 1640. This description does not in my view fit any known species satisfactorily; the lovebird in question does not have a yellow neck, but a grey head, and '*alouette*' was used by travellers as often in the sense of '*alouette de mer*' (waders: plovers, sandpipers, etc.) as in the sense of 'lark' (Alaudidae); birds as large as Grey

Plovers *Pluvialis squatarola* are referred to as *zalwet* in Mauritius to this day (Cheke 1982*b*). No other early traveller described birds that could be Grey-headed Lovebirds, and since I have found in writings unknown to the above authors clear evidence of its introduction in the 1730s, I believe there is no longer any reason to suppose there was any earlier occurrence. Cossigny (1732–55, 1764) stated explicitly that when he first arrived in Mauritius (in 1732) no such parakeet existed, but that it was brought in from Madagascar as a cage-bird and escaped; he gave (*ibid.* 1764) the date of introduction as "25 years ago", i.e. 1739. The population increased rapidly, as by 1755 (Cossigny 1732–55) there was already an "astonishing multitude". The species remained common until at least 1916 (Carié 1916), surviving in small numbers until perhaps 1950 (Guérin 1940–53). Rountree *et al.* (1952) pronounced the species extinct, wrongly attributing its decline to a cyclone in 1892 (a story since repeated: Staub 1973*a*, 1976).

The Ring-necked Parakeet *Psittacula krameri* was introduced to Mauritius after caged birds were accidentally released in about 1886 (Carié 1916), establishing themselves at first near Grand Port. By 1916 (*ibid.*) there was a second colony at Pamplemousses, but they were apparently still centred in those areas in 1940 (Guérin 1940–53) although the southern birds had spread across Mahébourg Bay to Pointe d'Esny and Beau Vallon. In 1952 (Rountree *et al.* 1952) they were still confined largely to the coastal plain in the south and east, Pamplemousses and also inland at Alma and Quartier Militaire. It is not clear when they reached Black River and penetrated the native forest where they now interact with the native *P. echo* (Chapter 5), but Newton (1958*a*) recorded them at Reduit and Vacoas before 1957, and considered the species much more widespread than did Rountree *et al.* (1952).

Other parrots of various kinds were evidently brought in regularly in the early nineteenth century. Holman, writing in 1830 (Barnwell 1948), stated that "a few black parrots [*Coracopsis* spp.] are brought from Madagascar as well as grey and green ones [*Agapornis cana*], and from the east coast of Africa. Cockatoos, lories, parrots and Java sparrows are sometimes brought from Batavia". Milbert (1812) also mentioned the presence of black parrots from "l'île Bonaparte" (= Réunion), presumably captive. Apart from the lovebird none of these appears to have established in the wild, though the risk still exists. The bird garden at Casela has cages full of miscellaneous parrots. These cages are built round trees and could well be broken open in a strong cyclone (pers. obs.).

Réunion

I have discussed above, under native birds, the probably introduced status of the unidentified black parrot (? *Coracopsis* sp.) in Réunion, first described by Bory (1804) and seen in captivity in Mauritius by Milbert (1812). There is no information on their date of introduction, and they appear to have died out in the 1860s. They were cited as very rare by Vinson (1861), Maillard (1862) and Coquerel (1864); Pollen, in the island in 1865 (Schlegel & Pollen 1868), clearly saw none. There are no subsequent reports and the only *Coracopsis* acquired by the St.-Denis museum in the period 1855–92 were two *C. vasa* from a sea captain (manuscript accessions book). Hartlaub (1877) considered it already extinct.

The Grey-headed Lovebird, although introduced very early into Mauritius (above), is not reported in Réunion until included in Maillard's list (1862) as "rare", though it may well have been present for a long time before that. It remained rare (Staub 1976), despite being supplemented by repeated introductions (Milon 1951), dying out around 1975 (Barré & Barau 1982). The Ring-necked Parakeet arrived from Mauritius in the early 1970s, cage-birds being released or escaping (Gruchet 1975, Barré & Barau 1982). They were rapidly placed on the list of pests (Servat 1974), and A. Barau (pers. comm. 1978), biological advisor to the *département*, claimed to have shot all of the incipient colony in 1976.

Rodrigues

Grey-headed Lovebirds were, like Zebra Doves, released from the ship *"Gemima"* (= *Teemayma*) in 1862 (Bertuchi 1923). E. Newton (1865*a*) shot two in 1864, though he was told their source was an American whaler. They were abundant (e.g. Slater 1875, Bertuchi 1923), even being exported in good numbers as cage-birds to Mauritius, until 1956, after which time they became very scarce, apparently due to a cyclone (C. M. Courtois *in litt.*), probably that of January 1957 (Davy 1971). Gill (1967) confirmed their scarcity in 1964, though a few persisted until at least September 1974 (Mr Tolvis pers. comm.). In addition to the possible effects of cyclones, they were heavily persecuted for their attacks on maize (various informants, pers. comm.; Staub 1973*b*, 1976).

The view that the lovebird was introduced in the early eighteenth century (A. Newton 1875*a*, Oustalet 1897, Staub 1973*b*) is discussed under Rodrigues native parrots (p. 47) and shown to be without foundation.

Bulbuls (Pycnonotidae)

Controversy raged in Mauritius for a long time as to who had released the new pest, the Red-whiskered Bulbul *Pycnonotus jocosus*, and whether it was deliberate (Carié 1910, 1916, Guérin 1940–53, d'Emmerez 1941). The consensus is that a number (one to six pairs, depending on the story) were released in April 1892 either intentionally by Gabriel Regnard or as a result of a cyclone destroying his aviary. However another story (Carié 1916, Guérin 1940–53) suggests there was (also?) an earlier release around 1880. Carié (1910) chronicled the birds' rapid spread across the island; it was already the most abundant bird on the island in 1916 (Carié 1916) and remains so today (pers. obs.).

It was reported in newspapers in 1972 (*fide* Staub 1973*a*, 1976) that Red-whiskered Bulbuls had been released in Réunion by tourists returning from Mauritius. It was rapidly declared a pest (Servat 1974), and though also subject to the attentions of bird trappers who find it attractive merchandise (T. Bègue and H. Gruchet pers. comm.), it seems slowly to be establishing itself (Barré & Barau 1982).

Rodrigues remains free of this bird, and special regulations exist prohibiting its importation (Lane 1946).

Shrikes (Laniidae)

In a well-known passage written in 1768 Bernardin (1773) referred to a recent introduction to Mauritius from the Cape of Good Hope of a bird called *"l'ami du jardinier"* (the gardener's friend), "brown and the size of a large sparrow". He described it as not only eating "worms, caterpillars and little snakes" but also sticking them on spines in hedges, adding he had seen a captive one suspending meat on the bars of its cage. Despite the typical laniid behaviour, some have taken this as a description of the native cuckoo-shrike *Coracina typica* (e.g. Oustalet 1897). However, true shrikes were introduced from South Africa in the eighteenth century. Oustalet himself cited Desjardins' notes referring to a newspaper article of January 1774

(reprinted in the *Revue Retrospective* 4: 339 (1953)): shrikes ("la vrai piegrièche") intended to "destroy little birds", were brought on the ships *Laverdy* and *Marquis de Talleyrand* and released in the Pamplemousses area. The plumage description given is inadequate to identify the species (black with some ash-grey and white patches); the birds were said to "hop more or less like a magpie and have a frequent movement of the tail".

Finches (Fringillidae)

The only true finches successfully introduced to the islands are two species of canary, *Serinus mozambicus* and *S. canicollis*.

Mauritius

Bernardin (1773) was the first to mention the *oiseau du Cap* (= *Serinus mozambicus*) in Mauritius, but left the history of its introduction and subsequent behaviour to Le Gentil (1779–81, Oustalet 1897), who wrote: "it is a species of *tarin* [greenfinch], a great destroyer; it comes from the Cape of Good Hope, whence it was brought during the last war [i.e. Seven Years' War, 1756–63] to the Isle-de-France as a curiosity and to give as gifts to ladies; (although this bird, which is a yellow and green species of *serin* [canary], is nothing unusual) but it is one of the most deadly presents given to the island." Le Gentil was on the island in 1760–1 and 1770, and it seems likely from his writing that the bird was already present in 1760. The identity of the bird is confirmed in drawings (*c.* 1770) in Commerson's manuscripts (Oustalet 1897) and in a specimen from the same source mentioned by Buffon (1770–83).

In contrast to the previous species there are no early descriptions of the Cape Canary *S. canicollis*, the two species probably being confused. The first record is a specimen collected by Dumont accessioned by the Paris Museum in 1804 (Oustalet 1897); Dumont was in Mauritius with Milbert in 1801 (Milbert 1812). It is not mentioned in the literature before the 1850s (Clark 1859).

Canaries were notorious pillagers of grain in the late eighteenth and early nineteenth centuries and a price was put on their heads in 1770 (Ly-Tio-Fane 1968, Bernardin and Le Gentil *locs. cit.*, Sonnerat 1782, etc.), but they ceased to be considered as pests when cereal growing declined, and by the 1850s were again regarded just as pretty songbirds (Clark 1859), although still very numerous (E. Newton 1861*b*). The Yellow-fronted Canary *S. mozambicus* remains very common on the island (Staub 1976) but the Cape Canary is extinct. According to Carié (1916) it was last seen in 1893 at the Trou aux Cerfs, and Meinertzhagen (1912) and d'Emmerez (1914) attributed its extinction to the cyclone of 1892. However I was told by Mr Paulo Jauffret Snr of Souillac that he used to catch these canaries at the Trou aux Cerfs in 1918–22; he attributed their final disappearance to their intense trapping for sale as cage-birds at this their last stronghold in Mauritius. Guérin (1940–53), writing in 1953, reported that attempts at re-introduction had failed. In Réunion the Cape Canary now favours high ground over 600 m (Staub 1976, Barré & Barau 1982), which is not available in Mauritius, and may explain why their last refuge in Mauritius was one of the highest points. However they had occurred in large numbers at low elevations in the past (E. Newton 1861*b*), suggesting perhaps a susceptibility to an introduced disease with vectors active at low altitudes (cf. Warner 1968 for Hawaii).

Some confusion exists in the literature about the local names of the two canaries in Mauritius. *S. mozambicus*, originally known as *oiseau du Cap* or *serin du Cap*, became the 'native canary' (*serin du pays*) after the introduction of *S. canicollis*, the epithet *'du Cap'* transferring to the newcomer from Africa. The date of this changeover is unknown but was complete by 1855 (Cheke 1982*b*).

Milbert (1812) reported that in 1801 canaries *Serinus canaria*, Bullfinches *Pyrrhula pyrrhula* and Goldfinches *Carduelis carduelis* were being brought into Mauritius from Europe as cage-birds, and that some had escaped. They did not establish; there were no further records.

Réunion

There is no published record of the introduction of the two canaries into Réunion; they are first listed by Maillard (1862), whose mistake in confusing the Cape Canary with the European Greenfinch *Carduelis chloris* was soon corrected by Coquerel (1867*b*) and Schlegel & Pollen (1868). Examination of archives of the second half of the eighteenth century for references to agriculture would no doubt reveal information on granivorous bird pests then present in Réunion. All I have found is that Pingré (1763) and the anonymous "Officer of the British Navy" quoted by Grant (1801) reported 'small birds' as pests in 1761 and 1763. Pingré

wrote that a tax of 100 birds' heads per slave (10 times the Mauritian figure) was levied. This was 9 years before it was introduced in Mauritius, no doubt because Réunion was the granary for both islands (Vinson 1868). The introduced seed-eaters may thus have reached Réunion sooner than Mauritius. Both canaries are still common today (Staub 1976, Barré & Barau 1982).

Rodrigues

Although E. Newton (1865*a*) was told in 1864 of a *serin* in Rodrigues this is more likely to have been, given local Creole usage (Cheke 1982*b*), the endemic fody *Foudia flavicans*, although Vinson (1964*b*) and Staub (1976) believed it was a canary. The first record I can find of Yellow-fronted Canaries on the island was in 1963 (Vinson 1964*b*), confirmed by Gill (1967) a year later; it is not a common bird there (pers. obs.).

Weaver-finches (Estrildidae)

Java Sparrows (*Padda oryzivora*), Spice Finches (*Lonchura* spp.) and waxbills (*Estrilda* spp.) have been introduced to the Mascarenes at various times.

Mauritius

Java Sparrows and other granivorous finches were already serious pests by 1765 (Le Gentil 1779–81). There is no earlier record. It seems unlikely that La Caille would have overlooked the birds had they been present in 1753. Already rare by 1859 (Clark 1859), they were last seen in 1892 (Carié 1916), apparently dying out because grain was no longer grown on the island. Their survival in the wild may have been prolonged by continued escapes from captivity; Holman (Barnwell 1948) referred to Java Sparrows being brought from the East Indies in 1830.

Avadavats *Estrilda amandava* were first recorded in 1753 by La Caille (1763: "*bengalis*"). Cossigny (1732–55) found none in 1732 but escaped cage-birds had become abundant by 1755. They apparently became rare after the introduction of *E. astrild* (see below); they were recorded by anon. (1873) but not by Clark (1859), E. Newton (1861*b*) or Hartlaub (1877). By 1900 they were so rare that captives were fetching high prices (Carié 1904), and they soon died out (Carié 1916: "about 20 years ago"). Sporadic reports up to the present are apparently due to re-releases (Guérin 1940–53, S. Temple pers. comm.).

Desjardins described a Common Waxbill *E.*

astrild in 1827 (Oustalet 1897). Although there were numerous references to "bengalis" before this date, the only descriptions clearly refer to *E. amandava*. Common Waxbills were introduced well after Avadavats, probably around 1800; they remain common today (Staub 1976).

The Spice Finch *Lonchura punctulata* was first certainly recorded by Desjardins in 1829 (Oustalet 1897), but in his memoir to Buffon, de Querhoënt (1773) named two birds as "*calfat*" (caulker), describing the Java Sparrow, but only mentioning another "less big, equally noted for the beauty of its plumage". Desjardins' use of the name "*marteau*" (hammer) and de Querhoënt's description of the birds' sounds as resembling the tapping of caulkers' mallets, suggest the second *calfat* was the Spice Finch. Le Gentil's (1779–81) "*moineaux de Chine*", present in 1765, may have been this species.

Réunion

The first estrildid recorded was a specimen of the White-rumped Munia *Lonchura striata* sent under the name *jacobin* by La Nux to Réaumur before 1757 (Brisson 1760). Various authors have considered this record to be a locality error (e.g. Mayr *et al.* 1968), but I follow Stresemann (1952) in accepting it as an introduction from Pondicherry, especially as Brisson gave the reliable La Nux as source. There is no further record of the species.

Unlike in Mauritius where the introduction of the different seed-eaters in the eighteenth century is well documented, we have no such records for Réunion. Java Sparrows, avadavats, waxbills and Spice Finches were well established when first listed by Maillard (1862). Morel (1861*b*) admitted to ignorance of the history of the Common Waxbill in Réunion. Java Sparrows, already rare in 1862 (Maillard 1862), were seen nesting with House Sparrows in St.-Denis in 1865 (Schlegel & Pollen 1868), but were not recorded as wild thereafter. Milon (1951) reported frequent but unsuccessful attempts to re-introduce them up to 1945. The other three species still occur on the island, only Common Waxbills being common (Milon 1951, Staub 1976, Barré & Barau 1982).

Vinson (1863) listed *Estrilda melpoda*, *Amadina fasciata* and "bengali bleu *Estrelda phoenicotis*" amongst birds he had "introduced", presumably only to aviaries.

Rodrigues

In Rodrigues *bengalis* (i.e. *Estrilda astrild*) were reported to E. Newton (1865*a*) in 1864; they may have arrived on the *Gemima/Teemayma* 2 years earlier. Although not seen by Slater (1875), Bertuchi (1923) confirmed their presence in 1916, and they remain common (pers. obs.). Slater (*ibid.*) saw a few Java Sparrows, but this must have been a short-lived introduction as no one else recorded them.

Sparrows (Passeridae): *Passer domesticus*
Mauritius

The dates given in the literature (Carié 1916, Guérin 1940–53, Rountree *et al.* 1952) for the introduction of the House Sparrow are based on an erroneous statement that the species was still absent in 1859. Dupont (1860) wrote in late 1859 that sparrows should be introduced from India, evidently unaware that the deed had already been done. According to Clark (1859), British soldiers had brought birds from India and released them in the barracks in Port Louis "two or three years ago", i.e. *c.* 1856; by 1859 the birds were already well established and "breeding fast". They were numerous all over the island by 1910 (Meinertzhagen 1912), and still are (Staub 1976, pers. obs.).

Réunion

Henri (1865) described the first feral breeding of escaped aviary House Sparrows in Réunion in June 1845. He did not (*contra* Schlegel & Pollen 1868) release the birds himself, but did encourage and feed them on his property. The birds spread throughout St.-Denis over the next 6–7 years. According to Henri the birds initially began breeding in the time of the northern summer, but "later" changed to nesting in the southern summer. Henri considered the sparrows to be of Indian origin, whereas other contemporary accounts (Maillard 1862, Coquerel 1865*b*) said they were from France; specimens from Réunion in the Paris museum are referable to *P. d. indicus* (C. Jouanin pers. comm., though Berlioz (1946) did not commit himself).

House Sparrows remain abundant throughout Réunion today (Staub 1976, Barré & Barau 1982).

Rodrigues

Bertuchi (1923) was the first to report sparrows (in 1916), although they could have been introduced somewhat earlier. They are common today (pers. obs.). Staub's view (1973*a*) that the *"moineaux"* seen by Marragon (1795, Dupon 1969) were *Passer domesticus* has already been discussed above under native fodies (p. 51).

Weaver birds (Ploceidae)
Mauritius

Although Bernardin (1773) referred to a *"cardinal"* as being present (as an introduced bird from Bengal!) in 1768, his description could as well apply to the native *Foudia rubra* as to the Madagascar Fody or Cardinal *F. madagascariensis*. However, in a memoir to Buffon soon afterwards, de Querhoënt (1773) distinguished clearly between the native species, the *"cardinal"*, and the *"foude"* introduced from Madagascar, already abundant and causing damage to crops. It was presumably brought in as a cage-bird on slave ships. The species has remained abundant ever since (Staub 1976, Chapter 4).

Village Weavers *Ploceus cucullatus* from South Africa were released around 1886 at Cap Malheureux, spreading rather slowly (Carié 1916); however they are now very widespread below about 1000 ft (305 m) (Newton 1958*a*; pers. obs.).

Réunion

As mentioned under the native fody, there are no definite records of Cardinals on Réunion before the mid-nineteenth century, when the species was listed by Maillard (1862) and all subsequent writers. I have already discussed why I consider Buffon's *mordoré* an aberrant Cardinal; if it did indeed come from Réunion, then this shows that Cardinals were present in the 1770s, but we know so little of Réunion birds (apart from certain specimens sent to Paris) between the 1730s and 1860s that there is no way of assessing a date of introduction. However, if the species was common in the 1750s, La Nux would surely have sent a specimen to Réaumur, and as it does not seem to have reached Mauritius before the 1760s the same probably applies to Réunion.

Although not mentioned in the literature before 1940 (Guérin 1940–53), Village Weavers (*oiseaux belliers* in Réunion) were introduced much earlier. Some escaped from a cage on a boat loading sugar at Bois Rouge, the property of Adrien Bellier (Albany 1974, A. Barau pers. comm.); this can have happened only during the short period when there was a jetty there,

around 1880 (A. Barau pers. comm.). Guérin (1940–53, followed by Staub 1976) was wrong in supposing that M. "Beylier" (*sic*) had released them. They are common today in lowland areas (Staub 1976, Barré & Barau 1982).

Rodrigues

In Rodrigues the Cardinal was not recorded before 1916 (Bertuchi 1923), but could have been introduced two or three decades earlier. It is abundant today (Gill 1967, Chapter 8). Village Weavers have not reached Rodrigues.

Starlings (Sturnidae): Common Mynah
Acridotheres tristis
Mauritius and Réunion

Various dates are quoted for the introduction of mynahs to Mauritius and Réunion; at least two independent accounts refer to an initial introduction, followed by the birds being killed, and then a second, successful release. Although Buffon (1770–83, vol. 3 (1775)) suggested that the first introduction to Réunion was a little over 20 years earlier (i.e. before 1755), he cited Desforges-Boucher as governor of the islands at the time, a post he assumed in 1759 (Toussaint 1972). Oustalet (1897) quoted an article abstracted by Desjardins from the *Annales maritimes et coloniales* of 1820, citing 1859 as the original introduction date to both islands. Foucher d'Obsonville (1783) stated that Desforges-Boucher organised the original release and then, after the colonists had wiped them out (they were thought to eat sown grain), wrote to the Comte de Maudave in Tranquebar for more; Foucher himself was delegated to obtain them, and did so. Foucher said this was "towards the end of the last war", i.e. the Seven Years' War, 1756–63; Céré, writing in 1777 (Pope-Hennessy 1889), gave the date as 1762. This coincides more or less with de Querhoënt's statement (1773) that in Mauritius the birds had "multiplied prodigiously in the 8 to 9 years since they were brought in" – giving a date of 1764–5; they were the subject of a special protection order in 1767 (Rouillard 1866–9, Ly-Tio-Fane 1968). Bernardin (1773) reported Mynahs as abundant in Mauritius in 1768–9. It is not clear whether the double introduction with the intervening extinction occurred on both islands or only on Réunion; it is more likely that it was on only one island and that the 1762 introduction on Mauritius was the first there, especially as neither Pingré (1763), Cossigny

(1764) nor Le Gentil (1779–81, on the island in 1760–1) mentioned the species. Vinson (1867) gave the date of the second introduction to Réunion as 1767.

The introduction of Mynahs was one of the first recorded attempts at biological control of an insect pest. The islands' agriculture had been persistently devastated since 1729 by grasshoppers introduced from Madagascar (Lougnon 1933–7, Ly-Tio-Fane 1968). According to Cossigny (1799) no grasshopper damage occurred after 1770, the disappearance of the pests being generally attributed to the Mynahs, though Cossigny himself was more cautious about the relationship. Manual destruction of grasshoppers by slaves was run like a military operation in Mauritius (Ly-Tio-Fane 1968), which may have contributed substantially; indeed d'Héguerty (1754) had claimed in 1751 that such methods had brought control in Réunion in the 1730s, long before the introduction of Mynahs, and the grasshoppers were certainly still under some control there before 1767 (Ly-Tio-Fane 1968).

Mynahs have remained abundant on both islands ever since (Staub 1976), rapidly becoming themselves a serious pest especially to fruit (Buffon 1770–86), to the extent that on Réunion an attempt was made in 1820 to persuade people to eat them as game birds (Vinson 1867); it failed because of the success of earlier propaganda that they were inedible (e.g. Buffon 1770–86).

Rodrigues

Contrary to Staub's (1973b, 1976) assertion, the Mynah was not introduced to Rodrigues by Marragon in the 1790s. Staub (1973b) misquoted Marragon (1795, Dupon 1969), who actually wrote, referring to grasshoppers and other destructive agents, "the introduction of the bird *martin would* [my italics] rid the colony of them" (my translation). He or others may have tried the introduction later, for Hoart (1825, Staub 1973b) was told that several unsuccessful attempts had been made to introduce *martins*. Corby (1845) suggested its introduction, again for pest control. However nothing seems to have happened until much later: E. Newton (1865a) did not record them in 1864, but they had been present for "some years" in 1874 (Slater 1875), although still rare; Slater collected specimens (Sharpe 1879). They are abundant today (pers. obs.).

Crows (Corvidae)

The introduction of crows from India to control rats(!) was suggested by Dupont (1860) but no action seems to have been taken. However around 1910 (Meinertzhagen 1912, Guérin 1940–53) a number of House Crows *Corvus splendens* became established, apparently derived from free-flying tame ones that travelled on ships from India; the birds were said to have been left behind when such a ship sailed (Guérin 1940–53). Numbers built up around the Roche Bois slaughterhouse (Rountree *et al.* 1952) and the Port Louis meat market (Guérin 1940–53), spreading to Case Noyale in the south-west, where there was a flock before the 1939–45 war (C. M. Courtois pers. comm.). Although numbers in Port Louis were controlled to prevent them stealing food in the bazaar (Guérin 1940–53), a flock of around 40 persisted at Roche Bois until decimated by the cyclone of 1945 (C. M. Courtois pers. comm.); at least one survived there until 1951 (Rountree *et al.* 1952). In 1950 two flew ashore in Port Louis off a ship from Ceylon (*ibid.*), and numbers in Roche Bois/Port Louis slowly built up to about 100 (Staub 1976). They have been seen in Beau Bassin (pers. obs. 1978), Mahébourg and Grand Baie (Staub 1976); Newton (1958a) recorded a pair at Cannoniers Point, near Grand Baie, as early as 1956.

Much earlier, various attempts were made to introduce Pied Crows *C. albus* from Madagascar to control rats and mice. Bernardin (1773) reported that in 1868 only three males were left of several pairs released, the rest having been killed for eating chicks. Oustalet (1897) gave details of two pairs introduced in 1831, and noted that Desjardins kept one for a while and occasionally saw them around the island (it is not clear whether this was before or after the 1831 release). They had certainly disappeared by the late 1850s, as Clark (1859) did not mention crows.

Vinson (1861) mentioned crows released in Réunion as breeding for a short time and then disappearing. He did not give any indication of species or date.

Miscellaneous proposed and unsuccessful introductions of birds and mammals
Mauritius

As if enough animals had not been introduced, with unfortunate effects, would-be biological controllers or aesthetes were always coming up with new ideas.

Commerson, writing to the younger Cossigny in 1770, exclaimed:

> What service would not be rendered to the colony if one could introduce here species of shrikes, *dominicans* [*Paroaria* spp.], tyrant-flycatchers, flycatchers, woodpeckers, *arniers* [?] and other insect-eaters that never attack cereals; little falcons, *bouchers* [? = shrikes?], night birds [? = owls], to balance the numbers of granivorous birdlings ["oisillons"] as also harmless snakes to destroy rats; we should go beyond the frogs it was thought useful to bring in to purge stagnant waters of the prodigious quantity of mosquito larvae that swarm therein. (My translation, from Cap (1861); see also Oliver (1909) who translated the passage somewhat differently)

This enthusiasm for biological control agents contrasts with Commerson's caution about some other introductions. According to Ly-Tio-Fane (1978), as a witness of some of these introductions Commerson had viewed the problem with perspicacity when he collected material for his great project on the *Histoire Naturelle*. He had noted the adverse effects of introduction in the case of the Java Sparrow, "which caused great depredations to crops".

Sonnerat (1782) suggested bringing in "large birds of prey" to help combat granivorous bird pests and Bory (1804) also recommended sparrow-hawks (*"eperviers"*) to the same end. Cossigny (1799) proposed introducing Siamese "bamboo rats" (? = *Rhysomys pruinosus*) "the size of cats, which make war on ordinary rats, and pass in that country for an exquisite dish"; he clearly did not think that rats that ate bamboo shoots might go for sugar cane or maize!

In 1878 an Acclimatization Society was established (Bouton 1883), but it occupied itself largely with introducing "useful" plants and game species (q.v.). Gleadow (1904, Brouard 1963), proposed introducing "a suitable woodpecker from India or Madagascar" (he suggested '*Picus chrysonotus*' = *Dinopium benghalense*) "to control a bark beetle"; fortunately this suggestion was not taken up.

Grant (1801, Pope-Hennessy 1886) reported the introduction from Senegal, release and subsequent near extermination by hunters of a "gazelle or antelope"; Milbert (1812) confirmed their extinction.

Clark (1859) remarked that Spotted Deer *Axis axis* were often kept and that "Numidian gazelles (*Antilope dorcas*)" were brought from Arabia but usually died of epileptic fits. Ranching of Eland *Taurotragus oryx* and Impala *Aepyceros melampus* has recently been proposed (Min. Econ. Pl. & Div. 1980).

Vinson (1963) suggested, at an ICBP meeting, importing endemic Réunion passerines to interbreed with the declining Mauritian species and thus provide 'hybrid vigour'; this suggestion was fortunately talked out.

Réunion
During the heyday of 'acclimatization societies' in the nineteenth century, there was also one founded in Réunion, which produced a *Bulletin* with details of numerous animals that were brought into the island by its members. Most of the birds appeared to be for aviary breeding, at least initially (cf. Vinson 1863: list of seven spp. comprising a pheasant, a quail, a parrot, three estrildines and ploceid). However Coquerel (1865b) proposed introducing tits (Paridae) to control the cane borer *Phytalus smithii*, and Nightingales *Luscinia megarhynchos* to liven up the "silent" forests. Vinson (1887) reported successfully breeding several species of foreign doves in aviaries. The fact that six Red-crested Cardinals *Paroaria* cf. *dominicana* are listed in the manuscript accessions book of the St.-Denis museum between 1855 and 1871 suggests a popular cage-bird at the time which may have been briefly feral; a population established on Agalega Is. probably around 1900 has survived ever since (Cheke & Lawley 1984).

Official plans to import birds for biological control of insects have surfaced from time to time. Leal (1878) wrote: "The administration is at this moment working on the introduction of certain insectivorous birds from Madagascar which have been designated by Mr Lantz" (my translation). The list consisted of two owls, a nightjar, a roller, a bee-eater, sunbirds, warblers, white-eyes, a drongo, vangas, a crow and coucals, couas and the Courol *Leptosomus discolor*!

Enthusiasm for introductions was not confined to the nineteenth century. Milon (1951) suggested releasing an egret, a snipe, a duck and the Hoopoe *Upupa epops*. He was motivated by a desire to improve the variety of game birds in the island and was careful to choose birds unlikely to compete with endemic forms or be injurious to agriculture. Such careful thinking seemed absent from suggestions aired anonymously in the French business magazine *Marchés Tropicaux* (anon. 1971*b*) to flood the island with pheasants, partridges, hares, antelopes, ibex, chamois, mouflons and goats. "In any case, populating the wild mountains of Réunion with wild animals is accepted as indispensable"! (my translation). The author was at least somewhat reticent about introducing monkeys – but the public consciousness of the dangers of such an introduction, to judge by the pet monkeys to be seen in St.-Denis, is much lower than it was in 1895 when Oliver (1896) reported that four *gendarmes* were sent to the site of a shipwreck to escort the captain's pet monkey to a cage in the Jardin Colonial (now the Jardin de l'Etat) in St.-Denis.

Rodrigues

The near introduction of house shrews and mongooses to Rodrigues has already been mentioned. The island was also threatened in 1858 with "weasels, ferrets or black snakes" which Magistrate Messiter proposed to bring in to fight rats (North-Coombes 1971).

REPTILES

Tortoises and terrapins (Chelonia)

Edwards (1872) and Vitry (1883) reported the presence of terrapins found in good numbers in the pond at Beau Plan in the Pamplemousses district of Mauritius. Edwards also described a specimen found in the sewers of Port Louis in 1871. From the descriptions given these terrapins were *Pelusios* sp., recently identified, from specimens collected at the same period, as *P. subniger* (Bour 1983, 1984*b*), probably introduced from Diego Garcia, the source of five received alive by Daruty in 1878 (note added to Vitry 1883, who had described them as "indigenous"). There are no recent records, Bour (1984*b*) believing it extinct, although the lake at Beau Plan has not been explored (Bour *in litt.*).

Another species, *Trionyx steindachneri*, which escaped at Moka after a flood in 1920 (Bour 1984*b*), had become widespread though not abundant by the 1950s (Vinson 1953*a*), and has remained so (Bour 1984*b*).

Amongst the bones collected by Thirioux early in the century on Le Pouce, a mountain behind Port Louis, are remains of *Geochelone (Asterochelys) radiata*

(Arnold 1980); bones have also been found in a cave in Réunion (Bour 1981). From 1830 onwards enormous numbers of this tortoise were imported to Réunion from the "inexhaustible" supply in the *Opuntia* thickets of Madagascar; they were sold for eating at 30 francs a dozen (Bour 1981; Vinson 1868). These shipments fell off somewhat in the 1880s (Vaillant & Grandidier 1910), but tortoises were still used as ballast on ships (!) in the 1920s (Mme Massé pers. comm.) and the trade persisted until the 1950s (Bour 1981). Given the very large numbers imported it is not surprising that some escaped to die in caves, but there is no evidence that feral populations ever became established. The species is still kept in captivity on both islands (Bour & Moutou 1982), but especially in Réunion where breeding Radiated Tortoises is a well-organised hobby (pers. obs.).

Aldabran Giant Tortoises *Geochelone (Dispsochelys) elephantina* were released on Flat Is. off the north coast of Mauritius in 1883 (Oliver 1891, vol. 2, p. 376n). They were thriving in 1910 when the island was visited by Meinertzhagen (1910–11), who assumed, from the numbers of animals of all sizes present, that they were breeding. The last record of these is a specimen in the Mauritius Institute collected in 1943 (Bour 1984*b*); feral cats were present in 1955 (Newton 1956). The species is widely bred in captivity in Mauritius and to a lesser extent in Réunion (Bour & Moutou 1982, pers. obs.), having been first brought in around 1766 (Bour in press).

Several other tortoises and terrapins have been cited by compilers as occurring in Mauritius (Bour 1984*b*), either in error or pertaining to escaped pets that never established; Bour (*ibid.*) recorded four species imported live since 1980.

Lizards (Squamata)

Geckos (Gekkonidae)

There is no account of the introduction of house geckos into the Mascarenes, but apart from *Hemidactylus frenatus* which I regard as possibly native (Cheke 1984, *contra* Vinson & Vinson 1969), they must all have arrived through human agency. The only one for which there is an early record is *Gehyra mutilata*, which was collected on Mauritius by Peron and Le Sueur (Dumeril & Bibron 1834–54, vol. 3) who were on the island between 1800 and 1804 (Toussaint 1972). However house geckos were seen in Mauritius in 1768 by Bernardin (1773), and were mentioned by

Milbert (1812), who got his details from Peron and La Sueur themselves. Bory (1804) also reported house geckos but it is not clear whether he saw these in Mauritius, Réunion or both. *G. mutilata* were presumably brought in on ships trading with the Far East, possibly by the Dutch, or by Poivre on his spice trips (Ly-Tio-Fane 1958), as it does not occur in continental India (Smith 1935). *Hemiphyllodactylus typus* and *Hemidactylus frenatus* have a similar natural distribution (though the latter is native also on the Chagos, St. Brandon and the Seychelles: Cheke 1984), while *Hemidactylus mercatorius* and *Ebenavia inunguis* are of Malagasy origin. Of these *Ebenavia* is not anthropophilous (*contra* Vinson & Vinson 1969: 206), though the consensus is nevertheless that it was introduced; it occurs only on Mauritius, while the other species are on all three islands (*ibid.*, Bour & Moutou 1982), except for *H. mercatorius* which has not reached Rodrigues.

Vinson & Vinson (1969) noted the existence of two old specimens in Paris of the large Indian gecko *Hemidactylus leschenaultii* collected on Réunion at an unknown date. The card file in Paris also lists another eight, from 'Ile de France' (Mauritius), seven collected by Marchal and one by Maillard; all are without date. I did not check these eight for correct identification, but Vinson & Vinson (1969) evidently did so for the Réunion two. It seems likely that this species was naturalised for a time in the early nineteenth century.

I have not been able to establish the identity of the '*Hemidactylus maculatus*' collected by Quoy and Gaimard in Mauritius (They were on the island in 1818; Desjardins 1830) and referred to in 1836 by Dumeril & Bibron (1834–54, vol. 3) as "juveniles". They are likely, as Dumeril & Bibron likened them to '*H. mabouya*', to have been *H. mercatorius* (see discussion in Loveridge 1947); these two forms are often treated as synonymous (e.g. Arnold 1980, Bour & Moutou 1982; see Cheke 1984).

Two day-geckos have been introduced to Réunion – *Phelsuma lineata* from Madagascar (1940s) and *P. cepediana* from Mauritius (1960s) – without, so far, having affected the native species (Cheke 1975g, 1982a, Bour & Moutou 1982).

Chameleons (Chamaeleonidae)

Desjardins (1837) stated that there were no chameleons in Mauritius, and Vinson & Vinson (1969), reviewing the evidence, decided that none had ever become naturalised. In view of the establishment at an unknown early date (before 1830; Bourgat 1967) of *Chamaeleo pardalis* in Réunion, it would be surprising if animals brought to Mauritius had not also established feral populations. Both *C. pardalis* and *C. verrucosus* were originally described from specimens allegedly collected in the Mascarenes, although the provenance of the latter's type, sent by Baron Milius, is certainly doubtful as he obtained many animals from Madagascar (see Geoffroy & Cuvier 1824–47). The Mauritius *C. pardalis* specimens studied by Cuvier (1829) and Dumeril & Bibron (1834–54) were collected by Lesson and Garnot and by Desjardins himself, and the Réunion ones by de Nivoy. In the early 1830s Charles Telfair of Mauritius presented a number of chameleons (three or four species) to the Zoological Society of London (Inventory of Reptiles, Amphibia and Fishes in the possession of the Zoological Society of London (manuscript accessions book)), but their origin could well have been Madagascar, in common with some of his other specimens. Vinson & Vinson (1969) thought there were no chameleon records in local journals, but overlooked two records of '*Chamaeleo verrucosus*' caught wild in Mauritius in 1882 and 1885 (*Trans. Soc. Roy. Arts Sci.*, Mauritius N.S. 17: 17(1885) and 18: 28(1886)), and also an unidentified chameleon seen in the 1840s (Blenkinsop 1851: 149). Both *C. pardalis* and *C. verrucosus* were probably naturalised on the island, the latter for nearly a century.

There are certainly no feral chameleons in Mauritius today (Vinson & Vinson 1969), but in Réunion *C. pardalis*, restricted in the 1960s to areas near St.-Paul (Bourgat 1967, 1970), has recently spread (anon. 1973, Bour & Moutou 1982) following protective measures. Bour & Moutou (1982) claimed that Cossigny (1732–55) alluded to this chameleon in Réunion in 1755, but he was referring to a dead specimen sent from Madagascar; the date of introduction is unknown.

Agamids (Agamidae)

According to Vinson (1870), *Calotes versicolor* was introduced to Réunion "about five years ago" (i.e. 1865) with sugar canes from Java unloaded off a ship, the *Saint Charles*. It was deliberately introduced to Mauritius from Réunion "some years later" by d'Emmerez de Charmoy (Koenig 1932; Vinson & Vinson 1969 suggesting *c.* 1900), and established by 1914 (d'Emmerez 1914). It is abundant on both islands today (*ibid.*, Bour & Moutou 1982).

Snakes (Serpentes)
Mauritius

The welcome absence, to travellers, of snakes on the Mauritian mainland was frequently commented on by seventeenth- and eighteenth-century visitors, and also the early nineteenth-century travellers de Guignes (1808) and Quoy & Gaimard (1824). Desjardins (1830) reported on a living snake ('*Coluber rufus*') found in Port Louis and commented that he knew of only one other such occurrence, presumably a reference to the "large boa" killed at Le Reduit in 1813 (Desjardins 1837). There were no further records for some time: Clark (1859) did not mention snakes, and Dumeril & Bibron (1834–54) listed no snakes from the Mauritian mainland. In 1879 Daruty (1883*a*), reporting on a specimen of '*Proepeditus lineatus*' (= *Lycodon aulicum*), stated that it had apparently been established for some time.

Günther (1869) identified a blind-snake sent from Mauritius earlier this year as '*Typhlops flavoterminatus*' (= *Typhlina bramina*); another was caught in 1871 (*Trans. Roy. Soc. Arts. Sci. Mauritius* N.S. 6:12(1872)). Daruty (1883*b*) commented that *T. bramina* had been established "for some years", and that two further boas had been caught since the 1830s; the blind-snake and the Wolf Snake *L. aulicum* were already very common, especially in cellars in Port Louis. It is no longer possible to identify Desjardins' '*Coluber rufus*', but the dearth of records between 1830 and 1879 suggests that *L. aulicum* was either introduced or became fully established in the 1870s. *T. bramina* could easily have been overlooked, and it is not possible to say when, or even whether, it was introduced. These snakes remain well established today (Vinson 1953*a*, Vinson & Vinson 1969).

Réunion

Early travellers to Réunion, as those to Mauritius, commented on the absence of snakes (accounts in Lougnon 1970), but after 1671 this appears to have been taken for granted, as there is no further mention of these reptiles until vol. 7 of Dumeril & Bibron's *Erpétologie Générale* appeared in 1854. Under *Lycodon aulicum* these authors quote Rousseau and Pervillez as saying that this snake was common on the island, having been brought from India in bales of rice (Rous-

seau was in Réunion in 1839: F. Roux-Estève pers. comm.). Maillard (1862) listed the same species, but Vinson (1868) referred to a remarkable multiplication of an unnamed snake accidentally brought in from Madagascar some 15 years earlier. Vinson attributed a decline in mice to this species, and blamed it for the near extinction of day-geckos *Phelsuma borbonica*. Guibé (1949, 1958; followed by Blanc 1972) listed the Malagasy snakes as "naturalised" in Réunion: one (*Acantophis dumerilii*) was a labelling error (Bour & Moutou 1982), but the colubrid *Liophidium vaillanti* may have been established for a while, though not present today (*ibid.*). However, the Wolf Snake eats geckos and was also blamed for their decline, so the status of *Liophidium* remains doubtful.

Maillard (1862) was the first to record the parthenogenic blind-snake *Typhlina bramina* which Vinson (1868) suggested was brought in in packages of imported plants. *Lycodon aulicum* and the blind-snake are still found quite commonly (Bour & Moutou 1982).

Rodrigues

The only snake known from Rodrigues is the blind-snake *T. bramina* (Vinson 1964*b*, Bour & Moutou 1982), presumably accidentally introduced.

AMPHIBIANS
Frogs and toads (Anura)
Mauritius

Early writers (Mundy 1608–67, Leguat 1708, Cossigny 1732–55 (in 1732)) commented on the absence of anurans in Mauritius, and the first positive mention is not until 1769 when Bernardin (1773) said that frogs had been introduced without success. Commerson in 1870 (Cap 1861) also referred to this introduction. Milbert (1812) said frogs had been imported from the Seychelles about 20 years previously; Clark (1859) named a Mr Genève as responsible and confirmed the date as 1792 but, like Chapotin (1812), gave the source as Madagascar. The species is the Malagasy *Ptychadena* (*Rana*) *mascareniensis*, which was certainly also present in the Seychelles in the early nineteenth century (Dumeril & Bibron 1834–54); it is common today (Vinson 1953*a*, pers. obs.).

Toads were first reported in Mauritius in 1837 when two were found in the Pamplemousses area (Desjardins 1837). Clark (1859) discussed these specimens, at the same time denying the existence of feral toads in 1859. Bouton (1875) implied that they were still found from time to time; he sent specimens (? the original 1837 ones) to Albert Günther who identified them as the Oriental *Bufo melanostictus* (Günther 1874). It was no doubt this that led Mertens (1934), in a major review of island herpetofaunas, to write that *B. melanostictus* occurred in Mauritius. However it seems that this species never became established, as there were no toads present in 1914 (d'Emmerez 1914) and the animal common today (Starmühlner 1976, 1979, pers. obs.) is *B. regularis* (*sensu lato*) introduced from South Africa in 1922 (Vinson 1953a). Vinson (1953a) also reported unsuccessful introductions in the 1930s and 1940s of the West Indian Giant Toad *B. marinus*; one shipment of 150 toads from Puerto Rico was refused entry (Easteal 1981).

Réunion

The original absence of frogs and toads in Réunion is attested to by François Martin in the 1660s (Lougnon 1970) and later by Cossigny (1732–55) in 1732. The first reference to the presence of frogs was in 1841 when Dumeril and & Bibron (1834–54, vol. 8) mentioned specimens of *Rana mascareniensis* collected by de Nivoy, presumably in the 1830s. Maillard (1862) listed two species of frog, *Ptychadena mascareniensis* and also *Rana cutipora*. The former is common today, occurring in large numbers at Grand Etang (pers. obs.) and elsewhere (Starmühlner 1979, Bour & Moutou 1982), but it appears that the reference to *R. cutipora* was an error (*ibid.*).

I have found no references in the literature to the introduction of the toad *Bufo regularis* ('*B. gutturalis*', Bour & Moutou 1982) to Réunion, but A. Barau (pers. comm. to H. Gruchet, *in litt.* Oct. 1979) says they were introduced from Mauritius in 1927, and Decary (1962; '*B. melanostictus*') found them well established in 1937. In the 1930s the toads were still spreading and did not reach Cilaos until after 1939 (H. Gruchet pers. comm.). These toads are abundant today in most parts of the island (Starmühlner 1979, Bour & Moutou 1982).

Rodrigues

No amphibians have reached Rodrigues.

APPENDIX
Table of mammals, birds and reptiles known to have occurred naturally in the Mascarene Islands (excluding migrants)

This table serves to update, for birds, that on p. 106 of Greenway (1967), and to provide similar information for mammals and reptiles. There is dispute (see text) about whether certain birds were native to the islands; these are omitted from the table, as are hypothetical species and ones for which the evidence is too tenuous to make an assignment to species; a list of rejected forms is given at the end of the table.

Abbreviations used are as follows:
+, extinct; (+), extinct, but breeding doubtful; [+], extinct, species not certainly confirmed
A, abundant; VC, very common; C, common; FC, fairly common; S, scarce; R, rare; VR, very rare; (V), vagrant only
⊙⊙, endemic genus and species; ⊙, endemic species; ×, endemic race; * (full) # (partial) entry in *Red Data Book* (Collar & Stuart 1985)

n, no historical records, perhaps extinct before first accounts; (n), no unequivocal historical record
o, offshore islets only; (i), introduced
The categories S and R overlap, 'scarce' being generally applied to small numbers widely spread, and 'rare' to similar numbers with a more restricted distribution (see Chapters 4, 5, 6 and 8).

Species	Réunion	Mauritius	Rodrigues
Landbirds (including freshwater species)			
Phalacrocorax africanus	[+]	+n	–
Nycticorax mauritianus ⊙	–	+(n)	–
megacephalus ⊙	–	–	+
sp. nov. ⊙	+	–	–
Butorides striatus	R	FC	FC
Egretta (garzetta) dimorpha	+	+	–
Phoenicopterus ruber	+	+(V)	–(V)
minor	?	?(V)	?(V)
Anas theodori ⊙	[+]	+	–
Alopochen mauritianus ⊙	–	+	–
sheldgoose, gen. & sp. nov. ⊙⊙	+	–	–
Circus maillardi ×/ '*alphonsi*' #	S	+	–
Falco punctatus ⊙ *	–	VR	–
sp. nov. ⊙	+	–	–

Appendix (continued)

Species	Réunion	Mauritius	Rodrigues
Aphanapteryx bonasia ⊙⊙	–	+	–
leguati ⊙⊙	–	–	+
Canirallus (Dryolimnas) cuvieri	–	+	–
Fulica newtoni ⊙	+	+(n)	–
Gallinula chloropus	S	FC	–
'Porphyrio caerulescens' ⊙	+	–	–
'Ornithaptera' solitaria ⊙⊙	+	–	–
Pezophaps solitarius ⊙⊙	–	–	+
Raphus cucullatus ⊙⊙	–	+	–
Alectroenas nitidissima ⊙	–	+	–
'A.' rodericana ⊙	–	–	+
sp. nov. (*'A. duboisi'*) ⊙	+	–	–
Nesoenas (Columba) mayeri (⊙)⊙*	[+]	VR	–
Lophopsittacus mauritianus ⊙⊙	–	+	–
'L.' bensoni ⊙⊙	[+]	+	–
'Necropsittacus' rodericanus ⊙⊙	–	–	+
[*'N. borbonicus'*] [⊙⊙]	[+]	–	–
Mascarinus mascarinus ⊙⊙	+	–	–
Psittacula eques/echo ⊙*	+	VR	–
exsul ⊙	–	–	+
'Scops' (?Otus) commersoni ⊙	–	+	–
owl, gen. nov. *sauzieri* ⊙⊙	+n	+(n)	–
murivora ⊙⊙	–	–	+
Collocalia francica ⊙ #	FC	C	–
Phedina borbonica ×	S	S	–
Coracina typica ⊙*	–	R	–
newtoni ⊙*	R	–	–
Hypsipetes borbonicus ⊙	C	–	–
olivaceus ⊙*	–	S	–
sp. nov. ⊙	–	–	+
Terpsiphone bourbonnensis ⊙	×C	×S	–
Saxicola tectes ⊙	VC	–	–
Timaliinae, gen. & sp. nov. ⊙⊙	–	–	+
Acrocephalus rodericanus ⊙*	–	–	VR
Zosterops borbonicus ⊙	×A	×VC	–
chloronothos ⊙*	–	S	–
olivaceus	C	–	–
Foudia flavicans ⊙*	–	–	R
rubra ⊙*	–	R	–
sp. ⊙	+	–	–

Species	Réunion	Mauritius	Rodrigues
Fregilupus varius ⊙⊙	+	–	–
Necropsar rodericanus ⊙⊙	–	–	+
Totals: species/endemic spp.	30/23	29/21	13/12
Extinct: all/endemic	19/16	16/11	10/10
% extinct: all/endemic	63/70	55/52	77/83

Seabirds

Species	Réunion	Mauritius	Rodrigues
Pterodroma arminjoniana #	–	oR	–
aterrima ⊙*	VR	–	+
baraui ⊙ #	FC	–(V)	VR[?+]
sp. nov. ⊙	–	–	+
Puffinus lherminieri ×	VC	+(V)	–
pacificus	oR	oC	oVR
Phaethon lepturus	S	FC	S
rubricauda	–	oR	R
Sula abbotti	–	+(n)	+
dactylatra	–	oVR	–
sula	–	–	+
Fregata ariel (& minor?)	(+)(V)	+(V)	+(V)
Anous stolidus	R	oA	oFC
tenuirostris	–	oA	oFC
Gygis alba	–	–	oVR
Sterna bergii	–	–(V)	(+)
dougallii	–(V)	–(V)	VR
fuscata	–(V)	oA	+(V)
Totals: species/endemic spp.	7/2	11/–	15/3
Extinct: all/endemic	1/–	3/–	7/2+
% extinct: all/endemic	14/–	27/–	47/67+

Mammals

Species	Réunion	Mauritius	Rodrigues
Dugong dugon	–	+	+
Pteropus niger ⊙	+	FC	+n
subniger ⊙	+	+	–
rodricensis ⊙	–	+(n)	R
Scotophilus leucogaster	?+	–	–
Tadarida acetabulosus	C	FC	–
Taphozous mauritianus	FC	FC	–
Totals: species/endemic spp.	5/2	6/3	3/2
Extinct: all/endemic	3/2	3/2	3/1
% extinct: all/endemic	60/100	50/67	67/50

Appendix (*continued*)

Species	Réunion	Mauritius mainland/ islets	Rodrigues
Reptiles			
Chelonia mydas	+(i)	VR/VR	VR
Eretmochelys imbricata	VR	VR/VR	R
Geochelone (Cylindraspis)			–
inepta ⊙	–	+/ ⎫	
triserrata ⊙	–	+/ ⎭ +	–
peltastes ⊙	–	–	+
vosmaeri ⊙	–	–	+
borbonica ⊙	+	+	+
sp. indet. ⊙	+	–	–
Cryptoblepharus boutonii ×	[?]	R/oC	–
Leiolopisma mauritiana ⊙	–	+/–	–
telfairii ⊙	+	+/oFC	–
Scelotes bojerii ⊙	+	VR/oC	–
Nactus serpensinsula ⊙	–	+/oS	–
Lepidodactylus lugubris	–	–	C
Phelsuma borbonica ⊙	xS	–	–
cepediana ⊙	–(i)	C/–	–
edwardnewtonii ⊙	–	–	+
gigantea ⊙	–	–	+
guentheri ⊙	–	+/oR	–
guimbeaui ⊙	–	S/–	–
ornata ⊙	xVR	xC/oC	–
Bolyeria multicarinata ⊙⊙	–	?/oVR[?+]	–
Casarea dussumieri ⊙⊙	–	+/oR	–
Typhlops cariei ⊙	–	+/–	–
Totals: species/endemic spp.	8/6	16/12/ 11+/8+	7/4
Extinct: all/endemic	5/4	8/8/ 1+/1+	4/4
% extinct: all/endemic	62/67	50/67/ 9+/12+	57/100

Notes:
1. Species omitted as probably not native:
   Birds: *Anas melleri*, *Streptopelia picturata* and the Réunion subfossil stork (see text). The rather ephemeral coloniser *Ardeola ibis* (Cattle Egret) is also excluded.
   Reptiles: Although *Hemidactylus frenatus* probably is native, it is omitted as this is in dispute, and its original status on Réunion is unknown.
2. Species omitted through lack of adequate material on which to assess them: Dubois (1670) reported quails and 'wood-rails' on Réunion which cannot be identified further, and Brown's (1773) colourful lizard is also too enigmatic to include. I have, however, treated three better-described species of Dubois' as full endemic species: the red and green parrot, the fody, and the grey pigeon (probably an *Alectroenas*) (see text).

3. Species excluded because they did not exist: Subfossil material described as '*Anhinga nana*' and '*Podiceps gadowi*' has been re-assessed (see text and Chapter 2). *Victoriornis imperialis*, *Kuina mundyi*, *Pezocrex herberti*, and *Testudophaga bicolor*, all from Hachisuka (1953), are imaginary (see text), *Leguatia gigantea, sensu* "géant", was a flamingo (Cheke 1983c and text), '*Necropsittacus francicus*' was an error of Rothschild's (1907a, b; see text), and Rothschild based '*Bubo leguati*' and '*Strix newtoni*' on single bones since re-assessed (text and Chapter 2). Finally '*Foudia bruante*' was an abnormal *F. madagascariensis*, although there had been a native fody on Réunion.
4. It is not known whether both species of Mauritian tortoise occurred on the northern islets. Horne (1887) reported a cave on Flat Is. that would repay excavation in its alluvial soil.
5. *Necropsar leguati* is omitted because there is no real evidence that it came from the Mascarenes (see text).

Acknowledgements

This study could not have been done without the hard-working staff of many libraries, who fetched dozens of dusty volumes from distant stacks. The Bodleian, Rhodes House and Radcliffe Science libraries in Oxford, the British Library and British Museum (Natural History) Library in London and the Mauritius Institute in Port Louis bore the brunt of my searches. I am also grateful to the Bibliothèque St. Geneviève, the Académie des Sciences and the various libraries of the Museum d'Histoire Naturelle in Paris for photocopies of manuscripts, and especially to Dr Yves Laissus of the latter's Bibliothèque Centrale for much useful information. Drs Kurt Bauer and Herbert Schifter kindly sent me details of unpublished Mascarene material in the Naturhistorisches Museum, Vienna. Many people in the islands provided useful unpublished information from their own experiences; I thank especially Messrs Armand Barau, Harry Gruchet and Auguste de Villèle for Réunion, Claude Courtois for Mauritius and Rodrigues and Alfred North-Coombes also for Rodrigues. Nicholas Barré, François Moutou and Roger Bour (Réunion) generously sent me information and papers in advance of publication, and Philip Baker (London) kindly copied relevant passages from an early eighteenth-century MS in Paris. I am also grateful to Drs Ian Atkinson and Malcolm Douglas for useful exchanges on, respectively, rats and deer, and to Nigel Collar for comments on a draft of the chapter. Finally I would like to thank Mary Hannah and Eivor Blake for typing the manuscript.

2

The fossil record

GRAHAM S.COWLES

Chapter 1 has outlined the extent to which many endemic Mascarene Island birds have become extinct, probably during the last 300 years since man arrived on the islands. Thirty extinct species are recognised today (Cowles in press), but of these only five are known from skins preserved in museums and institutions throughout the world. Four of these species, the Mauritian Blue Pigeon *Alectroenas nitidissima*, the Mascarene Parrot from Réunion *Mascarinus mascarinus*, the Rodrigues Parakeet *Psittacula exsul* and the contentious Leguat's Starling *Necropsar (Orphanopsar) leguati* of unknown locality, are represented by a total of only eight skins. The Réunion Crested Starling *Fregilupus varius* was better represented by 24–25 skins, all documented by Hachisuka (1953), although fewer survive today (Chapter 1). The remaining 25 extinct species are known only from fossil bones discovered in caverns and deposits on the three islands. In number these range to well over 200 elements for the better known Solitaire of Rodrigues *Pezophaps solitaria* and perhaps the Mauritius Dodo *Raphus cucullatus*, but the remaining species are unfortunately known from very few bones or bone fragments.

Identifications based on the osteological evidence are in some instances substantiated by field descriptions and illustrations in journals of seventeenth-century voyagers to the islands. The visitors who added notably to early ornithological history were Leguat (1708), on Rodrigues in 1691, Dubois (1674), on Réunion 1671–2, and Van Neck, on Mauritius 1598 or 1599 (see Strickland 1848). The other

voyagers were listed by Hachisuka (1953) and Cheke (Chapter 1). These brief accounts of the flora and fauna of the islands were compiled by people with little ornithological knowledge, but they do provide a most valuable and historical record of the lost avifauna of the islands.

The major descriptive osteological studies based on fossil bones were completed between the years 1848 and 1893, notably by Strickland & Melville (1848), Milne-Edwards (1867a, b, 1868, 1873), Owen (1866), A. & E. Newton (1870), Günther & E. Newton (1879) and E. Newton & Gadow (1893).

Recent research

At the invitation of the British Ornithologists' Union, I visited the Mascarene Islands in 1974 as part of the BOU research programme approved by the Government of Mauritius. Between September and December the islands of Mauritius, Rodrigues and Réunion were studied, the last after consultation with the French authorities. The sites of previous subfossil finds were investigated and several new areas explored in the attempt to obtain further bone material and thus increase our knowledge of the extinct avifauna of the islands.

In consequence of the fieldwork, a complete review of the Mascarene subfossil bird material has been made. This, together with additional material discovered in museums, and on the islands in 1974, has resulted in the re-identification of some older type material, and the description of seven new species of extinct birds. Included in the review are the previously unstudied collections of E. Thirioux from Mauritius about 1900, and the important collections of B. Kervazo made on Réunion in 1974.

The present chapter is based on the above data. The full review (Cowles in press) gives names and descriptions of the new species (referred to here as 'sp. nov.').

The following abbreviations are used:

BMNH(P), British Museum (Natural History) Department of Palaeontology;

BMNH(O) British Museum (Natural History) Sub-department of Ornithology;

UMZC, University Museum of Zoology, Cambridge;

MNHM, Muséum Nationale d'Histoire Naturelle, Paris.

History of sites and collectors
Mauritius

The Dodo, the most famous of the Mauritius endemics, is thought to have become extinct by about 1693 (although Cheke puts this at 1665: see Chapter 1). Strickland (1844) suggested that naturalists residing in the Mascarene Islands should search diligently in the alluvial soil and cave floors for evidence of the extinct avifauna. Twenty-one years later George Clark (1866), a school-teacher resident on Mauritius, discovered that a marshy area near the south-east coast of Mauritius called the Mare aux Songes contained bones of the Dodo and other animals. In September 1865 workmen digging peat for enriching the soil on the Plaisance Estate uncovered bones and a carapace of an extinct tortoise. The landowner allowed Clark to remove any bones that he might find in the marsh, and workmen were sent into deep water to feel with their feet for bones; this technique met with success and soon a few Dodo bones started to emerge. Clark then cleared some floating vegetation from the deepest part of the marsh, and trapped in the underlying mud he found large quantities of bones belonging to the Dodo and other birds. The material was sent to Richard Owen and Alfred Newton in England and Alphonse Milne-Edwards in Paris, for examination and identification.

In 1889, 24 years after Clark's great discovery, the Government of Mauritius appointed a Commission to enquire into the 'Souvenirs Historiques' of the island. Under the direction of Théodore Sauzier digging was resumed in the marsh. Many more bones were excavated, representing various extinct species which E. Newton & Gadow (1893) examined and described.

At the end of the nineteenth century the collector E. Thirioux made excursions to the mountain Le Pouce and ranges close to the capital of Port Louis and obtained many bones from various unnamed caves. There is in the Mauritius Institute a very fine and almost complete mounted skeleton of the extinct rail *Aphanapteryx bonasia* attributed to his collecting. The remainder of the Thirioux collection is in the UMZC.

It is extraordinary that nearly all the fossil evidence relating to the extinct avifauna of Mauritius has been obtained from the two sites, Mare aux Songes and Le Pouce. With the exception of one Dodo bone, no new avian material of any extinct species has been found on Mauritius in the last 80 years.

Réunion

There seems to be no previous published record of fossil bird bones having been found on Réunion. Berlioz (1946) and Barré & Barau (1982) have given details of the early visitors to the islands and discussed the birds seen and listed by them. Dubois (1674) mentioned many of the birds, now extinct, and much of his journal is now confirmed by the new fossil evidence found in 1974. I spent 10 days on the island in November of that year, investigating caves on the north-west coast, near the town of St.-Paul. It is known that the caves were inhabited in the seventeenth century by the first French settlers to reach the island, and in investigating these rock shelters it was hoped that bird bones would be found, perhaps in kitchen middens, which would reflect the early endemic avifauna. In April 1974, B. Kervazo had conducted excavations in two of the caves (Bour 1979), and through the kindness of Dr Dorst, Director of the Muséum National d'Histoire Naturelle, Paris, permission was granted for the examination and identification of the resulting material. The Kervazo collection has added much to our knowledge of the extinct avifauna of Réunion.

During November 1974 a visit was also made to a subterranean tunnel near the village of la Saline, West Réunion. The Caverne Vergoz, so named after M. Armand Vergoz who was first to discover bones in the cavern entrance, is situated on a seaward-facing slope called Bois de Nèfles. The tunnel, probably an old lava tube, descends quite steeply at an angle towards the sea. The interior is wet and receives the drainage from the soil above. Small stalactites hang from the roof. In the damp mud of the floor, bones were found of the extant Audubon's Shearwater *Puffinus lherminieri* and Wedge-tailed Shearwater *P. pacificus*, which still breed on Réunion. With these bones were found those of the giant tortoise *Geochelone* sp. which became extinct on Réunion in the nineteenth century (Arnold 1980, Bour 1981; Chapter 1). Tortoise bones were also found in the coastal caves near St.-Paul, together with those of a large extinct skink *Leiolopisma telfairii* not previously known from Réunion (Arnold 1980). It seems possible that the petrel bones are contemporary with the extinct reptile remains.

A 1-day excursion was made to an inland area called Plaine des Cafres, a large area of flat heathland about 1676 m above sea level. It was here that the flightless rail-like bird the *oiseau bleu* was said to have lived. The bird is known only from descriptions in the early literature of the island (Olson 1977; Chapter 1), but has been given the scientific name *Porphyrio caerulescens* (Sélys-Longchamps 1848). In the part of the plain which was visited there were no visible caves to explore and it was decided that a very careful study of the area would be necessary before any excavating is undertaken.

Rodrigues

François Leguat (1708), who landed on the uninhabited island in 1691, gave the earliest known account of the flora and fauna. He found the island fertile, with good vegetation and fresh-water streams. Rodrigues was rather different when visited in November 1974. There had been no rain for almost a year, the country was parched, the vegetation sparse, and water was in very short supply. Leguat described the birds of the island including the flightless Solitaire which, together with parrots, other birds and giant tortoises, his exiled Huguenot party captured for food. Leguat and his companions left the island after 2 years. Thirty-two years later, in 1725, Tafforet, a naval officer on board the sailing ship *La Ressource*, was marooned on Rodrigues with a survey party of four men for 9 months (North-Coombes 1971). An anonymous account of the now extinct avifauna described in the manuscript 'Relation de l'île Rodrigue' written about 1726, has been attributed to Tafforet by Dupon (1969). Some time after 1750 settlers came to the island; cats and other domestic animals were introduced, tortoises were exported from the island for food, and many of the endemic animals soon became extinct.

About 1789 (or 1786: see Chapter 1) bones heavily encrusted with a mineral deposit were discovered inside a cave situated on the Plaine Corail apparently by Mr Labistour, a resident of the island, and sent to the anatomist Cuvier in Paris (Strickland & Melville 1848). In 1832 Telfair (1833) announced that Colonel Dawkins, Secretary to the Governor of Mauritius, had visited the caves in which the bones had previously been found and had succeeded in finding more. At about the same time Mr Eudes, a resident of the island, was also successful in excavating bones "of a large bird which no longer exists on the island". The bones were presented to the Zoological Society of London and the Andersonian Museum,

Glasgow (Strickland & Melville 1848). In 1845 Captain Kelly of HMS *Conway* made an unsuccessful search for the exact locality of the excavation. The bones presented to Paris and Glasgow were examined and Strickland rightly concluded that they all belonged to the same species, the extinct Solitaire (*Pezophaps solitaria*) which Leguat had described.

Edward Newton (1865a) visited Rodrigues in 1864 in order to investigate the caves. In one of them he found remains of a giant tortoise and two bird bones. One of the bird bones, a tarsometatarsus, fitted the illustration of the Solitaire bones shown in Strickland & Melville's book. A third Solitaire bone was given to Newton by a Captain Barclay. Newton's hurried visit to Rodrigues was fortunately supplemented by the work of George Jenner, the magistrate of the island, who succeeded in uncovering bones which represented 16 individuals in the caves of Plaine Corail. The bones were despatched to Alfred Newton while Jenner continued to direct the digging inside the caves. In 1866 Newton sent four Indian labourers from Mauritius to work at the excavations under the supervision of police sergeant Morris. Nearly 2000 bones and fragments were excavated by this party and the results published by the Newton brothers in 1869, and Günther & E. Newton (in 1879).

H. Slater, one of the naturalists accompanying the Transit of Venus Expedition, visited Rodrigues in 1874 with special instructions to investigate the caves. With the assistance of nine men to dig, and a cook, he set out for the caves on the Plaine Corail in small boats from Port Mathurin on the north coast, the country being so rough that he would not have been able to carry his equipment overland without great expense (Slater 1879a). (In 1974, exactly 100 years after Slater's visit, the journey was made quickly by Land Rover from Port Mathurin to the caves with little trouble over fairly good roads.) Slater carried out his tasks in the caves thoroughly, although he remarked that he was surprised to find how much excavation had already been done: "Out of the first 13 caves, I found 12 to bear the unmistakeable signs of previous research". A report on Slater's important collection, combined with that of Jenner, was published by Günther & E. Newton (1879) and E. Newton & Clark (1879), adding greatly to the knowledge of the extinct avifauna of Rodrigues. In addition to the Solitaire, bones of other extinct birds had been collected, including an owl, pigeon, parrot, rail, heron and starling. A year after Slater's visit, more bones, including two almost complete Solitaire skeletons, were found in a cave by J. Caldwell, ably assisted by Sergeant Morris (Caldwell 1875).

The caves

Cheke (1974a) discussed cave nomenclature; here I use current local names. The caves are situated in the south-west of the island in the region named Plaine Corail, a flattish area sloping gently towards the sea. It is here that the island's small airfield is situated. Plaine Corail is composed mainly of calcarenite limestone (McDougall, Upton & Wadsworth 1965) with very little soil and a thin covering of sparse low vegetation. In places the surface limestone has been eroded into small pinnacles. With one exception (Caverne Tamarin) the caves visited lie beneath the surface of the limestone plain and might be described as 'pot-holes', as access is possible only by descent. Caverne Tamarin is situated on the northern edge of the limestone escarpment and can be entered from ground level. Below the surface of the flat plain is a labyrinth of tunnels and subterranean passageways which have been cut out of the limestone probably by the action of water. In 1974 water was found in two caves despite the island's drought. Balfour (1879a) suggested that the island was once at a lower level than at present, as he found evidence of raised beaches, so perhaps some caves were formed by sea action. The floor of most caves had a deposit of reddish soil brought down from the higher ground by the heavy seasonal rains, and ripple marks on this soil showed that at times water still flows through some of the tunnels. The cave entrances, with the exception of Caverne Tamarin, have been formed by the collapse of part of the tunnel roofs, and may be of any size from a small hole a metre or so wide, to a deep valley open to the sky with the original tunnel continuing as caves at either end. One of these valleys is perhaps 30 m deep and about 400 m in length and affords such good protection from wind and sun that quite large trees have grown in it; because the valleys are below the level of the plain they are not visible until one stands at the edge. Probably because of the difficulty of providing sufficient light further inside the caves, most of the digging by previous excavators took place just inside the cave entrances where the daylight penetrated, and in Caverne Tamarin there are still signs of the work which was carried out some years ago. Slater found the red cave soil soft and easy to dig, but the soil in Caverne Tamarin has

since been consolidated by the feet of cattle which are penned up in the cave during cyclones. In Grande Caverne, one of the larger caves, the soil of the main tunnel has been trampled hard by the feet of tourists who are at times taken through by guides. This cave is spectacular, with large stalagmites and stalactites, the roof being about 12 m high in places, but it was in the low-roofed side-tunnels that the new bone material was collected.

Caverne Tamarin was also explored, together with many smaller caves, where the roofs are so low that progress could be made only by crawling on hands and knees. The interiors of the caves are completely dark, and light for exploration was provided by a battery torch fitted on a head-band and by a hand-held battery lantern. Light for excavating was supplied by a small portable lamp which gave 75 watts of illumination for 5 hours. For digging a folding shovel and a small trowel were used.

Systematic list of Mascarene subfossil material
Procellariidae

Pterodroma sp. nov., Rodrigues. An extinct new species of large petrel is represented by bones obtained in the caves of Rodrigues. It may perhaps be the petrel referred to by Tafforet (c. 1726), which was said to nest in holes in the ground on the higher parts of the mountains.

Pterodroma aterrima, Rodrigues. The Mascarene Petrel *P. aterrima* is today confined to Réunion, but an upper jaw was correctly identified by Bourne (1968) from a nineteenth-century collection of bones from Rodrigues in the BMNH(P).

Podicipedidae

Podiceps gadowi, see *Numenius phaeopus* (p. 96).

Phaethontidae

Phaethon lepturus, Rodrigues. At least 25 skulls and many other bones of the White-tailed Tropic-bird have been found by the eighteenth-century and 1974 expeditions to Rodrigues. The species is common on Mauritius and occurs on Réunion; on Rodrigues it is now rare (Staub 1973, Cheke 1974a). From the subfossil evidence it would appear that *P. lepturus* was common, most probably breeding, on the main island of Rodrigues at some time in the past.

Phalacrocoracidae

Phalacrocorax africanus, Mauritius. The bones described as a darter *Anhinga (Plotus) nana* by E. Newton & Gadow (1893) were re-identified by Olson (1975) as the extant cormorant *Phalacrocorax africanus*. This is the first evidence of the Long-tailed Cormorant occurring in the Mascarene Islands.

Anhinga (Plotus) nana, see *Phalacrocorax africanus*.

Sulidae

Sula abbotti, Mauritius. A humerus and ulna excavated in the Mare aux Songes ("Gannet": E. Newton & Gadow 1893), were said by Bourne (1976) to resemble the bones of the extant Abbott's Booby *S. abbotti*. The species had not previously been positively identified from the islands.

Ardeidae

Nycticorax mauritianus, Mauritius. Seven fossil bones from a heron, found in the Mare aux Songes, were described as *Butorides mauritianus* by E. Newton & Gadow (1893). Of these bones only two are available for study today: the coracoid and a tarsometatarsus. The radius has been shown not to belong to the Ardeidae. Examination of the bones suggests a closer relationship to the night-herons *Nycticorax* than to the genus *Butorides* (*Ardeola*). Measurements taken from the extant species *N. nycticorax* confirm that the tarsometatarsus of *N. mauritianus* is more robust and longer, not shorter as might be suggested by the term "short footed heron" applied by E. Newton & Gadow and perhaps misinterpreted later by Hachisuka (1953) as being a very small bird. The tarsometatarsus of *N. mauritianus* is short only in comparison with species of genus *Ardea*, to which Rothschild (1907b) referred it.

Nycticorax megacephalus, Rodrigues. This was originally described by Milne-Edwards (1873) from several bones, including a complete sternum and skull, as *Ardea megacephala*. The recent discovery of a previously unknown cranium in the BMNH(P) confirms the opinion of Günther & E. Newton (1879) that the bones represent an extinct night-heron in the genus *Nycticorax*, and not *Ardea*.

Doubt must also be cast on the supposed total flightlessness of the bird (Hachisuka 1953). Milne-Edwards (1873) and Günther & E. Newton (1879) suggested the bird had reduced powers of flight and

indeed it was stated by an early voyager to Rodrigues that the bird flew only a little, and ran away when chased (Tafforet *c.* 1726). In their table of measurements Günther and E. Newton unknowingly used a skeleton of the large South American race *N. n. obscurus*. Comparison with this skeleton gave the impression that the wings of *N. megacephalus* were unusually small, but they are not small by comparison with *N. n. nycticorax*. The femur, tibiotarsus and tarsometatarsus of *N. megacephalus* are broader, longer and more robust than those of *N. n. nycticorax* and, like other species endemic to oceanic islands, the legs of *N. megacephalus* have become stronger as the need to fly decreased.

Nycticorax sp. nov., Réunion. The first osteological evidence to show that a species of *Nycticorax* also occurred on Réunion was obtained from the caves near St.-Paul in 1974. The incomplete tibiotarsus suggests that the extinct Réunion night-heron was probably longer in the leg than either *N. mauritianus* or *N. megacephalus*.

The historical record contains several references to 'bitterns' on all three islands, although there is no evidence that *Ixobrychus* or *Botaurus* has ever been found there. On Mauritius in 1696 Leguat stated "You shall see great flights of bitterns" (Oliver 1891). On Rodrigues Tafforet (*c.* 1726) wrote "There are plenty of bitterns . . . They are the size of an egret". Leguat also wrote of Rodrigues during 1691–3, "We had bitterns as big as fat capons" (Oliver 1891). On Réunion during 1671–2 Dubois (1674) described "Bitterns or Great Egrettes . . . they have grey plumage, each feather tipped with white . . . and the feet green". Berlioz (1946) thought Dubois' description reminiscent of a *Nycticorax* in immature plumage and it seems quite possible that all references to 'bitterns' were really night-herons. In stance and in immature plumage a night-heron appears very much like a bittern. The historical and osteological evidence confirms that an endemic species of *Nycticorax* occurred in each of the three Mascarene Islands, at least until the late seventeenth century.

Ciconiidae

Ciconia sp. nov., Réunion. A distal part of a tarsometatarsus found in the caves near St.-Paul in 1974 is the first positive evidence that a stork once inhabited the Mascarene Islands. This recent discovery may explain the identity of the bird recorded by Dubois who wrote (translated by Oliver 1897) "There are great birds of the height of a man because of their legs and neck which are very long. They have a body as large as the geese and plumage white and black at the point of the wings". Dubois referred to these birds as "*flamands*" and they may indeed have been flamingos *Phoenicopterus* sp., or perhaps the hypothetical giant bird of Mauritius *Leguatia gigantea* (see Chapter 1). The bone from Réunion is certainly that of a stork *Ciconia*, and Dubois' description could also fit a bird of this genus.

Anatidae

Sheldgoose sp. nov., Réunion. An upper jaw, sterna and other bones have been found in the cave near St.-Paul and represent a new extinct genus of short-billed Sheldgoose. Geese were said to be common on the fresh water near St.-Paul (Chapter 1) and Dubois (1674) described them as follows: "Wild Geese, a little smaller than the geese of Europe, they have the plumage the same and the beak and feet red". There are no wild species of geese living on Réunion today.

Alopochen mauritianus, Mauritius. This was described from a carpometacarpus by E. Newton & Gadow (1893) as *Sarcidiornis mauritianus*. However, Andrews (1897) showed that it was related not to the Comb Ducks *Sarcidiornis* but to the goose *Alopochen*. A second carpometacarpus from the Mare aux Songes recently discovered in the BMNH(P) confirms Andrews' conclusion.

Fossil bones of a duck *Aldabranas cabri*, described by Harrison & Walker (1978) from a femur and incomplete humerus collected on Aldabra Is., are said to resemble some of the characters found in the Sheldgoose *Alopochen*, but a direct relationship was not inferred.

Anas theodori, Mauritius. E. Newton & Gadow (1893) described the bones of a duck found in the Mare aux Songes and named it in honour of Théodore Sauzier. The incomplete sternum, coracoid, humeri and tibiotarsi are now supplemented by a previously uninvestigated cranium from the Thirioux collection. A recent study of the cranium confirms a dissimilarity to any other extant duck species already recorded from Mauritius. It has been concluded that it is the first known cranium of *Anas theodori* and differs significantly from those of other species.

Accipitridae

Accipiter alphonsi, see *Circus alphonsi* below.

Circus alphonsi, Mauritius. Milne-Edwards (1873) described two tibiotarsi from the Mare aux Songes as belonging to the genus *Astur* (= *Accipiter*). E. Newton & Gadow (1893), while describing further material obtained from the site, applied the name *A. alphonsi* to them although it appears from the rather contradictory text that they could find no difference between the excavated bones and those of *Accipiter melanoleucus*. However, recent comparisons made between the bones from Mauritius and those of the genera *Accipiter* and *Circus* show a quite obvious likeness to *Circus* rather than *Accipiter*. The tarsometatarsus, tibiotarsus and carpometacarpus are similar to bones of the extant Réunion Marsh Harrier *C. maillardi*, but until skeletal material of the latter is available the specific name of the harrier of Mauritius must remain in some doubt.

Falconidae

Falco sp. nov., Réunion. Bones from an extinct kestrel have been obtained from the caves near St.-Paul, and the bird called 'Merlin' of Réunion can now be identified. Dubois (1674, English translation by Oliver 1897) mentioned three different birds of prey on Réunion, of which only the Réunion Marsh Harrier or *papangue Circus maillardi* remains resident on the island. The identity of the *pieds jaunes* is explained by Verreaux (1863) who noted that the Réunion Marsh Harrier, when in immature plumage, is called by this name by the islanders. There seems little doubt that *papangues* and *pieds jaunes* were names given to the same bird and are still used (Barré and Barau 1982), though Cheke (1982b and Chapter 1) suggests *pieds jaunes* refers to the Sooty Falcon *Falco concolor*.

The third bird of prey referred to by Dubois can now also be identified, because confirmatory bone material has been found. Continuing his account of the birds of prey, Dubois wrote: "The third kind are Merlins [*emerillons*], which although small, still do not fail to carry off chickens and eat them". The word 'chickens' here means the young chicks of birds, not domestic fowl. (The original word is "poulet", normally used for domestic fowl, rather than "poussin" (baby bird) (ed.)). The Eurasian Merlin *Falco columbarius* is similar in size to a kestrel, of which several species are found on Indian Ocean islands (Chapter 5). It is not surprising that Réunion too should have supported an endemic kestrel, which was common enough in Dubois' time for him to comment on its habits. Since the seventeenth century it has become extinct, as have many of the other birds he listed.

Rallidae

Aphanapteryx bonasia, Mauritius. A large flightless rail with a long bill called *poule rouge* (red hen) has been known from the historical record since Cauche wrote of it in 1638 (published 1651). A painting by Hoefnagel made about 1610 and discovered by Frauenfeld (1868) depicts the rail with a decurved bill, long legs and reddish-brown plumage. The former existence of the bird was confirmed by Milne-Edwards (1868), who identified bones obtained from the Mare aux Songes, and used the name *Aphanapteryx broeckei* (Schlegel) for the rail. E. Newton & Gadow (1893) studied further bones from the site, and a cranium was described by Piveteau (1945). Hachisuka (1953) discussed the historical, pictorial and osteological evidence, and Olson (1977) produced a very comprehensive review of the species, giving the reasons for reverting to the name *A. bonasia*. It is thought that the bird became extinct around 1700 (Chapter 1).

Aphanapteryx leguati, Rodrigues. François Leguat (1708) saw this species alive on Rodrigues in 1691. He recorded the plumage as being bright grey, unlike *bonasia* of Mauritius which was reddish. The rail could not fly and was known to Leguat as a *Gellinote* (Woodhen). Tafforet (*c.* 1726) provided an additional record of the field characters and noted that the plumage was flecked white and grey, feet and bill red, and that it was similar in some ways to that of a Curlew; he also confirmed that it was unable to fly. Milne-Edwards (1874) provided confirmation of the sight records, when he recognised that the bones from the caves on Rodrigues belonged to a large rail which he named *Erythromachus leguati*. Günther & E. Newton (1879) examined additional material from the caves on Plaine Corail, and concluded that *leguati* should be placed in the genus *Aphanapteryx*. Olson (1977) thoroughly reviewed all the data relating to the species, which is now extinct. Further material was collected in 1974.

Dryolimnas cuvieri, Mauritius. Bones from the Mare aux Songes, now in the UMZC, were identified as "Water-hen" by E. Newton & Gadow (1893, p. 282)

and marked *Gallinula chloropus pyrrhorrhoa*, the Madagascar Moorhen, which is a common resident on Mauritius (Rountree *et al.* 1952) and also occurs on Réunion (Chapter 6). The bones are not, however, those of *Gallinula* but are *Dryolimnas cuvieri*, the Whitethroated Rail, a bird which no longer exists on Mauritius but still inhabits Madagascar and Aldabra and has been extirpated from Assumption Is. (Benson 1967). The only other evidence of *Dryolimnas cuvieri* occurring on Mauritius is the holotype study skin in the MNHM, collected in January 1809 from the Plain of St.-Martin in the west of the island.

It is possible that the bird illustrated in the journal of Thomas Herbert (1634), which has caused so much discussion amongst various authors over the years, and has been known as 'Herbert's Hen' (*Didus herberti*: Schlegel 1854), may be *Dryolimnas cuvieri*.

Fulica newtonii. Mauritius, Réunion. The first known osteological remains of a Coot in Réunion were recovered in 1974 from a cave near St.-Paul. *Poules d'eau* (water-hens) were recorded by Dubois (1674) as being present on Réunion in the early 1670s, but actual evidence of the coot's existence has, until now, been found only on Mauritius (Milne-Edwards 1867a). A comparative study of the bones has shown that the long- and strong-legged coot of Réunion is identical in anatomical details to *F. newtonii* of Mauritius, which may have become rare on that island by 1693 (Leguat 1708). The populations of the two islands form a monotypic extinct species. Olson (1977) suggested *F. newtonii* was probably derived from the extant *F. atra*.

Charadriidae

Numenius phaeopus, Mauritius. The single ulna said to be from the wing of an extinct grebe *Podicepes* [*sic*] sp. by E. Newton & Gadow (1893) (*Podiceps gadowi* of Hachisuka 1953), is in fact the ulna of a Whimbrel, which is a common migrant to the islands (Rountree *et al.* 1952).

Raphidae

Raphus cucullatus, Mauritius. Only the head and foot of the Dodo had been studied prior to the discovery by Clark (1866) that quantities of bones were deposited in the Mare aux Songes. Since 1866 the osteology of the bird has been the subject of much research. Hachisuka (1953) documented the data, extending the bibliography of Strickland & Melville (1848) up to 1941. Several museums and institutions throughout the world contain composite Dodo skeletons, constructed from the many bones found in the Mare aux Songes. It is generally accepted on osteological characters that the affinities of the Dodo lie close to the pigeons in the family Columbidae.

Pezophaps solitaria, Rodrigues. Since the discovery of the first Solitaire bones in the Rodrigues caves in about 1786, it has been stated that over 2000 bones and fragments have been collected, mostly in the years 1864 to 1875. E. Newton & Clark (1879) remarked: "We are not aware that the osteology of any vertebrate, other than man, has been studied with the same wealth of material as that of the Solitaire". There is very little fresh information at present that can be added to the early osteological studies (A. & E. Newton 1870, E. Newton & Clark 1879). The bone evidence confirmed the written account and drawings made by Leguat, who stated that there was a large flightless endemic bird which was good to eat and extremely fat, weighing as much as 45 pounds (20 kg), and described its habits and behaviour in some detail. It is still a matter of conjecture how so many Solitaires found their way into the Plaine Corail caves. Caldwell (1875) found two almost complete skeletons and suggested that the birds had sought shelter in the caves during a "hurricane" and died. The caves which Caldwell explored showed no signs of the action of water as did some of the caves which were visited in 1974. E. Newton thought that pigs might have destroyed the population; another theory (supported by North-Coombes 1971) is that a devastating fire may have driven the birds into the caves where they died.

Caldwell found gizzard stones *in situ* beneath the sternum and ribs of nearly complete Solitaire skeletons, confirming Leguat's statement that large stones, which he used for sharpening knives, could be found in the Solitaire's gizzard. Caldwell mentioned that the stones he found were basalt (dolerite: E. Newton & Clark 1879), a rock not seen by Caldwell within 2 miles (3 km) of the limestone plain. Temple (1977*b*) has suggested that the gizzard stones of the Mauritius Dodo *Raphus cucullatus* helped to remove the hard outer covering of large seeds, a process he believed was necessary to ensure the successful germination of the now nearly extinct tree *Calvaria major* on Mauritius. Wiehé (1949) mentioned that *Calvaria galeata* was probably a dominant tree in the original Rodrigues forest formation, but is now rare. Possibly trituration in the

gizzard of the Solitaire helped the seeds of some Rodrigues plants to germinate (but see Chapter 1 where Cheke argues against Temple's hypothesis).

The relationship between the Solitaire and the Dodo was discussed by Strickland & Melville (1848), and by Storer (1970), who argued that the two birds were derived independently from flying ancestors and should be placed in separate monotypic families, Pezophapidae and Raphidae. There is no osteological evidence supporting the suggestion that a Solitaire-like or a Dodo-like bird once existed on Réunion (Hachisuka 1937, 1953), although the bones may only await discovery.

A comprehensive bibliography of the family is given by Strickland & Melville (1848) and Hachisuka (1953).

Columbidae

Alectroenas rodericana, Rodrigues. Milne-Edwards (1874) identified a sternum and other bones from the caves on Rodrigues as a pigeon *Columba rodericana*. The sternum is, as Milne-Edwards pointed out, quite unlike that of any living genus known today; it may warrant separate generic status in the family Columbidae. Additional material collected in 1974 helps little to settle the affinities of the species. However, at present it is placed in *Alectroenas*, together with the extinct pigeon of Mauritius *A. nitidissima*, which is known only from three skins.

Leguat (1708) apparently saw *A. rodericana* on Rodrigues. It was somewhat smaller than "ours", he writes (presumably comparing it with *Columba livia*), grey in colour and very tame.

Psittacidae

Lophopsittacus mauritianus, Mauritius. There seems little doubt that flocks of parrots were present on Mauritius in the early seventeenth century. The eventual extermination of all endemic species, with the exception of the rare resident *Psittacula echo*, was probably caused or at least helped by the sailors and early visitors, who killed the parrots for food. A woodcut in the journal by Willem van West-Zanen (reproduced in Hachisuka 1953) depicts men catching large parrots in Mauritius about the year 1602.

Owen (1866) recognised and described a massive lower mandible and a tarsometatarsus as belonging to a large macaw-sized parrot. E. Newton & Gadow (1893) examined further material obtained by Sauzier from the Mare aux Songes, adding details of the femur, tibiotarsus and incomplete sternum to the record. The first known cranium of this extinct parrot has now been discovered in the collection of the BMNH(O).

Lophopsittacus bensoni, Mauritius. A second, and smaller species of *Lophopsittacus* was named by Holyoak (1973) from a lower mandible, palatine and tarsometatarsi collected by Thirioux in the caves near Port Louis. Mr Norman, late of the UMZC, has pointed out (pers. comm.) that the photographs of the holotype lower jaw, said to be (*a*) dorsal view and (*b*) ventral view, are in fact two quite separate specimens. Also, although marked 8/10 magnification, they are almost natural size. It is not at all certain that *bensoni* should have been placed in the genus *Lophopsittacus*.

Necropsittacus rodericanus, Rodrigues. Leguat (1708) and Tafforet (*c*. 1726) both provided details of the plumage and feeding habits (recounted in Hachisuka 1953 and Staub 1973*b*) of two Rodrigues parrots then common, but which no longer exist. From these field descriptions one species, the Rodrigues Parakeet *Psittacula exsul*, can easily be recognized as it is fortunately represented by two study skins. The other parrot, said to be larger than *P. exsul* and to have all green plumage, a long tail, large head and bill, is generally accepted to be *Necropsittacus rodericanus*, the fossil species named by Milne-Edwards (1867*b*) from an upper jaw discovered in the caves of Plaine Corail by Jenner. Günther and E. Newton (1879) illustrated a cranium and other bones of *N. rodericanus* from the collection of Jenner and Slater. This additional material helped maintain the theory of Milne-Edwards that the affinities of the parrot lie close to the genus *Psittacula*. The peculiarities of the cranium, however, keep *rodericanus* in the separate genus *Necropsittacus* as proposed by Milne-Edwards (1873). No further material of the parrot has since been collected but a comparative study of the recently found cranium of *Lophopsittacus mauritianus* and *N. rodericanus* shows that there is probably no very close relationship between the two island species of large-headed parrots from Mauritius and Rodrigues.

Psittacula exsul, Rodrigues. Parrots were apparently common on Rodrigues in the late seventeenth century. Two species were seen by Leguat in 1691, one of which, *Psittacula exsul*, was said to be in "abun-

dance". Both Jenner and Slater collected bones of *P. exsul* in the caves (Günther & E. Newton 1879), and a proximal end of a tibiotarsus collected in 1974 from the same area agrees well with the material in the BMNH(P). Two study skins of *P. exsul* are preserved, from which A. Newton (1872) first described the species, which was last seen in 1875 (Chapter 1).

Strigiformes

'*Tyto*' *sauzieri*, Mauritius, Réunion. New fossil evidence confirms that owls once inhabited all three islands of the Mascarene group. Bones collected in 1974 on Réunion are the first indication that a nocturnal bird of prey existed on this island. Previously, as Berlioz (1946) pointed out, no member of the Order Strigiformes had been recorded from Réunion (except for the rather prophetic misprint in a table of island species in Greenway 1967). Réunion owls were not mentioned by Dubois (1674), so the bird was either extinct before his visit in 1671–2, or was overlooked.

Comparison of the Réunion bones with those of *sauzieri*, named by E. Newton & Gadow (1893), shows them to be very similar, and there seems to be no definite osteological character on which to separate the two populations. E. Newton & Gadow placed *sauzieri* in the genus *Strix*, which is now *Tyto* (the barn owls) and not *Strix* as applied to the wood owls of today. Re-examination of the Mauritius material, however, shows clearly that *sauzieri* is not related to *Tyto*. Some osteological characters of the pelvis are not found in other extant genera and a new genus may have to be erected for the Mascarene owls.

'*Tyto*' *newtoni*, Mauritius. This was named by Rothschild (1907*b*), from two tarsometatarsi, smaller than those of *sauzieri*, but probably synonymous with *sauzieri*.

'*Tyto*', '*Ninox*' *murivora*, Rodrigues. In 1874 Milne-Edwards described as *Strix* (*Athene*) *murivora* an extinct species of owl from bones discovered in the caves of Rodrigues. Since that time, all subsequent authors have continued with the mistaken supposition that the Rodrigues owl *murivora* was a member of the genus *Athene* (or *Carine*: Günther & E. Newton 1879). Milne-Edwards used comparative osteological material named *Athene superiliaris* and *A. polleni*, but shortly after his publication, both species (now considered conspecific) were placed in the genus *Ninox*, where they have remained ever since. Overlooked too has been the greater emphasis placed by Milne-Edwards upon *Strix* (now Tyto) in his original description, and '*Strix murivora*' in the caption to his plate 11.

The remains of *murivora* show that the owl is not related to *Tyto* or to the present-day *Athene*. The pelves of *murivora* and *sauzieri* share the same unusual osteological characters not found in any other extant genera, and strongly suggest a close relationship between the two extinct species. The owl *sauzieri* of Mauritius and Réunion has longer and slimmer legs than those of the Rodrigues Owl *murivora*, and therefore the separate species are retained.

Timaliinae

Babbler, sp. nov., about thrush size, Rodrigues. An incomplete sternum collected from Rodrigues in 1974 is from a bird in the Order Passeriformes, about the size of a small thrush. The unique characters of the sternum, which differ greatly from those of other genera examined, have prompted the bird's assignment to a new genus, and represent a new species of extinct Mascarene Island passerine. The genus has provisionally been placed in the subfamily Timaliinae.

Pycnonotidae

Hypsipetes sp. nov., Rodrigues. Bones representing a new species of extinct bulbul have been identified from material collected in 1974.

Sturnidae

Necropsar rodericanus, Rodrigues. H. Slater was a member of the 1874 Transit of Venus Expedition and while on Rodrigues was first to discover and recognise bones of an extinct starling. Günther & E. Newton (1879) described, illustrated and named the material *Necropsar rodericanus*. Further material was collected in 1974 and examination of the remains confirms that they are from a bird in the family Sturnidae. It is possibly related to the extinct Crested Starling of Réunion *Fregilupus varius*, known from several skins and dissected specimens. Berger (1957) after studying the anatomy of *F. varius* concluded that it was not a starling and suggested a possible relationship to the family Prionopidae. *Necropsar rodericanus*, however, is closer osteologically to Sturnidae.

Discussion

The present evidence indicates that at least 30 endemic species of birds have been lost to the Mascarene Islands in the short period since records of the avifauna were first compiled, just over 300 years ago. Many different factors probably combined to cause the extinction of the birds; these are discussed in detail by Cheke (Chapter 1). Here I discuss possible reasons for the accumulation of bones in the few sites from which they are known.

It has been suggested by some authors that a severe fire once swept Rodrigues, causing the flightless Solitaires to seek shelter in the caves where some complete skeletons were later found. E. Newton (1865a), however, could find no evidence to support the fire theory. Prolonged drought, especially on Rodrigues where water is scarce in the dry season, must have resulted in a reduction of the bird populations, sometimes perhaps to dangerously low levels from which they never recovered. Birds unable to fly into the mountain range, where some springs continued to flow, may have entered the caves of Plaine Corail in a search for water, and died there. In the dry season of 1974 small streams and pools of water certainly did exist inside some caves. At other times of the year heavy rain in the wet season and high winds associated with the cyclonic period must also cause the death of some birds. Large flightless birds such as the rail *Aphanapteryx* and the Dodo would perhaps be most susceptible to waterlogging of plumage and death by hypothermia. There has been such a diversity of species recovered from the marsh, it would seem that a natural catastrophe such as severe weather was responsible for the bone deposits, although tortoises may have become trapped in the mud while at the edge of the marsh. Bones accumulated in the marsh because it received the surface water draining from the surrounding high country, and heavy rain may have carried down bones or carcasses, together with the alluvium seen by Clark (1866) which remained trapped in the marsh as the water drained out to sea. On Rodrigues, it is also possible that bones were deposited when birds fell into the underground tunnel system beneath Plaine Corail. Indeed, deep inside one cave in 1974 I saw the carcass of a goat which seemed to have fallen in through a hole when the tunnel roof collapsed.

On Réunion the bones found in the coastal caves may have been the result of early human occupation.

Bour (1979) suggested that a cyclone may have taken the birds to the vicinity. Petrel and shearwater bones were probably contained in burrows in the ground above the caves and as the cave roofs collapsed the soil containing the bones then fell to the floor. Many of the species found at St.-Paul were associated with water and some old maps show a lagoon close to the caves which would have attracted coots, herons and geese. These would have been easily captured by early settlers or mariners and St.-Paul is known to have been a port for sailing ships where fresh stores were taken aboard.

Acknowledgements

I am most grateful to Claude Michel, Director of the Mauritius Institute (now retired), to Mr Heseltine, Resident Commissioner of Rodrigues at the time of my visit, and to Mr Kwet-On, Agricultural Services Rodrigues, for their help. I am indebted to Mike Holland for his valuable local knowledge and hospitality while on Rodrigues. On Réunion, Harry Gruchet, Director Muséum d'Histoire Naturelle, and Auguste de Villele gave most generously their time and hospitality. I thank also Dr Jean Dorst, Director, and Christian Jouanin, Muséum National d'Histoire Naturelle, for allowing me access to reference material in Paris; and in the University Museum of Zoology, Cambridge, the late Con. Benson was always most helpful. My work was financed by the British Museum (Natural History). I am grateful to Dr D. W. Snow for kindly reading and criticising this chapter. A. S. Cheke was of assistance on Mauritius.

3

Vocalisations of the endemic land-birds of the Mascarene Islands

JENNIFER F.M.HORNE

One of the aims of the BOU Mascarene Islands Expedition was to obtain as complete a record as possible of the voices of the endemic birds. To this end I spent the breeding seasons of 1973 and 1974 in the islands, and recorded all the endemic land-birds. This report is concerned solely with vocalisations and associated behaviour. Some comparisons are made, where appropriate, between the voices of Mascarene birds and related birds elsewhere.

Itinerary and methods

The dates of my visits to each island, and the study areas I used, are listed in Table 1.

In both seasons the breeding cyle of all species had begun, so I could not study the initial period when territories were being set up and pairs formed. For some species, the time spent in the islands was not long enough for me to be able to determine the behavioural context of all the calls recorded. The passerine species were easier than the non-passerines to observe and tape-record, so that results for them are more complete.

Weather conditions were better for recording in 1974 than in 1973, when frequent high winds and cyclonic disturbance gave rise to background noise. I spent an average of 11 hours per day in the field. On return each night, whenever possible, a primary edit was done, the edit books written up, and the tapes re-spooled onto 7-inch reels. 'Playback' tapes also were prepared to attract and stimulate certain species,

especially *Zosterops chloronothos*, whose population density was so low in any given locality that playback proved very useful in helping to locate individuals and to hold them in an area for a short period.

Two full-track field machines were used: a Nagra IIIB with Beyer M66 dynamic microphone and a Stellavox SP7 with a Schoeps CMT 42 condenser microphone. In 1973 24-inch and 36-inch aluminium parabolas were used, but both proved to be very heavy and in 1974 were replaced by a 30-inch fibre-glass parabola. A tripod was used where possible. All recordings were made at $7\frac{1}{2}$ i.p.s. (19 cm/s). Fuller detailers of equipment used and operating conditions have already appeared (Horne 1979).

Sonagrams of all species (except *Collocalia francica* and the four *Zosterops* taxa) were kindly prepared by Professor Paul Bondesen of the Natural History Museum, University of Aarhus, Denmark, on a Kay Sonagraph 662-A. Those of *Collocalia francica* were prepared by Dr J. Gulledge at the Laboratory of Ornithology, Cornell University, on a Kay Electric Sonagraph 6061 B (0.085–16 kHz); and those of the four *Zosterops* taxa were prepared on a similar Kay Electric Sonagraph by the author in the Department of Ornithology, American Museum of Natural History, New York. The original tapes will be deposited with the British Institute of Recorded Sound, who supplied most of the tape for this work.

Table 1. *Dates and localities of fieldwork in the Mascarenes*

Island	Dates	Localities
Mauritius	22 Sept.–28 Oct., 12–29 Nov. 1973; 24 Sept.–30 Oct., 11 Nov.–1 Dec. 1974	Macabé Forest; Black River Gorges; Plaine Champagne; Plaine Paul; Rivière des Chevettes; Piton du Fouge; Baie du Cap
Réunion	29 Oct.–12 Nov. 1973; 12–23 Sept., 2–14 Dec. 1974	Forêt du Bébour; Plaine des Chicots; Plaine d'Affouches; Ravine de la Grande Chaloupe
Rodrigues	2–9 Nov. 1974	Solitude Forest Station; Cascade Pigeon; St.-Gabriel

Réunion Harrier *Circus maillardi maillardi*

The Réunion Harrier breeds much later in the season than the other land-birds of Réunion (M. Clouet pers. comm.), and the timing of my studies was such that I was able to obtain recordings only of vocalisations accompanying displays (except for a call of uncertain context heard twice in the forest). The study area at Ravine de la Grande Chaloupe had five resident pairs and at least one juvenile. Pairs displayed regularly each day, beginning at approximately 08.00 hours and terminating abruptly at about 09.00 hours, as was the case on the morning when I recorded their vocalisations. Males emit a series of distinctive calls during 'undulating' and 'rolling' displays, which involve long, steep ascents and descents; on one occasion I saw a male roll completely over.

Recorded vocalisations include (*a*) a 'chip', emitted at the top of the ascent of the rolling display (Fig. 1*a*); (*b*) the 'wail chatter' (transcribed by Clouet (1978) as *kaï pi-pi-pi-pi-pi* or *kaï ké-ké-ké-ké-ké*), emitted during the undulating display, the 'wail' commencing at the peak of the undulation and the 'chatter' continuing between peaks of undulations (Fig. 1*b*; cf. Fig. 1*c* showing the structurally comparable alarm call of Marsh Harrier *Circus aeruginosus*); and (*c*) a variable

'mew' call (two in Fig. 1*d*; cf. one of *C. aeruginosus* in Fig. 1*e*).

The 'chip' call is a short (0.075 s) figure with a frequency range of 0–6 kHz. Its fundamental tone is somewhat weaker than in the 'wail chatter' and 'mew' calls, with similarly strong lower harmonic tones. The 'wail' is a long (0.13 s) figure with the fundamental at 0.7 kHz. The 'chatter' is a series of short figures of variable frequency but regular time interval. The 'wail chatter' was uttered by either sex, but males seemed to employ it more frequently. This call resembles that described for the Hen Harrier *C. cyaneus* (Witherby *et al.* 1938–41, Peterson *et al.* 1974); possibly it is as variable in context as the descriptions of the call of *C. cyaneus* suggest. *C. aeruginosus* apparently omits the 'chatter' sequence from its display.

The 'mew' call was given by both sexes and was varied in context and in frequency, both individually and sexually (M. Clouet pers. comm.). Both sexes seemed to give it more continuously than does *C. aeruginosus*. The 'wail' of the 'wail chatter' could be derived from the 'mew' call, for it shows much the same pattern with strong fundamental tones and first and second harmonic tones.

A more detailed discussion of the vocal repertoire of this harrier is given by Clouet (1978).

Fig. 1. Vocalisations of *Circus maillardi* and *C. aeruginosus*. (*a*) The 'chip'. *Opposite:* (*b*) The 'wail chatter'. (*c*) Alarm chatter of *C. aeruginosus*, recorded by J.-C. Roché, France. (*d*) 'Mew' calls, uttered by a juvenile, possibly a male. (*e*) 'Mew' call of a female *C. aeruginosus*, recorded by J.-C. Roché, France.

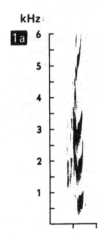

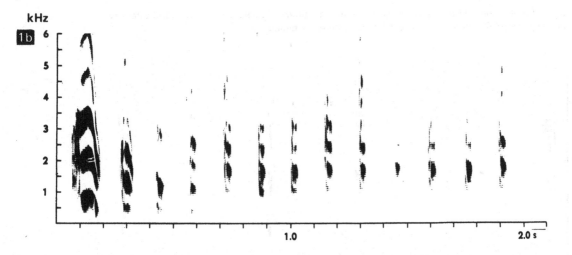

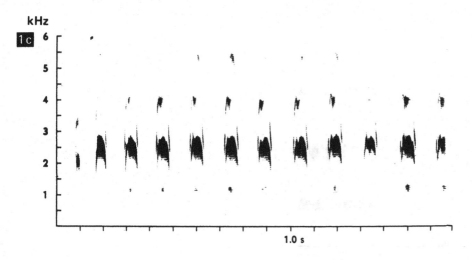

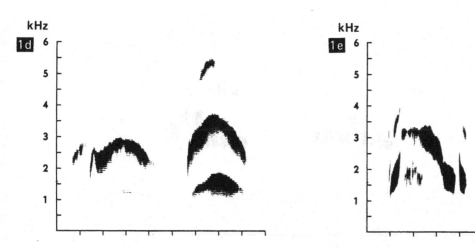

Fig. 2. Vocalisations of *Falco punctatus*. (*a*) The 'chip' call, uttered by a female, relatively low intensity. (*b*) Long sharp call, uttered by a male. (*c*) 'Wail', uttered by female.

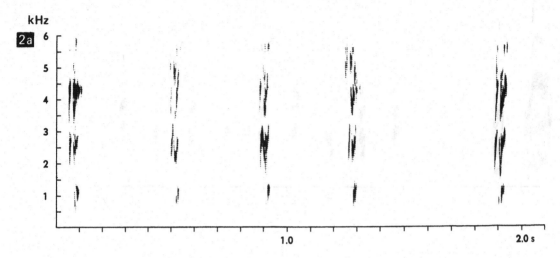

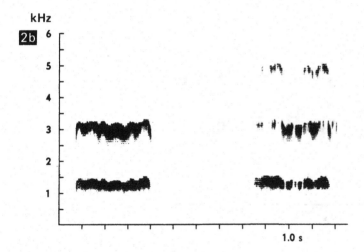

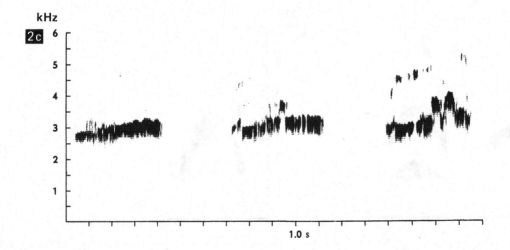

Mauritius Kestrel *Falco punctatus*

Recordings of the Mauritius Kestrel were obtained from the captive pair held at the aviaries of M. Louis Lenferna, La Balise, Black River. The vocalisations and behaviour are typical of the genus (S. A. Temple pers. comm.). As a general rule the female's calls are at a higher frequency than those of the male, and all the calls are variable. Both sexes give a 'chip' call, variable in context, and a call note which increases in intensity (on occasions) to develop into the 'wail' call.

The 'chip' call (Fig. 2*a*) is a short, sharp note with a wide frequency range. The fundamental tone at about 1 kHz is fairly weak and there are strong harmonic tones with the energy well distributed at about 2–3 kHz and again at 4–5 kHz. The tempo varied, accelerating for example during courtship displays. The call note (Fig. 2*b*) is a long, sharp note with the emphasis on the fundamental tone and first harmonic (alternating, or they may be codominant) at about 1.5 kHz and 3.0 kHz respectively, with variably intense harmonics above these.

The 'wail' (Fig. 2*c*) is a long note with a weak fundamental tone and a strong harmonic at 2.5 kHz, with other weak overtones. Comparison with Fig. 2(*b*) clearly shows the difference in frequency between the sexes (female higher, due to suppressed fundamental tone), and also illustrates similarities between the 'wail' and call note which suggest that the former is derived from the latter. During copulation both sexes simultaneously uttered another note somewhat similar to the 'wail' but of still higher intensity. This may correspond to the 'sex note' described for the Lesser Kestrel *F. naumanni* (Witherby *et al.* 1938–41, vol. 3). The call note of the Mauritius Kestrel is somewhat reminiscent of the larger species of *Falco*. It is vigorous, repetitive, lower in frequency and somewhat louder than, for example, this call of the American Kestrel *F. sparverius*.

Mauritius Pink Pigeon *Nesoenas mayeri*

I recorded the advertising call of the Mauritius Pink Pigeon, which consists of a short series of 'coo-phrases' (Fig. 3). I heard once, but did not manage to record, an alighting call of very low frequency.

The tempo of each phrase of the advertising call is slow, and there are long intervals between phrases in a series, up to 2 minutes or longer (maximum 2 min 25 s). The individual recorded usually gave three 'coos' in each 'coo-phrase', but four were occasionally heard. The introductory 'coos' in a phrase were always soft, short and low (about 300 Hz). The amplitude and frequency increased with each 'coo', the last 'coo(s)' commencing at 300 Hz, rising slowly to 400 Hz, then terminating sharply after a drop in frequency. In the four-'coo' phrase the energy is maintained at approximately the same level after the initial rise, with the emphasis between 380 Hz and 400 Hz.

A marked reaction to playback was observed in the pigeon whose advertising call is illustrated. As this bird, recorded in Macabé, approached me the interphrase interval between the 'coo-phrases' shortened to as little as 25 s (the mean of five was 45.4 s). On one occasion the playback was answered almost simultaneously by a 'coo-phrase'.

The species was seldom heard within the forests, possibly because the population surviving in the wild is very small. Nesting pairs found by S. A. Temple (pers. comm.) in March 1974 and S. D. McKelvey (pers. comm.) in late December 1975 and early January 1976 suggest that breeding takes place later in the season than in the other endemic land-birds. Presumably the pigeons bred after I left in November, which would explain why I did not hear the many vocalisations characteristic of the breeding season. Good descriptions of behaviour and vocalisations were given by McKelvey (1976).

Fig. 3. Vocalisations of *Nesoenas mayeri*. Four 'coo-phrases' from the same bird (not consecutive). The intervals between the phrases are not the actual intervals.

Echo Parakeet *Psittacula echo*

My many recordings of the Echo Parakeet include a variety of calls, though all have features in common. Both sexes gave the flight call (Fig. 4*a*) consisting of a noisy series of notes with wide-ranging frequencies, and a short, low-pitched fundamental tone with strong upper harmonic tones. In high-intensity situations such as that shown in the figure, the lower harmonics were emphasised. This call is variable in both context and structure. Fig. 4 (*c*) is a call of similar structure but longer duration, given by a perched bird.

Both sexes also gave the contact call (Fig. 4*b*), which is very similar to the flight call and is also variable. In one example illustrated the harmonics are closer, stronger and more distinctly formed than in the flight call; it was given by one of a pair sitting near a nest-hole which they had occupied but which had been taken over by a pair of Ring-necked Parakeets *P. krameri*. Both pairs were flying around and calling a good deal; both were seen to courtship-feed (by regurgitation), and *P. echo* to copulate. When the pair of *P. echo* were perched they repeatedly uttered the contact call and the 'yowl' call (Fig. 4*a, b, c*). This is a long, soft,

Fig. 4. Vocalisations of *Psittacula echo* and *P. krameri*. (*a*) *P. echo*, flight call. (*b*) *P. echo*, contact call, and 'yowl' call. *Opposite:* (*c*) *P. echo*, calls given by perched bird, two notes structurally similar to flight call and one with peaked elements similar to 'yowl' call. (*d*) *P. krameri*, calls of perched bird. (*e*) *P. krameri*, calls of same individual in flight.

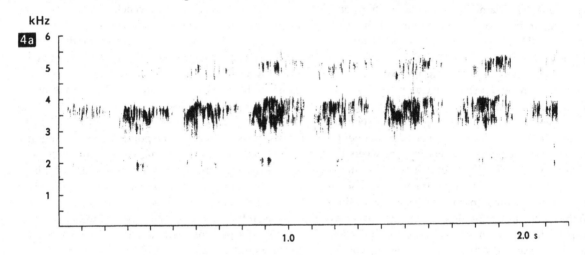

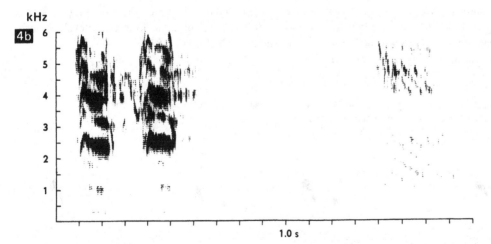

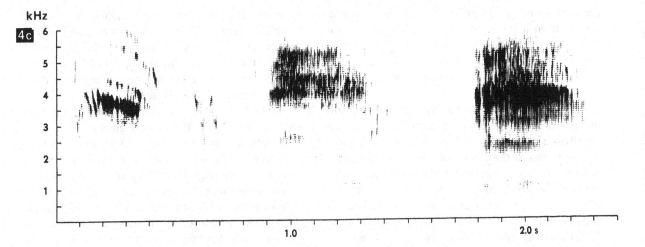

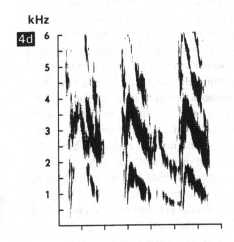

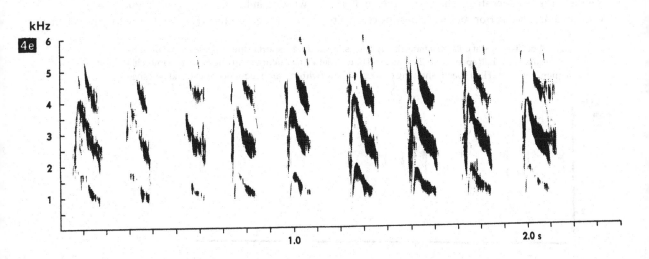

creaking sound, given probably by both sexes and always from a perch. It is a complex figure with numerous descending harmonics; this is well illustrated in the soft 'yowl' call shown last in Fig. 4(*b*). Peaked elements tend to appear in all the calls I recorded.

Two other vocalisations, the 'screech' call and the 'growl' (S. A. Temple pers. comm.), were apparently related to nesting activities and the young. I was unable to record these, as I observed no successful nesting; those nests that I watched were all usurped by either *P. krameri* or the Common Mynah *Acridotheres tristis*. The 'screech' call was uttered by a bird landing at its nest-hole, and is similar to a single figure of the flight call but has an ascending frequency. The 'growl', also given at the nest, can develop into a 'screech' call.

This species, like most parrots, is extremely vocal and its calls sound unmusical and harsh. There is a marked difference between its voice and that of the introduced *P. krameri*, so that it is impossible to confuse calling birds. The vocalisations of *P. krameri* (Fig. 4*d, e*) sound higher in tone, faster and more 'excited' than those of *P. echo*. The flight notes are shorter and more rapidly repeated (4–5 per second, as against 3–4 per second in *P. echo*). They show a more distinct form with a fundamental tone and close, clear harmonic tones; the upper harmonic tones attenuate above 6 kHz. There is fast frequency modulation with the energy more evenly distributed, and peaking is evident throughout.

Mascarene Swiftlet *Collocalia francica*

Recordings of two calls of the Mascarene Swiftlet, the echo-locating 'click' and the 'twitter', were obtained by day from birds of a nesting colony in a small semi-dark cave near Henrietta, Vacoas, Mauritius. Because of the small number of birds it was possible to record individuals as well as the more usual simultaneous calls of different birds. Observations were also made at a much larger, completely dark cave, with more birds, at Beau Bassin.

The Mascarene Swiftlet and some of its congeners (Medway & Pye 1977) can orient in darkness by means of echo-location, using a variable series of rapid loud 'click' calls which may increase in tempo to become a rattle (Fig. 5*a, b*). Structurally the clicking calls seem similar to those of *C. vanikorensis* (Griffin & Suthers 1970) and *C. hirundinacea* (Fenton 1975). Individual clicks form a double figure with a weak initial element not extending above 8 kHz, followed by a strong second element having frequencies apparently attenuating above 16 kHz. The double nature of the figure seems to be lost in fast rattles. Sound is emphasised at 5–6 kHz, and is variable and pulsed. The tempo of the calls increases to a rattle when a bird approaches its nest or other landing site, or moves into the cave away from the mouth, or approaches other birds, and decreases when a bird approaches the mouth of the cave from the inside. In the Henrietta cave birds apparently incubating at two nests began clicking as they left the nest. In returning birds, the tempo of the clicking quickened to a rattle as the bird approached its nest. One bird had difficulty landing on its nest; it made several false landings on ledges to either side before finally getting onto the nest. In full daylight at the mouth of the cave clicking calls began as or just before the birds entered the cave. On Réunion I heard clicking calls of this species while I was walking in forests enveloped in thick mist.

The 'twitter' (Fig. 5*c, d, e*) was emitted in the

Fig. 5. Vocalisations of *Collocalia francica*. (*a*) 'Clicks,' including a short rattle. *Opposite*: (*b*) Part of the same sequence at half speed. (*c*) 'Click' and call note ('twitter') sequence from one bird. (*d*) Same 'click' sequence as in (*c*) at half speed, showing echoes. (*e*) The 'twitter', same sequence as in (*c*), at half speed.

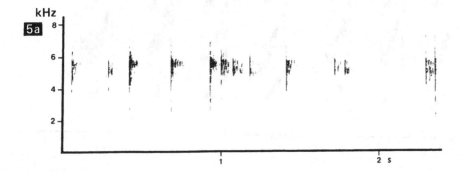

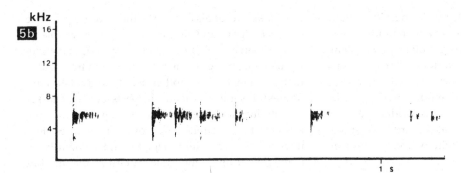

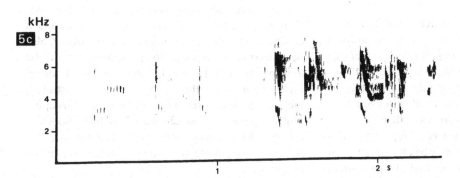

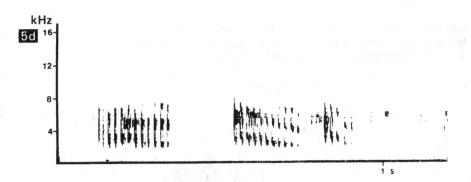

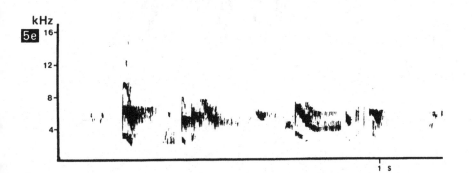

cave, alternating with bursts of clicking. Structurally this call has a weak, variable fundamental and stronger first harmonic in which pulsing is often very evident. The initial element exhibits overtones that apparently exceed 16 kHz. The frequency often drops during pulsations. The presence of another flying swiftlet elicited 'twitters', whereas lone individuals uttered only 'clicks'. Whenever numbers of birds flew simultaneously within the cave, 'twitters' were heard continuously along with 'click' calls.

Mascarene Swallow *Phedina borbonica borbonica*

Recordings of the Mascarene Swallow were made from two Mauritius breeding colonies, one on a cliff face at the Baie du Cap, the other under a bridge on the south coast road near Ste.-Marie. The cliff face colony was nesting and apparently had young, as begging calls were heard, and small flocks of adults could be seen in aerial interactions and displays.

The species has a varied vocal repertoire. The song, as in other hirundinids, may be uttered either on the wing or when a bird is perched. Three distinct motifs are recognisable from the recordings. One (Fig. 6a, right-hand figure) is a complex warble of approximately 0.85 s duration containing many elements with strong frequency modulation from 2 kHz to more than 8 kHz. Another (Fig. 6a and c, left-hand figures) is similar in many respects but is of lesser amplitude, has fewer modulated elements, and the major elements attenuate at lower frequencies, less than 8 kHz. The third (Fig. 6b and c, right-hand figures) is of much the same pattern as the first but ends with two distinctive glissando elements descending from 5.5 kHz down to 2 kHz.

A 'chip' call, sounding not unlike a sparrow's chirp (Fig. 6c, centre figure), appeared to be a contact call, possibly between members of a pair, but may also be interspersed among song phrases, as in the example illustrated. It may be uttered by birds either in flight or perched. Other vocalisations, not illustrated, include pre- and post-copulatory notes, various apparently aggressive notes heard when small flocks or groups met in aerial interactions, and the begging calls of young birds consisting of a fast, repetitive twittering.

Fig. 6. Vocalisations of *Phedina borbonica borbonica*, Mauritius. (*a*) Song phrase uttered in flight. *Opposite:* (*b*) Song phrase uttered by perched bird, ending in two glissando elements. (*c*) Sequence of three figures uttered by a perched bird, consisting of two song motifs and a 'chip' call.

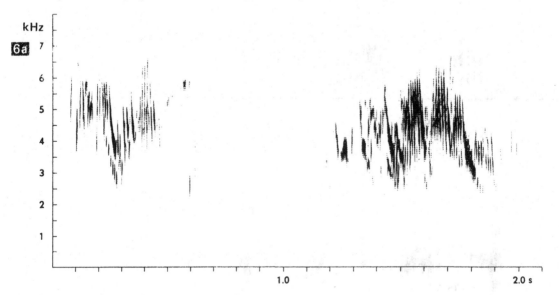

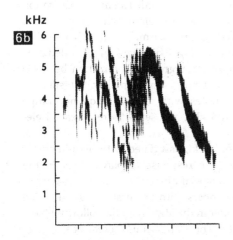

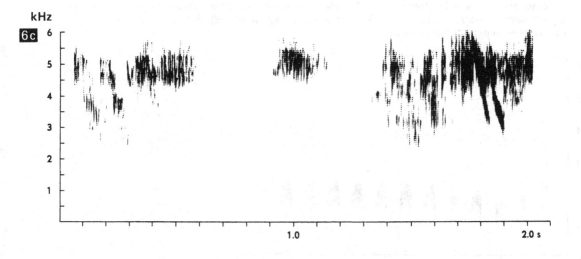

Mauritius Cuckoo-shrike *Coracina typica*

At least five calls of the cuckoo-shrike can be clearly distinguished, all of which are illustrated in Fig. 7 except for the 'rip' call, a sound like the ripping of cotton cloth which proved difficult to render satisfactorily in a sonagram, and the aggressive 'squeak' (Fig. 8*k*ii).

The 'whistle' call serves as territorial advertisement but is also used in other contexts. It consists of a series of short whistles preceded by a few introductory notes which are sharper and faster, with more and stronger overtones and is extremely variable. The variation is partly sexual but also individual and seasonal, and the intensity of the call varies with the situation. The tempo may be fast with a definite crescendo and ascending frequency (Fig. 7*b*), or slower and with no variation in frequency (Fig. 7*a*). The female 'whistle' (Fig. 7*c*) is shorter than the male's; the introductory part is longer but the whistled part much shorter, the first whistled note being twice the length of the second. The example shown was a soft reply to the male from a female on the nest, but females probably also give high-intensity whistles during territorial disputes.

Mated birds have a very strong pair bond, and employed a number of contact calls. The male vocalisation illustrated in Fig. 7(*d*) was related directly to the proximity of the female and was often the first indication of her presence. It is a short, melodious call with regular intervals between notes and variable frequency with the emphasis between 1.5 and 2.5 kHz. In

another contact call (Fig. 7*e*) the male's whistle, evident in the background, elicited from the female a 'buzz kek', a noisy buzzing note with the sound scattered over a wide range of frequencies with the emphasis between 3 and 5 kHz. Fig. 7(*f*) illustrates a female giving a series of soft 'buzz keks' increasing in tempo until the pair copulated. The male vocalisation in this case resembled the alarm whistle or squeak, which indicated not only alarm, but, at high intensity, also aggression (see Fig. 8*k*ii). Fig. 7(*g*), recorded during a high-intensity aggressive interaction between two males and a female, shows an alarm whistle, i.e. long irregular notes with a wide frequency range between 3 and 4 kHz and semi-discrete peaked elements.

The 'scold kek', associated with high-intensity aggression (e.g. in response to playback or in the presence of a conspecific in the nest area), is a series of loud, repeated notes with the first, or 'scold', note somewhat longer in duration than the following 'kek' notes (Fig. 7*h*). The same semi-discrete peaked elements are apparent as in Fig. 7(*f*) and (*g*). The notes and inter-note intervals are regular, with durations of 0.15 and 0.1 s respectively.

The 'rip' call (not illustrated) was probably associated with extreme aggression. I recorded it during territorial interactions and in answer to playback in the vicinity of the nest. S. A. Temple (pers. comm.) heard a male give this call when displaying at a Mauritius Kestrel.

Fig. 7. Vocalisation of *Coracina typica*. (*a*) 'Whistle' call of male. *Opposite:* (*b*) 'Whistle' call of male, crescendo and with ascending frequency. (*c*) 'Whistle' call of female, uttered by a bird on the nest in reply to the male. (*d*) Male's contact call. (*e*) Female's 'buzz kek' (right-hand end), with male's 'whistle' call in background (at *c.* 2.5 kHz). *Continued.*

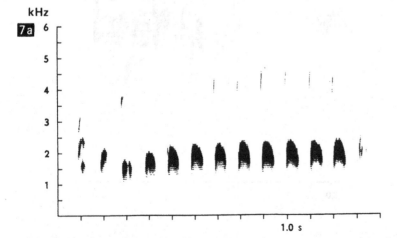

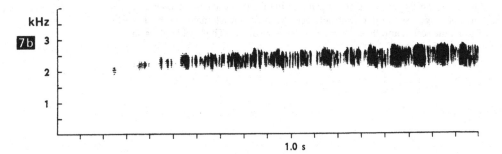

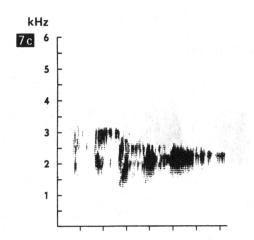

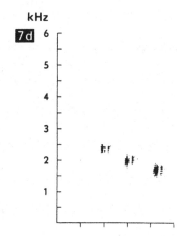

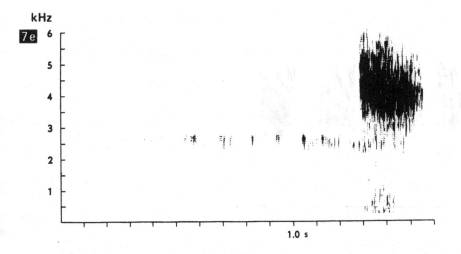

Continued overleaf.

Fig. 7 *continued.* (*f*) Sequence of calls preceding and during copulation (see text). (*g*) Sequence of calls accompanying high-intensity aggressive interaction between two males and a female: alarm whistle of male (from 1.0 to 1.5 s on time scale), and three 'buzz keks' from female. (*h*) The 'scold kek': a short sequence given in response to playback.

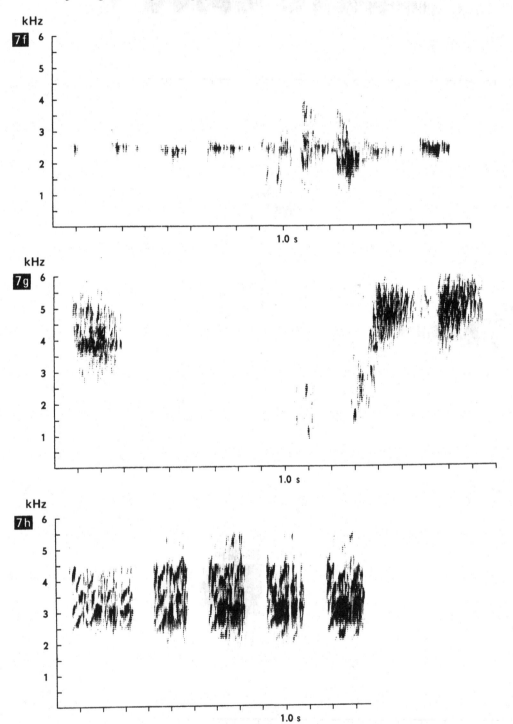

Réunion Cuckoo-shrike *Coracina newtoni*

My many recordings of the Réunion Cuckoo-shrike show that its vocal repertoire is similar to that of the Mauritius Cuckoo-shrike *C. typica*. *C. newtoni* has counterparts of the five distinct calls recorded in *C. typica*; a sixth, the soft begging calls of the young, was not heard from *C. typica* because I found no successful nest.

The 'whistle' call (Fig. 8*a–h*) is extremely variable, showing seasonal, contextual and probably individual variation. Although both male and female gave the call (M. Clouet pers. comm. and pers. obs.) I have no recordings positively ascribable to the female. Local inhabitants believe that the fast-tempo call is given by the male and the slow-tempo call by the female; but this is not so, as I several times saw and recorded a male as he uttered a fast-tempo call followed by a slow one. The 'whistle' calls of *C. newtoni* and *C. typica* were not as alike as their aggressive and contact calls, discussed below. The seasonal variation that was so marked in *C. newtoni* (compare Fig. 8*a–e* with *f–h*) was much less marked in *C. typica*, although there was some change in tempo. The 'whistle' calls of *C. typica* appeared to be more stable, varying in intensity and tempo but without the great shifts in form and frequency shown by *C. newtoni*.

Fig. 8. Vocalisations of *Coracina newtoni* and *C. typica*. (*a*)–(*e*) 'Whistle' calls of males, recorded in September, to show variation in tempo and in introductory notes. Overtones have been omitted from all except (*a*). *Continued overleaf.*

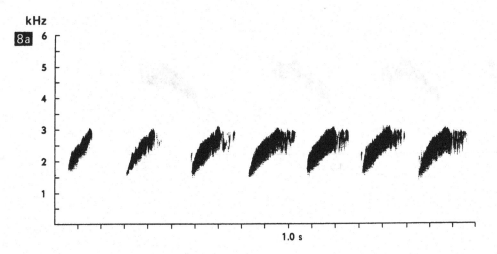

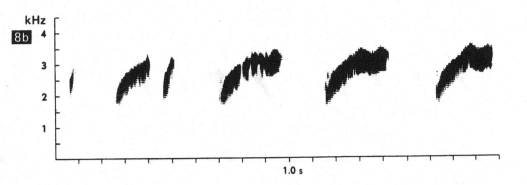

Fig. 8 *continued*

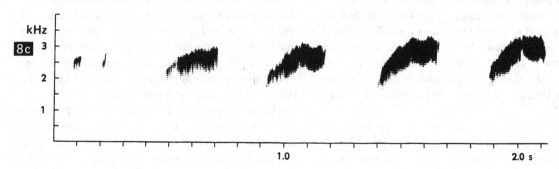

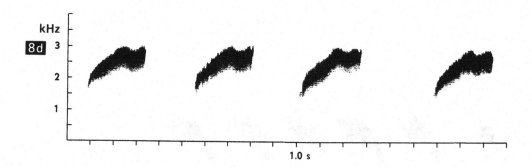

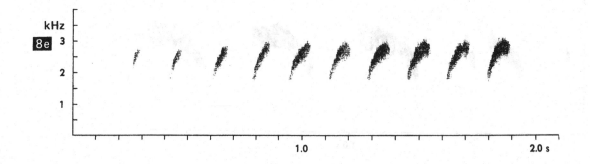

Fig. 8 *continued*. (*f*)–(*h*) 'Whistle' calls of one male, recorded near nest with young in December. Note tendency for the notes to be grouped in threes. These calls sound more melodious than those shown in (*a*)–(*e*). *Continued overleaf*.

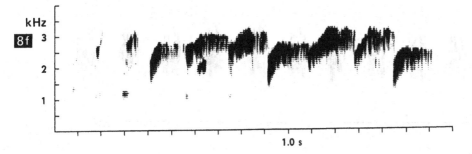

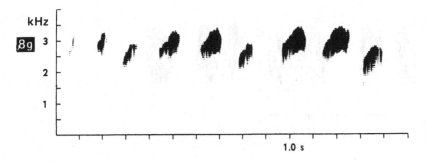

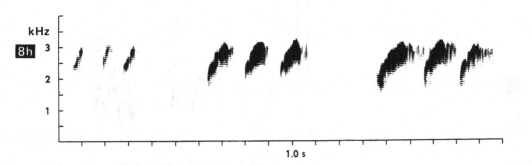

Fig. 8 *continued*. (*i*) Sequence of 'kek' notes. (*j*) Sequence of alternating 'scolds' and 'keks' (see text). (*k*i)
The aggressive 'squeak' of *C. newtoni*. (*k*ii) The aggressive 'squeak' of *C. typica*.

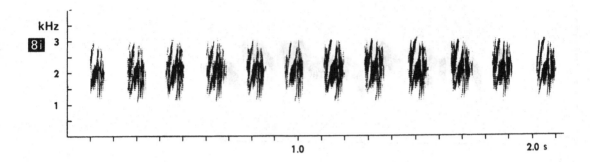

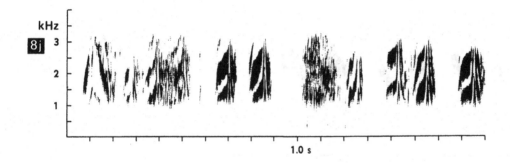

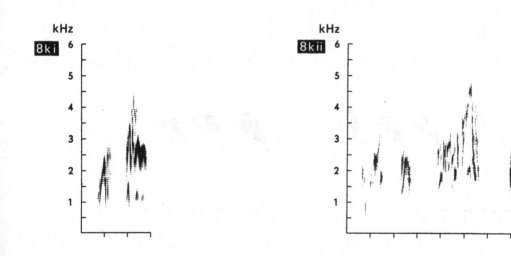

As in *C. typica*, the 'scold kek' (Fig. 8 *i–j*) seemed to be a high-intensity aggressive call and was given during territorial interactions and in response to playback. A recording made in September (Fig. 8 *i*) shows a succession of fast, regular but complex 'kek' notes, lacking the introductory scold note seen in the comparable phrase of *C. typica* (Fig. 7 *h*). The frequency is lower and its range less than in *C. typica*. Similar semi-discrete peaked elements are evident in both, those of *C. typica* being less clear, more noisy and more numerous. In an aggressive interaction between two birds, possibly both males (Fig. 8 *j*), one gave the 'kek' call (right and left-centre) and the other the 'scold' note (left and right-centre).

Fig. 8(*k*) shows the aggressive 'squeak' of both species of cuckoo-shrikes; (*ki*) is probably a high-intensity call given under stress, such as during interactions with an intruding male within the nest area, and as a response to playback. In December it was given repeatedly together with the 'rip' call whenever a Réunion Bulbul *Hypsipetes borbonicus* came near the nest; ten such instances on one day followed by the disappearance of the young cuckoo-shrikes by the next morning, coupled with frequent observations of mobbing of the bulbul by white-eyes *Zosterops olivaceus* and *Z. borbonicus*, suggest that *H. borbonicus* may have preyed upon the young *C. newtoni*. The agonistic calls of the two Cuckoo-shrikes were very similar (compare Figs. 8 *ki* and 8 *kii*); they have the same wide frequency range and peaked elements, with initial notes alike in duration, frequency modulation and amplitude. The long notes of *C. typica* were twice as long as those of *C. newtoni* (0.2 s and 0.1 s respectively), but the same peaked elements were apparent in both.

Mauritius Black Bulbul or Merle
Hypsipetes olivaceus

I recorded many of the Mauritius Black Bulbul's varied vocalisations, whose functions are not yet properly understood. These bulbuls apparently have a complex social organisation, but also a strong pair bond that may be maintained outside the breeding season. At least four types of vocalisation can be distinguished: the 'chuckle', the 'cat' call, the soft notes of pair contact, and the stress 'kek'. I have no information on the calls of the young.

The 'chuckle', or 'chuck chuck' (Newton 1958*a*), varies sexually, seasonally, according to context and in other ways (Fig. 9 *a–e*). It forms the basis of a simple song, and is used for recognition or greeting between members of a pair. Most of my observations on this call involved a pair in which the two birds were distinguishable, one having lost its tail. They had recently lost a nest with eggs, probably to a monkey, and were still very active around the nest-tree and adjacent area. The bird with the complete tail was apparently the male; he sang in the morning and evening, and during the day he chased other birds (conspecifics, *Pycnonotus jocosus* and *Acridotheres tristis*) whenever they appeared within the nest area. After such chases he returned to perch near the nest-tree and quivered his wings. He frequently gave an aerial display to his mate, using a rapid and exaggerated wing beat, with the tail raised and slightly spread, the head thrown up, and giving a series of rapid 'chuckle' notes.

Females may utter the 'chuckle' not as a song but in individual recognition after a call sequence when they rejoin their mate or foraging group.

The pair seemed to maintain close contact most of the time. While foraging, they would utter soft short 'chuckle' notes similar to those used by a group of birds feeding at a common food source, e.g. a fruiting tree. This suggests that pairs maintain contact within a group. When one of the pair rejoined its mate after a period of separation, even though they had been in vocal contact most of the time, they would 'chuckle' jointly. This was often followed by a 'chuckle' sequence from the male, sometimes culminating in an aerial display.

The 'cat' call (the 'mew' of Newton 1958*a*) consists of a rather long, clearly defined note with a weak fundamental, strong first harmonic and even stronger second harmonic. It attenuates above the 6 kHz shown (Fig. 9 *f*). It may be either a single note or preceded by some form of the 'chuckle'.

The stress 'kek' (not illustrated) is apparently given in situations of both fear and aggression, e.g. as a reaction to playback, and during inter- and intra-specific interactions.

There may be a shift in vocal behaviour at the end of the breeding season (S. A. Temple pers. comm.), the 'chuckle' being used mainly as a contact call within foraging groups, possibly as a means of recognition. The morning and evening singing ceases and individuals may be furtive and silent. Song also diminishes during nesting.

Fig. 9. Vocalisations of *Hypsipetes olivaceus*. (*a*) Single phrase of 'chuckle' notes given as dawn song, showing variation in length of notes and inter-note intervals; the last figure contains a buzzy sound which appears in most song phrases analysed. (*b*) 'Chuckle' song phrase recorded after playback, also with a buzzy sound near the end. (*c*) Series of 'chuckle' notes stimulated by playback. Note regularity of timing and similarity of alternate notes in the sequence. *Opposite:* (*d*) Series of 'chuckle' notes given by the presumed male of a pair before aerial display (see text). (*e*) 'Chuckle' series given by presumed male at dawn, with buzzing notes at end. (*f*) The 'cat' call, preceded by two 'chuckle' notes.

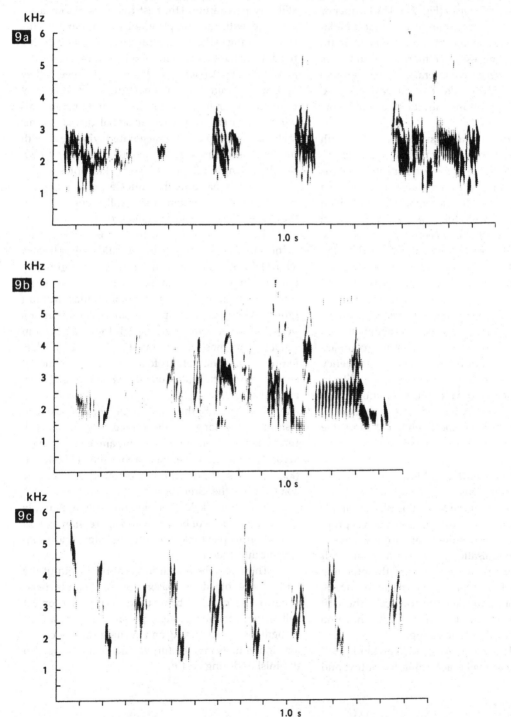

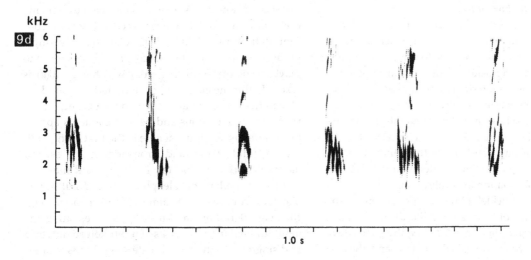

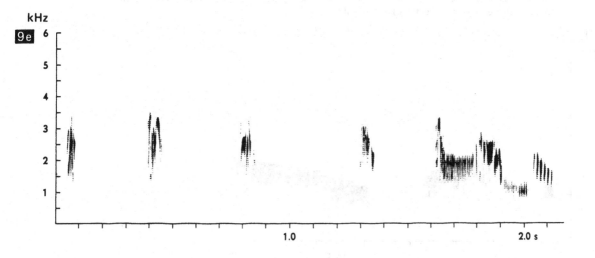

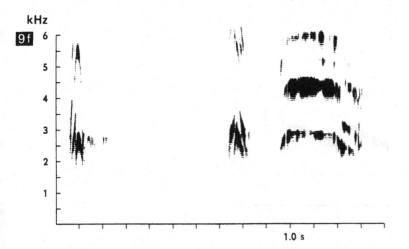

Réunion Black Bulbul or Merle
Hypsipetes borbonicus

Recordings were obtained from a number of different individuals of the Réunion Black Bulbul in two separate localities. The same four types of vocalisation can be distinguished as in the Mauritius Black Bulbul, and the same terminology is used. There is a general resemblance in the calls of the two species, but they are not closely similar. I have no information on the calls of the young. As in *H. olivaceus*, there is apparently a strong pair bond with a tendency to be social. I spent too little time on Réunion to be able to study seasonal variation in vocalisations.

Various 'chuckle' phrases were given in the continuous form of a song (Fig. 10*a–d*). In the most melodious form of the song (Fig. 10*a*) the short introductory note shows a peaked element and the fast frequency modulation that is a feature of the 'chuckle'; the second note also shows several peaked elements each with a marked downward accent. Energy is concentrated around 2 kHz. Fig. 10(*b*) shows a single phrase with two distinct motifs, given in response to playback of the recording from which Fig. 10(*a*) is taken. In this the energy is concentrated at around 2.5 kHz in the first motif and 2 kHz in the second. Rapid frequency modulations and peaked elements appear in both motifs; an unusual note in the first motif, with a weak fundamental and stronger first and second harmonics, is not repeated in the second motif. A somewhat similar 'chuckle' phrase (not stimulated by playback) consists of two almost identical motifs (Fig. 10*c*). Fig. 10(*d*) shows a 'chuckle' phrase repeating one motif, uttered in just over 1 s of unbroken sound, by a bird singing mainly the alternate song. Bouts of alter-

Fig. 10. Vocalisations of *Hypsipetes borbonicus*. (*a*)–(*d*) 'Chuckle' phrases given in the continuous form of song. For further details see text. (*e*) The 'cat' call.

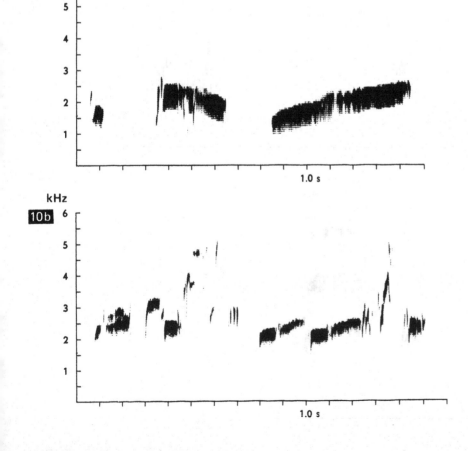

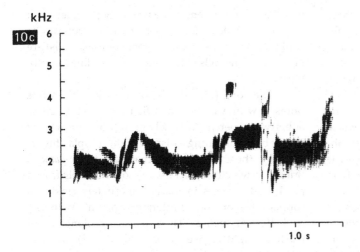

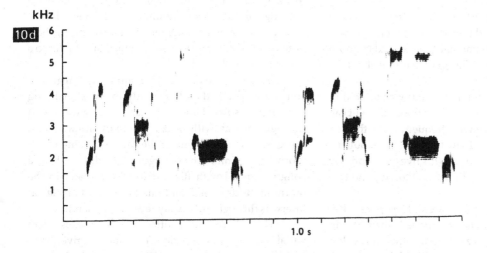

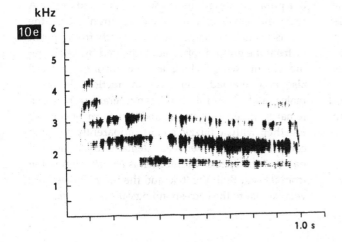

nate song, or alternate song and 'cat' calls, were commonly heard in the early morning and at intervals throughout the day. The two birds, probably a pair, could be in the same tree or on opposite sides of a valley. In the latter case, one of the calling birds usually flew to join the other.

One morning in the Forêt du Bébour birds were singing as they came separately past me down a mountain slope on their way down the valley. Six birds passed in this way, while I remained in one place for 10 minutes. In each case I could hear another bird approaching as I followed the first bird with my parabola down the slope. Probably they had roosted high up and were dispersing to forage at lower altitudes. Similar behaviour in *H. olivaceus* on Mauritius apparently varied from place to place; for example it was more noticeable in the Piton du Fouge than in the forest of Macabé.

The 'cat' call (Fig. 10e) is a long drawn-out note of slightly ascending and then descending frequency, with a very weak fundamental tone at 1 kHz and a series of harmonic tones. The emphasis is at about 2.3 kHz.

The stress 'kek' (not illustrated) was uttered in situations similar to those noted for the equivalent call in *H. olivaceus*. It was given frequently on the ten different occasions when I saw Réunion Black Bulbuls being mobbed by other birds (on five occasions by *Coracina newtoni*, three by *Zosterops olivaceus*, and two by *Z. borbonicus.*

The two species of Mascarene *Hypsipetes* differ vocally in many aspects, as indeed they do in behaviour, eye colour, size, and in other ways, but there are similar elements in the 'chuckle' and the call note. The Seychelles and Madagascar Bulbuls *H. crassirostris* and *H. madagascariensis* also have similar call notes (pers. obs.). The repertoire of the Seychelles Bulbul described by Greig-Smith (1979b) is very similar to those I describe for the two Mascarene species, but his terminology differs as do some of his interpretations (e.g. of the 'chuckle' in relation to song). It is possible that the call note and other acoustic displays I have described may show general similarities throughout the genus.

Réunion Stonechat *Saxicola tectes*

Recordings were obtained from different individuals of the Réunion Stonechat in two separate

localities intermittently between September and December. At least four types of vocalisation can be distinguished, two of which (song and 'tek' calls) are illustrated. The calls of juveniles and nestlings are not illustrated.

The song (Fig. 11a–c) was a series of discrete phrases with variable inter-figure intervals, given either while perched (usually on top of a prominent tree or bush) or during a display flight. Display flights resemble those of the Stonechat *Saxicola torquata* in Britain, in which the bird rises almost vertically from a perch and "dances in the air, alternately sinking a couple of feet or so and fluttering up again" (Witherby *et al.* 1938–41, vol. 2). These may be prolonged and reach a height of over 30 m (Milon 1951). *S. tectes* apparently gives these aerial displays more readily than *S. torquata*. Shorter display flights are also performed, from tree to tree within the forest, across a clearing, or out from a mountainside. These shorter displays are often accompanied by many sharp, short notes given in rapid succession, which are similar to a series of fast call notes.

In mid-September at high altitudes I heard no songs from perched birds and only an occasional song with flight display. Later in the season song was heard through the day with peaks of activity at dawn and dusk. This species is one of the earliest to sing in the morning, beginning before light, but I did not hear a song combined with flight display until later in the morning. Birds would continue to sing and display at intervals throughout the day, even in thick mist.

The 'tek' call notes, uttered by both sexes, were variable in form and function. They may be given with an initial plaintive 'hweet' (Fig. 11d), or as a single or double note. When a bird gave the 'hweet' at the start of a phrase it slowly raised and then lowered its tail. A male and female of a mated pair apparently used the call as a contact note, giving it frequently from a perch or from the ground, often accompanied by a bobbing movement, wing-flicking and tail-flirting. This call also accompanied aggressive interactions between males, and I heard it once in an interaction between females. Juveniles gave a modified form of the call together with a buzzy sound, apparently to keep contact with their parents and demand food. The local name of the species in Réunion is *tek-tek*, originating from its call. Both the 'tek' and the initial 'hweet' are very similar to the corresponding calls of *S. torquata*.

Fig. 11. Vocalisations of *Saxicola tectes*. (*a*) Song phrase from a perched bird. (*b*) Song phrase uttered during high display flight. Owing to the distance at which the recording was made the form of some weaker notes is not fully reproduced. *Continued overleaf.*

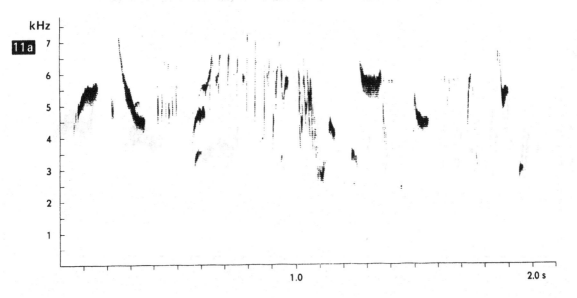

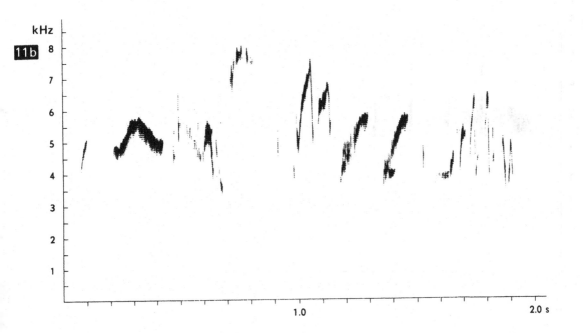

Fig. 11 *continued*. (*c*) Song phrase from near the termination of a display flight within a forest clearing. Note the double structure of the rapid short notes at the beginning (initial downstroke followed by a sharp click), which produces a harsh rattle. The three long notes at the end may be frequency modulated, as may the longer, strongly marked notes in (*a*) and (*b*). (*d*) 'Tek' calls with initial 'hweet'.

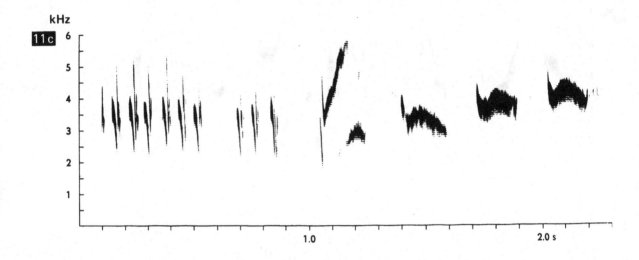

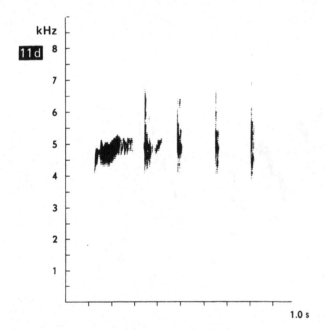

Mascarene Paradise Flycatcher
Terpsiphone bourbonnensis

Many recordings were made of the Mascarene Paradise Flycatcher at diverse localities in Réunion (*T. b. bourbonnensis*) and Mauritius (*T. b. desolata*). At least three types of vocalisation were distinguished: the song, which apparently is uttered only by the male; the 'buzz-song,' probably common to both sexes; and the 'buzz', certainly given by both sexes. No juvenile calls were recorded and the nestlings I saw called barely audibly.

Fig. 12(a) and (b) shows song phrases taken from a long sequence by a male *T. b. desolata* who was perched near his nest (which contained eggs) at nightfall. The notes are grouped in threes. In the first the sound is nearly continuous within each group; in the second there are distinct gaps between the notes, and the energy in the fundamental tone is concentrated at lower frequencies (1.5–3.0 kHz, compared with 2.0–3.2 kHz). Fig. 12 (c) shows a 'buzz-song' sequence from a male *T. b. bourbonnensis* given in response to playback (I found that the response was always greater to a 'buzz' or 'buzz-song' sequence than to song alone). The song notes are shorter than in the song, with longer inter-note intervals, the 'buzz' notes coming between the song notes. There are marked stroke elements at the beginning of the song notes, except for the last notes in each group. The 'buzz' notes are long and noisy, with sound scattered over a wide frequency range. Fig. 12(d) and (e) shows song phrases given by a male *T. b. bourbonnensis* during an interaction involving two males and two females, and probably represents high-intensity song. Much 'buzz' is audible in the recording, and the song notes tend to be fast in tempo with more frequency modulation than in typical song. This combination of 'buzz'-notes and song was often heard, especially in apparently territorial interactions. A section of this recording played back at the end of the interaction received an immediate response from a male and a female (a presumed pair); the male approached extremely close to the recorder and repeatedly raised and lowered his crest.

The 'buzz' (Fig. 12*f* and *g*) was probably the most widely used vocalisation. It was the only one that I could positively ascribe to a female, as I recorded it from both members of a pair during change-overs at the nest. It is variable in structure, tempo and context, and may occur as a single note or as a long sequence of notes. Response to playback of a 'buzz' sequence was immediate and intense. The bird, usually a male, would approach the recorder very closely. Playback of a 'buzz' sequence recorded on Mauritius elicited an immediate response from a male on Réunion.

There are differences in the songs and calls of the two races, but they tend to be obscured by the great individual variation within populations. Without more detailed analysis it is impossible to say precisely how different they are. A number of mainland African species sound, to my ears, similar to the island forms, and recordings from Japan are also similar. The 'buzz' and perhaps other notes may be characteristic of the entire genus.

Fig. 12. Vocalisations of *Terpsiphone bourbonnensis*. (a) Song phrases from a male *T. b. desolata*. *Continued overleaf.*

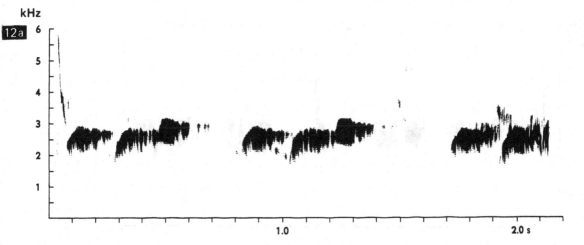

Fig. 12 *continued.* (*b*) Song phrases from a male *T. b. desdata.* (*c*) 'Buzz-song' of male *T. b. bourbonnensis* in response to playback. (*d*) Song phrase of male *T. b. bourbonnensis*; note that the second half repeats the first half almost exactly. *Opposite:* (*e*) Single section of a song phrase, from same recording as (*d*). (*f*) and (*g*) 'Buzz' sequences of two different males of *T. b. desolata.*

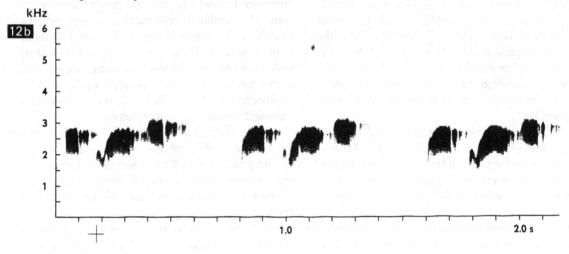

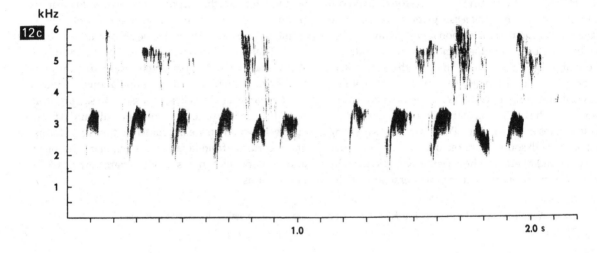

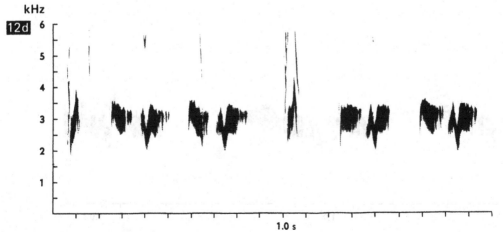

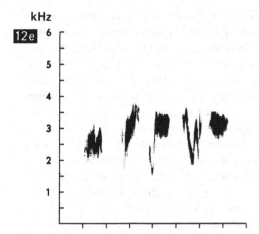

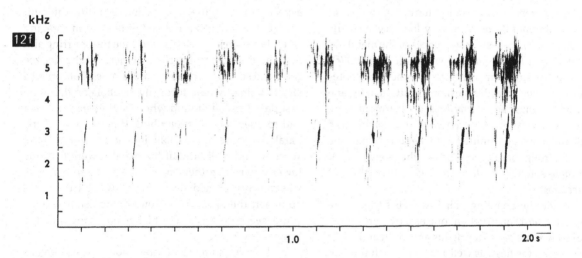

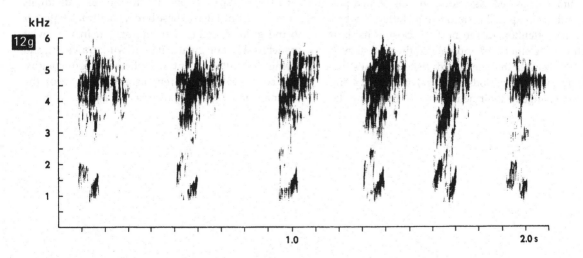

Rodrigues Brush Warbler
Acrocephalus rodericanus

A number of recordings of the Rodrigues Brush Warbler were obtained from the *Eugenia jambos* thickets at Cascade Pigeon. On 4 November 1974 I found a pair building a nest 4 m up in a *Eugenia* tree. Recordings were made in the immediate vicinity of this nest and within the territories of neighbouring individuals. I found the species extremely vocal, and distinguished at least four types of vocalisation: the song (Fig. 13*a–d*); the 'intimate notes', consisting of soft melodious notes or soft churring notes, given by members of a pair at close quarters; a distinctive 'zee zee zeit', apparently a contact note; and a squeaky note uttered during aggressive encounters.

Fig. 13 shows four variants of the song. Similar elements appear in some of them, though there are marked differences in overall pattern; e.g. the short trill near the middle of (*a*) is very like that near the beginning of (*c*). Two song sequences of the mainland African *A. gracilirostris* are shown in Fig. 13(*e*) and (*f*). The notes are similar in structure to some of the notes of *A. rodericanus*, but the harmonics, attenuating at 6 kHz, are stronger in *A. gracilirostris*. Harmonics are present but weak in *A. rodericanus* (attenuating at 5 kHz), and have been omitted from the sonagrams. The lack of strong harmonics makes the songs of *A. rodericanus* sound less full and melodious than those of *A. gracilirostris*.

While the breeding pair I was observing were completing nest construction, one bird (the presumed male) sometimes sang during his approach and while he worked at the nest, as well as when he left the site. In those visits which were accompanied by song he sometimes displayed, assuming an exaggerated posture with head up, tail up and wings fluttering which drew my attention to the nest, disclosing its exact location. The other bird approached the nest silently. On 7 November the silent bird visited the nest frequently and spent periods of several minutes at a time in it. A change in their behaviour the following day

suggested that the pair had begun incubating. I observed change-over between the parents at regular intervals. The commonest notes uttered when the two were together at the nest, both before and after incubation began were the 'intimate notes' which often alternated with song phrases. One bird gave song phrases and probably the soft churring notes; the other uttered soft melodious notes. Almost invariably a song phrase would be answered immediately by a similar song phrase from a bird in an adjoining territory.

Behaviour which I termed 'canopy singing' was heard at many times of day. Birds perched near the top of the *E. jambos* thickets were difficult to see through the thick understorey, and would sing in response to one another. Although the singing birds remained perched high in the thicket they were not necessarily immobile but might forage, or otherwise move about, especially later in the day, making recording difficult. I frequently noticed a marked difference in foraging levels between individuals who gave the churring and melodious 'intimate notes' and sometimes the 'zee zee zeit', and so were probably paired. When I observed a bird foraging I would focus my parabola on it; on the first day, I usually found when singing started that I had focused on the wrong bird. When I focused the parabola above the visible individual I had more success, and realised that the females were foraging lower down than their mates. Response to playback was always immediate; usually two birds would approach the speaker and on one occasion when I could see both birds, the higher one sang and the lower one did not.

Females sometimes sang under stress, such as when playback was used, or in highly aggressive interactions, such as one in which three individuals were involved. Two of these birds actually fell together to the ground, while the third stayed in the lower branches showing great agitation, wing-quivering and giving short intermittent song phrases together with a soft squeaky 'chirrup' reminiscent of some of the 'intimate notes' given by members of a pair.

Fig. 13. Vocalisations of *Acrocephalus rodericanus* and *A. gracilirostris*. (*a*)–(*d*) Four variants of the song of *A. rodericanus*. (*e*)–(*f*) Songs of *A. gracilirostris*. *Continued overleaf*

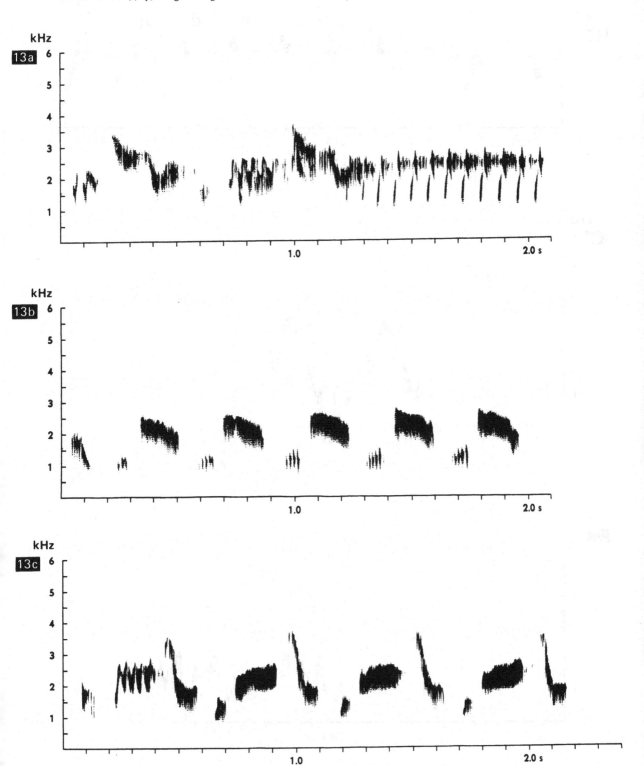

Fig. 13 *continued*

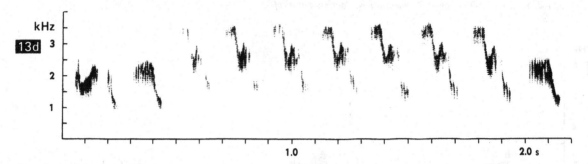

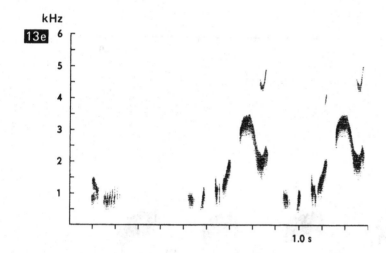

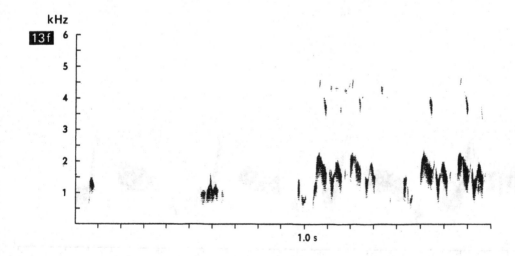

Mauritius Fody *Foudia rubra*

Both sexes of the Mauritius Fody were recorded at five different localities. At least five types of vocalisation can be distinguished: the 'plick' call, the territorial song, the 'chew' note, the 'rattle' (or 'buzz') note, and the flight call. Most were given by both sexes, some were very similar to the calls of *F. flavicans*, and certain of the female response notes had a marked structural affinity with those of *F. madagascariensis*. None of the nests that I watched was successful, apparently all being destroyed by monkeys; thus I recorded no nestlings or juveniles.

The 'plick' call was the most widely used vocalisation; both sexes gave it, and forms of it are heard throughout the year (S. A. Temple pers. comm.). It varied greatly in tempo, from very slow during foraging to a rapid 'splutter plick' (Fig. 14a) associated with the male's displays when he was close to the female during the breeding season. The female's 'plick' calls were similar to the male's but usually slower in tempo. During the breeding season a pair often gave 'plick' calls in unison; in all cases those birds that I recorded were near a nest-site and the calling in unison continued for many minutes, while the pair either foraged or gathered nest material. The male often terminated these sequences with a rapid 'splutter plick' as he flew either to a song perch or to the nest, or directly to join the female.

During the non-breeding season (May to September) the Mauritius Fody is apparently silent except for slow-tempo 'plick' calls and the flight call (S. A. Temple pers. comm.). At the start of the breeding season other calls, such as the 'chew' note, which was often uttered during chasing displays or near a nest-site, the 'buzz' note, and song phrases can be heard. Three series of 'chew' notes, all given by males, are shown in Fig. 14 (*b–d*). The 'chew' note was not heard from females except as part of a song phrase (see below). 'Buzz' notes, given by both sexes, are shown in Fig. 14(*d*)–(*f*).

A phrase from a sequence of the territorial song of a male (Fig. 14*g*) is a series of six almost evenly spaced disyllabic notes with wide-ranging frequencies (10 kHz to 4 kHz). Abrupt stroke elements of ascending and descending frequency terminate in a noisy or buzzy sound. Such phrases may be uttered by a male as a continuous song with pauses between the phrases, or may be delivered loosely with other vocalisations interspersed.

In a phrase from a sequence of female song, elicited by playback (Fig. 14*h*), the third and fourth notes are effectively 'chew' notes, while the last four notes are similar to those of the male's territorial song. Female songs reproduce less easily in sonagrams than those of males because of their lack of amplitude, although structurally they are very similar.

Fig. 14. Vocalisations of *Foudia rubra* (*a–h*) and *F. madagascariensis* (*i*). (*a*) A 'splutter-plick' uttered by a male during nest-building activities, preceded by an incomplete high-frequency female note. The upper frequencies of the 'splutter-plick' attenuate above the limit of the sonagram. The female note is of a kind usually uttered in response to the male's calls. (*b*) Fast-tempo 'chew' notes, uttered by a male in flight. *Continued overleaf.*

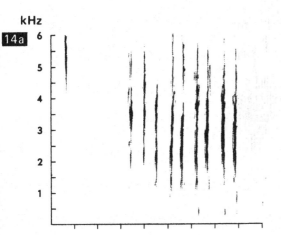

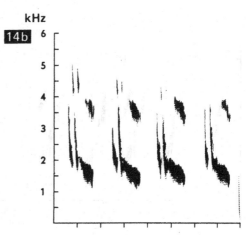

Fig. 14 *continued*. (*c*) Slow-tempo 'chew' notes, uttered by a male near a nest under construction, followed by two female response notes. (*d*) Five 'chew' notes of a male, followed by a female buzz, the 'chew' notes less clearly defined than those in (*b*) and (*c*), owing to the greater distance at which the recording was made. *Opposite:* (*e*) Series of 'rattle' (or 'buzz') notes uttered by a male and female during a territorial encounter with another male. The last four 'rattle' notes are those of the female, the two middle ones being synchronous with two fast-tempo male 'chew' notes. (*f*) Series of harsh 'rattle' notes given by both members of a pair who were building a nest. (*g*) Phrase from a sequence of male territorial song.

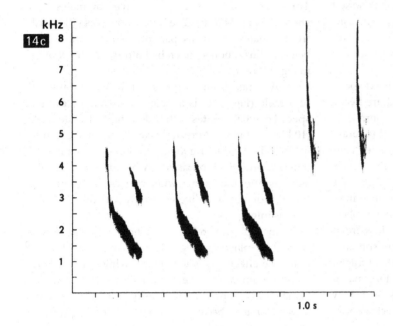

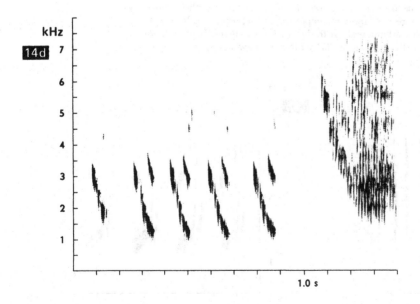

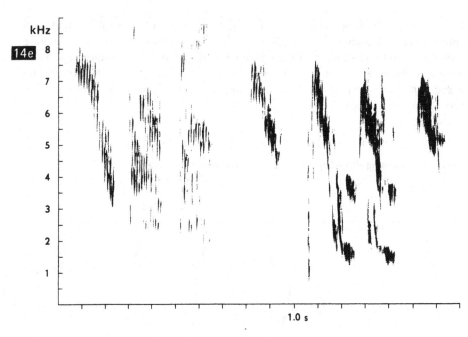

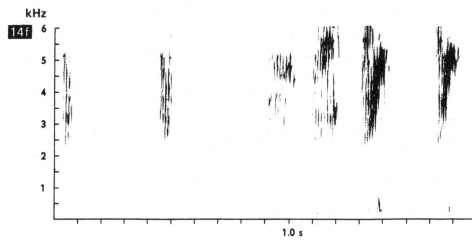

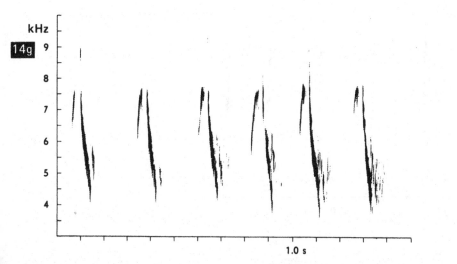

For comparison with *F. rubra* and *F. flavicans* (Fig. 15), Fig. 14(*i*) shows a phrase from a sequence of song from a male *F. madagascariensis*. The abrupt stroke elements are virtually identical with those in some of the vocalisations of *F. rubra* and *F. flavicans*. Displays and associated vocalisations of *F. rubra* and *F. flavicans* are compared under the latter species.

The flight call of *F. rubra* (not illustrated) is a single disyllabic, high-frequency note with a long inter-figure interval, reminiscent of some of the notes of *F. madagascariensis*. It was given only in flight, by both sexes, apparently throughout the year.

Fig. 14 *continued*. (*h*) Sequence of female song elicited by playback. (*i*) Phrase from a sequence of song of male *F. madagascariensis*.

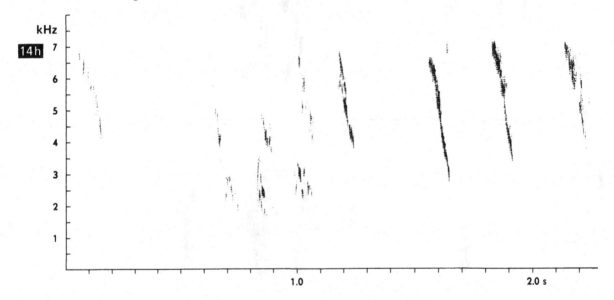

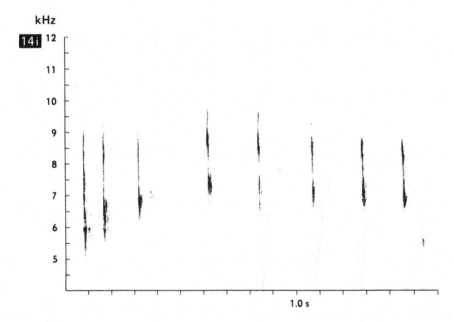

Rodrigues Fody *Foudia flavicans*

Many recordings of the Rodrigues Fody were obtained from both sexes. At least three types of vocalisation can be distinguished: the 'plick' call, the song (which includes 'rattle', or 'buzz', notes similar to those of *F. rubra*), and the 'chew' note. All were uttered by both sexes, either alone or in combination with other notes. I did not record any juvenile or nestling calls, nor did I detect a distinct flight call such as that of *F. rubra*.

The 'plick' call (Fig. 15*a*, *h*) was the most frequently heard vocalisation, and as in *F. rubra*, the slower-tempo forms are probably heard throughout the year. It was used in many situations, and accompanied by much wing-quivering. Interacting males gave the call, either in unison or separately in answer to each other. Males and females (presumed pairs) used the 'plick' call in establishing and maintaining contact, and often called for long periods of time. Frequently, I recorded a male and female 'plick'-calling in unison. When used in aggressive situations, the tempo increased; at the fastest tempo it became a splutter of sound that often signalled the departure of the calling bird. Playback of a sequence of 'plick' calls sometimes evoked a stronger response than playback of a song sequence.

Song (Fig. 15*b*–*g*, *i*) was related to breeding. Both sexes sang and both displayed by quivering their wings almost continuously while singing.

Under normal circumstances (i.e. before playback was used) during periods of singing the male spread and raised his wings to a point almost above his head, this movement always coinciding with the 'rattle' note of a song sequence. In stress situations, such as after playback, females may raise the wings almost to a vertical position over the head. In both sexes, the vertical wing-lift was preceded by a fast wing-quivering. After repeated playback both sexes quivered and raised their wings, without calling; this movement was not to the full, vertical position but was rather an upward flicking alternating with wing-quivering.

The female song illustrated (Fig. 15*f*, *g*) is much shorter in duration than the song phrase of the male, but there are many similarities in structure. The phrase illustrated in Fig. 15(*g*) begins with two melodious notes similar to those in Fig. 15(*d*) and (*e*), and ends with four striking, long noisy notes of high frequency.

'Chew' notes may be associated with 'plick' calls (female, Fig. 15*h*) or incorporated in a song phrase (male, Fig. 15*i*). 'Chew' notes unassociated with other notes were also given by the male, for example when two birds were chasing during courtship or as a form of contact call. During periods of singing they were heard especially in response to playback, suggesting they may have some aggressive connotation. The tempo of calling increased greatly in flight.

The displays of *F. rubra* and *F. flavicans* seen during the breeding season differed considerably, those of *F. flavicans* being apparently much more elaborate. Females of *F. flavicans* had almost the same aggressive territorial display as males; this was often demonstrated in their response to playback of their songs. They approached the sound source, looked about and listened, and followed this with further song sequences. Females of *F. rubra*, though they showed a positive response to playback, moved their wings very little in contrast to the continual wing-quivering and wing-lifting of females of *F. flavicans*. The songs of the female *F. rubra* were less audible than in female *F. flavicans* because they show less amplitude and seemed to be given only in the presence of her mate near the nest-site. Males and females of both species seem to respond more strongly to vocalisations of their own sex than to those of the opposite sex. Males of *F. rubra* display with their wings lowered, and can lower one wing at a time. This is in marked contrast to *F. flavicans*, in which males extend their wings high above the head. In *F. rubra* there seemed to be less synchrony between song and display than in *F. flavicans*, in which the wings were lifted in precise synchrony with the 'rattle' portion of each song.

One of the apparent displays of males of *F. rubra* was bill-wiping, which may represent a partly ritualised displacement activity. Repeated bill-wiping occurred in response to playback or during pair interactions at a nest-site. Sometimes it was a simple side-to-side movement, but it was sometimes accompanied by a slight sound that may have been vocal; it was repeated a number of times in succession, during either song or intervals of silence. Similar but silent and much less pronounced bill-wiping was observed in *F. flavicans*.

Fig. 15. Vocalisations of *Foudia flavicans*. (*a*) Series of 'plick' calls (double notes except for the first and last) terminating a male song phrase. (*b*) Sequence from one of the most commonly heard song phrases of males; this example was given apparently in response to slight playback stimulation. *Opposite:* (*c*) Song phrase of the same male as (*b*), given in response to playback of its own song. Note that the initial cluster of notes is incomplete. (*d*) End of a male song sequence terminating in two rattle notes and three typical song notes. At the end, one of the 'rattle' notes is analysed at half speed. *Continued.*

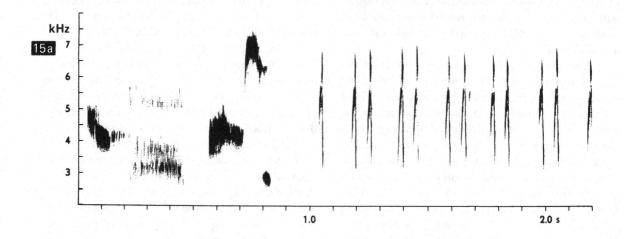

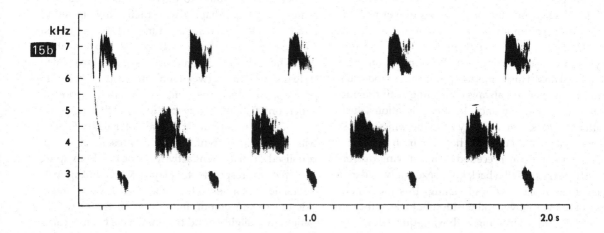

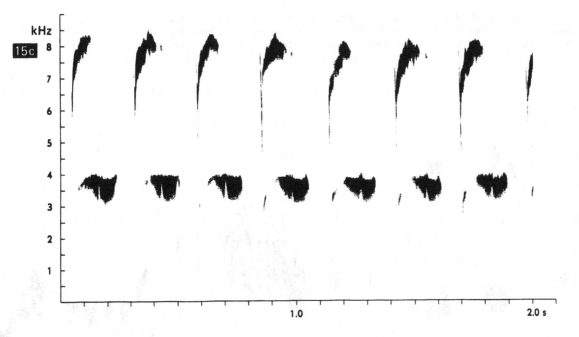

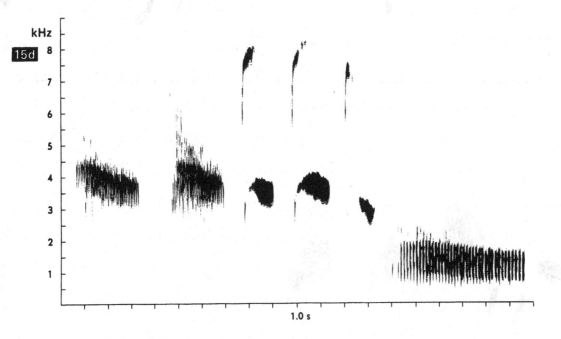

Continued overleaf.

Fig. 15 *continued*. (*e*) Complete male song-phrase consisting of typical song notes, 'rattle' and melodious notes. (*f*) Female song-phrase. *Opposite:* (*g*) Female song-phrase, incorporating 'chew' notes. (*h*) Female call consisting of three 'chew' notes followed by four 'plick' notes. *Continued.*

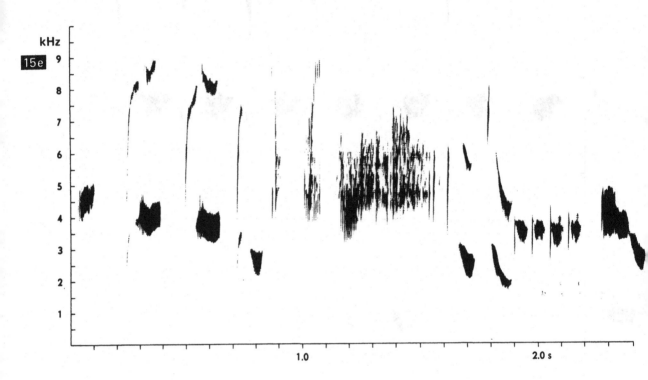

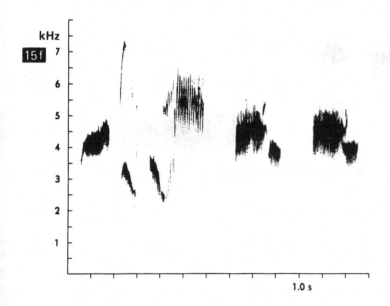

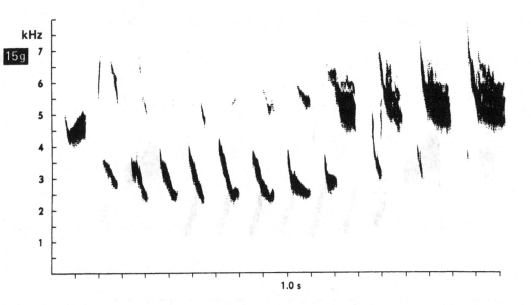

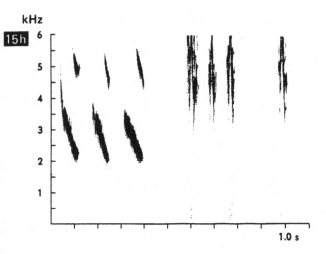

Continued overleaf.

Fig. 15 *continued*. (*i*) Phrase from a male song sequence, incorporating seven 'chew' notes.

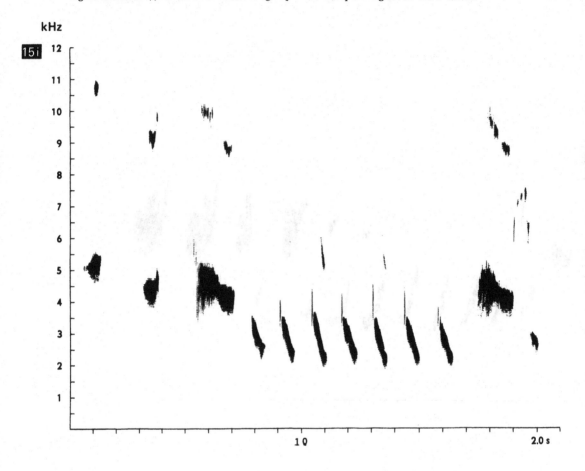

Mascarene Grey White-eye *Zosterops borbonicus*, Mauritius Olive White-eye *Z. chloronothos*, Réunion Olive White-eye *Z. olivaceus*

These species are considered together because of certain broad similarities in white-eye vocalisations and behaviour, and because the strong differences in structure of the vocalisations between the races *Z. b. borbonicus* and *Z. b. mauritianus* necessitate comparisons among all four taxa.

Fig. 16 illustrates representative sonagrams of vocalisations of *Z. olivaceus*, Fig. 17 of *Z. chloronothos*, Fig. 18 of *Z. b. borbonicus*, Fig. 19 of *Z. b. mauritianus*. Certain phrases from all taxa at a frequency range of 0–16 kHz to show the high frequencies attained are shown in Fig. 20. In the analysis of *Zosterops* vocalisations, even more than in any other case in this chapter, great care must be taken when comparing vocalisations because it is easy to be misled by poor-quality recordings. For the *Zosterops* comparisons I used good-quality equipment and recorded the birds from as close as possible. My tentative conclusions on this difficult group are based upon analysis of thirty 7-inch reels of taped *Zosterops* vocalisations, and over 300 sonagrams.

All four taxa have much the same basic repertoire in which at least five types of adult vocalisation can be distinguished, some previously described by Gill (1971a). The five types are (1) flight calls, (2) feeding (contact) and flocking calls, (3) agonistic calls, (4) songs and (5) roosting calls. The calls of nestlings and juveniles are important but not examined here. Comparison of the sonagrams shows clearly that all taxa differ from each other; interestingly the most marked differences seem to be between the two races of *Z. borbonicus* (e.g. see Figs. 18, 19, 20c–f).

There is an aptitude for mimicry in these four taxa that is typical of many species that I have recorded in mainland Africa and on other islands (e.g. *Z. modestus* of Seychelles). In all the Mascarene species mimicry occurred during periods of singing, at any time of day during the breeding season. The mimic phrases were often uttered more softly during elaborate 'warbled phrases' (Fig. 18g). All species gave soft and muted versions of their songs; these were uttered more commonly by races of *Z. borbonicus* and by *Z. chloronothos* than by *Z. olivaceus*.

The low population density of *Z. chloronothos* might account for the failure of past observers to report its vocalisations, other than the flight call. Where it is common and so has the necessary stimulation of sight and sound of conspecifics, it behaves like *Z. olivaceus*, defending a nectar source or nest-site against all other white-eyes of either species. On one occasion while near a *Z. chloronothos* pair who were constructing a nest, I played back a sequence of *Z. b. mauritianus* calls and one of the pair responded instantly by flying to the tree above me. In all of more than 150 cases response to playback of conspecific vocalisations was also immediate.

The songs (Figs. 16h, 17fi, g, 20bii) of the olive white-eyes *Z. chloronothos* and *Z. olivaceus* (both more solitary and aggressive species than *Z. borbonicus*), differ somewhat from those of *Z. borbonicus* (see below), as does their behaviour. Their songs consisted of fast warbling interspersed with abrupt flight call notes. During interactions at a nectar source both *Z. chloronothos* and *Z. olivaceus* sang with much bill-snapping (Figs. 16g and 17c). Birds sang in the early morning and late evening, but this singing was less sustained than in *Z. borbonicus*. When mimicking, both species of olive white-eyes gave long song sequences which sounded exactly like those of the species imitated; short mimic phrases were included by birds singing their own distinctive song.

The mobbing call and related behaviour of *Z. olivaceus* need special mention (Figs. 16b, 20g). On four separate occasions I recorded four or five *Z. olivaceus* mobbing a bulbul *Hypsipetes borbonicus*. Except when a mated pair were together, both *Z. olivaceus* and *Z. chloronothos* were usually solitary and aggressive; any conspecific within an area (especially near a nectar-bearing shrub or tree) was chased with much loud calling (Figs. 16a, 17a). When mobbing, several *Z. olivaceus* joined together in loud and strident calling. This call (Fig. 16b) was interesting in that one note seemed to be the usual aggressive 'pit' note and others, although sounding similar to the ear, were in fact structurally reminiscent of the 'Zosterops note' (see below). Both these notes showed a wide frequency range (Fig. 20g: 0–16 kHz) and were accompanied by considerable bill-snapping. The interactive calling of *Z. chloronothos* heard when two or more were together at a nectar source, as well as in response to playback (Fig. 17b), showed some similarity to the mobbing call of *Z. olivaceus*.

Fig. 16. Vocalisations of *Zosterops olivaceus*. (*a*) A series of notes, together with a number of bill-snaps, given by an individual defending *Hypericum* flowers. One loud 'chip' or 'pit' note uttered by a second bird is seen between notes 12 and 13. This sequence denotes high-intensity aggression typified by fast tempo and very short inter-note interval. It corresponds to the active 'tu-tu-tu' described by Gill (1971*d*: 45). (*b*) A high-intensity agonistic call uttered when a group of five birds mobbed a *Hypsipetes borbonicus*. Numerous bill-snaps (sharp, single, mechanical stroke elements) can be seen, and at least two birds utter notes which apparently have two distinct forms, one resembling the '*Zosterops* note' (see text) and the other the 'pit' note. (*c*) Two birds at a nectar source, one giving the foraging 'pit' notes, the other bill-snaps with some 'nectar defence notes', representing a low-intensity interaction. Similar vocalisations are uttered during aerial chases. (*d*) Non-aggressive, foraging 'pit' notes uttered by one individual feeding alone at a flowering *Hypericum* tree. This corresponds to the 'tchip' described by Gill (1971*a*: 45). (*e*) Flight call of one individual ('tchip-tchip-tchip', Gill 1971*a*: 45).
(*f*) Single individual uttering foraging 'pit' notes, giving bill-snaps and a warbled (frequency modulated) note; this forms a combined vocal-instrumental low-intensity agonistic call. (*g*) Bill-snaps and warbled note given by one bird at a nectar source; notes 13 and 14 resemble the plaintive agonistic call of *Z. b. borbonicus* ('eeeeeee', Gill 1971*a*: 43). The call is low-intensity agonistic. (*h*) A song phrase uttered without mimicry. Notes 8 and 9, the two '*Zosterops* notes' (see text), may represent mimicry, or more likely are notes common to *Zosterops* spp.

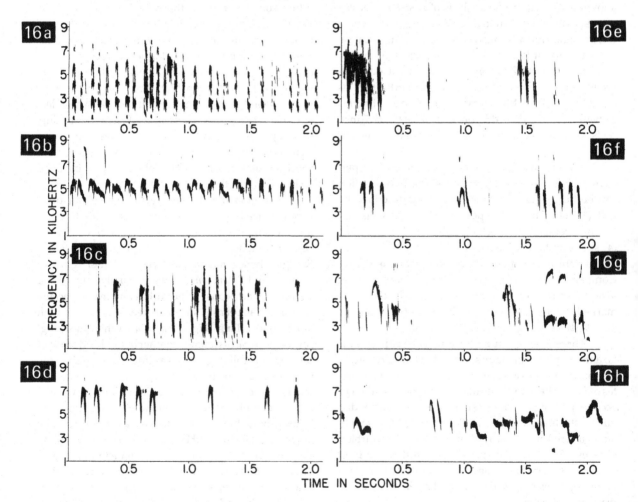

Fig. 17. Vocalisations of *Zosterops chloronothos*. (*a*i) A series of notes uttered by an individual defending *Eugenia dupontii* flowers. (*a*ii) Two birds taking nectar at a flower-head; one shows a tendency to sing warbled notes. (*b*) A sequence of frequency modulated warbled notes uttered by one (possibly both) of two interacting birds. This is the call repeatedly given in response to playback and is heard frequently about nectar sources. (*c*) A series of notes given by two birds at a nectar source, one uttering the 'pit' (notes 3 and 5), the other the bill-snaps and the agonistic notes. (In the background at the end is *Pycnonotus jocosus*.) (*d*i) Non-aggressive foraging (contact) 'pit' notes uttered by one individual foraging alone. (*d*ii) Two birds at a nest-site responding to playback with a 'pit' and two scold notes, the second of which is loud and buzzy. (*e*i) A typical sequence of a flight call showing a series of six 'pit' notes, a single 'pit' note and six more distant 'pit' notes (the flight pattern undulates somewhat in time to the call). (*e*ii) One phrase of a close flight call illustrating the structure of the 'pit' notes. (*f*i) A song phrase showing the fast warbling (frequency modulation) and strong lower harmonic tone typical of such a phrase. (*f*ii) A series of intense 'pit' notes uttered in response to playback of the phrase shown in (*b*) above, showing a wide frequency range in the initial stroke elements. (*g*) A typical song sequence depicting a series of notes with fast frequency modulation, strong fundamental tone and lower harmonic tone with an inter-figure interval approximately twice the length of the phrase. (*h*i) Roosting calls of two individuals uttered at 18.40 hours; they were situated low in the bushes of the dwarf forest at Plaine Champagne. Gill (1971*a*) described a similar call for *Z. b. borbonicus*. (*h*ii) One individual vocalising during nest construction; the mixture of 'pit' notes, bill-snaps and warbled notes closely resembles a low-intensity agonistic call sequence of *Z. olivaceus* in Fig. 16(*g*).

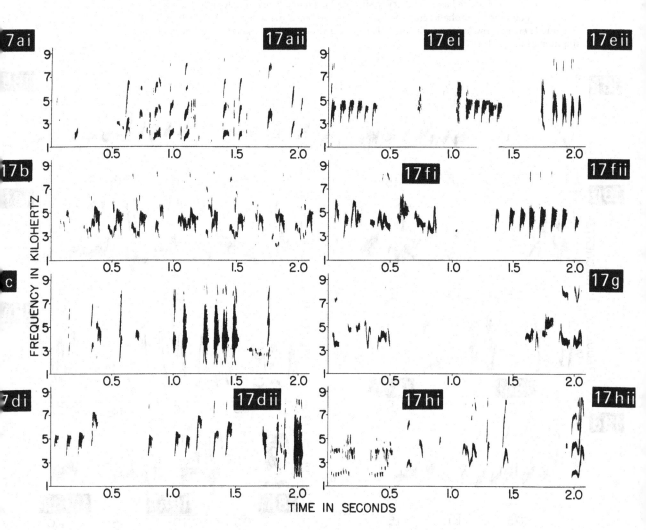

Fig. 18. Vocalisations of *Zosterops borbonicus borbonicus*. (*a*) A series of flocking (contact) notes given by a group of at least three birds, one of which is uttering an occasional warbled note. Shown are an initial downstroke, fast frequency modulation, a strong fundamental tone with (in some notes) lower harmonic tones, terminating with a downstroke. The whole strongly resembles in structure the 'Zosterops note' seen in the vocalisations of *Z. modestus* and many African mainland *Zosterops* spp. (*b*) A series of call notes and warbled (frequency modulated) notes uttered by two birds, possibly a mated pair. Notes 2 and 5 resemble the 'eeee' described by Gill (1971*a*: 43), as does the single note illustrated in the centre of this figure (see *hiii*); in notes 2 and 5 the lower harmonic tone is very evident, but is not seen in Gill's figure nor in (*hiii*) below. (*ci*) Five flocking (contact) notes uttered by two birds, possibly a pair. All the notes show the same noisy quality with the sound spread over a wide frequency range. Notes 4 and 5 show the initial 'chip' element which is so evident in many of the vocalisations of *Z. b. mauritianus* (e.g. Fig. 19*e,f*). (*cii*) The plaintive 'eeee' described by Gill, similar in structure to that shown in (*hi*), followed by two flight call notes from the same bird. (*d*) A series of flocking calls, many of the notes showing a strong lower harmonic tone. (*e*) A series of flight call notes that lack harmonic tones. (*f*) A randomly selected segment taken from a long recording of morning song. Note the strong harmonic tones. This type of song, seemingly derived from the flocking and contact calls (compare with *a*, *b* and *hii*) is interspersed with elaborate phrases containing mimicry, such as that shown in (*g*). (*g*) Portion of song showing an elaborate warbled phrase containing mimicry. (*hi*). The agitated 'eeee' call note (uttered here without an initial 'chip') has a buzzy, scolding quality with sound spread over a wide range of frequencies, and a strong lower harmonic tone. (*hii*). The contact notes of two possibly paired birds, one of which has a tendency to sing warbled notes. The last note (following the two 'Zosterops notes') is an emphatic 'chip'. (*hiii*) The plaintive 'eeee' described by Gill (1971*a*: 43) and included here to show how variable this note is in structure and the presence or absence of the harmonic tone (cf. *b*, *cii* and *hi*).

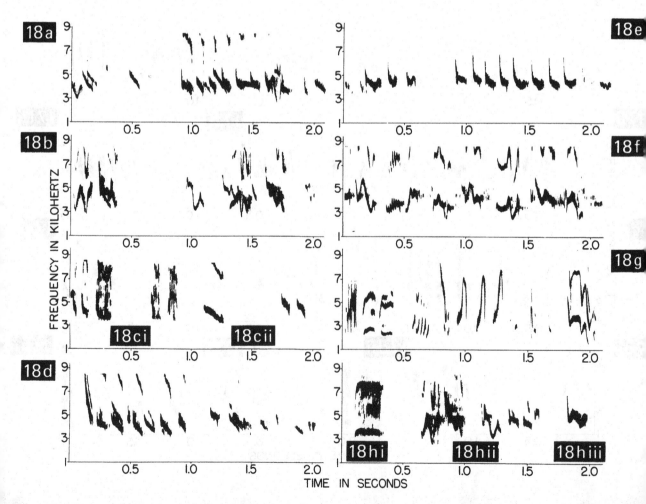

Fig. 19. Vocalisations of *Zosterops borbonicus mauritianus*. (*a*) Part of a sequence of agonistic 'squeak' call notes, showing an extremely complex tonal situation in which the bird is alternating the emphasis between the fundamental tone and the lower harmonic tone (see text for behaviour in response to playback of this call). (*b*) A high-intensity flight call uttered during an interaction with *Z. chloronothos*. The initial six notes were uttered by an approaching individual, and the recorded frequency range is apparently proportional to its varying distance. The abrupt 'pit' notes of *Z. chloronothos* overlap the sixth note, and the *Z. b. mauritianus* notes then fade abruptly. The initial sharp 'pit' element which is a marked feature of *Z. b. mauritianus* vocalisations (giving them an explosive quality) is clearly seen. This flight call was described by Gill (1971*a*: 43) as 'p-tree' or 'p-tee'. (*ci*) Contact (call) note between two birds, possibly a pair with a nearby juvenile. This note has a plaintive quality to it, and although it is similar to the 'eeee' of *Z. b. borbonicus* (Fig. 18*b*, *hiii*), the fundamental tone is suppressed, with the emphasis on the lower harmonic tone. The entire note exhibits complex, fast elements with close, wide-ranging harmonic tones. (*cii*) Contact notes uttered by one of two birds with a clear 'pit' (downstroke) element initiating the first note. The second note is a 'pit'. (*d*) A series of 'pit' notes uttered by a single foraging bird together with a contact note showing the usual initial 'pit' element. (*e*) Low-intensity flocking calls. The initial 'pit' element marks each note except the last, which very probably represents another bird responding to the first, nearer individual. (*fi*) Contact call of agitated individual showing 'pit' note and 'pit' element combined with complex harmonic tones. This may be the 'p-tree' described by Gill (1971*a*: 43). (*fii*) High-intensity flight or flocking call given by an abruptly departing individual. (*g*) Portion of a song sequence uttered by a bird near a nest-site. The tempo is slower, but the form of the notes is very similar to that of the other vocalisations illustrated. Apparently overtones are essentially lacking. (*h*) Section of a song following directly on that illustrated in (*g*), and showing a change to the more elaborate warbled phrases. These phrases, as in *Z. b. borbonicus*, often contain mimicry.

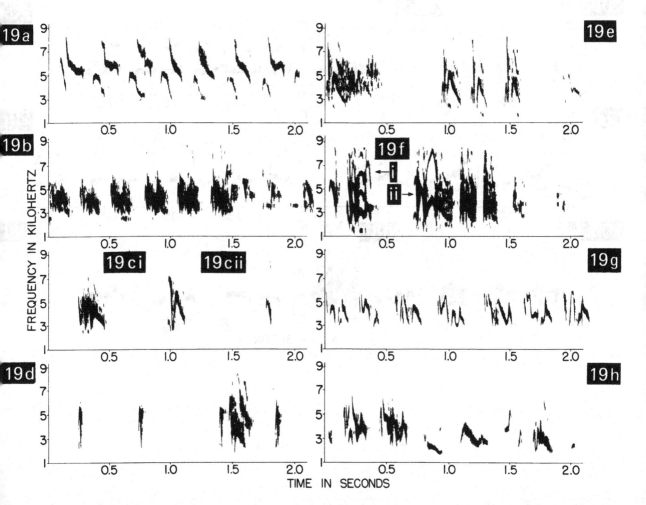

Fig. 20. Vocalisations of all four *Zosterops* taxa shown at 0–16 kHz, taken from selected phrases in order to illustrate the wide frequency range and strong harmonic tones occurring in a number of their calls. (*a*) *Z. chloronothos*. These two sonagrams are a 0–16 kHz version of the vocalisation shown in Fig. 17 (*h*ii), excluding the seventh and eighth notes (short 'pit' notes). (*b*i) *Z. chloronothos* flight call notes. (*b*ii) *Z. chloronothos* song phrase. (*c*) *Z. borbonicus mauritianus*, part of the agonistic 'squeak' call shown in Fig. 19(*a*). (*d*i) *Z. borbonicus mauritianus*, low-intensity flocking call; the initial 'pit' element is clearly illustrated, and resembles some of the structures seen in the vocalisations of *Z. chloronothos*. (*d*ii) *Z. borbonicus mauritianus*, detail of the agitated contact call ('p-tree'). (*e*i) *Z. borbonicus mauritianus*, another version of contact calls ('p-tree'). (*e*ii) *Z. borbonicus borbonicus*, the first six notes of the phrase shown in Fig. 18(*d*). (*f*) *Z. borbonicus borbonicus*, the last section (with the warbled note) of the phrase shown in Fig. 18(*a*). (*g*) *Z. olivaceus*, a section of the mobbing call shown in Fig. 16(*b*). (*h*) *Z. borbonicus borbonicus*, four notes of a portion of the morning song from which Fig. 18(*f*) is also taken.

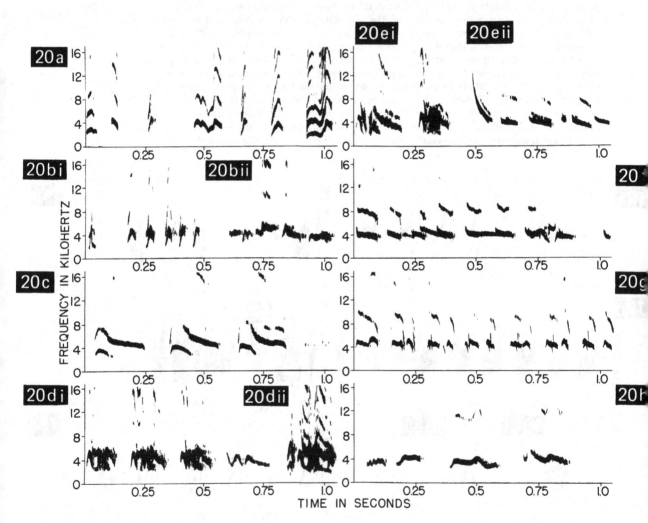

The vocalisations of the grey white-eyes *Z. b. borbonicus* and *Z. b. mauritianus* (Figs. 18, 19) show very marked differences together with certain structural similarities. Indeed, among the vocalisations of white-eyes known to me (i.e. the Mascarene and African mainland species, and *Z. modestus* of the Seychelles), those of *Z. b. mauritianus*, *Z. chloronothos* and *Z. olivaceus* are unique. The flight and flocking call of *Z. b. borbonicus* on Réunion differed considerably in structure from that of *Z. b. mauritianus* on Mauritius. The particular note structure seen in the flight call of *Z. b. borbonicus* (Fig. 18*e*) was similar to those of the African mainland and Seychelles species that I have analysed, and I have come to term this the '*Zosterops* note'. There was considerable variation in the use of the strong lower harmonic tone by both *Z. b. borbonicus* and *Z. b. mauritianus*. In both races it was most marked in certain agonistic, contact and flocking calls; it was also strongly evident in the songs of *Z. b. borbonicus*, but seemed to be absent from some of the latter's flight call notes (Fig. 18*e*). It was not characteristic of the songs of *Z. b. mauritianus*, but many of the latter's vocalisations had a strong harmonic tone with a suppressed fundamental tone. The frequency range was extremely wide (0–16 kHz: Fig. 20) in both races.

The agonistic 'squeak' call of *Z. b. mauritianus* (Fig. 19*a*) was heard at the beginning of the breeding season as pairs formed (and during interactions thereafter); it evoked a very intense response to playback (the birds returned repeatedly to the same place, tails erect, white upper-tail coverts and white rump patches fluffed up and white axillary feathers exposed). This call somewhat resembled the '*Zosterops* note', and showed a strong lower harmonic tone. A similar call was uttered by *Z. b. borbonicus* during intra- and interspecific interactions, and when mobbing *Hypsipetes borbonicus*.

The songs of *Z. b. borbonicus* and *Z. b. mauritianus* are apparently derived from flocking and flight calls, as seems to be the case in a number of mainland African species and *Z. modestus* of Seychelles (personal recordings) and possibly some Australian species such as *Z. lateralis gouldi* (Mees 1969). A bird would fly to a vantage point and then call while remaining stationary. This calling slowed in tempo and then continued into a song with intervals of soft and elaborate warbled phrases, many containing mimicry (Figs. 18*f,g* and 19*g,h*). The early morning and late evening songs ('challenge' songs of Gill 1971*a*) may simply consist of this slow-tempo flocking call modified by a little or much warbling (Fig. 19*g*), or they may be complex, containing mimicry and warbled phrases (Figs. 18*g* and 19*h*). It is possible that this variation is due to individual physiological conditions related to the reproductive cycle. In every case, the most elaborate songs that I recorded were within the nesting area of a pair; indeed this was so characteristic that the elaborate warbling could be used to locate the nest. The morning and evening singing seemed much louder than that heard during the day. However, I have found that individually the song amplitude is much the same, the difference being in the numbers of singing individuals involved. One notable difference between the songs of *Z. b. borbonicus* and *Z. b. mauritianus* (Figs. 18*f*, 20*h* and 19*g*, *h*) was the strong lower harmonic tone evident in the songs of *Z. b. borbonicus*. Although many vocalisations of *Z. b. mauritianus* showed strong harmonic tones, its song phrases did not.

On the strength of the strong vocal differences between *Z. b. borbonicus* and *Z. b. mauritianus* one could argue for greater, perhaps specific separation of these taxa. Alternatively with perhaps greater justification, it should be stressed that white-eyes are fundamentally a close-knit group, and that geographical isolation of conspecific populations can result in moderate vocal differentiation approaching, or even exceeding, that usually characterising distinct species.

Acknowledgements

I would like to thank the numerous people in the Mascarene Islands and elsewhere without whose help this study would not have been possible. I must mention especially the following: in Réunion, Messrs. A. Barau, H. Gruchet and Dr M. Clouet, and the many officials of the Office Nationale des Forêts for general assistance, transport and accommodation and the guides and porters which they so kindly made available to me; in Mauritius, Mr Robert Antoine, Messrs France Staub, A. Gardner, D. Ardill; and in Rodrigues the late Mr Lecordier and his family, as well as many other helpful officials of the Forestry Department. I owe much to Miss Annick Pelicier, my field assistant during 1974, and to Dr and Mrs S. A. Temple, my hosts during both field seasons. The late James Pass and Mrs Pass most generously donated the Stellavox recorder. I thank Dr Lester L. Short and the staff of the Ornithology Department of the American Museum of

Natural History for help in preparing the *Zosterops* species sonagrams, and the manuscript; Dr J. Gulledge and the Cornell Laboratory of Ornithology for the sonagrams of the swiftlet; and especially Professor Paul Bondesen for his truly tremendous effort in preparing all the remaining sonagrams. Dr David Snow's help in editing the manuscript has been invaluable. Mrs M. Lewin's timely loan of a typewriter was crucial. The British Institute of Recorded Sound (British Library of Wildlife Sounds) generously supplied most of the tape that I used. The British Ornithologists' Union substantially funded the project and provided other assistance. I am indebted to Sir Hugh Elliott and Mr Peter Hogg for their constant help and encouragement. Dr Claude Chappuis supplied many recordings of vocalisations of African and European species for comparison with those of the Mascarene birds. Mr H. O. Baker and Mr. C. Hardy of British Airways (Overseas Division) kindly arranged transport of my tapes to and from Mauritius. My thanks are due to Dr Lester Short who, owing to my many months in Africa, has had to do all the proofreading. Dr Diamond has been extremely helpful and I thank him.

4

The ecology of the smaller land-birds of Mauritius

Part II

A.S.CHEKE

THE SURVIVING NATIVE BIRDS OF MAURITIUS

Introduction
General

Already famous in the seventeenth century for the Dodo and the quality of its ebony, Mauritius is biologically the best known of the Mascarene Islands. However, ecological studies of the native biota date only from the 1930s, with Vaughan & Wiehé's pioneering work (1937, 1941) on the forests. The basic natural history of some of the birds has been known for two centuries, but no systematic ecological studies had been done before 1973 when Dr S. Temple and the BOU Expedition began work. Most of the native forest birds, all endemic or shared only with Réunion, had been considered endangered for over a century (E. Newton 1865b), so their study was overdue. The historical background of the abundance and distribution of Mauritian birds is given by Clark (1859), E. Newton (1861b), Carié (1904), Meinertzhagen (1912), Guérin (1940–53), Rountree et al. (1952) and Newton (1958a). Since 1973, numerous publications on the island birds and their conservation have appeared (e.g. Procter & Salm 1975, Cheke 1978a, Jones & Owadally 1982a, and others cited in appropriate places in the text), and F. Staub has written two books on the birds' natural history and identification (in French, 1973a; revised in English, 1976). This paper gives the detailed results of the BOU Expedition's observations.

Expedition personnel were based in Mauritius. I was on the island from late September 1973 to late February 1975 except for absences in December 1973, late April to mid-May, June, August and November

1974, and mid-February 1975. I also made short trips to Rodrigues on several occasions. I re-visited the island briefly in October 1978 under the auspices of the Jersey Wildlife Preservation Trust. Official expedition field assistants were J.-M. Vinson (JMV) (August 1973) and Ms J. Shopland (JS) (late September 1974 to early January 1975), and the following also visited Mauritius during the expedition and contributed substantially to the data presented here: Dr M. de L. Brooke (January 1974; MdeLB), Dr R. A. Cheke (August 1974; RAC), Dr R. E. Ashcroft (September–December 1974; REA) and M. A. Peirce (November 1974; MAP). A substantial amount of data that was to come from Dr S. A. Temple (SAT) was unfortunately never made available, though I have included such observations of his as I have to hand. J. F. M. Horne's observations on behaviour in connection with her sound recordings are given in Chapter 3. I have also been able to add unpublished observations from F. R. G. Rountree [1923–51; FRGR], F. B. Gill [1964; FBG], A. Forbes-Watson [1971; AFW], J. C. Lawley [1977–9; JCL] and C. G. Jones [1979–83; CGJ]. These sources are cited by their initials in the text to distinguish their contributions from published material.

An abridged version of the section on the Mauritius Fody has appeared elsewhere (Cheke 1983a).

The island

Mauritius (Maps 1 and 2) is an island of 720 square miles (1865 km²) (Ramdin 1969, Venkatasamy 1971), straddling latitude 20° S and longitude 57° 30′ E. It is 840 km east of central Madagascar, 164 km east-north-east of Réunion and 574 km west of Rodrigues (Admiralty Chart 4702). The relief is largely mild, the land rising from low plains in the north and east to a plateau around 610 m in the south-west. This smooth surface is broken in places by striking mountains, not high but precipitous, notably the Moka Range (Le Pouce, Pieter Both, etc.), Corps de Garde, Rempart–Trois Mamelles and the Montagnes Bambous. The Piton du Milieu is a classic exposed volcanic plug. In the south-west the plateau descends very steeply towards the sea on its western and southern flank, and includes a deeply dissected river bed, the Black River Gorge. Most of the remaining native forest is found in this south-western corner on the edges of the plateau, the gorge and along the southern scarp. The highest point (Black River Peak, 828 m) is still clothed in

indigenous forest and overlooks the gorge of the same name.

The island is largely surrounded by a fringing reef enclosing a lagoon mostly about ½–1 km wide, but extending out to 6 km or so in Grand Port bay (south-east). There are several included islets along the east coast, and also in the wide lagoon around Le Morne (south-west). Stretches of the west and south coasts lack any lagoon. A shelf 2–8 km wide surrounds the island, extending north-westwards from Cap Malheureux for up to 28 km, and includes the biologically important islets of Round Is. Serpent Is., Flat Is., and Gunner's Quoin (Coin de Mire).

The island is entirely volcanic (Simpson 1951, Arlidge & Wong 1975); the conspicuous, steep-sided mountains are the remains of a huge collapsed shield volcano and are the oldest rocks on the island (8–5.5 million years BP: McDougall & Chamalaun 1969). The rest of the island's surface is composed of more recent lavas, mostly degraded into soils of various kinds (Chazal & Baissac 1950, 1953, Simpson 1951, Ramdin 1969). In some areas there are more or less unaltered lavas; one such is the Plaine des Roches where the plantations at Bras d'Eau have proved a refuge for the native flycatcher *Terpsiphone bourbonnensis*. The most recent lavas date from between 700 000 and 179 000 years ago (McDougall & Chamalaun 1969, Arlidge & Wong 1975, Montaggioni 1975). Small areas of raised coral-sand beach of Pleistocene age exist in some coastal areas (Montaggioni 1975).

Climate

Mauritius lies near the southern limit of the tropics and has a seasonal climate dominated by the 'South-east Trades', more correctly east-south-east, which blow throughout the year (Padya 1972, 1984 from which the rest of this account of climate is condensed). The winds are stronger and more persistent in winter (May–August) than in summer, when they are interrupted during cyclonic weather.

Tropical depressions, often becoming violent storms (cyclones), develop during the summer in the oceanic belt between latitudes 5° and 30° S (Padya 1976). While only a few of these pass over the island, the passage of a depression within 300 km or so usually produces heavy rain; this component of summer rainfall is important but very variable from season to season (Vaughan & Wiehé 1937). Cyclones also very rarely develop in summer. Major cyclones are so

Map 1. Mauritius, showing the distribution of surviving native vegetation.
Notes:

1. The rectangle outlined by a dashed line in the south-west corner is shown in greater detail, enlarged, in Map 3.

2. The fine dotted line from Curepipe through Piton du Milieu and Montagne Lagrave and ending near Kanaka shows the approximate boundary of degraded native forest in 1950, before major clearances for tea and conifer plantations had taken place (based on the map Mauritius – Land Use DOS(Misc.)293 (1960), vegetation map in Vaughan & Wiehe (1937), and the literature (see text)).

3. Grid numbers from the 1 : 100 000 map of Mauritius DMS Y682 (1971), 10 km squares.

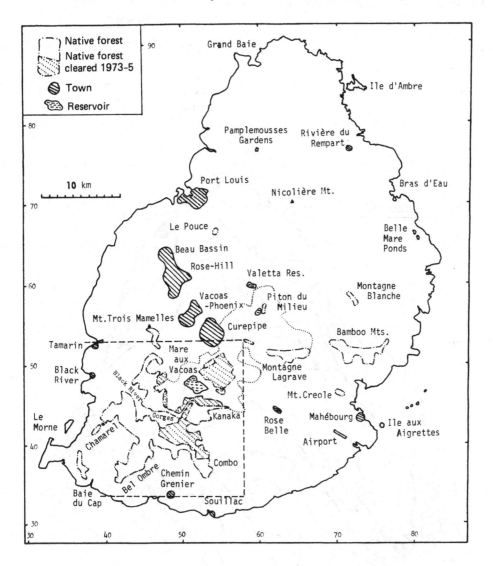

Map 2. Mauritius, showing topography, fringing reef, continental shelf, annual rainfall and mid-summer temperature (January).

Notes:

1. Contours in metres (......), isohyets in millimetres (— —), isobars in °C (— · —) and depth in fathoms (1.83 m) (— —). The 2200 feet (670 m) contour has been added to indicate the very high ground.

2. The rainfall and temperature stations used in Fig. 1 and Table 2 respectively are marked as follows. Rainfall: B, Bras D'Eau; P, Plaine Champagne (left dot); Petrin (right dot). Temperature: C, Curepipe Gardens; W, Fort William.

3. The mid-winter temperature can be reckoned approximately by taking the mapped 22 °C isotherm as 16 °C, 24 °C as 18 °C, and 26 °C as 20 °C.

4. Topography after Venkatasamy (1971) and DOS series map Y682 Mauritius, climatic data after Padya (1972) and sea depth after Admiralty Chart No. 711 (this last is incorrectly traced in Venkatasamy 1971). Open triangles denote mountain peaks.

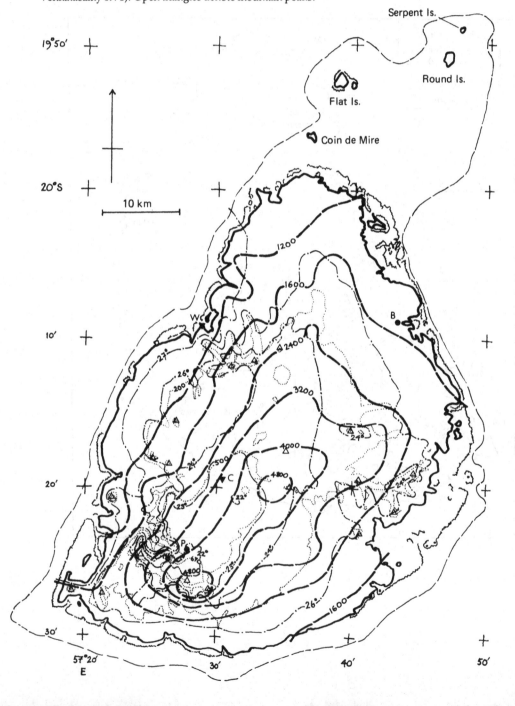

Fig. 1. Annual rainfall pattern in the mountain forests and the coastal lowlands of Mauritius.

Notes:

1. The localities (see Map 2 for their location) are:

	Symbol	Altitude	Annual rainfall (1931–60)	No. of rainy days (1931–60)
Petrin	Open bars	2150 ft (655 m)	3896 mm	230
Bras d'Eau	Stippled	30 ft (9 m)	1505 mm	180

2. To show annual variation the rainfall records for 1973 (triangles), 1974 (Squares) and 1975 (circles) are included. Bras d'Eau figures are the filled symbols, Plaine Champagne (Petrin not available) the open ones. The Plaine Champagne collector is at 2300 ft (701 m) (annual average not available). Total rainfall/no. of rainy days during 1973–5 were 82/91% (Plaine Champagne) and 77/79% (Bras d'Eau) of the 30-year average.

3. Data from Padya (1972; 1931–60 averages) and M. Lee Man Yan (*in litt.*; 1973–5 figures).

4. The diagram runs from June to May so as not to split the period of the principal breeding season of Mauritian birds, indicated by a horizontal bar at the top of the figure.

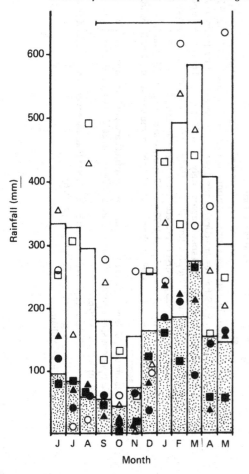

severe that the entire biota has had to adapt to them. The native forest is particularly resistant to leaf loss, branch loss and windthrow (Vaughan & Wiehé 1937, Brouard 1967, pers. obs.). Introduced plants are generally very badly affected; whole plantations of pines and eucalypts are destroyed in severe storms, though *Araucaria, Cryptomeria* and many exotic hardwoods are more resistant (Brouard 1963, 1967). The native fauna survives the actual storms much better than do introduced birds, but some species suffer from the aftermath, when flowers and fruit have been stripped from even native trees (Cheke 1975b, Jones 1980a). The severest cyclones of the past century are summarised in Table 1.

There is no truly dry season but little rain falls in the lowland coastal areas in winter. In the highlands there is still half as much rain in the winter months as in summer. Annual averages for Bras d'Eau (north-east coast) and Petrin (representing the native forest area) are given in Fig. 1, and the distribution of annual rainfall in Map 2. In no part of the island is a large part of every day spent in cloud as on Réunion (Chapter 6), but the Plaine Champagne–Les Mares–Montagne Cocotte area is often enveloped in cloud, and the upper parts of Montagne Cocotte (771 m) are enshrouded frequently enough to have developed a mossy forest (Vaughan & Wiehé 1937). The annual range of the daily mean temperature is 6–7 deg C, the daily range 7–8 deg C. The average temperature on the plateau in the summer at 2000 ft (610 m) is about 3 deg C cooler than on the coast. The monthly means for Curepipe Gardens (nearest the forest) and Fort William (nearest to Bras d'Eau in site and altitude) are given in Table 2, and the mid-summer isotherms on Map 2.

Relative humidity is very high on the south-western plateau, Curepipe Gardens recording over 90% daily average throughout the year, and over 95% during October–February. By contrast in the lowlands the annual range is up to 25%, from 69% (September) to 86% (April) at Tamarin. It is estimated (Padya 1972, maps 5: 1–4) that in the south-western forests the humidity never drops below 95%, though Vaughan & Wiehé (1941) gave a range of 80–90% with a mean of 87% for Macabé.

Climatic change over the last century in Mauritius and some other Indian Ocean islands was examined by Stoddart & Walsh (1979). Fluctuations in rainfall and patterns of cyclones were considerable, but no clear pattern emerged.

Table 1. *Severe cyclones in Mauritius since 1857*[a]

Date	Max. gust[b]	Max. SWS[c]	Intensity[d]	Duration (hours)	Source
11–16 February 1861	[98][e]	55	2	c. 144	Walter (1914a)[f], Roch & Newton (1862–3)
12 March 1868	[110]	62.5	2	60	Walter (1914a), Pike (1873)
28–29 April 1892	[121][g]	76[i]	3[h]	34	Walter (1914a), Jerningham (1892), Padya (1976)
5 March 1931	[90]	51[m]	2	70+	Padya (1972, 1976), Brouard (1967)
16–17 January 1945	[113]	63	2	c. 30	Padya (1976)
19 January 1960 (Alix)	120+	49[n]	2(–3)	c. 24	Brouard (1960), Padya (1976)
28 February 1960 (Carol)[j]	148	82	3	16	Brouard (1960), Padya (1972, 1976)
28 February 1962 (Jenny)[j]	122	80	3	3–4	Padya (1976)
6–7 February 1975 (Gervaise)	[115][k]	(65)[k]	2(–3)	20	Cheke (1975b, unpubl. obs.), Padya (1976)
22–23 January 1980 (Claudette)[l]	125	67	2	12	Valadon (1980), Jones (1980a)

[a] Only those cyclones particularly remembered for their damage are included; others of equivalent parameters but less damaging are omitted. The duration of the storm is one of the most important factors for small birds. To retain comparability all figures are from Pamplemousses; some others are added in notes.

[b] Wind speed is given in miles per hour, the form normally used in Mauritius, and that on which my cyclone intensity scale (Cheke 1979a) is based. To convert to metres per second multiply by 0.447.

[c] SWS, sustained wind speed over 1 hour.

[d]
Scale	Wind speed	Description (Padya 1972)
1	40–55	Moderate depression
2	55–70	Severe depression
3	70–85	Intense cyclone
4	85+	Intense cyclone (recorded only in Rodrigues)

[e] Figures in square brackets are estimated from the regression in Padya (1972, Fig. 8.7).

[f] Walters (1914a) SWS figures are consistently 1.36 times those given by Padya (1972), presumably scaled down on more recent information. I have recalculated the 1861 and 1868 speeds accordingly, though Padya (1976) gave 65 mph for the 1868 storm.

[g] Jerningham's figure; 132 mph estimated from Padya (1972), who later (1976) suggested c. 150 mph as more likely.

[h] The 1892 cyclone was the first to exceed scale 2 after wind speed records began in 1857; in the interval there were six more scale 2 storms: 1857, 1863 (2), 1874, 1876 and 1879. Unlisted post-1892 scale 2 storms were in: 1902, 1964 (and probably 1951, 1960 (Dec.), estimated from Davy (1971); information for 1914–39 is inadequate). Earlier in the nineteenth century the storms of 1806, 1818, 1824 and 1831 were very damaging (Padya 1976).

[i] 79 according to Padya (1976).

[j] Maximum gusts elsewhere in the island of 159 mph (Carol) and 162 mph (Jenny) were recorded (Padya 1972, 1976).

[k] Pamplemousses figures estimated (Padya 1976). SWS at Vacoas (max. gust 118) was 59 mph, but 82 was recorded at Mon Desert (Moka), with a gust of 174 mph (Padya 1976).

[l] Gusts of 137 mph at Fort William and SWS of 78 at Plaisance were also recorded (Valadon 1980).

[m] 62 at Vacoas, gusting to 115–120 (Padya 1976).

[n] 81 at Fort William.

Table 2. *Average monthly temperatures at two contrasting localities in Mauritius (from Padya 1972)*

	June	July	Aug.	Sept.	Oct.	Nov.	Dec.	Jan.	Feb.	Mar.	Apr.	May
Curepipe Gardens, 564 m: 1951–60												
Absolute maxima (°)	24.4	24.3	24.0	24.7	26.5	29.0	27.7	28.5	29.0	29.1	28.0	27.5
Mean maxima	21.9	20.1	20.7	21.6	22.3	24.4	25.7	26.0	26.3	26.3	25.3	24.3
Monthly means	17.9	16.8	16.9	17.5	18.3	19.7	21.1	22.1	22.1	22.1	21.1	20.1
Mean minima	13.8	13.5	13.2	13.5	14.4	15.1	16.6	18.2	18.0	17.8	17.0	16.0
Absolute minima	10.0	7.8	8.7	7.9	10.5	11.1	11.1	13.9	12.6	13.1	12.9	11.8
Mean daily range	8.1	6.6	7.5	8.1	7.9	9.3	9.1	7.8	8.3	8.5	8.3	8.3
Fort William, 6 m: 1961–70												
Absolute maxima (°)	30.5	29.9	29.1	29.9	31.4	33.0	34.1	33.2	35.1	35.6	34.3	32.8
Mean maxima	26.5	25.3	25.1	26.8	27.8	29.6	30.1	30.7	31.4	30.5	29.6	27.9
Monthly means	23.0	22.2	21.8	22.9	23.9	25.4	26.7	27.2	27.8	27.8	26.3	24.3
Mean minima	19.5	19.1	18.5	19.1	19.9	21.2	23.2	23.7	24.2	24.0	22.9	20.7
Absolute minima	12.9	12.2	12.5	12.7	14.1	15.0	15.8	15.5	18.5	18.0	17.3	12.5
Mean daily range	7.0	6.2	6.6	7.7	7.9	8.4	6.9	7.0	7.2	6.5	6.7	7.2

Note: The figures run from June to May to avoid having to split the period covered by the principal breeding season of Mauritian birds (September–March).

Vegetation

Of the original vegetation of the island, only remnants of the poorer upland forests and some at the middle levels remain; the lowland forests, the better highland ones, the dry palm savannahs of the northern plains are all long gone (Vaughan & Wiehé 1937, Brouard 1963; see also Chapter 1).

More or less degraded areas of climax forest, badly invaded by exotics, exist in the south-western uplands and the Lagrave and Bambous ranges (Maps 1 and 3). Even the best parts of this area (apart from parts of Bel Ombre) consist of the relatively low-canopied (20 m or less) forest that developed on the very poor soils of the south-western sector (Vaughan & Wiehé 1937), and are clearly a pale shadow of the magnificent forests that once clothed the central plains (Brouard 1963; Chapter 1).

A more or less unaltered part of this remnant forest was studied in some detail by Vaughan & Wiehé (1937, 1941). It is a rain forest with an emergent top canopy dominated by trees of the family Sapotaceae (*Mimusops, Labourdonnaisia* and *Sideroxylon* spp.), *Canarium paniculatum, Elaeodendron orientale* and *Nuxia verticillata*. It has a rich tree flora (40 spp. in 18 families) forming a dense secondary canopy, but a rather open, deeply shaded understorey now somewhat invaded by the exotic Strawberry Guava *Psidium cattleianum*. There are no saplings of some of the important canopy trees and inadequate regeneration of most species to maintain the forest in the long term. No long-term study has yet been done on the dynamics of this forest, nor of the reasons for its failure to regenerate, though there has been no shortage of theories: seed destruction by monkeys (Vaughan & Wiehé 1937, 1941, Owadally 1980), competitive exclusion by guava etc. (Owadally 1973, 1980), browsing by introduced herbivores (Thompson 1880, Cheke 1978a), grubbing by pigs (Owadally 1980), and the need for seeds to pass through the gut of now extinct animals before germination can take place (e.g. Temple 1977b; see Chapter 1). No doubt all of these factors, acting together, have been responsible for reducing regeneration well below replacement (Chapter 1).

The forest in some parts of the south-west is dominated by *Calophyllum tacamahaca* to the partial exclusion of the more mixed forest described above (Vaughan & Wiehé 1941); the southern scarp from Montagne Savane to upper Bel Ombre is typical. The dwarf forest at Petrin and Plaine Champagne,

regarded by Vaughan & Wiehé (1937, 1941) as an arrested successional stage (subclimax), is dominated by *Sideroxylon puberulum* (Sapotaceae). Native forest similar to parts of Macabé still occurs on the south-eastern slopes of the Montagnes Bambous, and regeneration and the general condition of the forest seemed much better there (pers. obs. in 1974). Most of these native forests are thoroughly penetrated by Strawberry Guava, privet *Ligustrum robustum* and the vigorous scrambling bramble *Rubus mollucanus* (= *R. roridus* auct.) (Owadally 1973, 1980). Where forest has been cut or blown down the guava forms extremely dense stands (with a canopy at 3–5 m) in which little or nothing else grows (Owadally 1973, pers. obs.). Other exotic trees that have penetrated the forest, such as *Litsea monopetala, L. glutinosa* and *Harungana madagascariensis* are non-pathogenic and indeed provide an extra food source for native birds, as does the ubiquitous shade-tolerant sub-shrub *Ardisia crenata*.

A characteristic of the plateau rain forests is a very high density of epiphytes (vascular, and mosses and lichens) which clothe trunks and branches at all levels (Vaughan & Wiehé 1937, pers. obs.). These provide habitats for large numbers of arthropods on which several of the birds feed. In contrast the guava has smooth flaky bark virtually without epiphytes, and its copious fruits appeal more to introduced mammals than to any of the surviving native birds.

Macabé–Brise Fer, the area chosen by botanists as the best remaining example of aboriginal upland forest (Vaughan & Wiehé 1937, Vaughan 1968), and popularly considered until recently the key nature reserve for wildlife, is devoid of Olive White-eyes *Zosterops chloronothos* and supports only low densities of the other native passerines apart from the Mauritius Cuckoo-shrike *Coracina typica* and the Grey White-eye *Z. borbonicus*.

Native forest of a drier type still exists on the western scarp slopes linking the Tamarin Gorge with Trois-Mamelles and Rempart mountains. This is dominated by *Elaeodendron orientale* and ebonies *Diospyros* spp. (pers. obs.), though some of the trees more typical of the plateau re-appear in the wetter patch under Trois Mamelles itself (Vaughan & Wiehé 1937). Similar forest occurs on scarp slopes from Tamarin Gorge to Chamarel and Piton du Fouge, and in the lowest part of Bel Ombre. It has a thinner canopy and fewer epiphytes than the rain forest, and is often badly invaded by the strangling vine *Hiptage benghalensis*.

Map 3. The forests of south-western Mauritius. The map is adapted using personal knowledge from the 1:100 000 map Mauritius Land Use December 1965, UNDP/FAO (1968). The kilometre grid is from the 1:000 000 map of Mauritius DMS Y682 (1971).

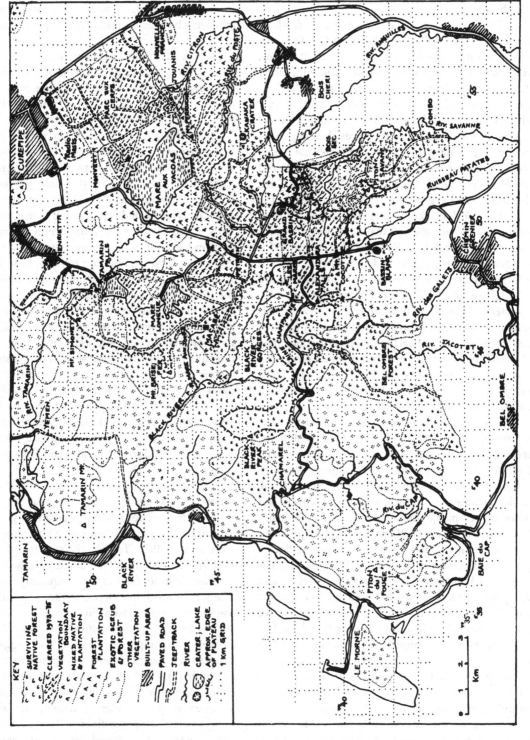

KEY

- SURVIVING NATIVE FOREST
- CLEARED 1975–'8?
- VEGETATION BOUNDARY
- MIXED NATIVE & PLANTATION
- FOREST PLANTATION
- EXOTIC SCRUB & FOREST
- OTHER VEGETATION
- BUILT-UP AREA
- PAVED ROAD
- JEEP TRACK
- RIVER
- CRATER; LAKE
- APPROX. EDGE OF PLATEAU
- 1 Km GRID

Km 0 1 2 3

Apart from the Grey White-eye, native birds are usually absent.

The logged-out plateau forests were left fallow for decades and until largely cleared in the 1960s and 1970s their mixed native and exotic composition provided acceptable habitat for the native passerine birds. Patches of the degraded forest remain at Monvert, Kanaka crater, Concession Jouanis, between Black River Peak and Chamarel (Map 3), and in small remnants elsewhere. In some places the logged-out forests (e.g. Kanaka and much of the Savane range) were interplanted with mixed exotic hardwoods and conifers, providing excellent bird habitat. This degraded forest vegetation was still extensive at the end of the 1939–45 war (Map 1) but was cleared from 1950 onwards for pine and tea plantations (Brouard 1963, 1966, Tilbrook 1968, Roy 1969). In the early 1970s a make-work programme independent of the old-established forestry service and financed by the World Bank (Owadally 1975, 1976, Procter & Salm 1975; Chapter 5) cleared the remaining large areas of this secondary forest, and also the unique marshy forest of Les Mares, making critical the already precarious state of native bird populations. The Kanaka block (the crater and some tracts by the Rivière du Poste excepted) was not finally lost until the late 1970s and Crown Land Gouly in 1980 (CGJ).

The vegetation of the rest of the island consists, in the uplands, of sugar, conifer and tea plantations, and scrubby areas dominated by guava and Travellers' Palm *Ravenala madagascariensis* with grassy clearings managed for deer hunting. Cane-fields dominate the lowland landscape, but there are some *Eucalyptus* and other timber plantations, scrubby grassland maintained for deer, exotic thorn scrub, dry grassland (around Port Louis) and the coastal 'pas géometriques' planted with *Casuarina equisetifolia*. The built-up areas of Plaines Wilhelms (Beau Bassin to Curepipe) and many smaller towns provide gardens well planted with fruit and ornamental trees, and in some places there are extensive Mango *Mangifera indica* orchards.

Of the native forest birds, only the abundant Grey White-eye, and to a much lesser extent the Paradise Flycatcher *Terpsiphone bourbonnensis*, have adapted to exotic vegetation; the rest are confined to shrunken remnants of the native upland forests. I believe some birds would adapt to exotic mixed hardwood plantations (if imaginatively managed), as they were beginning to do in the Kanaka area before it

was clear-felled (Cheke 1978a). A similar mixed forest still exists in the upper part of the Combo area (Map 3) and could be used for testing this hypothesis.

Biotic influences on the avifauna

The history and effects of introduced animals on the forests and the fauna have been considered in Chapter 1; here I will briefly summarise the direct effects on the avifauna.

Mauritian birds evolved in an environment where the only predators were two hawks (one now extinct), but are now subject to tree-climbing monkeys *Macaca fascicularis* and Ship Rats *Rattus rattus*, which both have a reputation for robbing nests (though the evidence is largely indirect). No native ground-nesting birds survived the early introduction of pigs; even most exotic ground-nesters were unable to resist the mongoose *Herpestes edwardsii* introduced in the early 1900s (Carié 1916). Nest-robbing is also attributed to the introduced Mynah *Acridotheres tristis* and bulbul *Pycnonotus jocosus*, the latter being blamed for heavy losses to white-eyes *Zosterops* spp. This bulbul, and the Cardinal *Foudia madagascariensis*, may compete directly with related native species and restrict their distribution, abundance or both. Mynahs and Asian parakeets *Pstittacula krameri* compete with the native *P. echo* for nest-sites (Chapter 5). Avian malaria and its vector (Peirce *et al.* 1977) and a disease resembling birdpox (see below; flycatcher and Grey White-eye) have been introduced, though their pathogenicity is not known.

The effect of man on the birds directly, in addition to his effect through habitat destruction, becomes more serious as bird populations get smaller. Two of the eight or nine remaining Mauritius Kestrels were shot in 1973 (Temple 1977a), I found traces of bird-lime on a Merle in 1974, and my study colonies of swiftlets *Collocalia francica* were frequently vandalised. The forests are subject to tremendous disturbance in the deer-hunting season (June–August; Owadally & Bützler 1972), from ride-cutting, clearing and burning as well as beating and shooting, and trigger-happy hunters in unguarded moments admit shooting passing birds (two Pink Pigeons *Nesoenas mayeri* in 1974, SAT; pigeons disturbed in 1975, Newlands 1975; Chapter 5). One endemic animal, the flying-fox *Pteropus niger*, already extinct on Réunion, is still subject to legal and organised shooting (Cheke & Dahl 1981). Finally the forests, reserves included, are daily penetrated by hundreds of woodcutters supposedly only cutting exotic weeds (i.e. privet and guava); however they not only take any other species in the 1–3 inch (25–75 mm) size range but also, by their noise and numbers, cause significant disturbance to nesting birds.

Species accounts
Mascarene Swiftlet *Collocalia (Aerodramus) francica (Gmelin 1788); zirondelle (zirondel)*
Affinities

Swiftlets of the genus *Collocalia* (*sensu lato*) are typical of the Oriental and Pacific Ocean biogeographical areas, and adept at colonising islands (Medway 1966, Brooke 1970). The Mascarene *francica*, and also *elaphra* of the Seychelles, are close to the South Indian and Ceylonese *C. unicolor*, the nearest species geographically (Medway 1966, Medway & Pye 1977). The Seychelles Swiftlet *elaphra* has been reduced to a race of *francica* (Proctor 1972), but differs sufficiently to be retained as a separate species (Table 3), especially in a subgenus of such critical specific characters as *Aerodramus*. Brooke (1972) upgraded *Aerodramus* to full generic level, a treatment followed by some recent authors (e.g. Medway & Pye 1977), but given the overall homogeneity of the group (Medway 1966) I prefer to treat all cave swiftlets as belonging to one genus, *Collocalia*.

General habits and behaviour

The Mascarene bird is a typical small swift, seen singly or in groups (rarely large assemblies) hawking over any kind of terrain. Other species are ignored, though one was once seen mobbing a kestrel (CGJ). The birds generally forage at elevations over 20 m above the ground, rarely feeding lower except over freshwater lakes. Feeding height is determined by the weather; birds often feed high (out of sight) when it is fine, but come down if it is cloudy, feeding only below the cloud-base. They land only for roosting and nesting, on inaccessible ledges on the roofs of caves and lava tunnels. The birds are noisy around nesting caves, especially in the evening, flying around in small parties, screaming. Inside the cave, when birds are actively flying around, there is the continuous rattle of their echolocation calls (Clark 1859, Chapter 3), which were also once heard out in the open (CGJ). The sexes are not externally distinguishable.

Distribution and numbers

Swiftlets can be seen anywhere on the island, and it is rare to spend a day in the field without seeing any. The birds are very mobile, and move around the island in response to weather conditions; only around roosting or nesting caves can one be certain of seeing birds.

Caves suitable for nesting are found over much of the island. They are generally lava tunnels, large gaseous bubbles in the original lava flows that formed cavities when the lava cooled. Such tunnels are particularly frequent in the youngest lavas (Chazal & Baissac 1950), which cover most of the island (map in Venkatasamy 1971). The entrance is typically a hole in the ground giving access through the roof of the cavity; these can often be detected by seeing swifts diving out of the sky apparently straight into the ground. In other cases a section of the roof has fallen in creating a trough with access to the cave at one or both ends.

All nesting caves I have been able to trace are marked on Map 4. Many of these are traditional sites: Clark (1859) singled out the sites at Roches Noires and Souffleur as particularly well known and containing very large numbers of birds; the former was certainly still active in 1973 (Staub 1973*a*, SAT). Likewise Pike (1873) visited the interconnecting cave system of Pointe aux Caves/Petite Rivière where Horne (Chapter 3) recorded echolocating swifts in 1976. I have not been able to locate the Caverne des Hirondelles at Souillac mentioned by Guérin (1940–53). The site referred to by Medway (1966) as that where Pike collected some nests, 'Black River', was presumably Pointe aux Caves which is in Black River administrative district. The locality where Edward Newton collected his clutches (preserved in Cambridge) is not known.

I made no attempt either to do a full survey of nesting sites or to census the birds. The well-known caves already mentioned have been said each to shelter thousands of birds, but recent counts at Petite Rivière revealed only between 5 and 30 nests and evidence of persistent vandalism (CGJ). The two caves that I studied in detail contained only very small colonies with a maximum of 10–15 active nests in each. As Staub (1973*a*) stated, swift caves are scattered throughout the island, but no one has drawn together local knowledge. Map 4 is only a bare beginning; a proper survey is required in view of increasing vandalism in nesting caves (see below). The total Swiftlet population in the mid-1970s was probably several

Table 3. *Characters of* Collocalia francica *and* C. elaphra

Character	C. francica (Mauritius and Réunion)[a]	C. elaphra (Seychelles)	Source
Wing length, mm (range; mean ± S.E.)	108–117; 112.09 ± 0.21 (86)[b]	119–124.5; 120.8 ± 0.32 (63)	Gill (1971b), Macdonald (1979)
Weight, g	7.8–13.5; 9.2–9.6[c]	9.5–14.3; 10.7 (13)	Gill (1971b)[d]
Wing shape (in flight)	Narrow at base; secondaries clearly shorter than tertials	Broad at base, more like a swallow (Hirundinidae); secondaries nearly as long as tertials	Pers. obs.
Eye (in life)	Small; relatively shallow lores 'mask'	Large; deeply recessed lores 'mask'	Pers. obs.
Activity period	Diurnal and crepuscular; activity stops within 35 min of darkness falling	Diurnal, crepuscular and active at night. Active until "long after the last light has disappeared"	Macdonald (1979)
Clutch	Two, rarely one	Invariably one	Proctor (1972), Macdonald (1979), Watson (1979), pers. obs.
Nest structure and dispersion	Nests spread out or clustered, infrequently sharing a common bracket. Attachment: vertical to horizontal surfaces[e]	Nests always tightly clumped, usually two or three sharing a common bracket. Attachment: horizontal to steeply angled surfaces (cave roofs)	Proctor (1972), Macdonald (1979), pers. obs.
Principal food brought to young[f]	Diptera (various families)	Flying ants (Hymenoptera: Formicoidea)	Macdonald (1979)

[a] Data from this study, in Mauritius.

[b] Gill (1971b) gave 108–116.5; 112.7 (13) for Mauritius and 108–112.5; 110.1 (8) for Réunion. See Chapter 9.

[c] Gill (1971b) gave 8.4–10.4; 9.0 (12) for Mauritius and 8.5–11.4; 9.0 (8) for Réunion. See Chapter 9.

[d] Macdonald (1979) only gave fledging weights of nestlings: 11–12 g.

[e] Clustered nests and attachment to horizontal or near-horizontal surfaces is more frequent in Réunion than Mauritius (Jadin & Billiet 1979; see Chapter 6).

[f] See Table 4.

thousand pairs, but this should not give rise to complacency as their nesting sites are so vulnerable.

Daily cycle

The Swiftlets return to the nesting caves every evening to roost in pairs at the nest-site whether or not they are breeding. Although they can navigate in the dark by echolocation, their evening return to the cave is closely related to the time of nightfall, usually within 20 minutes and never more than 35 minutes after dark (Fig. 2). By contrast the Seychelles Swiftlet is active long after dark (Macdonald 1979), and many Oriental

species fly throughout the night (Medway 1962*a*, 1969); the Seychelles bird has a markedly larger eye than the Mascarene Swiftlet, very noticeable when the birds are handled (pers. obs.).

Diet

The only information on diet comes from two food boluses I collected from birds bringing food to their young; these have been analysed by Dr John Farrow of the California State University, Long Beach, the psocids being identified by Turner (1979). Boluses have also been collected in Réunion (Jadin & Billiet

Map 4. Known breeding sites of the Mascarene Swiftlet and the Mascarene Swallow in Mauritius, 1973–5.

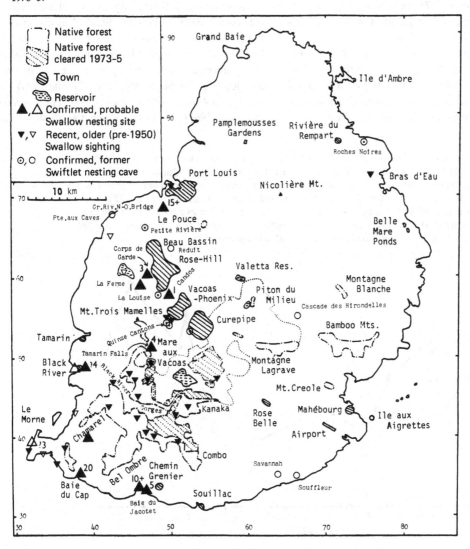

1977) and in the Seychelles (Macdonald 1979). The contents of boluses from these three sources (Table 4) show a striking similarity between the Mauritius and Réunion samples, and that the Seychelles Swiftlet provides its young with a quite different diet. The sizes of the prey items in Mauritius were remarkably constant: all were in the range 1.6–3.2 mm, except for a group of flying ants (5.2–8.2 mm) and a single anthomyid fly (4.2 mm). There is no information on the adults' own diet.

Breeding

The Swiftlet's breeding cycle was studied in two caves, about 250 m apart, at Quinze Cantons off the Vacoas–Mare aux Vacoas road. The caves were visited regularly between October 1973 and the end of January 1975, weekly where possible; gaps in the observations mark my absences from Mauritius.

Less frequently, birds were caught in mistnets set up in the caves, and ringed. This established that the colonies in each cave were independent; birds from one colony rarely visited the other. As will be seen, the two colonies, despite their proximity, were

never in synchrony with each other, although the birds within each cave nested more or less simultaneously. Synchrony is unusual in the genus (Medway 1969).

Unlike all the other Mauritian land-birds the Swiftlet cannot on present evidence (*contra* Staub 1973*a*) be said to have an annual cycle (Fig. 3). During the 16 months of observation there were no eggs or young in either cave only in April and May 1974, but even then nests were being built, so it seems likely that the birds might have bred had conditions been right. Newton (1958*a*) saw birds collecting nest material in June and July and found them breeding in October, and prematurely concluded that they nested all the year round. There is little other pertinent past material on seasonality; Edward Newton's clutches were all collected in early November, Staub's nesting observations (1973*a*) at La Louise cave were also made then, Pike (1873) found eggs at Pointe aux Caves in July, and FBG collected birds with gonads at all stages of development in September and October 1964. The south-east Asian species have long but regular seasons (Medway 1969), but in the Seychelles all stages from

Fig. 2. Roosting times of Swiftlets in relation to last light. Roosting time (black diamonds) is taken as the time the last bird was trapped flying in the cave (records for both caves at Quinze Cantons combined). Last light (open circles) is taken as the end of 'civil twilight' as defined in the *Smithsonian Meteorological Tables* (List 1951), from which it was calculated; at 20° S it is always between 22 and 25 minutes after sunset. Official Mauritius time (GMT + 4 hours) is 10 minutes ahead of local sun time for which the 'civil twilight' was calculated; however as subjective darkness (from my notes) coincided closely with the calculated figures, I have left them unadjusted in the graph.

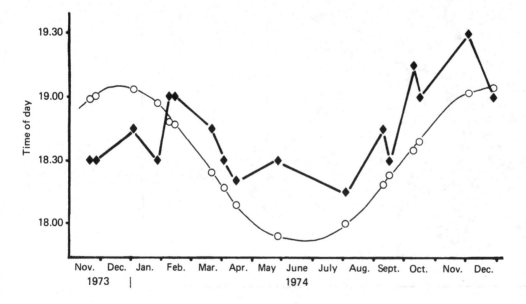

eggs to large young have been found in every month from June to November (pers. obs., Procter 1972, Macdonald 1979) and by implication (Watson 1979) throughout the year.

The nest is normally an individual bracket built on a more or less vertical part of a cave roof. The bulk of the nest is vegetable material – invariably *Usnea* lichen in the Quinze Cantons caves, though for nests in caves near the sea *Casuarina* 'needles' are used (Staub 1973*a*) and inland occasionally also conifer needles (Medway 1966). Staub (1976) gave a photograph of a nest with an adult sitting. Nest material is collected with and carried in the beak (cf. some Oriental species which use their feet: Medway & Pye 1977). The struc-

Table 4. *Food brought to nestling swiftlets in the Mascarenes and Seychelles*

| Food item identified | | Mauritius[a] 1st bolus | | | 2nd bolus | | | Réunion[b] | | Seychelles[c] |
Order	Family	A[d]	B	%	A	B	%	B	%	%
Psocoptera	Caeciliidae	2	5 } 7		1	1 } 20		} 1	1.5	} —
'	Hemipsocidae	1	2		1	1				
Hemiptera	Anthocoridae	1	1							
(Heteroptera)	Miridae	1	1							
(Homoptera)	Delphacidae	2	10 } 18		—	2 } 20		} 8	12.3	} 1
	Cicadellidae	2	4							
	Aphidae	—	3							
Coleoptera	Hydrophilidae	—	1							
	Staphylinidae	—	1							
	Nitidulidae	3	6 } 15			} —		} 8 }	12.3	} 0.5
	Chrysomelidae	—	2					2		
	Scolytidae	4	6							
Diptera	Chironomidae	—	19		—	3		10		
(Nematocera)	Simuliidae	—	—		—	—		1		
	Mycetophilidae	—	—		—	—		5		
(Cyclorrhapha)	Drosophilidae	—	23		—	3		3		
	Anthomyiidae	—	1		—	—		—		
	Pipunculidae	—	—		—	—		1		
	Lauxaniidae	—	— } 53		—	— } 60		5 }	57	} 4.5
	Phoridae	—	—		—	—		5		
	Muscidae	—	—		—	—		2		
	Acalyptera indet.	—	—		—	—		3		
(Brachycera)	Empididae	—	—		—	—		2		
(?)	Curtonotidae	—	11		—	—		—		
(indet.)	—	—	1		—	—		—		
Hymenoptera	Braconidae	—	1					6		
	Chalcidoidea	—	— } 7			} —		2 }	12.3	} 92
	Formicidae	2	6					—		++[e]
Araneae		—				—		2	3	<0.2
unidentified		—				—		1	1.5	—
Totals		—	104	100	—	10	100	65	99.9	98

[a] This study.
[b] Jadin & Billiet (1979); three boluses combined.
[c] Macdonald (1979); eight boluses combined.
[d] Columns marked A represent numbers of species, B numbers of individuals. Species were not distinguished by Jadin & Billiet (1979) on Réunion, and Macdonald (1979) only gave a bar chart, crudely drawn, for the diet of *C. elaphra*.
[e] The sign ++ indicates that the majority of the order were in the family indicated, but precise figures are not available.

Fig. 3. Numbers of active Swiftlet nests in two cave colonies at Quinze Cantons, near Vacoas.
Notes:

1. Open bars show eggs, dotted bars, young.

2. Estimated egg dates (and numbers of nests) are indicated by inverted 'v's at the top of the columns; the cave(s) were not visited on those dates, but the data were derived by calculating backwards from later visits.

3. Open triangles on occasions when no eggs or young were found indicate that active nest-building was observed. An underline with no other information indicates a visit during which no nesting activity was seen.

4. The simple horizontal bars above the nest data indicate periods of egg-laying. The 'interrupted tube' shows the period during which birds in each cave were moulting. The dashed line for September–November for cave F refers to a single bird only; no others were moulting there.

5. Capture–recapture population estimates (filled symbols) and the minimum number of birds known to be alive (dotted circles) are shown for each cave against a scale on the right-hand side (see also Table 5). The estimates for each cave have been corrected for the presence of birds known to be from the other cave.

6. Nest destruction by vandalism in the caves is shown by zigzag lines.

7. Cyclone Gervaise is indicated by 'C' with an arrow.

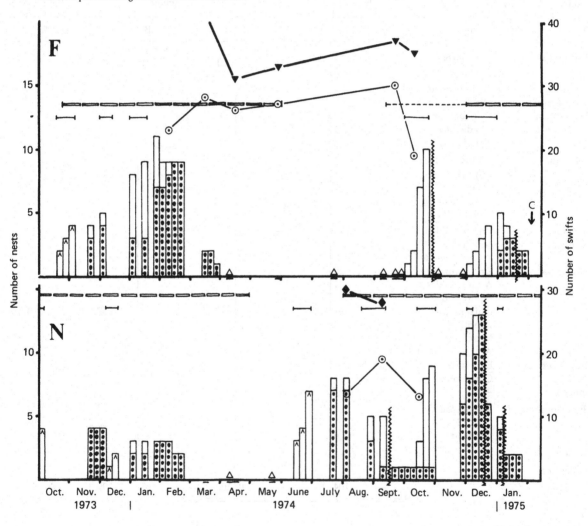

ture is attached to the rock, and the vegetable matter stuck together, with the salivary cement typical of the genus (Medway 1962*b*, 1966). Milne-Edwards & Grandidier (1879) described and figured the swollen salivary glands of 'Salanganes de l'Ile de France' collected in the breeding season, but as they considered various Asian forms to belong to the same species, it is far from certain that they obtained their material from Mauritius.

The birds take a minimum of 4 weeks from starting a fresh nest to laying; 4–6 weeks is the normal interval between layings after nests have been destroyed, but after a non-breeding interval the start-up is slower (Fig. 3). The re-laying interval (minimum 27 days, mean 35–50 days) is shorter than for Malaysian species (Medway 1969), where the minimum is 35 days and the mean as long as 87 days.

The clutch is usually two. Most clutches of one were suspect (incomplete, deserted) but three (6% of acceptable clutches) were genuine, and two more may have been. References in the literature to clutches of three or more (Milne-Edwards & Grandidier 1879 pl. 72, Guérin 1940–53, even Staub 1973*a*) stem from Pollen's misinformation (Schlegel & Pollen 1868; see Chapter 6). The eggs are chalky white (Staub 1976; pers. obs.).

I have approximate incubation periods from only three nests (21–23 days); it is certainly always more than 19 and less than 26 days (five more nests). This is shorter than the incubation period of the larger Seychelles Swiftlet (25–30 days: Macdonald 1979), but about the same as *C. esculenta* and *C. salangana* in Sarawak (Medway 1962*a*).

The young hatch more or less synchronously, and remain in the nest for at least 6 weeks. The shortest fledging period I recorded was between 35 and 43 days, the longest between 62 and 73, the uncertainty being due to the intervals between visits. The average is 45–55 days, about the same as *C. salangana* (45–48 days: Medway 1962*a*), but longer than *C. esculenta* (42 days: *ibid.*) and *C. elaphra* (42 days: Macdonald 1979). Often at the end of the nesting period only one young was present at the last visit prior to the nest being found empty, suggesting that fledging was not simultaneous. Also when young were very large the numbers in adjacent nests sometimes varied from visit to visit, suggesting trial flights after which birds returned to the wrong nest. The young are presumably fed by both parents, but I have

no direct evidence. Stages of growth were as described by Medway (1962*a*); Macdonald (1979) gave growth curves for the Seychelles Swiftlet.

When nests were not vandalised by humans, nesting success was very high: 84% of clutches hatched (38/44), and of those which hatched all but two (34/36) reared young to fledgling, though two nests still with young when cyclone Gervaise struck may not have survived. However a total of 24 nests were destroyed by vandals, 14 at the egg stage and 10 with young. Natural egg loss took the form of eggs vanishing between visits, presumably ejected by the parents (cf. Medway 1962*a*); there was no evidence of predation by cave-dwelling insects as occurs in Southeast Asia (*ibid.*), although Staub (1973*a*, 1976) recorded cockroaches as living in cracks near the nests. The only total failures during the nestling period were a case of young found dead in the nest and one of remains below an undamaged nest (possible vandalism), though in three other nests one of the two young died when at least half-grown. This level of nesting success is comparable to that recorded by Medway (1962*a*) in Sarawak, but in the Seychelles Macdonald (1979) found a much higher egg loss.

The interval between successive clutches in a nest, presumed to be laid by the same pair, was about 3 weeks (four nests). It seems that undisturbed pairs would normally rear two, and possibly sometimes three, broods in a cycle. Two broods would take 6½ months, three about 9½ months.

Population dynamics of the colonies at Quinze Cantons
Some idea of the dynamics of Mauritian Swiftlet colonies can be gained from considering together the trapping data and nest observations in the two caves at Quinze Cantons.

The cave further from the access road ('F') had a low ceiling only about 10 ft (3 m) high and it was possible to set a mistnet well within the chamber, in complete darkness, such that it lay directly across the path of birds flying between their nests and the only external opening. This resulted in a high trap rate and frequent recaptures of the same birds: of the 34 birds caught and ringed from November 1973 to May 1974, 29 (85%) were subsequently recaptured at least once before the end of October. After May, although catches in September and October were large, only seven new birds were caught. Two of the seven were birds of the year originally marked in the nearer cave

('N'), and it seems likely that by then all or nearly all the adults were marked. Capture–recapture analysis (Table 5) suggests a population increasing from 31 to 36–37 from April to September; 26, increasing to 30, birds were definitely known to be present during this period. There seems to have been a genuine decrease in the population (emigration?) between March and April (note the divergence between $\hat{M}_i$ and Σu_i at this date: Table 5), though the earlier population estimates are too high due to greater trappability of unmarked birds.

The other cave had three entrances all of which the birds used, was not totally dark inside, and had a much higher ceiling (18–20 ft; 5.5–6 m). Due to its height it was impossible to intercept the birds with a net in the main chamber of the cave, and the only place to trap them was just inside one of the entrances, the others being unsuitable for nets. The net was in a semi-lit place and the birds could often see it, avoid it, and use another entrance or exit. The recapture rate was poor in contrast with cave F (11 of 31 (35%) ringed November to May recaptured up to December) although the total number trapped was similar. Capture–recapture estimates of 30–31 birds were obtained during August and September, but the number of resident birds was certainly fewer as known cave F birds were also present on those days (Table 5).

Comparing the number of adults present with the numbers of active nests (Fig. 3) it is clear that not all birds necessarily breed in any given cycle. In cave N there were never more than four active nests between October 1973 and March 1974, although 16 adults were trapped and more were no doubt present; indeed there were about 15 additional nests under construction in early October, but they were apparently never used. In cave F, the first recorded activity (October–December 1973) was of only four nests, but by late January there were 11 active nests. Likewise in the following cycle in cave N only about three fifths (8) of the pairs present ($\leqslant 14$) nested in June–August, whereas 13 (all ?) did so later; in cave F the first attempt was by two thirds of the birds (10/15), but only five repeated after the vandalism in late October (though some birds are known to have been killed).

The recapture rate of birds ringed as juveniles and nestlings was about one third that of adults. Free-flying young (c. 8) were caught only in cave N and nestlings (12) ringed only in cave F. From November 1973 to early February 1974 juveniles were distinguished from adults by being in fresh plumage and not moulting; by mid-February, as the first adults completed the moult, they were no longer reliably separable. Two birds, one definitely the other probably juvenile, ringed in cave N moved to and settled in cave F; only one of the original eight was subsequently retrapped at N. Likewise of 12 nestlings ringed at F in February, two were later trapped at N, whereas only one was seen again at F. Four juveniles were caught at N in November before any young had fledged there (or at F). Evidently the majority of young move to other colonies after fledging. Only five adults moved between caves (four F to N, one N to F). All but one of these birds were subsequently retrapped at their original cave. The evidently more 'open' nature of the population at cave N partly explains the low recapture rate there, and this biases upwards the population estimate for residents.

The study did not last long enough to give accurate information on mortality, but did show that adult survival was at least 61% (17 alive in September 1974 of 22 alive 7 months earlier, cave F); the true figure is probably much higher (compare $\hat{M}_i$ and Σu_i in Table 5). Human vandalism is, however, a persistent and possibly serious threat to this species. While nest collection for soup, once a minor industry (Clark 1859), no longer takes place to any extent, the caves are a resort for all kinds of activity from illegal gambling to plain vandalism. The last takes the form of smoking birds out with burning tyres as well as direct destruction of nests. France Staub (pers. comm.) was once nearly trapped in the cave at La Louise when miscreants lit a fire in the entrance (since blocked off with rubbish: CGJ) while he and his party were far within the cave. Likewise I suspect my visits to the caves at Quinze Cantons stimulated interest in local youths potentially given to vandalism; this was despite excellent relations with people in the nearest houses. Some caves need to be secured against intruders (Cheke 1978a); until this is done the species should be considered 'vulnerable' in the sense used in the *Red Data Books* (Vincent 1966, King 1978–9). Mauritian Swiftlets also commonly harbour an unidentified non-protozoal blood parasite of unknown pathogenicity (Peirce *et al.* 1977).

According to local informants cave N filled up completely with water during cyclone Gervaise (6–7 February 1975: Cheke 1975b), during which 16 inches of rain (406 mm) fell in 24 hours (pers. obs.). The roofs

Table 5. *Trapping data and capture–recapture analysis for samples at two colonies of Xantus' auklets.* (title partly illegible)

Quantity	Notation^a	Cave	1973 Nov.	1973 Dec.	1974 Jan.	Feb.	Mar.	Apr.	May	June	July	Aug.	Sep.	Oct.	Nov.	Dec.
Total catch^{b,c}	n_i	F	5		11	13	15	10	11				23	17		4
		N	8		3	9		6	11			6	16	9		10
No. released	s_i	F	5		11	13	15	10	11				23	16		4
		N	8		3	9		5	11			6	16	9		10
No. of new birds^d	u_i	F	5		11	9	7	1	2				4	1		1
		N	8		3	9		5	2			4	10	7		5
No. of retraps	m_i	F	–		0	4	8	9	9				19	16		3
		N	–		0	0		0	9			2	6	2		5
No. of retraps also caught later	m_{it}^e	F	–		0	3	6	8	8				12	1		–
		N	–		0	0		0	8			1	3	0		–
No. of already marked birds not caught but seen subsequently	z_i	F	–		4	10	13	16	16				7	2		–
		N	–		4	7		7	16			7	3	4		–
No. of birds released that were subsequently recaptured	r_i	F	4		10	11	12	9	10				11	1		–
		N	4		2	1		2				2	3	1		–
Estimated population^{f,b} $(n_i+1)(m_{it}+1+z_i)/(m_{it}+1)$	$\hat{N}_i$	F	–		–	49	43	31	33				37	36		–
		N	–		–	–	–	–				31	30	–		–
Minimum known to be alive	n_i+z_i	F	5		15	23	28	26	27				30	19		4
		N	8		7	15		12	27			13	19	13		10
Estimated population of previously marked birds^g $(m_i+1)(m_{it}+1+z_i)/(m_{it}+1)$	$\hat{M}_i$	F	–		5	17	26	28	28				31	34		–
		N	–		–	–		–				13	12	–		–
Cumulative total of marked birds^b	Σu_i	F	–		5	16	25	32	33				36	40		40
		N	–		8	11		20	33			25	29	39		46

^{a} After Cormack (1968).

^{b} Cave F: includes 1 bird (Oct.) known to reside in cave N.
Cave N: includes 4 birds (Aug., 1; Sept., 2; Dec., 1) known to reside in cave F.

^{c} Omitting known juveniles (cave N: Nov., 4; Jan., 3; Feb., 1). These and the nestlings ringed in cave F are not scored until their first recapture, when they are classed as though unmarked. ^{d} Birds marked in one cave are counted as new when caught in the other. ^{e} Notation first used here.

^{f} By Manly & Parr's method (1968), modified for small samples on the principle suggested by Cormack (1968). Jolly's estimator (1965: (n_i/m_i) $[(z_i s_i/r_i) + m_i]$) was also calculated, giving very similar results for cave F, but excessively high estimates for cave N, even when corrected for small samples. The estimators for survival rate are not sufficiently sensitive to give meaningful results in a population of this size with very high survivorship between trapping occasions. ^{g} By extension from Manly & Parr's method (1968), using the small sample correction.

of both caves are below local ground level, over which a continuous sheet of water was flowing. Some 20 Swiftlets, were, however, flying around cave N in the evening of 9 February, so they must have survived submersion, presumably in air pockets in the cave roof. The cave was still flooded and impossible to explore; I left the island before I could make further observations.

Moult

As in other species of *Collocalia* (Medway 1962*b*, Macdonald 1979), moult in the Mascarene Swiftlet overlaps extensively with breeding. Moult was not synchronised in the two colonies (Fig. 4).

In 1973 most birds in cave N apparently began moulting in early October, though at least one must have started in August. The cave F birds were on average a month later, in keeping with their breeding cycle. In 1974 the cave N birds were poorly synchronised, some starting in August, others still not moulting in early December. No moulting birds were caught in the last big catch at cave F (9 October), though one bird, just beginning to moult in mid-September, had progressed rather slowly to one third complete on 24 December.

Moult normally takes 6–7 months, though it is possible that some birds complete it in a little over 5 months. This is comparable with the large Asian *C.*

Fig. 4. Primary moult in Mascarene and Seychelles Swiftlets.
Notes:
    1. Data are from this study (Mauritius) and, ringed with dotted line, from Macdonald (1979; Seychelles).
    2. The progress of moult is given as a 'moult score' (Newton 1966), where a value from 0 to 5 is given to each primary feather according to its stage of development (old = 0; new, fully grown = 5), and the figures for each feather summed. Lines join birds caught more than once during moult, dotted when one or other of the captures was just before moult began or just after it finished.
    3. Mean scores for each of the two caves for each trapping occasion are given by inverted triangles (F) and diamonds (N).

    4. Each bird is represented by a dot, except for non-moulting birds during September and October 1974. The actual figures were:
        F: Sept., 19; Oct., 14
        N: Sept., 12; Oct., 4
    5. The data for the Seychelles species *C. elaphra* are grouped by month and given for the highest number primary in moult, which I have converted to approximate equivalent moult scores. Two birds I caught on 20 November 1976 had scores of 0 and 3.

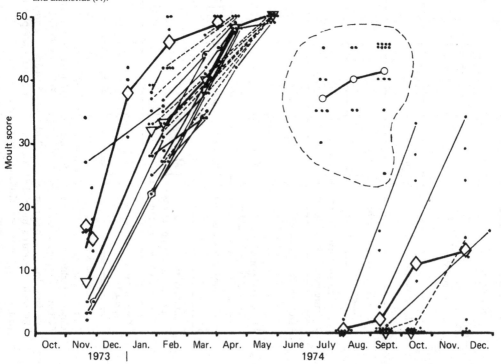

maxima (7 months: Medway 1962*b*); however, *C. elaphra* in the Seychelles appears to moult much more slowly, with only about half the birds in a colony in moult at any one time (Macdonald 1979).

Weights and measurements

No seasonal trend in weight was discernible. Males are not, *contra* Gill (1971*b*) longer-winged than females. Feather-wear shortened wing feathers by 1.2–1.6 mm (1.1–1.4%) annually. Full details are given in Chapter 9.

Mascarene Swallow *Phedina borbonica* (*Gmelin 1789*); 'zirondelle' (*zироñdel*)

Affinities

The nominate race of Mauritius and Réunion, *P. b. borbonica*, is larger and darker than the Malagasy *P. b. madagascariensis* (Hartlaub 1877, Berlioz 1946, Hall & Moreau 1970). Brooke (1972) considered *Phedina borbonica* a relict of an early radiation of swallows prior to the emergence of *Hirundo*, and that its nearest relative *P. brazzae*, confined to West-central Africa (Hall & Moreau 1970), differs sufficiently to be separated generically as *Phedinopsis*; Hall & Moreau (1970), on the other hand, saw no reason to keep either form out of *Hirundo*.

I prefer to use the English name 'swallow' rather than 'martin' to avoid the confusion that results when the latter name is used on islands where the same word is current as the local name of the mynah *Acridotheres tristis*.

General habits, habitat, behaviour and feeding

The Mascarene bird is a typical swallow, but has a somewhat heavy flight, and in the hand is a very solid bird compared with Palaearctic species. Except around breeding sites they are usually seen singly, hawking over reservoirs or along cliffs in the uplands, and, more commonly, amongst *Casuarina* trees or along the shore by the coast. Kestrels are often mobbed (Chapter 5), but other birds are ignored. Mascarene Swallows readily perch on the ground, especially on rocks by the shore (Staub 1973*a*); at Baie du Jacotet and Baie du Cap they sometimes congregated in little groups on the sand, and they are often seen on electric wires. There is no information on their diet; feeding activity peaks shortly before dusk (CGJ). The sexes are not externally distinguishable.

Distribution and numbers

Although Swallows can be seen hunting anywhere in the south-western part of the island, breeding colonies are concentrated on the coastal strip round the west and south from Port Louis to Souillac, and on inland cliffs in the same sector (pers. obs. and Staub 1973*a*, 1976; Map 4). Staub (*locs. cit.*) cited hearsay reports of nesting on fishing boats moored in the bay at Black River. The expedition has only one record outside the south and west: a single bird at Bras d'Eau on 27 October 1973. The distribution survey was not exhaustive and many more breeding sites must exist. The past distribution of the Swallow was apparently similar (Map 4); the colony at Baie du Cap was noted by Edward Newton (1861*b*), that at Beau Champ (= Baie du Jacotet) by Meinertzhagen (1912), and the Corps de Garde one by Carié (1904).

No attempt was made to census this species, but there is enough information to give some idea of the population. The biggest known colony (perhaps 20 pairs in 1973–4) is at Baie du Cap, and presumably provides the birds always to be seen at the summit of the Piton du Fouge. The colonies at Jacotet and under the nearby bridge over the Rivière des Galets contained 10–12 and 5–6 pairs respectively in 1974. There were about five pairs at Chamarel Falls and rather more at the Grand Rivière Nord-Ouest bridge, but all the other colonies in 1973–5 were of fewer pairs; the old Corps de Garde group numbered only three pairs in 1973–74, but Carié's remarks (1904) suggest it may once have been much bigger. The total of birds in known sites was only about 70–75 pairs; unrecorded sites might double this figure, but the total population is undoubtedly very small, as it is also in Réunion (Chapter 6).

Breeding

Staub (1973*a*) gave the breeding season as September to October, but at Baie du Cap birds were seen carrying nesting material on 22 November 1973; nesting had finished by early January in 1974 and 1975. Outside the breeding season the birds left the sites (except perhaps for roosting); from early January to mid-August the colonies were deserted. On 1 September 1974 the birds were back on site. A sample collected on 12 September 1964 in the Black River Gorges contained birds both with and without enlarged gonads (FBG). In 1974 breeding activity apparently did not start at the south coast sites until mid-October, though in the previous year there were young in the nest by 4 October.

The nest is an untidy pile of dry grass or *Casuarina* branchlets, with a few feathers, on a ledge on a cliff or building (Staub 1973*a*, 1976); Staub (*locs. cit.*) illustrated a bird nesting on an air-conditioner on a verandah. The clutch is recorded as two, the eggs being white spotted with brown (Staub 1973*a*). There is no information on incubation or fledging periods; according to Staub (1973*a*) both sexes feed the nestlings. The young are fed by parents after fledging (pers. obs.), but for how long is not known.

Moult

At Black River and Baie du Cap birds in early wing-moult have been seen in late December (1981, CGJ) and early January (1974, 1983); in 1974 birds at the Corps de Garde were two thirds moulted by 29 January. The birds at Tamarin Falls were half-way through moult on 5 January 1975, suggesting that they finished nesting earlier there than on the south coast. Two specimens in Cambridge collected in early April are in fresh plumage.

Mortality and disease

The population dynamics of the Swallow are unknown, but it appears to suffer heavy mortality during some cyclones. E. Newton (1863: 340, 1888; Roch & Newton 1862–3: 270) reported that a severe 6-day-long cyclone in February 1861 almost wiped the species out in Mauritius. After the storm he saw no birds until June 1862, and thereafter only a handful during the 16 years he remained on the island. By the time of the next report (Carié 1904) the bird was considered tolerably common but very local, notwithstanding the disastrous (but short) cyclone of 1892 (see Macmillan 1914). There have been no reports of this kind of mortality in recent cyclones; indeed Newlands (1975) reported up to 40 birds at Black River soon after Gervaise. Barré & Barau (1982) noted that the Réunion population was disturbed by cyclone Hyacinthe in January 1980 and by heavy rainfall in December the same year.

Swallows in Mauritius harbour an endemic trypanosome (Peirce *et al.* 1977) whose pathogenicity is unknown.

Mauritius Cuckoo-shrike *Coracina typica* (Hartlaub 1965); *merle cuisinier (merl kwizinye)*
Affinities

The Mauritius Cuckoo-shrike is very closely allied to *C. newtoni* of Réunion, but has no particular affinity to the Malagasy *C. cinerea* (Benson 1971). Benson suggested an Asian rather than African origin for both Malagasy and Mascarene Birds, but from different source species, and his conclusions from museum specimens are supported by the quite different voices of the Mascarene and Madagascar forms (pers. obs. on Mohéli, Comores).

General habits and behaviour

Mauritius Cuckoo-shrikes live solitarily or in pairs in the forest canopy. They appear to keep to well-defined territories throughout the year, although in winter and the early part of the breeding season they associate sporadically with parties of foraging Grey White-eyes passing through their range (Gill 1971*b*, JMV, CGJ; not seen by me). Grey White-eyes have also been seen mobbing a female cuckoo-shrike (JCL; January), and being chased off by one (JMV; August). I once saw a Cuckoo-shrike displaced by a Merle, and CGJ saw one chased by a mynah. Human intruders are sometimes mobbed, and JS saw an apparent distraction display on 31 December 1974: the male of a pair, the female having quietly vanished, hopped conspicuously around in a tree holding one wing stiff against the body, fluttering the other, spreading the tail sideways and chattering noisily. In August pairs of Cuckoo-shrikes start feeding together in very close association and interactions with birds from neighbouring territories are frequent. The arrival of a neighbouring male or pair is followed by all the birds chasing vigorously around the forest or above it. The residents then settle back to their feeding activities, and the visitor(s) return whence they came. Mynahs sometimes imitate the song, well enough to mislead an untrained observer (CGJ).

The species is strongly dimorphic, the males grey and black above, white below, the females rufous and about 2.5% longer in the wing (Benson 1971), but these features are not reflected in the behaviour of the two sexes.

Distribution, numbers and density

Past and present distribution. Cuckoo-shrikes were confined to the native forests in the south-west of Mauritius in 1973–5 (Map 5); since then, the Kanaka area birds have lost most of their habitat, and it is doubtful whether the now very isolated birds in Concession Jouanis could survive. CGJ has seen birds in lower Bel Ombre.

Information on past distribution is limited.

E. Newton's statement (in Pollen 1866) that the species was then "found in every part of the island where the original forest is left" is borne out by nineteenth-century specimen records outside the present range. Newton collected several birds at Vacoas (now a conurbation) and Bois Sec (now mostly under plantations) between 1860 and 1872 (specimens in Cambridge), while Bouton shot one at Montagne Blanche in 1829 (Oustalet 1897); this ridge is now capped with secondary scrub interspersed with a few native species. At the turn of the century the birds still occurred in the Forest Side suburb of Curepipe (Carié 1904). Guérin (1940–53) reported a bird heard singing at Chamarel in 1937, and, also in the 1930s, FRGR saw Cuckoo-shrikes in the Montagnes Bambous (inland of Anse Jonchée), down the Savane range to Jurançon (now under sugar) and in the Tamarin Gorge.

By the 1950s the species seems to have been restricted to the same general area it now occupies; Newton (1958a) "never saw it north of the Mare aux Vacoas" and FRGR's post-war sightings [1949–51] were all within the present range. However, the Piton du Milieu and Midlands areas may have held populations until they were cleared for tea in the 1950s and 1960s, as may the Parc-aux-Cerfs block until 1974. Apparently suitable habitat in the Montages Bambous seems now to lack Cuckoo-shrikes (CGJ; May 1984), though JMV heard what appeared to be a Cuckoo-shrike call there in 1973.

Population. The Cuckoo-shrike was written off as doomed earlier this century (Carié 1921, d'Emmerez 1928), but this gloom was misplaced. Although considered "local" and "rare" in the 1950s (Rountree *et al.*

Map 5. Distribution and density of the Mauritius Cuckoo-shrike, 1973–5. (The nature reserve boundaries on this and subsequent maps, from Procter & Salm (1975), are approximate.) Area covered by map is rectangle outlined by dashed line on Map 1.

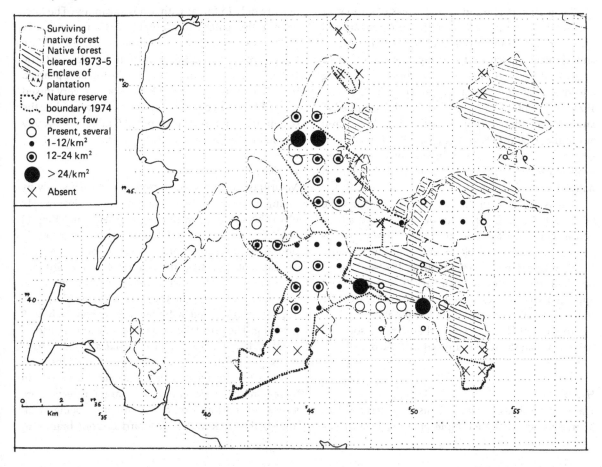

1952, Newton 1958*a*) no survey was done until 1973–5, by which time habitat destruction had further reduced its range. It was cited as "very rare" and given two stars (considerable anxiety as to its status) in the original bird *Red Data book* (Vincent 1966).

During 1973–5 the species was censused by the BOU Expedition in accessible areas of forest. Cuckoo-shrikes have a very distinctive song (Chapter 3) and are fairly vocal, especially in the breeding season, males being audible up to $\frac{1}{4}$ km away in the forest, or $\frac{1}{2}$ km across the canopy. Several visits are necessary to detect all the territories so some birds will have been missed in less well-covered areas. Preliminary results of these surveys given by Temple (1975, 1976*c*, King 1978–9) suggested a population of around 100 pairs, but full analysis shows it was nearer double this: 114 territories were located and allowing for inaccessible areas and those only partly covered, I estimate the total population in 1974–5 to have been about 210–220 pairs.

Density. Density varied from 2–4 territories/km^2 on Plaine Champagne and in other areas of scrubby or degraded forest, up to 28/km^2 in the best areas around Brise Fer, at Alexandra Falls and between Les Mares and Montagne Savane. The average density in native forest on the plateau was 16–22 territories/km^2 (giving a territory size of 4.5–6 ha; *cf.* 6–8 ha of Temple *et al.* 1974), but below about 1500 ft (457 m) this dropped to around 9/km^2 (Bel Ombre) before the birds' lower limit at around 1300 ft (396 m). The density in the mixed forests of Kanaka, where broadleaved evergreen exotics were interplanted with a selective logged-out native forest, was 7/km^2.

It is likely, as in the case of the Fody and the Olive White-eye, that the densities at Alexandra Falls and around Montagne Savane (Map 5) were abnormally high due to displacement of birds formerly living in Les Mares, clear-cut during 1972–4. When I revisited these sites in October 1978 the density appeared to be only about half that seen in 1974–5. The stable population in the now reduced habitat (including the loss of Kanaka) might therefore be some 30 pairs less than the total estimated in 1974–5, i.e. 180–190 pairs.

Habitat

The Cuckoo-shrike's habitat is usually described as 'native forest', though Temple (*in* King 1978–9) has

pointed out that they may also be seen in adjacent exotic vegetation. The expedition's surveys confirmed that the highest densities occurred only in fairly unaltered native forest, above about 1500 ft (457 m). Below that the forest is drier and warmer (Map 15) and the density falls off rapidly, no birds being found in 1973–5 below about 1300 ft, even in the well-preserved forest at Bel Ombre, although CGJ has seen them there as low as 800 ft (240 m). On the wetter high plateau, however, at over 1500 ft fairly heavily altered or degraded forest may be occupied. The area around the Kanaka crater held good numbers of birds in 1974–5. Commercial timber was selectively felled in this area in the early part of the century (Chapter 1) and the remnants of the native vegetation interplanted with commercial evergreens principally *Eugenia ventenatii*, *Cinnamomum camphora*, *Cryptomeria japonica* and *Eucalyptus* spp. This mixture clearly provided reasonable habitat for Cuckoo-shrikes and other endemic species, but the area has since been largely clear-cut (CGJ, 1982).

The Cuckoo-shrike is arboreal and reaches high density only in forest with a closed canopy. The low, open, rather patchy forest on Plaine Champagne and the 'guava with standards' between Macabé and Mare Longue reservoir held only low numbers (*contra* Carié 1921). The species composition of the forest appears to be unimportant – the Brise Fer forests are very mixed with a high component of sapotaceous trees, while the southern scarp forests (except Bel Ombre) are heavily dominated by *Calophyllum tacamahaca*.

Feeding ecology

Mauritius Cuckoo-shrikes feed almost exclusively by taking insects from twigs and foliage, by gleaning (18 obs.), hover-gleaning (2 obs.) and leap-snatching (4 obs.). They feed usually in the canopy but occasionally on undershrubs to within 2 ft (0.6 m) of the ground. Fly-catching was seen three times and one bird was seen to rob a spider's web. The only comment in the literature on feeding methods is by Gill (1971*b*) who apparently recorded a higher proportion of hover-gleaning. The prey taken was generally too small to be identified, but a half-grown *Phelsuma* gecko (Newlands per CGJ), 2–3 cm caterpillars (Lepidoptera; 4 obs.), a 6–8 cm stick-insect (Phasmida) and a 1 cm 'green insect' were noted. The larger insects were struck several times on a branch before being swallowed; the stick-insect was carried from branch to branch being struck frequently before being swallowed some 2–3 minutes after capture. Small insects were swallowed directly.

Beetles were recorded by E. Newton on the label of a male from Grand Bassin (specimen in Cambridge), d'Emmerez found mantids, beetles (*Cratopus*) and moths' wings (*in* Manders 1911) and Desjardins reported berries and particularly insects, especially weevils ("charançons"; Oustalet 1897). I have only one observation of a fruit being taken, from an *Elaeodendron orientale* in Macabé forest.

Cuckoo-shrikes also feed on Pink Pigeon eggs (McKelvey 1976, 1977a; Chapter 5). In both cases observed, a female attacked the eggs from underneath the nest while the pigeon was incubating; in one case the male Cuckoo-shrike 'distracted' the pigeon at the same time (McKelvey 1976).

Breeding
Pairs began to feed together in close association in August and may continue to do so throughout the season. Courtship feeding has not been seen but I once saw a female unsuccessfully begging. An attempted copulation was recorded on 20 August (JMV). Records of nests are few, the earliest one being built during September, the clutch being completed on 28 September (judged by incubation behaviour; SAT/JFMH, Macabé, 1973). The season may continue until late March; JMV saw a pair still incubating on 4 March 1967. Nest-building is slow and very persistent. A nest near Mare Longue was freshly started on 23 October 1974, had been destroyed by 5 November, was being rebuilt on 3, 7 and 9 December but was deserted by 29 December. Both sexes collect nest material and work on the nest, which is built on horizontal branches either at a fork or against the trunk, poorly concealed; several nests have been in trees adjacent to jeep-tracks. Recent nests have been around 20 ft (6 m) up in well-grown trees, usually somewhat beneath the canopy, but E. Newton (*in* Pollen 1866) found nests better concealed in smaller trees; the nest illustrated by Carié (1921) appears to have been in a small tree amongst vertical branchlets. Species chosen for nesting recently have been *Eugenia glomerata* (2), Ebony *Diospyros tessellaria* (2) and *Securinega durissima* (1); Newton (*loc. cit.*) recorded *Erythroxylum hypericifolium* (2) and *Eugenia glomerata* (1). W. Strahm (per CGJ) has suggested that Newton's 'bois balai' was not *E. hypericifolium* as he wrote, but *Grangeria borbonica* which has the same local name.

The nest is a shallow cup about 3 in. (76 mm) inside diameter, made of fine twigs decorated with lichen attached with spiders' webs (Newton *in* Pollen 1866; Carié 1921). The two recorded clutches are of two eggs (Newton *loc. cit.*, preserved in Cambridge; Carié 1921 (illustration; not mentioned in text)). The eggs are pale green much speckled with brown (*ibid.*). The only observation on incubation period is for the 1973 Macabé nest: approximately 3½ weeks (JFMH). Both sexes incubate in turn (JMV, JFMH). There is no information on either the nesting period or the number of young raised as all observed nests failed, only the one mentioned reaching hatching. Single recently fledged young accompanied by both parents were recorded on 13 November 1973 (SAT) and 13 January 1980 (Jones 1980b), and two juveniles were seen in late December 1981 (Staub per CGJ).

Nest predation has been seen only once but is apparently very severe. Staub (per CGJ) watched a Ship Rat take a nestling; the nests would also be easily accessible to monkeys, and may possibly be robbed by Red-Whiskered Bulbuls and Mynahs. Cyclones no doubt destroy nests of Cuckoo-shrikes as of all other species (Cheke 1975b, Jones 1980a), but the birds themselves seem unaffected (*ibid.*).

Moult
I saw a male with half-moulted primaries on 27 March 1974. The only two dated moulting museum specimens are from much later in the year: a male two thirds moulted from Vacoas, 6 June 1860 (Cambridge) and a male only a third moulted from Axenot (?), 20 July 1875 (Leiden; G. F. Mees *in litt.*). The time of the moulting must be very variable, no doubt depending on when individuals finish nesting.

Mauritius Black Bulbul or Merle *Hypsipetes olivaceus* Jardine & Selby 1836; 'merle' (merl)
Affinities
Hypsipetes is a genus of about 20 species, confined to the oriental region (Rand & Deignan 1960), except for one species-group which has radiated in the Indian Ocean (A. Newton 1876, Benson 1960). Most twentieth-century authors (e.g. Sclater 1924–30, Rountree *et al.* 1952, Rand & Deignan 1960) have considered the two Mascarene forms to be conspecific, but as Forbes-Watson (1973) has pointed out, they differ clearly in behaviour and voice in the field, and also have strikingly different iris colours (Réunion: white, Mauritius:

chestnut). The Mauritius birds are also markedly larger than their Réunion counterparts (Chapter 9), and the differences between them in the field are more striking than between the sympatric species-pair (*H. crassirostris* and *H. madagascariensis*) on Mohéli, Comores (pers. obs.). I follow recent authors with Mascarene experience in treating the Mauritian *H. olivaceus* as specifically distinct from *borbonicus* of Réunion.

Mees (1969: 302) has argued, *contra* Deignan (1942), that *Ixos* Temminck 1825 is the nomenclaturally correct generic name for the bulbuls usually placed in *Hypsipetes*. Only Benson *et al.* (1976–7) seem to have adopted this usage, and I prefer to continue using *Hypsipetes* Vigors 1831 for the Indian Ocean merles.

General habits and behaviour

Merles are noisy arboreal birds, usually seen in pairs or small parties. Spells of silent feeding when individuals are scattered in different trees are frequently interrupted by contact calls and chatter as the pair or group assembles again. Breeding pairs (sometimes groups) are wide-ranging and do not hold exclusive territories (Temple 1975, pers. obs.), though each has a centre of operations where the nest is built. Merles do not associate with other species, but do sometimes aggressively chase or displace them (Cuckoo-shrikes, Fodies, Red-Whiskered Bulbuls). They are inquisitive birds and will often investigate human intruders, but this trait is much less marked than it seems to have been in the past (e.g. Bernardin 1773, Clark 1859), when they would allow themselves to be knocked down with sticks and shot at point-blank range.

The sexes are similar in size and identical in plumage, so are not distinguishable in the field.

Map 6. Distribution and density of the Mauritius Merle in the south-western forests, 1973–5. Area covered by map as in Map 5.

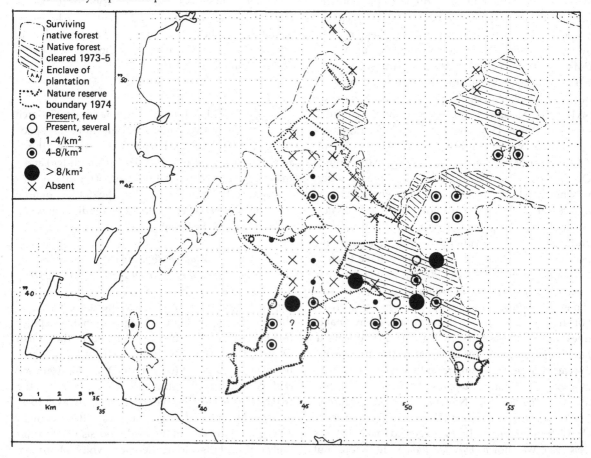

Distribution, numbers and density

Past and present distribution. The distribution during 1973–5 is plotted on Maps 6 and 7. The coverage of the Lagrave and Montagne Bambous forests was not complete and there were certainly Merles in more squares than indicated; a few may also exist on Montagne Blanche. Small numbers also occur in the Black River Gorges valley, on the slopes round to Tamarin Gorge (CGJ), and on 'Chamarel Peak' (= Piton Canot) (CGJ).

Nothing in the nineteenth-century literature gives any indication of the actual range of the Merle, the nearest being a remark of Clark's (1859) that it was "only found in the undisturbed parts of the forest". In addition to birds from within the present range Edward Newton collected birds in Vacoas, "near Curepipe", Bois Sec and Cachette (near Chamarel) in the 1860s (specimens in Cambridge); Aubebert collected a bird at Vacoas in 1875 (in Leiden; G. F. Mees *in litt.*). By 1910, according to Meinertzhagen (1912), the birds were "practically confined to the forest in the south-west of the island", an impression supported in general by Guérin (1940–53) who reported them in 1938–9 from the present range and Chamarel, but also noted them at Diamamou Falls at the western end of the Bambous range. FRGR saw birds in the

Map 7. Distribution of the Merle and Olive White-eye outside the south-western forests, 1973–5.

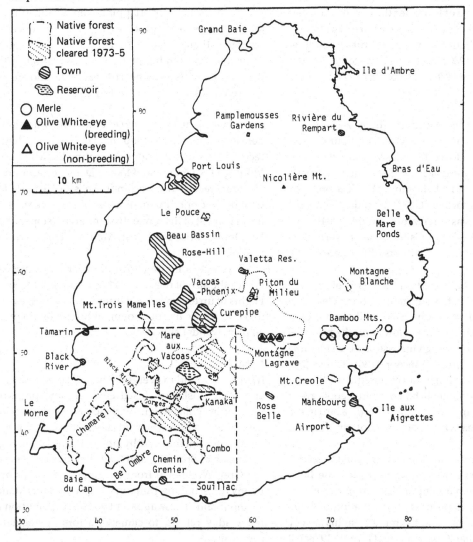

Bambous range in the 1930s, at Chamarel in 1949, and also outside the present range, in lower Bel Ombre (1950) and at Tamarin Falls (1930s). Newton (1958a) recorded them as far north as the Piton du Milieu, in areas since cleared for tea plantations. Since it can tolerate very degraded forest (see below) the Merle probably occurred until about 1950 throughout the forest area from the Piton du Fouge right across to the Piton du Milieu north-eastwards and through the Bambous mountains to the east. Since then the fragmentation and clearance of the old logged-out forests east and south of Curepipe, and further degradation around Chamarel, has reduced its range by about half. Kanaka Crater still supported two or three now isolated pairs in 1983 (CGJ).

A solitary bird, no doubt driven from its normal haunts by a shortage of fruit after the storm, was seen in a garden at Black River on the coast in early January 1980 shortly after Mauritius had been struck by cyclone Claudette (Jones 1980a). One was reported on the coast at Le Morne in 1976 (Steinbacher 1977), possibly in error.

Population. The Mauritius Merle was once a popular game bird, as its counterpart on Réunion still is (Chapter 6). Evidence of massive past slaughter (Clark 1859, Meinertzhagen 1912) and dramatic declines also from forest clearance (E. Newton *in* A. Newton 1876) led Hartlaub (1877) to predict imminent extinction. This was premature (Carié 1904), but Meinertzhagen (1912), a visitor, was nevertheless more pessimistic than local authors (d'Emmerez 1928, Guérin 1940–53).

Newton (1958a) stated that it was "not uncommon", but by 1964, after substantial forest clearance for tea (Chapter 1; also Roy 1969), Gill (1971b) considered the numbers "critically low" and the bird was rated as "very rare" and giving rise to "considerable anxiety" as to its future (Vincent 1966). Direct predation by man is no longer a serious threat, though some birds are undoubtedly taken where people are at work in the forest: in October 1974 I caught a bird in the Parc-aux-Cerfs area which had traces of bird-lime on its plumage.

Only that part of the Merle's range in the south-western forests was well surveyed during the BOU Expedition. Preliminary impressions from these data led Temple (1975) to postulate a population of "well over 1000 individuals", and to omit it from his survey of endangered Indian Ocean species (Temple 1976c); it

did not appear in the revised *Red Data Book* (King 1978–9). Analysis of the survey leads to a rather different conclusion. In the south-western forests (Map 6) only 72 pairs were located, with an estimate of about 140 in all. At least 10 more pairs were found on the Lagrave (7+) and Bambous ranges (3+ inland of Ferney) but the total in these hills must have been around 50. The estimated 1974–5 breeding population was thus only around 200 pairs. There is a strong suggestion (see below) that Merles may not breed until they are 2 years old, which would add a floating population of immatures; breeding success is, however, so low that there could hardly be more than 100 of these, making a total of only about 500 Merles altogether. Jones (1980a) recorded a sharp reduction in population after the severe cyclone Claudette of December 1979.

Merles have overlapping home ranges rather than discrete territories, so are harder to census accurately than Cuckoo-shrikes or Fodies. Our figures are estimated from the pattern of observations rather than head-counts as with territorial birds, and are thus inherently less exact than for those species.

Density and home range area. An indication of density is given in Map 6. The highest density recorded, around the ridge separating Les Mares from Montagne Savane, was 14/km². Densities around 12/km² were recorded at Alexandra Falls and in part of Bel Ombre, but only in small areas, and 4–6/km² is the norm in most of the range. In some areas (slopes above Bel Ombre, Brise Fer forest) Merles occur as sparsely as 1/km².

Temple's figure (1975) of "nearly 30 acres" (12 ha) for the size of the Merle's home range is too small. Flights of up to ½ km are not infrequent, so ranges of up to 30–40 ha are not unknown, although by no means all pairs wander so widely.

Habitat

Merles are found in evergreen broadleaved forest, but apparently dislike a closed even canopy, preferring trees of irregular height with some openings. However, that is not a sufficient description, as the birds are absent from the scrubby land of Plaine Champagne and north of Brise Fer, and (like all native birds except the Grey White-eye) from the drier and rather open Ebony forest linking Tamarin Gorge with Trois Mamelles mountain. These apparent gaps in distribution are not readily related to climatic factors or vegetation composition.

Merles were limited to forest containing at least some native vegetation, but were quite frequent in the partly exotic Kanaka area and on the degraded slopes of the Piton du Fouge and the Lagrave range, which suggests they might adapt to a forest of introduced trees if the composition was carefully chosen. The Kanaka area has already been described (above, under Cuckoo-shrike); the lower slopes of the Piton du Fouge and the north side of Montagne Lagrave are typical of logged-out areas of native forest – a few native trees and shrubs surviving in a thicket of guava, privet and bramble. There are, however, many such areas from which Merles are absent.

The clearance of the dwarf swampy forest of Les Mares, so serious for Pink Pigeons, Cuckoo-shrikes and Mauritius Fodies, has probably not seriously affected the Merle, which though recorded from there (Guérin 1940–53) does not occur in Pétrin or the swampy areas of Plaine Champagne and was thus unlikely to have been common. However, the degraded forest around Grand Bassin was a good site in the past (*ibid.*).

The Merle was apparently successfully, if only temporarily, introduced to a completely different *milieu*: Diego Garcia in the Chagos Islands. Hutson (1975) reported that the birds were well established and common in 1953 but disappeared shortly afterwards. About the same date Mynahs *Acridotheres tristis* were introduced (1954 or 1955 (*ibid.*), not 1953 as Bourne (1966) reported); by late 1955 Mynahs were conspicuously common (Scott 1961). Long (1981) attributed the demise of the Merles to competition from the Mynahs, which seems likely. Even in their home island Merles may be restricted by the presence of Mynahs; in Réunion, where Mynahs hardly penetrate the forests, Merles *H. borbonicus* are much more common than their Mauritian congeners (Chapter 6; Barré 1983).

Feeding ecology
The Merle is not wholly frugivorous as E. Newton insisted (1861b *in* A. Newton 1876) but also eats insects (Clark 1859, Guérin 1940–53, Newton 1958a, Temple 1973–4 [9 January 74]) and geckos. The BOU Expedition has seven observations of feeding on plant matter as against 13 on animal matter.

When seeking insects Merles fed by gleaning from leaves or leaf-bases (7 obs.), from twigs and branches (3 obs.) and by clumsily fly-catching (3 obs.). When gleaning, prey was seen from some distance away and stabbed at (if within range) or jumped towards. Fly-catching was by leaping from branch to branch catching flying insects *en route*, or by sallying out into mid-air (2 obs. of abortive attempts at butterflies (Lepidoptera)). Prey were rarely identified; JS saw "large black insects" taken from bark and Newton (1958a) recorded caterpillars (Lepidoptera). In Macabé forest I once saw a vigorous leap and chase in a tree after which the bird emerged with a *Phelsuma* gecko's green tail wriggling violently and promptly swallowed it: CGJ has twice seen adult day-geckos taken, once after a ground chase, and has had other observations reported to him; *Phelsuma* may be quite important in the diet. Insect-feeding was recorded on *Litsea* sp., *Lantana camara* and privet, as well as on a wide variety of native species.

Expedition observations of fruit-feeding were in October and November, but CGJ also saw it in March and May, as well as in January after a cyclone (Jones 1980a). Animal-feeding records are spread throughout the year. Berries up to 15 mm across are swallowed whole, but larger fruits are pecked at *in situ*. Species recorded during 1973–5 were *Aphloia theaeformis*, *Protium obtusifolium*, *Eugenia glomerata*, cf. *Gastonia* sp. and *Calophyllum tacamahaca* (all native), and *Ligustrum robustrum* and *Litsea monopetala* (introduced). The Mauritius Merle has not been seen eating the fruit of the introduced guava. E. Newton found *Aphloia* berries in birds he shot (Guérin 1940–53; labels on specimens in Cambridge: "gizzard filled with seeds of the fandaman"). Palm fruit (unspecified) and jackfruit *Artocarpus heterophyllus* were recorded by Newton (1958a), privet, raspberries *Rubus mollucanus* and (after a cyclone) papaya *Carica papaya* by Jones (1980a), and Guérin (1940–53), perhaps referring to captive birds, mentioned "mulberries [*Morus alba*], raspberries and cherries [*Prunus cerasus*]"; all these, except perhaps the unidentified palm, are introduced.

Fruits were mostly taken when still on the tree, merles being apparently very reluctant to visit the ground (E. Newton *in* A. Newton 1876; pers. obs.; cf. also *H. crassirostris* in the Seychelles (Grieg-Smith 1979b)). However, AFW saw a Merle pick up a round black fruit from the ground in Macabé forest in October 1971.

Temple (1975) mentioned flowers in the diet; I once saw a bird swallowing whole both open and shrivelled flowers of the large blue-flowered Morning Glory *Ipomoea congesta* and CGJ reports *Nuxia* flowers as frequently taken.

During the fruit-eating season Merles are vulnerable to severe cyclones, which defoliate trees and strip them of flowers and fruit. Jones (1980*a*) recorded a sharp reduction in numbers, with birds wandering well outside their normal range, in the aftermath of a particularly bad storm in December 1979. That it is not the storm itself that kills the birds is clear from my observations of plenty of living Merles 2 days after cyclones Gervaise in February 1975 (I then left the island, preventing further observation).

Breeding

There are too few observations to define the Merle's breeding pattern. Temple (1973–4) gave the season as September–January, though the observations available to me (mostly of birds carrying nest material) span November to late January, implying nesting through February. Specimens in Cambridge are noted as having enlarged testes on 2 February and 2 and 3 October 1860, and I caught a female with a brood patch on 27 January 1975.

The nest is a coarse structure usually in a low (6–8 ft; 2–2.5 m) bush (E. Newton in A. Newton 1876; pers. obs.), though I have a record of one 15–20 ft (4.5–6 m) up a tree. The cup is 8–9 in. (20–22.5 cm) across, lined with finer material than the twigs, coarse grass, leaves, roots, moss and lichens used to build the framework (E. Newton *loc. cit.*, Meinertzhagen 1912, pers. obs.)

The clutch, according to Guérin (1940–53), is "rarely more than two", but in the Cambridge collection there are groups of two, three and five eggs with the same collection number, though there is no proof that they constitute clutches. The eggs are pale pinkish spotted with brown, chiefly around the wide end (E. Newton *in* A. Newton 1876; pers. obs.) Guérin (1940–53) gave the incubation period as 14–16 days, both sexes incubating, the male only in the middle of the day. He added that the young, fed on insects, leave the nest after 3 weeks, and described the way the birds attack intruders, and perform a typical distraction display if the intruders get too close to the nest.

Guérin (1940–53) claimed that, after leaving the nest, the young remain with their parents for a few days only before being left to fend for themselves. But Merles, unlike other Mauritian passerines except the Grey White-eye, are seen throughout the year not only in pairs but also in small groups of three to five. This strongly suggests that families in fact stay together for a long time, as they do in the Seychelles species *H.*

crassirostris (Grieg-Smith 1979*b*). Supernumerary immatures are not infrequently associated with nesting pairs in the Seychelles (*ibid.*); their age is not known, but if, as the author implied, the birds bred only once a year, some at least must delay breeding until 2 years old. The colour difference between young and adult Mauritian Merles is less striking than in the Seychelles (pers. obs.), and the pale-brown juvenile flight feathers apparently do not survive the first moult, so it is not possible to be certain either that groups are in fact family parties, or whether some birds delay breeding for 2 years.

The two nests seen during 1973–5 both failed; there was no evidence of eggs being laid. One bird was seen carrying food (15 December 1974). Undoubted juveniles were seen on only two occasions: a single on 26 January 1974 and a family party (number uncertain) on 30 December 1974. Guérin (1940–53] commented that the Merle was "not very prolific". Monkeys *Macaca fascicularis* are usually blamed as the principal nest predator (E. Newton *in* A. Newton 1876, Temple 1975) but direct evidence is lacking. Newton (*loc. cit.*) noted an increase in Merles (and Cuckoo-shrikes) in an area where monkeys had been controlled 2 or 3 years previously. Rats may also be involved, and, as the nests are usually conspicuous, passing Mynahs may also account for any that are unattended.

Moult

Birds moulting tail feathers were seen from January to May, though non-moulting birds were also always present. A male specimen in Cambridge has wing-moult almost complete on 6 April (1860), but another in Paris at the same stage is dated 24 August (1872).

Guérin (1940–53) reported a post-juvenile moult beginning a few weeks after fledging. The juvenile specimen in Cambridge is in body-moult only, but another in Liverpool is moulting both body and wing. The Seychelles birds moult both body and flight feathers, though some juvenile feathers persist (Grieg-Smith 1979*b*); it seems likely that the Mauritian birds have a complete post-juvenile moult.

Paradise Flycatcher Terpsiphone *bourbonnensis desolata* (Salomonsen 1933) 'coq de bois' (kok debwa).

Affinities

This form is very close to the nominate race on Réunion, and is generally considered to be derived from the Malagasy *T. mutata*, principally on grounds of geographical proximity (Salomonsen 1933, Benson 1971). The voice and general behaviour of *mutata* on the Comores (pers. obs. 1975) and its small size closely resemble *T. bourbonnensis*, and are quite unlike the 'jizz' of the larger *T. corvina* (Seychelles) and African and Asian mainland species I am familiar with (*T. viridis*, *T. paradisi*, *T. atrocaudata*).

General habits and behaviour

This flycatcher generally keeps to the understorey of dense woodland, but sometimes forages in the canopy. It is generally solitary, but individuals occasionally join passing parties of Grey White-eyes (Guérin 1940–53, AFW, CGJ), though much less often than does the Réunion race (Chapter 6). Pairs are strongly territorial throughout the year, and neighbouring pairs are frequently seen chasing each other. An upright open-billed display by males accompanied by trilling seen occasionally (CGJ) was interpreted as a threat display by Barré & Barau (1982). The very distinctive song (Chapter 3) is heard all through the year, though rarely in mid-winter. Although it is most conspicuous at dawn and dusk (e.g. Desjardins *in* Oustalet 1897, Staub 1976) because other birds are quiet, singing can be heard throughout daylight hours, and males generally respond to a whistled imitation or playback.

This species is sexually dimorphic, the male having a glossy head which is replaced by dove-grey in the female.

Distribution, population and density

Distribution. Maps 8 and 9 show the distribution of the Flycatcher as recorded during the BOU Expedition. As the fieldwork necessarily concentrated in the south-west of the island where most of the rare birds live, no attempt was made to do a complete survey of the Flycatcher's distribution; there are no doubt many

Map 8. Distribution and density of the Mauritius race of the Mascarene Paradise Flycatcher in the south-western forests, 1973–5. Area covered by map as in Map 5.

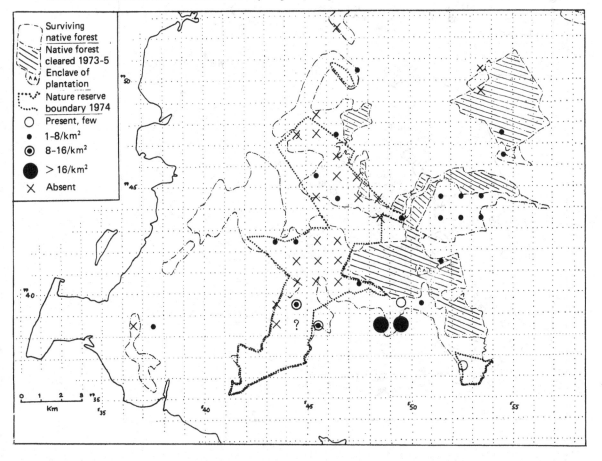

other places on the island supporting small numbers of this species. The distribution given by King (1978–9), suggests that Temple (his source) knew of further east coast localities, recorded the birds in the Montagnes Bambous (where I saw none), had seen birds in relict riverine forest in south coast valleys between Baie du Cap and Savane (not visited by the expedition, but two sites confirmed by P. Rountree, pers. comm.), and had found them in the forests between Tamarin Falls and Rempart Mountain (where neither I nor CGJ found any). Surveys in likely areas such as the mixed plantations at Belle Vue, L'Asile and Powder Mills, revealed nothing (pers. obs., CGJ), though SAT had earlier reported a few birds at the last.

In the past the species was evidently more widespread but no detailed information exists. In the 1830s birds were to be seen on Plaines Wilhelms (now the Rose-Hill–Vacoas–Curepipe conurbation) and generally in the Flacq area (Desjardins *in* Oustalet 1897). By the turn of the century they had all but vanished from new townships on Plaines Wilhelms, being little seen outside the forests (Carié 1904), though some birds survived in scrub around Candos hill (Quatres Bornes) until about 1940 (FRGR). Meinertzhagen (1912) recorded them as common in Pamplemousses Gardens, but by the 1950s they were gone (Newton 1958a), although there are occasional hearsay reports, and they were included by Owadally (1976a) in a list of the

Map 9. Distribution of the Paradise Flycatcher outside the south-western forests, 1973–5.

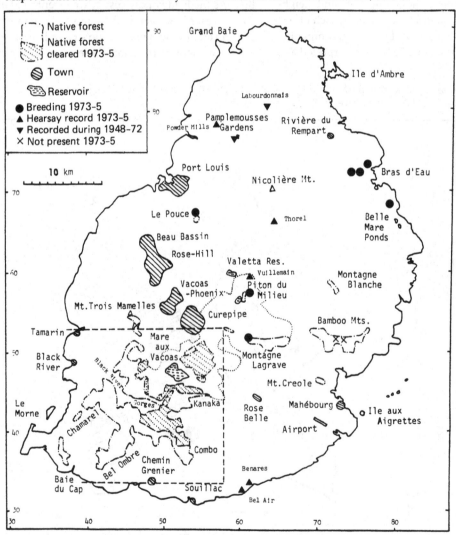

gardens' birds; I saw none there. Guérin (1940–53) reported a drastic decline by 1952 in the vicinity of Nicolière reservoir where they had once been common in *Eugenia jambos* thicket; the area is now largely planted to pine, though the birds do persist in small numbers in the plantations between Thorel and Nicoliére. Like several other species, Flycatchers were common in the Midlands area and also south of Curepipe (Pike 1873, Newton 1958*a*) prior to the big clearance for tea and pine plantations. Guérin also recorded them in the forests of Chamarel. On the other hand the thriving but far-flung population at Bras d'Eau on the north-east coast is a quite recent discovery (Staub 1971); indeed the plantations there would hardly have been suitable habitat before about 1940. Birds formerly resident in the gardens of the Labourdonnais estate were wiped out by the cyclones in 1960, and had not returned by 1974 (estate manager, per SAT).

Population. Flycatcher numbers have dropped dramatically as the distribution has contracted, but the species has always been considered, after the Grey White-eye, the least at risk of the native forest birds (e.g. Vincent 1966). Temple (1975) was the first to hazard a population figure, which he put at "somewhat over 1000 individuals", admitting that this was not based on an exact count; he later (1976*c*, King 1978–9) revised his figure upwards to "thousands".

Analysis of the BOU Expedition data suggests that these figures are over-optimistic. Only 60 territories of this species were located in the areas surveyed, a figure much lower than any of the other forest passerines! Despite its obvious song the Flycatcher is not always easily detected. In tall woodland with an open understorey the birds see an 'intruder', come over to investigate and are seen on about 40% of visits. In dense thickets, where visibility over a distance is very poor, the birds cannot respond in the same way, and the contact frequency drops to under one visit in three. Using these factors to adjust the figures still gives a total population in the areas covered, including Bras d'Eau, of only 125–150 pairs. The areas listed by Temple (*in* King 1978–9) and those yet undetected are unlikely to add more than about 100 pairs, as the sites are either very small or the birds are at such low density that I failed to find the birds Temple saw. This gives a maximum population for the island in 1974 of 250 pairs, and it was probably in fact somewhat less.

This puts the Flycatcher in the same numerical range as the other forest passerines, though its lower dependency on native forest makes it less vulnerable to degradation and damage of this habitat. On the other hand a much smaller proportion of its population than of any other endemic forest species is within current nature reserve boundaries.

Density, territory size and dispersion. In 1974–5 Flycatchers occurred in fewer 1 km squares in the native forests of south-western Mauritius than did any of the other forest passerines, and were at low density where they were present (Map 8). In the Macabé/Mare Longue area there were at best 1–2 pairs/km², 4–5 in Kanaka and 6–8 at Alexandra Falls. The density reached about 15/km² in parts of Bel Ombre and around 20 around Bassin Blanc. JCL confirms that in 1978–9 also, Bassin Blanc was the only native forest area where they were present in reasonable numbers. By contrast at Bras d'Eau, in mature exotic plantations, the birds averaged 50 territories/km².

At Bras d'Eau, where the most detailed work was done, the birds held territories of between 1 and 2 ha, normally contiguous with adjacent pairs (Maps 10–12). In mature forests, even in low-density areas (e.g. Macabé, Bel Ombre), pairs appeared to prefer to be adjacent to others. In scrub and immature forest the birds did not appear to be clumped to such an extent, though at Alexandra Falls and Montagne Savane the birds were present in only part of the area. During the clearance of the degraded forest in Parc-aux-Cerfs I recorded a 'clump' of four singing males in a very small area on 28 September 1974, but they must have subsequently moved on, as there were no further records despite intensive observation at the site. Territory size in the native forest is certainly greater than at Bras d'Eau, but is still very small in relation to density (except around Bassin Blanc and upper Bel Ombre), leaving large areas of forest with no birds at all.

Habitat

Undoubtedly the best habitat for Flycatchers, under present conditions on the island, is the plantation at Bras d'Eau. The birds favour stands of tall mature evergreen shade-bearers 50 ft (15 m) or more high, very open under the canopy, but with a fair amount of shrubby understorey, mostly *Cassia fistula*, *Leucena glauca* and saplings of the plantation trees (Cheke

1975c). The evergreens are principally *Araucaria cunninghamii* and *A. columnaris*, the former producing a more even shade, and an orchard of mature mango *Mangifera indica* interspersed with a variety of other fruit trees. Patches of broadleaved trees, deciduous in the dry season (*Tabebuia pallida, Vitex glabrata, Tectona grandis*), are utilised when they are in leaf but generally avoided when they are not; the few birds that have territories entirely in deciduous trees have presumably been excluded from the better habitat. Most of the forestry area at Bras d'Eau is planted to *Eucalyptus* spp., evergreen but not shade-bearing, which the Flycatchers avoid except where they are contiguous with better habitat. In addition the very abundant flying insects (especially mosquitos) and lack of human disturbance are beneficial to the birds (Cheke 1975c), as is the lack of the monkeys *Macaca fascicularis* so abundant in the native forest areas. The key structural factor determining the favourability of a habitat for the Flycatcher appears to be good visibility and still air under a closed canopy with ample, but not dense,

undergrowth on which to perch and rest. However there must be a climatic element also as there are structurally excellent areas of native forest with no birds at all (e.g. Brise Fer), and the distribution in the south-west suggests that the species is essentially a lowland one with a preference for forest that is drier or warmer, or both, than that found on the south-western plateau.

No Flycatchers were found in plantations structurally similar to Bras d'Eau (mostly without such dense shade and with too little undergrowth) explored at Belle Vue, L'Asile, Powder Mills and Balaclava. The first was also free of monkeys; the last two suffered considerable human disturbance. The only other exotic forest I found with a tall shady canopy was in the further parts of Le Pouce valley behind Port Louis; Flycatchers were present, so were monkeys, but there was little human disturbance.

The state of the undergrowth plays a very important part in whether Flycatchers can maximise their use of a habitat. Both for nesting and as feeding

Map 10. Paradise Flycatcher territories at Bras d'Eau October–March 1974. Individually marked birds are identified by number and unmarked ones by '+'; if one of a pair was not seen this is shown by a '?'. Base map from Cheke (1975c).

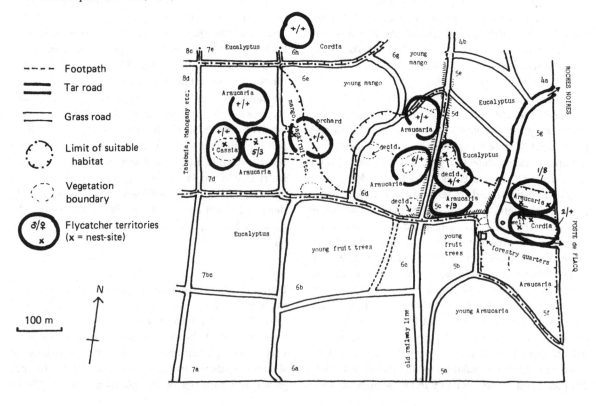

perches the *Cassia* understorey at Bras d'Eau seems to be best for the birds when between 1 and 3 years old (Cheke 1975c). In 1974 this undergrowth was coppiced at the beginning of the breeding season and as a result several pairs either did not nest or started very late, only to have their efforts destroyed by cyclone Gervaise in February 1975 (*ibid.*). Jones (1980a) reported that the coppicing is now being managed by the Forestry Services in favour of the birds along the lines I outlined in 1975 (Cheke 1975c). For several years the entire forestry area at Bras d'Eau has been under threat of being swept away to make way for a new international airport, though construction had not begun by late 1984 (CGJ).

Flycatchers are found in some but not all of the native forest of the structure described, and only in moderate density on the lower parts of the southern scarp of the south-western plateau (Map 8). Shorter forest with denser undergrowth (Alexandra Falls, Montagne Savane area, Black River Peak) is also inhabited to some extent. The birds are found too in low scrubby forest in a wide range of vegetation types from wet more or less native thickets (east Petrin, Rivière des Chevrettes, Piton du Milieu) to dry exotic scrub of *Eucalyptus*, *Schinus terebenthifolius* and *Furcraea foetida* (Belle Mare).

Feeding ecology
The Flycatcher feeds by fly-catching under the canopy and gleaning within it. Fly-catching is more commonly observed but it is a much more conspicuous behaviour so gleaning birds may have been overlooked. Also, Flycatchers are often so busy watching the observer that they stop feeding when being watched.

When fly-catching, the birds perch on shrubs or saplings in the understorey, or lower branches of trees, and sally out for insects, often sweeping downwards from the vantage point rather than upwards or outwards. When feeding in foliage birds either glean, like a warbler, or more vigorously by leap-snatching, fly-picking or rather clumsily hover-gleaning.

Prey is usually too small to identify, but larger items are brought to the nest or to dependent fledglings: I recorded a 3.5 cm moth (Lepidoptera), a lace-

Map 11. Paradise Flycatcher territories at Bras d'Eau September 1974–January 1975. Details as for Map 10.

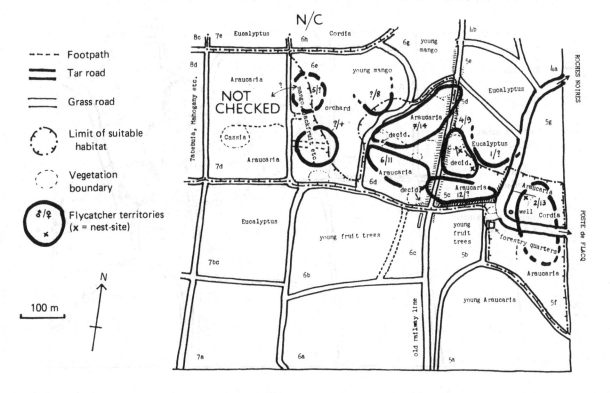

wing (Neuroptera) and a large fly (Diptera), and CGJ a large green grasshopper (Orthoptera). E. Newton recorded remains of beetles in a gizzard (label on specimen in Cambridge); Clark (1859) stressed Lepidoptera as important in the diet, and Guérin (1940–53) referred to moths and gnats (Diptera). Guérin (1940–53) also mentioned the birds adroitly robbing spiders' webs.

Breeding

Apart from one nest in November 1973 near Montagne Savane, all the expedition's breeding observations come from Bras d'Eau in the north-west of the island where the climate, and perhaps the birds' seasonality, differs from the south-western forests.

Breeding activity at Bras d'Eau began in August; a female was seen with nest material on 31 August 1974, and in 1973 a pair had fledged young on 30 September which indicates a similar timing. Clutches were recorded in September to December, and a nest nearly ready to lay in was found in late January 1975 (Table 6). There are nine clutches in Cambridge collected in the 1860s, all taken in November and December (Table 6). Three are from Black River and

Table 6. *Laying dates of Mauritius Paradise Flycatchers*

Source	Sept.	Oct.	Nov.	Dec.	Jan.	Total
Cambridge University Museum[a] (E. Newton & W. H. Power; eggs coll. 1863–5	—	—	5	4	—	9
Bras d'Eau, Mauritius[b] (this study)	2	8	1	2	1[c]	13

[a] Newton's eggs, from 6 clutches, are without locality data (his notebooks are lost: C. Bradshaw pers. comm.). Power's clutches (3) were collected at "Black River": this name could refer equally to the district as to the village, so the exact locality in south-west Mauritius is uncertain, but likely to be near the coast. A single egg from a clutch of Newton's dated 10 December is in the British Museum (Natural History) (Oates *et al.* 1901–2), presumably from the same clutch as the one in Cambridge dated 10.12.1864.
[b] Includes estimated laying dates for nests found with nestlings and for fledglings seen with adults. One nest from Belle Mare, about 4 km south-east, is included.
[c] The January nest was virtually complete, but eggs had not been laid before the study finished.

Map 12. Paradise Flycatcher territories at Bras d'Eau October 1978. Details as for Map 10.

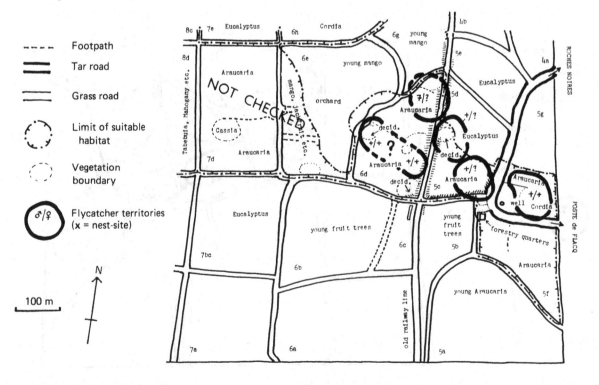

the others, collected by E. Newton, are likely to have been found in or near Vacoas where he lived. There is thus a suggestion that the season begins earlier at Bras d'Eau than in the south-west, although I can find no climatic factor that is out of phase in the two areas (Padya 1972; there are differences in magnitude of several parameters, but the annual cycle is the same). De Querhoënt (1773) recorded nesting in September, and Desjardins, who lived at Flacq (near Bras d'Eau), found nests in October, November and December (1830s; *in* Oustalet 1897, repeated by Guérin 1940–53), but there is nothing else on seasonality in the literature.

The nest, described by de Querhoënt and Desjardins (*locs. cit.*), and photographed by Staub (1976), is a cone usually placed in an upward fork of a sapling with a cup about 5 cm across and 4 cm deep. At Bras d'Eau the principal material used was green moss (not ferns as Staub stated) decorated with spiders' white egg cases, the lining being of fine grass inflorescence-stalks. Green (Desjardins) or dry (de Querhoënt; pers. obs.) leaves are used where there is no moss. De Querhoënt described a nest with a partial covering – I have seen one like this in Réunion (Chapter 6). The nest is usually 6–8 ft (1.8–2.4 m) from the ground, but may be as low as 3 ft (0.9 m). Any type of slender plant with a suitable fork may be used, but at Bras d'Eau *Cassia fistula* shoots are strongly favoured. If undergrowth is not available the nest is built at the end of a branch of a tree; such nests may be as low as undergrowth nests, but if the tree has no low branches can be much higher: one in an *Araucaria* at Bras d'Eau was 30 ft (9.1 m) high.

All six certain complete clutches held three eggs, as did the three clutches in Cambridge for which we have details and the one collected by Layard in 1856 (Layard 1867, Brooke 1976). Newton (*in* Hartlaub 1877) gave the usual clutch as two, three being less frequent, and several of the eggs in Cambridge collected by him are numbered in groups of two but have no data. Staub (1973a) and Temple (1973–4) quoted the clutch as two or three, the former later (1976) reverting to only two eggs. Newton's birds were probably collected around Vacoas; perhaps a lower clutch is normal in higher or wetter areas. Clark (1859) claimed four or five eggs, de Querhoënt four and Guérin (1940–53) three to five; in Réunion clutches over three have been ascribed to more than one bird laying in the nest (Chapter 6). The eggs are white densely spotted with

rusty brown (Layard 1867, Hartlaub 1877, Oates *et al.* 1901–12 (illustration vol. 3); Guérin 1940–53, pers. obs.; not greenish as de Querhoënt (1773) stated). The incubation period is 15–16 days according to Staub (1973a, 1976); one nest I observed took 15 days. Both sexes incubate (Guérin 1940–53, Staub 1976), but Staub (1973a) suggested the male sat more than the female; it is possible there is a diurnal rhythm, the female incubating at night and the male more during the day. Staub (1973a) gave the fledging period as 14 days. The juveniles are chestnut above with no grey or black (Benson 1971, pers. obs.), often obscurely spotted darker. At the time the young leave the nest the tail is only partly grown, but reaches full length after about 2 weeks. There then follows a rapid post-juvenile moult, completed in about 3 weeks, during which time the young are still fed by the parents. After this the young leave the parental territory very soon, all having left by 8–9 weeks after fledging. On 18 November 1973 I saw a Grey White-eye solicit a female Flycatcher with two recently fledged young. She tried to feed it a lacewing (too large!) realised her mistake and then successfully fed the insect to one of her own young.

Nesting success at Bras d'Eau is by Mauritian standards quite high: out of nine nests found with eggs, at least one young successfully fledged in five, and parties of fledged young with their parents were seen twice. Flycatcher nests are the most conspicuous of any native forest passerine and would be easily found by monkeys (absent at Bras d'Eau in 1973–5 but seen recently, CGJ), probably by both arboreal and ground-living rats and even, in some cases, mongooses *Herpestes edwardsii*.

The causes of nest failure at Bras d'Eau were not established. In two nests the clutch was never completed, in the other two eggs or young disappeared, together with, in one case, the female. Of those nests which were successful, four fledged all three young, the other only one. One of the birds that reared three was feeding only two about 12 days after fledging, and both the family parties seen without the nest being found had only two young. One pair saw all three young through the moult.

Although the success rate of observed nests was quite high, it was usual to find one or more (up to four) nests in a territory, especially towards the end of the season. This suggests a higher rate of failure, and a readiness to lay again. Clark (1859) and Guérin (1940–53) considered the Flycatcher to be double-brooded,

and though I have no unequivocal evidence of a successful second brood, one pair in 1973–4 built a nest
after having fledged a brood early in the season.
Repeats were frequent, always in a new nest.

Breeding may be delayed by several months if
conditions are unsuitable. At Bras d'Eau in 1974 the
Cassia was coppiced in early September and only one
of the pairs under observation nested before
December, although a different bird had been seen
building on 31 August before the undergrowth was
cut. The pair which laid in October were in a territory
where some undergrowth remained. By late January
the understorey was recovering well, but one pair
nested high up in an *Araucaria*. This pair was just
completing a nest on 27 January 1975, a date by which,
the previous season, all birds had long since finished
breeding and most were actively moulting.

Moult

The Bras d'Eau adults started moulting in mid-January
1974, when at least one pair still had three dependent
fledglings. This pair was half-way through wing-
moult by 12 February, and three others trapped on 28
March had nearly completed it. Moult at Bras d'Eau
must have been delayed at least a month the following
season because it had not started when I left the island
in February. However, a male caught in native forest
(Rivière des Chevrettes ridge) on 2 February 1975
was just beginning its moult. A pair at Bras d'Eau were
moulting in early March 1981 (CGJ). I have found
dated moulting birds in museums; a male from Vacoas
dated 2 January 1860 had not begun moulting and
another from Bel Ombre of 13 June 1862 had finished
(both in Cambridge).

The post-juvenile moult is of body-feathers only,
starts about 2 weeks after fledging and is completed in
about 3 weeks. The post-juvenile plumage of both
sexes resembles the adult female. The juvenile bas-
tard-wing and greater coverts are retained (sometimes
at least), and are tipped with chestnut unlike the plain
dull brown ones of the adult.

No information was obtained on age at first
breeding. However both birds of one pair during the
1973–4 season were in grey plumage so it is probable
that full plumage is not gained until the full moult of
the season following fledging, and that some birds
nevertheless breed before this happens. In Réunion
(Chapter 6) breeding males in grey plumage are much
more frequent.

Dispersal and mortality

All five nestlings and fledglings that I ringed in their
parental territories at Bras d'Eau disappeared within 2
months of leaving the nest (as did several more
unringed birds) and were never seen again. There may
be heavy mortality shortly after the parents stop feeding the young but the adult survival rate does not seem
to be high enough to support a loss rate of the 90–95%
implied if all the 'lost' birds had died. It seems more
likely that the birds disperse to a considerable distance;
this would also explain the very widely scattered
distribution of the species.

All six of the adults (four males, two females)
ringed within the main Bras d'Eau study area before
the end of March 1974 were alive the following
December, as was at least one of a pair on the fringe
(not thoroughly checked 1974–5), and a male ringed in
Macabé in October 1973. However none of the six was
to be found in October 1978, the only ringed bird seen
at Bras d'Eau then being a male first caught in October
1974 (of six more that season); all the other territories
were occupied by unringed birds (Map 12). The
Flycatcher's susceptibility to cyclones is not known but
but Gervaise, which hit the island just after my last
observations at Bras d'Eau in 1975, was particularly
severe.

Two birds caught in December 1974 and January
1975 showed symptoms resembling birdpox: each had
a scabby swollen foot and had lost at least one claw.
Another bird, healthy when caught on 24 October,
had a big warty excrescence on her face by 27 January,
just after rearing two young to independence. Several
Grey White-eyes caught over the same period showed
similar foot symptoms. By contrast only one of 19
Flycatchers tested for protozoan blood parasites was
infected, and that was a recently fledged juvenile with
an unidentified *Atoxoplasma*-like parasite (Peirce *et al.*
1977).

Measurements

Males are larger than females, with no overlap of wing
length in my samples of living birds (females, 66–71
mm; males, 72–77 mm); however, males of wing
length 70 mm exist in museums (Chapter 9). Skins of
grey-plumaged birds labelled 'female' but with wings
longer than 70 mm should be treated with caution
unless gonadally sexed; they are likely to be of immature males.

Mauritius Grey White-eye *Zosterops borbonicus mauritianus* (Gmelin 1789); 'pic-pic', 'zoiseau pic-pic' (pikpik, zwazo/zozo pikpik), oiseau manioc

Affinities

Apart from *Z. b. borbonicus* on Réunion, the Grey White-eye is not particularly close to others in the genus (Moreau 1957*a*). Vocally the Grey White-eyes, particularly *Z. b. borbonicus*, are closer to mainland African (and Seychelles) species than the Olive species (Chapter 3); I found *Z. maderaspatanus* on Mohéli (Comores) to have calls similar to *Z. borbonicus* but quite unlike *Z. chloronothos* or *olivaceus*.

General habits and behaviour

The Grey White-eye is a noisy, sociable species found in groups throughout the year (Gill 1971*a*), though less consistently so in the breeding season. Group size is usually three to six, though singles and pairs are also seen, but outside the breeding season flocks of 10 to 25 are not unusual, occasionally reaching 35 or more (FRGR, CGJ, Carié 1904); CGJ once counted 81. Parties of White-eyes moving through the forest sometimes attract other species to join them, particularly Mauritius Fodies (outside the breeding season), but also Cuckoo-shrikes, Flycatchers, Olive White-eyes and Red-Whiskered Bulbuls. These flocks often mob passing animals, including humans, hopping around in the vegetation very close to the intruder, screaming loudly; CGJ saw a mongoose attract attention, Clark (1859) noted hares and cats as particularly subject to mobbing, Meinertzhagen (1912) House Shrews *Suncus murinus*, and Desjardins (Oustalet 1897) rats. Conspecifics are sometimes displaced or chased during the breeding season; JCL saw a group mobbing a female Cuckoo-shrike in early January 1979, and CGJ has several times seen them mobbing perched Kestrels.

Breeding units are pairs or small groups (cf. Gill 1973*b* for *Z. b. borbonicus*); these have a very well-defined home range, within which they temporarily join passing groups. Pairs and trios clumping and allopreening have been seen in the breeding season (CGJ).

The sexes are identical in plumage and indistinguishable in the field.

Distribution, numbers and density

Past and present distribution. The Grey White-eye occurs throughout the island in any habitat with a reasonable number of trees or shrubs, including urban gardens. All commentators from Desjardins in the 1830s (Oustalet 1897) onwards have remarked on the catholicity of this species, and later writers (Carié 1904 onwards) usually contrasted this with the restricted distribution of the other native forest birds.

Population. No attempt was made to census this abundant species. Suitable habitat (the categories forest, scrub and built-up areas (half the last discarded) of the Mauritius land-use map UNDP/FAO 1968) covers some 68 000 ha, which at a density of about 0.5–1 pair/ha (see below) gives 34 000–68 000 pairs.

Density. In all the native forest habitats Grey White-eye density was substantially greater than that of all other native species combined. At Alexandra Falls, where most of the other species were at high density, Grey White-eyes were trapped four times as often as Olive White-eyes (the next most common species) and were at a density of around 150 'pairs'/km². The number of actual breeding units will be lower, some being groups of more than two (see below). No systematic censusing was done, but numbers seemed comparable in most other native forest areas, though certainly lower in the dwarf forest of Petrin and Plaine Champagne. In the degraded forest of Parc aux Cerfs, in an area still uncleared, Grey White-eyes were trapped over seven times as often as Fodies, giving an estimated density of 110–120 'pairs'/km². In non-native vegetation at Bras d'Eau the White-eyes outnumbered Flycatchers by 1.6:1 in trapping frequency. However, mistnetting there was deliberately selective for Flycatchers; visual observation suggested that the White-eyes were 2½–3 times commoner, giving a density of 125–150 'pairs'/km².

Since the density was 100–150 'pairs'/km² in each of three very disparate habitats, it is probably reasonable to use ½–1 'pair'/ha as an average over the Grey White-eye's range.

Breeding units appear to hold territories, although they do not always exclude passing groups of conspecifics. Territories appeared to be very small (a hectare or so), in keeping with the calculated density.

Habitat

The Grey White-eye is more catholic in its choice of habitat than any Mauritian bird apart from the introduced bulbul *Pycnonotus jocosus*, occupying most wooded vegetation from dense native evergreen forest through scrubby secondary growth with odd trees to gardens in towns. It occurs in pine and *Casuarina* plantations shunned by other native birds and is equally at home in the wet forests of Montagne Cocotte and the dry gardens of Grand'baie. Although trees or large shrubs are needed for nesting, the birds will feed in herbaceous vegetation (market gardens, wasteland etc.) and also visit tea plantations, when in flower, for the nectar.

Feeding ecology
Parties of Grey White-eyes are usually seen foraging warbler-like through the forest. They are principally insectivorous and feed mainly by gleaning, are adept acrobats and are capable of running up and down vertical tree trunks like a nuthatch (*Sitta*) (pers. obs., Gill 1971a).

Gleaning insects from foliage is the principal feeding technique. In addition, the BOU Expedition has 12 observations of other methods: foraging amidst flowers (7), along twigs (1), on bark of large branches and tree trunks (2), probing rotten wood (1) and feeding on the ground (1). When foraging in foliage the birds sometimes clung to the edge of a leaf to reach prey on the midrib. It was usually possible to distinguish insect-foraging in flowers from nectar-feeding; the former was generally amongst flowers (e.g. *Sideroxylon*, *Litsea*, *Psidium*) attracting tiny pollinating insects, though the birds may obtain both insects and nectar from the same species (*Grevillea*, *Terminalia belerica*, *Eugenia* spp.; *Bertieria*, *Gaertnera*). The birds mostly forage in the canopy but also often feed in the understorey, and the herb layer, and occasionally on the ground. Gill (1971a) recorded fly-catching on one occasion.

There is little information on prey. Gill (1971a) mentioned only the common moth-like hemipteran *Flatopsis nivea*, which CGJ also often saw taken. Only caterpillars (Lepidoptera) from gall-flowers of *Eugenia jambos*, a small butterfly, a 2 cm cricket (Orthoptera) from some rotten wood, and a 2.5 cm grasshopper (Orthoptera) were recorded by the BOU Expedition; the grasshopper was struck on a twig before being swallowed. I also twice saw grasshoppers being carried as food for the young.

The expedition has 10 records of nectar-feeding, principally on introduced species (unlike the Olive White-eye). Flowers are usually probed from the front, but a Morning Glory *Ipomoea congesta* was punctured at

the base, a technique also noted by FRGR for an unidentified garden climber. Species recorded are *Lantana camara*, *Holmskoldia sanguinea* (FRGR), tea *Camellia sinensis*, banana *Musa* sp., Bottlebrush *Callistemon citrinus* (often), *Erythrina variegata*, *Hiptage benghalensis* (all introduced), and *Eugenia mauritiana*, *Bertieria zaluziana* and *Gaertnera* sp. (native).

In Réunion the Grey White-eye feeds to a considerable extent on fruit (Gill 1971a, Chapter 6), but fruit-feeding seems to be irregular in Mauritius. Milbert (1812) accused them of damaging fruit, and Clark (1859) referred to a fondness for soft fruit, especially grapes. There are no other references, and even aviculturalists like Carié (1904) and Guérin (1940–53) made no mention of frugivory, but CGJ once saw a bird take, then reject, two berries of *Schinus terebenthifolius*.

Breeding
Breeding is spread over a very long period, and the species is likely to be multiple-brooded, although there is no firm evidence. The extreme dates are freshly built nests on 24 July 1981 (CGJ) and a fledged young being fed on 28 August 1949 (FRGR), to a pair collecting food on 4 April 1974. The pattern of breeding observations by the BOU Expedition is given in Table 7; there is some suggestion that birds in lowland areas start and finish earlier. Clutches in Cambridge collected for or by E. Newton are dated October (1), November (3) and December (3), while R. Newton (1958a) found nests in December, January and February.

As in Réunion (Gill 1973b; Chapter 6) this species in Mauritius also sometimes breeds in units of more than a pair. Most birds seen breeding appeared to be in pairs, but I have one definite and another possible case of group breeding and may have under-recorded it as

Table 7. *Breeding records of the Mauritius Grey White-eye 1973–5 (BOU) and 1979–83 (CGJ)*[a]

Part of island	July	Aug.	Sept.	Oct.	Nov.	Dec.	Jan.	Feb.	Mar.
Upland plateau	–	–	3*	9	1	2	*	1	2
Middle levels	1	–	–	1	1	1	–	–	–
Coastal areas	–	*	4	3	2	1	3	–	–

[a] All observations are corrected to the estimated month in which eggs were laid; thus an observation of young being fed in early April is tabulated as a breeding record for March. Three pertinent observations by Rountree (unpubl.) in 1949–50 are indicated by asterisks (*).

this species, being common, was not studied in such detail as the rarer endemics.

Nests are built from 5 ft (1.5 m) up in bushes to 27½ ft (8.4 m) up in trees (FRGR); tree nests are usually concealed in the foliage towards the end of a branch. The nest is hung hammock-like between fine twigs and made of moss, lichen, leaves or rootlets and dry grass (depending on locality) held together by spiders' web with which it is also attached to the twigs. The lining is of fine hairs loosely coiled round, and the whole structure is thin enough to see the eggs through, though structurally quite robust (FRGR, Guérin 1940–53 and BOU Expedition records).

The clutch is normally two, rarely three, rather elongated pale blue eggs (E. Newton *in* Hartlaub 1877, Guérin 1940–53, FRGR). I cannot trace any confirmed clutches of other than two. The incubation period is not recorded, but in one nest watched by FRGR the two young left the nest (though not yet able to fly) after about 10 days. The parents feed fledged young but for how long is not known. Single fledged young were seen being fed more often than two together, perhaps because each parent generally attends one of the young after it leaves the nest; juveniles were also recognised in white-eye parties. Freshly fledged young are recognisable by the yellow gape and partly grown tail, but very soon become indistinguishable in the field from adults.

Nesting success is unknown but fledged young and adults carrying food were seen more frequently than in any other endemic bird except the Paradise Flycatcher at Bras d'Eau. The nest-site would be relatively safe from arboreal predators such as monkeys and rats, but the Red-whiskered Bulbul *Pycnonotus jocosus* has been considered responsible for nest losses to the white-eyes. Carié (1916) collected but did not publish evidence that the Bulbul not only robbed eggs and young, but went on to refurbish and make use of the white-eyes' nests (species not specified, but probably mostly the Grey as at that time the bulbul had not spread to native forest areas). The BOU Expedition found only three Grey White-eye nests; the only one followed up was found destroyed 3 days after being discovered nearly complete with the birds actively building. This nest was low (5 ft; 1.7 m) in brambles and would have been easy for a predator to reach. No evidence of Bulbul predation was obtained.

Moult

Three out of nine adults trapped at Bras d'Eau on 3 December 1974 were one third, one third and half-way through wing-moult. None of three caught at Bras d'Eau on 11 January 1975 were moulting and indeed a bird was seen carrying nesting material there as late as 27 January. Birds caught in the plateau forests in mid-December and early February were not moulting. FRGR noted that about a third of Grey White-eyes were without tails in Vacoas in late March 1949; there is a specimen in Paris, taken in April 1825, just completing moult.

Moult presumably follows after any individual has finished breeding, which may be as early as December or not until mid-April.

Longevity, mortality and disease

The only longevity record is the sighting in Macabé on 3 October 1978 of a bird ringed on 20 November 1973. I ringed only two birds there, both on the same date; the surviving bird was in exactly the same place when I saw it again in 1978.

Some of the blood parasite data collected by the expedition (for the study by Peirce *et al.* 1977) bears further examination in the light of possible pathogenicity of the parasites. Only three species were sampled in large enough numbers to allow a breakdown by locality; one of those was the Grey White-eye (data provided by M. A. Peirce). Table 8 shows that avian malaria *Plasmodium* was not found in undisturbed native forest, whereas in the white-eyes *Leucocytozoon* was rife there but absent in the coastal plantations. In the disturbed upland forest both parasites occurred, and several white-eyes there had double infections. The differences in distribution are strongly significant (chi-square test; Table 8 notes). Both species of white-eye in Mauritius are parasitised by an endemic *Leucocytozoon*, *L. zosteropis*, presumably spread by the only local blackfly *Simulium ruficorne* (Peirce *et al.* 1977); the vector is probably scarce or absent from the coastal sampling localities as they lack open fresh water (see Mamet & Webb-Gebert 1980). Grey White-eyes are evidently extremely sedentary, whereas the Cardinals and Bulbuls move around more, so even coastal birds are infected, although they are usually parasitised by another species, *L. fringillarum* (Peirce *et al.* 1977, Peirce 1979), which may have a different distribution. The absence of malaria in the undisturbed forest agrees with the distribution of the typical vector *Culex* (*pipiens*) *quinquefasciatus* (= *fatigans*), which penetrates forests only where man is active (C. M. Courtois *in* Peirce *et al*. 1977, Mamet & Webb-Gebert 1980); the 'disturbed forest' was the Parc-aux-Cerfs area which at the time of sampling was full of people clear-felling it.

The pathogenicity of these parasites in Mauritius is not known (Peirce *et al*. 1977), but the effects of malaria and birdpox on bird life in Hawaii (Warner 1968, but see Atkinson 1977) are well attested and should be borne in mind. It is not impossible that some Mauritian birds are limited in their distribution by the presence of malaria outside the undisturbed forests.

Symptoms resembling birdpox (swollen scabby feet, loss of claws) were found on four Grey White-eyes at Bras d'Eau in October and December 1974 (out of a total of 18 trapped between October and January. Elsewhere only one bird with a claw missing was caught, at Parc-aux-Cerfs in September 1974. Similar symptoms were prevalent during the same months on Flycatchers at Bras d'Eau (q.v.).

Measurements

As large numbers of this species were caught it is possible to analyse the data for regional variation within the island. Wing lengths of upland plateau forest birds (altitude *c*. 600 m) averaged 1.04 mm longer than those of birds caught at Bras d'Eau (sea level); this difference is highly significant ($P < 0.002$; Table 9). There is thus in Mauritius, despite its smaller size and much milder topography, an echo of the within-island variation that is so striking in this species on Réunion (Gill 1973*b*; Chapter 6). Average weight increases about half a gram during the breeding season (Fig. 5). No birds were caught between March and July, but the fall in weight of Bras d'Eau birds in December and January before any similar drop in upland birds is additional evidence for a generally earlier end to the breeding season in the lowlands.

More details are given in Chapter 9.

Table 8. *Incidence of* Leucocytozoon (L) *and* Plasmodium (P) *infections in three Mauritian bird species in different habitats*[a]

Habitat	Parasite	Grey White-eye[b]	Cardinal	Red-whiskered Bulbul
Undisturbed upland	L	16/17	1/6	3/6
native forest[c]	P	0/17	0/6	0/6
Disturbed upland	L	19/22	2/9	2/15
native forest[d]	P	9/22	3/9	0/15
Lowland exotic	L	0/10	5/17	5/26
plantation[e]	P	3/10	6/17	0/26
		$n = 49$	$n = 32$	$n = 51$

[a] All samples taken between August 1974 and February 1975. A further small sample taken in October 1978 from Alexandra Falls was similar (Peirce 1979), but is excluded from the table to keep the data homogeneous.
[b] The distributions of *Leucocytozoon* and malaria show a strongly significant pattern using the χ^2 test on the following contingency tables ($\chi^2 = 153.4$ and 61.3 respectively, $P \ll 0.001$ for both).

	Leucocytozoon			*Plasmodium*		
	Infected	Not infected	Total	Infected	Not infected	Total
Upland undisturbed	16	1	17	0	17	17
Upland disturbed	19	3	22	9	13	22
Lowland exotic	0	10	10	3	7	10
	35	14	49	12	37	49

[c] Alexandra Falls and Kanaka.
[d] Parc-aux-Cerfs.
[e] Bras d'Eau and Flic-en-Flac.

Table 9. *Wing lengths of Grey White-eyes from upland forest and a lowland plantation (Bras d'Eau)*

	Wing length (mm)										
	50	51	52	53	54	55	56	57	58	59	Total
Upland forest, SW Mauritius	–	2	1	6	22	18	13	2	1	1	66
Lowland plantation, NE Mauritius	1	–	4	5	6	4	1	1	–	–	22

Fig. 5. Weights of Mauritius Grey White-eyes during the breeding season 1974–5. Upland and lowland samples are shown separately, with a bar indicating two standard errors on each side of the mean ($n = 4+$) or a line showing the range ($n + 3$ or less). Samples sizes are given in small figures at the foot of the diagram.

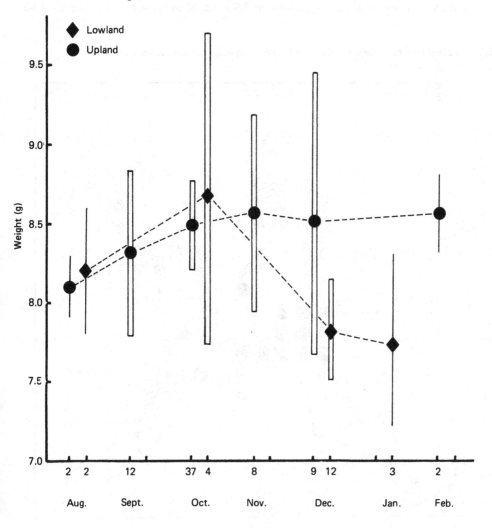

Mauritius Olive White-eye *Zosterops chloronothos* Vieillot 1817; 'oiseau pitpit' ['oiseau à lunettes']

Affinities

The Réunion and Mauritius Olive White-eye were treated as conspecific in Peters' *Checklist* (Moreau 1967; see also Moreau 1957a, b), but I follow Gill (1971a) in separating the two species. Gill considered the specialised Olive White-eyes to be earlier invaders of the Mascarenes than the generalised grey forms *Z. borbonicus* subspp. Moreau (1957a) assumed both invasions to have come from Madagascar on grounds of proximity, but there is no evidence to favour any particular source for the olive forms (Gill 1971a).

General habits and behaviour

The Olive White-eye's habits have been described by Gill (1971a). They live singly or in pairs, travelling quite long distances between favoured feeding sites (flowers), rarely associating with groups of the Grey White-eye. Both CGJ and I have seen birds displacing Red-whiskered Bulbuls. Like the Merle, the Olive White-eye has a large non-exclusive home range, but defends a small area around the nest-site or a favoured flower against conspecifics and Grey White-eyes (Chapter 3, CGJ). Pairs frequently indulge in mutual preening, and in sitting clumped together on branches, during the breeding season (see below). Olive White-eyes keep to their regular home range only during the breeding season; at other times many birds wander widely in search of flowers. The sexes are not distinguishable in the field.

Distribution, numbers and density

Past and present distribution. The distribution found in 1973–5 is shown on Maps 7 and 13. Past

Map 13. Distribution and density of the Mauritius Olive White-eye in the south-western forests, 1973–5. Area covered by map as in Map 5.

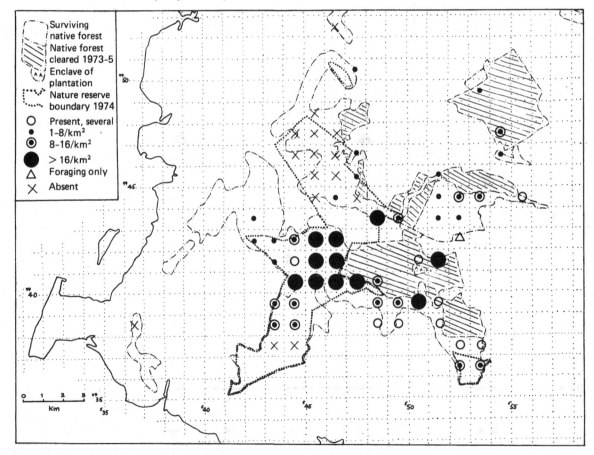

literature is little guide to the previous distribution of the species. In the 1860s it appears to have been largely confined to high ground except along the south coast (E. Newton 1861*b*), which is still the area where it is found at its lowest altitude (*c*. 1000 ft; 305 m). It was certainly resident as far north as Vacoas, as Newton collected six birds there between July and November (1860–66; specimens in Cambridge). By 1904 (Carié 1904) it was very rarely seen near the coast. Guérin (1940–53) evidently did not consider its distribution sufficiently restricted to require comment. J. Vinson (field notes, per CGJ) saw them on Le Pouce in the 1940s. In the 1930s FRGR saw Olive White-eyes regularly in Vacoas gardens, and also recorded them eastwards throughout the Montagnes Bambous and on Montagne Blanche. After the 1939–45 war he no longer saw them in Vacoas, though the range had not contracted as far as is suggested in his checklist (Rountree *et al.* 1952), as Newton (1958*a*) still found them in scrub and forest from the Piton du Milieu across to Black River Peak. Apart from the outlier at Montagne Lagrave, and occasional non-breeding birds seen around Piton du Milieu (1982: CGJ) and Le Pouce (1970s), the range has shrunk further with the clearance of the scrubby forests of the Midlands and Mare aux Vacoas areas from 1950 onwards (Map 1). Outside the breeding season the species was still sometimes seen in Curepipe gardens in the 1970s (Temple *in* King 1978–9), but no longer in Vacoas where odd ones were still to be seen in the 1950s and 1960s (Newton 1958*a*, Gill 1971*a*).

Population. Although several past writers have called the Olive White-eye rare, only d'Emmerez (1928) listed it as in danger of imminent extinction. It was not listed in the 1966 *Red Data Book* (Vincent 1966), but Temple (1974*a*) saw it as very rare and declining, and (1975, 1976*c*) gave a figure of around a thousand individuals for the population; his later figure of "probably over a thousand" was cited in the more recent *Red Data Book* (King 1978–9) with the prognosis 'vulnerable'. The names of the two White-eyes were accidentally reversed by Staub (1971), giving the impression that the Olive White-eye was common and the Grey rare.

This is the hardest of the endemic passerines to count accurately. Even during the breeding season when relatively sedentary, Olive White-eyes range over a wide area, but also remain very inconspicuous and silent unless moving around, when the flight call

draws attention to the birds. The first trait makes it difficult to know how many are in the area, the second makes a good proportion of the birds undetectable at any one time. Calculations from well-watched areas show that on average, apart from a few anomalous pairs that were always in evidence, only 32% of birds/pairs would be detected on any one visit. Underestimation may be partly offset by wandering birds being counted more than once, but as frequent visits reveal far more birds than one would suspect on any single visit this seems to be of only minor importance.

One hundred and twelve pairs were counted during 1974–5, but estimation from the one third sighting frequency, and areas of suitable habitat not visited, brings the likely population at this time to 340–350 pairs. However, I believe that, as for the Cuckooshrike and Fody, a good number of the living individuals in 1974–5 were birds displaced from Les Mares, Grand Bassin and Parc-aux-Cerfs/Mare aux Vacoas. The large numbers present on Plaine Champagne, Alexandra Falls and Montagne Savane range were probably above the long-term carrying capacity of these areas, and so the population is likely to stabilise at nearer 275 pairs in the reduced area now available (confirmed by CGJ in 1983).

Density. Density ranged from less than 1 (Macabé, Kanaka) to 37 pairs/km^2 (Alexandra Falls). High-density areas generally supported 15–20 pairs/km^2 (Petrin (west), Plaine Champagne, Montagne Savane, upper Bel Ombre), but all these areas, like the Alexandra Falls, were adjacent to recent clearances of good habitat, and were thus probably carrying above-normal numbers. Densities in more remote areas (Combo, lower Bel Ombre, Lagrave) were 10–15 pairs/km^2, which is probably a more usual long-term figure for good habitats. By October 1978 there had been a marked decline at Alexandra Falls, though no apparent change around Montagne Savane (pers. obs.); the Alexandra Falls and Plaine Champagne numbers have since stabilised at 10–12 pairs/km^2 (CGJ, 1983). In poorer habitats average densities were around 1–4 pairs/km^2. Home ranges in the breeding season appear to be up to 30 ha or more in low-density areas, and 20 ha or so in high-density conditions.

Habitat

On the plateau the Olive White-eye appears to prefer very wet areas, and rarely occurs in forests outside the 4000 mm isohyet (Brise Fer, Macabé, Mare Longue;

numbers decrease westwards across Plaine Champagne; see Map 13). This relation breaks down above the southern scarp; in Combo and parts of Bel Ombre the lower limit is at around 3200 mm. The area occupied immediately prior to 1950 (Midlands, Piton du Milieu, Parc-aux-Cerfs) is all within the 4000 mm zone, and this factor is perhaps connected with their presence on the Lagrave range but absence from the Montagnes Bambous; whether birds seen in these hills in the past bred there is not known. Within the above limits the birds are most abundant in native forest, much less so in degraded areas such as Kanaka. Low or open forest and tall forest (e.g. Bel Ombre) are all favoured (Newton 1958a, pers. obs.); the suggestion that the dwarf forest (often described as 'scrub') on Plaine Champagne is a particularly favoured habitat (Temple 1975, King 1978–9) is due largely to the exaggerated numbers there during 1973–5. Plaine Champagne is a good area, but it does not contain "most of the population" as Temple (1975) stated, but, even in 1974–5, only some 23%. FRGR commented that secondary forest dominated by Travellers' Palm *Ravenala madagascariensis* appeared to be suitable habitat in the 1930s, but that after 1945 the birds were much less often seen in it.

Feeding ecology

Olive White-eyes feed on both nectar and insects, and travel considerable distances to productive flowers (Gill 1970). The Mauritius species has much the longest beak in the family Zosteropidae, approached only by *Z. olivaceus* of Réunion (Moreau 1957a). Moreau was at that time unaware of its specialised nectar-feeding habits, although he was later to relate the species' tongue structure, the most divided and longest in its family, to nectarivory (Moreau *et al.* 1969). In fact Clark (1859) had pointed out that the bird was called *oiseau banane* because of its habit of visiting banana flowers for nectar (the name was shared with the Fody, also fond of nectar: see Cheke 1982b). Carié (1904), insisting that the bird was purely insectivorous, wondered why he had failed to keep them successfully in captivity. Guérin (1940–53) described the acrobatic method used for feeding on the nectar of hanging flowers, such as the *Trochetia* mentioned by Gill (1971a). The BOU Expedition has 14 observations of flower-feeding of which 13 are on nectar, seven of them on various species of *Eugenia* (*sensu lato*; some = *Syzygium* spp.). Mauritius supports at least 14 species of native *Eugenia*

and another, *E. jambos*, is widely naturalised in the forests (Vaughan & Wiehé 1937). Many bear nectariferous flowers, mostly during August–November (pers. obs.), which the white-eyes visit assiduously. Even a single isolated flower out of season will be found and used (Lagrave, April 1974). Species bearing flowers on the trunk or branches seemed to be more attractive in general than those with terminal inflorescences – the former appear to be adapted to pollination by lizards (e.g. *Phelsuma* spp.), the latter to insects (pers. obs.). No Olive White-eyes were seen on the large terminal inflorescences of the common tree *E. glomerata*. The taxonomy of the native *Eugenias* is confused, but those from which White-eyes have been seen feeding were identified tentatively (J. Guého pers. comm.) as *E. scandens* (a sparse-flowered vine), *E. venosa* (stem-flowered tree) and *E. obovata* (round-leafed shrub with large rounded leaves and terminal inflorescences; cited as *E. dupontii* in Chapter 3). The birds also feed from the introduced *E. jambos* (pers. obs.) and probably also from the related and widely planted *Eucalyptus* trees (*fide* Newton 1958a).

Use of other genera for nectar is more widely spread in time, but I have no records of flower-feeding during April–July apart from the single observation already noted. Species involved are *Gaertnera* spp., *Bertiera zaluziana* and *Erythospermum monticolum* (native) and tea *Camellia sinensis* and *Litsea* sp. (introduced); Jones (1980a) added *Callistemon citrinus* (introduced ornamental) and Rountree (unpubl. obs.) saw both White-eyes piercing the base of the tubular flowers of a garden climber (sp.?). *Trochetia uniflora*, cited by Gill (1971a), is now very rare; I have seen flowers pierced at the base by birds, clearly 'cheating' on this apparently bird-pollinated plant. *Bakarella hoyifolia*, also recorded by Gill, is now too rare to be an important food plant (CGJ). Many of the species mentioned, or close relatives, are illustrated by colour photographs in Cadet (1981).

I once saw an Olive White-eye prise open a bud of *Ochna mauritiana* and apparently eat something solid from inside it, perhaps stamens. Birds caught in August (1), October (1) and December (4) lacked the incrustation of pollen on the forehead so characteristic of *Z. olivaceus* in Réunion.

Flowers are in short supply after severe cyclones, but there is no evidence of white-eye numbers being seriously affected. Neither did I see any serious direct mortality during cyclone Gervaise.

The expedition has 21 observations of insect-feeding – by gleaning (20 obs.) or fly-catching (once) – suggesting this is a more important mode of feeding than Gill (1971a) believed. The birds usually gleaned from foliage, but there are two observations of gleaning from bark in the same way as the Grey White-eye frequently does. While gleaning in foliage the birds often investigate inflorescences, apparently for insects rather than nectar in the case of such species as *Sideroxylon cinereum*, *Mimusops maxima* and *Harungana madagascariensis*. Most insect-feeding observations were on native plants from shrubs to large trees, but exotic species that have invaded the native forest are also used: *Litsea* spp., *Ligustrum robustum*, *Harungana* and even *Citrus reticulata*.

Breeding

As Guérin (1940–53) noted, the Olive White-eye has very discreet nesting habits, and the BOU Expedition has very few observations on breeding activity. Carié (1904) gave the laying season as September to November, but we found preparatory activity beginning in early October: mutual preening by pairs, often accompanied by long periods of sitting clumped together, was the first sign. Birds carrying nest material were first recorded later in the month, (22 and 27 October 1974), but courtship feeding was not noted before 13 November. No nests were found, and there were no certain observations of juveniles (birds lacking the white eye-ring), but three groups of three, sometimes with one bird begging, were observed from 15 December through to early February. At other times of the year the birds were always single or in pairs.

E. Newton (*in* Hartlaub 1877 and Guérin 1940–53) and Carié (1904) described the nest as made of leaf veins dressed externally with spiders' webs and lined with fibres, animal hair, and especially feathers (which distinguish it from the nest of the Grey White-eye); it is also said to be more robust. According to Guérin (1940–53) the nest is completed in a few days. The clutch is two, rarely three (Newton *in* Hartlaub 1877; Carié 1904); there are two clutches of two in Cambridge, collected by Newton on 8 and 9 November 1863. The eggs are pale blue. During incubation the sitting bird, according to Guérin (1940–53), leaves the nest inconspicuously, flying low, but while there are young the adults give a distraction display feigning injury. Incubation and fledging periods are unknown, but Carié (1904) found that the "little groups of four or five" that are seen together after fledging very soon separate; we also saw very few family parties.

Moult

There are no dated moulting birds in museums, nor did the BOU Expedition trap any in moult. Birds at Alexandra Falls appeared to be in fresh plumage on 27 March 1974.

Mauritius Fody *Foudia rubra* (Gmelin 1789); 'zoiseau banane' (zwazo/zozo banan)
Affinities

The relationships of fodies were studied by Moreau (1960). The Mauritian bird, like the Rodrigues Fody *F. flavicans*, has a relatively fine 'insectivorous' beak and a well-developed brush tongue (Staub 1973b). Unlike *flavicans* and its ecological counterpart in the Seychelles *F. seychellarum* (see Chapter 8), the Mauritius Fody retains the bright red male coloration of its presumed Malagasy ancestors (Moreau 1960). The genus *Foudia* is endemic to the Malagasy Region and is closer to the African *Euplectes* than to any other ploceid genus (*ibid.*).

General habits and behaviour

Mauritius Fodies are inquisitive, solitary, arboreal birds, to be seen exploring foliage, branches, stumps and epiphytes from the canopy down to ground level. In the breeding season the males are highly territorial, and both sexes rather vocal; out of season they are much quieter (pers. obs., Newton 1959) and more secretive. Even during the breeding season behaviour is variable: a pair noisy and active one day can become impossible to locate the next.

Fodies often form temporary feeding associations with groups of Grey White-eyes passing through their territory, and outside the breeding season some appear to wander with White-eye flocks (pers. obs., Gill 1971b). CGJ saw a remarkable roosting movement of some 60 Fodies moving from the Plaine Champagne area to the south-west slopes of Montagne Cocotte, shortly before sunset on 16 May 1984.

Males are easier to see and catch than females, which accounts for the great disparity between the sexes in both museum collections and my trapping records (below). Males, as with the Rodrigues Fody (Cheke 1979a), respond well to playback and are easily mistnetted as they approach the tape-recorder. Females also respond to playback of male song but do not approach so close or impetuously. Outside the breeding season the response is more muted.

Distribution, numbers and density

Past and present distribution. In 1973–5 the Mauritius Fody was confined to the high plateau south-west of Curepipe and the upper parts of its south-facing slopes, with a tiny outlying group on the Piton du Fouge, an isolated peak of similar elevation (1956 ft; 594 m) in the extreme south-west (Map 14). Most of the present distribution is over 1500 ft (455 m), but descends to about 1000 ft (304 m) in Bel Ombre and Combo on the southern slopes. The birds seem also to be contained almost entirely within the zone where the mean January temperature is below 23 °C and the annual rainfall is above 2800 mm (Map 15). Native forest outside that zone contained few or no fodies (the summit climate of the Piton du Fouge seems to

have been ignored in the maps issued by the Mauritius Meteorological Services; all data from Padya 1972).

Past records indicate that the Fody was once more widespread in the uplands where rainfall is over 2000 mm and mid-summer temperature below 24 °C. There is nothing in E. Newton's writings (1865*b*, 1875, 1878*a*, *b*) to suggest that the bird was not widespread in the 1860s and 1870s, and they certainly occurred at lower elevations in the south then than now (Newton 1861*b*). In the 1890s Fodies could be seen in the gardens of Vacoas, Phoenix (both *c*. 1100 ft; 335 m) and Curepipe (1800 ft; 539 m), apparently retreating into the forests after 1895 (Carié 1904). Most subsequent writers apparently saw them only in the forests of the south-west, but Guérin (1940–53), writing in 1940, recorded their presence also near Curepipe (Forest Side) and in the Midlands area. Before the 1939–45 war FRGR saw Fodies between La Caverne and the Vacoas

Map 14. Distribution and density of the Mauritius Fody, 1973–5. Area covered by map as in Map 5.

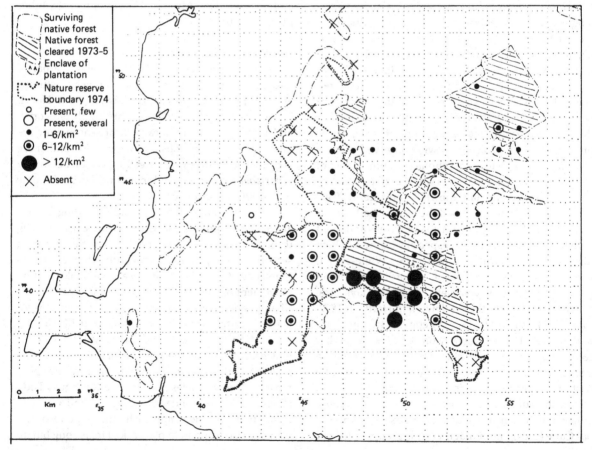

Ridges, at both ends of the Montagnes Bambous, and, more unexpectedly, at Nicolière in the north (Map 15; cf. Flycatcher); they also extended down the Savane range to Jurançon (below 800 ft; 244 m). The scrub and river reserves near Vacoas and at Jurançon, and the native forest at Nicolière, were lost during the war, and the Forest Side–Midlands area was cleared for pine and tea in the post-war period; a few birds still survived in 1974–5 in Crown Land Monvert, a small remnant of native forest just south of Curepipe (Map 14). They were then still widespread throughout the adjacent Parc-aux-Cerfs block which was being clear-felled at the time. Staub (1973a) reported occasional sightings in Vacoas and Curepipe, though regular watching in Vacoas during the expedition failed to reveal any. The distribution given by Newton (1959) accords closely with the present situation except that much habitat has since been lost. A published report of a bird on the coast of le Morne in 1976 (Steinbacher 1977) was presumably in error for a part-moulted Cardinal *F. madagascariensis*.

Map 15. Past and present distribution of the Mauritius Fody in relation to rainfall, temperature and the extent of native forest in the 1930s. Details of the vegetation in the 1930s are taken principally from Vaughan & Wiehé (1937). Climatic data from Padya (1972).

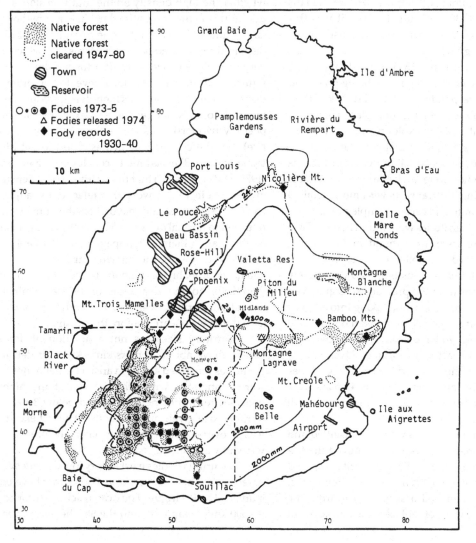

Population. The Mauritius Fody has long been a rare bird; fears for its future were already being expressed a hundred years ago (Slater *in* Hartlaub 1877). Carié (1904) observed a decrease (apparently unrelated to habitat destruction), and by 1910 Meinertzhagen (1912) considered it "a rare bird even in the south-west of the island". Newton (1959) called it "not uncommon" in the right habitat, though very local. By 1973 the new massive clearances of native vegetation on the plateau for forestry (Map 1) were causing alarm, and the threat to this species was emphasised by Temple (1974*a*, 1975) and myself (Cheke 1974 *b*, *c*).

Population figures given in 1974–5 (40–50 pairs, *ibid.*; 150–200 individuals, Temple 1975), before the census results were available, substantially underestimated the actual numbers. Temple (1976*c*) revised his estimate upwards (300 individuals), and it was this figure (equated to 100–120 pairs) that was used in the *Red Data Book* (King 1978–9).

Preliminary work in 1973–4 showed that Fodies were fairly easy to census, the males being noisy and conspicuous and responding well to playback of their calls. Repeated observations in one place suggested that two long visits (all day exploring a small area) or three line transects would normally reveal nearly all territories present. In 1974–5 all but one of the areas with resident Fodies were revisited, and accessible areas covered using line transects with more detailed coverage by exploration in certain sample plots.

A total of 121 territories were found in the 1974–5 season, and there were three found in 1973–4 that were not re-checked the following season (Cheke 1983*a*: Table 1). The 1974-5 census agreed very closely with the indications from the previous season. Given suitable habitat not censused, and taking into account the varying density (see below), I estimate the total breeding population to have been close to 250 pairs, about double those actually counted. Over half of these were along the forested south-facing slopes between Bel Ombre and Combo. Some 47 pairs were on land actually being cleared or threatened with clearance shortly.

Birds on land being cleared were unsettled and mobile. In the Parc-aux-Cerfs area I removed the territorial males from three adjacent territories, and found that all were replaced within a few days, although when I removed a second male from the central territory he was not replaced; the two adjacent birds expanded into the vacant area. The captured birds were later released elsewhere.

Returning in 1978 I noted a sharp decline in Fody numbers (Cheke 1979*a*; see also below); Jones (1980*a*) estimated the population as only 20–40 pairs but subsequently (*in litt.* 22 December 1980) considered this to have been an under-estimate. The Fody remains a priority species in the conservation programmes of ICBP (anon. 1981).

Density. The census revealed big variations in Fody density in different parts of the range, from a low of <1/km^2, to 27/km^2 on the southern scarp of the plateau between Alexandra Falls and Montagne Savane (Cheke 1983*a*, Map 14). It was not clear in 1974 whether the high density on the southern slopes was due to this being the optimal habitat or whether the large numbers were simply due to an influx of birds displaced from the adjacent Les Mares area which had very recently been clear-cut (Proctor & Salm 1975). A return visit in October 1978 revealed a dramatic decrease in fodies along this high-density slope. There was time for only one census, at Alexandra Falls (one, possibly two territories where there were previously five), but the change was just as clear around Montagne Savane. This led me to conclude (as Newlands (1975) had suggested) that the relatively high densities in this area in 1973–5 were an artefact of the adjacent habitat destruction, and that by 1978 the numbers had returned to normal densities (Cheke 1979*a*), i.e. nearer 6–7/km^2, as in Plaine Champagne and Bel Ombre in 1974–5. If this is so, the carrying capacity of existing habitat is probably only about 145 territories (rather than 250), so the clearances of 1972–4 eliminated habitat supporting over half the Fody's population (two thirds was cleared before the census).

Territory size was not a function of density except above about 20 pairs/km^2. At lower densities territories remained 4–8 ha in area whether or not they were contiguous with others. Even at the highest densities territories never dropped below 3 ha. The territory size given previously (8–20 ha, Cheke 1980) was an over-estimate based on density before the data were fully examined.

Mauritius Fodies exist at very low densities compared with other insectivorous Fodies that have been studied. The Rodrigues Fody can reach 5 territories/ha (= 500/km^2; Chapter 8) and the Seychelles species 37 individuals/ha (Diamond 1976). However, I found *F. eminentissima* at similarly low densities on Mohéli, Comores, in February 1975, and the situation is the same over much of Aldabra, though in certain parts of that island the species forms breeding colonies with defended territories at very high density (15/ha), the birds foraging beyond these limits (Frith 1976).

Habitat
Fodies occur in all types of native forest within the climatic limitations already described. The highest densities were in the forests along the southern scarp of the high plateau – very wet, 20–30 ft (7–10 m) tall, heavily dominated by *Calophyllum tacamahaca* (Vaughan & Wiehé 1941: 131) – but the concentration in this zone was probably an artefact of the destruction of the dwarf forest on the plateau above. Medium densities of 6–12 territories/km^2, probably the norm at present, occurred in the remaining dwarf forest (Plaine Champagne), high forest (but only in Bel Ombre) and in various degraded habitats invaded by exotics (especially guava and privet) and interplanted with exotic timber trees (notably *Eucalyptus robusta* and *Cryptomeria japonica*). Only one territory was found in almost entirely exotic vegetation (guava scrub with Travellers' Palm).

There is no reason to suppose that the Fody's feeding habits (see below) should confine it to native vegetation, provided appropriate substitutes are present, so it seems that the bird is limited by the structure of the forest and the nature of introduced species. A characteristic of the native dwarf forest (Procter & Salm 1975; = "*Sideroxylon* thicket" of Vaughan & Wiehé 1937) is a very open canopy, with bushy vegetation between the trees, themselves only 10–25 ft (3–8 m) tall. The structure of the degraded forest is usually similar: some tallish trees emerging from a low 6–10 ft (2–3 m) scrubby thicket. Territorial Fodies use the tall trees as vantage points for singing and as lookouts, often flying from point to point within the territory. However, Fodies were all but absent from structurally similar native habitat on the western edge of the plateau between Brise Fer and Simonet, and scarce around Mare Longue reservoir in the same area, suggesting that rainfall (or some consequence of it) is a more powerful limiting factor than structure. The typical upland climax forest (Vaughan & Wiehé 1937) on the north side of Black River Gorges supports very few

Fodies (Map 14) while similar forest in the upper part of Bel Ombre harbours good numbers; the latter area is mostly wetter but has a very similar temperature regime (Padya 1972).

In forests high humidity is usually associated with a good growth of epiphytes and rapid rotting of dead wood, both of which are necessary for the Fody's feeding habits. The absence of epiphytes and lack of dead wood may make forestry plantations within their range unsuitable for Fodies – monocultures of pine or eucalypts do not harbour the necessary insect fauna. Where pine plantations border native forest the birds sometimes used the pines *Pinus elliotti* for nesting but fed in the indigenous vegetation. Pines and *Cryptomerias* in mixed indigenous/exotic forest were also chosen for nest-sites, being possibly less vulnerable to arboreal predators. The dominant spontaneous exotics in the uplands, guava and privet, are smooth-barked and rarely support epiphytes, providing little forage for Fodies.

Typical suburban gardens, such as are plentiful in Curepipe and Vacoas, appear superficially suitable for Fodies, having a wide range of insectiferous exotics. Indeed these gardens were formerly inhabited, but the birds withdrew from these areas in the late 1890s for unknown reasons (Carié 1904). Excessive human and feline nest-predation is the most obvious cause, but an altitude-related disease is also a possibility; Fodies harbour several different blood parasites (Peirce *et al.* 1977). An artificial habitat suitable for Mauritius Fodies and other native birds could probably be created using suitable hardwood mixtures and tree-fruit plantations (Cheke 1978a).

The existence of large areas of native forest on Réunion similar in structure to other upland dwarf forest in Mauritius led me to suggest transferring some Fodies there to establish a second population (Cheke 1974b, c). The Réunion Fody has been extinct for 300 years, so there is no question of congeneric competition, and the island has no predatory monkeys (Chapter 1). This translocation was eventually agreed to by all the relevant authorities, and a trapping programme to remove birds from forests being clear-felled started in January 1975. Unfortunately little time remained before the expedition was due to end, and the programme was cut short by cyclone Gervaise, so only three birds (two males, one female) were translocated, being released at Bébour on 12 February 1975 (Cheke

1975*a*). A brief check in 1978 failed to reveal any birds (Cheke 1979*a*), nor have any been reported by Réunion birdwatchers (Barré & Barau 1982:78), but the area is very large and for the most part impenetrable. On 1 November 1974 some birds (two males, one female) were also translocated from Parc-aux-Cerfs to Montagne Lagrave (Map 15) in the hope of extending the birds' range in Mauritius; none was seen in a search of the immediate area on 29 November, but they could easily have moved further afield.

Feeding ecology

Mauritius Fodies feed on insects, fruit and flower nectar, from near the ground to the tops of trees. Insects are taken largely by probing or stripping dead wood and searching the bark of living trunks, branches and epiphytes like a nuthatch (*Sitta* sp.; Gill 1971*b*, Newton 1959, pers. obs.), though the birds also glean from foliage and flowers (Table 10). There is some suggestion of a sexual difference in feeding methods, females being more inclined to forage by leaf-gleaning than males.

All observations of these feeding techniques were on native species: *Sideroxylon puberulum, Labourdonnaisia glauca, Mimusops petiolaris, Calophyllum tacamahaca* and *Eugenia glomerata* were identified. It was rarely possible to identify insects captured but the following were observed: Orthoptera (both grasshoppers and crickets) (4), 'long grub' (? beetle larva) (1), 'small larvae' (1) and an entire small spider's web complete with attached insects (and ? the spider). Gill (1971*b*) saw a Fody eat a ½ in. long (12 mm) white grub, and d'Emmerez (in Manders 1911) reported small caterpillars.

One of the best ways of detecting Fodies in an area was to see whether the berries of the common small introduced undershrub *Ardisia crenata* had been 'treated'. Fodies feed on this fruit in a very characteristic manner, perching on the plant, peeling the fruit skin and then chiselling the flesh off the stone with the beak, leaving the stripped seeds still attached to the plant. The fruits are about 1 cm across, about half of which is the hard stone, presumably too large to swallow. Native *Eugenia* spp. berries are treated similarly but the birds swallow whole the little blue fruits of *Ossea marginata*, an introduced shrub. *Ardisia* is in fruit throughout the year, but native *Eugenias* are seasonal, and *Ossea* fruits only briefly in mid-summer (January to February). Guérin (1940–53), the only author to mention frugivory, stated that Fodies favoured Strawberry Guava and Rose-apple, both introduced. I never saw a Fody feeding on guava; the birds' range no longer extends into any significant areas of Rose-apple.

Fodies were seen taking nectar from the flowers of native *Eugenias*, *Aphloia theaeformis* and the introduced *Eucalyptus robusta* and *Callistemon citrinus*. All these flowers have shallow floral discs easily accessible to the bird's tongue. D'Emmerez (*in* Manders 1911) reported persistent feeding at banana inflorescences for "minute insects", but more probably in fact for nectar; bananas are absent from the Fody's present range (cf. the Créole name *oiseau banane*: Cheke 1982*b*).

Birds held in captivity prior to transfer to Montagne Lagrave and Réunion showed similar feeding patterns to those seen in the wild, readily stripping dead mossy branches, eating *Ardisia* and *Ossea* berries and taking nectar from Rose-apple flowers. It was,

Table 10. *Insect-feeding methods of Mauritius Fodies, 1973–5*

| Sex | Searching along, up or down trunks and branches[a] | | | Gleaning from leaves, leaf-bases and flowers[b] | | |
	Dead wood	Live wood	Unspecified	Methodical searching	Watch, see and jump	Totals
Male	11	11	2	3	1	28
Female	3	3	2	4	–	12
Totals	14	14	4	7	1	40

[a] One observation of foraging on dead wood on the ground is included in the table.
[b] All observations of males gleaning by searching on flowers or leaves were when they were feeding in close company with their mate. In one case the male kept with the female but did not feed, later going off to forage on trunks and branches.

however, clear that some birds were berry specialists; others rarely, if ever, fed on them. The incidence of 'treated' *Ardisia* berries in fody territories was likewise very variable, reflecting this tendency to specialise. Captive birds also took readily to banana (peeled; whole or crushed), and to an artificial nectar mixture. Captive birds fed on banana in two ways: some bit off chunks to swallow, others flicked their tongues in and out of the flesh as though feeding on nectar. They were not interested in free-moving insects offered to them, but readily took *Drosophila*, small moths, beetles and grasshoppers if immobilised in honey or by removal of limbs. Termites, offered in pieces of a nest, were taken at first (the bird breaking up the nest structure) accompanied by much drinking, but were later refused. Small earthworms were ignored.

In the wild more time is spent by the birds foraging for insects than feeding in other ways, but whether this is reflected in the quantity of food taken is impossible to say. Captive birds sometimes took as much as 5–10 ml of nectar mixture per day, and ate large amounts (several grams) of banana, yet would leave some of the insects offered. I suspect the ratio of insect to plant food is much higher in the wild.

When feeding in company with Grey White-eyes the foraging sites generally remained separate, the Fodies probing trunk and branches, the White-eyes gleaning on twigs and in the foliage.

Both in beak shape (Moreau 1960) and the development of a brush tongue (Staub 1973b) the Mauritius Fody is closest in the genus to the slightly smaller Rodrigues bird *F. flavicans*. Their feeding habits differ, however, in that *flavicans*, when feeding on insects, forages largely on foliage and twigs like a tit (*Parus* spp.) rather than concentrating on thicker wood, dead and alive, like a nuthatch (Cheke 1980, appendix; Chapter 8). I only once saw a Mauritius Fody perch on the edge of a large leaf to examine its under-surface, as *flavicans* commonly does (Chapter 8). The birds' tails reflect this difference: longest in the genus (relative to size) in *flavicans* (Moreau 1960), like a tit; shortest in *rubra* (*ibid.*), like a nuthatch. The other fodies are more generalist feeders, taking a high proportion of seeds (Benson 1960, Crook 1961, Frith 1976, Benson *et al.* 1976–7, Bathe & Bathe 1982b), though the Seychelles Fody has many feeding techniques in common with *flavicans* (Crook 1961, Lloyd 1973, pers. obs.). Amongst its other feeding methods, *F. eminentissima* probes bark and dead wood

(Frith 1976, Aldabra; pers. obs. on Mohéli, Comores), but it appears to obtain only a small proportion of its food in this way. The Cardinal *F. madagascariensis* very rarely forages in bark and epiphytes (pers. obs.; see below).

Breeding

Seasonality. Activity leading to breeding begins in August, male Fodies becoming vocal and territorial, and pairs associating closely. The earliest observations of nest-building are in the second week of October (9 October 1971, D. A. Turner per AFW; 12 October 1974, pers. obs.), though SAT heard birds with (? free-flying) young on 21 October 1973, which suggests egg-laying in mid to late September. Most recorded nests have been in October and December, but in some years nesting continues until late March: on 22 March 1974 two pairs were carrying food at Petrin. By 13 April there was no longer any nesting activity. The end of the season may depend on the incidence of cylones: in 1973–4 there were none, but in 1974–5 breeding was brought to an abrupt stop by the violent storm Gervaise on 6 February (see below). During 1979–83 CGJ observed territorial behaviour during August–February but none from March to July. Previously published dates for the breeding season only took in October to February (Newton 1959).

Nesting biology. Prior to nesting, pairs start associating very closely together, and are frequently seen feeding within a few inches of each other, both very vocal, the male singing and the female typically giving a churring call ('buzz-note'; Chapter 3), interspersed with rapid chipping notes in unison ('plick-call'; *ibid.*). I twice saw a feeding pair in which the female was continuously wing-shivering as though soliciting, although no copulation or courtship feeding took place. Wing-shivering is a highly developed behaviour in the Rodrigues Fody (Chapter 8), but not in other species of the genus; male Mauritius Fodies sometimes shiver their wings when singing.

The relative roles of the sexes in building the nest are unclear in this species and variable in the genus (Crook 1961 *contra* Newton 1959 for *madagascariensis*; Frith 1976; Chapter 8). In four cases the female alone was seen bringing nest material and working on the nest, the male actively fussing around and pursuing her but not helping, exactly as described by Newton (1959). However both birds were once seen working together on a nest, a single male was seen carrying nest material, and Gill (1971b) watched a nest where the male did most of the building. Whichever bird is actually building, the pair associate closely during nest construction, the male actively 'chipping' and singing, chasing the female as she works on the nest (Newton and Gill *locs. cit.*; pers. obs.) and, to any watching predator, generally advertising what is going on. Newton (1959) even saw a male mount a female collecting nest material. The male of the cooperating pair (above) was seen to bring material in his feet (JS).

The six nests observed in this study, and two more observed by D. A. Turner (per AFW), were all in trees, placed 7–30 ft (2–9 m) up, usually in the thickest foliage in the tree's canopy. Three were in indigenous trees (*Calophyllum tacamahaca*, 2; unspecified (DAT), 1), the others in introduced species (*Cryptomeria japonica*, 2; *Pinus* spp., 2; unspecified conifer (DAT), 1). The lowest nest (7 ft up) was in a *Cryptomeria* which had thick foliage at all levels; the pines and native trees usually had thick foliage only in the canopy. CGJ found a nest 1.5 m up in a guava, and another 17 ft (4.8 m) up a *Crytomeria*. Newton (1959) recorded nests from as low as 3 ft (1 m) in brambles, but usually 20–50 ft (7–16 m) "up in a tree towards the end of a branch". Ends of branches seem typical sites in conifers, but not in indigenous trees (pers. obs.). Hartlaub (1977, citing E. Newton as source) gave the height range as 1.5–5 m.

Five nests of which the construction material could be observed were made primarily of grass or grassheads, mixed with mosses, lichens or small twigs. In contrast Newton (1959) asserted that Fody nests were made of small roots and moss, and that this (in addition to greater robustness) distinguished them from the similar, but grass-built, nests of the Cardinal. Hartlaub (1877) described a similar mossy nest collected by E. Newton, adding that the lining was soft and embellished with down. Lafresnaye (1850) illustrated a nest made largely of grass stalks; de Querhoënts (1773) was of grass and leaves lined with feathers and "silk-cotton". Clearly materials used are variable; the nests I saw were made of the same materials as used by Cardinals in the same area.

The nest is domed with a lateral entrance like other *Foudia* nests and sometimes has a 'porch' extending the entrance somewhat, a feature which Lafresnaye (1850) considered diagnostic. However, I have found this detail to be variable in all fodies whose nests I have seen (*rubra*, *flavicans*, *seychellarum* and *madagascariensis*), and Frith (1976) noted the same thing on Aldabra (*eminentissima*).

No clutches or nestlings were observed in this study. Nests which reached the incubation stage were inaccessible, and all were destroyed before hatching. CGJ's guava nest contained three eggs, and Newton (1959) stated that "the usual number [of eggs] appears to be three", though he may (Newton 1958a) have seen only one clutch. There are four clutches in Cambridge collected by or for E. Newton in October and December 1865 – two C/3, and two C/2 – though one cannot be sure the latter were complete. Hartlaub (1877, Guérin 1940–53), citing E. Newton as source, gave the clutch as three. The brood size remains unknown. Staub (1976), not citing a source, gave incubation and nestling periods as 14 and 18 days respectively. The clutch is three, rarely two, in the Rodrigues Fody (Chapter 8), one or two in the Seychelles bird (Watson 1979), three (once four) in *F. eminentissima aldabrana* (Frith 1976), and two to four for the Cardinal in Mauritius (Meinertzhagen 1912, Carié 1916, Newton 1959), two or three in Rodrigues (pers. obs.). Mauritius Fody eggs, like those of Cardinals, are pale blue or blue-green (Newton 1959).

During incubation the male remains conspicuous and noisy, often diving from his high vantage point to mob a potential intruder. The female appears to do all the incubation (obs. on one nest only), as in other fodies (Frith 1976, Crook 1961; Chapter 8). Both sexes fed the young in one pair seen repeatedly ferrying food into a pinewood made impenetrable by brambles.

Guérin (1940–53) assumed that the Fody, unlike the Cardinal, nested only once a year. There is time to rear three or even four broods between mid-September and the end of March, but no evidence that this happens. Some late-nesting pairs appear to be young birds breeding for the first time: the male of the pair nesting in January and February 1975 whose nest was destroyed by the cyclone was still moulting into red plumage, and the female had a pale bill. This female could even have been a young fledged earlier the same season (see below).

Breeding success. Breeding success in this species seems to be extremely low. Nests were rarely found, and I have observations on only five for which the outcome was known: none progressed beyond incubation, four being destroyed by predators and one by cyclone Gervaise. Horne (Chapter 3) also failed to observe a successful nest. In each season there was only one observation of free-flying young, in October 1973 and January 1975.

The principal predator is probably the common introduced macaque, though Ship Rats are also likely to be very significant. No actual predation was observed, but monkeys were often seen 'bird-watching' and the fact that two of the predated nests were completely destroyed (one missing, the other torn to shreds) suggests monkeys rather than rats. The other two, less extensively damaged, might have been attacked by either predator. Newton (1959) and Guérin (1940–53) also implicated monkeys, the latter adding mongooses and lizards *Calotes versicolor*, both also introduced. However, the mongoose cannot climb trees, and the lizard prefers hotter, more open country than is favoured by Fodies (pers. obs.), so I doubt that either is much to blame.

I believe the low densities of Fodies are attributable to the very low nesting success, which explains why the distribution is often patchy: there are simply not enough Fodies to occupy available habitat, and those that there are retain a territory size related to a natural condition of higher density.

Arboreal predators were introduced so early into Mauritius (Chapter 1) that we have no pre-predator reports of the Fody, and there are no records of striking abundance such as in the now extinct Réunion Fody before the introduction of rats (Chapter 1), or the Seychelles Fody on predator-free islands (Penny 1974, Prys-Jones & Diamond 1984, pers. obs.) and the Aldabra form in places where nest-sites are least prone to rat predation (75% vs. 100%, Frith 1976). Mauritius is the only Indian Ocean island where monkeys have been let loose (Paulian 1961: 242), but Ship Rats are elsewhere the most serious predators on island fody populations, having wiped out the birds entirely on Réunion (Chapter 1) and most of the islands of the Seychelles (Diamond & Feare 1980). The Rodrigues Fody enjoyed high nesting success in 1974–8 as no Ship Rats seemed to be present on that island during the study (Cheke 1979*b*, Chapters 1 and 8). Outside the settlement (where they can nest with reasonable success) the Aldabran *F. eminentissima* is thinly distributed and has little or no success in breeding (Frith 1976), and the same species on Mohéli

(Comores) occurs at a low density comparable to the Mauritian bird (pers. obs.). In addition to rats, Mohéli supports lemurs *Lemur mongoz* (Tattersall 1977) and a tree-climbing skink *Mabuya maculilabris* (Blanc 1972, pers. obs.) which may also be nest predators. Amongst fodies, only the Cardinal appears to be resistant to rats and other mammalian predators, reaching very high densities in many areas (Réunion: Barré & Barau 1982, Barré 1983; Mauritius: FRGR (unpubl.); Rodrigues: Chapter 8).

Plumage sequences and moult

The adult male Mauritius Fody has a bright red head, throat and rump. The female is dull grey-brown, as are newly fledged young. As Staub (1976) has already pointed out, this species, like other fodies but *contra* Newton (1959) and Moreau (1960), has an eclipse plumage. In April 1974 I saw several males actively undergoing a complete moult (some had lost all tail feathers simultaneously), the red being replaced by brown. In April 1979 and late March 1983 all males were in eclipse (CGJ). Two birds on 20 May had only a few scattered red feathers; all others seen were entirely brown, their sex indeterminable. One dated museum specimen (in Cambridge) examined was in moult – a male one third through wing-moult on 15 March 1863. I was out of the island for most of June and July, but by early August many males were back in red plumage; the same was true in 1973 (JMV) and 1979–83 (CGJ). Throughout the breeding season some birds are to be seen moulting into red plumage, and I suspect these are young from the current or the previous season. In the hand, brown-plumaged birds can be sexed on wing length (see below). On 4 February 1975 I caught a brown male near Montagne Savane which had taken over the territory of a red male I had removed a few days earlier for transfer to Réunion. His plumage was very worn, suggesting a young bird from the previous season. I saw a similar bird in late March 1974 just beginning to acquire a few red feathers.

Freshly fledged young are recognisable by their pale horn-coloured bills (black in adults). The bill presumably normally darkens after a few months like that of the Rodrigues Fody (Chapter 8). It appears that, as with the Rodrigues Fody (*ibid.*), young males may moult directly into red plumage or spend a year in a second brown plumage. Two males in Bel Ombre in mid-summer (December–January) had pale bills but were acquiring red plumage (JS). Males still in brown plumage a year after hatching have black bills. Some females retain the pale beak until the next breeding season (CGJ, September and November). I saw a pale-billed female nesting in January 1975, paired with a half-red (? first-year) male; she may have been young of the year, as I had seen no pale-billed females earlier in the season.

The annual complete moult occurs at the end of a breeding season, males moulting only contour feathers to regain red plumage before the next season. The pattern is the same in the Rodrigues and Aldabra birds (Chapter 8, Frith 1976) and the Cardinal (pers. obs.), but the Seychelles Fody has only one moult per year (AWD). It is not yet known whether the Mauritian bird invariably has a complete post-juvenile moult or whether, as with *F. flavicans* (Chapter 8), this varies between individuals.

Behaviour during cyclones

The cyclones to which Mauritius is subject, and to which the native birds are presumably adapted, last from several hours to several days. During the storm normal activity is rendered impossible due to the high winds and torrential rain. It is not possible to watch birds in the field during cyclones, but I had three birds (two males, one female) in cages in my house during Gervaise (February 1975); they were clearly strongly affected by the storm. Although completely protected from any wind and rain, they were in a room with windows facing the full brunt of the storm. Throughout the day, although not inactive, they refused almost entirely to feed, two of them losing over a gram in weight. Both before and after the storm they fed readily. A visit to one of the best Fody sites (Montagne Savane area) 2 days after the cyclone revealed no apparent loss of birds, though a nest was completely destroyed. The behaviour of introduced birds visible around houses during cyclones was briefly discussed by Jones (1980a) and Cheke (1975b).

Relationship and interactions with the Cardinal Foudia madagascariensis and other species

Slater (*in* Hartlaub 1877) attributed the decline of the Mauritius Fody to the introduction and spread of the Cardinal, but as Carié (1904) soon pointed out, their habits are very different; Newton (1959) concluded that the rarity of the endemic form was due to predation and forest destruction as much as to any possible competition. I agree with Newton, but the ecological separation is not as clear-cut as he claimed.

Newton (1959) considered the Cardinal to be largely a ground-feeding granivore, occasionally taking insects. Such observations as I have of Cardinals feeding in Mauritius, which are mostly from upland forest habitat, suggest a much broader diet: grass and *Faujasia flexuosa* (Compositae) seeds (3), insect-gleaning in moss and lichens on branches (1), gleaning in flowers (*Mangifera indica*, 1), carrying insects to young (2) and obtaining nectar from flowers (3); Carié (1916) reported berries also. The nectar flowers were *Grevillea* spp. and *Lantana camara*, introduced species. The *Grevillea* was tackled from the front in the same way as would a Mauritius Fody, but the *Lantana* was plucked and nibbled in the way described by Melville (1979), which is widely used on Indian Ocean islands by Cardinals for small detachable flowers with corolla tubes (Chapter 8). I have only one record of a Mauritius Fody (a male) plucking *Lantana* flowers; the flowers were plucked and immediately thrown away, apparently as a displacement activity after he had failed to drive the observer (JS) away.

As Newton (1959) observed, Cardinals show a seasonal vertical migration, birds being absent or nearly so from the uplands from early May to August, many not returning until October. While on lower ground the birds congregate in flocks and are conspicuously granivorous. Males generally moult into eclipse in April or May, and remain so until August/September (lowlands) or October/December (uplands). However I have seen upland males moulting from worn red plumage directly into fresh red feathers (February and March 1974); these birds probably moult again in April–May into eclipse, thus having three moults in the year. At any time of year there were always some birds in full plumage. Breeding appears to be later than in the endemic Fody: November–April (uplands) and October–May (lowlands) (from extreme dates of nests seen; Newton (1959) reported late September–April, peaking December–March, altitudes combined). In the eighteenth century, when the Cardinal was newly established, the breeding season was reported as July and August (de Querhoënt 1773).

Although it occurs throughout the island, the Cardinal is, as Newton (1959) noted, at its lowest breeding density in upland forests; it was at a lower density in 1974 than the native bird along the southern scarp of the plateau. It is almost entirely absent from the Macabé area where the endemic Fody is also scarce. Few interactions with *F. rubra* were observed:

male Fodies displaced territorial Cardinals from song-posts or perches on a number of occasions. The reverse occurred at Montagne Lagrave in November 1974 when I liberated Fodies that had been caged for a few weeks and were disoriented. There, a male Cardinal was seen twice to displace a newly released male Fody, suggesting that the native species had been absent from this area for some time.

Given the catholicity of feeding and plasticity of habitat tolerance exhibited by the Cardinal, it is not impossible that over time it has been able, as in Rodrigues (Chapter 8), to displace the native bird from marginal habitat. However, it has been in the island for over 200 years (Chapter 1), so the situation has presumably stabilised and I doubt whether it is responsible for any of the endemic Fody's difficulties within its present distribution and habitat.

A male Fody was also seen twice to displace a male House Sparrow *Passer domesticus* (Parc-aux-Cerfs, September 1974), a species with which it is not normally sympatric. Other male Fodies tolerated less related species: on 6 December 1974 one was seen sitting quietly next to a singing Red-whiskered Bulbul (JS).

Despite the near-disappearance of the only resident raptor, *Falco punctatus*, a male Fody well outside the Kestrel's range, quietly preening, vanished tittering into the undergrowth when a dove *Streptopelia chinensis* glided over – a bird much the same size, shape and colour as the rare Kestrel (Parc-aux-Cerfs, 6 April 1974).

Weight and measurements
Females have shorter wings (<66 mm) than males (>66.5 mm), but are similar in weight (Chapter 9). A similar sex difference exists in Rodrigues Fodies, Aldabra Fodies and Cardinals (*ibid.*).

Acknowledgments
Thanks are first due to Wahab Owadally, Conservator of Forests, who facilitated the setting up and operation of the expedition and made the facilities of his department available to me. Tony Gardner, Assistant Conservator at the time, was also a valuable source of help and information. Stan and Barb Temple very kindly put me up for my first month in Mauritius, and later I was able to appreciate the friendly family atmosphere at the de Baritault's *pension* at Quatre Bornes. Stan introduced me to the birds; Claude Michel, France Staub and Jean-Michel Vinson, and the late Leo Edgerley gave freely of their knowledge of the island and its fauna. M. Lee Man Yan of the Mauritius Meteorological Services at Vacoas kindly sent me rainfall data for 1973–5, and Charlie Collins arranged for the identification of insects eaten by swiftlets. I am most grateful to Ruth Ashcroft, Mike Brooke, Bob Cheke, Tony Diamond, John Hartley, Yousoof Mungroo and Jennifer Shopland for help and companionship in the field at various times, to David Lorence and Joseph Guého for plant identifications, and to the Noordally family for regular hospitality after a hard evening's work in the swiftlet caves. Claude Michel, then Director, allowed me full access to the library collections of the Mauritius Institute, let the expedition use the Institute's address for mail, and kindly answered various queries after I left the island; Robert Antoine and Mlle M. Ly-Tio-Fane gave me access to the MSIRI library; my special thanks also to Dr Reginald Vaughan for lending me valuable books and giving of his time and knowledge. Back in England I have had useful correspondence and discussions on Mauritian birds with Frank and Paddy Rountree, Jonathan Lawley, Willie Newlands, John Hartley, Nigel Collar and Carl Jones. Shirley Bateman, for no tangible recompense, typed the manuscript – my thanks to her for taking on my handwriting and giving of her time.

Note added in proof
An update on the current status of Mauritian birds and conservation on the island will appear in *Oryx* in 1987.

5

The larger land-birds of Mauritius

C.G.JONES

Mauritius Kestrel *Falco punctatus* **(Temminck 1823;** *mangeur de poules (mañzer d'pul)*
Introduction
The Mauritius Kestrel was a little-known falcon until the current series of studies on its biology was started in January 1973 by Dr S. A. Temple (Temple 1977a, McKelvey 1977c, 1978, Jones 1980d, Jones *et al.* 1981, Jones & Owadally 1982b). I have drawn widely upon the unpublished manuscript of Temple (1978c) and have endeavoured to bring it up to date in the light of recent research. (Some of the following material also appears in Jones & Owadally (1985) (*Ed.*).)

Taxonomy
Brown & Amadon (1968) placed the Mauritius Kestrel in a super-species with the European Kestrel *F. tinnunculus*, Moluccan Kestrel *F. moluccensis*, Australian Kestrel *F. cenchroides*, Madagascar Kestrel *F. newtoni*, Seychelles Kestrel *F. araea*, and perhaps the American Kestrel *F. sparverius*. They also regarded the Lesser Kestrel *F. naumanni*, Greater Kestrel *F. rupicoloides* and the Fox Kestrel *F. alopex* as being typical kestrels. Fox (1977) kept the above grouping but gave them sub-generic status. Cade (1982) kept the sub-genus, calling it *tinnunculus*, but considered *F. rupicoloides* and *F. alopex* to have diverged earlier from the rest of the group.

Sub-generic status for the typical kestrels is more desirable than placing them in a single super-species. Within Brown & Amadon's super-species, *F. punctatus* and *F. araea* are too divergent to justify such a close

grouping. It would, however, be reasonable to place the kestrels which appear to be of recent radiation from *F. tinnunculus*, or a common ancestor, in a single super-species. These would be the allopatric *F. tinnunculus*, *F. sparverius*, *F. newtoni*, *F. cenchroides* and *F. moluccensis*.

The Mauritius Kestrel's nearest relative was regarded by Siegfried & Temple (1975) as *F. rupicoloides*, following Newton's (1958a) hint that the Mauritius Kestrel was "very reminiscent of the greater kestrel". Later, Temple (1977a) regarded the Mauritius Kestrel as being so specialised that its closest relative remained unclear, but suggested (1978c) that it belongs to the *tinnunculus* group, probably derived from an African source.

Benson & Penny (1971) and Jones & Owadally (1981) suggested that the kestrels of Seychelles and Mauritius are derived from *F. newtoni*, and that the Madagascar Kestrel is derived from the Kestrel *F. tinnunculus*, or a common ancestor. The radiation of the Indian Ocean kestrels from Madagascar seems most likely, although the exact relationship of the now extinct Réunion Kestrel with the others can only be inferred. Cowles (in press; Chapter 2) suggests that the Réunion bird and *F. punctatus* were closely related. A common origin is probable or perhaps one species evolved from the other.

Plumage
A good description of the Mauritius Kestrel is provided by Brown & Amadon (1968), and others by Guérin (1940–53), Staub (1976), McKelvey (1977c) and Cade (1982). The crown, nape, back, wing and tail coverts, and tail are similar to the plumage of the female kestrels in the *F. tinnunculus* super-species, being chestnut-brown, streaked or barred with black. The tail is barred with seven or eight black bands, the sub-terminal band being the broadest. The tail feathers are tipped with an off-white terminal band. The ventral surface is creamy-white spotted with distinct black heart-shaped spots which are rounder and smaller on the breast, more oval on the abdomen and broad arrowheads on the flanks giving an almost barred appearance in some individuals.

The plumages of the male, female and juvenile are similar with no readily detectable differences. Some males are not so heavily marked on the dorsal plumage with fainter streaking and spotting on the crown and nape and less barring on the back and wings. Brown & Amadon (1968) claim that in immature plumage the Mauritius Kestrel is paler chestnut above and more streaky below. I have carefully examined both live and dead Kestrels of known age and cannot confirm these differences. Juvenile Kestrels may be more spotted with less well developed arrow-head barring on the flanks but this is a minor plumage difference not easily detected. Field observations of recently fledged Kestrels give the impression that they are more streaked because they habitually sit with their feathers fluffed.

Since *F. punctatus* probably evolved from the *F. tinnunculus* super-species group, the lack of colour dimorphism between the sexes and the almost complete absence of a typical kestrel adult plumage pattern can be regarded as a degenerative feature. The evolution of adult plumage pattern in diurnal raptors usually involves an elaboration of the juvenile plumage, and the streaky and spotted plumages of the juvenile are probably closer to the ancestral plumage than is that of the adult (Johnson & Peters 1963). In the absence of some of the selective pressures common to continental areas the need for highly evolved plumage patterns has diminished. The Seychelles Kestrel *F. araea* has also lost the colour dimorphism between the sexes. But this species, which is also derived from *F. newtoni* of Madagascar, has lost the female plumage pattern; both sexes have the colouring of a male kestrel, and the juveniles have a distinct immature plumage (Brown & Amadon 1968).

Bare parts

The iris is dark brown, not dark ruby as described by Staub (1976); the culmen is blue-black shading to blue-grey at the base. Specimens collected by E. Newton (now in Cambridge) have the culmen described variously as grey, lead colour, bluish horn and horn colour. The cere is pale yellow and the periorbital region is yellow-grey or light yellow. The tarsus and toes vary from a pale yellow-grey to yellow or greenish-yellow. In captive birds fed a diet rich in carotene the cere, periorbital region, tarsus and toes may become a rich yellow. The claws are dark, almost black. The juveniles are similar to the adults although the feet, cere and periorbital region are a pale blue-grey becoming progressively yellower during their first year. See Growth, Development and Parental Care below for a description of the bare parts of nestlings.

Morphological features

The Mauritius Kestrel is a medium-sized kestrel slightly smaller and less heavy than the European Kestrel *Falco tinnunculus* (Tables 1 and 2; Chapter 9).

Proportions

The short rounded wings are an obvious morphological feature of the Mauritius Kestrel (Temple 1977a, 1978c, McKelvey 1977c, 1978, Jones 1980d). An adaptation for forest living, this shortening can also be seen in the New Zealand Falcon *Falco novaeseelandiae* and the Seychelles Kestrel *F. araea* (Fox 1977, Cade 1982) and among raptors is highly developed in the genus *Accipiter*. The convergence in wing shape between the Mauritius Kestrel and an *Accipiter* is shown in Fig. 1, where the wing outlines of a Hobby *F. subbuteo*, *F. tinnunculus*, *F. punctatus* and Sparrowhawk *Accipiter nisus* are all compared. The wing of the Mauritius Kestrel is relatively 17% shorter than that of a European Kestrel and 21% shorter than that of a Hobby. It is comparable to the Sparrowhawk which has only a 6% shorter wing (Table 1).

The tail is noticeably shorter than in the European Kestrel although measurements reveal that it is relatively 96% as long. The tarsus is long, in relative terms, some 32% longer than in the European Kestrel or Hobby and 66% longer than in the Eleonora's Falcon *F. eleonorae*, and only 6% shorter than the tarsus of a Sparrowhawk. A long tarsus is presumably an adaptation for reaching into vegetation to capture prey. The toes are also elongated, about 15 or 16% longer than those of *F. tinnunculus*. In absolute terms, however, the feet of the Mauritius Kestrel are virtually identical to those of the European Kestrel in both measurements and proportions and are typical for a kestrel, being thick-set and covered with scute-like scales.

The culmen is similar in length to that of a European Kestrel but is not so heavily built being neither as wide nor as deep.

Wing loadings

The size of the bird dictates the wing-loading value, the larger species having larger values (Table 2; Chapter 9). In a comparative study Fox (1977) has shown that in raptors of comparable size the highest values are a feature of the most rapacious attacking species that take fast agile prey. Attackers need to attain high speeds and this is achieved by a reduction of the wing

Table 1. *Body measurements and proportions of the Mauritius Kestrel and some selected raptors*

Species	Sex	Sample size	Total length (mm)	Body length, BL[a] (mm)	Culmen to skull (mm)	Wing (mm)	Wing % BL	Tail (mm)	Tail % BL	Tarsus (mm)	Tarsus % BL	Centre toe (minus claw) (mm)	Centre toe % BL	Tarsus + centre toe % BL
Eleonora's Falcon *Falco eleonorae*	?	1	385	196.5	27.5	310	157.8	161	81.9	35.5	18.1	35	17.8	35.9
Hobby *F. subbuteo*	M	1	308	158	23	252	159.5	127	80.4	36	22.8	32	20.3	43.1
European Kestrel *F. tinnunculus*	F	1	366	167	24	255	152.7	175	104.8	38	22.8	28	17.4	40.2
Mauritius Kestrel *F. punctatus*	M	3	292.3	134.5	22.5	170.6	126.8	135.3	100.5	40.5	30.1	27	20.1	50.2
Sparrowhawk *Accipiter nisus*	M	1	393.7	194.7	24	232	119.2	178	91.4	62.4	32.0	42.6	21.9	53.9

[a] Body length (BL) equals total length minus tail and culmen to skull.

Table 2. *Wing areas and wing loadings of the Mauritius Kestrel and some selected raptors*

Species	Sample size	Sex	Weight (g)	Wing area (cm^2)	Wing loading (g/cm^2)	Source of data
Greater Kestrel *Falco rupicoloides*	1	M	280	799	0.35	Cade (1982)
European Kestrel *F. tinnunculus*	2	M	181	704	0.26	This study
	2	F	219	788	0.28	This study
American Kestrel *F. sparverius*	11	F	113	305	0.37	Hartman, in Cade (1982)
Mauritius Kestrel *F. punctatus*	5	M	129	445	0.29	This study
	2	F	165	520	0.32	This study
Lesser Kestrel *F. naumanni*	1	?	147	546	0.27	
Sparrowhawk *Accipiter nisus*	8	M	177	552	0.32	
	9	F	295	737	0.40	Brown & Amadon (1968)
Sharp-shinned Hawk *A. striatus*	9	M	99	412	0.24	
	12	F	171	560	0.31	

area relative to weight, reducing profile drag and enabling a faster flapping rate (Cade 1982). By contrast the lowest wing loadings are possessed by species that feed on slow-moving prey. These raptors are termed 'searchers' and are typified by the kites and harriers. Intermediate between these two extremes are the 'typical' kestrels. My measurements of *F. punctatus* wing loadings show that they do not differ markedly from the values in other kestrels of similar size (Table 2). Cade (1982) gave the Mauritius Kestrel a higher value (0.43 for males and 0.50 for females) which places it among the 'attacking' falcons. This is due to his use of high body weights in his calculations (see Chapter 9). I believe these high values are misleading.

Sexual dimorphism

In common with most other raptors the female Mauritius Kestrel is larger than the male. Reversed sexual size dimorphism in raptors is related to diet: the faster the prey the greater the degree of dimorphism. In the genus *Falco* the lowest dimorphism is shown by the insect-eating species such as the Lesser Kestrel *F. naumanni*, and the greatest dimorphism is a feature of the bird-eating falcons such as the Orange-breasted Falcon *F. deiroleucus* and the Peregrine Falcon *F. peregrinus*. I. Newton (1979) showed this relationship using an index obtained by dividing the female wing length by that of the male. In the insectivorous Lesser Kestrel this index is 1.00, while in those species that prey on birds the index of dimorphism may be as much as 1.16. All the kestrels in or derived from the *tinnunculus* species grouping, including *F. punctatus* (1.06), have an index between 1.00 and 1.09 (Table 3). Temple (1978c) has suggested that island species of kestrels, including the Mauritius Kestrel, show more marked dimorphism than continental forms, but the available evidence does not support this. The degree of sexual dimorphism shown by this kestrel is consistent with the data on wing loading and body proportions, which all suggest that in common with the other kestrels it is an 'intermediate' hunter which preys mainly on moderately active prey such as insects and reptiles and occasionally on more agile prey like small mammals or birds.

Fig. 1. Wing outlines of some selected raptors.

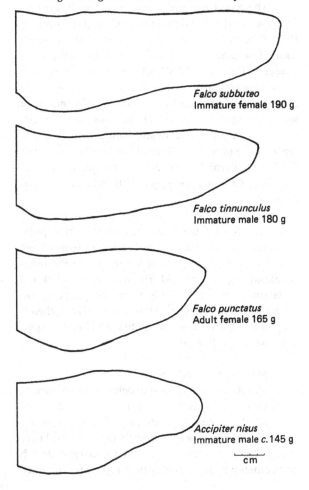

Falco subbuteo
Immature female 190 g

Falco tinnunculus
Immature male 180 g

Falco punctatus
Adult female 165 g

Accipiter nisus
Immature male *c.*145 g

cm

Table 3. *Sexual dimorphism index of kestrels*

Species	Index[a]
Kestrel *Falco tinnunculus*	1.04
Moluccan Kestrel *F. moluccensis*	1.04
Australian Kestrel *F. cenchroides*	1.08
American Kestrel *F. sparverius*	1.04
Madagascar Kestrel *F. newtoni*	1.09
Seychelles Kestrel *F. araea*	1.04
Mauritius Kestrel *F. punctatus*	1.06
Lesser Kestrel *F. naumanni*	1.00

Notes:
Wing length data from Brown & Amadon (1968), Cramp & Simmons (1980) and this study (*F. punctatus*).
[a] Female wing length/male wing length.

Flight characteristics

Mauritius Kestrels make much use of updraughts from cliffs and steep-sided gorges especially during courtship but also at other times of the year. While soaring the wings appear rounded and the tail is fanned giving the Kestrel the silhouette of a small *Accipiter*. Mauritius Kestrels often stoop headlong in angled dives with wings held nearly closed and close to the body, or occasionally when stooping steeply the wings may be held shut and against the body. It is a more direct and powerful flier than the *Falco tinnunculus*, *F. sparverius* and *F. naumanni* that I have observed in the wild, but is perhaps not as dashing as some of the small *Accipiters*. Below the forest canopy it is capable of rapid *Accipiter*-like flight, weaving between the trees with great agility. While in level flight it alternates between fast wing beats and glides, and the style of flight is intermediate between that of *F. tinnunculus* and *Accipiter nisus*.

Moult

Mauritius Kestrels moult during the summer. Temple (1978c) noted that individual birds could be identified by their distinctive moult patterns between October and March. These dates are earlier than other records available. McKelvey (1978) inferred that the adults are moulting their primaries and tails by mid-January. In a series of unpublished reports Newlands (Feb.–October 1975) recorded that in February 1975 after the passage of cyclone Gervaise the moult of the captive kestrels completely halted for 10 days, resuming in March and finishing in April.

All my observations suggest that the moult starts during December or more usually January and may not be complete until the beginning of June. At the end of January 1979 the male at Site 1 had its central tail feathers and also some primaries missing. The female had also moulted her central tail feathers, but her wing-moult was more advanced than that of the male. I could detect no moult in the Site 1 birds at the end of January 1981, but by mid-February the male at least, had moulted his central tail feathers. The Montagne Zaco pair moulted early at the end of 1981 and a secondary feather was found beneath a favourite perch on 3 December. In the middle of the month after the second nesting attempt by this pair had failed, I removed a single moulted primary feather from the nest cavity along with some body, breast and wing covert feathers. On 12 February 1982 in a roost-site used by the male, I found many moulted body feathers and an outermost tail feather. In April another moulted tail feather was found in this site along with a secondary feather.

I have examined museum skins at the BMNH and UMZC. Eight adult specimens collected between 20 June and 1 January show no evidence of moult. A specimen collected on 6 April 1860 was about half-way through the moult; most of the primaries were freshly grown except P 9 and 10 which were old and P 8 was still vascular at the base. The only new tail feathers were the two central ones. About half of the secondaries had been recently replaced.

Observations on the moult of five adult captive kestrels (three males, two females) confirm this timing. In four birds moult started in January and by the end of the month all had lost their central tail feathers. The other adult, a female, did not start her moult until March.

The moult has many similarities with those of *Falco tinnunculus* and *F. sparverius* (Cramp & Simmons 1980). The primaries moult ascendantly and descendantly from a moult centre at P 4 according to the usual sequence P 4,5,6,3,7,2,8,1,9,10 but with some variation between individuals and even between the wings of the same bird (see Table 4). The moult of secondary feathers is very variable, but usually progresses ascendantly and descendantly from a moult centre at S 5 or 6. The tail-moult follows the sequence T 1,2,6,3,4,5 which is identical to the sequence in *F. sparverius* (Cramp & Simmons 1980) and *F. tinnunculus* (pers. obs.).

In common with *F. tinnunculus* and *F. sparverius* the Mauritius Kestrel goes through a partial post-juvenile moult of ventral, dorsal and some wing covert feathers. This occurs in March–June when the birds are about 120–200 days old. There is strong individual variation in the extent of this moult. The markings on ventral feathers that are replaced are more arrowhead-shaped and barred than the spotted and heart-shaped markings of the juvenile.

Maintenance and comfort behaviours

Maintenance and comfort behaviours are similar to those of other members of the genus *Falco* (see Cramp & Simmons 1980). After a period of incubating the female perches nearby, usually defecates and may stretch her wings, tail and leg in a unilateral stretch and sometimes also raises both wings above her back in a two-wing stretch. This is accompanied by preening and vigorous body shaking (rousing). Kestrels often sit in the sun and recently fledged birds are fond of lying in patches of sun. I have not seen any specific sunning postures.

Both wild and captive birds bathe; it is usually infrequent among captive birds but may occur two to four times a week in some individuals (Jones *et al.* 1981). Dust-bathing has been recorded in wild and captive birds (Jones 1980*d*, Staub 1976, Temple 1978*c*, Jones *et al.* 1981). Our captive *F. punctatus* do not dust-bathe more frequently than captive *F. tinnunculus*.

Mauritius Kestrels sleep perching on one foot with their head sunk into their shoulders or with the head resting in the middle of the back, the bill under the scapular feathers. They roost at night in a specific roost-hole or a nesting-site. Captive birds roost in their nest-boxes or in the corner of the aviary (Jones *et al.* 1981). Roosting in sheltered areas and in cavities is probably a cyclone-avoiding strategy.

Distribution, status and population
Past status and distribution
It is not known how many Kestrels were on Mauritius during the early years of the island's colonisation, but Cade (1982) made a rough estimate. In 1753, 1644 km² of the island were probably suitable habitat, and allowing for one pair every 5–10 km² he estimated that there could have been between 164 and 328 pairs. Recent workers agree that in the past the population was probably well distributed over the island wherever

suitable tracts of evergreen forest existed (McKelvey 1977*c*, Temple 1977*a*, 1978*c*, Jones 1980, Jones & Owadally 1982*b*). In historical times, the contracting distribution of the kestrel paralleled the destruction of the native forest.

The earliest distribution records that I have traced come from the 1830s when birds were recorded from the dense forests in the centre of the island, Flacq and Camp de Masque in the north of the island, and from le Cap in the district of Savane in the south (Oustalet 1897, Guérin 1940–53) (Map 1).

It would seem probable that during the winter months unpaired birds would have wandered from the south into the forested areas of the northern plains, as suggested by the Kestrel that visited the property of Julien Desjardins in April 1830 (*ibid.*). However there may have been a permanent and possibly breeding population in the northern forested regions, since it is suggested that birds were collected there during the breeding season. In December 1829 and February 1830 Julien Desjardins obtained four Kestrels killed at Camp de Masque (*ibid.*).

During the 1860s and 1870s Edward Newton collected ten Kestrel specimens with locality data, all from the south-west corner of the island: six at Vacoas, one from near Mare aux Vacoas, two from Bois Sec, one from Baie du Cap, and a juvenile from Tamarin Bay. Temple (1978*c*) gave collection localities in the south-west and recorded a specimen from Mahébourg, without supporting details.

Temple (1978*c*) records sightings and specimens from Curepipe, Montagne Blanche and Montagne

Table 4. *Moult sequences in the Mauritius Kestrel*

	Primary number									
	(outer)									(inner)
	10	9	8	7	6	5	4	3	2	1
Usual moult sequence	10	9	7	5	3	2	1	4	6	8
Male kestrel left wing	10	9	8	6	3	2	1	4	7	5
Green 1979 right wing	10	9	8	6	3	2	1	4	5	7

	Tail feathers											
	6	5	4	3	2	1	1	2	3	4	5	6
	(outer)					(central)						(outer)
Usual moult sequence	3	6	5	4	2	1	1	2	4	5	6	3

Fayence, in the 1920s, and from Montagnes Bambous and near Montagne Lagrave in the 1940s. I have been unable to trace any of these specimens and the records must be regarded as doubtful. Guérin (1940–53) had, prior to 1940, seen Kestrels on the Montagne des Créoles. A label accompanying two Kestrel eggs at the Mauritius Institute and dated about 1928 states "bird with a tendency to disappear. It's with trouble if one

sees them along the mountain chain of Grandport [Montagnes Bambous]". Although we have no direct evidence it seems probable that Kestrels disappeared from these mountains in the early 1950s.

There is a questionable sight record of Kestrels from Montagne Longue in the Moka range above Port Louis in the 1940s (Temple 1978c) and I have a report from an elderly person who killed Kestrels on the

Map 1. Past and present distribution of the Mauritius Kestrel. The stippled area shows the Kestrel distribution 1976–83 and asterisks indicate isolated sightings or specimens. Other dates are for last records in areas of probable self-supporting populations.

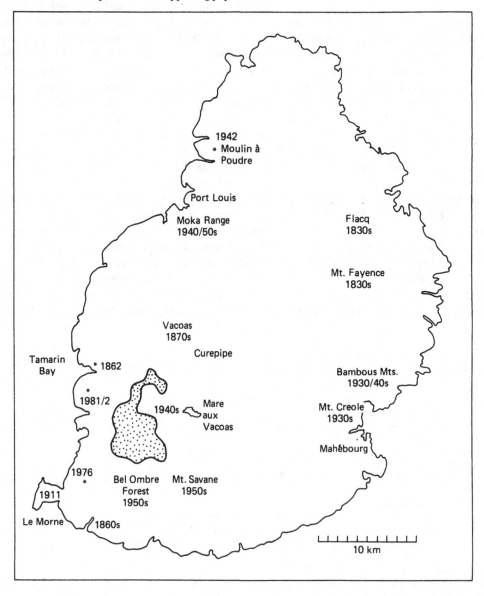

Map 2. Distribution of Mauritius Kestrels 1900–40.

Location	Probability	Location	Probability	Location	Probability
A Pieter Both	***	K Macabé	***	Q Le Morne	***
B Le Pouce	***	L Plaine Champagne–	***	R Bel Ombre Forest	***
C Montagne Ory	***	Les Mares–		S Montagne Savane	***
D Corps de Garde	*	Plaine Sophie		T Montagne Fayence	*
E Montagne du Rempart	**	M Black River	***	U Montagne Blanche	*
F Trois Mamelles	**	Peak		V Montagnes Bambous	***
G Tamarin Gorge	***	N Chamarel Gorge	***	W Montagne Lion	**
H Yemen	***	O Montagne Cocotte	***	X Montagne des Créoles	***
I Tourelle du Tamarin	**	P Piton du Fouge	***	Y Montagne Lagrave	*
J Brise Fer	***				

Probability of Kestrels occurring in specified areas:
*** Certainly/highly probably present   ** Probably present   * Possibly present

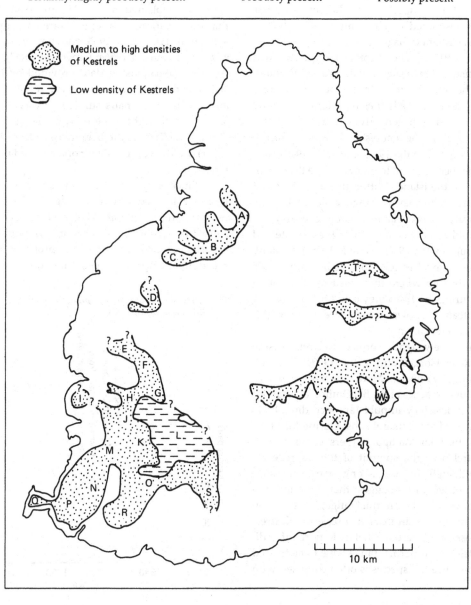

Moka range in his youth (1920s and 1930s). When the birds became extinct on this range is unknown, although there is a specimen in Geneva collected at nearby Moulin à Poudre in 1942 (MHNG 1291.62; C. Weber, pers. comm.). I have an unconfirmed report of a pair seen on Montagne Ory in the late 1950s or early 1960s (Mrs. K. Morgan, pers. comm.).

In the south-west there has been a gradual range contraction during the last century. In 1861 E. Newton (1861b) observed that in Savane district there were "A few, but not as many as elsewhere". ("Elsewhere" presumably meaning around Vacoas where Newton collected most of his specimens). This reference has been widely misquoted, e.g. by Temple (1977a, 1978c), Michel (1981). Meinertzhagen (1912) saw a Kestrel on Le Morne in 1911 and he claimed the species "was fairly well distributed throughout the island" although in his diary he noted Kestrels ("numerous"!) only from Chamarel, Alexandra Falls (nesting), and the Grand Bassin area (Meinertzhagen 1910–11 per A. S. Cheke; reported wrongly as 'field notes' by Temple 1977a).

Since the end of the 1950s and early 1960s Kestrels have not been proved to occur outside the south-west corner of the island. Within this area they had disappeared from the Savane range of mountains, the southern scarp below Plaine Champagne and the mountains of Piton du Canot, Montagne Laporte and Piton du Fouge by the 1960s (Temple 1978c). F. Staub (pers. comm.) saw Kestrels during the early 1950s above Baie du Cap where they used to take young chickens. During the 1940s they could be seen on the upland plateau in Gower near Mare Longue (M. d'Unienville, pers. comm.).

There are few sight records of Kestrels outside the Black River Gorges during the 1970s. F. Staub (pers. comm.) saw one near Montagne Savane in 1976 and in the upper reaches of the Bel Ombre forest about 1970. In 1976 (McKelvey: unpubl. tape recording, 1977) saw a single bird at Chamarel Falls. During July 1981 and August 1982, Jan Mainguard (pers. comm.) saw a single Kestrel near the summit of the Tourelle du Tamarin, although this area is only marginally suitable. These observations suggest that some individuals wander outside their normal range, but still keep to forested and mountain areas. In no case did Kestrels have to cross non-forested habitat. Reports of individuals outside their normal range have to be treated with caution since the species is often confused with

migrant falcons. The population trend is summarised in Fig. 2.

Claims that the Mauritius Kestrel occasionally wandered to Réunion (Meinertzhagen 1912, Guérin 1940–53, Watson *et al.* 1963, Staub 1973a, Temple 1977a) cannot be substantiated. A kestrel did occur on Réunion but it became extinct during the early years of the island's colonisation and is known only from one early account and subfossil remains (Chapters 1 and 2).

Recent status

Accounts published in the 1950s and 1960s suggested that the Mauritius Kestrel was rare and approaching extinction (Hachisuka 1953, Newton 1958a, Rountree *et al.* 1952, Vinson 1956a; Greenway 1967). Brown & Amadon (1968), using data provided by J. Vinson, claimed "this species is now on the verge of extinction and less than ten pairs survive". By 1970 the population was thought to be between six and ten pairs (Temple 1977a). At the beginning of the 1970s Staub (1971) could account only for one pair and two solitary birds.

Since early 1973 the Kestrels have been studied more or less continuously (Fig. 3). The numbers recorded in the wild are likely to be under-estimates since the non-breeding birds in the population are especially difficult to locate. It is probable that not all breeding attempts have been noted, especially those

Fig. 2. Tentative pattern of population decline of the Mauritius Kestrel 1930–80.

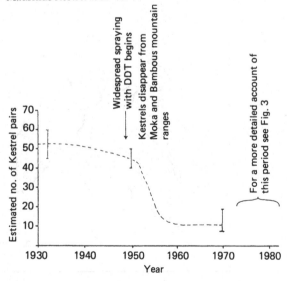

that failed early during the breeding cycle. Accounts that give the population of birds as one or two individuals (McKelvey 1977c, 1978, Temple 1977a, 1978c) are not realistic. The breeding sites discussed below are shown on Map 3.

1973/1974. In April 1973 the Kestrel population was first censused; only eight or nine individuals were located. In July a pair disappeared and it is suspected that they were shot. During the breeding season no young were reared although there were two pairs in the population. There were also two or three known non-breeding birds. In December a pair was trapped for a captive breeding programme (Temple 1977a, 1978c). The history of the captive breeding project (Jones 1980d, Jones *et al.* 1981) is beyond the scope of this paper.

1974/1975. The captive female died in March and an unmated female was captured in May. The remaining four birds, two pairs, survived the year and during the 1974 breeding season one pair nested in a cliff cavity at a site known as the 'Hole in the Wall' at the head of the Rivulet Roches Blanches in the Black River Gorges. This pair reared three young.

1975/1976. Between February and April 1975, following the passage of cyclone Gervaise on 6 February, there were no sightings of Kestrels in the wild (W. Newlands, Feb., March and Apr. 1975). In April Newlands wrote: "Most of the time spent in observation had been spent in the previous known territories,

but it seems that the birds have moved their ground, possibly as a result of a decline in the numbers of *Phelsuma* lizards in the high altitude forest following Gervaise". In May Newlands located two Kestrels in Macabé, two on Brise Fer and one in the Black River Gorges. Newlands also commented: "The birds seem to have been hunting under the forest canopy, making it extremely difficult to locate them." During September Newlands and Temple surveyed the population in detail and wrote:

> We found both of the known wild pairs – one in the Black River Gorges and the other on the Briz Fer–Tamarin Gorge escarpment. Their breeding timetable is running more than two weeks behind last year's schedule. The Gorges birds still have at least two of last year's three young birds within their territory . . . The cliff pothole from which the Gorges pair fledged their three young last season has not been reoccupied. This is unfortunate because the 'Hole in the Wall' was completely monkey proof. Instead, they have selected another cliff pothole, but this time in a lower rock which also contains caves used regularly by monkeys. [This site was above Rivulet Escalier.] . . . The Briz Fer nest site has not yet been pinpointed. (Newlands, October 1975)

Newlands left Mauritius at the beginning of the breeeding season and Temple (1978c) summarised the year's breeding results. Three pairs are alleged to have held territories. The Brise Fer pair was not detected breeding. One pair failed (presumably the pair above

Fig. 3. Mauritius Kestrel population 1973–83.

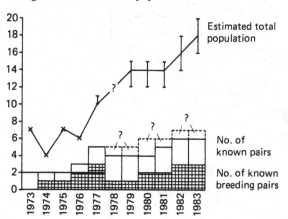

Map 3. Distribution of Mauritius Kestrel pairs and nest-sites 1975–83. The map shows all known nest-sites except the one at Tamarin Gorge. Letters denote Kestrel pairs/territories:

A   Site 1
B   Brise Fer
C   Le Bouton
  1   Hole in the wall
  2   Escalier cliffs
  3   Le Bouton cliffs

D   Site 3
E   Tatamaka cliffs
F   Morne Sèche
G   Montagne Zaco
H   Black River Peak

(Map drawn from Directorate for Overseas Surveys 1958 DOS 29 (Series Y881) and personal field observations.)

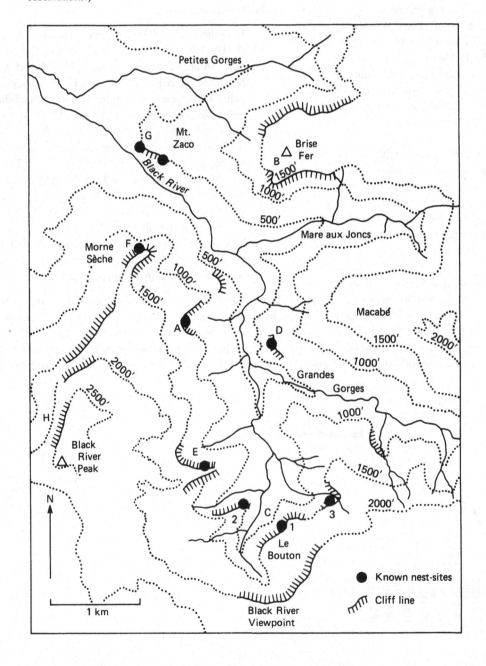

Rivulet Escalier), and a third (pair E, Temple 1978c) apparently did not attempt to breed. There is no exact record of this pair's location although Temple (1978c) implied that it was the same pair that produced young at Tamarin Gorge in 1976 and on Le Bouton cliffs in 1977 (these two sites are 8 km apart). At the end of 1976, the wild population was thought to be seven: three pairs and an unmated individual (Temple 1978c).

Temple (1978b) contradicted these observations in a discussion on nest-site traditions: "Being imprinted on a cliff-nesting site, two of these offspring paired and bred successfully on another cliff in 1975; meanwhile their parents also continued to breed successfully at cliff sites".

1976/1977. There were three pairs in the population. One pair nested in a cliff at the head of Feeder Jules Edouard and fledged three young (McKelvey 1977c, 1978). (This site is variously called Site 1, Roche Blanche and Bel-Ombre Cliff by McKelvey (1977c, 1978) and others.) Temple (1978c) recorded that this pair ('pair D') reared only two young; it is reputed to have nested at other localities in previous years (see account for 1975/1976). Another pair reared two young in the Tamarin Gorge (McKelvey 1977c, 1978). McKelvey (1977c) noted: "The one remaining wild pair has for the last five years been unsuccessful at nesting, due to choosing a tree cavity accessible to monkeys who prey on the eggs." This third pair was on Brise Fer (McKelvey, unpubl. tape recording). Since 1973 a pair has been seen frequently on Brise Fer; they were suspected to breed but this has not to my knowledge been proven. It has been stated that this pair were tree nesters (McKelvey 1977c, 1978, Temple 1977a, 1978c), although it has also been claimed that this nest-site was never found (Newlands Oct. 1975 and *in litt.* Jan. 1980; McKelvey unpubl. tape recording; T. A. M. Gardner *et al.* pers. comm.). In 1977 a pair in this area frequented a cliff site but did not breed (P. Trefry *in litt.* 1977, McKelvey 1978).

1977/1978. A pair was seen at Tamarin Gorge in July but during the breeding season only a single bird was seen. In the Black River Gorges three separate pairs were located. The Site 1 pair reared three young. A new pair, located at the head of Rivulet Roches Blanches in Le Bouton cliffs (Site 2), had two young, one of which was taken for the captive breeding programme. The third pair was found in low cliffs at the

junction of the Grande Rivière and Rivulet Roches Blanches (Site 3) by Jonathan Lawley. There were two young, both taken for captive breeding (Trefry unpubl. field notes, Jones *et al.* 1981).

1978/1979. In July 1978 Fay Steele captured an adult pair of Kestrels near Site 1 for the captive breeding programme. In the succeeding months another pair took over this nesting territory and reared three young (Jones *et al.* 1981.) During 1978 Kestrels were seen seldom at Site 2 and only once at Site 3. A pair was seen on Morne Sèche although they were not proved to breed (Steele 1979 and unpubl. field notes).

1979/1980. A pair nested successfully at Site 1 towards the end of 1979 and reared two young (Jones 1980d). During the year pairs were seen at Site 2, Morne Sèche and Brise Fer Mountain although no breeding was detected. In November a group of four Kestrels were seen from the Black River Viewpoint, flying over Le Bouton cliffs (Site 2).

1980/1981. Only two pairs are known to have attempted to nest. Site 1 was occupied and three young fledged, but one of these disappeared after fledging and is believed to have died. A new nest-site was discovered in the cliffs above Rivulet Tatamaka, but this pair reared no young. Birds were again seen on Brise Fer and Morne Sèche but no breeding was detected.

1981/1982. Again only two breeding pairs were located. The Site 1 cliff was again occupied, but a different cavity was used further down on the cliff face. I removed the eggs for artificial incubation and subsequent hand rearing. Two of the three eggs were fertile and hatched, one bird being reared. The pair is now known to have relaid. The second pair on Montagne Zaco laid three eggs which were taken for the captive breeding programme. All three hatched and two were subsequently raised. The pair moved to another cliff nearby and laid a replacement clutch. All three of these eggs hatched but they were not reared beyond day 1 (Jones 1982, Jones & Owadally 1981). Pairs of birds were seen or reported on Brise Fer, Morne Sèche and Tatamaka cliffs. A single Kestrel was seen in June 1981 and again in September 1982 on the southern slopes of Black River Peak. There are inaccessible cliffs in this area where Kestrels could easily nest quite undetected.

1982/1983. Early during the breeding season five pairs were observed displaying and holding territory and a further unconfirmed sighting of a pair was reported. Only three pairs attempted to breed.

In early October, the Montagne Zaco pair were displaying above the cliffs, but later in the month there was no sign of the male and the female was hunting on her own. It was suspected that the pair failed to breed because there was insufficient food available to support such an attempt. From 1 November the female was provided daily with food, and on 5 November the male was seen briefly in the area, but subsequently failed to appear. In mid-November the female laid an infertile clutch of two eggs which was removed and replaced by dummy eggs. She incubated until the very end of December after which the supplementary feeding was discontinued.

Nesting pairs were also located on the cliffs above Rivulet Tatamaka and in the cliffs of Morne Sèche. The pair in Tatamaka cliffs succeeded in fledging two young at the end of December. These disappeared soon after leaving the nest and they may have been killed by feral cats which are common in the area. The pair on Morne Sèche failed in their first nesting attempt, when it is thought they had young nestlings. A repeat clutch was laid in another nest-site lower down the cliff face in mid-December. The author removed the clutch of three eggs and successfully hatched and raised two males and a female for the captive breeding programme.

Displaying pairs were seen at Site 1 and in the valley of Rivulet Escalier, but these pairs progressed no further, and an unconfirmed report of a pair at Site 3 was obtained. A pair was seen on Brise Fer in August, but this area was not thoroughly checked subsequently. The cliffs on the southern slopes of Piton de la Rivière Noire were searched without finding any Kestrels.

1983/1984. Six or seven pairs were known in the population, but only three pairs attempted to breed.

A pair was present on Montagne Zaco – the same female that has been present since at least 1981/82, the male an unknown bird. They were provided with food daily from 26 September to encourage breeding in this fringe territory (see account for 1982/1983). In mid-October a clutch of three eggs was laid in the same cavity used the previous year. This clutch was taken for the captive breeding project. Only one egg

was fertile; it hatched and the bird was reared successfully. A replacement clutch was laid in mid-November in a cavity further down the valley that had previously been the male's roost-site. These eggs were removed sequentially and a dummy egg left in the nest scrape to keep the female loyal to the cavity. The female laid four eggs, but again only one was fertile; this was successfully hatched and reared at the captive breeding programme. The pair were given a well-incubated European Kestrel egg which they hatched on 27 November and fledged on 30 December. This cross-fostering was done to give the pair experience at rearing young (the female failed to rear three young hatched in late 1981). After fledging, the European Kestrel was almost totally dependent upon the male for food until it was caught on 6 February. On 7 January the pair started copulating again and a third clutch is believed to have been laid during the third week of the month out of sight over 1.5 m deep down a small hole at the back of the nest cavity. None of the eggs are known to have hatched.

The Site 1 pair reared one or two fledglings, a female of which was captured for captive breeding. Two young, a male and female, were reared by the pair on Tatamaka cliffs; this female was also caught for the breeding project. The Morne Sèche pair were seen courting from at least early October to early November, but did not progress further than this. Pairs of Kestrels were present on Black River Peak and Brise Fer mountain but their reproductive success is not known. Odd birds were reported from the Black River Viewpoint and Macabé Ridge during the year.

1984/1985. Data for this season were obtained too late to be incorporated elsewhere in the text or into any of the tables. Field coverage was good, perhaps the best ever, and eight or nine pairs were located and single birds were seen at other sites. Six pairs are known to have fledged 11 young, making this the most productive season on record.

All the known sites were occupied and the following pairs fledged young: Site 1 two young, Tatamaka cliffs three young, and Escalier cliffs two young. The pair at Montagne Zaco fledged one young and this pair is discussed in more detail below. Pairs were present at Morne Sèche and Brise Fer, the latter pair being located in Petites Gorges; both were observed nest prospecting and copulating but are not known to have progressed further than this (Fox *et al.*

1985). An unconfirmed sighting of a pair of Kestrels was made at Morne Sèche on the Magenta scarp in August but their breeding success is not known (A. Chevreau pers. comm.; see also reports in Fox *et al.* 1985).

A new pair was located at 'Stone Man', a site between Site 1 and Vanillerie valley. Kestrels had been seen on this ridge and in the valley since at least 1980 but were always assumed to have been Site 1 birds. One chick was successfully fledged although the remains of a second were found beneath the nest cliff where it is thought it had been predated by a mongoose. An unknown pair was discovered at the head of the Mare aux Joncs valley (Manava Gorge) and they fledged at least two young (Fox *et al.* 1985). This site had been checked in previous years without finding any sign of Kestrels.

Single birds were seen at five different sites: Le Bouton, Black River Peak, east of Brise Fer, mid-Manava Gorge and on Macabé Ridge. Fox *et al.* felt that these birds were from undetected pairs, although they could have been unpaired or floating birds or, in at least one or two cases, birds from known pairs. No birds were located or reported from outside the kestrel's known distribution.

The Montagne Zaco pair deserves separate treatment since it was subject to supplementary feeding and egg manipulation experiments which are summarised below. The pair was provided daily throughout the year with about 60–80 g of white mice.

The first egg of the season was laid on 12 September and the clutch of three eggs was removed from nest cavity C (see Fig. 5 for details of these cavities) on 23 September. One egg was fertile; it was hatched and reared at the captive breeding centre in Black River. The pair re-laid and on 7 October the first egg of the second clutch was laid in cavity D. Four eggs were removed on the days they were laid; after the removal of the fourth egg the female was allowed to complete her clutch and when the nest cavity was examined on 26 October there were four eggs which were removed for artificial incubation. The first four eggs that had been removed sequentially were placed under an incubating captive European Kestrel; three of the eggs were fertile but none hatched. Three of the four-egg clutch were fertile and one hatched on 22 November. This was later successfully fostered back under its parents. Meanwhile they had started their third clutch on 11 November in a new cavity (E). Three eggs were

taken sequentially, and following this a clutch of three eggs were laid and were removed on 1 December. Only two of these six were fertile and none hatched.

Radio telemetry studies had shown that the male would probably be able to support this female and a chick without supplemental feeding. To test the hunting efficiency of the male in this fringe territory additional food was reduced to one mouse (20 g) a day. On 2 December the chick from the second clutch, now 10 days old, was fostered under this pair and its development was uneventful. After this bird fledged but while it was still dependent upon its parents for food, the female again showed signs of wanting to lay and frequented a small, previously unused nest cavity (F). A smashed poorly calcified egg was found beneath the cavity on 15 January. During the following week the female was very sporadic in her attentions to the cavity and it was suggested that the male was unable to provide her with enough food, forcing her to do some of her own hunting. After about 10 days the female lost all interest in the cavity and when it was examined, a single apparently infertile egg was found.

The single young reared at this site was largely dependent upon its parents for food up until it was last seen on 7 March.

The sequence of the Montagne Zaco manipulations needs to be highlighted. Twenty eggs were laid in a total of four 'clutches'. Two of these, the second and third clutches, were extended to eight and six eggs respectively by egg-pulling. The first three clutches were laid while the food was being supplemented by white mice. The high productivity was undoubtedly due in part to the increased availability of food. The first clutch is the earliest recorded and the clutch of four eggs, with which the female completed her second series of eggs, is the largest recorded. This female also laid the largest number of eggs and clutches recorded in a season. Fertility was unfortunately low (47%) and hatchability was lower still (22%). Judging from other studies on falcons the low fertility is undoubtedly a feature of the pair rather than a product of the management. The low hatchability is, however, probably due in part to faulty incubation techniques. The possibilities of using supplemental feeding, egg-pulling and double-clutching to increase productivity are very real.

The increased number of birds found this season is undoubtedly in part due to the better field coverage, but also illustrates an increase in the actual population.

The growth in the number of birds located (Fig. 3) and the greater frequency with which they are being reported by gamekeepers, foresters and hunters is clear evidence of this. Kestrels are also being recorded at sites from which they were formerly absent and the last few years has seen a gradual filling-in of the patchy distribution which was evident in the 1970s (Table 12; Jones & Owadally 1982*b*). The larger number of single, unpaired or floating birds is also suggestive of a growing population. At the beginning of the 1984/85 breeding season the population probably stood at between 20 and 30 birds.

Temple (1978*c*) working on the assumption that a pair will defend 600 m of cliff and that home ranges can overlap by 50%, calculated the carrying capacity of the available habitat as follows: Black River Gorges could support about 18 breeding pairs with an additional population of 20–30 non-breeders; Montagnes Bambous region about 5 pairs plus an additional 10–15 non-breeders; Montagne Blanche 2 breeding pairs and 5 non-breeders; and the Moka range about 5 pairs and 5–10 non-breeders. Mauritius could thus support altogether 100–120 birds: 30 breeding pairs and 40–60 non-breeding individuals. Temple admitted this hypothetical number of birds is optimistic and he suggested that in reality only the Black River Gorges and Montagnes Bambous could support birds because of the likelihood of human disturbance. I believe Temple's estimates to be unrealistic since habitat quality is so poor that the actual carrying capacity of the available habitat must be low (see sections on Floristic and Faunistic Impoverishment and Food Availability, and Habitat Relations, and Home Range and Territory).

Habitat relations

The ecology of the Mauritius Kestrel is intimately tied to the native evergreen sub-tropical forests of Mauritius. These forests are divisible into three intergrading types: lowland, intermediate and upland (Vaughan & Wiehé 1937). The morphology of the bird is adapted to forest habitats (see Morphological Features). In historic times perhaps the highest densities of Kestrels were found in the upland forests where they may have hunted *Phelsuma* geckos among the decaying, cyclone-damaged extremities of the large emergent trees, as described by Vinson & Vinson (1969) for a bird seen on Macabé Ridge (see Food below). The mountains and cliff faces of the upland areas would also have provided updraughts upon

which the Kestrels could travel to hunting locations, or move around their territories with a minimum of energy expenditure as they still do today in the Black River Gorges (see Flight Characteristics above).

Kestrel distribution within the Black River Gorges is patchy and a reflection of habitat quality. What little of the forests have not been cleared, are unfortunately being degraded. As the older plants die or are destroyed by cyclones, they are increasingly being replaced by exotic species (Vaughan & Wiehé 1937, Vaughan 1968, Owadally 1980) and often the vigorous exotics outcompete and displace the native plants. This has had many direct and indirect effects upon the Kestrel population. On mountain slopes the Privet *Ligustrum robustum* is forming monotypic stands and in upland areas Strawberry Guava *Psidium cattleianum* is the dominant plant. Guava and privet not only change the floristic composition of the forest but also radically change its structure, forming a dense scrub layer. This affects the Kestrel's hunting strategies, since hunting beneath the canopy becomes increasingly difficult. Coupled with the degradation and simplification of native plant communities there has been a decline in the number of insects, geckos and passerines (see Floristic and Faunistic Impoverishment and Food Availability).

The nature and extent of the forest's degeneration is described in Chapter 1 and below under Pink Pigeon. Exposed slopes have become very degraded presumably due to the damaging effects of cyclones. In most of the Black River Gorges and surrounding areas the original forest canopy has been totally disrupted and all that remains are scattered emergent trees. Some good areas of native forest still exist but these are relatively small isolated patches in river valleys and in sheltered pockets on mountain slopes. The valley south of Morne Sèche, and that of Feeders Jules Edouard and Vanillerie are still well forested with pockets of lowland forest still keeping the integrity of the canopy. In the lower reaches there is bad degeneration on either side of the stream caused by the strangling creeper *Hiptage benghalensis*. The valleys of Rivulet Tatamaka and Rivulet Escalier show greater degeneration with 50–90% of the native forest replaced by guava and other exotics.

The main valley of Roches Blanches is almost completely degraded below Le Bouton cliffs at the south end of the gorge. The vegetation here is almost entirely guava with a few emergent Mango *Mangifera indica* and *Eugenia jambos.* Above these cliffs on Le Bouton, below Black River Viewpoint and on slopes leading up to Black River Peak there are still some good areas of native forest, some retaining 50–80% of the forest's integrity. On both slopes of the Grandes Gorges the forest has been badly invaded by guava and the upper slopes by privet. The canopy has been totally disrupted and all that remains are scattered emergent trees. Along some of the streams that drain into the Grandes Gorges River some good stands of native trees are still to be found. In most areas only about 5–10% of the native trees still exist and the large natural amphitheatre at the head of the gorge has been almost totally degraded by privet and guava.

Mare aux Joncs (also known as Manava Gorge) is virtually totally degraded on its northern slopes while the southern slopes have 10–50% of the canopy intact. The mouth of this valley is planted with *Eucalyptus tereticornis* and the northern slope has an area of self-seeding Tecoma *Tabebuia pallida*; further into the gorge there is a region of scattered Pencil Cedars.

The flood plain of the Black River from the mouth of the gorges all the way up to the tributary of Rivulet Roches Blanches is planted with Eucalyptus and on the banks of the river with a mixture of other exotic trees including *Terminalia arjuna, T. bellirica,* mahoganies *Swietenia mahagoni* and *S. macrophyla,* teak *Tectona grandis,* Yatis *Litsea monopetala, Vitex cuneata* and *Cordia myxa.* Smaller numbers of other exotic trees are also found including Mango, Badamier *Terminalia catappa,* and some old *Araucaria cunninghamii.* The understorey of this plantation is kept clear by forestry workers and flood action. There are also many open grassy areas which are kept clear by grazing deer. Kestrels are regularly seen in this exotic forest and I have observed one hunting for grasshoppers in the grass. Other local people have seen them hunting geckos, agamid lizards *Calotes versicolor* and House Shrews *Suncus murinus.*

Kestrels are occasionally seen outside the Black River Gorges along the west-facing slopes of Black River Peak, Magenta Scarp and the Vacoas Heights. All these areas are heavily degraded by privet and guava especially near the crest of the ridge, and at the base of the scarp by exotic acacias, Mauritius Hemp *Furcraea foetida* and in some areas dense monotypic plantations of Tecoma *Tabebuia* sp. The Tamarin Gorge and Petites Gorges are more protected and have some good stands of lowland forest on either side of the

streams and in protected hollows on the sides of the valleys. Tamarin Gorge is badly degraded in its upper reaches by privet, *Schinus terebinthifolius* and Tecoma. On rocky slopes there are dense *Furcraea* thickets.

Home range and territory

Mauritius Kestrels defend breeding territories (Newlands Oct. 1975, Temple 1978c) and have larger home ranges which may overlap with adjacent pairs by as much as 50% (Temple 1978c). During the winter months (April–August) the birds are less aerial and more difficult to locate; nevertheless it is suspected that during this period the home range is less stable and may include adjacent valleys and ridges, although a mated pair usually centre their activities around their nesting territory throughout the year.

With the approach of the breeding season Kestrels defend their territories. In September 1975 Newlands (Oct. 1975) observed: "The Gorges birds still have at least two of last year's three young birds within their territory and have been indulging in spectacular aerial disputes with them, sometimes ringing up beyond the range of 8× binoculars". Temple (1978c) referred to the same birds: "Often all four birds would soar together over the cliff, calling constantly and making repeated aggressive dives at each other." As a consequence of this conflict the resident pair moved from the 'Hole in the Wall' to the cliffs above Rivulet Escalier, 600 m away. McKelvey (1978) recorded the male displacing the previous year's young from the natal territory (presumably Site 1): "The following September, the adult male Kestrel disperses his remaining young with gradually intensifying hostility until they fear for their safety if they do not remain well out of the periphery of the home range. Quite spectacular flying contests occur between adults and young, accompanied by much diving, stooping and screaming, but always the young lose out and must disperse as the next breeding cycle is about to begin." I have not observed such overt aggression, but have seen additional birds tolerated in the vicinity of the breeding sites during all stages of the breeding cycle (Jones 1980d).

Temple (1978c) and L. F. Edgerley (pers. comm.) suggested that in winter there is a decline in the numbers of prey available in the upland forest and the kestrels shift to the lowlands. The birds are said to move back to the uplands during the breeding season, with a compression of the home range reflecting an abundance of prey in the montane forests during the summer months. There is little evidence to support this belief.

Kestrels exploit the area of their home range differentially, keeping to the ridges and slopes that are well forested with native vegetation and avoiding heavily degraded areas, presumably keeping to the areas of high gecko densities. During the winters of 1982 and 1983 it appeared very likely that the Montagne Zaco pair had largely independent home ranges, the male keeping to the slopes of Brise Fer Mountain north-east of the nest-cliff and the female keeping to the Montagne Zaco ridge and the valley bottom. One suspects that the birds remained independent of each other because of the paucity of food in this fringe area of suitable habitat.

Temple (1978c) suggested that the most intense territorial disputes occurred within 300 m of the nest-site and he speculated that the territory of a pair may be roughly a 600 m circle centred on the nest. Preliminary observations supported Temple's views on territory size although a more detailed analysis of my data suggests that this is an underestimate. Most Kestrels are to be found in a block of habitat of about 16.5 km² within and adjacent to the Black River Gorges. Since 1973 this area has supported eight different territories and now supports six or seven pairs (Table 12). Within this region, nest-sites are evenly distributed, being 0.93–1.65 km apart (mean distance 1.19 km). During the breeding season the male forages widely up to at least 0.5–0.7 km from the nest, suggesting a territory size of 79–138 ha, with a mean of probably about 100 ha. These data are consistent with the spacing between nests. Colour-ringed kestrels from established pairs have been seen up to 1 km from the nest cliff during the winter or when not breeding, suggesting a home range of potentially 315 ha or so. These figures are considerably more than demonstrated for the smaller Seychelles Kestrel, which has a mean territory size of only 40 ha (Watson, quoted by Collar & Stuart, 1985), but are consistent with a falcon living in now impoverished habitat that has to range widely to find food. A radio-telemetry study of four different territory-holding birds during the breeding season showed that they held self-contained territories of 44.6–74.5 ha with a mean of 59.8 ha (Fox *et al.* 1985). The lowest figure is believed to be an underestimate and year-round their requirements are likely to be larger than this.

Feeding ecology

Food

Until recently, accounts of the Mauritius Kestrel's diet were largely conjecture. Guérin (1940–53) assumed that the birds fed on small rodents and insects and occasionally lizards, frogs or young birds, and even mentioned such unlikely prey as partridge, quail and pigeon. Newton (1958a), on the basis of data from 'typical' kestrel species, provided a list of probable food items. He claimed that Kestrels "generally feed as much on insects, worms and beetles as on small birds or mammals and are adaptable". More recent accounts, undoubtedly echoing each other, claim that the Mauritius Kestrel feeds on insects, lizards and birds (Brown & Amadon 1968, Temple *et al.* 1974, Staub 1976, McKelvey 1977a, Temple 1977a, Jones 1980d) although Michel (1981) mentioned lizards, small birds and small mammals.

All recorded prey items, their weights and their relative frequency of capture are listed in Table 5. Some very unlikely items have been recorded; Guérin (1940–53) may have been merely suggesting what he thought to be likely prey, but others are based on sound evidence. E. Newton, for example, recorded freshwater shrimps on the label of an adult male Kestrel skin. I believe this bird was not healthy and was exploiting an easy food resource that it would not normally feed on. I have examined the skin, which is in very ragged plumage, and the information on the label supports my belief: "probably very old . . . T[estis] small. Gizzard filled with fresh water shrimps". It was collected on 1 January 1860 when the testis would have been enlarged if the bird had been in breeding condition.

Some authors have suggested the relative frequencies of different prey items in the diet. Temple (1977a, 1978c) claimed that insects, lizards and birds were taken in roughly equal numbers, but because of its calorific value the Grey White-eye *Zosterops borbonicus* was the most important source of energy; this bird has thus been claimed as the Kestrel's favourite prey (Staub 1976). McKelvey (1977d) noted that day-geckos *Phelsuma* spp. formed 50% of its diet, and Trefry (*in litt.* 1978) wrote "geckos seem to be the most frequently taken items in the wild", while his field notes record only one Grey White-eye delivered to the nest (22 Dec. 1977). Newlands (May 1975) described *Phelsuma* geckos as the Kestrel's main prey.

Table 5. *Recorded prey items of the Mauritius Kestrel*

Species	Approx wt prey (g)[a]	Frequency of capture[b]	Source[c]
Crustaceans			
Freshwater shrimps		*	11
Insects			
Insects, large insects			4,8,13,14,16,17
Flying insects			16
Dragonflies, large dragonflies		***	1,5,7,8,13,16
Cockroaches		* C	1
Grasshoppers, locusts, crickets		***	1,2,3,19
Cicadas		**	8
Beetles		**	3
Butterflies		*	16
Stick insects		*	19
Amphibians			
Frog *Rana (Ptychadena) mascareniensis*	6.0	* ?	4
Reptiles			
Geckos, lizards			3,4,7,8,10,14,16,17
Day-geckos *Phelsuma* spp.			1,5,7,13,16
Phelsuma g. guimbeaui	9.1/5.2	*****	1
Phelsuma g. rosagularis	6.5+/5.3	*****	7,18
Phelsuma cepediana	3.4/2.7	*****	1
Phelsuma ornata	3.4/3.0	*****	1
House gecko *Hemidactylus frenatus*	2.8	* C	1,5
House gecko *Gehyra mutilata*	3.8	C	5
Agama lizard *Calotes versicolor*	28/20	* C	1,5
Bojer's Skink *Scelotes (Gongylomorphus) bojeri fontenayi*		*	16
Birds			
Birds, small birds, small passerines			3,4,5,10,13,14,17
Young birds			4
Partridge ? *Francolinus pondicerianus*		+	4
Quail ? *Coturnix coturnix*	100	+	4
Young chickens *Gallus gallus domesticus*		* ?	4
Pigeons ? *Geopelia striata*	60	+	4
Madagascar Lovebird *Agapornis cana*	30	* ?	4
Red-whiskered Bulbul *Pycnonotus jocosus*	25	*	1,12,16
Zosterops sp.			7,8,12,16
Grey White-eye *Zosterops borbonicus*			
Adults	8.0	** ?	5,8,12
Chicks		*	16
Common Waxbill *Estrilda astrild*	8.0	*	7,8
Finch ? *Serinus mozambicus*	10	*	8
Mammals			
Small mammals, rodents			4,10,16
House Shrew *Suncus murinus*	45/24	** C	1,5,8
House Mouse *Mus musculus*	25	* C	5

[a] Figures before and after a solidus refer to male and female respectively. Single figures are unsexed samples.

[b] The frequency of each prey is estimated in five categories from very rare (*) to very frequent (*****). Prey items which have been caught by captive Mauritius Kestrels in their aviaries are marked C. Some records seem most unlikely; these are rejected and are marked +.

[c] 1, this study; 2, Bernardin de St. Pierre (1773); 3, Brown & Amadon (1968); 4, Guérin (1940–53); 5, Jones (1980d); 6, Jones & Owadally (1981); 7, McKelvey (1977c); 8, McKelvey (1978); 9, Meinertzhagen (1912); 10, Michel (1981); 11, E. Newton (1860), data on label of kestrel skin 1955.6.N.20.1362 BMNH; 12, Staub (1973a); 13, Staub (1976); 14, Temple (1977a); 15, Temple (1977b); 16, Temple (1978c); 17, Temple et al. (1974); 18, Vinson & Vinson (1969); 19, Manders (1911).

Table 6. *Prey items taken by the Mauritius Kestrel*

Species	Site 1 (1979–81)	Mt. Zaco (Oct.–Dec. 1981)	Tatamaka Cliffs (Nov.–Dec. 1982)	Morne Sèche (Nov. 1982–Jan. 1983)	Mt. Zaco female (non-breeding season 1980–3)	Various observations	Total no. of prey items
Dragonflies (Odonata)	–	2	–	–	–	–	2
Grasshoppers (Orthoptera)	–	1	–	–	–	–	1
Phelsuma day-geckos	11	205	13	14	1	4	248
Grey White-eye Zosterops borbonicus	–	1	–	–	–	1	2
House Shrew Suncus murinus	1	9	1	–	3	–	14
Totals	12	218	14	14	4	5	267

Note:
Most items were recorded as they were delivered to the nest-site by the male Kestrel. The 'various observations' are mainly from the non-breeding season, involve different birds, and include second-hand observations.

In the Macabé Forest, Vinson & Vinson (1969) and J.-M. Vinson (pers. comm. 1981) saw a Kestrel searching for food among the decaying extremities of indigenous trees. The bird was hopping from branch to branch and grabbed a lizard in one foot and swallowed it head-first. They suspected that this was an upland-forest day-gecko *Phelsuma guimbeaui rosagularis*. This lizard is probably a favoured food item of the Kestrel (Vinson & Vinson 1969; McKelvey 1977d); it is limited to native forest and has approximately the same range as the Kestrel.

My observations also suggest that the Kestrel feeds mainly on day-geckos, which made up 94% of 218 identified items delivered to the Montagne Zaco nest-site at the end of 1981 (Table 6). Less detailed studies on other breeding pairs and on birds outside the breeding season confirm that the Kestrel is almost totally dependent upon day-geckos. At the 1981 site, nine (4%) of the prey noted were introduced House Shrews *Suncus murinus*, which was also recorded as a food item by Meinertzhagen (1912) and Jones & Owadally (1981). The other prey were two dragonflies, a large grasshopper, and a Grey White-eye.

During the non-breeding season adult Kestrels, and especially young recently fledged birds, probably take more insects.

Historically it seems probable that the Kestrels would have fed on the skinks that were originally very common (Chapter 1). Some supportive evidence is provided by Temple (1978c) who found an osteoderm of the now almost extinct *Scelotes bojeri fontenayi* in a pellet; the Seychelles Kestrel feeds largely on skinks and geckos (Feare *et al.* 1974).

Hunting strategies

Several searching and hunting strategies are used by the Kestrels (Table 7). I describe these using the definitions of Fox (1977). Still-hunting is the most common searching technique; the birds sit quietly in a tree or on a rocky vantage point watching for prey. Other methods include hovering and poising (F. Staub and J. F. M. Horne pers. comm., Jones 1980d), but these are used less frequently than by kestrels, such as the European Kestrel, that live in open non-forested habitats. Oustalet (1897, Guérin 1940–53) reproduced

Table 7. *Mauritius Kestrel hunting strategies and success rate*

Species	Search technique Still-hunting	Direct flying attack	Dive/glide parachute	Sally	Ground	Rocks and cliffs	Trees and shrubs	Flying in air	Killed	Missed	Unknown
Phelsuma day-geckos	6	6				2			1	1	
							4			3	1
Dragonfly	2		1			1			1		
				1			1			1	
Grasshoppers	5		5			1			1		
					4					1	3
Insect spp.	4		2		2					1	1
				2			2			2	
Unidentified	27	10			5				1	1	3
							5				5
			17				13			5	8
					4					3	1
Totals	44	16	25	3	10	7	24	3	4	18	22
Per cent	100	36	57	7	23	16	55	7	18	82	–

Note:
Twenty-eight of the observations were made near nest-sites, and many of these were apparently low-intensity attacks. There are also a small number of observations of hunting by young inexperienced birds. Hunting success by adults is probably in the region of 25–30%.

an account of a Kestrel hovering over a poultry yard and taking chicks in 1830.

Prey is usually caught by direct flying attacks, and is snatched off a branch or caught after a brief pursuit, the Kestrel sometimes hopping and running among the branches. Staub (1976) noted that "Under [tree] cover the Kestrel flies from tree to tree, stalking mostly lizards, sometimes hopping after them along a branch with great rapidity." This hunting technique has been seen by many obervers including McKelvey (1978) who also described co-operative hunting: "Geckos are captured with incredibly swift dashes from a concealed position as they sun themselves on tree limbs. At times the male and female kestrel will hunt in unison, one pushing the gecko, who runs to the other side of the tree into the range of the second kestrel's talons". Jones (1980d) noted how very manoeuvrable the Kestrel is when in pursuit of geckos, and how it can turn in tight circles of less than 1 m while in flight.

Mauritius Kestrels are most active during the early morning and late afternoon (Staub 1973a, Temple 1978c, pers. obs.) when geckos are most available. Early in the morning, and during cool weather, *Phelsuma* geckos turn a dark, almost black, colour presumably to absorb heat, and are thus highly visible. When the sun is low and most of the trees are shady the geckos bask in the sun on exposed branches, increasing their vulnerability to predation. In the middle of the day the geckos are more active and difficult to detect; they are probably high in the canopy where their brighter green colour is highly cryptic. Temple (1978c) watched one pair of Kestrels that hunted in the late afternoon when almost all the geckos were on the west-facing side of tree limbs: "The kestrels would repeatedly fly directly to the western edge of their home range, which in this case was the tip of a ridge, and hunt their way slowly eastward with the geckos facing them and the sun at their backs. When they reached the eastern edge of their home range, they would return directly to the western edge and repeat the hunting sequence."

Prey that is relatively inactive, such as some insects, is captured by dropping from a vantage point or after a direct glide or flying attack (Jones 1980d). Active flying insects such as dragonflies are caught after a brief sally from the perch, or following a "truly skilled show of acrobatic flying" (McKelvey 1978).

I have little evidence that Kestrels feed on birds

and suspect that birds form a very small part of the total diet. Nevertheless several authors have recorded Kestrels hunting birds, usually the Grey White-eye (Staub 1973a, 1976, McKelvey 1977c, 1978, Temple 1977a, 1978c). McKelvey (1977c, 1978) and Temple (1978c) claimed that small passerines are usually caught in mid-air after a direct flying attack which may develop into a chase. Attacks upon birds, like those upon geckos, follow a period of still-hunting, although Guérin (1940–53) saw a Kestrel dive from quite high "with the speed of an arrow" and seize a Madagascar Lovebird *Agapornis cana* that had been feeding on maize. Temple (1978c) also described the Kestrel stooping after prey, in this case after a Red-whiskered Bulbul *Pycnonotus jocosus*: "The kestrel stooped and captured it as it dived headlong for the forest at the base of the cliff."

Food caching
Captive Mauritius Kestrels cache excess food and Jones *et al.* (1981) noted that the food is usually placed in the corner of a nest-box or where a branch makes an acute angle with an aviary wall. F. Staub (pers. comm.) has seen this behaviour in a wild Kestrel that cached a mouse or shrew in the crotch of a tree in Petites Gorges. At the Montagne Zaco nest-site in 1981, I often saw the male partly eat and then cache a House Shrew *Suncus murinus*, later retrieving it, feeding it to the female or eating it himself. Small prey items were usually passed to the female or eaten immediately, although I suspect that he occasionally cached one in the corner of the nest cavity for later retrieval by the female or himself. A hand-reared Kestrel first cached food, and later retrieved it the same day, when 60 days old. The three male Kestrels now in captivity are each fed about 25 g of mouse or beef every evening. This is partly eaten, then cached in the crotch of a branch or in the corner of a nest-box, and retrieved and eaten the following morning.

Relationship with non-prey species
In common with many island species the Mauritius Kestrel is very tame and frequently allows human approach to within 3–8 m (McKelvey 1977c, Temple 1978c, Jones 1980d), occasionally closer (especially in recently fledged birds). Most are indifferent to human presence although some are curious. A small male bird that fledged from Site 1 at the beginning of 1981 would fly up to within 1.5 m of me,

Table 8(*a*). *Species observed mobbing the Mauritius Kestrel*

Species	Reference
White-tailed Tropic-bird *Phaethon lepturus*	This study
Echo Parakeet *Psittacula echo*	Staub (1976), Jones (1980), this study
Mascarene Swiftlet *Collocalia francica*	This study
Merle *Hypsipetes olivaceus*	Temple (1978c)
Grey White-eye *Zosterops borbonicus*	Temple (1978c), this study
Mascarene Swallow *Phedina borbonica*	Temple (1978c), this study
Red-whiskered Bulbul *Pycnonotus jocosus*	This study
Common Mynah *Acridotheres tristis*	This study
Mauritius Fruit-bat *Pteropus niger*	Jones (1980b) this study

Table 8(*b*). *Species the Mauritius Kestrel may mob*

Species	Reference
White-tailed Tropic-bird *Phaethon lepturus*	Temple (1978c), Trefry (field notes), this study
Feral Pigeon *Columba livia*	This study
Pink Pigeon *Nesoenas mayeri*	Temple (1978c)
Madagascar Turtle Dove *Streptopelia picturata*	Temple (1978c)
Spotted Turtle Dove *S. chinensis*	Temple (1978c)
Eleonora's Falcon *Falco eleonorae*	Temple (1978c)
Echo Parakeet *Psittacula echo*	Temple (1978c)
Ring-necked Parakeet *P. krameri*	Temple (1978c), this study
Merle *Hypsipetes olivaceus*	Temple (1978c)
Red-whiskered Bulbul *Pycnonotus jocosus*	This study
Mauritius Cuckoo-shrike *Coracina typica*	Newton (1956), Temple (1978c)
Common Mynah *Acridotheres tristis*	Temple (1978c), this study
Mauritius Fruit-bat *Pteropus niger*	Jones (1980), this study
Crab-eating Macaque *Macaca fascicularis*	This study
Lesser Indian Mongoose *Herpestes edwardsi*	This study

sometimes perching about 4 m away. This may be the very tame male that is now at the Montagne Zaco site. Wild Kestrels usually appear to behave indifferently to deer *Cervus timorensis*, dogs and monkeys. (For the reaction of Kestrels to human intruders at the nest see Aggressive Behaviour below.)

Kestrels mock-attack birds and fruit-bats *Pteropus niger* and may in turn be mobbed by other species (Table 8). Mock attacks are usually silent. Temple (1978c) considered that these attacks are usually in play although he had seen aggressive interactions around the Kestrel nest-site, involving vocalisations, directed at tropic-birds and mynahs. All the mock-attacks that I have seen were in the nesting territory and were I believe agonistic. Birds and very occasionally fruit-bats mob Kestrels, which may be soaring, in direct flight or perched. Groups of about 8–15 Mascarene Swallows may briefly pursue soaring Kestrels, but Swiftlets mobbed singly. On the one occasion that I saw a fruit-bat mob a Kestrel it did so in level flight (Jones 1980d). White-tailed Tropic-birds stoop from above at soaring Kestrels (Jones 1980d). Echo Parakeets, Grey White-eyes and Red-whiskered Bulbuls mob perched birds while giving loud alarm calls (Staub 1976, Jones 1980d).

Aggressive behaviour

Aggressive behaviour is essentially the same as in other members of the genus such as *F. tinnunculus* (Cramp & Simmons 1980) and *F. sparverius* (Mueller 1971). It is characterised by increasing the body profile and direct staring with the beak pointing at the intruder. When an intruding tropic-bird, or other species, enters its nest territory the Kestrel stares at the intruder, lowers its head, fans its tail and launches an attack, flying at the intruder with a very strong rapid flight with deep wing beats.

Aggressive interactions between perched Kestrels involve staring with a 'wedge-head' posture (Fig. 4a). The nape feathers are flared and those at the rear of the crown raised, while the face feathers are flattened. This increases the head profile and accentuates the beak. While facing the intruder the aggressor bows forward, with a slight bowing of the head in mild conflict or a horizontal bow with the head, body and tail held in a straight line in a more intense conflict (Fig. 4b). The wedge-head posture is usually accompanied by a raising of the scapulars and back feathers, fanning of the tail and often a slight drooping of the wings. The bird may sleek its abdomen feathers and stand high on its legs. The loud '*kee-kee-kee*' threat call of other kestrels is either absent or very rarely used; aggressive interactions are usually silent.

An intruding Kestrel also adopts the wedge-head posture and bows or may remain upright in a 'griffon' posture with wings held out and half spread, beak open, and abdomen feathers fluffed out to increase profile and for protection. When intimidated by another Kestrel it may call '*ereep-ereep-ereep*' (*c.* two calls per second). Aggressive interactions may develop into ground grapplings with 'angel' (Fig. 4*c*) and 'fallen angel' postures as the birds fight with wings open. These displays have been well described and appear to be the same as those in the Buzzard *Buteo buteo* (Weir & Picozzi 1975) and the Peregrine Falcon *Falco peregrinus* (Nelson 1977).

Human intruders to the nest-site may be attacked by either sex. This behaviour is more highly developed in some birds than in others. While I was removing the eggs from Montagne Zaco (1981) neither bird attacked me although the female spent some of the time nearby in defensive wedge-head and 'griffon'

postures. At Site 1, while I was removing the eggs the female attacked repeatedly, flying very fast and silently, closing her wings as she shot past, striking hard and painfully at my scalp with her talons. While the female was incubating at Montagne Zaco (Site B) in 1981 the male would sometimes attack but not in such a determined fashion as the Site 1 female.

Breeding biology

The breeding biology of the Mauritius Kestrel was until recently unknown and the early accounts were based largely on assumptions. Brown & Amadon (1968) wrote "this little falcon will pass into extinction without anything significant being known of its breeding habits". Thankfully this has proved untrue.

Breeding is limited to the summer months and in common with other kestrels both sexes may reach sexual maturity in their first year (Jones *et al.* 1981).

Nest-sites and their selection

Guérin (1940–53) stated that the nest was made of sticks lined with roots, moss and hair in the tallest trees in the forest, and did *not* nest in cavities in rocks

Fig. 4. Aggressive and defensive postures. (*a*) Wedge-head posture with the Kestrel staring at the intruder with nape and crown feathers raised and cheek feathers flattened. (*b*) A more intensive threat: the body is bowed, wings held loosely, tail fanned, back and scapular feathers raised, and the bird stands on stiffened legs. (*c*) An intensive threat/defensive posture: the Kestrel spreads its wings in an 'angel' posture to maximise the body profile.

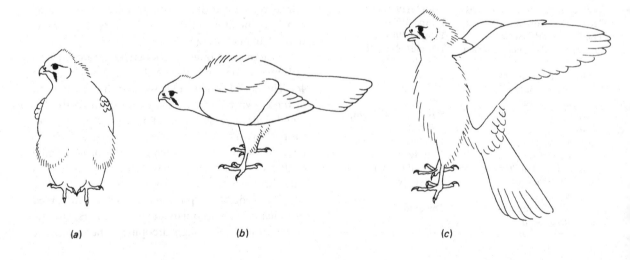

(a)   (b)   (c)

and trees. This was in part quoted by Staub (1973*a*); and as recently as 1981 Michel thought that the Kestrel made a "rough nest of twigs placed in holes e.g. those of a dead tree".

In common with other members of the genus this Kestrel does not make a stick nest but lays its eggs in a nest scrape. It was noted by Staub (1976) that they "breed in a hole in a cliff", and Temple (1977*a*) claimed that they "nest in tree cavities and holes in cliff faces". While the Mauritius Kestrel might occasionally nest in tree cavities, and in the past almost certainly did, I know of no authenticated cases (see Distribution

Fig. 5. Details of Mauritius Kestrel nest cavities.

1. Montagne Zaco cavity A, floor area 2.99 m². This cavity was used in 1981 for the first clutch of eggs.

2. Montagne Zaco cavity B, floor area 0.25 m². The second clutch by the above pair was laid in this cavity towards the end of 1981. The floor was strewn with loose rocks. A former tropic-bird nest hole.

3. Montagne Zaco cavity C, floor area 0.15 m². The nest-site for the only clutch laid by this female in 1982. It was again used for the first clutches in 1983 and 1984. The eggs were laid in an abandoned rat's nest at the rear of the cavity. This cavity had also been a tropic-bird nest-site. A former Kestrel roost site.

4. Montagne Zaco cavity D, floor area 0.50 m². Used for the second and third clutches laid in the 1983/4 season and for the second clutch laid in the 1984/5 season. The cavity had a rat's nest built at the rear, was a former tropic-bird nest-site and a former Kestrel roost site.

5. Montagne Zaco cavity E, floor area 0.67 m². The third clutch was laid in this cavity during the 1984/5 season.

6. Montagne Zaco cavity F, floor area 0.32 m². Only a single egg was laid in this cavity, the fourth clutch laid during the 1984/5 season. The cavity contained an old rat's nest.

7. Site 1, floor area 0.24 m². Cavity used in 1981, 1983 and 1984. It is a former tropic-bird nest hole and the eggs were laid among the voluminous feathers of at least two decomposed tropic-birds.

8. Morne Sèche, floor area 0.27 m². This cavity was used for the second clutch of eggs laid during the 1982/3 breeding season and was prospected in the 1983/4 breeding season. The eggs were laid on a bed of cockroach droppings among a mass of loose rocks.

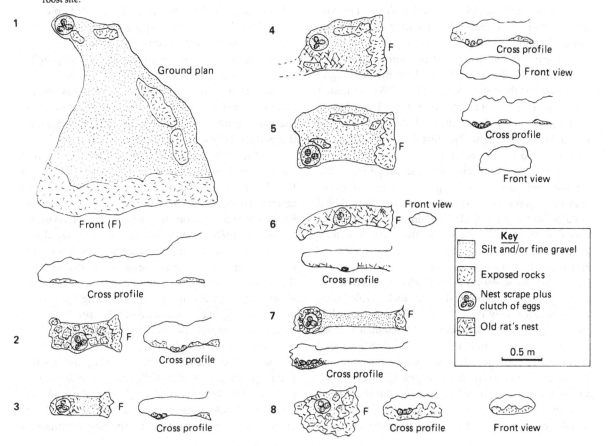

Status and Population above). All observed nests have been in cliff cavities, many quite deep (Fig. 5). Three sites that I examined were about 2 m, 1.2 m and 0.6 m into the cliff face and well sheltered. There has doubt-less been strong selection for this type of nest cavity which would provide shelter during cyclonic weather. Cade (1982) has suggested that prior to the intro-duction of monkeys the Mauritius Kestrel may have nested among the epiphytes in tree crotches as the Barred Kestrel *F. zoniventris* of Madagascar does. This is unlikely since the nest-site would be susceptible to destruction during cyclones.

Temple (1978*c*) gave the interior floor area of sites he examined as at least 300 cm^2, but they can be much more than this. A tree cavity prospected by Kestrels was in a vertical trunk, cylindrical in shape, about 35 cm deep, 20 cm in diameter and with a 20 cm diameter opening. Other tree cavities not examined closely had openings about 10 cm in diameter (Temple 1978*c*). Two of the ten sites examined by Temple (1978*c*), that were either occupied or prospected, had been used previously by other birds. One prospected site was probably an old White-tailed Tropic-bird site and another was apparently a roosting site for Mas-carene Swallows. Three of the eight nest-sites that I have examined had been used previously by White-tailed Tropic-birds. The site used by the Montagne Zaco pair for their second clutch in 1981 contained bones of a Tropic-bird and the pair at Site 1 in 1981 laid their eggs among the feathers of two decomposed Tropic-birds. Montagne Zaco Site C, used in 1983 and 1984, is an old Tropic-bird nest.

The substrate of nest-sites can range from fine silt to coarse gravel containing loose rocks 8–10 cm in diameter. Some sites are littered with Tropic-bird remains and McKelvey (1978) claimed "Feathers and bones of the prey carried into the nest hole may accidentally be incorporated into the nest scrape, as may molted plumage of the adults".

Nest-site selection is conducted by both sexes, although the male usually takes the leading role (Tem-ple 1978*c*, McKelvey 1977*c*, 1978*b*). When the male has found a suitable site, and provided the female is close by, he displays to her in a nest-site display, head lowered with beak pointed downwards, tail often half fanned and wings sometimes dropping. He calls with a high-pitched and sharp '*tink-tink-tink*' (in some birds this call is more of a '*chip*') or whines. The female usually reciprocates with similar appeasement pos-tures and a deeper '*chip*' or '*chup*' call and may also whine. McKelvey (1978) described this display, writ-ing the call as 'kek! kek! kek!' with the female also calling in a staccato 'kik kikit!' (= chitter-whining?).

Nest-scraping is done by both sexes (McKelvey 1977*c*, 1978, Temple 1978*c*, Jones *et al.*, 1981); typically, the scrape is a shallow depression about 16 cm in diameter.

The male's nest-advertising flight is described under Aerial Displays below.

Courtship displays
The courtship displays are very similar to those of other members of the genus. I describe these displays using current terminology (e.g. Cramp & Simmons 1980, Wrege & Cade 1977).

Aerial displays. With the approach of spring the kestrels become more aerial, both soaring over the nesting territories and calling frequently. Both sexes whine a great deal and 'prominent perch' on rocks and exposed branches. Courtship flights, often an elabor-ation of soaring, may begin in August or September (McKelvey 1977*c*, 1978, Temple 1978*c*, Jones 1980*d*) although low-level courtship flying may be seen as early as mid-July. The nature of the courtship flights (Fig. 6) is mainly self-assertive on the part of the male and submissive on the part of the female. This perhaps enables close integration of the pair and helps prevent the male being intimidated by the larger, potentially dangerous, female.

Mutual soaring often develops into chasing with the male flying rapidly across the cliff face after the female or stooping acutely at the female in mock attacks. The female sometimes rolls in mid-air onto her back or side to confront the male at the last moment. McKelvey (1978*a*) recorded that the female may also be the assertive sex during these aerial evolutions. The male indulges in power flying or power diving while the female soars nearby. He may fly rapidly making short stoops, throwing up and repeating the process in a zig-zag flight (= power diving, Nelson 1977). The male may fly rapidly out over the breeding territory in power flight and describe a large circle half a kilometre or more in diameter before rejoining the soaring female. During these flights the male sometimes calls with a loud chitter-whining. The female is usually silent or whines. A less overt form of zig-zag is shown by the male when he flies in shallow undulations

Fig. 6. Aerial displays. (*a*) Undulating displays (above) and zig-zag flights are prominent features of the male's courtship display, but they may also be expressed by the female. (*b*) Fluttergliding, a slow buoyant flight that the female uses to appease the male. She flies with quick shallow wing beats. This flight is identical to the begging flight of the juvenile. (*c*) Slow landing or 'V' flight, the nest advertising flight of the male, which may also be used by the female when she is trying to attract a male. The head is held low, tail fanned, feet held forward and wings held above the body in a 'V' shape. (*d*) Mutual soaring, a very common feature of the pre-incubation phase of the breeding cycle, which also serves to advertise the territory.

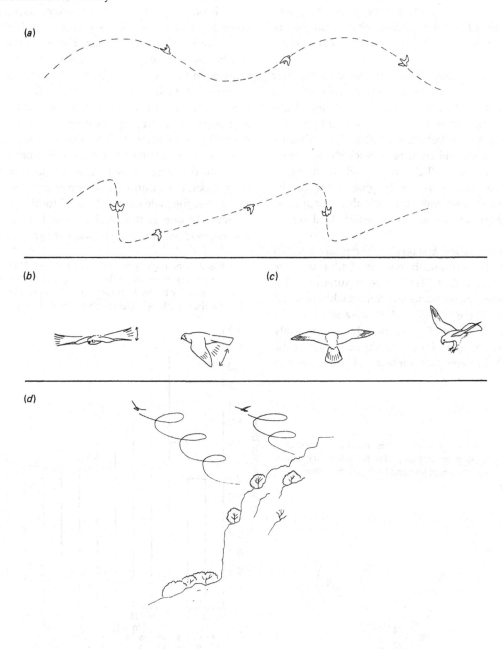

(*a*)

(*b*)   (*c*)

(*d*)

across the cliff face calling with repeated whining or *tick* vocalisations (each call about 0.5–0.7 seconds apart). When the male wishes to attract the female to the nest-site he flies in slow flapping 'butterfly flight' and glides to the nest-site with his wings held in a V above his back, his head low, tail fanned and depressed. This is a slow landing display. Sometimes he flies slowly from tree to tree 'prominent-perching' and calling with a loud high-pitched, sharp wail, before flying to the nest-site. Wailing may precede copulation.

The female takes on a more submissive role. Her vocalisations are mainly excitement and begging whines and she flies in a slow 'fluttergliding' flight (Jones 1980*d*). This is a slow buoyant flight with the wings fully spread but arched below the horizontal axis of the body and consists of quick shallow wing beats. She whines in flight and the whole display is indistinguishable from the begging flight of the recently fledged juvenile. It is probably an appeasement flight which encourages the male to feed her.

Pair integration, food passing and copulation. Much courtship occurs around the nest-site. Both sexes have nest-site displays (Fig. 7) and these postures may also be used when the pair come into contact while perched elsewhere. There is a great deal of variation in the amount and degree of posturing which is probably related to the degree of compatibility and length of the pair bond. In some poorly integrated pairs appease-

ment postures may be interspersed with aggressive displays. Compatible pairs nibble affectionately at each other's feet and beaks uttering soft chipping calls and often inverting or tilting their heads. These behaviours, which can also be seen in juvenile Kestrels, are probably important in pair integration and may occur well outside the breeding season.

Food passing is an important element of courtship that occurs before laying. The Montagne Zaco pair in 1981 were suspected to be food passing nearly 3 weeks before the first egg was laid. The female remains in the vicinity of the nest-site and the male returns with food, whining or chitter-whining as he flies in. The female sometimes solicits feeding with appeasement posturing (feathers fluffed, wings drooping, head and beak low) and whining vocalisations. Well integrated pairs pass food from beak to beak but the female may take or tear food from the male's foot. Early during courtship or in poorly integrated pairs the male may cache food for the female to retrieve. In one of the captive pairs I have seen a reverse food pass from the female to the male with no

Fig. 7. Nest-ledge display. This serves as a greeting and appeasement display and is characterised by bowing, with head held low and beak pointing downwards.

Fig. 8. Food passing in Mauritius Kestrels. Number of prey items delivered to the Montagne Zaco nest-site per hour by the male during the pre-incubation and incubation phases (October/November 1981).

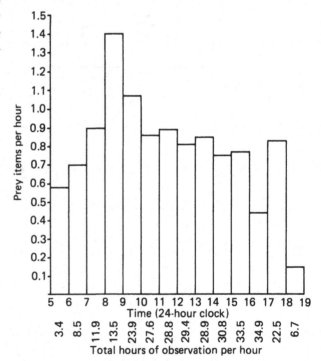

signs of aggression. Food passing occurs at or near the nest-site and continues through incubation and rearing until the young are nearly fledged. The daily pattern is shown in Fig. 8.

Copulation (Fig. 9) also occurs near the nest-site and may be solicited with overt posturing or by very subtle behaviour difficult to detect. The female may solicit silently or with shrill *'wear-wear-wear-wear'* calls (three or four per second), with abdomen feathers fluffed, head low, wings drooping slightly and tail held closed and above the horizontal. The male jumps onto the female's back and balances on his tarsi with his toes loosely clenched. His tail is placed under the female's and partly fanned; he keeps his balance by slowly beating his wings. During copulation, which usually lasts 5–10 seconds, the male is silent or may chip rapidly and the female whines or chitters. The number of copulations per day varies among both wild and captive birds from one to twenty a day (McKelvey 1977*c*, Temple 1978*c*, Jones 1980*d*).

Fig. 9. Copulation. (*a*) Female soliciting with tail raised slightly, vent feathers fluffed and wings drooping. (*b*) Copulation. The male balances on the female's back, his tail placed under the female's, and he beats his wings slowly to maintain balance.

(*a*)

(*b*)

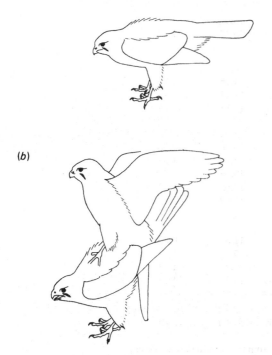

Eggs, egg-laying and incubation

The eggs are similar to those of other kestrels (*F. tinnunculus*, *F. sparverius* and *F. naumanni*) in shape and colour (Table 9). They are broad ovals, white or off-white, which may be spotted, blotched or speckled with pale or dark brown. They vary from plain white to so heavily marked that the ground colour is almost completely covered. Some are heavily marked at the broad end; most are heavily speckled all over with red-brown. Clutch size is usually three, though occasionally repeat clutches may be two. Accounts of more than three eggs in a clutch (Guérin 1940–53, Staub 1973*a*, McKelvey 1978) are probably speculation.

Egg-laying is recorded in September (Staub 1976), and Temple (1978*c*) gave dates of 18 September to 5 November. E. Newton (1861*b*) shot a male with enlarged testes at the beginning of October 1860, and he was given an egg (the only one he ever got) in September of that year (T. Stjernberg *in litt.*). An egg formerly in the Transvaal Museum (TM 43423) was collected in October 1920. Recent observations (Fig. 10) suggest that laying occurs in the wild from mid-September to early November with most eggs laid in October. In captivity eggs are laid at the same time but repeat clutches may be laid as late as early January (see Table 10). Meinertzhagen (1910–11, per A. S. Cheke)

Fig. 10. Dates of egg-laying in the Mauritius Kestrel 1971–83. Histogram calculated from Table 10.

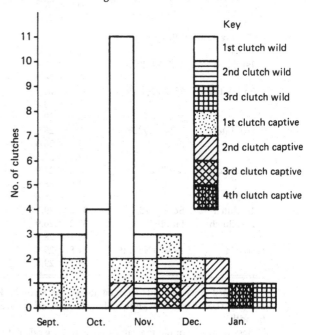

Key
- 1st clutch wild
- 2nd clutch wild
- 3rd clutch wild
- 1st clutch captive
- 2nd clutch captive
- 3rd clutch captive
- 4th clutch captive

Table 9. *Mauritius Kestrel egg data*

Date and collection details		Size (mm)	Fresh wt	Calculated fresh wt[a]	Ratcliffe shell index
(a) Wild Kestrels pre-1940					
MIPL	Pre-1940	36.9 × 31.6	—	19.34	1.30
		38.0 × 30.2	—	18.19	1.24
		38.0 × 31.1	—	19.29	—
		36.9 × 31.6	—	19.34	1.42
		38.8 × 31.0	—	19.57	1.27
		36.1 × 30.5	—	17.62	1.24
		36.2 × 30.6	—	17.79	1.40
		38.0 × 31.5	—	19.79	1.22
		35.8 × 30.85	—	17.88	1.25
		35.2 × 29.4	—	15.97	1.18
		34.7 × 30.6	—	17.05	1.20
BMNH	1919	37.7 × 30.6	—	18.53	1.24
SAM	1920?	35.7 × 30.6	—	17.64	1.34
ZMUH	1860	38.8 × 29.9	—	18.20	1.22
(b) Wild Kestrels 1981–3					
Mt. Zaco	Oct. 1981	35.9 × 29.9	—	16.84	—
		36.3 × 30.3	—	17.49	1.39[b]
		37.1 × 29.8	—	17.29	1.24[b]
Site 1	Nov. 1981	36.3 × 30.1	—	17.26	—
		39.2 × 29.6	—	18.02	1.21[b]
		38.0 × 29.7	—	17.59	1.16[b]
Mt. Zaco	Nov. 1982	37.0 × 27.8	—	15.01	1.06
		37.15 × 28.3	—	15.61	1.07
Morne Sèche, 2nd clutch	Jan. 1983	37.9 × 28.3	—	15.93	1.26[b]
		36.3 × 29.4	—	16.47	
		33.6 × 29.2	—	15.03	1.17[b]
Mt. Zaco, 1st clutch	1983	36.7 × 29.6	—	16.87	1.33
		36.2 × 29.6	—	16.65	1.23
		36.4 × 30.8	—	18.12	1.29[b]
Mt. Zaco, 2nd clutch	1983	36.3 × 29.2	16.70	16.24	1.37[b]
		37.8 × 29.6	17.16	17.38	1.26
		36.6 × 29.7	16.96	16.94	1.29
		34.9 × 28.8	14.83	15.19	1.20
(c) Captive Kestrels 1978–83					
Female 4		32.8 × 29.8	—	15.29	1.06
1st clutch	1–7 Oct. 1978	34.4 × 28.4	—	14.56	1.28
		33.8 × 28.8	—	14.71	1.18
2nd clutch	18–22 Oct. 1978	34.7 × 28.1	—	14.38	1.14
		33.8 × 28.8	—	14.71	0.77
		34.6 × 29.7	—	16.02	1.23
3rd clutch	7–14 Nov. 1978	32.0 × 28.1	—	13.26	1.22
		33.1 × 28.3	—	13.91	1.09
		33.4 × 29.5	—	15.25	1.20
4th clutch	Jan. 1979	35.8 × 27.9	—	14.62	1.27
		34.0 × 28.0	—	13.99	1.21
Female 8		35.1 × 28.8	—	15.28	—
1st clutch	Sept. 1978	36.2 × 29.1	—	16.09	1.09
2nd clutch	Jan. 1979	33.2 × 26.9	—	12.61	0.80
		35.1 × 28.9	—	15.38	1.07
Female 14	19 & 23 Nov. 1983	34.8 × 29.9	16.40	16.33	1.23
		35.1 × 29.8	16.40	16.36	1.19

[a] Calculated fresh weights derived using the formula $W = Kw(LB^2)$ with the constant $Kw = 0.0005248$ based on known Mauritius Kestrel egg fresh weights. L is egg length and B breadth.

[b] These represent minimum values since they are calculated from the weight of hatched egg shells.

Table 10. *Dates of egg-laying and clutch size in the Mauritius Kestrel*

| | Wild Kestrels | | | | | | Captive Kestrels | | | |
| | Tamarin Gorge | | | | | | | | | |
Year	Le Bouton	Site 1	Site 3	Tatamaka cliffs	Mt. Zaco	Morne Sèche	Female B	Silver	Yellow/blue	Blue
1974/5	1st W/10						18–24/9 (3)			
1975/6							Mid/11 (3); 1st W/12 (3)			
1976/7	Mid/9	Mid/9					28/10 (1)			
1977/8	4th W/10	Late/9–early/10	3rd–4th W/10							
1978/9		Late/10–early/11						Mid/9 (3); 31/12–2/1 (2)	24/9 (1)[a]; 1–7/10 (3); 15–22/10 (3); 7–14/11 (3); Early/1 (2)	
1979/80		1st–2nd W/10							2/12 (1)[b]	
1980/1		4th W/10		/10 or /11						
1981/2		1st–2nd W/10 (3)			3rd W/10 (3) 2nd W/11 (3)					
1982/3				Mid/10	11, 13/11 (2)	? mid/10 (3); mid/12 (3)				
1983/4		3rd W/10		2nd W/10	3rd W/10 (3) 14,16,18,23/10 (4)[c] 3rd W/1					11,17,23/11 (3)

Notes:

Laying dates are recorded as accurately as possible. When exact dates are not known they have been calculated from hatching and fledging times, assuming a laying period of 6 days, an incubation of 32 days and a nestling period of 38 days. When laying cannot be calculated to the day it is expressed to the week, part of the month or whole month. Days are expressed before the month and the latter are expressed as plain numbers: thus Mid/10 is mid October, 1–7/11 denotes that the eggs were laid between 1 and 7 November and 2nd W/9 shows the clutch was laid during the second week of September. Figures in brackets at the end of an entry are the clutch size.

[a] This egg was laid off the perch and does not represent a full clutch.
[b] The female died of egg peritonitis while laying this incompletely formed egg.
[c] This clutch was extended by 'egg pulling'.

watched a pair feeding young on 23 February 1911. Captive birds usually lay eggs every other day. Temple (1978c) gave the inter-laying interval as 2–4 days.

The female does most of the incubation. McKelvey (1978) thought that she did it all although the male would enter the nest hole during a brief absence of the female. Temple (1978c) noted that both sexes contributed to incubation and during a 2 day watch during the middle of the incubation period: "The female incubated 78% of the observation time with attentive periods that averaged 143 minutes. The male incubated 16% of the observation time with average attentive periods of 17 minutes. The eggs were unattended 4% of the time with the longest inattentive period lasting nine minutes." In the captive pair that reared a fledgling in 1979, the male did not take any part in the incubation of the first clutch. During the incubation of the second clutch the male would enter the nest box, presumably to incubate, for 5–10 minutes while the female was out of the box feeding. In both clutches inattentive periods were about 5–15 minutes (F. N. Steele pers. comm.; Jones *et al.* 1981). At the Montagne Zaco sites in 1981 the female usually remained in the nest cavity 90–96% of the time during day-time watches. The male would sometimes enter the cavity when the female was not in the nest hole, and he spent about 1–4% of the daytime watch in the cavity. The female always incubated at night.

Leading up to full incubation (>85% of the time in the nest cavity), the female spends an increasing amount of time in the nest hole. At the two Montagne Zaco sites in 1981 in the week prior to full incubation, and for the two days after this, she spent an increasing proportion of the observation time in the nest cavity (28%, 46%, 37%, 63%, 48%, 24%, 76%, 87%, 92%, 94% per day). Full incubation probably does not start until the laying of the third egg, which results in staggered hatching. A clutch taken from the wild after 6 days of full incubation subsequently hatched in an incubator over a 3 day period after incubation periods of 31, 32 and 33 days.

Published incubation periods are variable: 30–35 days (Staub 1976), 28 days (McKelvey 1977c, 1978), and 30 days for an egg laid in captivity that hatched in an incubator after 6 days of natural incubation (Temple 1978c). The captive Kestrel bred in 1978 hatched after an estimated 32 days of incubation (Jones *et al.* 1981). McKelvey (1977c, 1978) maintained that the hatching chick first cracks (pips) the shell on the twenty-sixth day before finally hatching on the twenty-eighth day.

Three artificially incubated eggs took about 32 hours 15 minutes, 45 hours and 53 hours from pip to hatch. After hatching the chick is wet, helpless and exhausted and sleeps for several hours with its head tucked between its legs. The occipital muscles that aid in hatching are swollen and very obvious, but atrophy after about 48 hours.

Growth, development and parental care

There is little published information on the fledging period. The available information is conflicting: 30–35 days (Temple 1978c), 35–42 days (Staub 1976), 49–56 days (McKelvey 1978) and 63–70 days (McKelvey 1977c). The following observations are taken from hand-reared Kestrels and my unpublished field notes, unless stated otherwise. I believe that the development of hand-reared chicks does not differ significantly from that of wild birds. The day of hatching is taken as day 0.

Day 0. Weight *c.* 12 g. The young, in common with other falcons, can be classed as semi-altricial (Thomson 1964) but they nevertheless show many embryonic features and are helpless. The eyes are large and bulbous, the head is large and the legs and wings are poorly developed. The eyes may be partly open at hatching but close a few hours later. The chicks are covered by white down, long (5–10 mm) on the back, shoulders, femur and wings, shorter (5 mm) on the crown, thorax, inside of femur and around the abdomen. The abdomen is largely bare. The bare parts are pink and the toenails and beak a clear white. After hatching the chicks are very weak and can raise their heads only unsteadily and for short periods of time. They have two calls: a whine '*vree-vree-vree*' with which they beg for food, and a cheep when they are in discomfort. Most of the time they spend sprawled out asleep. The female broods almost continuously and leaves the nest scrape only to fetch food from the male three or more times a day.

Day 2. Stronger and can gape for food proficiently. Begin to practise behavioural thermoregulation and huddle together if cold, or after feeding, and sprawl if temperature is adequate or too hot. Eyes open by about 48 hours. Dark pigmentation at base of toe nails. No noticeable crop.

Day 4. Weight about 20 g. Noticeably stronger and more co-ordinated, sitting upright on tarsi. Rudimentary preening and allopreening.

Day 6. Dark pigmentation in skin of scapular and wing tracts. Ejecting faeces well clear of nest scrape. Can see food 5 cm away.

Day. 7. A few dark specks of second down follicles on dorsal tract. Can detect food from 30 cm away.

Day. 8. Weight *c.* 40 g. Grey-brown second down erupting on ventral tract, scapulars, dorsal tract, wings and tail. Can detect food 45 cm away.

Day 9. Cere, feet and periorbital region a pale cobalt blue.

Day 12. Weight *c.* 65 g (males) or 80 g (females). Oesophagus expanding at base into a crop. Grey second down well developed on major tracts but poorly developed on head. Tail, primary and secondary feathers in pin. Female may leave chicks unattended for up to 15 minutes at a time while she takes food from male and eats (Jones *et al.* 1981).

Day 13. Pin feathers breaking through on scapulars, wing coverts, nape and ventral tracts.

Day 15. Double wing stretch over back.

Day 16. Weight *c.* 95 g (males) or 120 g (females). More conscious of outside world, focusing on objects and head-bobbing. Move around on tarsi and strike at objects with feet in play. May sleep with head tucked into scapulars.

Day 17. Primaries, secondaries and their coverts begin to split their sheaths; other wing feathers still in pin. Stand correctly with some difficulty. Snap with beak at insects around nest scrape. Female leaves chicks unattended for 20–25 minutes at a time (Jones *et al.* 1981).

Days 18, 19. Feathers unfurling and beginning to show through down on all major body tracts. Tail feathers 6 mm out of sheaths. Stretching on extended legs and double wing stretch downwards.

Day 20. Weight *c.* 120 g (males) or 150 g (females). Standing for long periods. Males noticeably more proficient at walking than females.

Day 21. Unilateral stretch of wing and leg while lying on side.

Day 22. Female leaves chicks unattended for most of the day but still spends night in nest cavity (Jones *et al.* 1981).

Day 23. Feet developing a greenish-yellow tinge; cere and periorbital region still cobalt blue. Males are 1 or 2 days more advanced than females.

Day 24. Weight *c.* 135 g (males) or 178 g (females). Have more structured play, pulling at and playing with pieces of grass and moulted feathers, gently nibbling their own and their siblings' toes.

Day 25. 'Chupping' and 'chipping' vocalisations used after feeding, when under mild stress and during low levels of excitement.

Day 26. Losing down rapidly; large patch of down still present on wing coverts and centre of dorsal tract. Feathers coming through on crown. During 26 hours of observations Temple (1978c) recorded the male and the female bringing food to the young 32 times.

Day 27. Weight *c.* 140 g (males). Wander around in nest cavity and stand at the entrance of the nest hole looking out. Greet parents returning with prey by whining loudly and rapid wing trembling movements.

Day 28. Weight *c.* 135 g (males). Rapidly losing down from head and wings; rest of body almost clear of all signs of down except for odd flecks adhering to tips of feathers on breast and back.

Days 32–35. During this period McKelvey (1978) recorded the female helping the male to find food for the young (cf. observations of Temple when chicks 26 days old). A captive bird took its first bath at 35 days old.

Day 36. Weight *c.* 135 g (males).

Days 36–38. Very little down visible except for odd flecks on back and crown. Leave the nest cavity for the first time; males more precocious than females. Clamber around on cliff face or among branches of nearby tree. Incapable of sustained flight. At night return to nest cavity. Sleep in prone position.

Day 40. Weight *c.* 134 g (males).

Days 40–42. Some primaries and tail feathers still vascularised at base, secondaries fully formed. Practising wing flapping exercises while perched near nest-site. Capable of flying 2–3 m.

Day 44. Weight *c.* 132 g (males). During the following week the weight stabilises at about 5 g below this.

Days 45–54. Rapidly becoming increasingly volant, flying on exercise sorties 30–50 m to nearby bushes and back to the area of the nest-site. As the days progress they fly more and may soar briefly with the adults. The young stay within about half a kilometre of the nest-site, return to the site often and at night roost in the nest hole. Fledglings are fed by the adults at or near the nest cliff and greet adults with much excited wing-trembling and loud food-begging whines. The young snatch food from the adults, mantling over it with wings and tail spread and their back to the other siblings to prevent them seeing and stealing the food. The food is quickly consumed. The young may cache excess food in the crotch of a tree or in the nest cavity. While not eating or flying they spend much time preening or may lie prone, dozing in the sun. By the end of this period the young are indulging in much aerial play – soaring, chasing and mock-attacking each other, often chasing and weaving between the trees of the canopy. When hungry the young have a begging flight 'fluttergliding' (described under Aerial Displays above). Much of this play is reminiscent of the aerial courtship displays of the adults.

Days 55–80. The young begin following the adults to more distant areas. At the end of this period they are still receiving some food from their parents (Temple 1978c). McKelvey (1978) noted that the parents return almost every two hours with prey and may drop a lizard or wounded bird on rocks near to the fledglings for them to grab. McKelvey elaborates: "There is a stage when the parents actually seem to fear the aggressive, demanding young and they then deliver the prey to a nearby ledge or drop it out of the

air to be captured by the pursuing young on the wing. After 70 days the parents only deliver food twice a day to encourage the young to hunt on their own." At Site 1 in February 1981 when the young were about 75 days old, I watched the adult male, with I believe a food item in his claws, circling rapidly upwards being pursued by two young birds whining as they flew. The male then stooped and climbed again above the valley, all the time being followed by the young birds. I watched them for about 5 minutes and had the impression that he was encouraging them to chase him.

Days 84–98. The young become independent of their parents (McKelvey 1978). This requires confirmation since they stay within the parents' home range for several months (see below).

Days 100–300. The young stay around the natal cliff site and in adjacent areas, coexisting with the parents without conflict until the approach of the following breeding season (Staub 1976, McKelvey 1977c, 1978, Temple 1978c, Jones 1980d). They continue to indulge in playful antics and occasionally beg food. I have seen the 'fluttergliding' begging flight by a male as late as about 190 days old, although the adults did not feed him.

Breeding rate and success

The mean production of Mauritius Kestrels per year is about 0.84 fledged young per territorial pair (Table 11), and since there is a small floating population the annual production is probably only 0.3–0.4 fledged young per adult. I. Newton (1979) has shown that among small temperate falcons 1.7–3.2 young are

Table 11. *Mauritius Kestrel breeding success 1973/4–1983/4*

No. of territory-holding pairs (years combined)	32
% of pairs attempting to nest (eggs laid)	53 (17/32)
% of nests from which young flew	75 (12/16)
Mean clutch size	2.8(17/6)
% of eggs fertile	68 (13/19)
Mean brood size at fledging	2.3 (27/12)
Mean no. of young per clutch started	1.7 (27/16)
Mean no. of young per territory-holding pair	0.84 (27/32)

Note:
Data from manipulated nesting attempts are not included.

Table 12. *Mauritius Kestrel breeding results 1973/4–1983/4*

Year	Unknown tree cavity	Brise Fer	Le Bouton (Site 2)	Site 1	Tamarin Gorge	Site 3	Morne Sèche	Montagne Zaco	Tatamaka Cliffs	Black River Peak	No. of known pairs	No. known fledged	Estimated pre-breeding population
1973/4	0	?									2	0	7
1974/5		?	3								2	3	4
1975/6		?	e								2	0	7
1976/7		?		3	2						3	5	6
1977/8		?	2*	3	0	2**					5	7	10–11
1978/9		?	0?	3	(?)		0?	0?			4 or 5	3	?
1979/80		?	0?	2			0?	(?)			4 or 5	2	12–15
1980/1		?	(?)	3			0?	(?)			4 to 6	3	12–15
1981/2		?	(?)	e×			0?	e×/e	e	(?)	5 to 7	0	12–15
1982/3		?	0?	0			e/e×	e	2	?	6 or 7	2	14–18
1983/4		?	(?)	1* or 2			0	e×/e×/e	2*	?	6 or 7	3 or 4	16–20

(?) Possible pair present (single birds seen).
? Pair present but breeding success not known.
0 Pair present but not breeding.
e Eggs laid but no young produced.
e/e Two clutches laid but no young produced.
2 Number of young fledged.
× Clutch taken for captive breeding.
* One young taken for captive breeding.
** Two young taken for captive breeding.

usually produced per pair per year. A lower annual production than for temperate regions is expected from *F. punctatus* because of the smaller clutch size and the probable greater longevity of tropical species. Kemp (1984), in the only other study on the population biology of a small tropical falcon, recorded 1.3 fledged young per adult per year in the Greater Kestrel *F. rupicoloides*.

The most significant finding is the low number of pairs (53%) that breed, or attempt to breed, in any year (Table 11). This is considerably lower than in any of the small- or medium-sized diurnal raptors from healthy populations listed by I. Newton (1979), and may be a consequence of poor territorial quality.

Fertility is rather low (68%) but the sample size is too small to come to any conclusions. Of all nesting attempts 75% are successful in rearing at least one fledgling and 56% of eggs result in flying young (Table 11).

The low numbers of breeding birds per year could be a consequence of poor habitat quality and an impoverished food supply. Populations are likely to fluctuate more in poor habitats since their breeding density will be more directly dependent upon food availability. During years of food abundance normally poor-quality marginal territories will be used, while in years of food shortage competition for good-quality territories will be high. In a raptor population such as the Mauritius Kestrel that is heavily dependent upon one food resource, in this case *Phelsuma* geckos, small fluctuations in food availability may affect the breeding density.

Good-quality territories seem to be in great demand. A breeding site occupied every year since 1976 (Site 1) is often frequented by non-breeding adults which I have seen present during courtship, incubation and rearing of the young. In July 1978, when the breeding pair were trapped from this site for the captive breeding programme, they were replaced by another pair that successfully bred in the succeeding months (Jones *et al.* 1981).

The longevity of the Kestrel is not precisely known but is thought to be high (Temple 1978c). If we consider the high level of mortality in the first year to be only 50% and the annual production only 0.3–0.4 young per adult, then the surviving young must live on average 5.0–6.7 years to replace themselves. Longevity may even be longer than this considering the upward trend in the population.

Causes of decline and rarity

The primary reason for the rarity of the Kestrel is habitat destruction and degradation. The species is adapted to a forest existence and cannot live in non-forested habitats, so the fragmented and impoverished forests of Mauritius have become increasingly unsuitable for it. While this has been the main reason for the Kestrel's decline there are several important secondary factors which are discussed below.

Floristic and faunistic impoverishment and food availability

Degraded areas of forest with thickets of exotic vegetation are largely unsuitable for Kestrels. They are structurally and floristically simpler than less invaded areas of native forest and are less capable of supporting the diversity and density of animals. An obvious consequence of this impoverishment is a decline in the amount of food available to the Kestrel. Low food availability is probably the main reason for the low population density of Kestrels. Previous workers do not agree with this: McKelvey (1978) claimed "Only a physically or mentally defective young bird could fail to find sufficient food among the warm tropical forests and gorges of Mauritius". Additional ecological data collected since McKelvey worked with the Kestrel (December 1975–August 1977) support the hypothesis of poor food availability.

In the native forest the density of *Phelsuma* day-geckos, the Kestrel's main prey item, is related to the density of native trees and shrubs. In the upland forest there is a direct correlation between the numbers of *P. guimbeaui rosagularis*, the most common gecko in this forest, and the native plants (Vinson 1976b). In areas lacking many native trees this gecko was either rare or absent. The other species and sub-species of *Phelsuma* that are found on Mauritius are unable to maintain populations in areas heavily degraded by privet and are found only in small numbers in areas degraded by guava (J.-M. Vinson pers. comm.). Healthy populations of geckos can be found in some mixed exotic forests, but these areas are largely unsuitable for Kestrels.

Competition with other species

Competition for food with introduced birds or mammals is unlikely to be great; there are no other resident raptors. Other raptors appear in small numbers during the northern winter: Eleonora's Falcon *F. eleonorae*, Sooty Falcon *F. concolor* and Peregrine Falcon *F. peregrinus* (Staub 1976); of these, only Eleonora's Falcon is an annual visitor (pers. obs.). None of these species competes directly with the Kestrel because they typically restrict their hunting to open areas. *Circus maillardi*, the endemic harrier from nearby Réunion, is an accidental visitor (Staub 1976) and the Madagascar Sea Eagle *Haliaeetus vociferoides* has been recorded only once (Rountree *et al.* 1952, Benson 1970–1).

Fossil remains and early records from Mauritius include two species of owls and a harrier (Chapters 1 and 2). The competitive relationships between these and the Kestrel are unknown and there can be little point in speculation.

Gecko numbers are probably depressed by predation by Common Mynahs *Acridotheres tristis* and Red-whiskered Bulbuls *Pycnonotus jocosus*, both introduced. Mynahs frequently eat lizards including an agamid *Calotes versicolor* (pers. obs.) and house geckos *Hemidactylus* and *Gehyra* spp. (J.-M. Vinson and S. McKeown pers. comm.) and it is reasonable to suggest that they will also take *Phelsuma* geckos. The Red-whiskered Bulbul is probably a major predator of young geckos (McKeown pers. comm.).

Competition for nest-sites has been thought to be significant. McKelvey (1977c, 1978) claimed that Kestrels had to compete with White-tailed Tropic-birds *Phaethon lepturus*, Common Mynahs and feral pigeons *Columba livia* for nest cavities. Tropic-birds are, according to McKelvey, competitively superior and "mynahs and feral pigeons will, at times occupy holes suitable for kestrels and, by sheer perseverance and superior numbers, discourage the kestrels". Tropic-birds, Mynahs and Feral Pigeons are present at most of the cliff sites frequented by Kestrels, and during the 1981 breeding season two of the three nest-sites used by Kestrels had been previously occupied by White-tailed Tropic-birds. A cavity that a male Kestrel prospected, made a nest scrape in and subsequently used as a roosting hole was also an old Tropic-bird nesting cavity. The cliffs are honeycombed with holes and there are abundant suitable cavities, so competition for nest-sites is probably insignificant.

Human persecution

Locally the Kestrel is known as *mangeur de poules* ('chicken eater'), and this led to unnecessary persecution in the past (Guérin, 1940–53, Staub 1973a, McKelvey 1977c) and may be a reason for its disappearance from some of its former range. This name has been used since at least 1763 when it was reported by l'Abbé de la Caille (Cheke 1982b). It is believed it was applied to the Kestrel by the early settlers who used the name for any predatory bird. Some elderly people claim to have killed Kestrels in their youth "because they killed chickens"! In both 1971 and 1973 a pair is believed to have been shot (Temple 1977a, 1978c). Today persecution of the birds is rare due to conservation education on the island (Jones & Owadally 1982b). Hunters who shoot deer in the Black River Gorges and the surrounding areas frequently keep us informed regarding the location of Kestrels they have seen, and are eager that they be conserved. Some persecution may still occur; in June 1982 I was asked to examine a falcon shot by a hunting party in February 1980 at Albion on the east coast. The owners believed it was a Mauritius Kestrel but it proved to be an immature Eleonora's Falcon.

Egg predation by monkeys

The role of the Crab-eating Macaque *Macaca fascicularis* as a predator of birds' eggs and young on Mauritius is a controversial subject. I examine these views together with data on the ecology and population of these monkeys more thoroughly in the section on the Pink Pigeon. Here I consider only the evidence that is directly applicable to the Kestrel. Temple (*et al.* 1974, Temple 1978b) stated that monkeys take Kestrel eggs, though later (1978c) he was more cautious, saying that such predation was only "strongly suspected". Temple also claimed that in some years monkeys may be 100% efficient in locating and destroying the nests of Mauritius Kestrels and Pink Pigeons, but gave little evidence. It would, however, be unreasonable to believe the monkeys do not take and eat eggs and nestlings that they find; also they can and do reach some of the cliff nest-sites that Kestrels choose to nest in. The Montagne Zaco nest-site that I climbed to in 1981 was littered with old dry droppings of both monkeys and Ship Rats *Rattus rattus*.

Temple (1978b, c) suggested that monkey predation of Kestrels nesting in trees was high enough to have reduced the population. He further attributed the upward trend in the population from 6 in 1973 to 14 or 15 in 1977 to a change in the choice of nesting-sites (Temple 1976b, 1977a, 1978b, c). Temple (1978c) summarised his view: "In 1974 a tradition shift occurred in the population, and one pair of birds nested on a cliff rather than in a tree cavity. The cliff site was safe from predators, and the pair raised young which were imprinted on cliff nesting-sites and bred on cliffs when they matured. By 1977 the population had tripled because of the reproductive success which resulted from the new tradition"; this view has been widely quoted (e.g. Spencer & Everett 1982, Cade 1982).

Temple (1978c) suggested that the former nesting habits of the Kestrels can be judged by examining the tail wear of museum specimens, on the basis that the rectrices of cliff-nesting birds should be more worn than those of birds nesting in tree cavities because of the repeated abrasion against coarse basalt; he claimed that as much as 20 mm of the tail can be worn off. After examining museum study skins (location of specimens and sample size not given) Temple suggested that prior to 1900 about half of the Kestrels nested in cliff cavities and about half in tree holes. After this date the cliff nesting tradition is thought to have died out, and the population became locked in a tradition of nesting in tree cavities.

The belief that the population remained locked in a tradition of tree nesting for three-quarters of a century does not stand up to critical examination. While it is plausible that nesting traditions are a feature of the Mauritius Kestrel, as they are of other falcons, it is unlikely that the more adaptive tradition of cliff nesting would have died out in favour of the less adaptive tree-nesting tradition. Temple (1978c) himself provided evidence suggesting that nesting traditions in this Kestrel are not very conservative. In 1974 a pair was observed prospecting 17 different possible nest-sites, 11 of which were tree cavities and 6 in cliffs. This suggests that the actual location of the nest-site, in a tree or a cliff, may not be a rigid feature of the bird's behaviour. While it can be inferred that Kestrels did once, or do occasionally, nest in trees (see Nest-sites and their Selection) there is little evidence to support this. The nest-site of the birds on Brise Fer assumed to nest in trees was, to the best of my knowledge, never found (see Distribution, Status and Population above).

Temple's evaluation of the museum material is less than adequate. A large percentage of museum skins show tail wear, which is to be expected from a cavity-roosting and cavity-breeding species. However, only seven of the 50 measured skins show tail

wear approaching the 20 mm which Temple claims is characteristic of cliff-nesting birds. The majority of skins show wear of only about 1–4 mm. Also, retrospectively judging the former nesting habits of a Kestrel by examining tail wear of a study skin must be very unreliable since feather wear is just as likely to be related to the size and shape of the nest cavity as it is to the nature of the cavity walls. Some closely observed wild Kestrels have badly abraded tails while others that are known to nest and roost in cliffs do not. A female nesting in a cliff site at the head of Feeder Jules Edouard, Black River Gorges, in 1981 had an intact tail with the pale terminal band clearly visible. Some captive Mauritius Kestrels that roosted and nested in wooden nest boxes developed very worn tails.

Temple's hypothesis does not explain either why Kestrels disappeared from areas with safe nesting cliffs such as the Moka Range in the north-west, Corps de Garde in the west and Montagne Bambous in the east where the habitat quality is poor.

Finally, the Kestrel population cannot have been locked in a tree-hole nesting tradition after 1900 since there was until recently a Mauritius Kestrel's egg in the Transvaal Museum, South Africa, that was taken from a cliff nest-site near Grand Bassin in 1920 by Mr George Antelme (Miss T. Salinger *in litt.* March 1982).

Pesticide contamination

In the past organochlorine pesticides were used extensively on Mauritius for agricultural purposes (Ricaud 1975) and malaria control (Mamet 1979). Seabirds around Mauritius have been shown to carry pesticides that result in the typical eggshell thinning syndrome (Temple 1976c). Temple (1978c) had the contents of a single egg produced by the captive female Kestrel in 1974 analysed. He stated that although detectable levels of several toxic chemicals were present, they were below the level normally associated with reproduction dysfunction in raptors. Similarly, eggs laid by wild Kestrels (1982–3) contained low levels of pesticides but these cannot be implicated in the low breeding success of this species.

Eggs laid by captive Kestrels 1978–9 did, however, contain appreciable quantities of the organochlorines DDE and dieldrin (Table 13). It is suspected that the captive birds picked this up from wild-caught lizards, mice and birds that were used to supplement their diet in 1977–8 and these probably

Table 13. *Pesticide contamination of captive-laid Mauritius Kestrel eggs 1978/9*

Female reference	Date egg laid	Clutch number	DDT ppm lipid wt	DDE ppm wet wt	DDE ppm lipid wt	HEOD ppm wet wt	HEOD ppm lipid wt	% thinner than normal
Silver	Sept. 1978	1st	—	71.4	500	14.2	100	—
Silver	Jan. 1978	2nd	—	250	1750	76.1	533	28.8
Yellow/blue	7–14 Nov. 1978	3rd	8	—	38	—	—	3.6
			32	—	104	—	—	13.9
			24	—	52	—	—	5.1

Notes:
DDE is pp'DDE, the principle metabolite of DDT; HEOD is the active ingredient in the insecticide dieldrin and a metabolite of the active ingredient in the insecticide aldrin.
Analysts: the first and second egg by Ian Newton, Institute of Terrestrial Ecology, England. The other three eggs by Dr David B. Peakall, Toxic Chemicals Division, Canadian Wildlife Service.

Table 14. *Mauritius Kestrel egg shell thickness indices*

Egg sample	n	Thickness index Range	Mean	Standard deviation	Coefficient of variation
Pre-1940 (wild)	13	1.18–1.42	1.27	0.07	5.5
1978–9 (captive)	14	0.77–1.28	1.12	0.15	13.4
1981–3 (wild)	15	1.06–1.39	1.24	0.10	8.1

Notes:
The thickness index is derived by the formula: weight/length × breadth.
There are no significant differences in egg shell thicknesses between recent or pre-1940 wild Kestrel eggs. Captive-laid eggs are, however, significantly thinner than wild eggs ($P < 0.01$).

Table 15. *Calculated weights of Mauritius Kestrel eggs*

Egg sample	n	Calculated weight (g) Range	Mean	Standard deviation	Coefficient of variation
Pre-1940 (wild)	14	15.97–19.79	18.30	1.09	5.9%
1978–9 (captive)	15	12.61–16.09	14.67	0.96	6.5%
1981–3 (wild)	18	15.01–18.21	16.66	0.98	5.9%

Notes:
Egg weights are calculated from dimensions using the formula $W = Kw(LB^2)$, where Kw is the constant 0.0005248, derived from known weights and measurements of fresh eggs, L is egg length and B the breadth.
The eggs laid by the wild Kestrels 1981–3 are significantly lighter than the eggs laid pre-1940 ($P < 0.01$), and the eggs laid by captive birds 1978–9 are significantly lighter than those laid by wild birds 1981–3 ($P < 0.01$).

became contaminated from the regular sprayings with organochlorines that are carried out in and around Black River to control malaria-carrying mosquitoes. The small captive population at this time was showing many signs of probable organochlorine poisoning. The eggs were thinner-shelled than eggs laid by wild birds (Table 14) and two of the three eggs laid by a female in her first clutch in 1978 failed to hatch despite being fertile. Retarded gonadal development was also evident in 1979: the birds did not come into breeding condition until late, the first and only egg being laid on 2 December (the latest first clutch recorded, see Table 10). Some of the captive birds that died in 1979–80 showed clinical signs of poisoning (Cooper *et al.* 1981). That all these birds died between 26 September 1979 and 13 January 1980, during the breeding season, is probably significant.

Pesticide contamination has not been demonstrated to have affected the wild population although the disappearance of Kestrels from the Grand Port and Moka Ranges in the 1940s and 1950s may well have been due to this cause.

Examination of eggs laid by wild birds before 1940 and those laid 1981–3 shows that there are no differences in egg shell thickness (Table 14), but surprisingly the recent eggs are significantly smaller, and those laid by captive birds known to be contaminated by organochlorines are smaller still (Table 15). While contamination by pesticides can be implicated as a cause of the lighter eggs laid by the captive birds the reason for the small eggs laid by the wild birds is not clear. No eggs laid between 1940 and 1981 are available for examination.

Pink Pigeon *Nesoenas mayeri* Prévost 1843; *pigeon des mares (pizoñ demar)*
Introduction
The behaviour and ecology of the Pink Pigeon *Columba mayeri*,* the rarest and most endangered pigeon in the world (King 1981), was virtually unknown until the work of McKelvey (1976, 1977a), Hartley (1978), Jeggo (1978a, b, 1979) and Temple (1978c). The main focus of all these studies has been

---

* The author's preference for *Columba* over *Nesoenas* (see under Taxonomy) has been overridden here in favour of consistency with usage in the current Mascarene literature including the other chapters of this book (*Ed.*).

conservation, and in particular captive breeding; much work on the bird in the wild and in captivity remains unpublished.

There is now a large self-supporting captive population of Pink Pigeons giving some security for the future of the species. The wild population has declined due to habitat destruction. Whether the small remnant population can survive in the reduced area of suitable habitat left is a matter of debate.

Taxonomy
The Pink Pigeon was originally placed in the genus *Columba* (Prévost 1843) until Salvadori (1893) created the monotypic *Nesoenas*. Salvadori considered the rounded wings and rufous tail as distinct features worthy of generic separation from *Columba*. Most subsequent authors recognised *Nesoenas*.

Goodwin (1959, 1967) returned the Pink Pigeon to the genus *Columba*, but kept it within the sub-genus *Nesoenas*. He initially (1959) regarded the species as a possible early offshoot of the *oenas* species-group but later (1967) placed *mayeri* as an early offshoot of *Columba*. More recently Staub (1973a, 1976) resurrected *Nesoenas* and was subsequently supported by McKelvey (1976, 1977a), Temple (1978c) and King (1981), but Hartley (1978) and Jeggo (1978a, b, 1979) used *Columba*. Goodwin himself (1983) has, following McKelvey (1976), reverted to *Nesoenas*.

In support of Goodwin's earlier thinking, *mayeri* shows some resemblance to the *oenas* species-group in colour and colour pattern. Within this assemblage its closest affinities are probably with the *palumbus* subgroup and its closest living relative may be *Columba unicincta*. *Mayeri* shows many similarities to *palumbus* pigeons in ecology, behaviour and morphometrics which are discussed elsewhere in this paper.

The neighbouring island of Réunion may have had the same or a closely related species '*Nesoenas duboisi*' named by Rothschild (1907b) from an early description given by Dubois (1674).

Plumage, moult and bare parts
Good plumage descriptions can be found in Rothschild (1907b), Guérin (1940–53), Hachisuka (1953), McKelvey (1976) and Goodwin (1983). The mantle and wings are dark brown, the primaries and to a lesser extent the secondaries are edged on the outer margin with beige. The back is a dusky pink and in some individuals strongly suffused with grey on the

rump. The tail and tail coverts are russet-brown except for the outermost feathers which are off-white tinged with russet. The ventral surface is a pale pink becoming paler on the head where it is off-white, usually clearer in the male. The bare areas, feet, cere and ocular region are coral red. The iris is pale straw colour or off-white.

Juvenile plumage has been described by McKelvey (1976), Jeggo (1978) and Temple (1978c). It is not as striking as the plumage of the adult. The wings and upper mantle are chestnut to chocolate brown. Individual feathers may be bordered with a lighter shade of brown. Body feathers are a light chestnut shading to apricot on the rump. The tail is similar to that of the adults. The cere, feet and ocular region are all a dusky purplish-grey. The iris is dark brown.

McKelvey (1977a) claimed that "The moult process seems to be quite ungoverned by season or sexual activity, and birds in all stages of the moult have been seen courting, feeding young, or foraging". Careful examination of the available data does not entirely support this. Observations on wild and captive birds suggest that moult usually starts at the beginning of summer between October and January and continues until about May or June, with few or no birds moulting during the period of greatest food shortage at the end of winter. Two birds in Cambridge collected in July are in fresh plumage. It should be stressed, however, that there is considerable variation between birds in the dates of moulting.

Primary moult starts at the innermost feather (P 1) and finishes at the outermost (P 10). Secondary moult is irregular in some birds, but others moult ascendantly from the first (distal) feather. There is often considerable difference in the timing and sequence of the moult of the wings of the same bird. Tail moult sequence is irregular with no clear pattern obvious from the limited data (see Table 16).

Table 16. *Moult in the Pink Pigeon*

(a) Secondaries

Ref.	Date	Sex	Wing	1	2	3	4	5	6	7	8	9	10	11	12
1[a]	08/02/79	M	Left	O	M	O	0.9	O	N	O	N	O	O	O	O
			Right	O	O	O	0.2	O	O	O	O	O	O	O	O
19	31/03/79	F	Left	N	O	O	O	O	O	O	O	O	O	O	O
			Right	N	N	0.8	O	O	O	F	F	F	F	F	F
21	19/05/79	F	Left	N	N	N	N	0.9	0.8	O	O	O	O	O	O
			Right	N	N	N	N	0.8	O	O	O	O	O	O	O

(b) Rectrices

			Outer					Centre					Outer
Ref.	Date	Sex	6	5	4	3	2	1	1	2	3	4	5	6
1[b]	08/02/79	M	O	0.6	0.5	O	0.9	O	0.8	0.8	0.4	N	O	O
16	30/04/79	M	N	N	N	N	0.6	N	0.6	N	0.6	N	M	N
17	01/05/79	F	N	N	N	N	N	0.5	N	M	N	N	N	N
19[c]	01/03/79	F	N	0.6	N	0.7	N	0.5	N	N	N	0.8	0.8	0.8
45 (juv.)	04/05/80	?	O	O	0.3	0.3	0.3	0.3	0.3	O	O	0.3	0.3	0.3
17/Col/35/2/4	<1876	F	N	N	0.7	N	N	N	N	N	N	N	0.7	N

Notes:
F, fresh; M, missing; N, new; O, old. Feather growth shown as estimated tenths of full length.
All pigeons were captive adults except 45 which was moulting from juvenile plumage and 17/Col/35/2/4 which is a skin at the UMZC.
[a] On 16/05/79 all secondaries on both wings of bird 1 were new.
[b] On 16/05/79 all of the tail feathers of bird 1 were new.
[c] On 31/03/79 all of the tail feathers of bird 19 were new.

Moult from juvenile to adult plumage starts at about 2 months and is usually finished by 6 months. McKelvey (1977a) gave a longer period, suggesting that the last primary is moulted at about 8 months and the last tail feather at 9 months. The pattern of juvenile moult is similar if not identical to that of the adults. Body moult starts before the change of the primaries and tail feathers and the secondaries are not usually moulted until primary moult is nearly complete. Juvenile moult starts with the wing coverts (?) and head plumage (McKelvey 1977a) before progressing to the ventral tracts, the plumage first being replaced in the centre of the tracts. The moult is often irregular in pattern giving the bird a patchy appearance. By $3\frac{1}{2}$ months the bird is beginning to look like an adult, but the colours are not so bright and the head, neck and ventral tract are not so light.

General habits and behaviour
Maintenance and comfort behaviour

Most of these behaviours are typical for a *Columba* pigeon and McKelvey (1976) described and illustrated some of them. Bathing and rain-bathing are common; the former is described adequately by McKelvey (1976) and is no different from bathing by other members of the order (Goodwin 1983). Heavy rains after a period of drought will send all the captive Pink Pigeons into spontaneous rain-bathing. The bird lies on its side, fans its tail and spreads one wing, stretching it over its back. The feathers are ruffled to allow the rain to penetrate, and when the plumage is sufficiently wet, the pigeon stands, shakes itself and repeats the posture on the same or the other side. Sun-bathing behaviour involves a similar posture to rain-bathing. The bird lies on a branch or the ground and spreads one wing and half of the tail on the same side (McKelvey 1976, Temple 1978c, Staub 1976, pers. obs.). Both Staub (1976) and Temple (1978c) describe dust-bathing but subsequent studies have failed to confirm this; it seems likely they were confusing it with sunning. Dust-bathing in an arboreal *Columba* would be most unusual (D. Goodwin *in litt.*). Pink Pigeons drink standing water in the usual pigeon manner, but they also drink rain droplets and dew that accumulates on leaves (McKelvey 1976, pers. obs.). The expression of other comfort and maintenance behaviours (e.g. preening, sleeping) is typical of most if not all other pigeons (see Goodwin 1983).

Roosting

Pink Pigeons roost on their own or in groups (Temple 1978c) at what may well be traditional sites. The roost is usually in a valley or on the steep side of a hill or ravine. In 1974 the entire population was distributed among five known sites: four along the south-facing scarp and one on the Macabé Ridge. Each site would contain between 1 and 13 birds every night, but the usual number was 5–10, suggesting some movement between roosts. By December 1975 the whole population roosted in one flock using the *Cryptomeria* grove below Plaine Paul (McKelvey 1976). During the breeding season many of the birds still use this grove, roosting within their nesting territories. For the rest of the year I suspect they roost singly, in pairs, or in small groups along the south-facing scarp.

Reactions to other animals

Few species mob Pink Pigeons and they do not often react to other animals, although McKelvey (1976) cited an impressive list of interactions. The typical sleeking of feathers which pigeons do at the approach of a predator has been observed by McKelvey in Pink Pigeons when they see distant flying Echo Parakeets, Ring-necked Parakeets *Psittacula krameri*, White-tailed Tropic-birds, Mauritius Kestrels and Peregrine Falcons, although upon identifying these they ignore them, excepting the Kestrel and the potentially dangerous Peregrine Falcon (see Predation on Adults). I have seen a Red-whiskered Bulbul make a pass at a perched Pigeon and Temple (1978c) saw a Kestrel make a mock attack on one in flight.

McKelvey (1976) claimed that monkeys elicit a staring and neck-craning response from perched Pigeons, and the Pigeons also fan their tails and sleek their plumage. I have been unable to substantiate these observations. Captive Pigeons ignore tame monkeys even if they run across the top of their aviary. I have also watched a group of three Pink Pigeons foraging among some low scrub vegetation in close proximity to a group of monkeys. The monkeys and Pigeons ignored each other, and the Pigeons only moved when the monkeys came within 2–3 m.

Pink Pigeons are naturally very tame and will allow close approach by humans to 3 m or less; while sitting on an observation platform I have had them land in full view only 1.5 m away.

Habitat

During the last century Pink Pigeons frequented all the native vegetation communities in the south-west of the island that lie above 300 m. The upland marshes of Les Mares, *Sideroxylon–Helichrysum* scrub of Plaine Champagne, upland forests of Macabé, Brise Fer and Bel Ombre, pole forests of the logged-out upland plateau and the transitional forests of the south-facing scarp were all used by Pigeons until the 1960s to mid 1970s, during which time some of these areas were exploited and destroyed. Temple (1978c) believed that a mosaic of habitats was essential to meet all the needs of the Pink Pigeon, although it is probable that very little of their food (by biomass) is (or was) obtained from any one particular plant community (see Feeding Ecology and Table 17). It is likely that the Pigeons have been restricted to this one area of the island because it is the only large tract of native vegetation left supporting high enough densities of suitable food plants.

The core area in which Pink Pigeons have been found for over a hundred years lies within the 4.5 m isohyet. Centred on Mount Cocotte and Les Mares (hence the bird's vernacular name *pigeon des mares*) this region included Mount Savane, Bassin Blanc and Plaine Paul. This area was probably favoured because of its high density and diversity of plants (J. Guého pers. comm.). The most important areas of Les Mares and Plaine Paul were destroyed in the 1970s when 1512 acres (612 ha) were cleared and planted with soft-woods (R. T. Hanoomansing *in litt.*, Chapters 1 and 4).

Today the best place to see Pink Pigeons is in the grove of mature self-seeding *Cryptomeria japonica* on the scarp below Plaine Paul, where the birds nest and often roost. The *Cryptomeria* is interspersed with native vegetation and Pigeons forage in and around the grove. Further west than the breeding site, including the ridge of Bassin Blanc and the southern slopes of Mount Cocotte, Crown Land St.-Marie as far as Cascade Cinq Cent Pieds and Alexandra Falls, the scarp is convoluted with spurs and holds pockets of native vegetation. The area has a high species diversity. The dominant trees include *Calophyllum* spp., *Sideroxylon puberulum*, *Labourdonnaisia glauca* and *Syzygium* spp. Other common species are *Tambourissa* spp., *Nuxia verticillata*, *Erythrospermum monticolum*, *Aphloia theiformis* and the non-pathogenic exotic *Litsea monopetala* (pers. obs.). Patches of this forest and the north-facing slope of Mount Cocotte are heavily degraded by exotics (pers. obs.).

Pink Pigeon habitat extends into the low montane wet forest of Bel Ombre, across the dwarf (sub-climax) forest of Plaine Champagne up to the low-canopied upland forest on the northern slope of Black River Peak. These areas still have high plant species diversity and *Nuxia*, *Erythrospermum* and *Aphloia*, the three most important food plant species, are abundant.

Habitat destruction, population status and distribution

The primary reason for the Pink Pigeon's present low population has been habitat destruction and impoverishment. The destruction of native plant communities on the south-west plateau during the last 50 years has been considerable. Relatively large tracts of land suitable for most of the native birds have been cleared for tea and forestry plantations. The upland plateau stretching from Piton du Milieu to Black River Peak including the forests of the south-facing scarp and the Black River Gorges provided a contiguous area of native forest until the 1940s. Within this area the Midlands and Kanaka/Grand Bassin forests had been selectively logged in the nineteenth century and the turn of this century (Chapter 1; Brouard 1963) and most of the large canopy trees removed. However, detailed information on the exploitation of the native forests is lacking (Edgerley 1957). Old plantations existed in the Grand Bassin and Midlands Ranges and in the Sophie and Bel Rive–Malherbes sections; these had been developed from small haphazard plantations of exotic trees, and were interspersed with large areas of native vegetation (Edgerley 1957, 1961, King 1945; R. E. Vaughan pers. comm.). The pole forests of the upland plateau, although badly degraded in areas, still supported a good diversity of plant species until they were cleared (R. E. Vaughan pers. comm.), and all the endemic birds had been recorded there.

The destruction and fragmentation of this area accelerated in the 1940s and 1950s. The tea industry obtained a new lease of life during the Second World War and there followed an expansion in the areas planted under tea (Roy 1969). This development occurred on the south-west plateau, using areas above 300 m with more than 3 m of rain a year, and was centred on the Midlands area.

In 1950 a White Paper on Crown Forest Land and Forestry (Allan & Edgerley 1950) reserved 5925 acres (2370 ha) in the central plateau for tea-planting. By 1972 the area under tea, most of which had less than half a century before been utilised by the Pink Pigeon, Echo Parakeet and other native birds, totalled 13 500 acres (5400 ha).

The insidious encroachment upon the Pigeon's habitat continued with the clearing of the upland scrub forests of Quartier Militaire and Midlands Ranges in the 1950s (Edgerley 1957) and large areas south of Curepipe in the 1960s. These were planted with pine. Between 1967 and 1971 the Forestry Service concentrated on clearing and then planting with softwoods the area around the Mare aux Vacoas reservoir, Grand Bassin, Pétrin, Mare Longue and Rivière du Poste (Brouard 1969, 1970, Owadally 1971a, b, 1973). In the early 1970s most of the clearing and planting in this region was taken over by other Government bodies, but some smaller-scale development of this area by the Forestry Service has continued to this day.

The Government development programmes Travail pour Tous and the Rural Development Project operated by the Development Works Corporation and financed by the World Bank greatly accelerated the destruction of this area. Between 1973 and 1981 7563 acres (3025 ha), much of it key Pink Pigeon and Echo Parakeet habitat, were cleared and planted with softwoods. The consequence of this destruction was the dramatic population crashes of Pink Pigeons and Echo Parakeets.

In late 1977 work started on a new road passing from Les Mares to the south coast. The road passes close to the Pink Pigeon's main breeding and roosting site and by Bassin Blanc. The increased number of visitors and disturbance that the road will bring does not augur well for the future of Pink Pigeons in this area. By early 1984 the road was virtually complete.

The area of contiguous native vegetation and pole forests available to the Pink Pigeon in 1940 was in excess of 40 000 acres (16 188 ha). Today this area has been reduced to only 9000 acres (3642 ha) of which only about 5700 acres (2307 ha) are utilised by the Pink Pigeon.

Distribution and status up until the 1950s

Before Mauritius was colonised by man, the Pink Pigeon may have ranged widely over the whole island.

Subfossil bones found in the Mare-aux-Songes (Newton & Gadow 1893) give some support to the notion that the birds were to be found in the lowlands, although McKelvey (1977a) considered their distribution to have been the whole of upland Mauritius. Temple (1978c) claimed that the Pink Pigeon ranged widely over the upland plateau during the 1800s and quotes specimens collected from the "regions around Curepipe, from Candos Hill near Quatre Bornes, and apparently from mountains above Mahébourg". Temple does not provide the present locations of these specimens and the records need to be substantiated. Nineteenth-century specimens for which data have been obtained were all collected on the south-west plateau. Four of these are at the UMZC: an adult female collected by J. Caldwell at Vacoas prior to 1876 and three Pink Pigeons collected for, or by, Edward Newton in the 1860s, two from Grand Bassin and the other from Bois Sec. Written accounts from this period (E. Newton 1861, 1865, 1875, 1878) testify that the pigeon was rare and restricted to forests in the island's south-west. All the evidence suggests that it was localised although it probably ranged widely over the scrublands and forest between Black River Peak and Piton du Milieu in the area roughly circumscribed by the 4 m isohyet (see Padya 1972).

Information from the turn of this century confirms the Pink Pigeon's rare status and limited distribution. A specimen in the BMNH obtained in 1909 has written on its label "About to become extinct, indigenous species special to Mauritius" and another skin obtained at the same time and collected by Maurice Ulcroy has on its label "They are found only from the area of Savanne to Les Mares".

In 1912 Meinertzhagen wrote that "A few still remain in the south-west corner of the island and are said to be increasing", although he does not provide any supporting evidence for the latter statement. Guérin (1940–53) noted in 1940 that some Pink Pigeons survived but that they might become extinct in a few years. He gave their distribution as Les Mares, Montagne Cocotte, Montagne Perruche and occasionally as far as Bel Ombre and Chamarel. Information from local hunters suggests that in the 1940s Pink Pigeons were commoner than Guérin suggests. E. Closel (pers. comm. to J.-F. Martial) claimed that up to 25 could be seen on every visit to Bassin Blanc which was, and still is, one of the best places to see them. Their distribution included much of the south-western plateau and they could occasionally be seen as far north-west as Tamarin Falls (Map 4).

In the 1950s the accounts support the population and distribution data from the previous half-century and it seems likely that at this period the population was stable. Rountree *et al.* (1952) reported it "Local – confined to the remote forests of the south-western plateau. Rare." Newton (1958a) observed that the Pigeon was "confined to the high ground in the south of the island roughly from Mare aux Vacoas southwards to the Piton de la Rivière Noire." Greenway (1967), using data provided by Georges Antelme (for the 1958 edition), stated that they are rare and then later writes "they are found commonly enough, although localized, in the forests near Savan[ne] and Rivière Noire", the latter statement probably referring to the forests of Black River Peak. Benedict (1957) described them as extremely rare. The different magnitudes of rarity are probably due to different terms of reference used by these naturalists to describe what was then a small, stable and probably healthy population of a relatively easy-to-see species.

Map 4. Distribution of Pink Pigeons in the 1940s, with a list of locations mentioned in the text.

1	Black River Peak	16	Kanaka
2	Macabé	17	Mt. Perruche
3	Plaine Champagne	18	Mare aux Vacoas
4	Plaine Paul	19	Plaine Sophie
5	Les Mares	20	Mare Longue
6	*Cryptomeria* grove	21	Tamarin Falls
7	Bassin Blanc	22	Black River Gorges
8	Mt. Cocotte	23	Grandes Gorges
9	Bel Ombre Forest	24	Le Bouton
10	Mt. Savane	25	Black River Viewpoint
11	Mt. Capote	26	Brise Fer
12	Bois Sec	27	Chamarel
13	Combo	28	Chamarel Gorge
14	Gouly	29	Monvert
15	Grand Bassin		

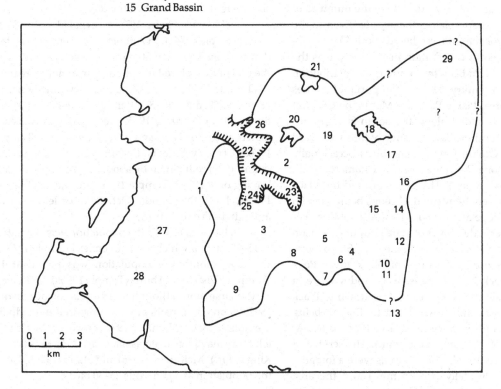

The actual numbers of Pigeons in the 1950s have been variously estimated by local naturalists with a good degree of agreement. The population probably lay between 40 and 60 birds and observers respectively suggested that there were less than 50 (F. Staub pers. comm.), more than 50 (L. F. Edgerley pers. comm.) and 50–60 (J. Guého pers. comm.). These estimates are regarded as realistic by others familiar with the pigeons in the wild (J.-A. Lalouette, M. d'Unienville, J.-M. Vinson and others pers. comm.).

Distribution and status 1960–83

Cyclone Carol hit Mauritius in late February 1960 causing unprecedented damage (Padya 1976); the Pink Pigeon population was severely affected and may have been reduced by 50% (F. Staub pers. comm.) or more. The storm caused irreversible damage to the native forest and it seems probable that the population did not recover its former status. By the mid 1960s the Pigeon could again be found regularly in its favoured haunts (J.-M. Vinson, J. Guého pers. comm.) and could be observed on virtually every outing to the area around Grand Bassin and on most day trips to the Macabé Ridge. However, following the encroachment upon their habitat by forestry plantations during the 1960s, there was probably a decline in the numbers of birds during the latter half of the decade. Even so the population probably still numbered about 40.

The Pigeon's distribution in the 1960s was the same area that had been the centre of its distribution for well over a century. The most important localities were Les Mares, Plaine Paul and Montagne Cocotte. Their range extended along the south-facing scarp from Bois Sec and the Savane range of hills westward into the Bel Ombre forest, some birds occasionally wandering along the ridge towards Chamarel (L. F. Edgerley pers. comm.). The ridges around the Black River Gorges were frequented, Pigeons being seen on the Macabé Ridge and even in the forests of Brise Fer extending over to the shores of Mare Longue (J. Vinson unpubl. field notes; L. F. Edgerley, F. Staub pers. comm.). The northern slopes of Black River Peak above the Black River Gorges, were and still are, a favoured locality, this area extending down to Black River Viewpoint and Crown Land Le Bouton being described as a likely place to observe Pigeons (G. A. d'Argent, J.-M. Vinson pers. comm.). The scrub and pole forests around Mare aux Vacoas were a foraging area (A. W. Owadally pers. comm.) and the area extending north towards Curepipe and east to

Nouvelle France was suitable Pigeon habitat (A. S. Cheke pers. comm.), but I have been unable to confirm the species' occurrence in this region. The Crown Lands south of Mare aux Vacoas, around Kanaka Crater and towards Grand Bassin were regularly frequented (J. Vinson unpubl. field notes; J.-M. Vinson pers. comm.). Their range also extended south to Les Mares and Plaine Champagne and south-east along the tributaries of Rivière du Poste and Rivière des Anguilles down into the valley of Bois Sec. The total range of the Pink Pigeon in the 1950s and early 1960s was probably between 130 and 160 km^2.

The population of Pigeons probably remained stable or declined slightly in the early 1970s, but in the middle of the decade there was a severe crash in the population which was exacerbated by cyclone Gervaise which struck in February 1975. The population has failed to recover fully from this decline due to the destruction of large tracts of habitat during the period 1973–81. All local naturalists to whom I have spoken concur unanimously with this assessment of the Pigeon's population trend during the decade. There is some variation in the year-to-year estimates which almost certainly reflects changes in the population (Fig. 11), although some observer bias is probable with six different workers providing data.

Field research in 1973–4 revealed only 25–30 birds (Temple 1974a). In August 1974, Temple *et al.* felt that perhaps less than 30 birds survived. A survey at the end of 1974 placed the population at not more than 28 (Temple 1975), only 29 (Temple 1976a) and between 27 and 38 (Temple 1978c). Before cyclone Gervaise hit the island in February 1975, the population was estimated at 25–30 (Newlands 1975) and 28–30 (Temple 1976b). The cyclone badly affected the population (Staub 1976) reducing it by about 50% (Temple 1978c), leaving only 10–12 (Temple 1975), around 12 (Temple 1976a), 12–15 (Newlands, Oct. 1975) or less than 20 individuals (Temple 1976b).

In November 1975 the population was estimated at 15–22, two of which were juveniles (Temple 1978c). The next month the population was placed at 19 (Temple 1976a) and 24 birds (Temple 1976b). McKelvey for the same month thought the population numbered 18–20: 6 breeding pairs and 6–8 unpaired individuals (Temple 1978c). In January 1976, however, McKelvey felt there may have been as many as 24 Pigeons (W. B. King *in litt.*). McKelvey (1976) and Staub (1976) placed the population at a generous 25–30 birds.

Map 5. Distribution of Pink Pigeons 1970–5 (based on several sources of information: see text).

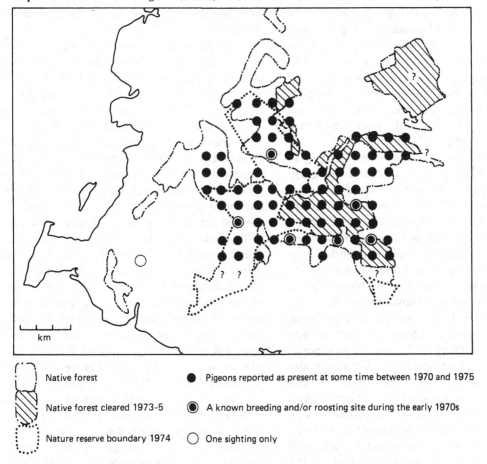

Native forest

Native forest cleared 1973-5

Nature reserve boundary 1974

● Pigeons reported as present at some time between 1970 and 1975

◉ A known breeding and/or roosting site during the early 1970s

○ One sighting only

Fig. 11. Population trend of the Pink Pigeon 1955–83. The continuous line represents the mean population estimate 1955–83. The vertical bars are probable extremes.

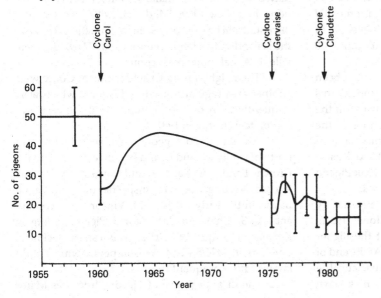

At the beginning of 1976 10 pairs were found breeding and an eleventh pair was reported to be the unlikely mixed pairing of a male Pink Pigeon and a female Madagascar Turtle Dove *Streptopelia picturata* (McKelvey 1976). At least 5 young were reared to independence during the year but this productivity was offset by the removal of some adults, a fledgling and eggs for the captive breeding programme (see Table 24). The year closed with an estimated Pink Pigeon population of 27 birds.

The 1977 population estimates are conflicting. In April Hartley (1978) captured 8 adults and a fledgling, and took 4 eggs for captive breeding at the Jersey Wildlife Preservation Trust and also at the Government Aviary in Mauritius (Durrell 1978) (see Table 23). Before these birds were trapped Temple (1978c), using information provided by McKelvey, claimed the population was at least 6 pairs and 6 or 7 unmated individuals (this makes at least 18 or 19 birds which is substantially less than the estimate made only weeks before in late 1976). Temple was probably confusing McKelvey's pre- and post-trapping estimates (see below). McKelvey (King 1977) gave the wild population as 6 or 7 adult pairs and 6 or 7 unpaired males (this predominance of males seems most unlikely and I regard them as non-breeding birds of both sexes). McKelvey mentioned 4 young successfully reared during the year, but Temple (1978c) quoted McKelvey as saying only 2 young were produced. In his general account of the Pink Pigeon Hartley (1978) placed the population generously at 20–30 individuals. Temple (1978c) was less optimistic and gave a low estimate of only 10–12 birds at the end of the year (this is probably based on his misinterpretation of the year's events). In late 1977 the population probably stood at around 20 birds.

In 1978 the population was thought to have been approaching 30 individuals (Steele 1979) although this seems very generous. During an all-day watch at the *Cryptomeria* grove in late May, when all or most of the birds could be expected to be in the vicinity, an estimated 16 birds were seen (Steele 1979). At least 3 young fledged during the year. The 1978 Pink Pigeon population was probably in the low twenties.

The 1979 population estimates are variously given as 12–20 (Jones 1979) and 20–30 (Jones 1980a; Pasquier 1980), although with hindsight the higher figures are undoubtedly over-optimistic. At the end of December cyclone Claudette hit Mauritius, and the storm and post-cyclone food shortages may have greatly reduced the population: only 6–10 Pigeons were subsequently seen in the *Cryptomeria* grove (Jones 1981).

No thorough survey of the population has been conducted in the early 1980s; however, the frequency of sightings and distribution of the birds suggests that the population has probably stabilised between 10 and 20, perhaps the carrying capacity of the now reduced area of suitable habitat available to the Pigeons.

The contracting range of the Pink Pigeon in the 1970s and its distribution in the early 1980s can best be illustrated by summarising some of the field observations during this period. The Pigeons have been forced into the small remaining areas of native vegetation by the destruction of large tracts of good habitat in the 1970s. In addition further suitable areas are now essentially unavailable to the remaining birds due to habitat fragmentation. The south-facing scarp between Montagne Savane and Bel Ombre forest is the stronghold for the species today (Map 6). The gamekeeper and foresters for the southern part of the Bel Ombre forest have, however, never seen any there. East of their present distribution Bois Sec was a favoured area until recently. In this locality a small grove of *Cryptomeria* below the ridge on the Rivière Chevrettes was a breeding site. A. S. Cheke (*in litt.*) reported pairs building nests here on 21 February and 20 April 1974 and saw a fledged young on the latter date. The following year in the same area, a pair were frightened away by trail-cutting in early June (Newlands, June 1975). The Pink Pigeons there started to decline with the clearing of native vegetation, to make way for forestry plantations, in the late 1960s (M. d'Unienville pers. comm.), but could still be regularly seen up until the mid-1970s, the last individual being seen about 1977 (M. d'Unienville, J.-A. Lalouette pers. comm.).

The neighbouring Crown Lands of Gouly were another area regularly used by Pigeons, which were frequently seen by the guardian until the area was completely cleared in 1981.

A decade ago Pigeons were seen in the area north of Gouly around Grand Bassin, extending to the Crown Lands of Kanaka and towards Mare aux Vacoas. Nesting was even suspected on Piton Grand Bassin in the early 1970s (J.-M. Vinson pers. comm.) and A. S. Cheke (*in litt.*) saw a Pigeon calling in Kanaka on 13 April 1974. There was a sharp decline in numbers in 1974/5 (J.-M. Vinson pers. comm.) and I know of no subsequent records.

The disappearance of Pigeons from Les Mares

and Plaine Paul occurred in the 1970s when this very important area was cleared. Further west, Plaine Champagne, which is still covered by native vegetation, is apparently under-utilised by Pigeons. This area lies between the south-facing scarp above Bel Ombre forest and the edge of the gorges. Some birds do fly over this low forest on their way to and from the forest on the north side of Black River Peak. In the early 1980s I recorded single Pigeons flying over Plaine Champagne about once every 30–50 hours of field work.

Montagne Cocotte has always been one of the best places to see Pink Pigeons, but the frequency with which they can be seen has declined. J. Vinson (unpubl. field notes), who did not always record what birds he saw, noted Pink Pigeons on five out of ten visits between 1942 and 1962. Today they are seen on only about one visit in four or five.

I can find no records of Pink Pigeons in the Combo region even though this is a large area of seemingly suitable habitat and Pigeons are to be found only 5 km to the north-east. The gamekeeper who lives in the forest has never seen any.

The ridges around the Black River Gorges used to be important areas. Today the only place around the Gorges where Pigeons can be seen is on the northern slope of Black River Peak which is probably frequented throughout the year. Formerly Pigeons used to forage further into the gorge at the foot of Black River Viewpoint and Crown Land Le Bouton. Newlands (Oct. 1975) recorded them here in September 1975, but there are no subsequent records.

Macabé Ridge, Brise Fer and the forest stretching over to Mare Longue were areas where Pigeons could be regularly observed. A. S. Cheke (*in litt.*) described them as occasional in 1974 but after this date there was

Map 6. Distribution of Pink Pigeons 1980–4.

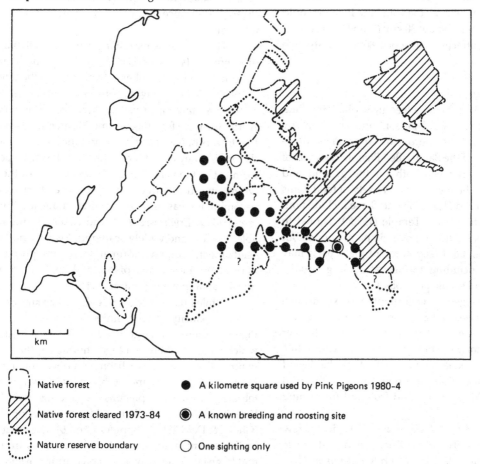

Native forest	● A kilometre square used by Pink Pigeons 1980-4
Native forest cleared 1973-84	◉ A known breeding and roosting site
Nature reserve boundary	○ One sighting only

a sharp decline in the number of sightings. J. Hartley (pers. comm.) saw a single Pink Pigeon in late January or early February 1977 and F. N. Steele (unpubl. field notes) saw one at the entrance to Macabé on 6 September 1978. This was the last sighting in this area. Pink Pigeons no longer use this forest probably because access would have to be across Les Mares/Pétrin, now pine-covered, or across Grandes Gorges, an area not utilised by the Pigeons. There are few reports of Pink Pigeons ever being found down in the Black River Gorges. Hunters and gamekeepers familiar with the bird and with an intimate knowledge of the gorges, extending back to 1951, have never seen it at such low elevations. The only record of a Pink Pigeon in the gorges is a single bird seen at the conjunction of the Grandes Gorges and Rivulet Roches Blanches on 3 January 1981 (A. Gardner pers. comm.).

Pink Pigeons rarely wander outside their usual range although Temple (pers. comm. to A. S. Cheke) saw one in Chamarel Gorge below Petite Moka in 1974.

In the early 1980s all the wild Pink Pigeons could be found in an area of about 30 km² (Map 6), representing a range contraction of about 80% in only two decades.

Feeding ecology

In many aspects of its feeding ecology the Pink Pigeon is a typical 'wood pigeon' and shows many similarities to both the European Wood Pigeon *Columba palumbus* and the Band-tailed Pigeon *C. fasciata* of North America. It feeds both on the ground and among trees and bushes. A wide variety of food items have been recorded (Table 17) and these include both native and exotic plants. Temple (1978c) claimed to have seen birds feeding on over 50 plant species but did not list them all. Many different plant parts may be consumed including the leaves, young shoots, flowers, fruit and seeds (pers. obs.).

The Pink Pigeon is well adapted to exploit such a wide food base. It has short rounded wings, a longish broad tail and strong feet enabling it to clamber and balance among the flimsy ends of branches while feeding. The beak is similar to that of the Wood Pigeon, moderately long and hooked at the tip, well suited for plucking flowers and fruit and for ripping leaves.

The bulk of the diet is provided by just a few species of plants. Temple (1978c) stated that *Nuxia verticillata, Aphloia theiformis, Erythrospermum monticolum, Lantana camara, Pittosporum senacia* and *Dio-*

spyros sp. accounted for 38% of observed feedings in his study but the importance of the first three species was greatly understated and on a year-round basis they may provide most of the diet by biomass (pers. obs.; see Table 18). These three food plants are widely distributed and common over the Pink Pigeon's habitat, being found in nearly all the vegetation communities. They were even to be found in the badly degraded, logged-out pole forests of the upland plateau, where they formed low shrubs in the secondary thickets (Vaughan & Wiehé, 1937). The common and widely distributed exotics *Lantana camara, Ligustrum robustum* var. *walkeri* and *Litsea glutinosa* are often taken by wild birds (McKelvey 1976, Temple *et al.* 1974, pers. obs.) but do not provide a very large part of the diet by biomass and are largely ignored by captive birds in favour of domestic grains.

Some of the plants included in Table 17 (e.g. *Diospyros tesselaria, Hornea mauritiana*, and *Euphorbia pyrifolia*) must form only an incidental part of the Pigeon's diet since they are very rare or absent within its range.

There are seasonal changes in the availability of different foods. Temple (1978c) saw Pigeons foraging in Kanaka from September to December on the flowers (?) of *Antidesma madagascariense* and the leaves (?) of *Tabernaemontana mauritiana*, while from January to April they were foraging around Montagne Savane on the flowers of *Nuxia verticillata* and the fruits of *Aphloia theiformis*. They in fact forage for these last two species for most of the year (Table 17). At the end of the dry winter months (August to mid-November) food is becoming increasingly scarce and fruiting and flowering *Aphloia, Erythrospermum* and *Nuxia*, if still available, are few and widely scattered. Leaves and fallen seeds must, I assume, form an increasing proportion of the diet. The shortage of food at this time of the year may be an important limiting factor.

While I have little direct evidence to show that a seasonal paucity of food limits the numbers of Pink Pigeons, there is a large body of historical evidence which suggests there is a food shortage at the end of winter. Fat cycles have been described for a wide range of frugivorous and herbivorous Mascarene vertebrates: tortoises, pigeons (including the Pink Pigeon), Dodos and Solitaires, Merles and flying-foxes (Cheke & Dahl 1981; Chapters 1 and 6). Observations in the seventeenth to nineteenth centuries indicate that large fat deposits were put on during the summer and autumn months when there was an abundance of

Table 17 continued

Species	Status[a]	Range[b]	Part eaten[c]	Importance of food items[d]	J	F	M	A	M	J	J	A	S	O	N	D	Refs.[e]
Ebenaceae																	
Diospyros spp.		M	Lv	**													E,F,I
D. tesselaria Poiret	V	M	Fr,Sd	***													E
Oleaceae																	
Ligustrum robustum Blume var. *walkeri* (Decaisne) Mansf.		I	Lv / Fr	*? / ***													A
Apocynaceae																	
Tabernaemontana mauritiana Poiret	Nt	M,R	Lv? Fl?	***	x												I
Loganiaceae																	
Nuxia verticillata Lam.	V	M,R	Yg,Lv / Fl	* / *****	xxxx	xxxx	xxxx	xxx									A,B,H,I,J
Convolvulaceae																	
Ipomoea sp.		I	Fl	**													A,E,F
I. purpurea Roth.		I	Bd,Lv,Fl	C													F
Verbenaceae																	
Lantana camara L.		I	Fl / Fr	*** / ****		x											A,B,E,F,I
Lauraceae																	
Litsea glutinosa (Lour.) C.B. Robinson		I	Fl? Fr	**	x												J
Loranthaceae																	
Korthalsella opuntia (Thunb.) Merrill var. *richardii* (Van Tieghem) Danser		N	Lv	**		xx											A
Euphorbiaceae																	
Antidesma madagascariense Lam.	Nt	N	Lv?	***									xxxx	xxxx	xxxx	xxxx	J
Cordemoya integrifolia (Willd.) Pax	V	M,R	Fl?	**		x							x	x	x		I
Euphorbia pyrifolia Lam. Arg.	V	N	Fr	*?													C,H
Homalanthus populifolius Graham		I	Fr	**													A
Stillingia lineata (Lam.) Muell. Arg.	Nt	N	Fr,Sd	***													E,F,I,J
Casuarinaceae																	
Casuarina equisetifolia L.		I	Lv	C												x	A
Musaceae																	
Ravenala madagascariensis Sonn.		I	Nr	?													F
Palmae																	
Tectiphiala ferox H.E. Moore	E	M	Sd	*?													B
Graminae																	
Axonopus compressus (Swartz.) P. Beauv.		I	Sd	*													I
Ischaemum aristatum L.		I	Sd	*													A
Panicum maximum Jacq.		I	Lv,Sd	C													A
Stenotaphrum dimidiatum (L.) Brongn.	Nt	N	Lv	C													I
Animals																	
Tadpoles (*Bufo regularis*)				?													A
Pond snails				*?													E
Small land snails				*?													E
Small black beetle				?													F

Table 17. Food items taken by Pink Pigeons

Species	Status[a]	Range[b]	Part eaten[c]	Importance of food items[d]	Dates of feeding observations												Refs.[e]
					J	F	M	A	M	J	J	A	S	O	N	D	
Taxodiaceae																	
Cryptomeria japonica (L.f.) Don	I		Sd	*													A
Flacourtiaceae																	
Aphloia theiformis (Vahl) Benn.	Nt	N	Fr	*****	xxxx	xxxx	xxxx	xx									A
Erythrospermum monticolum Thouars	Nt	M	Fr	**				x			x	x	x		xx	x	A,B,C,D,G,I
Pittosporaceae																	
Pittosporum senacia Putterl.	Nt	M,R	Lv	***													E
subsp. *senacia*			Fl? / Fr	****													E,I
Guttiferae																	
Harungana madagascariensis Choisy	Nt	N	Fl	***													A
Sterculiaceae																	
Trochetia sp.	V	M	Fl	**													F
Rutaceae																	
Murraya paniculata (L.) Jack	I		Lv,Fl	C													A
Meliaceae																	
Melia azedarach L.	I		Lv	C													A
Malvaceae																	
Hibiscus rosa-sinensis L.	I		Lv	C													A
Sapindaceae																	
Hornea mauritiana Baker	V	M	Fl	?													F
Leguminosae																	
Glyricidium sepium (Jacq.) Kunth ex Griseb.	I		Lv,Fl	C													A
Leucaena leucocephala (Lam.) Dewitt	I		Lv	C													A
Pisum sativum L.	I		Lv,Fl	C													F
Phaseolus sp.	I		Lv,Fl	C													F
Pongamia pinnata W. F. Wight	I		Lv,Fl	C													A,B,C,D,G,H,I
Myrtaceae																	
Eugenia sp.			Lv	**													E,F
Psiloxylon mauritianum (Bouton ex Hooker f.) Baillon	V	M,R	Fr	*				x									A
Melastomataceae																	
Ossaea marginata (Desv.) Triana	I		Fr	*													A
Passifloraceae																	
Carica papaya L.	I		Bk,Lv,Fl	C													A
Rubiaceae																	
Paederia foetida L.	I		Lv,Fl	C													A

continued

fruit and flowers, to be used up during the cool dry winter when food was scarce. Today these seasonal food shortages may be even more severe due to forest degradation and competition for flowers and fruits from introduced species.

In early summer (December and January) following the onset of the rains the Pigeons feed a great deal on flowers. Leaves are eaten throughout the year in moderate amounts, and those of several species are taken by both wild and captive birds (McKelvey 1976, pers. obs.; see Table 17).

Most feeding episodes are arboreal (c. 60–75%), and the Pigeons can move around the extremities of trees and bushes with agility. They can, by gripping with their strong feet, hang virtually upside down and when necessary they spread their wings and tail for support. McKelvey (1976) provides an illustration of a Pink Pigeon feeding on berries, and Goodwin (1955) gave a series of sketches of Wood Pigeons feeding arboreally that could equally apply to Pink Pigeons. Flowers, fruit or leaves are grasped firmly in the beak and pulled or torn off the stalk by a tugging and twisting motion.

When food is abundant Pink Pigeons are capable of eating a large percentage of their daily food requirements in one feeding episode (McKelvey 1976, Temple 1978c). During a single feeding episode of 21 minutes David Todd and I observed a Pigeon eat an estimated 60+ fruits of *Homalanthus populifolius*.

Pigeons search the ground for fallen fruits and seeds and feed on the leaves and seeds of grasses (McKelvey 1976, Temple 1978c, pers. obs.). They regularly pick up small pebbles and also crushed snail shells (McKelvey 1976, Table 18, pers. obs.). Pink

Table 18. *Crop and gizzard contents of Pink Pigeons*

Species	1 Crop	Gizzard 12/07/1864	2 Crop	Gizzard 23/04/1974	3 Crop 28/12/1981	4 Crop 18/09/1983
Aphloia theiformis seeds	Full	Full	—	—	—	—
Erythrospermum monticolum seeds	—	—	23	Bolus	11	23
Psiloxylon mauritianum fruit	—	—	1	—	—	—
Ligustrum robustum var. *walkeri* berries	—	—	—	—	2	—
Nuxia verticillata flowers	—	—	17	—	—	43
Hard unidentified seeds c. 2 mm in diameter	—	—	—	c. 20	—	—
'Crop 'milk'	—	—	—	—	c. 50% by vol.	
Small pebbles (3.0 × 1.5 mm to 7.0 × 2.8 mm)	—	—	25	Several	—	—

Crop and gizzard contents data taken from: 1, specimen label 17/Col/35/2/2 at the UMZC; 2, data provided by J.-M. Vinson (*in litt.*) from a male pigeon poached on the Macabé ridge and later confiscated; 3 and 4, partial crop contents removed by author from two c. 6-day-old squabs and a fledgling respectively.

Notes to Table 17:

[a] Based on IUCN *Red Data Book* categories (Anon 1980). Nt, not threatened; V, vulnerable; E, endangered.

[b] M, Mauritius; R, Réunion; N, native; I, introduced.

[c] Bd, buds; Fl, flowers; Fr, fruit; Lv, leaves; Sd, seeds, Yg, young.

[d] The importance of the food items is estimated in five categories from only one feeding record (*) to frequently consumed with several independent records (*****). C, plants growing in their aviaries that are readily eaten by captive pigeons. Banana *Musa* sp. is the only plant growing in the aviaries that is not eaten by the Pink Pigeons. Not recorded are the domestic seeds, grains and pulses that are readily eaten by the captive pigeons. These include: sunflower, millet, rice, wheat, barley, oats, maize, beans, mung beans, peas, lentils and peanuts.

[e] A, this study, including unpublished observations by P. Ahimaz, A. S. Cheke and D. M. Todd; B, unconfirmed records by hunters, foresters and naturalists; C, Guého & Staub (1979); D, Guérin (1940–53); E, McKelvey (1976); F, McKelvey (1977a); G, Staub (1974); H, Staub (1976); I, Temple (1978c); J, Temple *et al.* (1974).

Pigeons very occasionally eat invertebrates (McKelvey 1976, 1977a) and McKelvey (1976) recorded one trying to catch tadpoles. They search among the leaf litter flicking loose particles aside with sideways movements of the bill. All of the *Erythrospermum* seeds removed from the crops of two squabs, and 21 out of 23 seeds taken from the crop of a fledgling Pigeon (Table 18), had lost their red epicarp and had presumably been found in this condition beneath the parent tree. While on the ground Pink Pigeons show no reticence about going under dense bushes to look for food.

The birds usually forage alone (McKelvey 1976, pers. obs.), in pairs or in family groups (McKelvey 1976, Temple 1978c). An area is searched by flying from tree to tree or by flying directly to known food sources (McKelvey 1976, Temple 1978c, pers. obs.). During most of the day individuals are widely distributed throughout their range, regularly travelling up to 6 or 8 km from their roosting sites (Temple 1978c, pers. obs.). Loose concentrations of Pigeons occasionally gather at a food source and Temple (1978c) recorded nine birds foraging within a radius of about 5 m. Flocking appears to be rare (see The Effect of Cyclones) although Temple (1978c) noted: "flocks of several dozen Pink Pigeons were often seen as recently as during the 1930s at Les Mares . . . The birds are said to have flown down from nearby Montagne Cocotte to feed in the marsh, and the owner of the hunting lodge formerly located at Les Mares reportedly entertained his guests by attracting the birds to grain spread on the lawn".

Pink Pigeons feed most frequently during the early morning and late afternoon, although these peaks are less developed during the winter. In October and November, before the onset of the summer rains, the Pigeons may forage during most hours of the day.

An aspect of the Pink Pigeon's feeding biology that is well known on Mauritius is its supposed fondness for a fruit that has a narcotic or intoxicating effect upon the bird, and renders its flesh toxic, causing anyone eating it to have convulsions. This story stretches back to the earliest days of the island's colonisation (see Chapter 1).

There has been considerable confusion in the literature regarding the identity of the offending fruit, but most local informants agree that the plant involved is the *fandamane*, i.e. *Aphloia theiformis* (G. A.

d'Argent, M. d'Unienville, E. Closel via J.-F. Martial, *et al.* pers. comm.). Several other plants have been put forward as contenders; Temple (1978c) believed there had been a confusion over the local names and the plant responsible was *fangamme*, *Stillingia lineata*. He experimentally force-fed some domestic pigeons *Columba livia* fruits of *Erythrospermum monticolum*, *Aphloia theiformis* and *Stillingia lineata*. Only the *Stillingia* produced any ill-effects: "The pigeon became violently ill about 20 minutes after ingesting the fruit. Initially its rate of respiration increased; it then suffered convulsions; it lost coordination; and finally entered a coma. After three hours, the bird recovered its coordination slowly, and after five hours, it was almost fully recovered." McKelvey (1976) also thought that *Stillingia* was the offending plant, but it seems most unlikely that the pigeons would habitually consume a fruit that has such a drastic result (see Goodwin, 1967, p. 20). In his experiment Temple fed a whole fruit and not the dehisced seed that the pigeons are more likely to eat. I fed six seeds to a captive Pink Pigeon without any ill effects; a fruit that affects domestic pigeons need not necessarily affect Pink Pigeons (cf. Goodwin 1967). Guérin (1940–53) and later Staub (1973a) blamed the common food plants *Aphloia* and *Erythrospermum* but gave no supporting evidence.

Descriptions of intoxicated Pigeons have been given by various authors. Guérin (1940–53) claimed that the bird becomes drowsy and stays motionless for hours on a branch with eyes half-closed. The Pigeons can then be easily caught with a noose on the end of a stick. Donald Taylor (*in litt.* to Jean Vinson 24 Aug. 1961) wrote that he had been assured by *guardiens* at Les Mares and Montagne Cocotte that Pink Pigeons eat *Aphloia* berries and become so drunk that they can be picked up off the branches. The only account that gives details that in any way suggest some advanced state of intoxication is from Temple (1978c) who gave the following description of a single bird that he saw. The Pigeon "was staggering on the ground with its wings drooping and was barely able to fly. It flew awkwardly to a perch. After perching for about 60 minutes, it was able to fly away, but it was still obviously suffering from incoordination." Since Temple did not see the cause of this bird's condition there is no evidence to suggest that the Pigeon was anything other than very ill, from whatever cause. Several informants have claimed to have seen *fandamane*-intoxicated Pigeons, but on close questioning it was clear that they were describing unaffected birds but could not believe that they were naturally so tame.

The poisonous nature of the Pigeon's flesh has also given rise to many reports. Staub (1976) and Guého & Staub (1979) suggested that the flesh becomes toxic after they eat the narcotic (?) *Euphorbia pyrifolia*. This plant is unlikely to be an offender since it is very rare or even absent from the Pink Pigeon's range. Hachisuka (1953), quoting Emmerez de Charmoy, noted that the flesh was bitter but that no ill effects are experienced from eating it. Desjardins (*in* Oustalet 1897), quoting local informants, claimed that the flesh is poisonous unless the Pigeon is beheaded immediately after death. More recently E. Closel, who has eaten several dozen roasted Pink Pigeons without ill effects, asserted that the flesh is bitter unless the birds are decapitated to bleed them (J.-F. Martial pers. comm.). Two birds shot and decapitated near Macabé forest in July or August about 1943 were roasted and a small piece given to each of the 15 family members present; the flesh tasted good but all subsequently suffered nausea and vertigo (M. Koenig pers. comm.). In 1977 a Pink Pigeon that had been accidentally killed in April, not decapitated and subsequently deep-frozen, was curried and consumed at a dinner party. None of the guests developed any signs of toxaemia (S. Ward pers. comm.).

Breeding biology
Breeding season

Published accounts of the Pink Pigeon's breeding season are not consistent. Guérin (1940–53) suspected that they nested between July and December, as did Staub (1973*a*). Newton (1958*a*) found hatched egg shells at the end of April and a nest with an egg in May, but believed that their normal breeding season was October to February. Year-round breeding was stated to occur by Staub (1976), with a peak from December to March. Hartley (1978) suggested October to May with a peak in egg-laying between November and March. Temple (1978*c*) speculated that there are two general peaks, April to May and December to January. Using his own field observations McKelvey (1976, 1977*e*) noted the start of the breeding season in late December 1975 and the fledging of the last young on 1 November 1976. This would give the period of egg-laying from early January to early September. Field data and pub-

lished observations of breeding activity show that wild Pigeons start breeding in December, reach a peak in January to June, with the number breeding decreasing during the drier winter months, and few or no birds breeding from August to November (Fig. 12). Captive birds on Mauritius, by comparison, are capable of breeding throughout the year with no marked peak in breeding activity.

There is considerable variation in the onset and length of the breeding season from year to year which is dictated by weather conditions and consequent food availability. Cyclones strip trees of leaves, flowers and fruit causing severe food shortages and abruptly halting any breeding in progress (see The Effect of Cyclones). Following cyclone Gervaise on 2 February 1975 no breeding activity was recorded until May (Temple 1978*c*) or early June (Newlands 1975). In 1977 bad weather and high winds delayed most breeding until late March and early April (Hartley 1978, Durrell 1978), although McKelvey (*in litt.*) collected an egg on 28 February. Following cyclone Claudette on 23 December 1979 I was impressed by the absence of breeding activity. The first courting pair was not seen until 29 May, 6 months after the storm. Heavy rains on their own may cause new plant growth and hasten flowering, which encourages early breeding.

The breeding season coincides with the possibility of cyclonic weather (December–April) when all nests, eggs and squabs may be destroyed in a single storm. With such high selection against breeding at this time of the year, reproduction would be expected to be outside this period. That breeding coincides with the greatest likelihood of cyclonic disturbance suggests that the abundance of food available during the summer months is a proximate factor governing the breeding of this species.

The breeding cycle of captive Pink Pigeons is very sensitive to food availability. Five pairs of Pigeons that were being kept on a maintenance diet and not breeding, all laid eggs between 17 and 26 days after being given an *ad libitum* food supply. Similarly the wild Pink Pigeons start breeding the month after the onset of the summer rains (Fig. 12) when food becomes plentiful.

Captive Pink Pigeons can lay several clutches in succession and will lay a repeat clutch after successful breeding. Eggs laid in captivity are removed for foster rearing and the birds usually re-lay. A maximum of 28,

Fig. 12. Pink Pigeon breeding seasonality. (*a*) Number of clutches recorded per month. Dates are calculated from observations of courtship behaviour and fledged young as well as clutches found. Data include all records, this study, published and unpublished sources. (*b*) Rainfall per month at Bois Sec (from Padya 1972). (*c*) Number of fledglings recorded per month. Data from this study, published and unpublished sources.

(*a*)

(*b*)

(*c*)

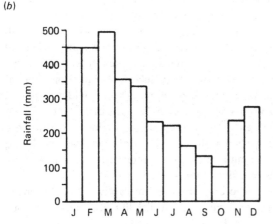

31 and 33 eggs have been laid in a year by productive pairs, representing 18, 23 and 18 clutches respectively. Wild Pigeons will also re-lay after predation of their nests, and McKelvey (1976) noted three or four clutches from productive pairs.

Territorial behaviour

In the early morning, evening, and to a lesser extent at other times of the day, birds holding nesting territories call with a low monotone cooing *coo-coo-cooooo-oooo*. They usually call from exposed branches where they are highly visible, the pale head and breast being particularly conspicuous. Calling birds are usually males although in captivity unpaired females will also call. Territories are also advertised by display flights which are similar if not identical to those of the Wood Pigeon. The bird leaves its perch and climbs rapidly and steeply above the forest and claps its wings in rapid succession usually three to five times, and then glides back into the forest with wings held horizontally and the tail fully spread. The tail is particularly conspicuous during this display, the bright chestnut being enhanced by the pale outer feathers.

Wing-clapping display flights are used in a number of contexts when the male wishes to advertise his territory. A displaying male may prompt wing-clapping in adjacent pairs. Trespassing birds may be pursued out of the territory after a wing-clapping display. In a typical example of this behaviour an intruding bird was flying just above the canopy towards a territory. The resident male left his perch and rose quickly and steeply above the canopy, wing-clapped three or four times and then glided back down to the level of the canopy before pursuing the intruding bird out of the territory, flying with deep purposeful wing beats. Both birds were uttering deep, grunting *wah-wah* anger notes.

A relatively slow flight with deep wing beats is often used when Pigeons are flying over their territories.

A Pigeon landing in another Pink Pigeon's territory is repelled by one or both birds. The resident Pigeon may fly and land near the intruder and supplant it, chasing it out of the area. If the intruder does not leave immediately it will be attacked with quick downward strokes of a partly opened wing. The birds fight by striking each other with sharp blows from the carpus, or jabbing with pecks aimed at the head and neck.

Nesting territories in the *Cryptomeria* grove are all between about 0.2 acres (0.08 ha) and 0.4 acres (0.16 ha) in area. Temple (1978c) suggested that only an area within 5 m of the nest is defended and claimed to have found adjacent nests as close as 10 m. Such social behaviour must be very rare and is not supported by my observations.

Courtship

The bowing display is the typical sexual advance of the male and is common to some extent to all pigeons (Murton & Isaacson 1962). This display of the Pink Pigeon has been briefly described by McKelvey (1976) as the 'step-bow display'. It is always used in the context of courtship, and often precedes copulation.

During the bowing display the male calls *coo-woo-wah, coo-woo-wah*. With his crop inflated, his tail closed and held either horizontal or down, and his irides dilated, he bows, his head pointed to the ground, and squats slightly. He utters *coo-woo*, then stands upright and utters *wah*, completing the vocalisation. The whole vocalisation is rendered in one continuous sequence usually three to six times. If the female is responsive she may solicit copulation.

As a precursor to the bowing display or between displays, the male pursues (drives) the female on foot with his head held close between his shoulders and crop inflated. The bowing display may also be preceded by slow silent head-nodding on the part of the male. This often occurs when a pair make initial contact and are perched near one another.

Caressing and courtship-feeding may also occur prior to, or between, bowing displays. Courtship-feeding may develop from caressing, especially if the caressing is directed around the eyes and cere. The female usually takes the initiative and inserts her beak into the male's gape; regurgitatory movements then take place. Solicitation by the female often follows courtship-feeding and copulation may follow.

When the male has 'paired', and if a suitable nesting location is available, he squats and calls from the site with a deep quiet *cooo-ooo-oo*. His head is occasionally nodded and held low and his tail is inclined upwards. While calling the male quivers his wings and sometimes his tail. This may attract the female and while he is calling she often caresses the male's neck and head feathers. Nest-advertising may be repeated at more than one site and if nest-building is already advanced the birds may arrange and re-arrange the twigs in the nest. Nest-calling is to my ears identical to territorial calling although McKelvey (1976) regarded it as quieter and lower pitched.

Nests and nest-building

In common with other pigeons the male collects the nesting material and the female constructs the nest. I have examined about 20 nests of wild Pink Pigeons and all were typical pigeon nests: a loose weave of sticks lined with finer twigs. The nest material was mainly privet *Ligustrum robustum* but also included twigs from native trees. These are collected on the ground or may be broken from the ends of branches; the Pigeon grasps a suitable twig in its beak and breaks it with a quick twisting movement of the head. In the wild and also in captivity, an old nest may be dismantled to provide material for a new nest. Hartley (1978) stated that privet twigs were used for nest-building, but both McKelvey (1976) and Temple (1978c) claimed that the nests were usually made from *Cryptomeria japonica* twigs. Nest-building among wild birds is completed in 1 (McKelvey 1976, 1977a) or 2 days (Hartley 1978). Among captive birds it is more protracted and may continue for several days. Twigs are often added to the nest during incubation. Many nests of wild Pigeons may be quite substantial, about 30–35 cm in diameter and up to 8 cm deep. McKelvey (1977a) by contrast reported nests to be less substantial and noted that "fifty or sixty twigs constitute a well made nest; nests containing eggs have been found consisting of 23 twigs".

All the nests of wild Pink Pigeons that I have examined, and also about another 20 examined by Hartley (pers. comm.), were between 4 m and 15 m above ground, the average height being about 10 m. Most of the nests, certainly all those that I have examined, have been in *Cryptomeria japonica*. This exotic was first planted within the Pink Pigeon's range around 1918–20 (Edgerley 1962) and has been widely used for nesting since at least the 1940s or 1950s (Edgerley pers. comm., Newton 1958a). A few nests have been found in *Eucalyptus robusta* and single nests in *Diospyros* sp., *Labourdonnaisia* sp. and *Sideroxylon bojerianum* (McKelvey 1976, Hartley 1978, Temple 1978c).

Guérin (1940–53) claimed that the Pink Pigeon nested in holes in trees. Unlikely as this may seem, the claim cannot be entirely dismissed. Wild Pigeons usually choose nest-sites next to the trunk in a *Cryptomeria* tree, often surrounded by dense foliage in locations not unlike open cavities. In the pristine forests of Mauritius suitable open tree holes would have been far commoner than they are today and nesting in holes would give the Pigeons far greater shelter during cyclonic weather than the nest-sites that are now used. Captive Pink Pigeons often preferentially nest in a nest box rather than in an open basket.

Today, most, if not all, of the Pink Pigeons nest in a 2.4 ha (6 acre) grove of *Cryptomeria japonica* on the south-facing scarp between Bassin Blanc and Montagne Savane. Between 1971 and 1976 nests were also recorded on Montagne Cocotte, Piton Grand Bassin and Bois Sec (McKelvey 1976). In addition in early 1977 McKelvey saw a pair with two fledglings on Black River Peak and he suspected that they had nested nearby (King 1977). This seems most unlikely; the young had probably just followed the adults to a favoured foraging area. The Pink Pigeon used to nest widely within its range; Newton (1958a) found one nesting in a *Cryptomeria* on Piton Capote in 1954 and Edgerley (pers. comm.) found a nest and eggs in an ebony in lower Bel Ombre forest in the mid-1950s.

The present semi-colonial nesting of the Pink Pigeons is probably an artifact of many suitable nest trees being in one small area rather than an inclination of the species to be colonial.

Eggs and incubation

The first egg is laid one or sometimes two days after the completion of the nest (McKelvey 1976, 1977a), though in captivity this gap may be as long as a week. In captivity the female may spend 2 or 3 or sometimes more days sitting on the nest before the first egg is laid (Jeggo 1978, pers. obs.). The Pigeons lay one, or more usually two, eggs with an interval of 1 or 2 days between successive eggs – a time period agreed by McKelvey (1976, 1977a). The eggs are clear dull white, not as glossy as those of the domestic pigeon.

Incubation starts with the first egg, although eggs removed from the nest during the first 48 hours after laying often feel only slightly warm to the touch and I suspect that incubation proceeds gradually during the first days after laying. Incubation follows the usual pigeon pattern. The male incubates during the day from about 2 hours after sunrise until about 1 or 2 hours before sunset. The female incubates for the rest of the time (McKelvey 1976, 1977a, Durrell 1978, pers.

obs.). There is some variation (Jeggo 1978, pers. obs.), and during early incubation the female may spend a large portion of the day on the nest.

The incubation change-over between the pair may involve nest-relief displays which include submissive and appeasement posturing. Prior to changing over the partner may stand or squat beside the incubating bird and caress it around the head and neck. The incubating bird will then usually get up and leave the nest. However not all change-overs involve caressing and the partner landing on a perch next to the nest-site may be sufficient to encourage the incubating bird to leave.

Most of the eggs laid in captivity hatch after 13–15 days (Table 19). Incubation periods in excess of this have resulted from sporadic early incubation. McKelvey (1977a) gave the incubation period as 18 days, a mistake copied by Temple (1978c); Staub (1976) also gave the incubation as 18–20 days. Jeggo (1978a, b) recorded incubation periods as 13–14 and 14 days for captive birds.

Development of squabs

Soon after the eggs hatch the empty shells are removed by the parents and dropped away from the nest (Jeggo 1978a, pers. obs.) or alternatively just get pushed over the side. The newly hatched squab is covered with sparse down which is off-white or very rarely (*c.* 3%) dirty yellow in colour (not golden-yellow as described by McKelvey 1976). In common with all altricial young they have many embryonic characteristics including a relatively large head, closed bulbous eyes and poorly developed limbs (Table 20). I cannot agree with McKelvey (1977a) who claimed that they have a strong grasping reflex from the "very moment of hatching". The grasping reflex does not become functional until the third or fourth day and then it is still weak.

The squabs are fed by both parents on regurgitated 'crop milk'. After the first few days other foods are added to the milk; I have noticed leaf fragments on

Table 19. *The incubation period of Pink Pigeon eggs incubated in captivity by Barbary Doves*

Incubation period (days)	13	14	15	16	17
No. of eggs hatching	25	86	39	5	1

Note:
n = 156; mean = 14.2 days.

the fourth day after hatching. Jeggo (1978*a*) saw seeds being included on the fifth day and McKelvey (1977*a*) noted: "The young are fed a crop milk for six days, after which the curd-like substance is enriched with fragments of tender leaves, flowers and fruits". Two squabs about a week old, taken from the *Cryptomeria* grove, had their crops filled with about 50% milk and 50% whole *Erythrospermum* seeds and privet berries (Table 18).

Days 1–3. Continuously brooded, eyes closed, spend most of time sleeping and unable to move around; can raise head only for brief periods.

Days 3–7. Often left unbrooded for increasingly longer periods during the day, still being brooded at night. Eyes beginning to open. Grasping reflex develops. Rudimentary first preening seen. Moves to edge of nest to defecate. Feather papillae visible as faint specks of pigmentation below skin.

Days 8–12. Unbrooded at night. At beginning of this period sit up on tarsi; stand towards end of this period and will preen while standing. Start of wing-flapping exercises. Leg and wing stretch well developed. Call with high pitched *shree-shree* food-begging call. Feathers present as intact quills.

Days 13–17. Threaten intruders by puffing out feathers and raising wings vertically over back and may lunge at intruder with open beak. Covered with fringing feathers.

Days 18–22. Leave nest: at first only walk around on surrounding branches but may return to roost in

nest at night. Start making exploratory pecks at objects. When hungry squeak at approaching adults and excitedly flap wings. While being fed close eyes and raise wings above the body. Feathers fully developed on most tracts.

Days 23–27. Perch for night and only occasionally return to nest. Start to feed themselves and fly short distances. Sun-bathe by raising outstretched wing above back annd head and partly lying on other wing.

Days 28+. Independent of parents for crop food, but roost next to them. Follow parents for several days (weeks?) learning best feeding areas.

In captivity young Pink Pigeons are independent of their parents in 4 weeks, but McKelvey (1977*a*) claimed "the parents continue to feed the squabs until they are six weeks of age". McKelvey also noted that the young first leave the nest at 28 days which is a week later than my observations on wild and captive birds and also later than birds bred at the Jersey Wildlife Preservation Trust (Jeggo 1979).

Breeding success
The Pink Pigeon's breeding success is low and data collected in 1976 and 1977 suggest that only about 70% (32/46) of the adult population were breeding (McKelvey 1976, 1977*a*, Temple 1978*c*). In the mid 1970s it is probable that there was a surplus of birds following the destruction of large areas of Pink Pigeon habitat during the previous decade. Today I suggest that most, if not all, adult pairs attempt to breed per year.

In 1976 only five young were raised from at least 48 nesting attempts (0.10 per attempt) (McKelvey 1976, 1977*a*). Temple (1978*c*), using an incomplete set of data for 1974–6 (including some of McKelvey's results), recorded seven young from 34 breeding attempts (0.21 per attempt). Since 1978 there have been 68 documented breeding attempts (not all necessarily representing clutches laid) and only eight squabs/fledglings recorded (0.12 per attempt) (Table 21). Brood size at fledging has been recorded on 15 different occasions. Seventeen young were reared, giving a mean brood size of 1.1 young per successful nest.

Breeding success is greater at the beginning and end of the breeding season than during the peak months (Table 22). The Pigeons are about four times more likely to rear a brood per nesting attempt

Table 20. *Weights and measurements of newly hatched Pink Pigeon squabs*

	Number	Range	Mean
Weight (g)	18	8.3–13	10.5
Wing length (mm)	6	11–13	12.1
Tarsus (diagonal)	6	9.8–11	10.5
Culmen to skull	6	12–13	12.4
Head length excluding culmen	6	13–14.5	13.8
Head width	6	10–11	10.6
Centre toe	6	9–9.5	9.2
Centre claw	6	2–2.5	2.3

between June and December, when not only is disturbance by cyclones unlikely but breeding is sporadic
and therefore less prone to predation. During the
months of peak breeding activity when several pairs
may be breeding in the *Cryptomeria* grove it becomes
more worthwhile for a predator to search systematically for eggs.

Breeding success is adequate to maintain the
population and even to allow the Pigeons quickly to
recoup losses caused by cyclones (see Distribution and
Status 1960–83).

Factors limiting the population
Predation on adults

The adult Pigeons would have had few predators
before the island was colonised by man and the exotic
mammals that he introduced. Human predation has
never been very serious due to the assumed inedibility
of Pink Pigeon flesh (see Feeding Ecology) and very
few individuals have ever been taken for aviculture
(see Conservation). There has, however, been some
human predation in recent years. In 1973 a single bird
was shot and the following year two birds were killed
by a deer-hunter (Temple 1978c).

Table 21. *Pink Pigeon productivity data*

% of pairs attempting to nest 1976/7 (wild)	69.6 (32/46)
Mean clutch size (captive)	1.76 (92/52) (Mauritius)
	1.65 (86/52) (Jersey)
% of eggs fertile (captive)	64.2 (267/416) (Mauritius)
	38.1 (67/176) (Jersey)
% of fertile eggs hatching (captive)	72.7 (216/297) (Mauritius)
	58.2 (39/67) (Jersey)
% of young reaching 30 days (captive)	52.37 (113/216) (Mauritius)
	43.6 (17/39) (Jersey)
Young produced per nesting attempt (wild: see Table 22)	0.10 (5/48) (McKelvey)
	0.21 (7/34) (Temple)
	0.12 (8/68) (this study)
Number of young reared per egg laid	0.08–0.15 (wild)
	0.19 (113/593) (captive, Mauritius)
	0.06 (17/262) (captive, Jersey)
Mean brood size at fledging (wild)	1.1 (7/15)

Note:
Data from captive birds from the Government Aviary, Black
River, Mauritius, and the Jersey Wildlife Preservation Trust,
England.

Cat predation is likely to be a significant but as
yet undetected source of mortality among adult
Pigeons feeding on the ground. McKelvey (1976)
records cats killing Madagascar Turtle Doves *Streptopelia picturata* and I have found unidentified dove
feathers in cat droppings. Feral cats are widespread
and common in the native forest and live completely
independently of man; they are very shy and seen
only about once in every 150 hours of fieldwork.

The mongoose *Herpestes edwardsi* has been variously regarded as an incidental predator (McKelvey
1976) and a serious threat (Temple 1978c). It is a very
common and generalised predator that eats household
scraps, insects, toads *Bufo regularis*, rodents and birds
up to the size of domestic chickens (pers. obs.) and is
probably capable of killing a Pigeon; one is suspected
of killing a captive bird. I did, however, watch a group
of six apparently adult mongooses, that were foraging
for scraps left by tourists at the Alexandra Falls viewpoint, completely ignore a Spotted Dove *Streptopelia
chinensis* feeding in the same area. McKelvey (1976)
recorded a Pink Pigeon being attacked by a Peregrine
Falcon. The Pigeon was returning to the *Cryptomeria*
grove when it was attacked; it veered, began to retrace
its course, was stooped upon again but managed to
zig-zag and avoid the Falcon, and then it flew hard for
cover and escaped. Attacks by Peregrine Falcons must
be very rare since the above is only the third Peregrine
recorded on Mauritius (Jones unpubl.).

Nest predation

Nest predation is known to be very high and there is
considerable debate concerning the identity of the
offending predators. Most people consider the monkey to be the main culprit (discussed in next section),
but there are other possible contenders. Perhaps the
most dangerous of these is the Ship Rat *Rattus rattus*,
which according to McKelvey (1976) has been seen
raiding nests of the Mascarene Paradise Flycatcher,
Red-whiskered Bulbul and Spotted Dove and which
he suspected of preying on Pink Pigeon nests. Studies
elsewhere have shown that the Ship Rat will eat the
eggs and young of forest species nesting in trees,
taking eggs up to 61 mm in length (Atkinson 1978).

Cuckoo-shrikes *Coracina typica* occasionally rob
Pink Pigeon nests and accounted for the failure of four
out of 48 nestings in 1976 (McKelvey 1977a). In the two
cases observed the female attacked the eggs from

beneath while the Pigeon was still incubating, and on one of these occasions the male tried to distract the Pigeon while the female was trying to get to the eggs (McKelvey 1976). Mynahs are regarded as potential nest predators; McKelvey (1976) recorded them stealing domestic pigeon eggs placed in an artificial nest, and later dismantling the nest. The density of Mynahs is, however, too low in the native forest for them to be regarded as a serious pest.

Nest predation by humans has never been significant although in recent years a few eggs and young have been taken for the captive breeding programme (Table 24). Surprisingly the Pink Pigeon has a nest distraction display (McKelvey 1976) which was very efficient in luring me away from the area the first time I was party to examining a nest containing a squab. The bird left the nest and fluttered into some low vegetation flapping its wings in a distressed manner and keeping its head low and tail partly fanned. When I ran forward to capture it, it fluttered further into the undergrowth until I had been lured perhaps 50 m from the nest; then it flew away. This distraction display is well developed in some individuals but poorly in others, and some birds do not show it at all. The display may be used by both males and females when there are eggs or young in the nest. One of our captive birds will even try to lure a human intruder from the nest before she has laid any eggs. It is curious that such behaviour should have been retained by the species when historically there were only avian nest-predators on the island, against which the display would have been useless.

Nest predation by macaques.

An exotic monkey, the Crab-eating Macaque *Macaca fascicularis* is very common in areas of native and secondary forest. Their total population has been variously estimated at 12 000 (McKelvey 1977a) and 15 000 (Owadally 1980) with about 3500–4500 in native forest (Temple 1978c). The Macaques move around as individuals or more usually in troops of 80 or more. They have long had a reputation of being competitors and nest predators. As early as 1741 Grant (1801) stated "the birds much diminish in the woods, as the monkeys, which are in great numbers devour their eggs". The Macaques have been accused of robbing the nests of many native birds (see individual species accounts). These predations have been considered to be most serious upon Pink Pigeons (Temple 1974, 1976b, 1978c, Temple *et al.* 1974, McKelvey 1976, 1977a, Staub 1976). The importance of monkeys as nest predators of pigeons has long been recognised. Wallace (1865), using biogeographical evidence, pointed out that the most striking abundance of pigeons

is confined to the Austro-Malayan subregion, . . . yet all forest-haunting and fruit-eating mammals, such as Monkeys and Squirrels, are totally absent. But Monkeys, besides consuming vast quantities of fruit, are exceedingly destructive to eggs and young birds; and Pigeons, which build rude, open nests, and whose young are a long time helpless, must be more particularly exposed to their attacks. This is no doubt the reason why, in the dense forests of the

Table 22. *Pink Pigeon breeding success during the peak breeding season and the rest of the year*

% fledged young produced from breeding attempts during the peak breeding season Jan.–May	% fledged young produced from breeding attempts during the rest of the year June–Dec.	Level of significance between both sets of data (Student's t test)	Source of data
0.03 (1/35)	0.31 (4/13)	$P < 0.05$	McKelvey 1976, 1977a, in litt.)
0.09 (4/47)	0.19 (4/21)	N.S.	This study
0.06 (5/82)	0.24 (8/34)	$P < 0.05$	All data combined

Amazon, where Monkeys are most abundant, Pigeons are scarce or almost entirely absent; and in South America generally it is to be observed that by far the larger number of the Pigeons inhabit the districts where Monkeys are almost or quite wanting.

Predation by exotics upon island species that have evolved without many predators and competitors can be expected to be severe. Perhaps the extinction of the Blue Pigeon *Alectroenas nitidissima* in about 1830 was the result of nest predation by monkeys? (See also Chapter 1.)

McKelvey (1976) claimed that monkeys, which are very common in the Pink Pigeon nesting areas, accounted for the destruction of most of the nests under observation during the 1975/6 breeding season. Elaborating, he noted that 42 eggs "were observed to have been destroyed by predation" – mostly by monkeys – and he witnessed the total failure of the birds to hatch and rear young (an exaggeration since he caught a juvenile on 2 March 1976: McKelvey 1976). Temple (1978c) provided similarly depressing figures; between 1974 and 1976 only seven young fledged from a minimum of 34 nesting attempts. He further claimed that in some years monkeys may be 100% effective in locating and destroying Pink Pigeon nests.

These studies support beliefs widely held among Mauritian naturalists. Most of the evidence appears to be circumstantial as no one has recorded seeing a monkey robbing a Pigeon's nest. There are, however, some supporting anecdotal observations. On 17 April 1978 Fay Steele (pers. comm., unpubl. field notes), who had been watching a Pink Pigeon incubating tightly for 3 days, returned to continue his observations. The incubating bird was absent and there were six to eight monkeys in and around the nest tree. On closer examination he discovered the nest had been pulled apart and thrown out of the tree with no sign of any eggs. Similarly a nest that I discovered with one egg on 26 February 1981 was later robbed with no sign of the egg and the centre of the nest pulled out.

In March/April 1976 while catching Pigeons for the captive breeding programme, a fully fledged squab was removed from a nest and replaced with a "rammier pigeon" (*Streptopelia picturata*) which was accepted by its foster parents. The following day it was discovered that "monkeys had found the nest. It had

been destroyed and the rammier chick devoured" (Durrell 1978). The nest structure was completely dismantled (J. Hartley pers. comm.) – a sure sign of Macaque predation. Sussman & Tattersall (1980) in their study of the Macaque initially supported the above views, stating that it was "likely that the macaques are the most dangerous predator of some of the most highly endangered bird species of Mauritius." Subsequently they questioned the role of monkeys in preying on birds' eggs and believed that it had been exaggerated, the Ship Rat being the real culprit (R. W. Sussman, R. Jamieson & I. Tattersall pers. comm.). Macaques in an area of lowland exotic forest and savannah were almost entirely vegetarian, and in two years of study no evidence of nest predation was detected. Ninety-three per cent of their feeding time was spent on fruits, stems, leaves and flowers, 5% on invertebrates and the remaining 2% was undetermined (Sussman & Tattersall 1980). Unfortunately a comparable study on the monkeys has yet to be done in the native forest where the endangered birds are to be found. It is quite possible that different populations of macaque will have different feeding preferences. In the closely related Japanese Macaque *Macaca fuscata* feeding on eggs has been shown to be the result of a feeding tradition found in some groups but absent in others (Miyadi 1965, 1967).

To test the rate and extent of monkey predation on eggs McFarling (1982) set up a series of experiments. The study covered 28 days during July and August 1982 and involved 460 egg-days. Chicken eggs were placed randomly on the ground in an area of exotic lowland open savannah covering about 10 000 m^2; daily predation rate was recorded and predated eggs were replaced, but never in the immediate locality. During the first 5 days egg predation was patchy probably due to the chance finding of eggs by monkeys and other predators. Following this initial period there was an increase in predation rate and by the twentieth day predation was consistently high and approaching 100%. Monkeys were thought to have accounted for 90% of the predation, the other predators being mongooses and rats.

In a second experiment 24 eggs were individually placed in old weaver-bird *Ploceus cucullatus* nests which were secured in well-spaced trees. The eggs were all taken by monkeys within 24 hours and nine of these predations were observed through a telescope.

Monkeys are clearly capable of exerting a significant influence upon the breeding success of the Pink Pigeon and it seems probable that they are a major nest predator of this species.

Inbreeding

Inbreeding depression may be a contributory reason for the Pigeon's low population. There is no direct evidence of this from the wild birds, but captive birds after only one or two generations in captivity show a number of congenital and developmental problems. Relatively high numbers of infertile eggs, embryonic and squab deaths, inclined feet, kinky sternums, neurological disorders and split tail feathers have been recorded. Some, if not all, of these conditions can be congenital in the domestic pigeon (e.g. Levi 1941). A detailed analysis of the effects of inbreeding in the Pink Pigeon is in preparation. (See the account of the Mauritius Kestrel for a discussion on the subject.)

The effect of cyclones

Cyclonic weather may affect the Mascarenes at any time during the southern summer (November–April) when winds of 80–160 mph may be experienced. Severe cyclones may cause direct mortality and some birds may be blown out to sea (McKelvey 1977a). The high winds strip trees of flowers, fruit and most of their leaves, bringing food shortages and starvation for some of the birds. Any breeding in progress will be abruptly halted, with the probability of all the nests being destroyed in the storm. In the week following cyclone Carol in 1960 F. Staub (pers. comm.) was shown ten dead pigeons on Mount Cocotte by the guardian and he saw a pair, much weakened and hungry, feeding ravenously on some of the few leaves available. The cyclone did unprecedented damage (Padya 1976) and killed off a high proportion of all the birds. Two weeks after the cyclone, food was clearly still in short supply. Most, if not all, of the remaining Pink Pigeons were apparently foraging in a large group, and a flock of about 20 was seen on Mount Cocotte (M. d'Unienville, pers. comm.).

In March and April 1975, following the passage of cyclone Gervaise, starving Pink Pigeons were seen walking on the ground, so weak they could not fly (F. Staub *in litt.* to S. A. Temple). Newlands (Feb. 1975) also records an observation of a "sick or injured' Pigeon being seen after the cyclone.

Severe cyclones may halve the Pigeon population, as demonstrated by cyclones Carol (Feb. 1960), Gervaise (Feb. 1975) and Claudette (Dec. 1979) (see Distribution and Status 1960–83). In the months following a cyclone, when food is once again abundant, depletions in the population can be quickly recouped by rapid breeding (see Breeding Season). The ability to lay multiple clutches is probably an adaptive strategy which helps offset losses caused by cyclones, and now possibly acts to compensate for the high rate of nest predation as well.

The Pink Pigeon does not appear to have any highly developed strategy that helps it to cope directly with cyclones. Hole-nesting would confer some protection and has already been discussed (see Nests and Nest-building). During a storm Pigeons may shelter on the ground beneath bushes as I have observed in Zebra Doves *Geopelia striata* and Spotted Doves. Pink Pigeons are well able to withstand heavy rain and unlike Zebra Doves and Spotted Doves do not become wet or waterlogged in torrential downpours.

Exotic columbiforms and other birds and their competitive relationship with the Pink Pigeon

There are four exotic columbiforms on Mauritius: the Feral Pigeon *Columbia livia*, Madagascar Turtle Dove *Streptopelia picturata*, Spotted Dove *S. chinensis* and Zebra Dove *Geopelia striata*. All were introduced in the eighteenth century (Chapter 1) and are now well established and common. Recent attempts to introduce other species – the Diamond Dove *Geopelia cuneata* in the mid 1970s and sporadic attempts with the Barbary Dove *Streptopelia 'risoria'*, a common cage bird on the island – failed probably due to adverse competition with congeners.

The established exotics are widely distributed and are found at all altitudes, but do show specific habitat preferences. The Feral Pigeon is commensal with man and feeds in and around villages and towns, occasionally feeding in flocks on cleared sugarcane fields. Most remain in and around built-up areas to breed, and some of the pigeons in Black River nest in the axils of coconut palm fronds. Many of the mountains and cliffs have Feral Pigeons breeding on them even though they may be several kilometres from the birds' feeding areas. I have seen Feral Pigeons on Signal Mountain, the cliffs of Tamarin Gorge and Vacoas Heights, Takamaka cliffs and Morne Sèche in the Black River Gorges, Corps de Garde, Trois Mamelles and Tourelle du Tamarin.

Spotted and Zebra Doves are commonest in the drier lowlands and prefer open and fringe habitats: gardens, roadsides, cleared areas and lowland exotic savannah. They can be found in native forest but only where there are open clearings and roads. These doves are seen singly, in pairs or in small groups. Larger parties are not infrequent, especially around abundant food sources where the birds may occur in mixed groups. The largest flock I have seen comprised 45 Zebra Doves and 15 Spotted Doves.

Road counts (Fig. 13) show that Zebra Doves are

Fig. 13. The number of doves seen per car journey between Chamarel and Plaine Champagne, May–November 1979. The distance between Chamarel and Plaine Champagne is 6.9 km and the ground rises from *c.* 290 m to *c.* 580 m above sea level, passing through badly degraded and mid-altitude forest.

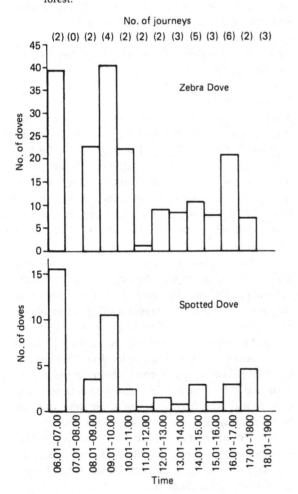

most active in the morning and to a lesser extent in the late afternoon. They exploit roadside habitats more than Spotted Doves, no doubt due to their preference for the finer seeds of grasses and herbaceous plants that grow there.

The Madagascar Turtle Dove prefers dense wooded areas and is most common in the native forests of the upland plateau. It frequently enters well-wooded gardens in Curepipe and can be seen in the botanical gardens of Curepipe and Pamplemousses. In the lowland exotic savannah it is seen only about half as frequently as it is in the upland forests (Table 23).

The competitive relations between the exotic pigeon and doves and the Pink Pigeon are not clear. The most common species in Pink Pigeon habitat is the Madagascar Turtle Dove. This is, however, very different from the Pink Pigeon in feeding ecology since it (like the other exotics) feeds on the ground largely on seeds (Goodwin 1983). The dietary overlap between the exotics and the Pink Pigeon is not great but their presence may prevent the Pink Pigeon from greater exploitation of terrestrial food sources. This competition is, however, unlikely to be very significant.

The very common Red-whiskered Bulbul competes with the Pigeons for *Nuxia* flowers. The high densities of this exotic bulbul undoubtedly contribute to the premature exhaustion of this food source.

Conservation

The conservation of the Pink Pigeon has been the major motivation behind the present study, and has been discussed in several reports and publications (e.g. Temple 1976*b*, McKelvey 1976, Durrell 1978). Captive breeding has been the main approach and all the captive birds are derived from 13+ eggs, 5 young and 11 adults (see Table 24). Up until the end of 1983 111 young had been bred and reared to fledging at the Government Aviary on Mauritius and a further 24 had been bred in Europe at the Jersey Wildlife Preservation Trust. Other captive populations are held at the Casela Bird Park (Mauritius), the Vogelpark Walsrode (West Germany), New York Zoological Society and Rio Grande Zoological Park (Albuquerque, New Mexico) (Hartley 1984).

These captive populations offer hope for the continued survival of the species. The survival of the species in the wild is regarded as a very high priority, and plans for managing the population are under way.

Since it is clear that the wild birds suffer from food shortages at the end of the winter months (August to mid-November) a programme of supplemental feeding at this time could help the population and probably increase the carrying capacity of the habitat. Available information augurs well for the success of artificial feeding since in the 1930s hunters used to feed Pink Pigeons at Les Mares on grain spread on the lawn of the hunting lodge (see Feeding Ecology).

The release of captive-bred birds back into the wild cannot be justified while there are still Pigeons surviving in suitable habitat. There is some hope in releasing Pigeons into the area of forest between Brise Fer, Mare Longue and Macabé where Pigeons were formerly found but are now absent, because access across Les Mares has been prevented by the planting of pines.

A project to release Pigeons into the 60 acre (24 ha) Royal Botanical Garden at Pamplemousses has recently begun (Durrell 1984), and it is hoped that these birds, which will be supplementally fed, will form a self-supporting, semi-wild population (Pasquier & Jones 1982).

The protection of nest-sites and the control of nest predators has been suggested as a potential conservation measure (McKelvey 1976, Temple 1978, King 1978–9). It is unlikely that these measures would, in the long term, increase the population since the pre-sent level of breeding success is sufficient to maintain the population at the carrying capacity of the habitat.

The translocation of captive-bred Pink Pigeons onto Réunion has been suggested (Cheke 1974b, 1978a, Temple 1976, 1981). The rationale behind this is sound since Réunion has suitable habitat and once provided a home for the now extinct 'Nesoenas duboisi' (Hachisuka 1953). Hunting pressure on Réunion is, however, severe and a large tame pigeon would be at risk without effective conservation propaganda.

On Mauritius the Pink Pigeon and most of its habitat is protected by law, but it remains a matter of speculation whether these measures alone can save the wild population.

Echo Parakeet *Psittacula echo* A. & E. Newton 1876; *Cateau (vert) (kato, katover)*
Introduction

The Echo Parakeet is the rarest and most endangered bird in the Mascarenes. It is likely that this, the last surviving endemic parakeet from a western Indian Ocean island, will soon become extinct. There have been only two studies of its biology (A. & E. Newton 1876, Jones 1980c) and a few brief accounts of its life history (Guérin 1940–53, Staub 1973a, 1976, Forshaw & Cooper 1978). This report summarises my fieldwork, Temple's unpublished manuscript (1978c), and published material.

Table 23. *Pigeon and dove observations per hour in different habitats*

Locality (feet above sea level)	Hours	Observations per hour[a]				
		G.s.	S.c.	S.p.	C.l.	N.m.
Lowland exotic savannah (200–250)	10.0	10.6	4.5	0.8	0	0
Black River Village (sea level)	2.0	1.0	1.5	0	5.5	0
Cryptomeria grove (1750)	11.3	0.1	0	1.3	0	5.5
Macabé Ridge (2000')	16.3	2.0	1.0	1.7	0	0

Notes:
The observations of pigeons and doves per hour were collected in 1979 by systematically searching in different habitats during early to mid-morning and noting down all sightings. The data from the *Cryptomeria* grove also include observations collected from observation platforms. The results give an index of density in different habitats.

[a] G.s., *Geopelia striata*; S.c., *Streptopelia chinensis*; S.p., *S. picturata*; C.l., *Columba livia*; N. m., *Nesoenas mayeri*.

Table 24. *Numbers of eggs, young and adult Pink Pigeons taken from the wild for captive breeding*

	Eggs	Young	Adults
1976	5	1	3
1977	7+	1	8
1978	1	0	0
1979	0	1	0
1980	0	0	0
1981	0	2	0
Totals	13+	5	11

Description

Bare parts

The bare parts have been described by Hachisuka (1953) and Forshaw & Cooper (1978) and I have taken additional information from museum labels. The iris has been variously described as pale yellow, yellow or greenish-yellow, although A. S. Cheke (*in litt.*) noted that some birds have pink and others white irides. The skin around the eye is orange and the legs and feet are greenish-grey, grey or dark grey.

In common with many species of *Psittacula* the adult male has a red, and the adult female a dark, almost black, upper mandible (A. & E. Newton 1876, Staub 1976, Smith 1979). The beak colour of the recently fledged bird has been described or implied to be black (A. & E. Newton 1876, Staub 1976, Forshaw & Cooper 1978) or red (Temple 1978c, Smith 1979). The source of this confusion is the description given by A. & E. Newton (1876). They correctly described the adults but described a young male as "hardly distinguishable from the female, except that the bill is blackish brown" (i.e. not almost black as in the female). They described another young "male" in which the "red is beginning to show itself at the base", implying incorrectly that the bill changes from black-brown to red.

I suggest that the original immature birds were incorrectly sexed. In the Newton Collection in Cambridge there are two skins labelled as juvenile males (18/Psi/67/k/1 and 18/Psi/67/k/2). The former is clearly an adult female collected at a time (12 April 1860) when all the birds would be showing adult bill colours. The other specimen, collected on 31 December 1860, has a red bill in the process of turning dark (the original label notes "bill reddish brown"). I believe this is an immature female in the process of changing its bill colour from red to black. George A. Smith has examined these specimens and agrees with my assessment.

Forshaw & Cooper (1978) described the immature as being "similar to the female" but do not elaborate. They referred to specimen USNM 487076, which is labelled male but has a black beak and is certainly an adult female. Temple (1978c) noted that recently fledged birds have red beaks, but in immature females this colour is soon masked by darker pigmentation. A captive female fledged in December had gained a fully dark beak by the end of the following April. A captive male of the same age developed a dusky cast on his beak, but it remained redder than the female's as is the case in many *Psittacula* parakeets (Smith 1979). A young male seen by A. S. Cheke (*in litt.*) on 20 May 1974 had a dull pink bill. The young taken for the captive breeding programme in 1975 had fledged by the end of the year, were noted to be females in early March, and by 23 March had had ashen grey beaks for a week (S. D. McKelvey *in litt.*). The red (or in some species horn) bill colour is a primitive feature shown by all young parrots. The darker bill colour shown by females of some *Psittacula* species or by both sexes in *P. k. krameri* is caused by a secondary deposition of melanin (G. A. Smith *in litt.* 1975).

Plumage

Good plumage descriptions are found in Rothschild (1907b), and also in Hachisuka (1953) and Forshaw & Cooper (1978) who both illustrated the adult male. The colour is similar to *P. krameri* but is a darker, richer green. The male has an incomplete black collar and an incomplete pink collar both of which fail to meet at the hind-neck. The occiput is tinged with a powdery blue. The female has an indistinct, incomplete black collar, and a green collar, dark green across the cheeks shading to yellow-green on the hind-neck. The immature plumage is apparently similar to that of the female.

Moult

Moulting occurs mainly during the summer months although there is considerable variation in its timing between individual birds and probably also between years. Body-moult apparently starts in late winter and continues for several months. Between November and January the primary moult starts and is well advanced before the start of the tail-moult in March or April. By the end of June most if not all birds are probably fully moulted (see Fig. 14).

An adult female collected on 17 August 1964 was in body-moult (USNM 487076) and a female, one of three birds seen on Plaine Champagne on 6 December 1974, was in ragged moult (A. S. Cheke *in litt.*). Some Parakeets are still moulting their body feathers in March; a freshly moulted body feather was found on 6 March 1984 and I saw a bird in deep body-moult in late March 1980.

Two adult specimens collected during October and a single adult collected in April (all in Cambridge) show no signs of moult although an adult female (in New York) collected on 4 November 1907 had just started the primary moult (Table 25). Cheke (*in litt.*)

Table 25. *Moult in the Echo Parakeet*

(a) *Primaries*

AMNH No.	Date	Sex	Wing	Proximal 1	2	3	4	5	6	7	8	9	Distal 10
9531		F	Right	O	O	O	O	0.5	B	N	M	O	O
			Left	O	O	O	O	0.7	0.3	N	0.9	O	O
621311	4/11/1907	F	Right	F	F	F	O	O	0.2	O	O	O	O
621309			Left	F	F	O	O	O	M?	O	O	O	O
621309		M	Right	F	F	F	F	F	M	O	O	O	F
			Left	F	F	F	F	F	F	0.3	O	F	F
621310		F	Right	F	F	F	M	M	N	N	N	0.8	O
			Left	F	F	F	0.1	0.3	N	N	N	0.7	O
448948		M	Right	F	F	0.7	N	N	N	N	N	B	0.3
			Left	F	F	0.7	N	N	N	N	N	0.9	0.2

(b) *Tail*

AMNH No.	Date	Sex	Outer 6	5	4	3	2	1	1	2	3	4	5	Outer 6
9531		F	M	F	F	F	F	O	O	O	O	O	O	O
621311	4/11/1907	F	O	O	O	O	O	O	O	O	O	O	O	O
621309		M	O	O	O	O	O	O	O	O	O	O	O	O
621310		F	N	N	0.8	0.9	O	M	M	O	0.9	0.8	N	N
448948		M	N	N	N	N	0.5	0.5	0.5	0.5	N	N	N	N

Note:
B, broken; F, fresh; M, missing; N, new; O, old. Feather growth shown as estimated tenths of full length.
Data provided by Mary LeCroy, American Museum of Natural History (*in litt.*) and pers. obs.

Fig. 14. Moult timetable of the Echo Parakeet.

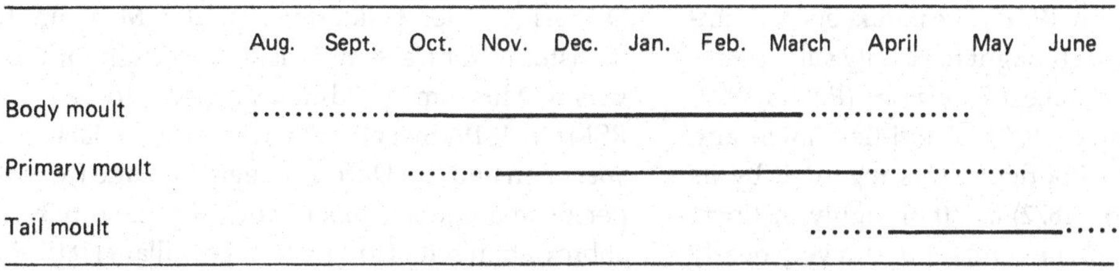

	Aug.	Sept.	Oct.	Nov.	Dec.	Jan.	Feb.	March	April	May	June
Body moult											
Primary moult											
Tail moult											

saw a single bird on 10 January 1974 that was moulting its inner primaries.

At the end of March 1980 there were Parakeets deep in moult; one had some tail feathers missing and another had a very short tail with the new feathers only part-grown. In mid-April 1979 I saw a Parakeet on Macabé Ridge missing its centre two tail feathers and the following month another bird (or perhaps the same one?) was in tail-moult. On 4 June 1974 two males in a group of four were still growing their tails (Cheke *in litt.*).

Primary moult starts with P6 or P7 and moves in both directions, the inner primaries probably being moulted last. It is difficult to age the inner primaries on museum skins since they seldom look old and are accordingly marked 'fresh' in Table 25. In specimen AMNH 448948 the inner primaries are duller than the new ones, suggesting they they were yet to be moulted. The primary moult is usually regular and can best be designated

$$\frac{(7),8,9,10}{(6),5,4,3,2,1}$$

This sequence is typical of most parrots although the moult centre is usually 6 (Forshaw & Cooper 1978). Some modification of the 6,5,4,3,2,1 loss of primaries is not unusual in parrots (Forshaw & Cooper 1978) and *P. echo* shows some inconsistencies (e.g. 5,6,4,3,2,1). Tail-moult appears to be consistently bilaterally symmetrical, starting with the outer feathers and progressing inwards (i.e. T6,5,3,4,1,2).

Taxonomy

The taxonomic relations of the western Indian Ocean parakeets of the genus *Psittacula* have aroused much debate, but the birds are generally considered to have mixed antecedents. The extinct Seychelles Parakeet *P. wardi* is usually considered to have been derived from the Alexandrine Parakeet *P. eupatria* (Peters 1937) or to be only sub-specifically distinct (Greenway 1967, Penny 1974). The parakeets of Réunion and Mauritius were thought to be only sub-species of the Ring-necked Parakeet *P. krameri* (Peters 1937, Hachisuka 1953, Smith 1979). The little-known and extinct *P. exsul* from Rodrigues was regarded by its describer (A. Newton 1872) as "thoroughly distinct" from other members of the genus and "not very nearly

allied" to either *P. echo* or *P. wardi*. Subsequently little has been written on its taxonomic position although Smith (1979) regarded it as a race of *P. krameri*.

Recent studies have shown that *P. echo* of Mauritius is clearly specifically distinct from *P. krameri* from which it differs in ecology and behaviour (Gill *in* Forshaw & Cooper 1978, Temple 1978c, Jones 1980). The resemblance that these species show in plumage is largely fortuitous. The parakeets of the western Indian Ocean are probably all derived from the Alexandrine Parakeet of India (G. A. Smith *in litt.*). Their most likely dispersal route from Asia was to the Seychelles and then to Rodrigues before spreading to Mauritius and Réunion. The further the radiation from the source the greater the loss of *eupatria* characters.

Early descriptions of *P. exsul* are ambiguous but it seems quite likely that both a blue and a green morph existed (see also Chapter 1, (*Ed.*)), the green form having, in the male at least, a red alar patch (see Tafforet 1726, Leguat 1708) – a character shared with *P. wardi* and *P. eupatria*. The only surviving specimens are two skins of the blue form and close examination of these suggests that they were a simple blue mutation (G. A. Smith *in litt.*) caused by the loss of psittacin (the yellow and red pigments) (Smith 1979); consequently the blue morph could not have a red alar patch. The beak was sexually dichromic as in *echo* with a red upper mandible in the male and black in the female (Newton 1872, 1875, A. & E. Newton 1876, Rothschild 1907b, Forshaw & Cooper 1978).

Most controversy has arisen over the relations between the parakeets of Mauritius and Réunion. No Réunion Parakeet *P. eques* specimens were thought to have survived to allow comparison with *echo* (Hachisuka 1953), although from the descriptions of Brisson (1760), Buffon (1770–83) and Levaillant (1801–5) these forms are inseparable.

There is an early parakeet specimen which may be from Réunion in the Royal Scottish Museum in Edinburgh. This skin, together with the rest of a natural history collection formerly owned by Louis Dufresne, chief taxidermist in the Muséum Royal d'Histoire Naturelle in Paris, was bought for the University Museum of Edinburgh (later to become the RSM) in 1819 (Sweet 1970). The original label on this specimen and the Dufresne register name the bird "la perruche à double collier" and add "Lev. p 39". The abbreviation refers to a plate in Levaillant (1801–5) and both this and the name are referable to *P. eques* from Réunion. However, since the early collectors were not very careful about noting the true origin or identity of their specimens the possibility remains that this specimen could be an early *echo* from Mauritius, the source of all but one of the other Dufresne specimens from the Mascarenes (R. McGowan *in litt.*).

The Edinburgh skin does not differ in any significant way from the description given by Brisson, Buffon or Levaillant, other than being tailless. The specimen is similar to a male *echo* and in comparing the skin with the description and illustration of the latter species in Forshaw & Cooper (1978), R. McGowan (*in litt.*) notes that the pink collar is "not nearly as broad as that illustrated in [the] plate" and "the outer upper webs of the outer two primaries on each wing have a blue tinge"; this last feature is not mentioned in the descriptions of *echo* in Forshaw & Cooper (1978) or Hachisuka (1953). I have compared detailed photographs of the Edinburgh specimen with *echo* skins at the Mauritius Institute and do not think that these differences are significant. The width of the pink collar is variable and both the adult male specimens that I examined had a blue tinge to their outer primaries. The wing measurement of the Edinburgh specimen (193 mm) is within the range for male *echo* but longer than average; the tail measurement given by Brisson (196 mm) is about average for *echo*, but too short for *krameri*.

Previous authors have claimed or alluded to the conspecificity of the two populations. In their original description of *echo* A. & E. Newton (1876) suggested only that the two populations may have been distinct, solely on the basis of the general dissimilarity of the avifaunas. They agreed that *echo* "no doubt, answers in nearly all particulars to the true *eques*". Later Edward Newton (1888) treated the populations as conspecific stating in a footnote only that "they may perhaps be distinct". Salvadori (1891) listed them as conspecific but unaccountably gave only Mauritius as their country of origin. Other authors split them (e.g. Rothschild 1907b, Hachisuka 1953) but based their accounts only on secondary sources of information. Most recently Cheke (Chapter 1) after considering the available evidence considered them to be conspecific, which is the only pertinent conclusion to be drawn from the available evidence.

The specific name *echo* has been retained in this book for reasons of familiarity, but I suggest that in future both forms be united under the prior name *eques* Bodd. Given that neither females nor juveniles were described from Réunion it is appropriate to retain nominal distinction between the islands at the subspecific level, the Mauritian bird becoming *P. eques echo* Newton.

General habits, behaviour and field characters

The Echo Parakeet is strictly arboreal and usually keeps to the forest canopy while feeding and resting. It is less gregarious than the Ring-necked Parakeet *P. krameri* and usually moves around singly or in small groups. Although very vocal it is less so than *P. krameri*. Vocalisations can be heard throughout the year, but the birds are considerably more vocal during the breeding season (September–December). They will often approach to within 10–15 m in response to the playback of contact calls.

In common with most Mauritian birds the Parakeet is tame and shows little fear of man. During the late winter (June–November) when food is scarce the birds are tame and can often be approached to within 3–5 m; however during the height of summer when blossoms and fruit are readily available they are noticeably more wary and become alarmed when approached to within 10 or 15 m. Temple (1978c) claimed that breeding birds were not disturbed by cars passing along a track 10 m from the nest and adults are not alarmed when their nests are examined, but merely sit in the tree a few metres away.

The Echo Parakeet can be difficult to separate from the Ring-necked Parakeet in the field if vocalisations are not heard. The more intense green of *P. echo* is a useful feature in good light conditions as is the black beak in the female. These features are not usually obvious and proportions and flight style are probably the most reliable characters. Echo Parakeets have relatively shorter tails and are stockier birds. In flight *P. echo* are usually slower fliers with slower wing beats. They are quite adept at exploiting updraughts while traversing ridges and flying up over cliffs within the Black River Gorges. The birds may rise rapidly by circling in the updraught slowly beating their wings to gain the desired height.

Vocalisations

The species has a wide vocal repertoire (Chapter 3) being most vocal late in the evening prior to roosting. The contact call is the one most commonly heard, a low nasal *chaa-chaa, chaa-chaa* (about two calls per second). The alarm or excitement call is a higher-pitched *chēē-chēē-chēē-chēē* (three or four calls a second). It is most usually heard when the Parakeet is in flight, and is accompanied by shallow rapid wing beats, the wings being kept below the horizontal axis of the body. If suddenly disturbed or frightened the bird may call with a short sharp *ark*. A. & E. Newton (1876) noted that when sitting on the tops of trees the parakeet not infrequently "whistles melodiously". On two occasions I have heard a deep quiet purring *werr-werr* and *prr-rr-rr* (5 May and 24 July 1979) from an adult female after landing in a tree. The courtship calls

(see Chapter 3) are most frequently heard between September and December although I once heard a male calling on 22 May (1979). In common with other parrots they have a growling 'anger' vocalisation (P. Ahimaz pers. comm., pers. obs.).

Activity cycles and roosting

The species is in most respects a typical *Psittacula* parakeet. The birds forage mainly during mid-morning and especially mid- and late afternoon (Fig. 15). Bad weather depresses but does not necessarily disrupt feeding activity (see Feeding Ecology). During the middle of the day, and occasionally at other times, groups of Parakeets gather on large emergent trees where they rest and preen; these are not usually roost trees.

Activity is greatest during the afternoon, when the Parakeets are travelling to and from feeding areas. This is also the time when the birds are most vocal. For over an hour before dusk they are excitable, flying around in groups calling frequently and briefly stopping to perch prominently on the tops of trees before continuing to circle in the area. Similar pre-roost behaviour has been noted in *P. krameri* in the Black River Gorges (pers. obs.) and in the Emerald-collared Parakeet *P. calthorpae* (Forshaw & Cooper 1978).

My observations suggest that the Parakeets do not settle for the night until shortly before dusk. Temple (1978c) saw them at roost sites between 3 hours and 25 minutes before sunset (with a mean arrival time at the roost of 78 minutes before dusk). The pre-dusk activity that is so obvious in this species was not mentioned.

Parakeets roost in sheltered locations on hillsides or in ravines, usually in dense mature stands of trees. They perch close to the trunk or inside cavities, and preferred trees are those with dense foliage such as *Eugenia*, *Erythroxylum* or *Labourdonnaisia* spp. (Temple 1978c). Most of the birds roost on the Macabé Ridge with some on the south-facing scarp. An adult male was roosting in a group of large native trees on the eastern slopes of Mount Cocotte during July and August 1979 and a male (presumably the same one) was roosting there in October 1983 (P. Ahimaz pers. comm.). In August 1984 two or more birds were roosting on the scarp off Plaine Champagne.

Temple (1978c) located 17 different roosting and nesting areas. Numbers at any one site varied between

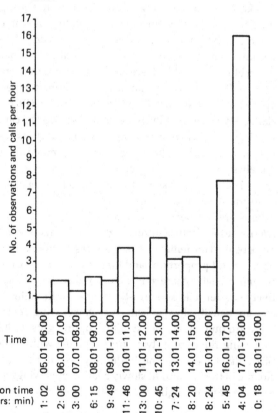

Fig. 15. Number of observations and vocalisations per hour of *P. echo* on the Macabé Ridge, 3 May–28 August 1979.

one and 11, with some shifting between sites. During May 1979 Parakeets were watched going to roost on the Macabé Ridge, the six birds then present all choosing one site on the south side of the ridge among some old native trees above Plateau Colophane. In August 1984 three birds were roosting on the same slope, or on the ridge, but about 1.5 km further up the valley. In the morning the birds usually quietly leave the roost in the hour after dawn, although Temple (1978c) saw some remaining at the roost for several hours.

Territorial and mobbing behaviour

Echo Parakeets are territorial only during the breeding season when they defend an area around their nest tree. Territoriality is poorly-developed and is neither consistent nor vigorously expressed (Temple 1978c). Outside the breeding season the birds remain in loose association with their breeding area.

Temple (1978c) described the range of territorial behaviour shown by *P. echo* and the following is summarised from his work. Territorial aggression may be directed at conspecifics, *P. krameri* and other species. Most encounters are subtle and hardly recognisable as territoriality. Before laying both sexes take part in territorial defence, but afterwards the male takes the dominant role. The first reaction to an intruder is vocal, and persistent calling may be adequate to discourage a trespasser. More intense interactions involve one or both members of the pair approaching an intruder, cautiously moving closer by jumping from limb to limb. As the intruder tries to get to the nest cavity, the territory-holders and the intruder circle slowly around the trees in the vicinity of the nest cavity. Temple recorded one such chase that lasted 90 minutes before the intruder finally left. An example of territorial conflict between Echo Parakeets and Mynahs is given under Competition for Nest-sites. Actual fights are rare although Temple recorded one between two males in low bushes in which both ended on the ground before one broke free and flew off. Neither appeared to be seriously injured.

Mobbing behaviour is well developed in the genus and groups cluster and noisily scold strange or threatening animals (Smith 1979). In *P. echo*, mobbing may stem from territorial conflicts, or flight may be diverted to chase a passing bird. Both *P. echo* and *P. krameri* have been seen to chase Mauritius Kestrels, White-tailed Tropic-birds *Phaethon lepturus*, and each

other (Temple 1978); M. d'Unienville (pers. comm.) recalls seeing groups of two or three *P. echo* mobbing flying fruit-bats south-west of Piton du Milieu in the early 1950s. Kestrels are regularly mobbed, as first noted by Abbé de la Caille in 1753 (Staub 1976). The following description is probably a typical interaction between these species. On 3 May 1979 on the Macabé Ridge I watched two Echo Parakeets mob a circling Kestrel in flight. The Kestrel alighted on a prominent dead branch of an emergent tree. The Parakeets were calling wildly with alarm calls *chēē-chēē-chēē* (c. 1.5–2 calls per second). Other Parakeets in the area joined in until there may have been as many as six all calling excitedly, flying around and landing in surrounding trees. The Kestrel remained calmly perched for 5 minutes, spending some of the time preening, before leaving.

Both species of parakeet respond to the presence of monkeys with loud calls (Temple 1978c), though I have also seen monkeys foraging near to Echo Parakeets being totally ignored.

Habitat

Echo Parakeets restrict all their activities to areas of native vegetation and there are no observations outside this area. The importance of native forest to the parakeet was already recognised by A. & E. Newton (1876) who said that it "frequents the forest only, retiring before cultivation".

The remaining birds centre their activities on the upland forest of the Macabé Ridge, an area they have favoured since at least the 1930s (R. E. Vaughan pers. comm.). Along this ridge the large trees east of Plateau Colophane at and below the corner of Brouard's Folly are their favourite haunt. These mature trees (e.g. *Canarium paniculatum, Syzygium contractum, Mimusops maxima, Labourdonnaisia* sp.) are some of the largest left on Mauritius.

The lowland, intermediate and scrub forests are also very important as feeding areas for the Parakeets (see Feeding Ecology). The importance of the upland scrub and dwarf forest was clearly demonstrated by the catastrophic decline in the population following the clearing of large areas of this habitat in the 1970s (see Habitat Destruction and Degeneration).

The relative frequency with which Parakeets are seen in these different habitats is illustrated in Fig. 16.

The annual fluctuations shown in the histograms reflect changes in population density (see Status and Distribution) although the inflated number of observations on Macabé during 1983 (Fig. 16a) is due to increased coverage of this area in the breeding season when the Parakeets are very active and vocal. Con-trastingly the high number of observations in lowland forest in the same year is due to a single male Parakeet that used to forage in one of my study areas at the base of Morne Sèche at the mouth of the Black River Gorges.

In 1979 the frequency of Echo Parakeet observations and/or vocalisations seen and/or heard in dif-

Fig. 16. Parakeet observations and vocalisations per hour in different habitats 1979–83.

The histograms illustrate the number of separate observations plus contact calls detected in the different habitats per hour. In quantifying my results, a single bird seen flying past vocalising in flight is recorded once although a single bird that is seen flying on three separate occasions and is heard, but not seen, on another two will be recorded five times. Individual birds are not always easy to keep in sight and it may not be clear whether ten observations or vocalisations represent ten different birds or the same bird ten times. Despite these difficulties the histograms are a good index of activity but not necessarily of parakeet numbers.

(*a*) Number of observations and vocalisations per hour of *P. echo* and *P. krameri* on the Macabé Ridge 1979–83. The dashed section for *P. echo* for 1979 includes the total number of observations plus vocalisations if the evening period of high activity is included. This period of activity is excluded from the subsequent year's results.

(*b*) Number of observations and vocalisations per hour of *P. echo* on Plaine Champagne in July and August 1979 and October 1983.

(*c*) Number of observations and vocalisations per hour of *P. echo* in the lowland and mid-altitude forest (below 365 m), along the west scarp of the Black River Gorges above Rivulet Roches Blanches and in the forest north and north-west of Morne Sèche, 1979–83.

(*d*) Number of observations and vocalisations per hour of *P. krameri* in lowland forest and along river valleys in the Black River Gorges, October and November 1979 and 1980. The dashed section includes the period of high activity in the morning up to 09.00 hours (see Fig. 17).

(*e*) Number of observations and vocalisations per hour of *P. krameri* and *P. echo* from Black River Viewpoint in November 1979 and October/November 1983.

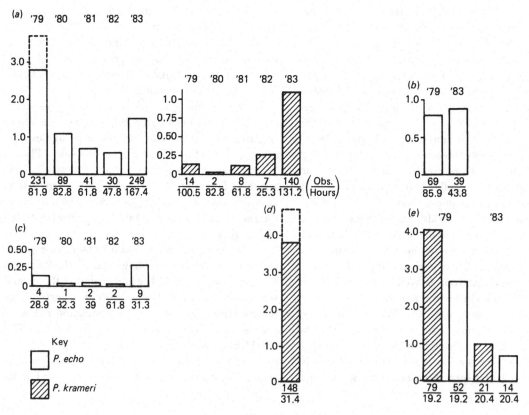

ferent habitats was two or three per hour in upland forest on Plaine Champagne and about one every 8 hours in lowland forest.

Status and distribution

The distribution of the Echo Parakeet has always been closely tied to native forest. The decline in numbers and contracting distribution are concomitant with the destruction of native vegetation. Many of the early accounts suggest that the parakeet was always thinly distributed.

Milbert (1812), who was on Mauritius 1801–4, described Parakeets as being fairly rare, although he did note: "Among the birds which swarm in the leafy trees, . . . the large and the small parakeet . . . are there in great number." These birds were seen on the upland plateau near what is now Quatre Bornes and Vacoas. From his descriptions (expanded elsewhere in his account) it is clear that he was referring to the introduced Madagascar Lovebird *Agapornis cana* and the larger *P. echo*.

In the 1830s the Parakeets were described as "quite common" (Oustalet 1897). W. Bojer killed some in the forests of Black River in October 1829 and in May 1836 E. Dupré killed one in Flacq (Oustalet 1897, Guérin 1940–53). A. & E. Newton (1876) stated that "its numbers are gradually failing" and indicated that the species was still to be found in the eastern forests

Fig. 17. Number of Ring-necked Parakeets seen or heard per hour in the lowland exotic forest of the Black River Gorges, October/November 1979. The birds are most obvious in early morning and late afternoon. Many birds leave the gorges during the day and return in the late afternoon to roost. In this histogram the late afternoon observations are probably under-recorded. All observations were taken away from the main roost site and the data were recorded in the same manner as for Fig. 16.

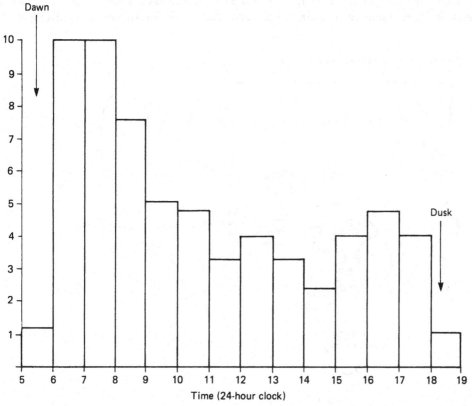

1: 20　2: 00　2: 00　2: 00　3: 00　3: 00　4: 15　4: 00　3: 00　1: 40　2: 00　1: 30　1: 00　0: 40

Observation time (hours: mins)

in the district of Grand Port. I have located six speci- mens that formerly belonged to Edward Newton, of which four have collection locality and date; of these two were collected in October 1873 at Bois Sec and two from Vacoas, one on 12 April 1860 and the other in 1897. Carié (1904) suggested that the population was reasonably large, although Rothschild (1907*b*) called the species "rare and apparently on the verge of extinc- tion". In 1911 Meinertzhagen (1912) noted that the Parakeet was common near Mount Cocotte and Grand Bassin, and that he had seen a pair near Curepipe in April.

Temple (1978*c*) (quoting personal communi- cation with several Mauritian naturalists) noted that in the 1920s and 1930s Parakeets were found in the upland areas south and east of Curepipe including around Montagne Perruche (Map 7). This is confirmed by eggs collected in this area by G. Antelme (see Eggs and Incubation). Although more widely distributed than today, the birds were seen only occasionally; Guérin (1940–53) saw this species only once (two birds in March 1939 near Plaine Champagne). In the early 1950s Echo Parakeets were quite commonly encoun-

tered on the Crown Lands of Merlo and Petite Merlo, south-west of Piton du Milieu (M. d'Unienville pers. comm.). This suggests that they would have ranged over the area between Mare aux Vacoas, Nouvelle France and Piton du Milieu (Map 7) until this area was cleared for tea and forestry between the 1950s and early 1970s. Rountree (*et al.* 1952) considered the species rare and confined to remote scrublands and forests on the south-west plateau – the only place he had seen them (Rountree MS field notes, per A. S. Cheke). R. Newton (1958*a*) saw the birds only once: a pair in the Black River Gorges in April 1954. Newton did by comparison see the elusive Mauritius Kestrel five times in three years and the highly localised Pink Pigeon "perhaps a dozen times" and heard its cooing more often. Gill (quoted by Forshaw & Cooper 1978), who was on Mauritius from August to October 1964, regarded the Echo Parakeet as a quiet unobtrusive bird and collected a specimen (USNM 487076) near Mount Cocotte in August.

During the late 1960s and 1970s F. Staub (*in litt.*) noted that the Parakeets centred their activities on Macabé, and were not known to roost elsewhere.

Map 7. Probable distribution of Echo Parakeets in the 1940s.

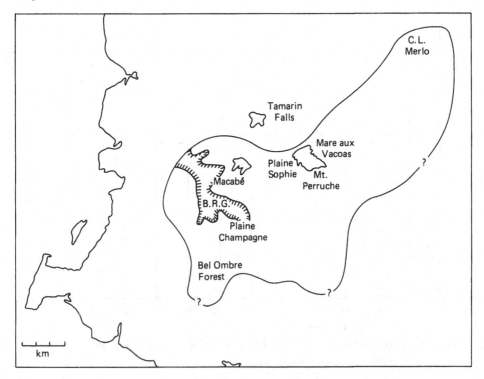

During the day they foraged, travelling widely as far as Grand Bassin and Kanaka Crater, and were seen flying towards Bois Cheri and over towards Mount Cocotte. Staub once heard a Parakeet flying above the Bel Ombre forest and saw one on the flanks of Savane Peak above Bois Sec resting in a tree. He has never encountered them in Combo forest or anywhere in this region and feels that this area, which faces south-east, is too exposed and windy for them.

Around 1970 Staub thought there were roughly 50 pairs left although I suspect this was an over-estimate. At this time Echo Parakeets were easy to see and locate in and around the Black River Gorges because due to the cutting of the upland scrub and forest on Plaine Champagne, Les Mares and elsewhere on the south-west plateau (see Habitat Destruction, Population Status and Distribution in Pink Pigeon section) all the birds were forced into the remaining areas of native forest in the south-west. The numbers recorded in the early and mid 1970s must therefore have been at an elevated population density; with some of their best foraging areas destroyed, the birds were forced to travel intensively within what suitable habitat was left to find sufficient food.

Between August 1973 and December 1974 Temple (1978c) censused the population and estimated that between 32 and 58 birds survived; this and later census data is summarised in Fig. 18. In 1974 Temple noted that the population numbered about 50 and that "the recent downward trend . . . is ominous". In February 1975 Newlands (1975) noted that there were about 40 birds before cyclone Gervaise struck. In May he claimed "It is possible to see more than twenty individuals during a morning spent on the forested Macabé Ridge", and later in the same report he noted "the population probably totals no more than 50". At the end of the year McKelvey (*in litt.* to S. A. Temple) wrote "The parakeets at Macabé Ridge are 14 in number with no young of the year in evidence but a yearling male is among them". Temple (1976b) reported "The Mauritius Parakeet population has remained relatively stable over the past three years. All censuses have revealed between 50 and 60 individuals". At the end of 1976 McKelvey thought there were about 45 birds. Temple (1978c), quoting McKelvey (*in litt.*), however felt the population had dropped noticeably in 1976 and 1977. No detailed census of the birds was made in 1977 although McKel-

Fig. 18. Population trend of the Echo Parakeet 1973–83. The continuous line represents the probable population trend and the vertical bars are the different estimates per year (see text for further details).

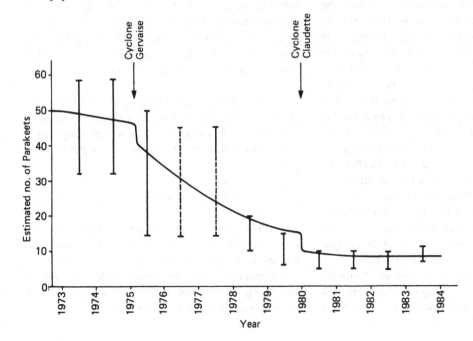

vey and Staub both felt that the population was between 35 and 45 individuals (King 1977). The estimates for 1976 and 1977 were not I believe based on an actual head count and I regard McKelvey's count of 14 for December 1975 as a realistic minimum for this period. In 1978 when Steele censused the Parakeets he felt certain there were less than 20 birds and probably less than 10 (F. N. Steele pers. comm.). In June 1978 he recorded "Dr F. Staub feels as I do, that *P. echo* could be in more danger than has been realised. My observations have been very limited but I have found this bird somewhat difficult to locate. Dr Staub has been watching them for many years and in his opinion the population has declined rapidly the past 4 or 5 years". Steele (1978, 1979) further expressed his concern about the species in reports of August 1978 and January 1979, in both of which he mentioned seeing a flock of six birds on Macabé Ridge.

During 1979 I could definitely account for only six or seven birds (three or four males and three females), although there could have been as many as 15 individuals (Jones 1980*a*; see also Map 8). In 1980 only six known different birds were seen; they could be located on Macabé Ridge and were seen occasionally up until 3 June 1981 when a group of five or six birds was seen.

During 1982 fewer sightings were made and it was felt that the population had declined further. For 3 months at the end of 1983 an intensive study of the remaining Parakeets was made (anon. 1984, Ahimaz 1984) and all but one observation suggested that only four birds remained (three males and one female). On 15 December 1983, however, on Macabé Ridge, a group of six vocalising Echo Parakeets flew into some nearby trees. They were approached to about 20 m when six flew away followed by a further two, leaving another two birds perched in the tree. At this time a different Parakeet was heard on the south-facing slope of the ridge. The total count was 11 birds although none of the four observers could be absolutely certain that there were no Ring-necked Parakeets intermingled in the flock and the most pertinent conclusion is that there were between seven and 11 birds with the probability of only two females. The flock is most likely to have represented the total population.

An examination of the number of Parakeets located per hour on Macabé Ridge 1979–83 (Fig. 16) reveals that a major drop took place between 1979 and

1980. This was probably the result of mortality caused by cyclone Claudette (23 December 1979).

Breeding success 1973–83

The breeding results of the Parakeet are shown in Table 26. Below I provide a year-by-year summary of all known breeding attempts.

1973. Information for this year is contradictory. Temple *et al.* (1974) claim that seven nests were found, of which two were taken over by Common Mynahs, two by Ring-necked Parakeets and one was destroyed by monkeys. The other two nests were successful. Temple (1978*c*) gives an entirely different set of results for the year. Eleven breeding pairs were known and seven were successful in raising a total of 11 young. This account is probably in error for 1974.

1974. At least six pairs were successful in producing 11 young. A pair of these young were taken for the captive breeding programme (Temple 1975, 1976*b*). Temple (1978*c*) gives conflicting information claiming there were 11 known pairs, of which three failed, four young fledged and four pairs are assumed not to have bred.

1975. In September Temple (1975) noted that there were six active nest-sites. Only two were still active in December, each with two young. The young from one nest were taken for captive breeding (Temple 1978*c*). The other young evidently did not survive since McKelvey (*in litt.* to S. A. Temple) could find no young of the year in evidence at the end of December.

1976. Temple (1978*c*) noted that "In 1976 S. D. McKelvey (*in litt.*) suspected that no parakeet nesting attempts were successful". However King (1977) quoted McKelvey as having seen three pairs with a total of five young.

1977. No data available.

1978. Three pairs were located. In October there was a pair apparently prospecting in the trees below the tourist lookout at Alexandra Falls. Another pair was regularly seen on the Brise Fer side of Mare aux Joncs gorge (A. S. Cheke *in litt.*). The only pair known to have nested used a tree on Macabé Ridge. They

Table 26. *Echo Parakeet breeding data 1973–83*

Year	No. of known pairs	No. of pairs attempting to breed	No. of pairs failed	No. of pairs rearing young	No. of young reared	Reference
1973		7	5	2	2–4	Temple *et al.* (1974)
		11	4	7	11	Temple (1978c)
1974	–	–	–	6	11	Temple (1975)
	11	7	3	4	5	Temple (1978c)
1975	–	6	5	1	2	Temple (1975)
	–	6?	4?	2	4	Temple (1978c)
1976	6–11	–	–	3	5	King (1977)
	–	–	–	0	0	Temple (1978c)
1977	–	–	–	–	–	No data available
1978	3	1	1	0	0	Steele (1979)
1979	3	0	0	0	0	Jones (1980a)
1980	2	1	1	0	0	Jones (1982)
1981	2	1	0	0	0	Jones (1982)
1982	1	0	0	0	0	This study
1983	1 or 2	0	0	0	0	Ahimaz (1984); this study

Map 8. Distribution of Echo Parakeets and observations per hour, calculated from data collected in 1979. In early 1984 the distribution remained unchanged but the frequency of observations had dropped by about 40–50%.

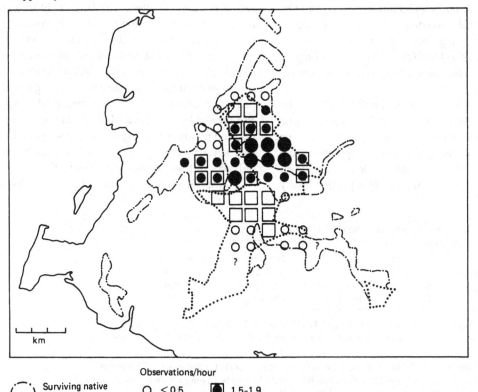

km

	Observations/hour	
Surviving native forest	○ < 0.5	◉ 1.5–1.9
Nature reserve boundary 1979	☐ 0.5–0.9	● > 2
	● 1.0–1.4	

failed due to interference by non-breeding males (Steele 1979) (see Helpers and Disrupters at the Nest).

1979. Two pairs were seen prospecting, one below Black River Viewpoint and the other on Macabé Ridge. There is no evidence that either pair progressed further than this (Jones 1980*a*, *c*).

1980 and 1981. Both years one pair tried to breed on Macabé Ridge but were thwarted by Mynahs that took over the nesting cavity (Jones 1982) (see Competition for Nest-sites).

1982. No evidence of any breeding activity.

1983. On 26 October a pair were seen prospecting on Macabé Ridge in the same area inspected/used 1978–81. Copulation was observed once elsewhere on the ridge late in the afternoon of 1 November. There is no evidence that the birds progressed further than this (P. Ahimaz pers. comm.).

The possibility cannot be ruled out that there could be some undetected breeding in the population. Parakeets frequenting the scarp above Bel Ombre forest are difficult to locate and a low level of reproduction here or elsewhere could have been entirely undetected.

During this study less than half (47.6%) of the detected pairs tried to nest per year. This figure may be an over-estimate since non-breeding birds may be overlooked. Considering the whole population, just over a third (36.5%) of the birds tried to breed annually. The proportion of nesting attempts that finally fledged young is low (46.4%), due in part to nest competitors and predators, and possibly to an impoverished food supply. Echo Parakeets usually rear one (Staub 1973*a*) or two young (Staub 1976) per successful nesting attempt; no brood of more than two has ever been recorded (Temple 1978*c*).

There have been few studies of the population dynamics of parrots although the recent study by Saunders (1982) of White-tailed Black Cockatoos *Calyptorhynchus funerus* allows some tentative comparisons with *P. echo*. This black cockatoo is a large, potentially long-lived species; it does not start to breed until 4 or 5 years old, lays only a one or two egg clutch, and when successful rears just a single youngster (*ibid.*). By contrast *Psittacula* spp. are probably shorter lived, can in most species breed when 2 years old, and lay larger clutches (two to four in *P. echo*) (Forshaw & Cooper 1978). Annual production and population turnover would be expected to be higher in *P. echo* than in *C. funerus*.

The production of Echo Parakeets is below replacement, with an average of only 0.6–0.7 young reared per nesting attempt and only 0.29–0.34 young per pair in the population annually. By contrast black cockatoos maintain a stable population with a production of 0.65 young per breeding pair which, accompanied by a high proportion of pairs breeding (84%), results in an annual average of 0.55 young reared per pair. In another locality where habitat quality is poorer and there is less food available the black cockatoos are declining: only 0.2 young are produced per adult per year since an average of only 56% of pairs breed and only 35% of these are successful (Saunders 1982).

The survival rate and longevity of the Echo

Table 27. *Echo Parakeet breeding success 1973–83*

% known pairs attempting to nest	% of nests from which fledglings flew	Mean brood size at fledging	Mean no. of young per attempted breeding	Mean no. of young per pair in population
42–43 (10/23 or 24)	40 (16/40)	1.6–1.7 (40–42/25)	0.6–0.7 (24–26/40)	0.29–0.34 (10/29–35)

Note:
Data summarised from Table 26. Since there are conflicting reports for the years 1973–6 some of these data must be in error. The errors are, however, likely to be in magnitude rather than in the levels of productivity per pair; the data probably give a true index of breeding success in the population.

Parakeet are unknown but some evidently live for several years. The surviving birds in the population must be at least 6–8 years old. To achieve replacement in the population at its present rate of production of young, fledged birds must live an average of 7.2–7.6 years.

Feeding ecology

The Echo Parakeet feeds mainly on native plants (A. & E. Newton 1876, Staub 1976, Jones 1980a), taking a wide range of parts including buds, young shoots, leaves, flowers, fruits, seeds, twigs and even bark and sap (Staub 1973, Temple 1978c, pers. obs.). Some exotic plants are eaten but they form only a very small part of the diet. *P. echo* feeds arboreally and never, or rarely, descends to the ground (A. & E. Newton 1876, Temple 1978c, pers. obs.). Of 95 feeding episodes observed during this study, 53% were on fruits, 12% on flowers, 31% on leaves and 4% on bark and/or sap. Temple (1978c) saw 239 feeding episodes on over 30 different plant species. Over 25% of these were on *Calophyllum tacamahaca, Canarium paniculatum, Tabernaemontana mauritiana, Diospyros* sp., *Erythrospermum monticolum, Eugenia* sp., *Labourdonnaisia* sp., *Mimusops maxima, M. petiolaris, Nuxia verticillata* and *Protium obtusifolium* (Table 28).

Some species are more important food sources than others. In July and August 1979 and up until October in subsequent years Parakeets feeding on *Calophyllum parviflorum* fruit on Plaine Champagne flew directly to and from fruiting bushes. In 28 feeding episodes that I watched they ignored other trees in fruit including *Syzygium contractum* and *Sideroxylon cinereum*. The only feeding observation not on *Calophyllum* fruit was of a female that sampled a *Sideroxylon cinereum* leaf. Many species, now rare, may formerly have been favoured food items. The *bois perroquets* (parakeet tree) *Olax psittacorum* probably acquired its name from the Parakeets' fondness for it. Some fruits that are easily available are not exploited; there are no observations of Echo Parakeets feeding on the very common Strawberry Guava *Psidium cattleianum*.

In pristine Mauritius parrots must have had a significant impact upon the seed production of some favoured food plants. Some fruits have a very hard epicarp which is resistant to attack by parrots and it seems likely that the woody stone evolved as pro-

tection against seed predators. *Sideroxylon grandiflorum, S. sessiliflorum* and *Canarium paniculatum* have a very hard epicarp surrounded by a fleshy pericarp favoured (in two of the species at least) by Echo Parakeets. After feeding on the edible exterior, the hard stone is rejected (Temple 1983), and this undoubtedly aids in the trees' dispersal.

Parakeets forage in different areas at different seasons. From November to March they have been noted foraging in the *Sideroxylon–Helichrysum* scrub (Temple 1978c). In October (1975) Newlands (unpublished field notes) watched a Parakeet feeding for 45 minutes on the terminal buds of the emergent bushes of the dwarf forest of Pétrin. F. Staub (*in litt.*) has also frequently seen Parakeets in this area and in April has noted them feeding on *Ochna mauritiana* seeds. The dwarf forest and scrublands of the south-west plateau are important to the Parakeet as a foraging area throughout the year, different species being utilised when flowers, fruits or other edible plant parts become available to them. Food does not, however, always become available on a regular seasonal basis since the flowering, and hence fruiting, times of many food plants is irregular (J. Guého pers. comm.). In addition many of the species on which the birds depend are now rare (Table 28). During the winter and early spring when fruits are scarce the Parakeets eat considerably more leaves (Fig. 19) and spend a greater proportion of the day foraging. The birds that were feeding on *Tabernaemontana mauritiana* on Macabé Ridge in November 1983 would eat the leaves during the late afternoon, presumably after failing to find more substantial food earlier in the day.

Parakeets usually forage singly or in small groups (L. F. Edgerley pers. comm.). Temple (1978c) felt the species while feeding was "usually more or less solitary" though he did see groups of up to 13 sometimes feeding in the same group of trees, ignoring each other with no aggressive interactions. It is difficult to estimate how social or gregarious this species is since so few now survive. I have, however, seen a group of four birds feeding in October (1980) on the blossoms of Black Ebony *Diospyros tesselaria* deep in the valley of Grandes Gorges River (Jones 1981) and several groups of three (two males, one female) feeding on a variety of food plants. Pairs usually keep in loose association throughout the year and can be seen foraging together.

Table 28. *Food items of the Echo Parakeet*

Species	Status[a]	Range[b]	Part eaten[c]	No. of feeding observations[d]	J	F	M	A	M	J	J	A	S	O	N	D	Refs.[e]
Flacourtiaceae																	
Erythrospermum monticolum Thouars	Nt	M	Lv / Fl? / Fr?		x												A,C,D,E
Clusiaceae																	
Calophyllum parviflorum Bojer ex Baker	Nt	M	Lv / Fr	1 / 30							xxx	xxxx	xxxx	x			A / A,G
C. tacamahaca Willd.	V	M,R	Lv / Fr	1 / c. 12	x	x								xx		x	A / A,D
Ochnaceae																	
Ochna mauritiana Lam.	Nt	M	Sd					x									B,C
Burseraceae																	
Canarium paniculatum (Lam.) Benth. ex. Engl.	E	M	Lv / Bk / Fl	1 / 3											x x		C,D / A
Protium obtusifolium (Lam.) Marchand	Nt	M	Fr													x	D
Celastraceae																	
Elaeodendron orientale Jacq.	Nt	Masc.	Yglv / Fr	1 / 2											x		A
Pleurostylia leucocarpa Baker	R	M	Fr	1								x					A
Myrtaceae																	
Eugenia spp.																	
Syzygium glomeratum (Lam.) DC.	Nt	N	Fr	3											xx		A,D / A,G
Sapotaceae																	
Labourdonnaisia glauca Bojer	Nt	M	Yglv / Bk													x	C,E
L. revoluta Bojer	R	M	Yglv / Fr	2													A,C
Mimusops maxima (Poiret) Vaughan	R	M,R	Fr	2	x				x								D,E
M. petiolaris (DC) Dubard	V	M	Fr	1													D

Species	Category[a]	Island[b]	Part[c]	No.[d]	Feeding observations	Source[e]
Sideroxylon puberulum DC.	Nt	M	Lv	1	x	A,B,E
S. cinereum Lam.	E	M	Fr	1		
S. grandiflorum DC.			Lv	1	x	A,G
			Fr	1		F
Ebenaceae						
Diospyros spp.	Nt	M	Fr	5	x	A
D. tessellaria Poiret			Fl		x x	
Oleaceae						
Ligustrum robustum Blume var. *walkeri* (Decaisne) Mansf.	I		Fl	4	x x	A,G
			Fr		x	C
Lauraceae						
Litsea monopetala (Roxb.) Pers.	I		Fr	2	x	A
Apocynaceae						
Tabernaemontana mauritiana Poiret	Nt	M,R	Lv	50+	xx xxxx xxxx xxx	A,G
			Fr	20+	xxx xxxx xxx	A
Loganiaceae						
Nuxia verticillata Lam.	V	M,R	Yg lv	1	x	A,C,D,G
			Fl			
Euphorbiaceae						
Cordemoya integrifolia (Willd.) Pax	Nt	M	Lv	1	x	A
			Fr?			
Securinega durissima Gmel.	Nt	Masc.	Fl	1		A
			Lv	2	x	
Stillingia lineata Mull.	Nt	M	Fr	2		A

[a] Based on IUCN *Red Data Book* categories (Anon 1980). Nt, not threatened; R, rare; V, vulnerable; E, endangered.

[b] M, Mauritius; R, Réunion; Masc., all three Mascarene Islands; I, introduced.

[c] Bk, bark; Fl, flowers; Fr, fruit; Lv, leaves; Sd, seeds; Yg, young.

[d] All the numbered, and most of the dated, feeding observations were collected during this study and include observations by P. Ahimaz, A. S. Cheke, J. F. Martial and D. V. Merton. Include data collected 1974–84.

[e] A, this study; B, Staub (1973); C, Staub (*in litt.*); D, Temple (1978c); E, Temple *et al.* (1974); F, Temple (1983); G, Ahimaz (1984).

The Echo Parakeets wander in search of food and travel several kilometres to and from feeding areas (Temple 1978c, pers. obs.). Birds from Macabé Ridge wander as far as Bel Ombre forest, the west-facing scarp above Petite Rivière Noire and the Petites Gorges.

While feeding the Parakeets clamber silently around removing fruits or flowers with the beak, and may hang upside down to reach a particular item. The food is then transferred to a foot in which it is held while being eaten. Birds feeding on *T. mauritiana* leaves would typically scoop out the mesophyll with the bottom mandible leaving the cellulose in place. When the mesophyll has been largely removed the petiole and midrib are discarded. Many leaves are, however, only partly eaten and discarded after one or two beakfuls. In a sample of 152 leaves 49 (32%) were bitten off and dropped, 34 (22%) were partly eaten and in 69 (46%) most of the mesophyll was eaten. Similarly several fruits may be sampled and dropped before the bird settles to eat one.

While eating the Parakeet takes a bite which may be masticated for several minutes before swallowing. In a typical feeding episode a male flew from the *Calophyllum* bush where it had been foraging and landed on an exposed branch in full view of me. It then transferred the fruit from its beak to its foot, removed the kernel and masticated it for 10 minutes before swallowing. After feeding the bird wiped its beak on the branch, defecated, scratched its head and left the area, calling in flight with the monotone contact call.

Most foraging occurs during the morning and late afternoon (Fig. 20). Feeding activity is depressed by high winds and heavy rain (Temple 1978c), although I have seen Parakeets foraging during torrential downpours and in windy weather.

Breeding biology

In its breeding biology *P. echo* appears to be similar to other species in the genus. The breeding season usually starts in August or September. Temple (1978c) observed that between 1973 and 1975 breeding activity "commenced rather synchronously in September".

Nest-site selection and characteristics

Nest-site selection occurs early during the breeding season. In August or September 1971 F. Staub (*in litt.*) watched a group of two males and a female moving along Macabé Ridge flying from tree to tree, clamber-

Fig. 19. Food parts consumed at different times of the year by the Echo Parakeet.

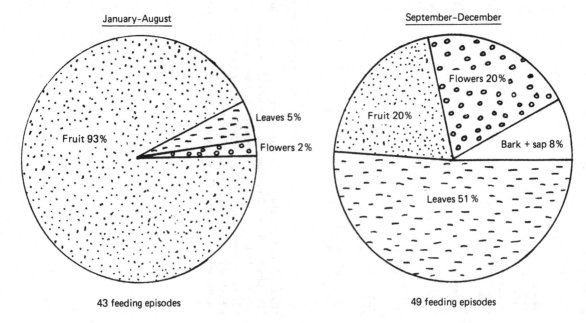

January–August

Fruit 93%
Leaves 5%
Flowers 2%

43 feeding episodes

September–December

Flowers 20%
Fruit 20%
Bark + sap 8%
Leaves 51%

49 feeding episodes

ing among the branches and inspecting cavities for about 20 minutes; he suspected that they were prospecting for a nest-site. Steele (1978) saw a pair chewing at a suitable hole in a tree in September. Tree cavities may also be inspected outside the breeding season; I observed a female examining a hole on 5 May 1979.

The characteristics of a typical nest-site have been listed by Temple (1978c), who inspected 24 different cavities. Thirteen of these were used while he was on Mauritius, four had been used during the recent past and seven had been prospected with a great deal of interest. They were all in large living emergent native trees, usually *Calophyllum*, *Canarium*, *Mimusops* or *Sideroxylon*. The cavities were all at least 10 m above the ground with exposures other than south-easterly, which are affected by the South-East Trade winds. F. Staub (*in litt.*) watched two nests: both were in emergent *Mimusops erythroxylon*, one in a limb and the other in the main trunk, between 6 and 7.7 m, and about 7.7 m above the ground, respectively. Both nests had easy access and one faced north-west.

Nest cavities are at least 50 cm deep and 20 cm wide, with entrance holes 10–15 cm in diameter, and with overhangs or other features to prevent flooding (Temple 1978c). These dimensions are much larger than the cavities I have been shown. Anti-flooding features are not common to all nest holes since Temple (1976b) lists "flooding of nest holes" as one of the natural disasters that takes a toll. Nest cavities are usually in horizontal branches rather than vertical trunks (Temple 1978c).

Courtship and copulation

The entire courtship sequence has yet to be described although some elements have been observed. Pair maintenance behaviour (regurgitatory feeding, billing and preening of the female's nape feathers by the male) can be observed throughout the year in compatible pairs. The male is usually the assertive sex in these behaviours.

Copulation has been seen during September and October. The sequence leading to copulation is similar to that in *P. krameri* and *P. eupatria* (Smith 1979). Before approaching the female the male wipes his

Fig. 20. Foraging and feeding activity of Echo Parakeets on Tatamaka bushes *Calophyllum parviflorum* on the scrublands west of Alexandra Falls, Plaine Champagne (12 July–15 August 1979).

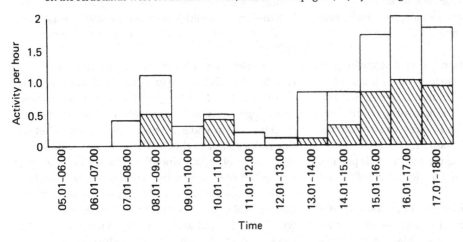

0: 10   2: 30   5: 00   6: 25   7: 40   8: 00   8: 30   9: 00   9: 30   9: 45   8: 00   8: 00   3: 21

Total observation time (hours: mins)

Echo Parakeets heard or seen flying in area per hour

Echo Parakeets seen feeding in Tatamaka bushes *Calophyllum parviflorum* per hour

beak, then moves towards her in a slow stately walk and preens her nape feathers. The female solicits copulation by squatting with her body horizontal. The male mounts her back and during copulation raises and drops his head in a ritualised hammering. The male may mount the female for up to 5 minutes and frequently changes the side from which he copulates. After copulation the male may feed the female and they both preen.

The following sequence, modified from the field notes of Yousoof Mungroo (28 October 1980) is probably typical.

06.35   Three Parakeets, two males and a female, land near the road on Macabé Ridge. A pair perch in one tree and the third in an adjacent tree.
06.54   The pair fly to a former nest tree, *Labourdonnaisia glauca*, used in 1978. Female disappears, probably into nest hole; female reappears, approaches male and solicits copulation. The male attempts to mount female but is unsuccessful.
07.25   The male chases away the other male and returns to nest tree.
07.28   Male and female both preening.
07.35   Preening stops. Female, then male, flies to branch near nest hole. Male preens female's nape feathers.
07.39   Female solicits copulation, crouching on branch with body held horizontally. Male mounts, successful coition, tail held to left and sometimes to right. During mating (which lasts 4½ minutes) male's head moves up and down in a regular hammering motion.
07.43   The pair disengage and the male perches about 0.3 m from the female. Hear soft, quiet *cheek* call.
07.45   Male wipes bill a few times, moves closer to female and touches her beak seven or eight times before they lock bills. Male feeds female for a few seconds before parting; this is repeated five or six times and is followed by preening.
08.05   Male flies to neighbouring tree. Preens body, wing and tail feathers. Female joins male. Male continues preening for 8–10 minutes.
08.17   Both birds leave area and shrill *cheeek* vocalisation heard.

Eggs and incubation

It was suspected that during the years 1973–5 most pairs began incubating "around mid-September" (Temple 1978c). In 1978 at a single nest watched by Steele (1979) eggs were believed to have been laid at the beginning of October (see Helpers and Disrupters at the Nest). Back-dating from fledging dates provided by Temple (1978c) (allowing 86 days between the first egg and fledging), a range of laying dates from 20 August to 10 October is derived with most in late September and early October. The actual range must be greater since an immature female collected on 31 December 1860 is advanced in the process of changing its bill colour. Temple (1978c) noted that the dark mandible is not completely acquired until 6 months of age. If this immature female is aged as 4 months old, the eggs must have been laid at the beginning of August. Data accompanying an egg in the Transvaal Museum give the date of collection as November (1921).

Late clutches may be the result of repeat layings. Staub (1973a) suggested that "it is not impossible that the Echo Parakeet raises two broods per year". On 17 December 1971 Staub (*in litt.*, see below) watched a female that was still incubating eggs or brooding young chicks, suggesting that the eggs had not been laid until at least mid-November; he also saw a pair of Parakeets using the same nest-site on 27 February 1972.

Clutch size has been variously assumed to be two (Staub 1973a) or two or three (Staub 1976, Temple 1978c). A. & E. Newton (1876) said they were given "two eggs . . . which were taken with the mother from the nest in a hole of a tree." It is not clear whether this was a complete clutch. These eggs were probably taken in October 1873 since this is the date on the only labelled adult female in the Newton collection in Cambridge. Three eggs, probably a clutch, were given to George Antelme by a keeper on his father's estate which was located at Forest Side (Viader 1939). The label accompanying the Transvaal Museum egg, quoting G. Antelme, gives the clutch size as three or four. The eggs are typical parrot eggs, rounded and white; measurement details are given in Chapter 9.

The incubation period is recorded by Staub as a month (Staub 1976), although he does not mention how he obtained such a figure. As in other members of the genus, incubation is usually entirely by the female

(F. Staub and F. N. Steele pers. comm.). Temple (1978*c*) watched a nest for two consecutive days in 1974. The eggs were only attended for 68% of the time and the female was attentive for periods that averaged 110 minutes. The male also spent some time in the nest hole (average 11 minutes). The nest was left totally unattended for periods that averaged 26 minutes, and the longest period of absence was 64 minutes. Temple did not give the date or the stage of incubation; his data suggest that incubation had not started properly or there was a shortage of food and the female had to forage on her own for some of her nourishment.

While the female is incubating she is fed by the male four or five times a day (Staub 1973*a*) or about once every 2 hours (Staub 1976). Typical feeding episodes were described by Staub (1973*a*, 1976) and are best illustrated by his field notes (pers. comm.) for 17 December 1971.

> 09.20 Male flies back to nesting area calling in flight, perches near to nest tree in a *Diospyros tesselaria*.
> 09.25 Female comes out of nest hole, scratches head, flies to neighbouring tree, defecates, preens, flies to the ebony in which the male is perching.
> 09.30 Male feeds female, bills placed at right angles, male feeds female by regurgitation, male leaves area.
> 09.40 Female flies directly back to nest hole in a *Mimusops erythroxylon*.
> 10.40 Male returns to repeat the feeding episode.

At the same nest on 24 November 1971 Staub saw the male return to the nesting area calling in flight; he alighted in the nest tree and climbed down to the nest entrance, head first, knocking on the branch with his bill as he did so. The female came out and flew to the ebony tree where the male fed her.

Post-hatching care and development
After hatching the young are brooded by the female; by 2 weeks old they are left unattended for most of the day. Temple (1978*c*) watched a nest that contained young about a fortnight old. They were fed at intervals averaging 79 minutes and the parent delivering food remained for an average of only 11 minutes. Temple implied that at this stage both parents fed the nestlings, although Staub (1973*a*, 1976) said that only the

female feeds the young for the entire nestling period. In common with other parrots it is probable that the female cares for them for the first few days, she in turn being fed by the male. Later when the young are homeothermic, and the female forages for food also, the male assists in the care of the nestlings, feeding them directly (Forshaw & Cooper 1978).

Well-developed nestlings threaten intruders with loud menacing growls and lunge and bite at them. If the intruder persists in trying to enter the cavity, the birds retreat into nooks and remain perfectly still (Temple 1978*c*). The young are estimated to remain in the nest for 8–9 weeks (Staub 1976, Temple 1978*c*). This is longer than expected and may be an over-estimate. Most members of the genus are nest-bound for only about 7 weeks (Forshaw & Cooper 1978, Smith 1979, Low 1980).

Fledging dates were recorded between 14 November and 4 January, most in the second half of December during 1973 to 1975 (Temple 1978*c*). Calculating from laying dates, the range of fledging can be at least from the end of October to the beginning of February or later. When they leave the nest fledglings remain near the entrance "for some time" (Staub 1973*a*). As soon as they can fly the young accompany their parents foraging (Staub 1973*a*). On 27 February 1972 Staub (*in litt.*) saw a female Parakeet and two young feeding on privet *Ligustrum robustum* berries; the young were imitating the adult which was carefully selecting the fruits. Fledgling Parakeets are reputed to remain with their parents, being fed by them, for at least 2–3 months after leaving the nest. Immature birds have been observed being fed by adults as late as March (Temple 1978*c*).

Helpers and disrupters at the nest
The presence of intraspecific 'helpers' at the nests of Echo Parakeets was first observed by Temple (1978*c*) and later by McKelvey (King 1977) and others. The 'helpers' are adult males that are reputed to be active only after the young have hatched (Temple 1978*c*), although Steele (1979) recorded them at a nest with eggs (see below). These additional adult males attempt to feed the adult female and her nestlings but are usually rebuffed by one or both members of the pair. Between 1973 and 1978 most of the observed nests were attended by additional males. Temple (1978*c*) saw 'helpers' at all but one of the unstated number of nests he watched. The number of additional males

was variable; as many as five (Temple 1978c) or six (King 1977) were recorded. After the young fledged, 'helping' waned appreciably (Temple 1978c).

Steele (1979) watched a nest in 1978 that eventually failed due to the disruptive influence of 'helpers':

> Eggs were apparently laid about the the 1st or 2nd of October and for the first week things went well. Then another male made his appearance at the tree. This new male spent his time sitting in the entrance to the nest while the male to the pair was away feeding, invariably up the hill, across Macabé road and into the forest about one quarter mile away. The interloper was often seen entering the nest, possibly trying to feed the female. When the original male returned from his feeding grounds the visiting male would fly to a spot about 200 yards down the valley from the nest and would feed there.
>
> Eventually a third male appeared at the tree and there was considerable conflict among the three males. The next day there was only one male present and this male spent his time sitting in the nest entrance. He fed about 200 yards down the valley. The obvious conclusion was that the interloper had won out over the original male. When the hen emerged to be fed this male would feed her just a little but apparently not enough to satisfy her hunger. She began to chase after him and he would then fly to his feeding grounds often followed by the female. She would usually return to the nest within 10 minutes but after a week or so of this the nest tree was deserted.

'Helpers' at the nest are, in some species, associated with a skewed sex ratio (Skutch 1976). The ratio in the *P. echo* population 1974/5 appeared to be around 5 males to 1 female due to the predominance of males at the nest-sites, but counts at roost-sites showed the ratio to be nearer 15:11 (Temple 1978). A. S. Cheke (*in litt.*) also noted a preponderance of males in the population in late 1973 and Temple's roost counts may be misleading. Since 1979 the sex ratio has been about 3 or 4:1. Captive parrots often succeed in rearing far more males than females (R. Low *in litt.*); if this is reflected in the wild, skewed sex ratios may not be unusual in small populations of parrots.

The male 'helpers' at the nest formed part of a floating population and Temple (1978c) counted 10–15 non-breeding birds during the 1974 breeding season. In subsequent years the population numbers decreased and became even more skewed. During 1976/7 McKelvey felt the entire population was 35–45 with a great preponderance of males. He had seen 18 single unpaired males and only three pairs with five young (King 1977). Historically the sex ratio was probably even since I have traced in museums a total of 15 male and 17 female specimens.

Nest 'helping' in *P. echo* is poorly developed and may be a recent phenomenon. F. Staub (pers. comm.) who watched this species in the 1960s and carefully observed two nests in 1971, never saw this behaviour. With the destruction of many of the Parakeet's foraging areas on or near Plaine Champagne in the early 1970s (see Habitat Destruction, Degradation and Food Availability), and the consequent increase in Parakeet density in the other areas of native vegetation, for a few years there was a population far in excess of the normal carrying capacity of the habitat with a consequently large floating population. The nest helping/disrupting behaviour was, I believe, caused by the large number of displaced non-breeding birds.

Notes on the Ring-necked Parakeet Psittacula krameri and competition with the Echo Parakeet Psittacula echo

The Ring-necked Parakeet was introduced into Mauritius in about 1886 (Chapter 1). Today there is a flourishing population which competes with the Echo Parakeet for nest-sites and quite probably for some food resources. In view of their close taxonomic relationship and similar morphology (see Description) they provide a good example of the competitive relationship between two congeneric species.

Distribution

Soon after Ring-necked Parakeets were introduced they established themselves near Grand Port and by 1916 they had formed a second colony at Pamplemousses (Carié 1916, Guérin 1940–53) where they are still found. Rountree *et al.* (1952) claimed they were confined to the coastal plain in the south and east, Pamplemousses, Alma and Quartier Militaire, although in May 1949 and 1950 G. de Roubillard shot three individuals in Flacq (specimens in Mauritius Institute). R. Newton (1958a) considered *P. krameri* more widespread and common and recorded them in the hill forests at Vacoas and Le Reduit, noting that their movements were influenced by food supplies. Today this parakeet is widespread and may be found within the Black River Gorges where there is a large roost of several hundred. These birds travel out of the gorges and along the base of the north-west-facing scarp to Le Morne and up along the Magenta scarp to Yemen. Other birds fly over the sides of the gorges and I have watched Ring-necked Parakeets flying up and presumably over the north-east-facing slopes of Black River Peak. While standing on the top of this peak (826 m) I watched one fly past and continue on out and over towards La Gaulette and Case Noyale. I suspect that some of the birds that feed on the maize fields of Chamarel may also leave the gorges by this route.

Habitat requirements

Disturbed areas of exotic vegetation are preferred by the Ring-necked Parakeet and I saw them most often in the mixed exotic forests and savannah at the base of mountain ridges and in the plantations at the mouth of the Black River Gorges. They also forage in the mixed stands of exotic trees that flank many of the river and stream valleys in the Gorges and elsewhere. Many authorities claim that they live in close association with man (Forshaw & Cooper 1978, Ali & Ripley 1968–74, vol. 2). On Mauritius they are wary of people and do not usually come close to human habitations or into small back gardens, but they do forage in the Royal Botanical Gardens, Pamplemousses, and in the orchards of Bras d'Eau in the north of the island.

Feeding ecology

The feeding requirements of *P. krameri* are much broader than those of *P. echo*. They "appear to eat every vegetable substance including the bud, the flower, the unripe and ripe fruit, the sown seed, the growing crop, ripening ear and the harvested corn" (Smith 1979). Similar sentiments have been expressed by others familiar with their ways (Ali & Ripley 1968–74, vol. 2 1969, Forshaw & Cooper 1978). On Mauritius their diet is similarly eclectic; Table 29 includes food recorded by observers in other parts of their range, but since these plants commonly occur on Mauritius it seems probable that they are also consumed there.

The birds feed both in trees and on the ground (Smith 1979); *P. echo* never or rarely descends to the ground (A. & E. Newton 1876). Ring-necked Parakeets forage singly or in large flocks. They are a major pest on maize, often appearing in hundreds as soon as the crop is ripe, even though they may rarely be seen in the area at other times. Some birds forage in the Black River Gorges and I have seen them regularly in small groups of about three to eight feeding on the flowers and fruits of Yatis (*Litsea monopetala*) and *Clerodendron serratum*. Single birds feed on *Eucalyptus tereticornis* and *Tabebuia pallida* flowers. The latter flowers are pulled off and held in one foot while the nectar is removed by chewing the calyx.

Breeding biology

The breeding biology of *P. krameri* on Mauritius appears to be similar to that elsewhere in its range (Smith 1975, 1979, Forshaw & Cooper 1978). Eggs are laid probably from June to mid-October (Newton 1958a, pers. obs.).

Competition and interactions between P. krameri and P. echo

The feeding habits of the two species are different (Tables 28 and 29), although their bill lengths are almost the same. *P. echo* is, however, a larger and heavier bird (Chapter 9). Temple (1978c) observed no competition for food but saw both species feeding in the same group of trees. Some of the interactions between the two species are undoubtedly subtle and I suspect that they are ecologically incompatible. They use different areas of the Black River Gorges, suggesting that there is some degree of competitive exclusion. *P. echo* most frequently keeps to the ridges, forested slopes, the head of stream valleys and the native scrub-forest of the plateau. By contrast *P. krameri* is found in disturbed areas and plantations at the mouth of the gorges and along the river and stream valleys. This separation may, in part, be related to the population density of *P. krameri*, since in the early 1970s they were nesting in trees on Macabé Ridge. It is probable that the forests on the ridge are sub-optimal habitat and are used by *P. krameri* only when population densities are high. Temple (1978c) gave some support to this notion since he saw "members of a pair of *P. krameri* that nested on the Macabé ridge fly distances of 4 km or more to feed in forestry plantations in the lowlands".

Interactions between *P. echo* and *P. krameri* are usually passive outside the breeding season, although Temple (1978c) recorded both species pursuing each other. I once saw two Echo Parakeets pursue a single Ring-necked Parakeet (June 1981). The few brief associations that I have observed between the two species have not been overtly aggressive. I have seen a group of parakeets, two *P. echo* and one *P. krameri*, flying together below Black River Viewpoint, although the *P. krameri* soon separated from the others. On another occasion I was watching a group of three Ring-necked Parakeets feeding in a tree at the head of the Grandes Gorges River; they were joined by two Echo Parakeets, which stayed close to them for about

Table 29. *Food taken by the Ring-necked Parakeet* Psittacula krameri

Plant species	Part eaten	Reference
Meliaceae		
Melia azederach L.	Fruits	This study
Anacardiaceae		
Mangifera indica L.	Flower and fruits	This study
Leguminosae		
Acacia sp.	Fruits	This study
Albizzia lebbek Benth.	Fruits	Toor & Ramzam (1974)
Arachis hypogaea L.	Fruits	Ali & Ripley (1968–74)
Cicer arietinum L.	Gram	Ali & Ripley (1968–74)
Erythrina indica DC	Flowers	Ali & Abdulali (1938)
Combretaceae		
Terminalia catappa L.	Fruits	This study
Myrtaceae		
Eucalyptus tereticornis Smith	Flowers	This study
Psidium guajava L.	Fruits	This study
Lythraceae		
Punica granatum L.	Fruits	This study
Rubiaceae		
Coffea arabica L.	Berries	Ali & Ripley (1968–74)
Compositae		
Xanthium sp.	Fruits	Ali & Ripley (1968–74)
Solanaceae		
Capsicum sp.	Chillies	Ali & Ripley (1968–74)
Bignoniaceae		
Tabebuia pallida Hemsl.	Flowers	This study
Verbenaceae		
Clerodendron serratum Spreng.	Flowers and fruits	This study
Lauraceae		
Litsea monopetala (Lour.) C. B. Robins.	Flowers and fruits	This study
Moraceae		
Ficus sp.	Fruits	Forshaw & Cooper (1978)
Casuarinaceae		
Casuarina equisetifolia Forst.	Seeds	Ali & Ripley (1968–74)
Palmae		
Livistona chinensis (Jacq.) R. Br. ex Mart.	Fruits	This study
Graminae		
Zea mays L.	Ripe corn	This study

1 minute before they left, calling in flight. F. N. Steele (pers. comm.) saw a similar interaction in May 1978.

It is tempting to speculate that there has been competitive exclusion between the two species of parakeets. *P. echo* has possibly undergone niche contraction as a result of competition from its more generalised congener. It is probable that the Ring-necked Parakeet is preventing the Echo Parakeet from modifying its feeding ecology and adapting to an increasingly altered forest. The Echo Parakeet's arboreal nature and dependence upon mature forest preclude it from entirely occupying the niche now occupied by *P. krameri*.

Probably the severest form of competition between these parakeets is over nest-sites. Cavities used by *P. echo* have in subsequent years been occupied by *P. krameri*. F. Staub (pers. comm.) knew of one such site in a *Mimusops petiolaris*, and noted (Staub 1976) that *P. krameri* often competes with *P. echo* for nest holes. Temple (1978c) recorded that Echo Parakeets are easily frustrated in defence of their nesting territory and that two pairs "passively relinquished their cavities to the intruders without really engaging in physical defensive behaviour". Temple (1978c) also claimed that between 1973 and 1975 he "documented three instances in which *krameri* displaced *echo* from nesting cavities". Temple *et al.* (1974) noted that two out of seven *P. echo* nest cavities were taken over by Ring-necked Parakeets in 1973. In 1974 five nesting pairs of *P. krameri* were known in the region of Macabé Ridge; but between 1979 and 1982 no pairs of *P. krameri* have been found nesting in, or near, this area, although in 1983 when the *P. krameri* population had noticeably increased a single pair nested in a large tree on the ridge next to the road (P. Ahimaz pers. comm.).

Causes of decline and rarity

In common with the other rare endemic birds of Mauritius the primary reason for the Echo Parakeets' decline is undoubtedly habitat destruction and alteration. There are, however, important secondary causes of rarefaction which are depressing the population and accelerating its extinction.

Habitat destruction and degeneration

The Echo Parakeet is apparently tied to native forests for its food and nesting requirements. The decline in population and contracting distribution of this parakeet is largely mirrored by destruction of the native forest (see Status and Distribution). The reasons for the present rarity and low density in the areas of native vegetation are probably related to its dependence on native plants. The degradation of the forest will continue to dilute its food supply and the birds will have to travel further to find food. The disappearance or decline of key food species may have resulted in the birds suffering from a severe food shortage at certain times of the year.

The dramatic population crash that this species showed in the 1970s is almost certainly due to the destruction of large portions of feeding areas. Between 1971 and 1974 about half (2800 ha) of the *Sideroxylon–Helichrysum* scrub on Plaine Champagne, Les Mares, Kanaka Crater, Grand Bassin and Plaine Paul was cleared for commercial forestry plantations by the Development Works Corporation, funded by the World Bank (Procter & Salm 1975, Temple 1976b, Chapter 1) (see Distribution, Status and Population). There are many field observations which suggest that the Parakeet has suffered from food shortages following the cutting of these upland scrub-lands. The remaining birds forage widely for food, travelling several kilometres to specific feeding areas (see Feeding Ecology). The pattern of breeding success shows similarities with other parrots known to be suffering from impoverished food availability (see Breeding Success 1973–83). At a nest watched in 1974 the incubating female was absent for 32% of the time for periods that averaged 26 minutes (Temple 1978c). This is reminiscent of the behaviour shown by the White-tailed Black Cockatoos studied by Saunders (1982); in an area of poor food supply the female often left the nest to forage for herself, while in areas of better habitat the male was able to meet her food demands. At the Echo Parakeet nest watched by Steele (1979) in 1978 the female would leave her eggs and pursue the male who was not feeding her adequately; eventually she deserted the eggs.

Nest-site availability

The availability of suitable nest cavities for Parakeets is considered by some to be the factor limiting the number of breeding pairs (L. F. Edgerly pers. comm., W. A. Newlands *in litt.*). This is unlikely to be true since the population density has apparently always been low even when there were greater areas of less degraded upland forest available to provide suitable nest-sites. The forests used to, and still do, support several hole-dependent species (*Pteropus subniger, Phaethon lepturus, Acridotheres tristris* and bees *Apis* sp.).

Temple (1976*b*, 1978*c*) suggested that there are adequate cavities available, but Parakeets have difficulty in locating another suitable nest-site if their nest hole is destroyed or taken by a competitor. In 1974 Temple (1978*c*) recorded that two pairs lost their nest holes to competitors and a third lost their eggs to a predator. None of these pairs attempted to find another nest-site despite there being suitable unoccupied tree cavities nearby.

Many of the large old emergent trees used by the Parakeets for nest-sites are damaged by cyclones. Cyclone Gervaise in 1975 destroyed 8 out of 14 known nest trees (Temple 1976*b*) (9 out of 14 according to Temple 1978*c*). Seven of the cavities used in 1974 were lost in Gervaise and in the 1975 breeding season "at least 4 pairs (presumed to be the same birds that nested in 1974) remained in the vicinity of their former nesting cavities and never even attempted to breed" (Temple 1978*c*).

Nest predation by monkeys

A. & E. Newton (1876) implied that the monkey *Macaca fascicularis* was a nest predator of the Parakeet. They noted that the numbers of *P. echo* were failing, but that in the district of Grand Port where the monkey density had been thinned the species was "enjoying a transient prosperity". Further evidence is provided by Temple *et al.* (1974) who recorded one nest destroyed by monkeys. In a discussion of the conservation of this Parakeet Temple (1976*b*) wrote: "Monkeys manage to destroy the eggs or young of a small number of pairs." Temple (1978*c*) recorded eye-witness reports of nest predation on six species including *P. echo*. Later in the same manuscript Temple contradicted this and his earlier reports by stating that he had "no positive evidence of nest predation", but "found two female parakeets with missing tails during the 1973 and 1974 nesting seasons". He suggested that monkeys pulled out the tail feathers while attempting to take eggs from the nest cavities. I suggest that perhaps these tailless females were in moult. I have no direct evidence of monkeys raiding Parakeet nest holes and neither do any of the Mauritian naturalists to whom I have spoken (L. F. Edgerly, A. W. Owadally, F. Staub *et al.*).

Competition with other species

The role of the Ring-necked Parakeets is discussed under Notes on the Ring-necked Parakeet *Psittacula krameri* and competition with the Echo Parakeet *Psittacula echo*. Here I discuss species other than *P. krameri*.

Competition for food resources. Dietary overlap with several native species does occur, e.g. with the Pink Pigeon, Merle, and fruit-bat *Pteropus niger*, although the density of the first two species is so low that they are unlikely to have a significant competitive effect. In the original unmodified forests of Mauritius competition between *P. echo* and *Pteropus niger* (and also perhaps *P. subniger*) would have been negligible due to the greater density of fruiting trees and the nocturnal foraging habits of the bats. However today, with a paucity of fruiting trees in the forest and a great demand for the available fruit, the fruit-bat may have some impact upon the amount of food available to *P. echo*.

The most destructive and important competitor for food is the monkey *Macaca fascicularis*, which frequently rips off all the fruit on a tree before it is ripe. A carpet of leaves, branches, flowers and fruit regularly develops on the ground beneath endemic trees that monkeys have been feeding on. Ripe fruits are rarely produced by some species since the monkeys get them before they are fully developed. *Canarium paniculatum* fruits have never been found by foresters in the native forest, although some have been collected along the Rivière des Papayes on the Henrietta Sugar Estate where monkeys have been eliminated. Ship Rats *Rattus rattus* also damage fruit although the impact of this species on the native forest has yet to be evaluated.

Competition for nest-sites. Cavities suitable for nesting are in great demand by hole-nesting species. The most severe nest competitors that *P. echo* has to cope with are bees, White-tailed Tropic-birds, Ring-necked Parakeets, Common Mynahs and Ship Rats. Artificial nest boxes placed in the forest for *P. echo* have been used by some of these species (Table 30). During a total of forty nest-box years, 22 (55%) were occupied by alien animals, most (82%) by bees and wasps. Ship Rats and Mynahs occupied the other boxes, but since the boxes were checked only from the ground these species were probably under-recorded. Some boxes were used sequentially by different hole-dwelling animals: nest-box 7 was occupied by bees in 1979 and

1980, was empty in 1981 and was used by Mynahs in 1982; box 15 was occupied by Mynahs in 1979 and bees in 1981.

Temple (1978c) saw Tropic-birds and bees take possession of former Parakeet nest-holes. He also saw Common Mynahs displace Parakeets from nesting cavities on two occasions; one of these cavities had been used for four successive years by *P. echo* before the Mynahs took possession. Mynahs appear to be competitively superior to Parakeets (Temple 1978c) but are a serious threat only when the forest has been opened up by roads and clearings; they are rare in undisturbed areas of native forest (Watling 1975, pers. obs.).

At the end of October 1980 Y. Mungroo (pers. comm.) watched a pair of Echo Parakeets inspecting a nest cavity and later copulating. Two days later the birds were absent from the area and a pair of Mynahs had taken possession of the tree hole. The following year a similar but more elaborate sequence of events was again observed in the same place. Mungroo watched the interactions between a pair of *P. echo* and a pair of Mynahs on 29 September 1981:

10.22   Male *P. echo* arrives and perches in a tree adjacent to the 'nest tree'.

10.32   Female *P. echo* comes out of the nest hole and chases away a pair of mynahs that were in the tree. The mynahs charge back. The male parakeet flies to the aid of the female, but the mynahs chase him away.

10.34   Female parakeet re-enters the nest hole. The male parakeet stays in the vicinity, perching or flying around but avoiding close contact with the mynahs.

10.40   Female parakeet comes out of the nest hole. A mynah lands in the tree but is chased away by the female. A second mynah attempts to land in the tree but is also chased away.

10.43   Both mynahs try and land in the tree but are chased away by the female.

10.45   Female parakeet re-enters nest hole.

10.46   Two male *P. echo* seen flying along the ridge but when they approach the parakeet nesting territory they are chased by a single mynah. The parakeets offer no resistance.

10.55   No mynahs to be seen in the area.

On subsequent days no Parakeets were seen in the area but a pair of Mynahs were occupying the nest-hole.

Temple (1978c) discovered evidence of rats having gnawed the entrance to one Parakeet nesting cavity, and found a rat nest in another cavity apparently never used by Parakeets.

Cyclone-related mortality

Cyclones probably kill birds directly by damaging the hollow limbs in which they shelter, or by post-cyclone starvation. Cyclones strip nearly all the fruit from trees although many trees flower soon after a cyclone has passed, and food would soon be available again. Fol-

Table 30. *Occupancy of artificial nest boxes erected for Echo Parakeets on Macabé Ridge*

Date	Orientation	No.	Empty	Bees	Wasps	Mynahs	Ship Rats	Not checked	Damaged
								Contents	
July 1979	Vertical	19	8	2	—	1	1	6	1
June 1981	Vertical	23	7	13	—	2	—	—	1
June 1981	Horizontal	6	3	2	1	—	—	—	—

Note:
Twenty vertical boxes were erected in the forest during 1977. In August and September 1978 all were checked, and bees removed from several of them (Steele 1979). Seventeen further boxes, six of them horizontal, were added in October 1979. Observations on occupants since 1978 have been from the ground; the boxes have not been climbed to, nor any occupants removed. Some recorded as empty might therefore have held undetected occupants. By July 1979 one of the original 20 boxes was missing, and a further seven had fallen or were missing by June 1981.

lowing cyclone Gervaise in 1975 Temple (1978c) felt certain that some Parakeets had been killed and Newlands (Feb. 1975) noted that they were "wandering further than usual in search of food", although in March he noted "this species seems to have recovered rapidly". After cyclone Claudette I noted that Parakeets were seen less frequently and that they were wandering further in search of food, and in the year following the cyclone the number of Parakeet observations per hour was more than halved (Fig. 16).

Disease

The population crash of *P. echo* in the 1970s has been attributed to disease (S. A. Temple pers. comm. to R. F. Pasquier) although there is no evidence to support this. Temple (1978c) found that droppings from wild birds were free of evidence of parasites. Four nestling Parakeets taken into captivity were free of external parasites and examination of droppings and blood smears for parasites proved negative. A faecal sample that I collected in August 1979 from an adult male showed no evidence of parasites. Bacteriological culture subsequently yielded a *Proteus* sp., *Escherichia coli* and *Enterococcus* but these could well be part of the normal gut flora of this species (J. E. Cooper *in litt.*).

Captive birds have died from a variety of causes. A young male (5 months old) died on 10 April 1975 from visceral gout. Bacteriological culture of the intestinal contents yielded a *Citrobacter* sp., and of lung, liver and kidney, *Klebsiella pneumoniae* and *K. ozanae*; these are probably part of the usual bacterial flora of the bird (J. E. Cooper *in litt.*). A year-old female died on 25 December 1975 from traumatic injuries including bone fractures after the bird panicked in the aviary (J. E. Cooper *in litt.*). On 4 June 1976 a female died due to a sudden and severe infection of *Staphylococcus aureus* and *E. coli* (S. D. McKelvey in litt.). The fourth parakeet, a female, died on 14 July 1975. The only lesions found *post mortem* were some haemorrhagic and inflammatory foci in the kidney (J. E. Cooper *in litt.*).

Human persecution

Persecution has played only a very minor role in the decline of this species. Small numbers have been collected for museums or for trophies. In 1972 two birds were shot by a hunter who had them mounted. Two years later when Temple (1978c) examined a previously well-used nesting cavity, not far from a

forest road, he discovered two small-calibre bullets lodged in the tree near the cavity entrance. Guérin (1940–53) claimed that Parakeets were commonly killed for food, being regarded as a delicacy by the island's rich inhabitants. In the eighteenth century the extinct grey parrot *Lophopsittacus bensoni* was considered better eating (Chapter 1). Very few Parakeets have been taken for the avicultural or pet trade; the species is unknown to modern aviculturalists (e.g. Low 1980). Seth-Smith (1903) knew of one that belonged to a Dr Russ in Germany which "became extremely tame and affectionate, and learnt to speak well".

Conservation

The conservation of the Echo Parakeet has been the subject of much discussion and has been covered in several documents (Temple 1976a, Steele 1979, Jones 1980b, 1980c). Several approaches have been tried but none has been successful.

In the belief that nest-sites were limiting the population, nest-boxes have been placed in the forest. Two were erected in 1974 (30 cm × 30 cm × 75 cm with a 12 cm diameter entrance hole); one was prospected but not used (Temple 1978c). In 1975 three were placed in trees on Macabé Ridge (Newlands *in litt.*) but they were removed at the end of the year before Newlands left the country since it was thought they would encourage nest robbers if left unobserved and unguarded. Trefry (*in litt.*) placed 20 nest-boxes in the forest during August 1977 (30 cm × 30 cm × 72 cm with a 9 cm entrance hole) and in October 1979, 17 boxes (23 cm × 23 cm × 80 cm with an 8 cm diameter entrance hole) were placed on the south-west side of Macabé Ridge. Parakeets have failed to use any of these, although some have been occupied by other hole-dwelling and potentially competing species (Table 30).

A nest-site used in 1978 was 'monkey-proofed' by covering the main trunk with a thick layer of grease to a height of several feet. The nest failed but not due to monkey interference (Steele 1979).

A feeding platform was built on Macabé Ridge and baited with cultivated fruits, sunflower seeds and maize (June and July 1979). It was hoped that Echo Parakeets would be attracted to it and the population could be provided with additional food. Unfortunately no Parakeets visited the platform and it is now clear that the baiting was done at the wrong time of year and in the wrong manner. Chillies *Capsicum* sp. threaded onto cotton and suspended from favoured feeding trees were sampled by Parakeet(s) in August 1984. Supplemental feeding during the months of greatest food shortage (September–November) could well encourage some of the remaining birds to breed.

The control of nest competitors and predators around nest-sites has been suggested (Temple 1976*b*, King 1978–9) but never been implemented.

The translocation of Parakeets to Réunion has been suggested (Cheke 1974*b*, 1978*a*, Temple 1976*b*, 1981). The rationale behind this is ecologically sound since Réunion has more forest, which is less degraded than that remaining on Mauritius, and does not have *P. krameri*. It is unfortunately too late to consider such a move since the population is now too low to risk such a procedure, even if the remaining birds could be caught.

Captive breeding of Echo Parakeets has been attempted on two occasions. Two young birds were taken from the wild in 1974 and another two in 1975. None of these birds lived to maturity (see Disease).

A resolution passed at the ICBP World Conference on Parrots held in St. Lucia in 1980 recommended that all the remaining Echo Parakeets be captured for captive breeding (Pasquier 1982). It was further suggested that they be sent to a collection specialising in the breeding of parrots (Pasquier *in litt.*). The adults have proved difficult to capture and, on Mauritius, it was considered that the best chance for the future survival of the species would be to take eggs or young for a captive breeding programme should the wild birds nest.

Captive breeding is still regarded as the best chance for saving *P. echo*. At the end of 1983 Vogelpark, Walsrode, West Germany, financed a 3-month study and the attempted, but in the end unsuccessful, capture of Parakeets (Ahimaz 1984). During late July to early September 1984 a further unsuccessful attempt to capture the birds was made by Don Merton of the New Zealand Wildlife Service financed by the International Council for Bird Preservation and the Jersey Wildlife Preservation Trust.

Acknowledgements

I arrived on Mauritius in January 1979 to work on the International Council for Bird Preservation's (ICBP) bird and bat conservation project (now called the Mauritius Wildlife Research and Conservation Programme). Since then I have worked with the Mauritius Forestry Service and am grateful for the help and support they have given. The Conservator of Forests, A. Wahab Owadally, has been a close colleague helping in all aspects of the work. When I arrived the then Assistant Conservator of Forests, Tony Gardner, and his wife Jean made me welcome and introduced me to the island and its wildlife. Many Mauritians have given freely of their knowledge and time to help in many ways and I have particular pleasure in thanking G. A. d'Argent, the late L. F. Edgerley, Joseph Guého, J. A. Lalouette, J.-F. Martial, France Staub, M. d'Unienville and J.-M. Vinson. R. E. Vaughan, whose pioneer work on plant communities is a major cornerstone for all those working on Mascarene ecology, has provided much useful advice and has freely loaned valuable references.

Visiting and corresponding biologists have provided valuable guidance and support, especially T. J. Cade, A. S. Cheke, D. Goodwin, J. F. M. Horne, R. Low, W. A. Newlands, I. Newton, G. A. Smith, F. N. Steele and S. McKeown. The staff of the Jersey Wildlife Preservation Trust (JWPT) have been a constant source of advice and encouragement and special thanks must be given to John Hartley and Lee and Gerald Durrell. John E. Cooper from the Royal College of Surgeons (England) has been the veterinary advisor to the conservation project since its inception and has helped in many aspects of the work.

Much of the work reported here would have been impossible without the enthusiastic and dedicated assistance of the following who have helped with the care of the captive animals and with fieldwork, often under difficult conditions: Julian P. L. Davis (May–October 1979), Richard E. Lewis (October 1979–January 1980), Fabien Kendall (September 1981–June 1982), Harriet Rathborn (1982), Wendy Strahm (January 1982–continuing), Willard Heck (October–December 1982), Preston Ahimaz (October–December 1983), and David M. Todd (January 1984–continuing).

Government personnel who have been of considerable help are Yousoof Mungroo (Scientific Officer, Wildlife) and Inderdeo Khelawon (Forest Officer, Black River).

Several conservation organisations have supported my work: ICBP (US national section), World Wildlife Fund (WWF) (US appeal) and the New York Zoological Society (1979–March 1981). From April 1981 to March 1983 funding came from WWF (International) while still under ICBP (US) administration. Since April 1983 the management of the project has been taken over by the ICBP (International) and funding has come from the JWPT, and Ian Pollard who provided help at a difficult time. Funding was provided to Willard Heck by the Peregrine Fund and to Preston Ahimaz by the W. W. Brehm Fund. My friends and colleagues at the ICBP have taken a special interest in the Mascarenes and have worked tirelessly to secure funding and provide help. In particular I need to mention Warren King, Roger Pasquier, Christoph Imboden, Paul Goriup and Jane Fenton.

Writing the various drafts of the manuscript was tedious but was made easier by the good humour and constructive comments of Paul F. Brain, Anthony S. Cheke, Nick C. Fox, Wendy A. Strahm and David M. Todd. Harriet Rathborn helped bring order to the first draft of the manuscript and typed it with characteristic efficiency.

6

The ecology of the surviving native land-birds of Réunion

Part III

THE SURVIVING NATIVE BIRDS OF RÉUNION AND RODRIGUES

A.S.CHEKE

Introduction

Réunion is the largest and least ecologically disturbed of the Mascarene Islands, but has nevertheless lost more of its original avifauna than Mauritius (Chapter 1). Only the Réunion Cuckoo-shrike *Coracina newtoni* is rare and at risk, and for that reason more time was devoted to it than to the other species. Gill (1971*a*, 1973*b*) has considered the ecology of the two native white-eyes *Zosterops* spp., and Clouet (1976, 1978) has studied the harrier *Circus maillardi*, but little other ecological work has been published since that of Pollen (Schlegel & Pollen 1868). Berlioz (1946) compiled a check-list, which Milon (1951) corrected and augmented, and more recently an identification guide has appeared (Barré & Barau 1982). Barré's important paper (1983) appeared too late for full consideration here.

I visited Réunion five times during 1973–5, for a total of 12 weeks: 29 October to 12 November 1973, 25 April to 26 May, 7–27 August (with my brother, Dr R. A. Cheke (RAC)), and 1–26 November (with Ms R. E. Ashcroft (REA)) 1974, with a very brief stopover, 12–14 February 1975. I made a further short visit from 10 to 12 October 1978. Dr A. W. Diamond (AWD) was on the island in January 1975, and has kindly allowed me to use his notes on his visit to the Plaine des Chicots; I have also been able to use information from M. Harry Gruchet (HG), curator of the natural history museum in St.-Denis, M. Charles Armand Barau (CAB), M. Théophane Bègue (TB), gamekeeper at the Plaine des Chicots, Dr M. de L. Brooke (MdeLB), M. Nicolas

Barré (NB), Dr Michel Clouet (MC), Mr M. J. E. Coode (MJEC), M. Emile Hugot (EH), M. Bertrand Trolliet (BT) and Messrs A. Forbes-Watson, S. Keith and D. A. Turner (cited as AFW). Dr F. B. Gill (FBG) has kindly passed me typescript notes on his visit in 1964 (MS 1964), and his field notebook in which he recorded details of specimens collected in 1967; the 1964 typescript has been supplemented by details from the Smithsonian Institution accessions book, where the specimens collected during the International Indian Ocean Expedition are deposited. These colleagues are referred to by their initials (above). Mrs J. F. M. Horne visited Réunion in November 1973 and September and December 1974 to record songs and calls (Chapter 3).

The island

Réunion is a volcanic island of 2512 km², 665 km south-east of the nearest point on the coast of Madagascar and 164 km west-south-west of Mauritius. It is an oval in shape (Map 1), the longer axis, 70 km in length, lying north-west to south-east; at its widest the island is 50 km across. The relief is remarkable, the highest point being 3069 m above sea level, the volcanic massifs dissected by deep erosion ravines with cliffs commonly over 1000 m high. Geologically there are two overlapping shield volcanos: an older extinct one, centred on the Piton des Neiges, much dissected into gorges and erosion basins surrounded by ramparts known as cirques, and the still active Piton de la

Map 1. The island of Réunion.

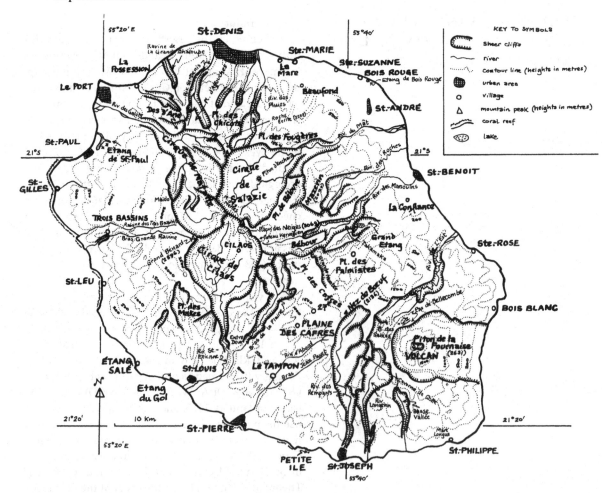

Fournaise ('the Volcan') which is still an almost complete dome, cut into by only two major ravines in the older part. The coastal plain is narrow, seldom more than 4 km wide, much less around the fringe of the Volcan; in the north-west there is a stretch of tall sea cliffs for about 10 km. Around most of the island the land shelves steeply into the sea with no fringing reef, but there is a narrow lagoon up to a few hundred metres wide along parts of the west and south-west coasts.

An important difference from the other Mascarene Islands is the large proportion of the island which is at a considerable altitude; 61% of the surface is above 1000 m, a height not reached by even the highest point in Mauritius (2717 ft; 827 m). The altitude and the ruggedness of the terrain has meant that large areas of upland vegetation have remained more or less untouched by man's influence, but the relatively small area of lowland forest has been all but totally destroyed, together with the species that inhabited it (Chapter 1).

Réunion is politically an overseas *département* of France. The standard work on the island is Defos du Rau (1960), from which I have taken geographical information, with additional geological data from Upton & Wadsworth (1966). Anon (1962) is a useful compendium of geographical information in map form.

The climate

Like the other Mascarenes, Réunion's climate is dominated in winter by the South-east Trades and in summer by the western Indian Ocean cyclone system. Due perhaps to its larger size, and certainly to its much greater height, there is much more contrast between the windward and leeward sides of the island than in Mauritius or Rodrigues. Data for this section are from cyclostyled maps and leaflets produced by the Météorologie Nationale, Service de la Réunion, which are more up to date than information published in standard works. Some details are personal communications from M. Lenombre; data for individual years are contained in the *Bulletin Climatologique*, published annually by the Service. Cadet (1977) and Robert (1980) discussed the island's climatology in greater detail.

The distribution and frequency of annual rainfall is shown in Maps 2 and 3, monthly figures for selected places being given in Fig. 1. The rainfall contours in Map 2, and isotherms in Map 4, are approximate due to the shortage of recording stations away from the coast; the rainfall at the Plaine des Chicots, home of the Réunion Cuckoo-shrike, is certainly higher than the map shows. Prodigious falls are often recorded during cyclones; over a metre in 24 hours is not

Map 3. Distribution of number of rainy days in a year in Réunion (1951–69). (After cyclostyled map by the Météorologie Nationale, Service de la Réunion.)

Map 2. Distribution of annual rainfall in Réunion (1951–74). Isohyets in metres. (After Moutou 1980.)

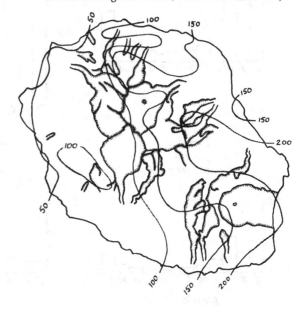

unusual, and the record is 1.84 m in 24 hours in January 1966. Connected with the high rainfall on the mountains and on the windward side is extensive cloud cover, almost invariable in the afternoons but sometimes persisting for days; the cloud-base is usually around 1500 m, so many upland areas are shrouded in mist, only the highest points (over about 2500 m) remaining clear. This massive cloudbank reduces afternoon sunshine considerably even along the dry east coast, where the sky directly overhead is usually clear. In the three cirques the cloud forms within the enclosure formed by the rock walls and sweeps powerfully up them in late afternoon, taking a 'fountain' of cloud up well beyond the (often clear) rim of the encircling wall. This updraught is used by the endemic Barau's Petrel *Pterodroma baraui* to save energy in reaching its nesting places (Brooke 1978).

Temperature isotherms are shown on Map 4, and details of monthly variation at selected sites in Fig. 2. Temperature varies fairly regularly with altitude; around the coast 35 °C is occasionally reached in summer, while frosts are regular on winter nights above about 1000 m; temperatures down to −5 °C are frequently recorded at the hunting lodge at the Plaine des Chicots (1834 m) (T. Bègue pers. comm.), and the effects can be seen as a severe burning of young leaves on the trees in this area. Snow falls very occasionally on the summit of the Piton des Neiges (3069 m). More information on temperatures at high altitudes is given by Cadet (1974). The hottest month of the year is February, and the coolest August.

Cyclones are less frequent than in the other two Mascarene Islands; the last really severe one was in 1948 (Milon 1951). The steep topography and the very

Fig. 1. Monthly rainfall at four selected stations. (Data from cyclostyled tables issued by the Météorologie Nationale, Service de la Réunion, 1973 edition; dates and number of years not given.)

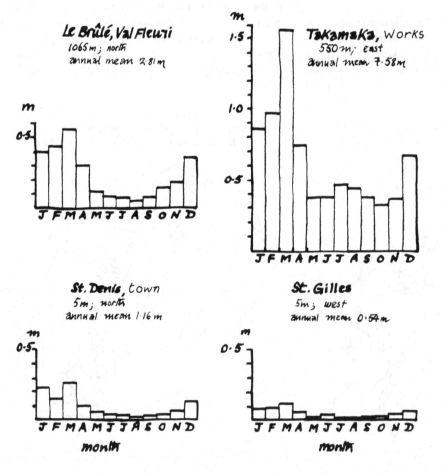

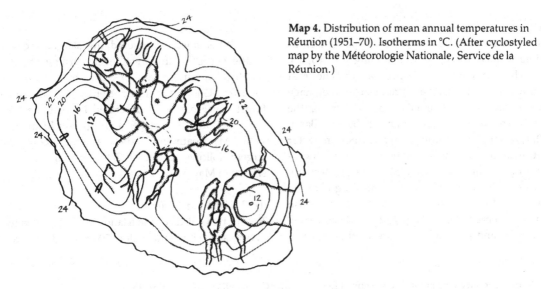

Map 4. Distribution of mean annual temperatures in Réunion (1951–70). Isotherms in °C. (After cyclostyled map by the Météorologie Nationale, Service de la Réunion.)

Fig. 2. Monthly temperatures at four selected localities. Graphs show mean, mean maximum, mean minimum, and the high and low extremes recorded for each month. (Data from manuscript records of the Météorologie Nationale, Service de la Réunion (M. Malick *in litt.*); Bellecombe data from their *Bulletin Climatologique, année 1973*)

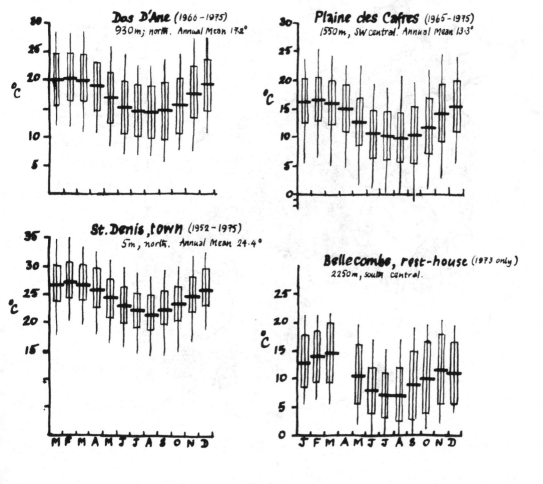

high rainfall during cyclones make for very powerful run-off which can cause major landslips as well as flooding. Hyacinthe, an extremely long wet tropical storm in January 1980 (e.g. 1.74 m of rain in 24 hours at Grand Ilet, Cilaos) caused enormous erosion damage (*Bull Climatol.* Jan. 1980), killed up to 50% of the endemic birds and destroyed all nests (Barré & Barau 1982). Cyclones have clearly been an important factor in the evolution of the vegetation and the physiognomy of the forest (Rivals 1951). The greatest frequency of cyclones is in January, coinciding with the latter part of the birds' breeding season; they are also frequent in February and March but, apart from a few in December and April, are very rare at other times.

The forests

When it was discovered, the island was covered to the shores with forest, even earning the name 'English Forrest' [*sic*] amongst British navigators (Tatton 1625) (Chapter 1). The present state of the island's forest cover has been described by Antoine (1973), Baumer (1981), Cadet (1974, 1977) and Rivals (1952, 1968). My treatment of the major vegetation zones largely follows Cadet (1977); their distribution is shown in Map 5.

The lowlands (up to 500 m)

Round two-thirds of the island, from St.-Denis to Les Avirons, the lowlands below 500 m are dominated by

Map 5. The surviving natural vegetation of Réunion, after Cadet (1977). The dividing line between tropical and subtropical rain forest is somewhat arbitrary as there is a fairly wide zone of intergradation; both types are known locally by the term *bois de couleurs* (mixed evergreen forest). *Tamarin* (*Acacia heterophylla*) forest is generally interpenetrated by *bois de couleurs* formations.

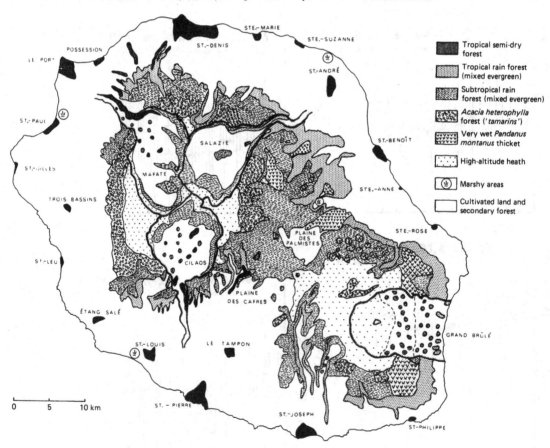

sugar cane *Saccharum officinarum* cultivation, except for the area of young lavas between Ste.-Rose and St. Philippe where there is forest. On the dry west coast a belt of rough grazing occupies most of the land under 500 m.

Amid the sugar in the east and south are pockets of other crops, principally maize *Zea mays*, wooded ravines (mostly exotic species, notably Cinnamon *Cinnamomum camphora*, Mango *Mangifera indica*, Rose-apple *Eugenia* (*Syzygium*) *jambos*, *Litsea chinensis*, *Schinus terebrinthifolius*, bamboos etc.), and villages with gardens (many fruit trees). There is a mere and reedbed (of *Typha javanica*) at Bois Rouge (Cambuston), and large expanses of gravel, scrub and trickles of water where the large rivers reach the coast; these fill their beds only during cyclones.

On the young lavas of the south-east corner there is forest in various stages of succession down to the coast, except for unvegetated places where the eruptions of 1961, 1976 and 1977 (Cadet 1977, Tricot 1978) broke through to the sea. Much of this forest is native; Cadet (1973, 1977) has described the development of the indigenous forest of *bois de couleurs* ('wood of many colours') on young lavas. Over quite large areas the Office Nationale des Forêts (ONF) is also practising so-called 'natural regeneration', which consists in practice of raising saplings of selected native timber trees after clear-felling (Miguet 1980). The principal trees favoured are *Elaeodendron orientale*, *Mimusops maxima*, *Labourdonnaisia calophylloides* and *Terminalia bentzoe*. Although a few hectares of tall forest have been set aside as 'biological reserves' (Baumer 1981) most of the lowland forest is earmarked by the ONF for this form of regeneration (Moulin & Miguet 1968), thus removing the island's last surviving area of this rich and diverse habitat. The dominant trees in the native lowland rain forest include the four mentioned above, and also *Cordemoya* (= *Mallotus*) *integrifolia*, *Calophyllum tacamahaca*, *Homalium paniculatum*, *Eugenia* spp., *Diospyros melania*, *Nuxia verticillata*, *Ochrosia borbonica* and *Sideroxylon majus* interspersed with tall palms *Dictyosperma album* and *Acanthophoenix rubra*, which are now mostly gone, poached for the palm cabbage, a popular and expensive delicacy. The mature forest extends from Tremblet to Basse Vallée, but most of it below 300 m is badly degraded or cleared. Along the foreshore itself and a little inland there are plantations of *Casuarina equisetifolia* and *Pandanus utilis*.

The west coast is very different. Between St.-Pierre and La Possession there is a coastal strip, widening northwards, in which there is little cultivation, the very dry stony land being occupied by scrub and rough grazing; the scrub is dominated by *Pithecellobium dulce* and *Schinus terebenthifolius*, *Leucena glauca* and *Furcraea foetida*. A scrubby woodland of the same species composition is found in the ravines reaching the coast between St.-Leu and St.-Gilles. Swampy land at St.-Paul and St.-Louis (Etang du Gol), now partly drained, is under sugar, but areas of open water exist in both places, with reedbeds of *Typha javanica* and, at the Etang de St.-Paul, *Cyperus papyrus*. At Etang Salé and along the coast south of St.-Gilles are extensive plantations of *Casuarina equisetifolia* and at Etang Salé, other exotics planted to stabilise dunes on beach-sand. The only visible relics of the original vegetation are occasional tall *Latania commersonii* palms in copses and gardens. The bed of the aptly named Rivière des Galets ('river of boulders') is a wide stony area virtually free of vegetation – over time this river has built up the barren shingle spur of the Pointe des Galets.

Between La Possession and St.-Denis sea cliffs up to 200 m high line the shore, broken by the mouths of the Ravine de la Grande Chaloupe and the Ravine Jacques. The lower slopes of the high land behind the cliffs are cultivated, mainly with maize and sugar. The Ravine de la Grande Chaloupe preserved the best-surviving relics of the semi-dry tropical forest that was formerly an important vegetation zone on the leeward (west) side of the island.

The uplands (500–1500 m)
From 500 m to the average cloud-base (around 1500 m), there is an almost complete band of mixed evergreen forest (*bois de couleurs*) all the way round the island. However, below about 1000 m the forest on the windward side is largely secondary and heavily invaded by Rose-apple and Strawberry Guava *Psidium cattleianum*; on the leeward side this level is the sugar cane zone with, in the upper part, food crops, and the essential oil crop Scented Geranium *Pelargonium graveolens*. The forest is interrupted over several kilometres at the Plaine des Cafres by cultivation, and, higher up, by pasture. Over much of the slopes of the Grand Bénard the mixed forest has been largely cleared for cultivation, but *Acacia heterophylla* forest survives higher up.

The lowland tropical forest already described grades into the subtropical upland mixed forest dominated by Sterculiaceae (*Dombeya* spp.) at about 700 m in the east. In the west the 'lowland' forest is shifted uphill by the dry climate to a few pockets between 700 and 1100 m, notably the Plaine des Makes and Grand Montagne. The sterculiaceous forest, so characteristic of the Réunion uplands, is rich and varied, differing markedly in composition depending on slope, aspect and rainfall; in sheltered places a tall forest develops (Petite Plaine, Plaine d'Affouches), while stunted shrubs may be all that can survive on exposed or precipitous places (Grand Montée, cliff below Coteau Kerveguen). One of the features of Réunion forests is that many species of tree (and tree-ferns *Cyathea* spp.) occur in high proportions at all levels from near the coast to the tree-line. Important amongst these are: *Antirrhaea verticillata*, *Aphloia theiformis*, strangling figs *Ficus* spp., *Molinaea alternifolia*, *Nuxia verticillata* and *Weinmannia tinctoria*. The figs are indicative of well-developed forest, while more marginal areas are characterised by the shrub *Forgesia borbonica*. The mixed forests grade into the next type of vegetation.

In the wetter parts of the windward side, on badly drained slopes, a cloud forest in a stage of arrested succession (Cadet 1977) may occur as low as about 800 m. This is a dwarf forest rich in tree-ferns *Cyathea glauca*, *Pandanus montanus*, *Acanthophoenix crinita* (much poached), *Phillipia montana*, with rather few real trees, and with *Machaerina iridifolia*, *Sticherus* (= *Gleichenia sens. lat.*) *flagellans* and *Sphagnum* as characteristic ground-cover.

Above 1200 m on the main massif, and on well-drained lavas protected from the prevailing winds on the Volcan, there are zones dominated by a large endemic tree, the *tamarin des hauts* (*Acacia heterophylla*), which contrasts in its height and often huge girth with the mixed evergreens with which it grows. There is some controversy over the status of the *tamarin* in the forest succession. Under natural *tamarin* forest, in places with an almost impenetrable understorey of the endemic bamboo *Nastus borbonicus*, there is very little regeneration, whereas after a fire, or a clearance, the seeds lying in the soil germinate in thousands (Rebeyrol 1966). This suggests that the *tamarin* forest is a fire- or catastrophe-climax, perhaps maintained naturally by volcanic eruptions (causing fires) and cyclones (Rivals 1952, Cadet 1977, pers. obs.). Cadet (1977)

neatly related the soil and forest types on the Plaine des Chicots to the age of the lavas there, showing that the *tamarin* grows naturally only on the most recent lavas. The other view, held by foresters in the ONF, and current since the 1880s (Miguet 1957, 1973, 1980), is that the *tamarin* is the climax, and in principle an invader of mixed evergreen forest, but that in the last century or so the forests of old *tamarins* have somehow become 'decadent' and fail to regenerate so that man needs to step in to 'restore' the *tamarin* to its 'natural dominance' by artificial management (Miguet 1957). Given the remarkable and extended enforced dormancy, under standing forest, of *tamarin* but not of mixed evergreen seeds, which is not disputed by the foresters, it is difficult to see how the *tamarin* could ever have regenerated to any extent within its own stands. The physiological reasons for the seed dormancy are not understood (Cadet 1977), but it is clear that light and soil aeration play an important part in its interruption (Rebeyrol 1966, Cadet 1977; cf. also the ecology of the very similar *A. koa* in Hawaii (Scowcroft & Wood 1976)).

This question has an important bearing on the management of forests for bird conservation, in particular of the Plaine des Chicots, home of the endangered Cuckoo-shrike. At present the ONF regards this area as degenerate *tamarin* forest in need of rejuvenation, whereas Rivals, Cadet and I regard it as in an advanced stage of natural succession from *tamarin* forest to mixed evergreen – a difference which strongly colours views on how the area should be managed in the long term. A complicating factor is the deer *Cervus timorensis*, which are overstocked and eat regenerating trees of most species (see below under Conservation: also Cheke 1975f, 1976, Moutou 1979, Barré & Barau 1982).

In the same altitude zone there are also plantations of introduced trees, principally *Cryptomeria japonica*, and in some places woods of naturally regenerating exotics other than the invasive species mentioned already. At the upper end of the zone there are managed forests of *tamarins*, covering large areas of Bélouve and around the Maïdo on the slopes of the Grand Bénard. *Tamarins* cast a weak shade and plantations are often heavily invaded by an introduced climbing bramble, the *vigne marronne Rubus mollucanus*, which requires considerable maintenance to keep at bay (Miguet 1980). Elsewhere open ground, especially on landslips, is colonised by this bramble, preventing a natural succession back to forest; some forest areas, such as the mixed *tamarin* and mixed evergreen forest at the west end of the Plaine des Fougères, are also heavily penetrated, and many trees die as a result.

Many introduced plants, some of them very pernicious from the point of view of preserving the native vegetation, are much used by endemic birds. Strawberry Guava fruits are fed on by Merles *Hypsipetes borbonicus*, and flowers of several other species provide nectar for the white-eyes *Zosterops* spp. *Fuchsia magellanica* ('*F. coccinea*'; Gill 1971*a*) occurs very commonly as a non-pathogenic invader at the Plaine des Palmistes and Bébour, and is evidently a very important source of nectar for the Olive White-eye *Z. olivaceus* in that area.

The highlands (1500 m upwards)
The zone from about 1500 m to 2500 m is within the clouds most afternoons of the year, although the average cloud-base varies seasonally and in different locations, being lower on the sheltered west than on the windward side; at some places, such as the Plaine des Chicots, there is often a clear zone at 1600–1800 m between layers of cloud.

In the west, from the middle of the Plaine des Fougères round to the Rivière Langevin, a dwarf cloud forest, similar to that already described but dominated by heather *Philippia montana* and with less *Pandanus*, extends upwards to about 2000 m in places (Mazerin, Plaine des Salazes, Volcan (south)), petering out lower (around 1800 m) elsewhere. In the north and east, and in some well-drained areas around the Volcan, the *tamarin* forest extends to the tree-line, 1800 m at the Plaine des Chicots, but over 2000 m in the east. At these altitudes the small tree *Sophora denudata* is sometimes an important component of the vegetation. In the Cirque de Cilaos there is no cloud-forest zone, and the mixed evergreen forest merges gradually into the heath vegetation at around 2100 m.

Above the tree-line is a wide zone dominated by *Philippia montana* and *P. galioides*, often associated with *Senecio hubertia*, *Phylica leucocephala* and *Stoebe passerinoides*. Just above or within the forest *Philippia* becomes very large, reaching 6–8 m, but the height of the plants falls off rapidly with increasing altitude, becoming a low heath under a metre high by 2200 m in

most areas, though bushes over 2 m high persist to 2500 m in sheltered places. Around 2700 m *Philippia* gives way to *Stoebe* which in turn fades to only occasional plants above 2800 m. At the summit of the Piton des Neiges there are only a few grasses growing very sparsely out of the pumice-like rock.

Distribution of native birds in relation to vegetation
This section is a summary of the distribution of the native forest passerines; details of their ecology are deferred to the species accounts.

In the man-made environments round the coast only one species of native bird occurs regularly: the Grey White-eye *Zosterops borbonicus*. Anywhere there is scrub, gardens or wooded stream-edges etc. the bird can be found, though it is scarce in the dry scrub of the eastern coastal strip, and may be absent from some places (e.g. the slopes of La Saline, and all cane-fields); it is, however, common in the *Casuarina* plantations of the same coast.

In forest, even if secondary and rich in introduced species, all the five common native birds are found: Merle, Flycatcher *Terpsiphone bourbonnensis*, Stonechat *Saxicola tectes* and the two white-eyes. However the total bird density and the relative proportions of species vary a great deal between different types of forest and with altitude.

As my visits to Réunion were irregular, and my sorties to the various parts of the island (apart from the Plaine des Chicots) were exploratory rather than systematic, my observations on bird density are necessarily subjective. However, certain patterns have emerged from the data (Appendix 1), and are discussed as a basis for further studies (see also Barré 1983).

The richest habitat, in terms of bird density, was the mixed evergreen forest, though a substantial admixture of *tamarins* or *Philippia* apparently made little difference. Within a forest type, altitude seemed to make little difference to the overall density of birds, although the species composition varied. At low altitudes, up to 600 m, Grey White-eyes were particularly common and Olive White-eyes and Stonechats scarce or absent; with increasing altitude Grey White-eye density declined, the decrease being compensated by the high density of the othe two. Flycatchers were most common at low altitudes, while Merles were generally found higher up, though this probably reflects hunting pressure on Merles rather than their natural distribution. (The tremendous drop in Merle numbers since the early 1800s is described below in the species account).

The overall picture of bird density in the mixed forest was complicated by a well-marked vertical migration shown by the Merle and the Olive White-eye. Both descended to lower elevations after the breeding season (January), returning to the uplands in winter (May to July, depending on locality). The Merles are said to follow the fruiting of the Strawberry Guava in different places and elevations, returning to higher parts when fruiting is over, but this piece of folk-lore is an oversimplification (see below). The Olive White-eyes were probably forced down by lack of suitable nectar-flowers in late summer. The movements of these two species greatly altered the usual density and relative frequency of species: the enormous numbers of Olive White-eye at the Ilet à Guillaume in early May 1974, outnumbering Greys by at least five to one, are an example (even the Merles were commoner than Grey White-eyes on that occasion). At the same time there were rather few of both on the Plaines des Chicots and d'Affouches just above. I recorded no Olive White-eyes at all above the deer fence (1450 m) at the Plaine des Chicots in February 1975.

The mixed forests of the Plaines des Chicots and d'Affouches are the only home of the Réunion Cuckoo-shrike, but it is far from clear why it is confined to this area (see species account below).

By contrast the cloud forest supported very low numbers and diversity, Olive White-eyes and Stonechats being scarce or absent, the worst places lacking Merles and Flycatchers as well (e.g. top of the Rivière de l'Est gorge; trail above Bélouve).

Pure or nearly pure *tamarin* forest was also poor in birds. The only natural *tamarin* forest with a low mixed evergreen component that I visited was on the slopes of Bellecombe near the Volcan, where the bird density resembled that in the plantation of pure *tamarin* at Bélouve. There were no Merles in either area, but the relative frequencies of the other species differed between them. Stonechats and Flycatchers were the commonest birds at Bélouve, but were rare and absent respectively at Bellcombe. Probably the ground-cover and understorey is important – certainly there is very little ground-layer on the lava slopes at Bellecombe

and this no doubt makes it less suitable for Stonechats. Densities of both white-eyes were low in both places.

There is considerable controversy on the island as to the effects of *Cryptomeria* in monoculture plantations on the wildlife of the island. The forestry authorities stoutly maintain, against the protests of conservationists, that birds are happy in *Cryptomerias* (e.g. anon 1974b). My observations (Appendix 1) are to the contrary – birds were very scarce indeed in *Cryptomeria* plantations, in which there was clearly little or no suitable food. White-eyes were rarely seen feeding; usually they were simply travelling through. Some birds nested in the conifers, but generally only close to mixed forest in which they could feed.

There is no doubt, however, that the total area under tree monocultures is too small to have any significant effect on bird populations, and this applies equally to the plantations planned for the future (Moulin & Miguet 1968, anon. 1974b, Bourgenot 1974), except that an extension of plantations in the Plaines d'Affouches and des Chicots would gravely affect the survival of the Cuckoo-shrike (Cheke 1976, and see discussion on conservation below). The botanical implications of further plantations on the best-developed upland forests (i.e. those on flat land, which are easiest to work) are more serious (Baumer 1981), but outside the scope of the present discussion.

In heath vegetation above the tree-line, bird density was generally fairly low, though higher near the edge of the forest especially where there was plenty of *Hypericum* amongst the *Philippia* (e.g. Plaine des Chicots). Where there was *Hypericum* Olive White-eyes were present when it was in flower, but otherwise only Grey White-eyes and Stonechats were found. These became progressively less frequent with altitude, a few white-eyes reaching 2750, around the upper limit of *Philippia*. Stonechats were seen at the summit of the Piton des Neiges (3069 m).

Distribution of introduced birds in relation to vegetation

Many fewer introduced species have penetrated the forest in Réunion than in Mauritius (Chapters 4, 5). Only the Cardinal *Foudia madagascariensis* and the Madagascar Turtle-dove *Streptopelia picturata* occurred regularly in the forests; the latter was generally extremely scarce, and the Cardinal was seldom plentiful either, occurring scattered through forest up to

1800 m. Cardinals were much scarcer than in native forests on Mauritius and indeed, except at Bébour, were fewer than the thinly dispersed Mauritius Fody *F. rubra* in those forests (Cheke 1983*a*, Chapter 4). Cardinals had a marked vertical migration, most remaining at low altitudes (below 500 m) throughout the winter, apparently from April to September, then coming up presumably to breed. The majority of the population was distributed around the coast at low elevations remaining there all year, reaching very high densities in places: I estimated five to eight territorial males per hectare in sparse *Pithecellobium dulce* scrub at La Saline les Bas (50–150 m) on 18 November 1974.

In heath at high altitudes the Cape Canary *Serinus canicollis* was widespread though nowhere really common; it also occurred at lower elevations in cultivated areas and in exotic scrub. The Malagasy Partridge *Margaroperdix madagascariensis* also occurred in *Philippia* heath.

Other introduced species occurred only casually in native vegetation, even the Common Mynah *Acridotheres tristis* being only very rarely seen, except on the much-cleared heathland at the Plaine des Cafres, and in the forests of Grand Bassin (where there were no Merles). The waxbill *Estrilda astrild* was occasional in forest edge up to 1300 m. Both these species were very common in man-made habitats at low elevations, as was the Yellow-fronted Canary *Serinus mozambicus*, which also penetrated the forest edges around St.-Philippe and Basse Vallée.

Ecology and distribution of native land-birds
Little Green Heron Butorides striatus (Linn.); crabier (krabye)

This species appears to suffer periodic extinctions and recolonisations on Réunion. In the mid-nineteenth century a 'grey egret' was reported to live and breed in the island's rivers (Vinson 1861, Coquerel 1864), and Sclater (1924–30) listed this species, but by 1948 it had recently become extinct (Milon 1951). The birds were present in 1963 (Jouanin 1964*b*), and CAB found two nests along the Rivière St.-Jean in both 1966 and 1967, but when I visited the area with him in 1974 he told me he had seen no birds since October 1967 and all four nests had failed. I searched all suitable areas in 1974 but saw no birds. However in 1979 NB saw birds at the Rivière St.-Jean and also at the Etang du Gol (Barré & Moutou 1979), where they later bred. In 1981

NB found two pairs at the former and ten birds at the latter site (Barré & Barau 1982).

Salomonsen (1934) and White (1951) included Réunion in their reassignment of Mascarene *Butorides striatus* from the Malagasy subspecies *rutenbergi* to the Asian race *javanicus*, but neither in fact examined any Réunion specimens (C. M. N. White *in litt.*, Jan. 1977) as none were available (Berlioz 1946). Jouanin (1964*b*) assigned the 1963 Réunion bird to *rutenbergi*, which suggests the recolonisation was from Madagascar and not from the neighbouring island of Mauritius.

These herons are confined in Réunion to fresh-water meres and sluggish rivers with heavily vegetated banks. There are no mudflats for shore-feeding.

The two 1966 nests had newly hatched young on 15 and 24 December respectively (CAB), representing eggs laid in late November/early December assuming the same incubation period as elsewhere (21–23 days, Hancock & Elliott 1978). Both 1967 nests had eggs in early October. Laying dates in nests found by NB in the 1979–80 season were November, December and January (*in litt.*), but in 1980 he found one with small young in late November (Barré & Barau 1982, where the season is given as September to February). The clutch is three (*ibid.*). Nesting success appears to be low. All four 1966–7 nests failed after hatching, apparently due to interference from children destroying weaver nests along the river (CAB). Two of NB's nests in the same area in 1979–80 were abandoned before hatching, though he recorded a successful nest at the Etang du Gol in 1980. He frequently saw young there but never at the Rivière St.-Jean (*in litt.*, 1982). Due to an editorial mistake (F. Moutou *in litt.*), *Ardeola idae* instead of the Little Green Heron was listed as a native Réunion bird in Moutou (1984).

Réunion Harrier Circus maillardi Verreaux 1863; papangue (papañg)
Taxonomy

This is an endemic race, *C. m. maillardi*, of the Malagasy Harrier *C. m. macrosceles*, the type of the species being from Réunion. I prefer to consider these forms as a full species, rather than submerge them in *C. aeruginosus* as is commonly done (Brown & Amadon 1968, Nieboer 1973, Clouet 1976, 1978), as retaining the distinction at the specific level between the African and Malagasy forms emphasises their different origins and habits. The Malagasy birds resemble, and are presumably derived from (Berlioz 1946, Clouet 1976, 1978), the Asian *C.* (*a.*) *spilonotus*, rather than the African *C.* (*a.*) *ranivorus* or the European Marsh Harrier *C. a. aeruginosus*, which migrates to Africa. Berlioz (1946) apparently mistook a young male for a female in his description; Clouet's descriptions (1976, 1978) are correct.

The Réunion Harrier is smaller and more markedly dimorphic sexually (in both colour and size) than other members of the marsh harrier complex (*ibid.*). Clouet also showed that it has a shorter tarsus, longer claws and longer middle toes than its presumed Malagasy ancestor – adaptations to larger and more predominantly avian prey. The island was originally well supplied with pigeons and parrots (Chapter 1), but no mammals apart from bats. The wing is also shorter and more rounded, an adaptation to flying in forest, a feature also found in the Mauritius Kestrel *Falco punctatus* (*ibid.*, Chapter 5).

Status and distribution

The Harrier, which Milon (1951) said "could not be called rare" in 1948, had become scarce enough by the middle 1960s to require some kind of protection. Until then it was regarded as vermin, but it was withdrawn from this category in 1966 (Gruchet 1975), although it was not granted explicit protection until 1974 (Servat 1974). Since 1966 the numbers have apparently increased noticeably (CAB, HG).

The species is found throughout the island above the zone of sugar cultivation, coming down to lower levels in wooded ravines and in the south-east where there is little cultivation. There is some vertical migration, CAB reporting birds coming down to take lizards in fields after the cane is cut (late winter), but the phenomenon is certainly not as marked now as it was in the mid-nineteenth century (Schlegel & Pollen 1868). Young birds may be seen over the marshes at Bois Rouge and St.-Paul at any time of year, behaving like typical marsh harriers. Birds only rarely visit areas over 2000 m altitude. Map 6 shows the breeding distribution in 1973–5. Since then some of the vacant parts of the Grand Brûlé near the coast have been colonised (pers. obs. 1978).

The density of breeding pairs varied a great deal depending apparently on altitude and human disturbance. In the lower part of the range at Grande Chaloupe and lower Cirque de Salazie, between 400 and 800 m, the density was high (3 km^2 per pair)

(Clouet 1978), whereas at the Plaines des Chicots and d'Affouches (1100–2000 m) I estimated 4–6 km^2 per territory, but I might have overlooked some birds due to the terrain. Some suitable areas, such as the Cirque de Cilaos, had very few birds because they were still shot. Taking a conservative figure of 10 km^2 per territory over the whole area of the breeding distribution (1300 km^2) gives an estimate of 130 pairs; the 50 or so actually observed by MC and myself are marked on the map, and their distribution suggests the total population in 1974–5 may well have exceeded 200 pairs, the figure now generally quoted (Clouet 1978, Barré & Barau 1982).

Breeding and ecology

Little was known of the bird's ecology and nothing of its breeding before Clouet's (1976, 1978) studies from September 1974 to early 1976. Here I give those observations I made myself, and the information on nesting and habits gleaned from local inhabitants before Clouet started work, which supplement his results; the local information proved to be very accurate, and the common practice of ignoring such sources can waste useful knowledge.

Informants in different parts of the island reported that nests were on the ground or in low bushes, that two or three eggs were laid and that two young were reared. Clouet found nests only on the ground, but confirmed that two or three eggs were laid of which one failed to hatch. However in November 1974 I saw what appeared to be a single family with three young, so all the eggs may sometimes hatch. TB's assistant reported a nest with four young and two eggs, presumably laid by two different females. Clouet only found clutches laid from mid-January to late May, but in March 1970 CAB was brought a large nestling which must have come from an egg laid in December. The incubation period is 33–36 days, the fledging period 45–50 days; the young are fed by the male for at least 2 months after fledging, and remain in the parents' territory until December (*ibid.*).

In several hours watching Harriers, I never once saw a bird attack or catch any prey when hunting from the air. My only feeding observation was of a family party at Grand Etang on 14 November 1974: the young were walking over the cracked drying mud of the lake bed, eating small unidentified items, taken after a hop, skip and a jump from where the bird first saw them. The only prey I could find were dead and dying dragonfly nymphs (Odonata: Anisoptera) and frogs *Ptychadena mascareniensis*, but the frogs were very agile and could probably have avoided the clumsy antics of the birds.

My informants all said that the main food was rats *Rattus* spp. and Tenrecs *Tenrec ecaudatus*. TB added that they would take anything they could get, alive or dead, and claimed that they would imitate Merles and chickens (wails and cluckings respectively) to attract those prey into the open. He once found a feral cat, a Harrier and a Tenrec, all dead, locked together by their claws, and another time surprised a Harrier gripping a young deer fawn which it failed to kill.

Other prey species recorded, or reported to me, were: small birds and their nest contents; chickens; lizards *Calotes versicolor*; incubating Malagasy Partridge (TB per BT); Feral Pigeons *Columba livia*, mice *Mus musculus* (Schlegel & Pollen 1868); grey grasshoppers (Orthoptera: Acrididae), principally in summer, on grassy deer pastures above the tree-line on the Plaine des Chicots (TB). I found such grasshoppers at the Plaine des Cafres in May and November 1974; in May there were many Harriers there, but in November there were none. Feral Pigeons, considered the Harrier's favourite food by Pollen (Schlegel & Pollen 1868), are now more or less confined to towns, whereas in the mid-nineteenth century they were widespread on cliffs and ravines (Coquerel 1864).

FBG dismissed the idea of the Harriers feeding on small birds (Gill 1973b: 12) on the grounds that they

Map 6. The distribution of the Réunion Harrier *Circus maillardi* in 1973–5. The approximate lower limit of the breeding distribution is given by the line of crosses (+ + + +) and the upper limit by inverted 'V's (∧∧∧); see text for details. Observations of particular birds are indicated by circles: ⊙, pairs; ○, single birds. The two localities where single birds were seen near the coast are both marshy land surrounding meres. Observations by the author and M. Clouet (pers. comm.) combined.

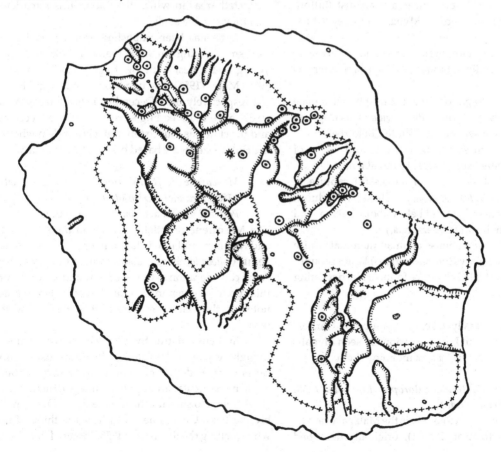

are too large and clumsy (*in litt.*, Jan. 1974); TB commented that they caught white-eyes when the birds were "otherwise occupied, e.g. with their heads buried in *Hypericum* flowers". The Harriers do a lot of hunting over forest, either beating low over the canopy, sometimes briefly hovering, or circling much higher, watching; if they are after birds, their kill rate is certainly very low. Both types of 'easy' prey (nests and tenrecs) are seasonal: most nests are available from November to January (at least in the areas where Harriers are found), and Tenrecs hibernate underground from May to October, so the birds would need to rely on other more difficult prey during the winter. MC has confirmed that Harriers do take birds' nests (immature seen at Stonechat's nest on 14 Nov. 1974), and also listed (Clouet 1976, 1978) several species of small birds as important components of the diet, although introduced mammals make up half the prey taken. He did not publish details of his feeding observations, but has sent me the following (*in litt.*, 24 Apr. 1976):

1. Failed attacks seen once each against Button-quail *Turnix nigricollis*, Mynah and Grey White-eye.
2. A small grey bird, the size of a female sparrow or female Cardinal, brought to the nest by a female Harrier.
3. A toad or frog brought to the nest by the female (probably the former, *Bufo regularis*, as frogs are scarce or absent at the Grande Chaloupe).
4. Skulls of three young Tenrecs, and bones of unidentified small birds, found at a nest.
5. Several observations of successful capture of lizards *Calotes versicolor*. These introduced lizards are only found below about 1000 m, and never in the forest (pers. obs.)
6. Several observationss of small mammals (mice *Mus* and shrews *Suncus murinus*) being brought to the nest (FBG also once found a Harrier eating a *Suncus*).

Other birds generally take little notice of Harriers (pers. obs. FBG), but I have occasionally seen Mynahs and Swiftlets *Collocalia francica* mobbing them.

Moorhen Gallinula chloropus Linnaeus 1758; poule d'eau (puldo)

The Moorhen in Réunion is the Malagasy race *G. c. pyrrhorrhoa* (Benson 1970–1), originally described

from Mauritius (A. Newton 1861*b*). Maillard (1862) considered it very rare, though only a few years later Schlegel & Pollen (1868) called it common and noted that its principal locality was the Etang de St.-Paul; Coquerel (1864) also commented that it could occasionally be seen there, and Vinson (1868) called it "very common in the meres". Despite the above and later records (Hartlaub 1877), Berlioz (1946) claimed he knew of no authentic record for Réunion, an oversight quickly corrected, however, by Milon (1951), who saw some at St.-Paul and reported them common enough on other meres.

I found them still common at the Etang de St.-Paul, and also saw some at the Etang du Gol and at the Etang de Bois Rouge. These three sites are the only suitable areas of reed-fringed meres on the island. Barré & Barau (1982) reported Moorhens along some of the more permanent rivers of the east coast but I saw none myself. The birds at the three meres are extremely shy (though easily heard), presumably as a result of heavy hunting pressure; considering the restricted areas in which they occur it is remarkable that they survive at all.

Little has been published on their habits. NB saw small young on 20 September 1980 and small young and a nest with eight eggs on 27 November 1979. Milon (1951) saw two broods of 2–3 and 8 days old respectively on 5 December 1948. Barré & Barau (1982) gave the breeding season as July to December and the clutch as four to ten, but it is not clear whether the clutch sizes were based only on Réunion data.

Mascarene Swiftlet Collocalia francica (Gmelin 1788); zirondelle (zirondel)

Past and present distribution and numbers

The Mascarene Swiftlet is found on both Mauritius and Réunion, and is a distant outlier of the large Asian genus *Collocalia*; there is also a form in the Seychelles. A full consideration of the taxonomy and synonymy of this species is given in Chapter 4, which also includes notes on the field characters of this species and the next.

In Réunion the Swiftlet was fairly common, though very much less so than in Mauritius. Nesting caves were evidently present in many ravines, but I made no particular attempt to enumerate them. Milon (1951) refers to sites up the Rivière de St.-Denis, notably the Ilet à Guillaume, but I heard nothing of this when visiting the îlet in May 1975, though I did hear of other caves along this river. Pollen mentioned especially the Rivière des Galets (Schlegel & Pollen 1868) and Coquerel (1867*a*) added also the Rivières St.-Etienne, Ste.-Suzanne, and two sites in the Rivière de Mât (L'Escalier and La Fontaine Petrifiante). I can confirm none of these sites, and as there seems to be no active tradition of these localities, I suspect many colonies then extant are now extinct. The "thousands of nests" at La Fontaine Petrifiante (*ibid.*) in the 1860s also suggests a substantial decline since then; I neither saw nor heard of any concentrations which would suggest such numbers exist today. I visited two colonies myself: a small cave at the head of a waterfall on the Bras Chansons (Bébour) about 200 m east of the forest road (*c.* 15 nests), and (with HG) a cavern formed by a vast boulder at La Chapelle (Cilaos; < 10 nests). Jadin & Billiet (1979) found respectively 60 and 30 active nests in these caves (or nearby ones; their names and descriptions differ somewhat) in late 1978, and over 100 in a third in the Bras Patience (Takamaka). Gruchet (1976) reported on a cave at St.-Gilles (65+ nests) and has visited another active one near St.-Joseph (pers. comm.); to judge by the number of birds around it in November 1974, there were nests in the tower of the church at Ste.-Anne. HG remembered very many more nests at La Chapelle in the 1940s; Jadin and Billiet (1979) reported that this site was decimated after access became easy following the construction of a new road. At no time in Réunion during 1973–5 did I see more than about 50 Swiftlets together, and then only twice – over the drying lake-bed at Grand Etang on 14 November 1974 and two-thirds of the way up the path to the Plaine des Chicots on 14 February 1975. On 11 October 1978, however, there were 100+ birds over a wooded hillock by the sea near Ste.-Rose. Normally birds were seen in ones and twos almost anywhere on the island up to about 2000 m elevation, but it was possible to spend a day in the field and see no Swiftlets at all, which never happened in Mauritius. Barré & Barau (1982) suggested a population of 2500–5000 pairs (12 times the number of Mascarene Swallows) but this may well prove an over-estimate.

Ecology

I studied the Swiftlet in some detail in Mauritius (Chapter 4); there is no reason to suppose that its habits on Réunion are any different.

Barré & Barau (1982) tentatively gave the breeding season as September to January, but the season in Mauritius is very long and probably not regularly annual, so the few breeding records for Réunion may not mean very much. Pollen found eggs and young (Schlegel & Pollen 1868) but gave no date, so the only records available are the following:

1. Nests with small young (? eggs also), St.-Gilles, 8 October 1975 (Gruchet 1976).
2. FBG collected a female carrying an oviduct egg in November 1964; 8 others (both sexes) collected had only slightly enlarged gonads.
3. Birds apparently incubating, Bébour, 1 November 1973; others seen collecting nesting material (*Usnea*) at nearby Bélouve, 2 November 1973. Young in nests, Bébour, late November 1977 and 1978, but some birds also incubating (Jadin & Billiet 1979).
4. Several birds building, and one nest with an egg, Rivière de St.-Denis, early December 1948 (Milon 1951).
5. Eggs (possibly also young) at Cilaos, early December 1978 (Jadin & Billiet 1979).
6. Eggs, and young at all stages of growth, Takamaka, early December 1978 (*ibid.*).
7. Nests with young, St.-Joseph, 3 January 1972 (HG).
8. Birds definitely not nesting, Cilaos, 2 May 1974 (pers. obs.) and early May 1980 (NB).

The eggs are plain ivory white (Jadin & Billiet 1979). The clutch size is apparently always two (Gruchet 1976, Jadin & Billiet 1979, Barré & Barau 1982); Roussin's (1860–7, vol. 4) lithograph of a nest shows two eggs. Pollen, however, described the clutch as three to six and referred to nests containing both freshly laid and well-incubated eggs, and young at all stages from newly hatched to fledging (Schlegel & Pollen 1868). I am at a loss to account for this from so careful and critical an observer as Pollen; he implied elsewhere that he saw at least some nests *in situ*, though many were procured for him by collectors who could have mixed up the contents.

Nests are suspended from the roofs or sides of caves, clustered together; Gruchet (1976) and Jadin & Billiet (1979) have published photographs of groups of nests. The nests are composed of *Usnea* lichen or moss (*Pilotrichella* sp., *Aerobropsis* sp.) glued together with varying amounts of saliva (Milon 1951, Gruchet 1976, Jadin & Billiet 1979).

The type of site used for breeding in Réunion

differed from that in Mauritius. Réunion largely lacks the huge underground tunnels and caves which are so characteristic of Mauritius (Chapter 4), and the 'caves' are mostly no more than large rock fissures, though the St.-Gilles site is in a lava tunnel (Gruchet 1976). In Réunion Swiftlets thus usually nest in very deep shade but not totally in the dark, as the cave is rarely more than a few metres deep.

Food boluses were collected from three Swiftlets by Jadin & Billet (1979). They contained Diptera (11 families), micro-Lepidoptera, Hymenoptera (2 families), Homoptera, Coleoptera, Psocoptera, spiders and plant and feather debris. These data are tabulated with the Mauritian material in Chapter 4. Coquerel (1867a) found largely Lepidoptera and Diptera amongst the insect remains below a colony in the Rivière des Galets.

Maillard (1862) and Coquerel (1864) both listed two species for the island: *C. francica* and *C. esculenta*. One might dismiss this as confusion due to these authors believing Buffon's bird to differ from that used by the local Chinese for birds'-nest soup; but in November 1973 J.-M. Vinson (pers. comm.) saw a number of Swiftlets at Takamaka that were dark all over, lacking the pale rump and underparts of *C. francica*. During 1974 I kept a careful watch for such birds, visiting Takamaka twice, but saw none there or anywhere else; NB also confirms the ordinariness of the Takamaka birds. Coquerel (1867) later denied the existence of a second species, while admitting two "varieties", but the question is worth remarking on in case a second form is ever found on the island.

In the late 1940s Chinese residents still used the nests for soup and Milon (1951) gave some details of the trade. This could be advanced as a reason for the decline over the past century, but the evidence is insufficient, and Gruchet (1976) stated that nests are no longer taken for food. Jadin & Billiet (1979) reported continuing nest destruction (pure vandalism) at Cilaos. Coquerel (1867a) reported that people with batons killed large numbers of the birds for food as they entered or left caves; this slaughter certainly no longer occurs.

Malagasy Swallow *Phedina borbonica* (Gmelin 1789); zirondelle (ziroñdel)

P. b. borbonica is found in both Mauritius and Réunion (Berlioz 1946), differing in bill size (pers. obs.) and colouring from the Malagasy race which is partly migratory (Moreau 1966, Clancey *et al.* 1969) and thus might possibly turn up in Réunion. Coquerel (1864, 1867a) described a "very remarkable" paler, larger form with a larger beak occurring along the shore. The Madagascar bird is paler, but not larger, and has a smaller bill (Hartlaub 1877, pers. obs.), so if Coquerel found some he confused the details; the migrations of the species seem to be sporadic and unpredictable (Moreau 1966), and although no birds have been reported recently, it is possible that they occurred in Réunion in the mid-nineteenth century.

The Swallow may once have been much commoner in Réunion than it is now; Maillard (1862) called it abundant and Coquerel (1864, 1867a) fairly common, though Pollen on the contrary considered the bird as "getting rarer day by day" and restricted to certain localities only (Schlegel & Pollen 1868). Pollen apparently saw them only between La Possession and St.-Paul. Coquerel (1867a) noted that they occurred principally "dans les hauts": a vague term, but it may well account for Pollen's failure to find them, as he seems to have spent his time only in the region between the Rivière des Pluies and St.-Paul, where the Swallows are certainly rare at all elevations today. Milon (1951) called it "common", but noted it was less so than the Swiftlet.

The species is easily missed by the casual visitor to Réunion, as the breeding distribution is mostly in a narrow band at 200–500 m altitude along the east side of the island from St.-Denis to the Rivière de l'Est; and again in the vicinity of St.-Joseph (Rivière des Remparts; Manapany) and (*fide* Horne 1975) at the Grande Chaloupe. This zone is inland of the main road, but below the forest zone and so outside the places usually visited by naturalists. Within their breeding range the birds were local; for instance I never saw any around La Confiance or Grand Etang. There were one or two pairs at Cilaos, at 1200 m, apparently resident.

Feeding birds could be met with sporadically anywhere up to 1500 m, but were absent from the dry west coast except around the meres at St.-Paul and Le Gol where one or two could usually be found; odd birds were often seen from the main road between St.-Denis and St.-Benoît.

All the colonies appear to be very small; the largest recorded is of "several" nests at Takamaka (Jadin & Billiet 1979). The total population in 1974 seemed unlikely to have exceeded 200 pairs, but Barré

& Barau (1982) allowed 200–400 pairs. Barré (*ibid.*) saw 190 birds together on wires at the Rivière des Pluies after an overnight deluge (31 December 1980).

Feeding behaviour is described in Chapter 4; there is nothing to add for Réunion.

As early as the mid-eighteenth century de Querhoënt (quoted by Buffon 1770–83) described the breeding cycle on the basis of information from an elderly inhabitant who told him that laying "occurred in the months of September and October; that he had several times taken nests in caves, holes in rocks etc.; that they were composed of straw and a few feathers, and that he had never seen anything therein but two grey eggs finely spotted with brown" (my translation). Jadin & Billiet (1979) and Barré & Barau (1982) confirmed this description, including the egg colour, the former also finding broods of three at Takamaka (November or December 1977 or 1978). Barré & Barau (1982) gave the clutch as two or three, and the breeding season as September to December. Confirmatory observations on seasonality are few: CAB collected a female with an enlarged ovary (3 mm follicles) on 29 October 1967 at Bois Rouge (FBG's notes), birds in the Rivière des Remparts on 22 November 1974 (about three pairs) showed evidence of nesting behaviour (pers. obs.), and Horne (1975) saw active nests on 14 December 1974 at the Grande Chaloupe.

In addition to open nest-sites (rock ledges and buildings) similar to those used in Mauritius, the swallow in Réunion sometimes nests in much darker and more secluded sites, in caves together with swiftlets. Milon (1951) saw a nest in the swift cave he visited in December 1948, and HG has also seen both species together in caves.

Réunion Cuckoo-shrike *Coracina newtoni* (Pollen 1866); *tuit-tuit* (*twit-twit*)

Introduction

The Réunion Cuckoo-shrike first appeared in print in Maillard's list (1862), under '*Oxynotus ferrugineus*' (the then current name of the Mauritius Cuckoo-shrike). More detailed notes by Pollen (1865*a*) led A. Newton (1865*b*) to suggest that the forms on the two islands differed, and Pollen described the Réunion bird as a new species in 1866. Apart from two more articles by Pollen (1865*b*, Schlegel & Pollen 1868) there was no new information until Milon (1951), who saw but one in 1948. On the basis of information from C. Jouanin, Vincent (1966) included the species in the *Red Data*

Book, giving it three-star status, indicating 'very grave anxiety' as to its prospects for survival. As the primary brief of the BOU Mascarene Expedition was to study the distribution and ecology of endangered species, I directed most of my efforts in Réunion to the Cuckoo-shrike, the only member of the avifauna thought to be at risk. A special report on the conservation of this species has already been produced by the BOU (Cheke 1976), and the account which follows is largely an English version of that paper.

Past and present distribution

Pollen by his own admission appears not to have travelled widely in Réunion, but his description of the Cuckoo-shrike's distribution (Pollen 1865*b*) applies equally well today: "I have found the *Tuit-tuit* to be very abundant in the heights (inland) of La Possession, principally in the forests of the Dos d'Ane. Mr Lantz and I have found them in the heights of St.-Denis and in the mountains, near to the Ravine le Frais and the Camp Rattaire" (my translation).

It is difficult to know whether the apparent restriction of the bird's distribution to the area around headwaters of the Rivière St.-Denis is real or due to a shortage of information. Apart from Milon's observation of a female in the "hauts de St.-Benoît" in 1948, there is no conclusive evidence of the birds having occurred elsewhere. A pair collected by de l'Isle in August 1875 (Jouanin 1962) is labelled "près du Grand Brûlé", meaning, in principle, the lower eastern slopes of the Volcan; however, since the region behind St.-Denis where the birds have always occurred is known as Le Brûlé it is possible that de l'Isle confused the localities.

The only other evidence for a wider distribution is equally tenuous. Mr A. Hoareau told me in 1973 that there had been Cuckoo-shrikes in the forests near St. Philippe (near the Grand Brûlé!) "about 10–15 years ago" (i.e. about 1960) when he used to work there; EH reported similar comments from forestry staff. M. Hoareau was familiar with the species at the Plaine des Chicots, and was careful to distinguish the Cuckoo-shrike from albino Merles, both of which are sometimes known locally as *merle blanc*.

TB told me that his father used to talk of a *gros zoiseau blanc* that he had known 30 or 40 years before (1930s) in the forest at Bellecombe, and which TB now thinks was probably the Cuckoo-shrike. Three ONF labourers claimed to recognise a whistled imitation of the Cuckoo-shrike's call and told me they had heard it at le Cratère (behind St.-Benoît; Milon's locality?), the Rivière d'Abord (Plaine des Cafres) and the Ravine St.-François (inland of Ste.-Anne). TB also remembers having heard a call in the forests behind Ste.-Anne which in retrospect, now knowing the bird, he thinks was the Cuckoo-shrike.

I surveyed Cuckoo-shrike populations at the Plaines des Chicots and d'Affouches, and also visited many other parts of the island trying to find the birds. Of those places mentioned above as having once perhaps held Cuckoo-shrikes, I visited the following without success.

St. Philippe: Forêt de Mare Longue, 8.11.73; Basse Vallée, 8.11.73 and 11.11.74

Ste.-Anne: along the Ligne Domaniale to the Bras Felix, 17–18.8.74

Volcan: Forêt de Bellecombe, 25.8.74 and 10.11.74

Plaine des Cafres: headwaters of the Rivière d'Abord, 3.11.74

I was unable to visit the Massif du Cratère, and I suspect the Ligne above Ste.-Anne is too high up – the cloud forest there certainly seemed most unsuitable for Cuckoo-shrikes.

The place I considered most likely to harbour undiscovered Cuckoo-shrikes was the Plaine des Fougères, east across the canyon of the Rivière des Pluies from the Plaine des Chicots. The eastern end near the Rivière du Mât, visited on 17 August 1974, proved to have cloud forest, but the western part, above the Rivière des Pluies, has forest essentially similar to that at the Plaine des Chicots. RAC and I spent 19–21 August 1974 up there exploring separately along and above the Ligne Domaniale to 1550 m, but found nothing. EH told me that he too had never found Cuckoo-shrikes on the Plaine des Fougères.

The other most probable place is the forest seawards of the crest that bounds the Plaine d'Affouches on the west. I explored the Ligne Domaniale from the north end as far as the Ravine Mal Coté on 15 November 1974, again without success. I was unable to visit the other, Dos d'Ane, end of this forest.

Other forested areas which I visited but where there was no sign of Cuckoo-shrikes, were as follows.

Ilet à Guillaume (Rivière de St.-Denis), 4–6.5.74

Takamaka, 10.5.74; Grand Etang, 11.11.73, 14.11.74

Bébour–Bélouve, 31.10–5.11.73

Cilaos (Grand Matarum etc.), 26.4–2.5.74

Plaine des Makes (Forêt du Bon Accueil), 16–17.11.74

FBG (*in litt.*, Gill 1972, 1973*a*) encountered Cuckoo-shrikes only at Le Brûlé–Plaine des Chicots in his 9 months of explorations all over the island in 1967, and the same applies to all other ornithologists who have been there recently (C. Jouanin pers. comm.; Horne 1975; AFW, BT, NB). EH, who knows the island as well as anyone, and who is particularly interested in the Cuckoo-shrike, has never met with it anywhere else.

Nevertheless the possibility still exists that there are Cuckoo-shrikes in other places, though it is unlikely there are many. Many of my visits to prospective sites were necessarily rather brief, and the species is not easy to find, especially when it is not being vocal (see below); indeed on my first visit to the Plaine des Chicots (6–7 November 1973) I failed to discover any at all. I suggest that future searchers concentrate their efforts in the following areas (in order of likelihood of positive results):

Western slopes of the Piton Ravine à Malheur, Dos d'Ane

Massif du Cratère, St.-Benoît

Heights behind Ste.-Anne, below the Ligne Domaniale

Heights of St.-Philippe–Tremblet

Plaine des Merles, Salazie

Ravines of the Rivière d'Abord, Ravine du Tampon and Bras Jean Payet, Plaine des Cafres

Forest along the Ligne Domaniale on the slopes of the Grand Bénard.

The present confirmed distribution of the Cuckoo-shrike does not differ from that given by Pollen (1865*b*). His "forêts du Dos d'Ane" are adjacent to the Plaine d'Affouches (the little cirque of Grand Coin was not fully settled and deforested till 1880 (Defos du Rau 1960)), and the "hauts de St.-Denis" etc. clearly represent the Brûle–Plaine des Chicots area. The Ravine le Frais is marked on the IGN 1:50 000 map (see Introduction), though I have been unable to find the Camp Rattaire. An interesting difference between Pollen's data and my own is that he gives the bird's altitudinal range as 800–1400 m, whereas the lowest I found a bird was about 1300 m, and they now occur up to the tree-line around 1800 m. Possible reasons for this change will be discussed below, but I will add here that it may be relatively recent as EH remembered a

Cuckoo-shrike being killed at around 900 m above Le Brûlé in his youth (1915–1920); Milon (1951) claimed to have heard one at 600 m at the same place, but this altitude is below the village, where there is unlikely to have been undisturbed native forest in 1948, so I suspect a printing error or mistake in estimating altitude.

The birds are spread throughout the forests of the Plaines d'Affouches and des Chicots; their distribution as I found it is shown on Map 7; detailed discussion of the census methods and results is given in the next section.

The Cuckoo-shrike population: past estimates and the 1974 census
The first person to estimate the Cuckoo-shrike population was Jouanin in March 1965: numbers "not known precisely but from widespread enquiries pursued during a two months stay it seems unlikely that more than about 10 pairs of the birds survive" (Vincent 1966). Gill (1972, 1973a) stressed the extreme scarcity of the species and this very pessimistic view was generally held until I was able to census the species in 1974. While the 1974 population was 120 pairs or more (Cheke 1976, and below), it is difficult to know whether this represents a substantial recent increase or whether earlier impressions were wrong. As early as the 1880s Lantz (1887) was predicting imminent extinction, but sweeping statements of impending extinction were not unusual at the time and are difficult to assess. To judge by EH's remarks the Cuckoo-shrike was not considered by the locals as particularly rare, within its range, in the 1920s and 1930s, and it seems to have been Berlioz's statement

Map 7. Plaines des Chicots and d'Affouches. Distribution of Réunion Cuckoo-shrikes censused in 1974–5. Each territory is marked by a ring, within which is indicated the numbers of times birds were noted at that site in relation to the number of visits (2/14 = 2 contacts in 14 visits). Other symbols; N, nest; TB (broken rings), additional record from T. Begue; PO, many observers. In some territories the positions of the birds on different occasions are shown, giving an indication of territory size. Scale as Map 8. (From Cheke 1976.)

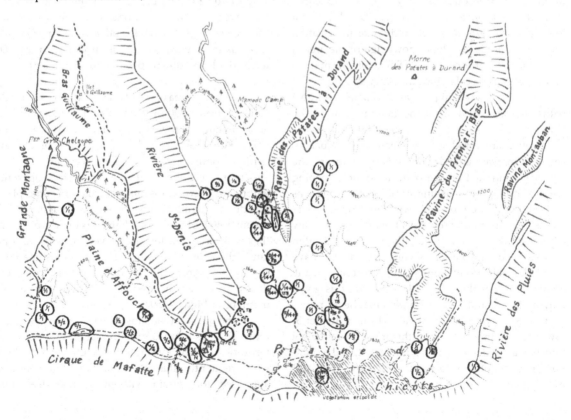

(1946), unsupported by evidence, which established it as "virtually extinct". In his brief visit in December 1948 Milon (1951) encountered only two birds, though he reported hearing of others; he also said it was on the way to extinction. I believe that the Cuckoo-shrike population was lower in the 1940s to 1960s than it is now, possibly due to poaching and to the disastrous cyclone of January 1948 (Defos du Rau 1960, Davy 1971). Dispersion of birds by the storm might account for Milon's sighting at low altitude (300 m) inland of St.-Benoît.

Since becoming gamekeeper of the Plaine des Chicots in 1969, TB says he has noticed a steady increase in Cuckoo-shrikes, which he ascribes in part to his own efforts to reduce poaching of all kinds. This coincides with Jouanin's low estimate in 1964, though as the bird is very easy to miss and as he did not attempt a census, I suspect he underestimated the numbers four- or five-fold.

From 9 to 16 August 1974 I tried to census Cuckoo-shrikes in the accessible parts of the Plaines d'Affouches and des Chicots, assisted by RAC. To the results of this survey I have added observations made during visits to these areas in May (7–9) and November (5–8, 18–20) 1974, and January (AWD, 21–22) and February (14) 1975, and also additional information from TB from his excellent knowledge of the area. I chose August for the census as TB had told me that the birds were most vocal in July and August.

Our census technique was to walk along forest paths listening, and whistling an imitation of the Cuckoo-shrike's song call at frequent intervals; silent males often answered such a whistle. Each Cuckoo-shrike heard was marked on the 1:50 000 map, and the observer continued on the path, if possible keeping the bird calling by whistling until the next one was encountered. Where possible we followed the movements of individuals within their home range. We saw Cuckoo-shrikes very infrequently indeed, and often they stopped calling after two or three bursts. I have based the estimate of numbers of territories on the assumption that the song call *tui-tui-tui-tui* is given only by males, and that each individual heard represents a territory and hence a pair. Horne (Chapter 3) saw a female produce it, but this must be extremely unusual, as TB denies ever having seen it, and nor did I or any of my other informants (REA, MC, RAC and BT).

The results are shown on Map 7. The total is an estimate, as it is impossible to be certain that two birds, recorded on different days, say 300 m apart, are really two different individuals. I have chosen the ones which seem most reasonable on the information I have, a total of 49; eliminating the most doubtful of these would still leave 43, but the total could be as many as 53. In addition there are many places where TB has found Cuckoo-shrikes; I have not marked these, as his description was not precise enough to be mapped, except for a concentration around the head of the Rivière de St.-Denis (not included in the total). Each ring on the map includes a figure for the number of times the bird was recorded, and the number of times the home-range was crossed by an observer (all visits included); the frequencies on the main path from le Brûlé show that any given bird has a very low probability of being recorded especially if, as was often the case on this path, the observer passes through quickly without stopping to whistle every 100 m or so. The success rate during the census itself was much higher.

An indication of territory size is given in four cases, two near the upper altitude limit and two near the lower. The two upper ones were the best known; each covered about 8/ha, and this can perhaps be taken as the average size of the species' territory. The territories lower down appeared to be smaller (about 6 ha).

The total area of the Plaines d'Affouches and des Chicots, from 1300 m to the tree-line (*c.* 1850 m), is about 16 km². Of the parts covered (see Map 8), about 42½%, although forest, lacked evidence of Cuckoo-shrikes (compare Maps 7 and 8). According to TB, in 1974 there were birds in at least some of the apparently empty areas, so an absence of records in some places no doubt signifies insufficient coverage. If one assumes 40% of the 'likely' habitat is unoccupied, and that each occupied territory covers 8 ha (= 12.5 per km²), then the 1974 population can be estimated at $16 \times 12.5 \times 0.6 = 120$ territories; if the birds lower down do in fact have smaller territories, the total might be as high as 150.

The question remains as to how many pairs this represents, as there is no information on how many territories are held by single males. One sees the birds so infrequently that this question could not be answered for certain without a great deal more research. In species in which males are the more vocal sex, they are always seen more often than females (Yapp 1962), and this holds for the Cuckoo-shrikes, although the total number of sightings by myself and collaborating observers in 1974–5 was only 12 (repeated observations and those at nests excluded). The observations were: male seen alone five times, pair together, five times, and female alone twice. Horne (1975) considered that a territory whose incumbents she was recording contained three birds, two males and a female, on the basis of having seen a male twice as often as a female (16:7); her ratio agrees with mine and I consider this to be the normal frequency of sightings of the different sexes for a pair of this species. I conclude that there is no reason for supposing there to be many unpaired territorial males; I did however find one bird which was almost certainly unpaired. At a time (November) when other birds were relatively quiet (and nesting) he was exceptionally vocal when

aroused by whistling and could be followed visually with ease; at no time did I see a female in his territory. He had a brownish wing and nape and a weak ear-patch, and was probably an immature.

Summing up, I estimate that the total population in 1974 was about 120 territories (= pairs), and can be certain that it lay somewhere between 100 and 150 pairs. Since then there appears to have been little change. In 1978 TB reported that the Cuckoo-shrikes and Merles had disappeared from the trail to La Bretagne where he frequently found evidence of poachers (lime-sticks etc.), but that by contrast he had noted an increase around his house and along the edge of the Rivière St.-Denis gorge.

Habits and ecology
I have already described the Cuckoo-shrike as difficult to find and observe. In behaviour it closely resembles the Mauritian *C. typica* (Chapter 4) but is much more

Map 8. Plaines des Chicots and d'Affouches. Coverage of the area during censuses of the Réunion Cuckoo-shrike in 1974–5. Shading indicates area effectively covered, including auditory range on either side of paths. Superimposed grid is of kilometre squares. (Adapted from Cheke 1976.)

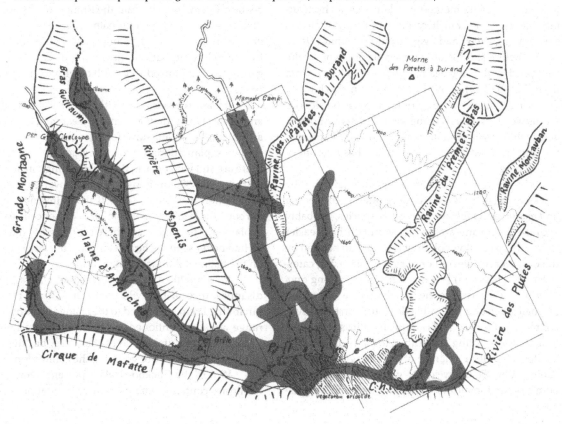

silent, and tends to keep more exclusively within the canopy, neither descending below it nor flying over it at all frequently. Population density is lower, and there are relatively fewer paths through its habitat, both of which contribute to less frequent encounters.

Cuckoo-shrikes live and feed solitarily, though sometimes both members of a pair feed together. Interactions with other species are apparently rare; TB reported that they sometimes associate with feeding flocks of Grey White-eyes. He also said that Merles returning to higher areas after their downward migration chase and disturb the resident Cuckoo-shrikes, and suggested that this contributed to their increased noisiness in July and August, separated pairs calling to maintain contact. I saw no such aggression in August, but in May I watched two Merles persistently chasing a male Cuckoo-shrike, which called vigorously, at the Plaine d'Affouches. Horne (Chapter 3) reported Cuckoo-shrike mobbing Merles. It may be relevant that one of the Merle's contact calls is very similar to a single 'tweet' of the Cuckoo-shrike's song.

When feeding the Cuckoo-shrike was very silent and inconspicuous, moving slowly through the tree with long waits between short flights from branch to branch. Its feeding method was largely search-gleaning, sitting on a branch carefully surveying all the surrounding foliage and twigs (including tufts of *Usnea* lichen) before striking at an insect or flying on to the next perch and trying again; prey were detected from up to 2 ft (0.61 m) away. I once saw a male clumsily 'hovering' (his feet were gripping a twig too slender for his weight) while reaching for a food item; another, also male, once spread his wings and hissed at something in an *Usnea* tuft on a *tamarin*. BT has a record of fly-catching.

I have very few observations on food taken by Cuckoo-shrikes. They were more often seen feeding on *tamarins* and *Philippia* than other species, probably because these plants are less dense and the birds easier to see. They usually took prey too small to be seen by the observer. However I once saw a pair taking caterpillars (Lepidoptera) ½–1½ in. (1.25–3.75 cm) long on a *tamarin*, and TB assured me that they were "very fond of small green and yellow beetles found on tamarins"; NB recorded spiders and caterpillars. Birds at the nest brought to the young, in addition to many unidentifiable items, adults and larvae of large longicorn beetles (Coleoptera: Cerambycidae), green caterpillars and both large and small moths (Lepidoptera) and, surpris-

ingly, earthworms (TB, MC, AWD, BT). Cuckoo-shrikes also foraged in mixed evergreen trees, and TB has seen them searching for food in the branch-whorls of the endemic bamboo *Nastus borbonicus*.

Pollen (1865b) was very specific about what the Cuckoo-shrike then ate: "The bird I'm discussing lives principally on beetles and their larvae. It particularly likes the larvae of the *Oryctes* which live in the palm trees that give us the palm-cabbage. In the stomachs of 14 individuals of *Oxynotus ferrugineus* I found nothing but remains of beetles and *Oryctes* larvae" (my translation). Pollen collected his specimens in January, May and June (labels on specimens in Cambridge, Leiden and in the British Museum).

The remarkable thing about Pollen's observations is that *Oryctes* beetles lived in palms *Acanthophoenix rubra* which have since disappeared from within the birds' range (due to wholesale cutting for the palm-cabbage), and the endemic *Oryctes* species appear to have become extinct. Gomy (1973a and *in litt.*) could not find these insects after several years of beetle-hunting on the island; although the endemic palms are now rare, they are not extinct, so there seems no obvious reason for the disappearance of the beetles. Gomy (*loc. cit.*) said the Rhinoceros Beetle *O. rhinoceros*, a pest of coconut palms *Cocos nucifera*, had not been introduced to Réunion, and added (*in litt.*) that therefore no parasite (including disease?) of the genus has been introduced for biological control purposes.

It was the association with *Acanthophoenix* palms and their fauna that led me to search the forests around the Ilet à Guillaume for 3 days in May 1974. The area has plantations of mature *Acanthophoenix* and there are still wild ones in the surrounding forests; the altitude, 700–800 m, is right for the lower end of the distribution in Pollen's time. If the palms no longer harbour the beetles, then the birds' absence is easily explained. The watchman there confirmed to me that he had never seen the bird in the area, but knew it well on the Plaine d'Affouches higher up.

It seems likely that the combined effects of the disappearance of the wild palms and of the beetles inhabiting them are related to the change in altitudinal range already mentioned, and also to the decline in Cuckoo-shrike numbers since the 1860s, but there is now no possible way of verifying the connection.

Within the Cuckoo-shrike's present range most territories contain elements of three vegetation types:

Philippia heath, *tamarin* forest and mixed evergreen forest. None are without *Philippia* or mixed evergreens, but some of the lower territories lack *tamarins*. Although the birds make plenty of use of the *tamarins*, it appears that the mixed evergreens must make up a substantial proportion of the area for it to be suitable for the birds. The area of forest with the highest proportion of *tamarin* is that immediately around the rest-house and gamekeeper's house at the Plaine des Chicots, and the first kilometre of the path towards the Dos d'Ane. There were no Cuckoo-shrikes there in 1974–5, although there had been a pair in previous years (TB, HG) and also subsequently (TB); whether or not this particular pair was killed by poachers (as TB thinks), there have certainly never been the five or six pairs one might expect from the extent of the area on the map, so it is presumably fairly unsuitable.

Breeding
Until 1974 nothing was known of the Cuckoo-shrike's nesting behaviour. I found the first nest in November, AWD found another in January 1975, and several more have been observed since (TB, MC, BT). Nest-building can begin as early as August, and the birds take up to a month to complete a nest (TB), which is composed of lichens, especially *Usnea* and a whitish foliose species providing an effective speckled camouflage (MC and BT photographs, *in litt.*); spiders' egg-cases may also be used (TB, per AWD). The nest is

small and neat, 10–15 cm across, placed on a branch of similar thickness, usually at a fork near the top of a tree. In the final stages the female sits in the nest and lines it with fine grass inflorescences brought by the male (pers. obs.) The clutch is two blue-green eggs with brown spots; incubation begins as soon as the first egg is laid (TB), taking 3–3½ weeks (no exact record). Both sexes incubate, the female apparently more than the male (REA and pers. obs.). After hatching, both sexes feed the young, bringing items every 5–10 minutes (MC, TB, BT), also removing faecal sacs and taking turns to shade the young from the hot sun in the middle of the day (BT). The fledging period seems to be about 20 days, possibly a little less. The parents continue feeding the young after they have left the nest (TB, BT), but the period is not known. The laying season continues until at least January, with young sometimes in the nest until late February. TB believes that some pairs are double-brooded: at the Piton Grêle in the 1977–8 season, he observed two successive successful nests which he thought were produced by the same pair. Details of nests are given in Table 1.

Three of the five nests of which the outcome is known fledged young successfully, in one case despite the collapse of the actual nest structure when the nestlings were only part-grown (Table 1). One nest was destroyed by wind, and in another, smallish young disappeared overnight, suggesting predation

Table 1. *Details of recorded nests of the Réunion Cuckoo-shrike* Coracina newtoni. *All were at or near the Piton Grêle, Plaine des Chicots*

Season	Nest no.	Eggs laid (month)	Result	Remarks	Observers (see Introduction)
1974–5	1	Nov.	Hatched early Dec. Young vanished overnight 12–13 Dec.	Predated at night, possibly by rats	ASC, REA, J. Horne (per MC)
	2	Dec.	2 young fledged early Feb.	Nest partly collapsed when young half-grown, after which they sat on branch above	AWD, MC, TB
1977–8	3	Sept.	Blown down by gust before hatching		TB
	4	Oct.	2 young fledged late Nov.	Apparently the same pair	BT, TB
	5	Jan.	2 young fledged late Feb.		TB
1979–80	6	Sept.	(Not known)		TB (per NB)
	7	Sept.	(Not known)		TB (per NB)

by rats *Rattus rattus*, frequently seen in trees around the Piton Grêle (TB, RAC, pers. obs.), although Horne (Chapter 3) blamed Merles. BT reported the birds being very nervous when Harriers were around (the exposed nests would be rather visible to them), but the Cuckoo-shrikes were never seen to mob the raptors. After fledging the risk from the weather is still high: in 1975 cyclone Gervaise narrowly missed the island just after nest 2 (Table 1) had fledged, and when I visited the area on 14 February, a week after the storm, there was no trace of parents or young; the trees on the crest where the nest had been had lost half their leaves and many branches had broken off. NB noted, however, that after the severe storm Hyacinthe in January 1980, the population of Cuckoo-shrikes (and Merles) appeared unaffected (Barré & Barau 1982).

On 22 January 1975 AWD saw Cuckoo-shrikes carrying food in two other parts of the Plaine des Chicots; TB saw two fledged young on 30 January that year and (with NB) a pair feeding a single fledgling on 17 December 1979. Juveniles are recognisable by having pale tips to the feathers that give a barred appearance on the back (as well as the breast); this distinguishes them from females (MC, TB).

These few observations suggest that the nesting success of the Réunion bird is better than its counterpart in Mauritius. There birds are very rarely seen carrying food, and no recently observed nest has been successful, although a few juveniles have been seen (Chapter 4).

Moult

The only information on moult comes from five specimens in Leiden collected by Pollen and van Dam on 4 and 5 January 1865 (G. F. Mees *in litt.*). Of these, two adult males are in early wing-moult, tails old, while two adult females are not visibly moulting. This suggests that males moult sooner than females, possibly before the young are independent. The fifth bird is a juvenile, but we do not know whether the male and female collected on the same day were its parents. Another juvenile, collected on 30 May, is moulting its wing coverts but no flight feathers. No other museum specimens are in moult, but one in Cambridge, collected by Pollen on 24 May 1865, is in very fresh plumage.

Merle (Réunion Black Bulbul) *Hypsipetes borbonicus* (Gmelin) 1789; merle (merl)

Taxonomy

The Réunion Merle is treated here as a full species distinct from its Mauritian counterpart. *H. olivaceus*; the reason for this decision is given in Chapter 4, as it is the Mauritian form whose nomenclature is thus changed from current usage.

Status and distribution

The Merle is common in the mixed evergreen forests, occurring more sparsely in degraded forest and *tamarins*, and exhibiting a marked annual cycle of vertical migration. The species is classed officially as game, and the open season lasts for about 2 months, always in the period May–August. The actual dates vary from year to year and are given annually in a special decree. Merles are widely shot outside the official season (pers. obs.), and are also limed after being attracted to call-birds (Barré & Barau 1982).

The Merle was formerly extraordinarily abundant. In the mid-1700s even a "mediocre hunter" could kill 12–15 dozen in a morning (La Motte 1754–7), and when Bory de St. Vincent (1804) explored the Plaine des Chicots in 1801, he and his porters took only rice, relying for protein on Merles and for vegetables on palm-cabbage. Merles were so abundant, tame and inquisitive, that a few minutes' work with a baton would easily supply the day's need! By the 1860s Pollen (Schlegel & Pollen 1868) wrote that Merles were formerly very common but were now limited to an altitude of 800–1200 m in the mountains, where they were shot and limed in hundreds; he feared for the species' survival. Maillard (1862) noted it as scarce and Coquerel (1864) as "beginning to become quite rare". This impending extinction appears to have been illusory, as EH told me that his father and contemporaries used to have no difficulty in bringing back 150 Merles a day from the forests around the Volcan in the period from about 1900 to the 1920s; he had, however, noticed a very spectacular decline in numbers since about 1915. Milon (1951) saw only one, and was told they were much more numerous before the cyclone in 1948; he also gave details of persistent hunting. By 1964 there had been a good recovery, Merles being common in all inland forested localities (Gill 1964, Jouanin *in* Vincent 1966), and there has been no obvious change since (Gill 1973*a* and pers. obs.).

My interpretation of the apparent conflict between the mid-nineteenth century and early-twentieth century observations is that throughout the nineteenth-century there was probably a steady decline, taking the form of a diminution of numbers in lower, more accessible and more disturbed areas, while birds in inaccessible areas remained at a high density. By the early years of the century it was necessary to go to the forests around the Volcan to find large numbers, and by the middle or late twenties these last strongholds of abundance began to fail. Until 1948 there was a steady decline in density in the areas where Merles were still found, and the violence of the cyclone devastated the remainder. The recovery since then has only been partial, and the Merles, though common by Réunion standards, today are obviously at least an order of magnitude rarer than they were in the best areas only 50 years ago. Even the most assiduous hunter would be lucky to bag 15–20 in a day nowadays, and the average density is certainly a great deal less than that of *Hypsipetes* bulbuls in the Seychelles and the Comores (pers. obs.). I saw Merles in numbers equivalent to those on the Seychelles or the Comores only in May 1974 at the Ilet à Guillaume, when a large proportion of the population from the higher levels had moved down and concentrated in the forested depths of the Rivière de St. Denis.

To assess density in a favourable area, I mapped encounters with Merles during the Cuckoo-shrike census at the Plaine des Chicots in August 1974. Merles are very mobile, only sporadically vocal, and one individual can contrive to sound like several; my records in no way constituted a census, but suggested a density of about 25–40 pairs per km^2. At Bébour, in November 1973, I noted the encounter frequency as "*c.* a pair every 100 yards", which gives a density of about 115 per km^2, or two to three times as common as at the Plaine des Chicots. By contrast the numbers were much lower in cloud forest, and in the rare places where the birds occurred at low altitudes (Forêt de Mare Longue (St.-Philippe); Rivière des Remparts). From the breeding distribution described below (about 1000 km^2), and taking 25 pairs per km^2 as an average density, the total island population in 1974 can be estimated at roughly 25 000 pairs.

Merles were almost ubiquitous in forested land, other than plantations, but totally absent from heath vegetation and cultivated areas. They occurred below 1000 m only in patches of mixed evergreen forest, such as on the slopes of the ravines and canyons, and in the St.-Philippe area. They were totally absent from the occasional wooded ravines on the lower slopes in the west, where all the vegetation was exotic. They were also noticeably lacking from the *tamarin* forest at Belle-combe, and the mixed forests on the slopes of the Bras de la Plaine (above Grand Bassin), and the headwaters of the Rivière d'Abord (see Table 1 in the Appendix). It may be no coincidence that these last two areas are well within the sphere of influence of the notorious bird-catchers of the Plaine des Cafres. The breeding distribution of the Merle closely resembles that of the Harrier (Map 6), except that they do occur in Cilaos, but not above the tree-line (see Map 5), or in the other places mentioned above.

General habits and ecology
Once absurdly confiding, Merles no longer tolerate close approach. "Dr. J. C." (1861) wrote that when they came down to lower levels with their newly fledged young, they would come inquisitively to a call-bird and allow themselves to be shot one after another without taking any notice of the repeated bangs from the gun. Gill (1964) wrote, by contrast: "exceedingly shy and difficult to collect". There would appear to have been strong selection for caution over the past century.

In general behaviour the Réunion Merle closely resembles other black bulbuls on Indian Ocean islands (Rand 1936, Benson 1960, Grieg-Smith 1979b, Chapter 4). The birds were usually not seen until disturbed, when they flew off calling noisily, often stopping briefly on a vantage point chattering before vanishing over a ridge, usually flying several hundred metres before settling. The flight call is very distinctive, and Merles rarely flew without using it, providing the best way of detecting them. The 'cat' call was also frequently given, and otherwise silent Merles often responded to an imitation Cuckoo-shrike whistle with this call before bursting into a bout of chatter calls, often given by a pair together. The birds generally fed silently, with intervening spells of interaction between individuals or pairs. More than one pair sometimes congregated at a good food source, and during the period when the birds had moved down to lower elevations, they kept together in groups of four to ten, usually three to six, perhaps family parties. These groups were often noisy, unlike the silent pairs.

Interactions between Merles and other species

were infrequent, not I suspect because the Merles are docile but because they are so aggressive that all the other passerines usually keep well clear of them. The nearest in size to the Merle is the Cuckoo-shrike; in aggressive interactions with this species (see Cuckoo-shrike account) Merles were dominant. Horne (Chapter 3) also saw both species of white-eye mobbing Merles.

Food and feeding

Merles are largely frugivorous. As they usually fed silently, and flushed readily if disturbed, it was quite difficult to watch them feeding; I have only nine feeding observations, six being of fruits.

Aphloia theiformis seems to be one of the most favoured foods, and I saw Merles feeding on the white berries in May, August and November, which indicates the long season during which the fruits were available, though they were rather scarce in August. One bird swallowed three berries – each 1–1.5 cm across – in quick succession. I twice saw Merles feeding on figs *Ficus* spp.; I sampled the fruit of one of the fig trees and found it sweet and palatable (by contrast *Aphloia* berries are revoltingly bitter to the human tongue). My other observation was of a bird carrying an unidentified berry, presumably to a nest.

TB told me he had seen Merles feeding on the fruits of *Aphloia* (most often), *Sideroxylon borbonicum* and mistletoe (Loranthaceae spp.), and the flowers of *Nuxia verticillata*. Although I often flushed Merles from *Sideroxylon* bushes, I never actually saw them eating the small (0.75 cm) black berries. *Sideroxylon* was fruiting at the Plaine des Chicots in August and November, and according to TB the mistletoe has fruits all the year round.

'Dr. J.C.' (1861) reported the Merle to be entirely frugivorous, mentioning *Schmidelia* [= *Allophyllus*] *integrifolia* and *Prockia* [= *Aphloia*] *theiformis* as much sought after; the former is known locally as *bois de merle*. Pollen (Schlegel & Pollen 1868) repeated these observations, but added that occasionally the Merle also ate insects. Cadet (1977) considered the Merle to be the principal dispersal agent of the fruits of the two *Acanthophoenix* palms, and Lavergne (1978) blamed the species for the spread of the introduced arborescent weed *Solanum auriculatum*.

Several people told me that the introduced Strawberry Guava was much eaten by Merles, and in local bird-lore the Merles' vertical migration is closely associated with the fruiting of wild guavas (see below).

Merles also take nectar from flowers – though I saw them doing so only on introduced *Eucalyptus* (*robusta* and cf. *citriodora*). I also twice saw birds eating whole flower parts of *E*. cf. *citriodora*, apparently buds but possibly young fruit.

Apart from Pollen's remark the only mention of Merles eating insects is a passing reference by Jacob de Cordemoy (*in litt.* to Manders 1911). I saw insectivorous behaviour only twice, which suggests that it is less prevalent in this species than in other Indian Ocean merles (see e.g. Chapter 4, Benson 1960, Benson & Penny 1971, Frith 1979, Grieg-Smith 1979b). One bird was clumsily fly-catching unidentified insects from a tree perch; the others, in a group, were 'leap-snatching', also clumsily, in Camphor trees *Cinnamomum camphora* (both obs. 5 May 1974).

Breeding

'Dr. J.C.' (1861) gave the breeding season as July to December and rhetorically asked whether the bird was single- or double-brooded, while inclining towards the latter. Barré & Barau (1982) gave the same dates without saying on what they were based. TB commented that Merles and Olive White-eyes generally started nesting earlier than other species.

FBG collected a male and a female on 16 November 1964, both in breeding condition, the male with large testes and the female with a brood patch. He also found an empty nest, though does not record whether it was fresh or old, on 17 June 1967. A juvenile in the Paris museum, halfway through its wing-moult, was collected by Carié in March 1925. I saw a bird carrying a berry at Bébour on 31 October 1973, and BT saw a recently fledged brood there in early February 1978. I saw no other breeding behaviour, nor did I see any birds in juvenile plumage, though the groups of three to six in late April and early May may have been family parties. Unmoulted juveniles of all Indian Ocean merles are easily recognisable by paler body plumage and rusty-coloured primaries and tail feathers (A. Newton 1876, pers. obs.).

'Dr. J.C.' (1861) wrote that the clutch was normally of "two eggs the size of a hazel-nut and of a pale blue". Barré & Barau (1982) agreed that the clutch is two, but describe the eggs as pinkish-white with dark spots, like those of all other Indian Ocean merles

(Bendire 1894, Benson 1960, Penny 1974, Chapter 4), so it seems that 'J.C.' mistook the eggs of another species (Guérin (1940–53) suggested the Mynah) for those of the Merle. He described the nest as made principally of moss, which though differing from the materials used by merles in Mauritius, the Seychelles and Aldabra (Bendire 1894, Grieg-Smith 1979*b*, Chapter 4), agrees with Benson's observations in the Comores (1960), and FBG's nest in 1967. Barré & Barau (1982) noted also stems and leaves of ferns, and roots, as a framework under a mossy covering.

Moult

The only dated adult specimen which I have examined that was in moult was collected in March 1925 (Paris: 1964–313). It had lost all its old feathers, but the outermost two primaries were in pin; a juvenile, less advanced in wing-moult, was collected at the same time. Two late May specimens in Leiden are in fresh plumage (G. F. Mees *in litt.*), while one in Liverpool collected by H. H. Slater (presumably in January 1875) is very worn.

Several Merles I saw at the Ilet à Guillaume and at Cilaos in late April and early May 1974 were moulting their tails, and one at Cilaos was also halfway through wing-moult. Only a minority of birds were moulting, the rest having perhaps finished by then; certainly there were no birds showing juvenile plumage or dark bills.

Altitudinal migration and a fat cycle

A seasonal cycle of fat was reported by La Motte (1754–7), and this was associated by "Dr. J.C." (1861) with an altitudinal migration. Merles ascended high into the mountains in July and August prior to breeding, at which time they were so fat and heavy that hunters described them as 'burying themselves' when they fell. La Motte (1754–7) gave the fat months as July to September. Whether the fat cycle still exists I cannot say, having caught only one bird. However I never heard it referred to by the many people with whom I discussed the movements and poaching of Merles, so it may no longer be as marked as it once was. The vertical migration still occurs, though it is difficult to get a clear picture of its extent and timing over the island.

At the Plaine des Chicots some Merles were present at all times, but there were certainly fewer in

February and May than in August and November; there were unusually large numbers in the bottom of the adjacent gorge, at the Ilet à Guillaume, in May, though there had been 'normal' numbers in November 1973. TB told me they returned in numbers to the higher parts of the Plaine des Chicots in July and August.

Two other people gave me the following information. M. Antoine of the ONF St.-Benoît [November 1973] told me that Merles were at Grand Etang in June–July, went up to Bébour following the guava crop in July–August, then re-descended in September–October, possibly nesting at Grand Etang. An anonymous mayoral chauffeur from St.-Denis [November 1973] said that Merles descend towards the coast in large numbers when guava is in fruit (December–February at St. Philippe, February–March at St.-Denis (Le Brûlé)).

These accounts are incomplete and inadequate: some Merles certainly remain to nest at Bébour (I saw one carrying food), and, while guavas were fruiting heavily at Le Brûlé at 800–900 m in early May 1974, there were no guavas to speak of at the Ilet à Guillaume, but plenty of Merles. There remains (second informant) the interesting possibility that the timing of the migration may not be the same in different parts of the island.

Réunion Stonechat *Saxicola tectes* (Gmelin) 1879; *tec-tec* (tektek)

Taxonomy

Most authors have treated the Réunion Stonechat as a full species, but Ripley (1964 *in* Peters *et al.* 1931–, vol. 10) considered it a race of the widespread *S. torquata*. While it is clearly close to *torquata*, it differs in that the male always has a white (instead of black) throat and often also a white eye-stripe. The habitat range occupied by the Réunion birds is also greater than is usual in *torquata*. Schlegel & Pollen (1868) briefly surveyed the various forms of Stonechat and also concluded that Réunion birds were distinct from all the rest, although at the time they preferred to keep them within *torquata*.

Status and distribution

All previous authors have considered Stonechats common or abundant, and the same is true today. Pollen

(Schlegel & Pollen 1868) noted that they were commoner above 1200 m than below, which is still the case. Stonechats were evidently resident at sea level at that time, if only sparsely, whereas now they are unusual at very low altitudes: Pollen (1868) saw them along the road from St.-Denis to Ste.-Marie where they would never be seen today, and Régnaud (1878) reported them from sea level to the highest parts of the island. Milon recorded the birds as "very common everywhere", so the change may be recent. There seems to be no obvious reason for this retreat from low altitudes, but it parallels that seen in the Olive White-eye (q.v.).

The Stonechat's present distribution closely resembles that of the Harrier (Map 6), but unlike the Harrier it is found around the Volcan and at Cilaos. The lower limit of distribution is 800–900 m in the north and north-east, declining southwards to only 300 m at Mare Longue (St. Philippe) and the Rivière des Remparts. Further north it occurs at lower altitude in secluded places (Ilet à Guillaume, 700 m; Grand Etang, 500 m). My information for the west is rather sparse, but suggests that north of St.-Pierre, it is very scarce below 950 m, except in the beds of major rivers (Bras de la Plaine, Rivière St.-Etienne). There is no upper altitude limit; I saw Stonechats on the top of the Piton des Neiges in May, although I doubt whether they breed as high as that (MB did not see any above 2900 m in January). The birds do still sometimes come down to sea level, especially outside the breeding season; one was hawking insects over the water from a clump of Papyrus *Cyperus papyrus* at the Etang de St.-Paul on 11 May 1974, Barré & Barau (1982) gave two records (canefields, west coast scrub), and I found a singing male at La Vièrge au Parasol (Grand Brûlé) on 11 October 1978.

Habitat and abundance
The Réunion Stonechat occurs in a very wide range of habitats, though it is not equally abundant in each. In addition to typical stonechat habitat (open scrub), the Réunion bird is also found in 'wheatear' (*Oenanthe* spp.) habitat (open ground) and 'robin' (*Erithacus* spp.) habitat (forest), the one species taking advantage of all the 'small thrush' niches open to it. It appeared to be most common in *tamarin* and mixed evergreen forest with openings and clearings, though it occurred in good numbers even under closed canopy as tall as 20

m or more, as at the Forêt de Mare Longue (600 m). It occurred in any kind of secondary vegetation at the right altitude and was the commonest bird in *tamarin* plantations, but occurred only at very low density in *Cryptomerias*. In *Philippia* heath it was common, but less so than in mixed forest, and its density declined upwards as the heath got shorter and gave way to grass (Appendix); the sparse vegetation around the Volcan supports only a small population, very thinly spread. Stonechat density was quite low in the introduced gorse *Ulex europaeus* scrub on the Plaine des Cafres. At Grand Etang, at about 500 m, the birds were very scarce, as they were in the cloud forest along the east side of the island. In apparently suitable habitat at the Bois Bon Accueil (Plaine des Makes, 970–1200 m) there were very few Stonechats. Human activity probably reduces the bird's density, but I saw them in gardens at the Plaine des Palmistes and at Cilaos.

In November 1974 I counted nests in an area of *tamarins* with little undergrowth around the rest-house at the Plaine des Chicots (1834 m) (Map 9). In the six 1-ha grid-squares I found eight nests, and suspected the existence of another three – an average of nearly two per hectare. I have no reason to suppose this density to be unusual, and it corresponds with my subjective impressions throughout the Plaines d'Affouches and des Chicots, Bébour, the forest up to the Petit Matarum at Cilaos, and elsewhere. The densities in the cloud forest are at least an order of magnitude lower. I have not attempted to estimate a total population for the island, but it must be well over a hundred thousand pairs.

General habits and behaviour
Réunion Stonechats are conspicuous birds, tame and inquisitive. They draw attention to themselves by perching near an intruder in their territory, and calling vigorously a repeated *tek tek* (hence their local name, which was already known to Brisson in 1760, and from which derives Gmelin's specific name *tectes*). The song, performed in flight or from a perch, is loud and distinctive.

The birds live solitarily, though interactions between adjacent territorial males or brief encounters between members of a pair were frequent when at high density. In the breeding season pairs were strictly territorial, but I do not know whether the same birds remain in the territories throughout the year. There must be some movement to account for the variable occupation of some places, such as the summit of the Piton des Neiges, but this could be due to dispersing juveniles.

Food and feeding

The birds invariably fed singly, and almost entirely on invertebrates. Most food was taken on the ground: the bird perched on an eminence, or branch, surveyed the surrounding area until it saw a prey, then dashed over and snatched it up. Food was spotted up to 3 m away. If the item was large the bird often landed beside it before attempting to take it. On most occasions the bird returned to a perch (usually a new one) to eat the prey, but if there were many items together it remained at the site to eat them. Large insects were beaten on branches before being swallowed. Food was also taken in flight from foliage, walls, rocks, etc., and in vegetation the feeding method occasionally turned into leap-snatching or search-gleaning; they also often fly-catch. Stonechats quite often fed on the ground like Robins *Erithacus rubecula*, watching for a movement and hopping or flying forward to pick up the prey, but clearly preferred a vantage point from which to survey the area. HG and I once watched for over 20 minutes a bird feeding by simply picking small items from among the gravel on a wet road.

It may well be the need to feed largely on the ground which limits the stonechat's numbers in cloud forest. In this habitat the close and rampant undergrowth of *Machaerina* and ferns, and the thick moss and fallen debris on the ground, make the normal feeding methods of the birds impossible, and confine them to foraging on vegetation, which, being dense, prevents surveying from a distance. An open field layer with little ground vegetation seems to be necessary for them to make full use of a habitat.

The food taken appears to be almost exclusively insects and other invertebrates. Most items are too small to identify in the field, but occasionally larger items such as caterpillars (Lepidoptera) up to 4.5 cm long were taken. Like Robins in Britain, the Stonechats were always on the alert for unexpected sources of food; TB used to break open stumps of dead wood and

Map 9. Sketch map of the area around the rest-house (*gîte*) at the Plaine des Chicots, showing distribution of Stonechat nests in November 1974.

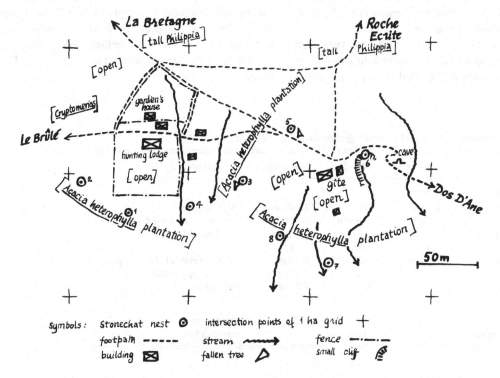

symbols: Stonechat nest ◉   intersection points of 1 ha grid ✛
footpath -----   stream ➤➤   fence —·—·—
building ✉   fallen tree △   small cliff ▨

the watching birds would come and feed on the large (5–7 cm) cerambycid beetle larvae thus exposed. The bird feeding on the wet road was apparently eating tiny larvae, probably dipterans.

The guts of two nestlings from the Plaine des Chicots who died fully fed when their nest was flattened (probably by a mule) contained a wide range of insects in both size and type (Table 2) and a very small amount of vegetable matter. Food brought to nestlings must vary with availability: MB saw only small items (<0.75 cm) brought to a nest in Cilaos he watched for 1¼ hours in January 1974. Although faecal matter from the same dead nestlings contained easily identifiable remains, an adult sample contained such tiny fragments that only a few caterpillar spiracles could be identified.

There is very little about the bird's food in the literature; Pollen (Schlegel & Pollen 1868) mentioned Diptera caught by fly-catching, and Guérin (1940–53) added "grasshoppers, larvae, ants, and flies picked up from the surface of the ground".

Breeding

Stonechats' nests were the easiest to find of any Mascarene endemic bird, so a quantity of data is available, although none of my visits, nor those of any colleague, spanned a full breeding season.

The breeding season seems to be very well defined, being from mid-October to late January. In 1973 and 1974 the most advanced nests I found at the beginning of November had hatched on 31 October (1974, Plaine des Chicots; i.e. first egg c. 15 October) or 1 November (1973, Bébour). TB reported a bird incubating on 12 October 1978. Similarly the last nests were on very similar dates in 1974 and 1975; MB watched a nest at Cilaos with young still only 4 days old on 21 January 1974, and AWD found one or two nests with large young at the Plaine des Chicots on 22 January 1975; by 14 February, when I was there again, there were certainly no active nests, but a cyclone had struck in the interval.

Past records are similar: Carié found eggs in November and January (Berlioz 1946), Milon (1951) found eggs and young in December, FBG likewise in November, and F-W, K & T saw a female carrying nesting material on 17 October 1971. BT and Barré & Barau (1982) confirmed these dates. Regnaud (1878) reported the laying season as starting in September; the birds then present at low altitudes may have begun earlier.

The start of breeding was not well synchronised between individuals; in 1979 there was at least 3½ weeks between the first eggs of the earliest and latest nests in eight nests at the Plaine des Chicots. Although there would be time for the early starters to have two broods by the end of January, I suspect that they do not do so. The sparse data available imply that pairs, with their young, become nomadic for a while after nesting; certainly the birds I was familiar with around the rest-house in November seemed to have gone in February.

Table 2. *Stomach contents of two nestling stonechats* Saxicola tectes

	Life-stage	Number	Size range	Comments
Arthropods				
Lepidoptera	Larvae	6+	15–25 mm	At least 2 spp.
	Adults	Several	?	Fragments and wing scales
Coloeoptera	Larvae	9+	5–10 mm	All Melidridae
	Adults	1+	3 mm	
Diptera	Adults	5	4–7 mm	Some referable to Dolichopodidae and Trypetitidae
Spider		1	4–5 mm	Salticidae
Plants				
Seeds		2	2–3 mm	Unidentified
Plant fragments		Several		Strips of fibre

Nests were easily found once incubation had started, as females sat fairly tight until approached within 2 or 3 m, when they rapidly left the nest. During the nestling period parents bringing food could be followed to the nest quite easily.

Nests were on the ground or in a tuft of ground vegetation, although a common variant was a vegetated bank where a path had been cut out of a slope. Nests were built of moss and leaves, lined with feathers and fine grass; Milon (1951) gave a detailed description with measurements. The nest was usually well concealed, although some on banks were exposed; it was the birds' behaviour which betrayed the site. I have only one observation on nest-building: by the female, as in *S. torquata* (Johnson 1971).

The clutch is two or three (Barré & Barau (1982) claimed it may reach four). Milon (1951) found 11 clutches of three to 1 of two, but my ratio was 8:6 (1973 and 1974 combined; all but one in November). All my clutches of two were being incubated, and only one was not confirmed by a subsequent visit. In addition I found two broods of three and two of two after the young had hatched. Since in seven nests I followed up all the eggs hatched, these late-found nests can legitimately be included in the ratio. The average clutch size probably varies from year to year and possibly also with date, altitude or locality.

The eggs are generally a pale, dull blue-green (Berlioz 1946, Barré & Barau 1982), rarely cream (pers. obs.), finely spotted with brown, chiefly at the large end. I have seen eggs of both ground-colours in the same clutch.

I was unable to measure the incubation period exactly, but it is probably around 14 days, as four nests had intervals of 12, 13, 13 and 14 days from first being found (all clutches complete) to date of hatching. Only one nest gave a reasonable estimate of the nestling period, large young still being present at 13 days old. In both cases the true figure is probably 13–15 days, which corresponds very closely with that for European Stonechats (13–14 days for incubation and 12–16, usually 13–14, for fledging: Johnson 1971).

Only females incubated, but males helped to feed the young, both in the nest and after they fledged. MB watched a nest with three 4-day-old young in it for 75 minutes on 21 January 1974: the male brought food five times, the female six; the female also brooded the young for four periods (1½, 2½, 4 and 7¼ minutes). I do not know for how long the young are dependent on their parents after leaving the nest.

Nesting success appeared to be high, in spite of the ease with which nests could be found, probably because ground predators are quite rare. MC saw an immature male Harrier approach and inspect a Stonechat nest, but was too far away to see exactly what happened. Of the nests I followed to beyond half-grown young, six were still successful when last seen, two had been trodden on (there were two mules and a bull at the Plaine des Chicots; this particular problem must be rare elsewhere!), one was probably raided by humans, one lost eggs to an unknown predator, and one may have fledged, but there were rodent droppings in it as well as feather-dust. BT found an incubating female eaten on the nest.

In January and February there were young Stonechats in evidence everywhere (MB, ASC, AWD), but the young had moulted by April and were then no longer distinguishable from adults. Presumably the adults moult at the same time. The juveniles are typical spotty young thrushes. I made the following notes in February 1975: Juvenile plumage more or less as female, but orangey-buff breast feathers have brown centres; throat white; rump slightly rusty; head and back more or less spotty, the feathers darker brown than in the female but with pale tips; a palish bar on the greater coverts (formed by pale-tipped feathers).

Plumage variation

In November 1973 I noticed apparently female Stonechats singing and acting territorially; FBG (1964) noted the same. It seems that such 'females' are in fact males in female-type plumage; indeed among males a complete range of plumages exists from female-type through to fully black and white. In addition there are several phases of black and white plumage.

There are essentially two types of variation: the first is in the dominant colour of the head and back, which varies from dull grey-brown ('female') to black ('male'), and the second is in the amount of white around the eye, in the wing and on the rump. The presence, absence and extent of white in these parts appears to vary independently, so that birds can be found with all possible combinations. The extreme cases of 'black' and 'pied' morphs are illustrated in Fig. 3, together with an inset showing an intermediate head pattern (see also Cheke 1975d, Barré & Barau

1982). In addition to a larger or smaller area of white, or none at all, the rump may be greyish, contrasting somewhat with the back. The rusty breast patch varies considerably in size and intensity.

A typical female is dull grey-brown on the back with darker wings and tail, with no trace of pale on wing or rump. She has a pale eye-stripe encircling the ear-coverts and is pale underneath, often with a slight rusty or fawn wash on the breast and underparts.

At Bébour and Bélouve in November 1973 female-type birds holding territories often apparently lacked a mate. In 1974, however, at the Plaine des Chicots, one breeding male (nest 8) was extremely like his mate, though distinguishable by a trace of white on his rump, and darker ear-coverts. The other breeding males ranged from fairly female-like to clearly black and white birds. The pair at nest 8 started breeding rather late, in keeping with the possibility that female-type plumage indicates immaturity.

To investigate whether or not this plumage variation is age-related, I ringed 21 nestlings and wrote an article (Cheke 1975*d*) in the local natural history journal in the hope that bird-watchers might report on the appearance of ringed birds over the following years. Unfortunately I had no reports, but neither AWD nor I could find any ringed young in January or February, only 2 months after I had marked them.

Fig. 3. Colour phases of male Réunion Stonechats: (*a*) full dark phase; (*b*) full pale phase; (*c*) intermediate phase, head only, showing partial eye-stripe. The shaded areas on the breast are rusty orange. Note that the presence and amount of white on face, wing and rump vary independently, and individual birds may have any combination of dark and pale in those areas. (From the original drawing used for Cheke 1975*d*).

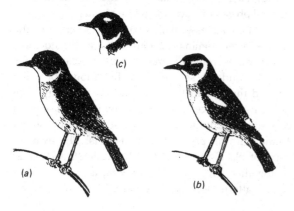

On present evidence I suggest that the brown to black variation may be age-related, but that the degree of white in the final plumage may be polymorphic.

Mascarene Paradise Flycatcher *Terpsiphone bourbonnensis (Müller) 1776; çakoit (sakwat), "oiseau de la Vièrge"*

Taxonomy and nomenclature

The nominate Réunion form *T. b. bourbonnensis* differs slightly from the form in Mauritius, and the two are generally considered subspecifically distinct (Berlioz 1946, Benson 1971).

The local name *oiseau de la Vièrge* is well established in the literature but is nearly extinct as a Créole (as opposed to French) name (Cheke 1982*b*). The term *zoiseau la vièrze* is now applied widely to small birds in general, especially the white-eyes, and is interchangeable with *zoiseau bon-Dié*, although it still applies specifically to Flycatchers in the Cirque de Salazie. *Çakoit* is usually pronounced *tchakoit* by French-speakers.

Status and distribution

Nineteenth-century authors considered the Flycatcher to be common and widespread, inhabiting "not only the forests which cover the mountains, but also wooded areas near the coast, and even sometimes town gardens" (Schlegel & Pollen 1868: my translation). The situation is much the same today, although there has undoubtedly been some retreat from the lowlands, and the birds have gone from the larger towns, though Gill (1964) saw them in gardens in Hellbourg.

The present distribution is similar to that of the Merle, but in addition they reach the coast around the Volcan, and occur in secondary woodlands mostly above about 200 m on the east side of the island (Milon 1951, Barré & Barau 1982). Flycatchers do not appear to occur in the scrubby woodlands along the west coast, although they are seen in similar habitat in Mauritius (Chapter 4); however Gill (1964) and BT found them not uncommon in the *Casuarina* forests at Etang Salé. Like the Merle, they are absent from the forest at Bellecombe (Volcan) and very scarce in the copses on the upper part of the Plaine des Cafres, but are not uncommon at Grand Bassin. The tree-line is the upper limit of their distribution, and they are seen only in the edges of *Philippia* heath.

Habitat and abundance

The Flycatcher was confined to woodland, but was very catholic in its choice of forest type. In general it was most abundant in low and mid-altitude mixed forest – e.g. Tremblet (more or less sea level), Mare Longue (600+ m), Grand Etang (500–600 m) – but occurred regularly through all types of forest, maintaining reasonable densities even in cloud forest where other species were scarce (Appendix). Its absence from forests near the Volcan, noted above, remains to be explained, and I found it relatively scarce at Bébour and Cilaos, in apparently suitable habitat, although Gill (1964) considered Cilaos to be the area of greatest abundance. It was fairly common in 15–25-year-old *tamarin* plantations at Bélouve, but rarely occurred in *Cryptomeria*.

I noted how often I met Flycatchers at the Plaine des Chicots, an area of only middling density for the species. The observed density proved to be very low – only about five pairs per km², which is considerably fewer than the Cuckoo-shrike. I consider that this figure underestimates the real numbers of Flycatchers, because even in low densities Flycatchers seem to hold only very small territories (2–3 ha in Mauritius: Chapter 4), so that in any transect an observer could miss many of the territories. A figure of 10 pairs per km² is more reasonable for the Plaine des Chicots. The density was certainly much higher than this in mixed evergreen forest at lower altitudes: Barré & Barau (1982) cited one pair per hectare at the Tremblet (Grand Brûlé) in June 1980.

Habits and behaviour

The habits of the Flycatcher in Réunion are similar to those of the same species in Mauritius (Chapter 4). On November 1973 at Grand Etang I saw two Flycatchers chasing each other in flight at a considerable altitude (estimated at 200 ft; 60 m) above the forest. They disappeared before I was able to see what sex they were. At no other time either in Réunion or Mauritius did I see Flycatchers outside the canopy.

Food and feeding

All observers agree that the Flycatcher is exclusively insectivorous, though no analysis has been attempted of what exactly is eaten. Pollen, who may have examined stomach contents, mentioned small flies and dragonflies (Schlegel & Pollen 1868). I have no observations on food in Réunion, but BT noted moths,

Diptera and mayflies (Ephemeroptera) being fed to nestlings in December 1977.

Flycatchers feed by fly-catching, fly-picking, leap-snatching or gleaning, and my impression was that in Réunion they gleaned more frequently than in Mauritius. Fly-catching is the primary mode of feeding, but the others are important and occasionally practised continuously over long periods if there is a suitable food source. Flycatchers often associate with passing parties of Grey White-eyes, and glean or fly-pick with them.

Breeding

There is little information on breeding in the literature, and I have only a few observations to add. Focard (1861) wrote that he thought the nests were started in September, and that young fledged in late November or early December; Barré & Barau (1982) also gave the season as September to December. Milon (1951) found two nests in mid-December 1948, in one of which the male was incubating eggs near to hatching. FBG collected ten birds (five of each sex) in November 1964 and found them all in breeding condition, while a male (in grey plumage) collected on 14 September 1967 had very small testes. CAB found a nest with eggs being incubated at Cambuston on 27 October 1964 and another with one egg on 10 October 1976. BT watched a nest with young at the Plaine d'Affouches in mid-December 1977, and saw a freshly fledged brood there in late January 1978.

I saw a pair feeding young at Bélouve on 3 November 1973, and found a nearly completed nest at Bébour 2 days earlier. Although TB saw a bird carrying food at the Plaine des Chicots on 16 November 1974, there was little apparent nesting activity there during that month, though flying young were vociferously present by 22 January 1975 (AWD) and again in mid-February (ASC). Some of the February young were in full juvenile plumage while others had nearly finished the post-juvenile moult. In Mauritius this moult was complete within 5 weeks of fledging (Chapter 4), so even the most advanced birds could have fledged as late as early January.

These records suggest that the peak of breeding may be later at high altitude than lower down (nights are still cold at the Plaine des Chicots in November). Focard's, Milon's and Gill's nests were at relatively low altitudes, and Cambuston is only 20 m above sea level. Bélouve is, however, at 1500 m, and MB saw fledged and moulted young being fed on 16 January 1974 at 1400 m at Cilaos, so that situation is not clear-cut. Late breeding is suggested, though, by MJEC's observation of a family party still together on 30 March 1974 at the Plaine des Cafres (1650 m). In Mauritius the breeding season extended from August to January and February at sea level, usually with a peak in October and November, but there is little information for upland areas (Chapter 14).

Focard (1861) and CAB (*in litt.* 1976) gave the clutch size as two or three, though Focard noted hearsay records of four or five eggs which he ascribed to two females laying in the same nests. I was unable to reach the nest in which birds were feeding young at Bélouve, but the five records of fledged young in 1974–5 refer to only one or two being seen, although three fledged from BT's nest in 1977. In Mauritius the clutch is apparently always three, all of which sometimes fledge (Chapter 4). CAB (*in litt.*, Barré & Barau 1982) described the egg as pinkish white, spotted with irregular rusty spots which form a dense band around the broad end; Focard (1861) claimed that the ground-colour was greenish-blue.

The nest-site was described by Focard (1861) as being in forks of branches, 4–5 (French) ft (1.2–1.5 m) up in "cultivated trees"; Milon's nests (1951) were on shrubs, 1.3 m from the ground. The nests I saw ranged from 5 to 12 ft (1.5 to 3.7 m) above ground, and CAB also recorded a nest at 3 m. Mauritian birds generally chose very slender supports for their nests (Chapter 4), but while some Réunion birds do so, others built on stouter material: two on *tamarins* in a plantation at Bélouve were on wood up to 1½ in. (37 mm) thick. Focard mentioned "bois de gaulettes" as a favourite site for forest-nesting birds; this is presumably the small tree *Doratoxylon apetalum* (Rivals 1952: 62, Cadet 1977: 310). The only forest nest I found was in a large old *tamarin*, at the end of an overhanging branch instead of the more usual upright fork. BT's nest was in cut branches of *Cryptomeria* piled on the ground around the trees.

The shape of the nest is usually conical, fitting into the fork. The outer part is of soft green moss with pale patches of frondose lichens (spiders' 'egg-bags' are used instead in Mauritius (Chapter 4)), the cup being of fine fibres. The nest was described in detail by Focard, Pollen and Milon (*locs. cit.*); all referred to the mossy outside being held on by finely woven individual threads of spider's web; these are not visible on superficial examination of nests *in situ*, but CAB (*in litt.*), describing a nest before him, confirmed this delicate use of web threads. In two nests measured by Milon the cups were 48–52 mm across and 39 mm deep; CAB's nest was 45 mm across and 30 mm deep. The external depth depends on the shape of the fork or support in which the nest is placed. One nest I found, not quite finished, was roofed over from one side with fine fibres, somewhat like a half-built Cardinal's nest.

Plumage variation

Paradise Flycatchers in general are notorious for polymorphism (Chapin 1948, Hall & Moreau 1970), and for the males taking a long time to reach full plumage (Ali & Ripley 1968–74, vol. 7). The Mascarene species is superficially monomorphic, but, particularly in Réunion, shows plumage variation that might be partly polymorphic, although certainly also related to age.

The plumage acquired during the post-juvenile moult in both sexes resembles that of the adult females, and after an unknown interval the males acquire the glossy black head of the typical adult plumage (Chapter 4). Whereas in Mauritius it was rare to see territorial pairs with a 'grey' male, in Réunion it was much more common, though variable from place to place. The ratio of black to grey males varied from mostly grey (1:9, Bébour, November 1973) to all black (Ilet à Guillaume, May 1974), with all combinations between.

A priori there are three possible explanations for the situation in Réunion, any of which could be operating in combination (cf. the Stonechat).

1. There is a very high juvenile survival rate, so that a high proportion of the population is young. (There are many indications that, in general, nesting success in all species is higher in Réunion than Mauritius.)
2. The acquisition of full male plumage is delayed in some or all individuals.
3. Some males remain permanently in 'post-juvenile' plumage.

Without further study I do not feel justified in suggesting which of these processes might provide answers in this case; however, in view of the well-established habitat-related polymorphism in the Grey White-eye (Gill 1973b), and the situation in the Stonechat (q.v.), the Flycatcher would be an additional and interesting subject for further investigation. My observations are summarised in Table 3.

Table 3. *Proportions of black- and grey-headed male Paradise Flycatchers* Terpsiphone bourbonnensis *in Réunion*

Place and altitude	Date	Habitat	Pairs seen		Individuals seen (includes pairs in col. to left)		Subjective estimate (no actual count) of ratio black:grey
			Male black	Male grey	Black males	Grey birds	
Mare Longue, 600 m	8.11.73	Tall mixed evergreen forest					4:1
Ilet à Guillaume, 700 m	4–6.5.74						No grey-headed males seen, although flycatchers common
Cilaos, 1200–1500 m	26.4–2.5 1974	Mixed evergreens			3	5	
		Secondary scrub and woodland	5+	2			
Pl. des Chicots, 1400–1800 m	7.11.73	Mixed evergreens and *tamarins*					1:3
	12 and 13.8.74		4	4	5	19	
	5–6.11.74		3	1			
Bélouve 1400–1600 m	2–4.11.73	*Tamarin* plantation	1	4+			1:9 (based on pairs and individuals not all separately noted at the time. Only 1 black-headed male seen)

Mascarene Grey White-eye Zosterops borbonicus (Boddaert 1783); zoiseau blanc (swazo blañ)

Gill (1971a, 1973a) studied both Réunion white-eyes in detail; here I summarise relevant parts of his work, and expand only on points he did not cover or on which I can provide new or conflicting information.

Taxonomy

The Grey White-eye on Réunion exhibits great variation, which Gill (1973b) has shown to be partly clinal and partly polymorphic. Gill concluded that earlier attempts to classify the morphs as subspecies could not be upheld, although there was a substantial geographical element in the variation. The names proposed by Hartlaub (1877) and Storer & Gill (1966) can be considered synonyms of *Z. b. borbonicus*, for which Boddaert, not Gmelin, is the earliest authority (Cheke 1973b). There is a distinct race in Mauritius (see Gill (1971a) for details of the differences).

The distribution of the Réunion colour morphs is shown in Map 10. The morphs can be summarised as follows. (*a*) Birds with brown backs: brown-headed (C on Map 10), western and central; grey-headed with brown nape (B), southern; grey-headed with grey nape (A), northern and eastern. (*b*) Birds with grey head and back: rare below 700 m in the north and east, 1200 m in the west, rising to 65–100% incidence above 1300 m (north and east), 1600 m (west). Occurs in conjunction with all three brown morphs. Within and between each morph type there is a complete range of colours, but the proportion of intermediates is low.

Status and distribution

The Grey White-eye is, and has always been, an abundant bird found almost throughout the island. A good summary of its habits was given as early as 1778 by de Querhoënt (*in* Buffon 1770–83), who added that it was "found everywhere in Bourbon". The species can still be seen everywhere on the island except where woody vegetation is very sparse (e.g. around the Volcan, above 2750 m, some areas of very dry scrubby savannah on the west coast). It is the only native bird (apart from the Flycatchers at Etang Salé) likely to be seen below about 800–1000 m on the west coast from Les Avirons north to La Possession.

Gill's (1971a) estimate of the total population in 1967 as upwards of 556 000 agrees with my aggregate estimate of bird density (Appendix), i.e. about 2.8 birds per hectare of occupied habitat (about 2000 km^2) over the whole island. Although of the right order of magnitude, I regard this estimate as a little high for this species alone in 1974–5. EH considered the Grey White-eye to have declined by about 80% in his lifetime.

Habitat

Gill (1973*b*) defined the habitat preferences of the Grey White-eye as follows:

> Unlike most remaining endemic Réunion birds, which depend on the remnants of indigenous forest, *Z. borbonica* [*sic*] is common in most habitats, including gardens, evergreen forests, *Casuarina* forest, and highland heath. But despite its seemingly broad habitat tolerances, it is primarily a species of open and disturbed areas rather than intact forests; its present distribution reflects the widespread destruction of original habitats. In tracts of undisturbed interior forests it is decidedly scarce except at the edges of large clearings and along streambeds.

Although the first part of the statement accords well with my observations, I disagree that the species prefers disturbed areas and edges to true forest; Gill's conclusion may result from correlating scarcity with 'remote' forests without taking into account the kind of forest which has been preserved in large tracts. The largest areas of "undisturbed interior forests" are cloud forest on the eastern slopes of the island which are intact largely because they are biologically very unproductive – there is no useful timber and the slope, rainfall and soil conditions are unsuitable for cultivation. I suspect that the places where Gill found Grey White-eyes scarce are precisely those areas where they are scarce for reasons quite other than lack of disturbance, i.e. where bird density is very low due to the poor quality of the habitat. My surmise is supported by a comment by FBG (*in litt.*) that "if you hike back into the Bébour forests the absence of white-eyes, except near streambeds and old clearings, is striking". One of the characteristics of that forest is that stream beds are the places where the vegetation is at its richest and most developed, a healthy mixed evergreen community thriving in the crevices, compared with the poorer forest of the *plaine* on either side; hence the white-eyes (and other birds) tend to concentrate there. I never found an area of biologically rich, undisturbed forest where Grey White-eyes were scarce. They were particularly common in the untouched forest around 600 m at Mare Longue (St.-Philippe), and were plentiful in the Bois Bon Accueil

Map 10. Distribution of brown-backed morphs of the Réunion Grey White-eye. A, grey-headed; B, grey-headed with brown nape; C, brown-headed. Zones of contact and overlap are indicated by dots within the cross-hatching. In the sketches of the birds, cross-hatching represents brown, stipple represents grey. (From Gill 1973*b*; 30 (Fig. 19).)

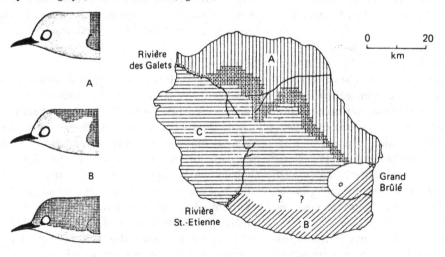

(Plaine des Makes), and the quiet parts of the Plaine des Chicots. The only remote, mixed forest with rather few Grey White-eyes was the western part of the Plaine des Fougéres, where all bird species were at a relatively low density (Appendix).

I conclude that the Grey White-eye's original optimal habitat was lowland mixed evergreen forest (as at Mare Longue) but that it had also penetrated to a greater or lesser degree all the other habitats originally available on the island, major physical boundaries and wide climatic range allowing the development of variation in this very sedentary species. Since man has modified the island it has successfully adapted to forest-like habitats (orchards, treed gardens, plantations, secondary forest) wherever food is available. This modification has undoubtedly disturbed the balance of the polymorphism, as Gill (1973b) pointed out, by "altering the parameters of optimal adaptation". The Grey White-eye has certainly suffered little from man's intrusion, but I cannot support the view that it was 'pre-adapted' to disturbed habitats, especially as these would have been of very limited extent (e.g. landslips, fires due to vulcanism) before man's arrival (see also Barré 1983).

Apart from the sparsely vegetated natural habitats the only types of vegetation not or scarcely penetrated by Grey White-eyes were cane-fields, geranium fields, *Cryptomeria* plantations, and the sparse dry thorn scrub on the west coast. The birds were, however, relatively scarce in cloud forest, most areas of *Philippia* heath, and *tamarin* plantations. Their abundance in secondary forest and *Casuarina* plantations was certainly equal to that in most mixed evergreen forests, though not as high as at Mare Longue.

General habits and behaviour

Réunion Grey White-eyes are very small brownish or grey birds, often paler below, with a white rump. They foraged noisily in parties usually of four to ten birds, but sometimes up to 15–20; at good feeding sites birds would be coming and going all the time, and several groups were sometimes present at once. The birds were sociable and rarely aggressive to each other or to other species: Gill (1973b) noted frequent huddling and mutual preening. In the breeding season the striking white axillaries and rump were displayed to good effect.

Gill (1973b) saw several colour-ringed birds in the same place over several months, and concluded that, while very mobile over a very small area, at the island level the birds were extremely sedentary. There was, however, some vertical migration up and down the 1000 m cliffs of the Rivière des Remparts to take advantage of flowering *Sophora "nitida"* (= *denudata*). Gill also found that in the breeding season birds would form pairs or groups with restricted ranges and that these birds would join passing parties temporarily. Only detailed study of individuals over a long time in one place would reveal the true mobility and range of these parties.

De Querhoënt (*in* Buffon 1770–83) described Grey White-eyes mobbing ground animals (partridges, hares, cats), a habit known to hunters and used by them to help find game. Apparently the birds danced about the animal making a characteristic call. Grey White-eyes have a very piercing mobbing call, often addressed to human intruders, but I never saw or heard tell of the type of mobbing described by de Querhoënt. Perhaps the habit has died out – the ground animals referred to were relatively recent introductions in the mid-1700s and might then have been more stimulating to the birds (but see Chapter 4).

Food and feeding

Grey White-eyes glean insects on twigs, flowers and leaves, and by bark-creeping, but also eat fruit and (rarely) buds, and take nectar from flowers and sweet sap oozing from trees (Gill 1971 & below). They are extremely agile, capable of probing while hanging upside-down from the bark of large branches, as well as hanging tit-like from twigs. The species exploits niches occupied elsewhere by warblers (Sylviinae), tits (Paridae) and creepers (Certhiidae and Sittidae) as well as being a 'normal' white-eye (fruit-eating) and extending frequently to nectar feeding; in short it is an extremely versatile bird. As Gill (1971) noted, flycatching was very infrequent; we each saw it only once.

I have re-analysed Gill's data on stomach contents (extracted from his manuscript collection notebook) to investigate seasonal changes in feeding habits, and possible different feeding behaviour in different habitats. Gill's (1971a) suggestion that the birds ate more fruit during the winter than during the breeding season is confirmed by Fig. 4, which shows a peak of fruit-eating in June and July, declining to a low in November. The graph is almost an exact inverse of that for testis size over the same period (Fig. 5).

Further examination of the full data suggests that birds living at high altitudes (over 1500 m) rarely ate fruit; when this was identifiable it was usually the white berries of *Aphloia theiformis*. Although these were available from April to November at least (pers. obs.), they were not found in stomachs after July. Birds living over 2000 m sporadically ate buds of *Philippia montana*, and in the only sample taken at 2500 m (Caverne Dufour) all ten stomachs collected contained only *Philippia*. Heathers are poor foods for most herbivores (Watson *et al. c.* 1970), so it is remarkable that a generalised white-eye should be able to feed on it. The Caverne Dufour sample was taken in June, during the peak of the fruit-eating period, and suggests that at very high altitudes there is a shortage of insects at this time (there are no fruit-bearing plants in the *Philippia* zone). The ability to eat *Philippia* must very considerably increase the carrying capacity of the island for Grey White-eyes during lean times of the year.

Fruit-eating from October to December was confined almost entirely to the lowlands (below 500 m), and the middle elevations (500–1800 m) on the leeward side of the island. Fruits identified were *Lantana camara* and *Schinus terebenthifolius*, common introduced shrubs. Both species fruit throughout the year, but the samples show a sharp peak of *Lantana* consumption in July.

In contrast to the evidence of these stomachs, I only once saw a Grey White-eye eating fruit (*Schinus* berries near St.-Gilles in August). This is in part a reflection of the small amount of time I spent at low elevations.

Gill (1971a) referred to a shortage of insects in the winter, but until seasonal changes in insect abundance and in fruiting of plants have been studied one cannot conclude that fruit-eating is a substitute for depleted insects rather than the reverse, or usefully

Fig. 5. Gonad size and breeding condition in Réunion Grey White-eyes, 1967. Symbols: Birds with fully ossified skulls (○), incompletely ossified skulls (△) and 1967 juveniles (◇). Mean (position of symbol) and range (vertical line) of each sample are given; crosses mark single observations, or cases where one bird falls well outside the range of the rest (these are included in the mean). During the breeding season information for females is presented as percentage in breeding condition (taken in presence of enlarged follicles, oviduct eggs or corpora lutea); hatched columns show birds with fully ossified skulls, open columns, incompletely ossified. (All data from F. B. Gill's manuscript notebook.)

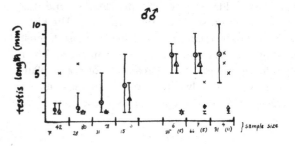

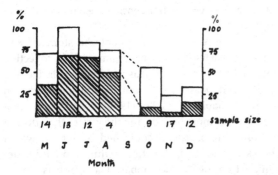

Fig. 4. Seasonal variation in fruit-eating by Réunion Grey White-eyes, 1967. Upper (unshaded) part of graph shows percentage of samples in which at least one bird had eaten fruit, lower (shaded) part shows percentage of samples in which half or more of the birds had been eating fruit. A sample is one day's collection from any locality; only those with two or more birds are included (only two samples were of two birds, and only six of three). There were no observations in September. (All data from F. B. Gill's manuscript notebook.)

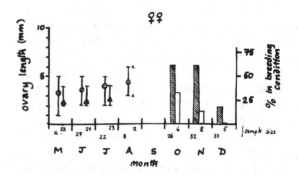

consider other hypotheses such as a response to a need for more moisture in the drier months, or simply an opportunistic reaction to a possible abundance of fruit in the winter. In this context it is perhaps of interest that the consumption of caterpillars follows a reverse pattern to that of fruit, reaching a low in July (zero) and a peak in November.

Nectar feeding cannot easily be studied from stomach analyses. Gill's (1971*a*) field observations suggested that it was more or less opportunistic, occurring when and where flowers of suitable kinds were abundant. I have no records of nectar-feeding in April–May, the time when nectar-feeding was at its most varied in the Olive White-eye (q.v.). To Gill's list of flowers used I can add *Aphloia theiformis*, *Eugenia* (*Syzygium*) *cymosa*, *Pittosporum senacia*, *Rubus mollucanus*, *Weinmannia tinctoria*, and an un-identified rubiaceous bush. I did not see Grey White-eyes feeding on *Eugenia jambos* which Gill reported as much favoured at Beaufond (Ste.-Marie), nor on *Dombeya* spp. (see under Olive White-eye). Grey White-eyes did not actively seek out rare flowers out of season as did Olive White-eyes. Gill (1971*b*) also recorded occasional feeding on sweet sap oozing from trees.

Breeding and the annual cycle

Apart from Milon's nest (1951), Gill's (1973*b*: 10–11) observations are the only information on breeding in the literature. His field observations and notes on gonad condition of specimens give a good idea of the extent of the season (Fig. 5): gonad size and courtship behaviour increased from early August and nesting began in mid-September to mid-October depending on locality, continuing through December "perhaps into January and February". However NB saw adults feeding a fledged brood in August, and MB and AWD both saw flying young being fed in mid-January, but no evidence of young in the nest. I saw no evidence of breeding in my brief February visit. My own obser-vations on breeding behaviour, nest-building, carry-ing food and feeding flying young, were all in the month of November. I saw far more nesting activity in November 1973 than in November 1974, which suggests that the peak time may vary from year to year; it was presumably later in 1974 than 1973, as there was no evidence of juveniles in the population in November 1974. NB has found nests in October and November, and Milon (1951) found ten nests, mostly

with eggs, in early December 1948. It remains unknown whether the species is double-brooded.

Gill saw 'helpers' feeding young, which I did not, but I did once see a third bird (apparently an adult) soliciting food from a pair which were carrying it to a nest. The third bird accompanied the pair begging actively, but was ignored. 'Helpers' have also been noted at nests of Seychelles White-eyes *Z. modestus* (Grieg-Smith 1979*a*).

I re-examined the data from Gill's specimens to investigate longevity, age of first breeding, and maturation period. Longevity may be assessed roughly from the proportion of immatures (i.e. birds with incompletely ossified skulls) in the population. Fig. 6 shows the percentage of each sex with partly ossified skulls during 1967; apart from the 86% for males in May, which must be a sampling error, the inference is that 50% of the population is of young birds during the winter, which implies an adult annual mortality of around 50% (assuming a stable population and that 1967 was a normal year). In August the percentage of unossified skulls dropped rapidly, indi-cating that full skull ossification took about 9 months. About 10% were still incompletely ossified by December – if they remained so for life this would of course lower the mortality figure calculated above. Fig. 5 shows that first-year birds of both sexes bred later than adults; indeed it seems possible that some of the females do not come into breeding condition at all in their first year.

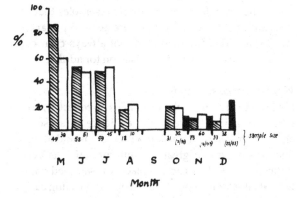

Fig. 6. Percentage of the population of Réunion Grey White-eyes with incompletely ossified skulls, 1967. Hatched columns, males; open columns, females; solid black, new (1967) juveniles (total percentage in sample). (All data from F. B. Gill's manuscript notebook.)

de Querhoënt (*in* Buffon 1770–83) gave the clutch as "commonly three", and Milon (1951) recorded six of 2, three of 3 and one of 4. NB found single clutches of two and three. The pale blue eggs closely resemble those of the Olive White-eye (Berlioz 1946).

Milon (1951) described the nest as a small cup attached to twigs 1–5 m up in trees, covered in green moss when in foliage (e.g. *Casuarina* branchlets), or grey-white material (lichen?), sometimes with dead leaves attached, in more exposed sites. Milon noted that the latter type was less conspicuous than green in the bare branches. The cup is of fine stems delicately woven.

Gill (1973*b*) quoted local hearsay of nest predation by rats (*Rattus rattus*), and introduced reptiles *Lycodon aulis* (sic; = *aulicum*, a snake) and *Calotes versicolor* (lizard); the nest appears to be camouflaged rather against avian predators, probably the now extinct kestrel (Chapters 1 and 2) and possibly also the Harrier.

Moult

Birds collected by Gill in May were completing body-moult and all were in fresh plumage, while a few birds in December were beginning to moult body feathers. None of the dated museum specimens I have examined was in moult, and none of the birds I trapped in August and November was moulting, although nearly half of Gill's August and September birds were moulting head feathers (some of my Olive White-eyes were doing this, but none of the Greys). Several of the juveniles collected by Gill from October to December were undergoing body-moult (FBG manuscript notes). If body-moult starts in October, the birds must start to moult soon after fledging. It is not clear whether the wing and tail feathers are also moulted then. The juvenile plumage closely resembles that of the adult, but the body feathers are generally sparser and looser. The differences are not always clear-cut and some juveniles could be mistaken for adults.

Weights and measurements

FBG weighed many of the specimens he collected, and I have examined his data for evidence of seasonal changes in weight. Variation within the island at any one season was so great that interpretation is very difficult, especially as no one place was sampled regularly at all seasons. FBG did most of his collecting early

in the morning when weights are at their least consistent (one bird weighing 7.7 g at 06.45 had already put on 0.5 g by 08.30, whereas four birds I retrapped between 10.00 and 18.00 changed by only 0.1 or 0.2 g: data for the two white-eye species combined). The overnight weight loss in near-freezing conditions in August was 1.0 and 0.8 g from 8.7 and 8.4 g respectively for two Grey White-eyes.

FBG's weight data and my August and November samples are summarised in Fig. 7. During the winter (May–August) males (8.18 g, $n = 131$) average 0.37 g more than females (7.81 g, $n = 104$; $P < 0.001$, *t*-test), but there is considerable overlap. Occasional samples were abnormally light while some very high weights were reached by pregnant females in November (the maximum is 11.0 g: Storer & Gill 1966). The samples are so variable and their origins so diverse that I hesitate to draw conclusions from the graph. However, male weights appear to rise from May to a June–August winter peak, declining to a low in October and rising again through November to December. Samples from the same locality in successive months parallel this trend, but there are only two such sites and three intervals. High weights and variance of females in November were due to pregnancy, while the December values were all low, perhaps indicating that the breeding season was much harder on females than males. The highest October weights for females were from the same locality as the lowest November ones, suggesting that breeding may have begun and ended early in that region (Le Brûlé, behind St.-Denis) that year.

Immature birds were lighter than older birds. Table 4 compares weights of birds with ossified and unossified skulls in the period May to July, and shows that older birds were heavier more often than the reverse. The overall means for the months considered were: males: adult (ad.) 8.21 g ($n = 60$), immature (imm.) 8.11 g (65); females: adult 7.95 g (47), immature 7.68 g (54). Of the four relevant comparisons (ad. males:imm. females, ad. males:ad. females, imm. males:imm. females, and ad. females:imm. females) only the means of the first are not significantly different (others $P < 0.02$, *t*-tests).

Gill (1973*b*) examined the influence of altitude on weight, but found a significant relationship only in the brown-headed brown morph, the prevalent form in the west and centre of the island. In his raw data the birds of the western lowlands were consistently lighter

Table 4. *Comparison of the weights of adult (skull ossified = so) and immature (skull incompletely ossified = sio) Grey White-eyes during May, June and July 1967*
F. B. Gill's data, lumped into 13 geographical zones. The table shows the number of samples in each category: samples containing fewer than three of either adults or juveniles excluded.

	Males	Females	Totals
Mean *so* heavier than mean *sio*	7	7	14
Mean *so* within ½ g of mean *sio*	1	2	3
Mean *so* lighter than mean *sio*	3	1	4

Note:
Combining males and females and omitting the *so = sio* class, the suggestion that adults are heavier than immatures is supported by a Sign Test ($P = 0.03$).

than the others, which may account for the significant regression. In grey-headed brown morphs and grey morphs the relationship was negative rather than positive, which would require complex interpretation had those regressions been significant.

Wing length varies with morph and altitude. As a general guide Gill (1971a) gave an average of 55 mm in the lowlands, "rising with altitude to an average of about 58 mm". Later, Gill (1973b) gave 54.6 mm for brown morphs (all head colours combined), rising 0.76 mm per 1000 m altitude – i.e. never reaching 58 mm within the island's altitudinal limits (3000 m); presumably this latter figure applies to grey morphs (which predominate higher up), 12 males of this type averaging 57.7 mm (Storer & Gill 1966: '*Z. edward-newtoni*': females, however, averaged only 56.1 mm so the combined value for both sexes is 57.06 mm).

Fig. 7. Seasonal variation in weight of Réunion Grey White-eyes, 1967 and 1974. Only samples with five or more weights are included (range 5–13), except for one of three in November 1974 (indicated). Code at bottom of graph gives geographical area/morph type and altitude. A, B and C refer to Gill's map (1973, p. 30 = Map 10 here); A/C indicates a border zone. Altitudes as follows: l (low = 0–500 m), m (medium = 500–1500 m) and h (high = 1500+ m). (Data from F. B. Gill's manuscript notebook, with some weights of live birds in 1974 (ASC etc.) added.)

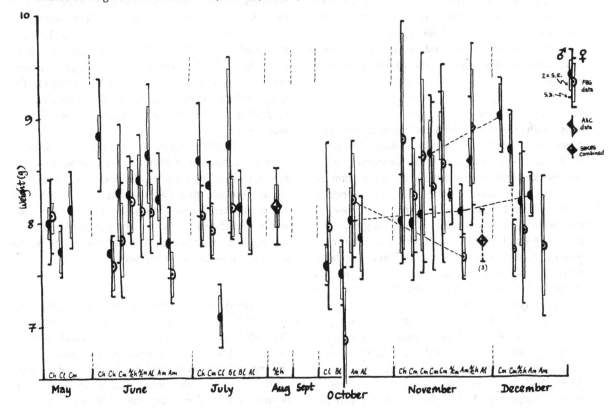

My own measurements at the Plaine des Chicots (1800 m) confirmed the longer wing of the grey morph: 12 greys averaged 56.9 mm (53–60.5) while 10 browns averaged only 55.4 mm (54–56.5), a statistically significant difference ($P < 0.02$, t-test).

Gill (1973b) found that most populations showed a sex difference in wing length, males averaging 1–1.5 mm longer than females. This difference is clearly shown in Storer & Gill's measurement table (1966).

Tail length correlates closely with wing length, increasing with altitude, and shows similar sexual variation to the wing. Tarsal and bill lengths increase, decrease or show curvilinear altitudinal relationships depending on the morph, and the bill (but not the tarsus) also differs between the sexes in some morphs, males having longer bills in grey-headed brown (0.5 mm) and highland grey (0.9 mm) forms.

Réunion Olive White-eye Zosterops olivaceus (Linnaeus 1766); zoiseau vert (zwazo ver)
Taxonomy and nomenclature

The Olive White-eye on Réunion differs from its counterpart on Mauritius in being larger and having a blackish rather than a greyish face. In treating them both as full species I follow Gill (1970), who gave full supporting details in his paper. Moreau (1957b) discussed the nomenclature, and discarded Hartlaub's name *haesitata* often used for the species as a junior synonym.

Status and distribution

Nineteenth-century accounts are vague on the abundance and distribution of the Olive White-eye; Coquerel (1864) called it "très commun", while Schlegel & Pollen (1868) noted that it was less common than the Grey White-eye. Schlegel & Pollen appear to have muddled the distribution of the two white-eyes: they referred to *Z. borbonicus* as occurring preferentially high up (1200–1500 m), rarer around the coast, whereas *Z. olivaceus* was said to occur "in the lowlands rather than high in the mountains". This is the reverse of the present position; there is no evidence to support such a change of distribution.

There has, however, undoubtedly been a general retreat of Olive White-eyes away from the coast since the 1860s. Pollen (1868) shot one in St.-Denis in March 1864, and Maillard obtained one in the Rivière de St.-Denis in May of the same year (specimen in Paris). This suggests that such occurrences were not unusual, and EH confirmed that in the 1920s Olive White-eyes were regular, though few, in lowland areas, but have since become very rare, and have also declined in the mountains, though much less so. Milon (1951) recorded Olive White-eyes in December 1948 only above 500–600 m, which agrees with my observations, although as recently as 1967 Gill (1971a) reported them as plentiful around 200 m in August to November in vegetated ravines in the north and east. I saw no Olive White-eye below 500 m even in the east, and they were generally scarce at this altitude even in native forest (see Appendix); J.-M. Vinson (pers. comm.) saw some at Bois Blanc on the east coast in November 1973. Olive White-eyes descend to lower elevations between March and May, when Pollen and Maillard were collecting (see below), but they certainly do not now roam widely in the lowlands. This contraction in range can be compared to that of the Stonechat and perhaps the Flycatcher (q.v.). After cyclone Hyacinthe in January 1980, Olive White-eyes frequented the botanic garden in St.-Denis and gardens in Ste.-Marie for 15 days or so (NB).

Olive White-eyes, like Merles, were present throughout the forest. They differed in being absent from the cloud forest in the west, but present in *tamarin* plantations. They also occurred in the Forêt de Bellecombe, the Plaine des Cafres and the upper ravine forests to the west and south, though they were absent (at least in November) from the bottom of the gorge of the Rivière des Remparts. The birds range up to 2300 m where flowering *Hypericum* and *Sophora* were found.

Gill (1971a) estimated the total Olive White-eye population as around 154 000, based on mark–recapture at Cilaos. The breeding distribution extends over about 1500 km^2, giving a calculated density of about 1 bird per hectare. This is of the same order of magnitude as my density estimates in many places (Appendix), but may be on the high side for the whole island population, as Gill also suggested.

Habitat

The Olive White-eye is by preference a nectar-feeder (Gill 1971a, and below), and thus its distribution depends largely on that of the flowers which provide its food. The most favoured native plants were *Hypericum lanceolatum*, *Sophora denudata* and *Forgesia borbonica*, while *Fuchsia magellanica*, abundant only at Bébour and the Plaine des Palmistes/Grand Montée, was the most important exotic. These are all plants of the middle to high altitude evergreen forest and the *tamarins* (*Sophora*), rather than the wettest and driest zones. The flowering of most food plants is seasonal, leading to considerable movement and vertical migration of the birds as they seek alternatives, largely within the mixed forest zone, including secondary forest in the east.

Although gardens contain many nectariferous flowers which are potential food plants for the white-eyes, the inhabited zone around the coast is nowadays avoided by the birds. The same is true in Mauritius (Chapter 4), and it is difficult to explain, especially as garden flowers were visited where available in an enclave within the forest (see below). In view of a similar and modern avoidance of the lowlands by the Stonechat and, partially so, by the Flycatcher, I am inclined to attribute it to some common factor; possible explanations are discussed in the context of conservation below.

General habits

The Olive White-eye is a small, dull, dark olive-green bird with a blackish head against which the white eye-ring makes a striking contrast. The rump is a brighter yellow-green, and the underparts grey, browner on the flanks, fading to yellowish under the tail.

Gill (1971a) summarised the habits of the Olive White-eye as follows: "*Z. olivacea* [*sic*] is an aggressive asocial white-eye that occurs in isolated pairs throughout the year rather than in small flocks. Rarely are several individuals together without some interaction; in areas with many flowers much of their time is spent in aggressive chasing and chattering, for they are highly territorial throughout the year in relation to flowers." They are also usually very noisy, their staccato calls being given in flight and especially in interactions around flowers (Chapter 3). Gill (1971a) described the inter- and intra-specific interactions at defended flowers: Olive White-eyes defend flowering bushes against their own species and Grey White-eyes, but are less aggressive towards the latter, and tolerate them if they fail to go away. This behaviour appears to be adapted to the defence of *Hypericum* bushes, which are usually scattered as isolated plants in forest or heathland. At large aggregations of flowers (e.g. *Sophora* groves, some *Hypericum* clumps) the territorial defence seemed to break down and the plants were fed on by numerous white-eyes of both

species in succession, though rivals were still chased from the immediate vicinity of a feeding bird. At Bébour, where *Fuchsia magellanica* was evenly spread in the forest, it seemed not to be actively defended, unlike *Hypericum* bushes in the same area. Study of marked birds would be needed to discover whether a defended bush was owned by a single individual, or whether a succession of white-eyes defend it for shorter periods.

When the white-eyes descended to lower elevations they could become concentrated in small areas, and then seemed to be less aggressive. At Ilet à Guillaume in early May 1974 I noted: "basically in pairs, but aggregate in good trees, sometimes indulging in the loud aggressive calls and chasing, often all peacefully together". The "good trees" were ones with an apparently good source of insect food; the birds were gleaning, not flower-feeding. At the same time one individual was seen persistently driving off a Grey White-eye (apparently for no reason as both were gleaning insects in Camphors and Gums).

Food, feeding and vertical migration

Zosterops olivaceus has a remarkably long beak for a white-eye (Moreau 1957a) and as early as 1860 Morel (1861a) noted that it fed on flower nectar, mentioning also insects and sweet juices of soft fruit. The same author also remarked on the laciniate tongue ("pencil-lée"), which Moreau *et al.* (1969) found was the most deeply cleft of any white-eye's.

Although birds may be seen gleaning insects from foliage at any time of the year, the striking feature of the feeding ecology of the Olive White-eye is undoubtedly its attentiveness to flowers. *Hypericum* was the most popular species and in the off-season birds would appear from nowhere to feed on the very rare individual flowers, each often several hundred metres from the next. *Hypericum* was also the most abundant and widespread nectariferous flower in the upper parts of the bird's range (1500+ m) and its marked seasonality of flowering was obviously very important in the White-eye's biology. *Sophora denudata* was the other principal high-altitude nectar flower, but it has a restricted range and flowers only in August.

Several trees in the mixed forests attracted white-eyes, especially *Forgesia borbonica* which has a wide distribution and a long flowering season. The introduced *Fuchsias* are also established in this zone, and at the lower end of the zone in the west are areas dominated by *Eugenia jambos* which is also favoured when in flower (Gill 1971*a*).

The flowering seasons of any species differ somewhat in different parts of the island and at different altitudes, so the birds' movements were not synchronised throughout the island; this also accounts for differences in dates given here and in the literature. My observations were made primarily in the Plaines des Chicots and d'Affouches and the valley of the Rivière de St.-Denis between, and apparently general remarks below should be taken to refer to that area unless otherwise stated.

At the Plaine des Chicots *Hypericum* flowered from mid-winter (exact month uncertain) to January. In August flowering was confined to above 1700 m, but was more generalised in November; *Hypericum* grew as low as about 1300 m, but plants below 1500 m rarely flower. During the non-flowering season Olive White-eyes descended to lower elevations; none were seen above the deer fence (*c.* 1450 m) in February and only small numbers in May. Most of these birds seemed to concentrate in the forested gorge of the Rivière de St.-Denis, where I recorded extraordinary densities in May at the Ilet à Guillaume; Olive White-eyes were commoner than all other species together, and about five times as numerous as Grey White-eyes (Appendix). During this period, when good sources of nectar were not flowering, almost any flower was visited; I recorded nearly twice as many species visited in May as in any other month (Table 5). About 60% of visits were to flowers of the introduced bramble *Rubus mollucanus*; I could detect only a trace of nectar in either unopened or fully open flowers, though White-eyes visited both. The flower was abundant but the yield to visiting birds must have been very low. The birds regularly visited flowers in the keeper's garden and often fed from the flowers of *Eriobotrya japonica*, a tree they ignored in the forest.

There seems to be local variation in the species of flowers visited. Gill (1971*a*) found that in the wet forests *Dombeya* trees, some of which flower throughout the year, were visited regularly. Although these trees, especially the white-flowered *D. umbellata* (*sensu* Rivals 1952), are conspicuous and common, I recorded Olive White-eyes visiting them for nectar only once, at the Plaine d'Affouches in May; at Cilaos (April), Béfour (November) and at other times at

the Plaines d'Affouches and des Chicots, *Dombeya* flowers were ignored. On the other hand Gill did not record feeding on *Rubus* or *Eucalyptus*, on both of which I found it regular and seasonally frequent.

The off-season alternatives to *Hypericum* varied with altitude and area. In May *Rubus* was predominant at the Ilet à Guillaume, but a few hundred metres higher up, at the Plaine d'Affouches, the White-eyes were almost all feeding on *Aphloia theiformis*. At this altitude the *Rubus* was hardly in flower, while in turn *Aphloia* was rare at the Ilet à Guillaume. *Eugenia jambos* is probably important in some areas, although I frequently saw White-eyes ignoring it; Gill (1971*a*) found it much favoured in the ravines inland of Ste.-Marie. In August at the Plaine des Chicots, *Forgesia* was little visited in the *Hypericum/Sophora* zone (1700–1900 m), but was the most popular nectar source at lower altitudes.

Table 5 summarises my observations of nectar-feeding by Olive White-eyes. When feeding on nectar Olive White-eyes generally perched near the flower and probed into it from the front with the bill. *Fuchsia* flowers were often found pierced at the base, although white-eyes I actually saw feeding (both species) always probed the flower from the front, often hanging to do so. Gill (1971*a*) described how Grey White-eyes pierce *Fuchsia* and *Kalanchoe* flowers, sometimes also holding them with a foot, but did not say whether he saw *Z. olivaceus* doing so. The corollas of *F. boliviana* are so long that piercing is the only possible way of reaching the nectar, but although I again found many pierced flowers I never saw the birds at work. The third introduced *Fuchsia*, *F. hybrida*, has small flowers but very long slender peduncles, too long for the birds to reach when perched on the flowering branch; this species, though widespread, was ignored. Olive White-eyes sometimes hovered briefly in front of *Hypericum* flowers, which were often quite difficult to get at from a perched position.

Gill (1971*a*) found stamens and pollen in stomachs of Olive White-eyes and implied that they were eaten deliberately. This may be so, but I suspect they are often ingested by mistake; some stamens (e.g. *Hypericum*) detach very easily, and the amount of pollen in some flowers is such that it would be impossible to avoid swallowing some. In August and November both species of white-eye in areas with flowering *Hypericum* were often plastered with yellow pollen on their faces.

Table 5. *Nectar-feeding by Olive White-eyes*

Species	a	b	c	Flowering season and months in which nectar-feeding recorded[a]												Notes and observations
				Jan.	Feb.	Mar.	Apr.	May	June	July	Aug.	Sept.	Oct.	Nov.	Dec.	
Agave americana	*															Cilaos
Aphloia theiformis		G	Zb	○		○	●		- - -			- - -				Not favoured when *Hypericum* and *Forgesia* available. Not flowering in Aug.
Dombeya 'umbellata'		G	Zb				○		- - -			- - -				See text. Not flowering in Aug. This is Rival's (1952) "*mahot blanc*", illustrated, apparently wrongly, under the name *D. pilosa* by Cadet (1981)
Eriobotrya japonica	*															
Eucalyptus cf. *citriodora*	*						○	○								
E. robusta	*															
Eugenia (*Syzygium*) *jambos*	*	G	Zb		○					○	○					Records at Ravine Grande Chaloupe. Not flowering in Nov. at Grand Etang, though plenty at Basse Vallée
E. paniculata								○								Ilet à Guillaume; only fed on in garden, forest trees ignored
Forgesia borbonica		G		○				○			●			●		Very few flowers in May; none in Feb. at the Pl. des Chicots
Fuchsia boliviana	*						○							○		?Flowers all the year round; seen at Bélouve and Cilaos. Flowers found pierced, feeding not actually observed
F. magellanica	*	G	Zb	○				●						●		Local; Pl. des Palmistes area and Bébour mainly. ?Flowers all the year. Similar but long-peduncled *F. hybrida* is barely used as flowers cannot easily be reached by the birds

continued

Table 5 (*continued*)

Species	a	b	c	Flowering season and months in which nectar-feeding recorded[d]												Notes and observations
				Jan.	Feb.	Mar.	Apr.	May	June	July	Aug.	Sept.	Oct.	Nov.	Dec.	
Hypericum lanceolatum		G	Zb	●				○			●		○	●		Virtually no flowers Feb.–May; see text
Labourdonnaisia sp.								○								Ilet à Guillaume
Quivisia ovata								○								Cilaos
Rosa sp. (cultivar)	*							○			○					Ilet à Guillaume, garden
Rubiaceous shrub (? genus)			Zb								○					Pl. des Palmistes. Specimen lost before being identified
Rubus mollucanus	*		Zb				●	○								Ilet à Guillaume; see text
Salvia sp. (cultivar)	*		Zb					○								Ilet à Guillaume, garden
Solanum auriculatum	*													○		Bébour; although common, apparently not fed on elsewhere
Sophora denudata		G	Zb								●					Very short flowering season
Weinmannia tinctoria			Zb								○					Not flowering Feb. at Pl. des Chicots. Rivals (1952) commented that bees are very attracted to this species and that it produced the best honey

Notes:

[a] *, introduced species.

[b] G, also recorded by Gill (1971a).

[c] Zb, also fed on by Grey White-eyes (this study and Gill 1971a)

[d] The flowering season is indicated by a continuous line, broken where flowering is sparse; details are given only where data are adequate. ○, a few observations; ●, many observations, species very important.

Gill (1971a) also recorded *Camellia sinensis* (tea), *Geniostoma* sp., *Melia azerdach*, *Musa* cf. *sapientum* (banana) and *Pelargonium* "*capitatum*" (it is in fact *P. graveolens*). The tea plantations have been removed since 1967. BT adds *Callistemon* cf. *speciosus* (*, Zb) and *Grevillea robusta* (*). I also found pierced flowers of *Cuphea ignea* (*, Zb) an undershrub common at Mamode Camp (above Le Brûlé).

Olive White-eyes ate insects commonly and regularly. Gill (1971*a*) recorded insect-feeding only 20 times in 9 months, whereas I have one or more observations on 15 days in a period of 12 weeks, and was observing white-eyes much less assiduously than he was. Seven of my observations were in April/May, four each in August and November, reflecting the poor supply of flowers in May; two of the November observations were of birds feeding young. I found that insect-feeding was much more persistent and continuous in May than at other times, and it was also the only time I recorded Olive White-eyes congregating to seek insects – I several times recorded up to eight birds in a single tree at the Ilet à Guillaume.

Insects were sought mostly by gleaning in foliage, though flowers and flower-buds were also probed apparently for insects rather than nectar. White-eyes gleaned in many kinds of vegetation from shrubs to trees, native and exotic, sometimes in the same trees in which they were also nectar-feeding (*Eriobotrya, Weinmannia, Eucalyptus*). *Dombeya umbellata* was a popular foraging tree, but as noted above the flowers were usually ignored. I once saw a bird attempting to forage in a *Cryptomeria*, but noted that it "soon gave up". In heathland birds occasionally foraged in foliage between seeking out *Hypericum* bushes. Like Gill (1971*a*) I never saw any bark-creeping or probing by the Olive White-eyes, but unlike him I never saw them fly-catching.

I did not identify any insects taken, but could see that birds feeding young were collecting caterpillars (Lepidoptera) amongst other unidentifiable items (Bébour and Plaine des Makes, November). Gill (1971*a*) found insects in almost all stomachs he examined, but did not identify them.

Birds kept in captivity by Morel (1861*a*) fed readily on sweet fruits such as bananas, custard-apples, guavas and rose-apples, and he saw wild birds attracted to split lychees. Milon (1951) recorded Olive White-eyes attacking fruit. On the other hand Gill (1971*a*) commented that in captivity they ignored various fruits that *Z. borbonicus* took readily, and that he had never seen them feeding on fruit in the wild nor had he found any in their stomachs. It seems strange that a nectar-feeding bird should not accept sweet juicy fruits, and Gill did record sap-feeding by *Z. olivaceus* (and *Z. borbonicus*; from *Claoxylon glandulosum*). Perhaps the present lack of interest in fruit by Olive White-eyes stems from a lack of it in their current range, and that in the last century when the species was common in the lowlands the birds were familiar with and made use of any deliquescent fruits available in gardens and orchards.

Breeding

Previous records of breeding are confined to Milon (1951). Recent observations suggest a season extending from June to January. TB has seen a nest begun in June which fledged young on 15 July (per NB). FBG (1967 MS notes) recorded ovaries enlarged, though not yet ready for laying, on 26 August and 5 September in localities as widely separated as Beaufond and Nez de Boeuf. TB saw a newly fledged brood of two in mid-September 1978.

In 1973 at Bébour I saw birds feeding a fledgling on 1 November, and others collecting food and nest material around the same date. In November 1974 I saw nest material being carried (8 November, Plaine des Chicots), food being collected (16 November, Plaine des Makes), and a nest at the Plaine des Chicots on the 19th, the clutch being completed next day. Milon (1951) found two nests with eggs on 12 December 1948 and Carié collected eggs in December 1911 (Berlioz 1946); FBG shot a juvenile and a pregnant female at Cilaos on 2 December 1967. In mid-January 1974 MB saw a juvenile at Cilaos, and in January 1975 AWD found birds in groups, presumably families, on the path to the Plaine des Chicots, and saw young begging from adults.

The clutch size is two or three: Milon's nests had completed (part-incubated) clutches of two and three, and mine contained two eggs. The eggs are pale blue-green (ASC) or blue (Berlioz 1946), without spots. There is no information on incubation or fledging periods.

The fledgling lacks the white eye-ring of the adult, and the blackish facial feathers are replaced by green, uniform with the rest of the head and back. The eye-ring is gained very rapidly (to judge by the very few birds seen without it, and the fact that birds with eye-rings have been seen soliciting food from adults (AWD)). The green facial feathers are moulted rather slowly; several birds caught in August at the Plaine des Chicots still had green or largely green faces, some of them actively moulting to new blackish feathers. FBG recorded body-moult on a juvenile collected as early as 2 December.

The nest structure was described by Milon

(1951), whose description and measurements agree closely with mine. The nests are of moss and white pappus material on the outside, lined with fine dry grass, the cup being 40–45 mm across. All three were 1.5–2 m from the ground; Milon's were both in forks of branches, mine on a horizontal branch (of *Stoebe passerinoides*).

Moult
The complete annual moult presumably takes place mostly between January and March. The only specimen in wing-moult that I know of is one in Leiden collected on 31 May 1865 that was still moulting its secondaries, the primaries all being new (G. F. Mees *in litt.*). Two more Leiden specimens were in complete fresh plumage by 21 March (1865), and another seven May and one June specimen in this and other museums are all in fresh plumage. None of my birds caught in August or November was moulting wing or tail, though, as mentioned above, some birds of the year had minor head-moult in August. Likewise FBG collected birds in head- and body-moult in August and September, and a juvenile in similar condition in December.

The survival and conservation of native Réunion birds
Forest clearance
The rather wide ecological tolerances of most of the endemic species remove them from any serious risk of extinction; even the most active destruction of forests would spare the steepest areas, the cloud forests of the east and heath vegetation, allowing much reduced but probably viable populations to survive. Destruction of mid-altitude forests would most affect the Olive White-eyes and Merles during the descent phase of their vertical migration, and would eliminate a high proportion of the Flycatchers. While the birds are relatively immune from any likely 'development' of forest land, this in no way detracts from the botanical and ecological necessity of preserving as much as possible of Réunion's vegetational heritage (Rivals 1968, Melville 1973, Baumer 1981).

Introductions
More serious for the endemic birds would be introductions of predators, competitors or diseases. So far Réunion has escaped the influence of monkeys, so destructive in Mauritius (Chapters 1 and 4). Feral cats and dogs are scarce, and mainly affect mammals (Moutou 1979). Only the rats *Rattus rattus* and *R. norvegicus* are widespread potential predators (of nests), but as one of these (*R. rattus*) has been in the island since the late 1600s (Moutou 1980, Chapter 1), they are presumably not a threat to the surviving birds. Although the import of any animals other than domestic ones is strictly prohibited, pet monkeys *Macaca fascicularis* are frequently smuggled in from Mauritius and there is a serious risk that this pest might become established.

Competitors are already well established. How much the existing introduced birds have affected the native forms is hard to say, although the Mynah may compete with the Merle at lower elevations, and may also predate the nests of other species; fortunately it is very unusual in native forest. More serious is the recent (1972) introduction of the Red-whiskered Bulbul *Pycnonotus jocosus* (Staub 1973a, Barré & Barau 1982), which Carié (1916) found to be a major predator on white-eyes and their nests in Mauritius, and considered to have been a major cause of their decline. The scarcity of the Mauritian Merle compared with the Réunion one may also be related to the presence of the bulbul.

The introduction of alien plants and herbivores has a more insidious impact, by degradation of the habitat. Several species of plants are competitively superior to native forest, and invade cleared areas preventing regeneration of native vegetation; a few species can also penetrate standing forest.

The bramble *Rubus mollucanus* is the most vigorous of the woody weed species, and very invasive of stands of native forest up to at least 1900 m altitude (Lavergne 1978, pers. obs.). The raspberries it produces are bird-dispersed. It can stifle plantations of *tamarins* unless controlled (Miguet 1980, pers. obs. at Bélouve), and can similarly invade natural forest. Landslips on steep slopes where previously native trees grew are rapidly colonised. Only the driest areas are unsuitable for it. The long-term consequences of this plant progressing unchecked are probably very serious, and measures to control it seem appropriate and urgent (Lavergne 1978).

In the east where forest is cleared it is spontaneously replaced by Rose-apple *Eugenia jambos* and the Strawberry Guava *Psidium cattleianum*, and, lower down, *Schinus terebenthifolius*; this plant is also the most important exotic in the drier western side. None of these actively invades native forest, though given bulbuls and monkeys to disperse the seeds the guava would surely do so as it does in Mauritius.

There are numerous other undesirable species in Réunion forests (Lavergne 1978), but the four mentioned pose the most serious long-term threats. The effect of feral herbivores is discussed below in relation to the Cuckoo-shrike's habitat. The grazing of sheep and cattle under forest in some areas is resulting in debarking of certain species and puts a complete stop to regeneration.

The introduction of pathogens is perhaps the most insidious and least observable risk to bird populations (Warner 1968); the rarefaction or extinction of birds for 'unknown reasons' may often, in fact, be due to epidemic fatal diseases. Brasil (1912) suggested that disease was the only plausible explanation for the disappearance of the Réunion Starling *Fregilupus varius*, and it may also be connected with the retreat from the lowlands of the Olive White-eye and Stonechat, and possibly also the Merle and Flycatcher. The general reduction in Grey White-eye numbers during this century also requires an explanation; reduction in Merles may also not be due solely to human persecution.

Blood samples collected in 1974 were reported on by Peirce *et al.* (1977). Avian malaria was found in only one bird, a Cardinal, trapped at 120 m altitude. Birds trapped at the Plaine des Chicots (1800 m) carried only haematozoa thought to be non-pathogenic. *Culex (pipiens) quinquefasciatus* (= *fatigans*), the malaria vector in Hawaii (Warner 1968), is reported to occur up to 1600 m in Réunion (1000 m in Hawaii), but only in association with human habitation. Proximity to humans may perhaps carry the risk of disease as well as danger of predation, and this could explain why Olive White-eyes do not now frequent apparently attractive garden and orchard flowers whereas in the nineteenth century they did so. Likewise the ground-feeding insectivore niche formerly occupied in the lowlands by the Stonechat is now vacant.

Direct human pressure

Although long forbidden, the catching of small birds, largely white-eyes, for food, still continues. They are caught on lime-sticks and sold as a delicacy. The trade is much reduced nowadays, due not so much to a strict enforcement of the law, as a shortage of birds making the practice uneconomic. M. Picard

assured me that up until 6 or 7 years previously (i.e. 1967–8) birds were a great deal more abundant at the Plaine des Cafres and the Rivière de l'Est, and there were many more full-time bird-catchers at the Plaine des Cafres than the two or three operating there in 1974. Similarly 10 years previously forestry labourers took only rice with them into the forest and used to catch their meat: it only took half an hour or so at lunchtime to catch a dozen White-eyes, but by 1974 this was no longer possible. Picard also reported that "in his father's time" (*c.* 1920s) there was one bird-catcher who lived on the gizzards of the birds he caught, selling the true meat; as a white-eye's gizzard probably weighs about 0.2 g, (cf. the similar-sized Bluetit *Parus caeruleus* (O'Connor 1976)) this man must have been catching hundreds of birds a day. Apart from indicating human pressure these stories also underline the decline in bird numbers, evidently rather recent in the Plaine des Cafres area. Picard suggested disease rather than poaching as a cause of the decline, on the grounds that the latter had been going on for many decades, whereas the decline was sudden and recent. Poaching is also prevalent in the Brûlé area, and I twice found fresh lime-sticks at the Plaine des Chicots. TB reported frequently encountering poachers of various kinds (after deer and Tenrecs as well as birds); he believed there was some decline in poaching as a result of his own wardening efforts. It seems improbable that poachers are now contributing substantially to the mortality of any species except perhaps the Merle (q.v.), which is shot as well as limed. Illegal shooting of Harriers and Moorhens is probably locally important.

The inhabitants of Mauritius and Réunion are inclined to be destructive of wildlife in general, and survival in populated areas of species with easily found nests is likely to be hazardous: the Flycatcher and Stonechat are probably major sufferers, though if the Olive White-eye always nests as low as in the three nests recorded then it too would be very vulnerable. Catapults are also commonly used against birds. The enormous increase in the human population during this century (173 000 in 1902 (Toussaint 1972); 511 000 in 1976 (anon 1976)) may well have influenced bird numbers directly.

Cyclones

No discussion of influences on bird populations in Réunion would be complete without some consideration of the effects of cyclones. The last major storm to hit Réunion directly was in January 1948, just before Milon's visit (Milon 1951), but near-misses did considerable damage in 1962 (Jenny), 1966 (Denise), 1975 (Gervaise) and 1980 (Hyacinthe) (Davy 1971, pers. obs., Barré & Barau 1982). Milon (1951) collected many local comments on bird populations before and after the storm; he reported that in particular Merles and Olive White-eyes had been seriously depleted. Barré & Barau (1982) considered that populations of "small endemics" (i.e. white-eyes, Flycatchers, Stonechats) were halved by Hyacinthe on the exposed side of the island, Merles and Cuckoo-shrikes being relatively little affected. Given the disastrous effect of 10 days' incessant rain on a colony of bats (*Tadarida acetabulosus*: 90% dead), the effect on swifts and swallows could also have been very severe (*ibid.*). Severe cyclones defoliate trees extensively and although native species are more resistant (Brouard 1967, Cheke 1975b), they lose fruits and flowers readily. It is probably the loss of fruit and flowers which affects the Merle and Olive White-eye so badly, rather than the force of the storm itself. Flowers take weeks to regrow (pers. obs.) and fruit obviously longer still; January, the worst month for cyclones, is in any case a bad month for flowers for Olive White-eyes.

Although the populations of native birds have survived countless cyclones over the centuries, any species that suffers a high mortality is in danger of extinction if the population drops to the point where any survivors would be too scattered or otherwise too few to recover. If mortality was 99%, as may have been the case for the Merle in 1948, a population of 10 000 might not be sufficient buffer against a cyclone if the 100 remaining individuals could not find each other in the rugged topography of the island. This kind of calculation puts a new perspective on the level of minimum populations judged necessary for long-term survival in places subject to severe natural catastrophes. Like the Rodrigues Fody (Cheke 1974a, 1979a, 1980) the Cuckoo-shrike might be brought below the safe limit if the long-planned afforestation of much of its habitat with conifers and *tamarins* were to take place (Cheke 1976); to judge by its scarcity in the 1950s and early 1960s it may well not be secure in its present area.

The special case of the Cuckoo-shrike

The Cuckoo-shrike is confined to one restricted area of mixed forest, which is currently subject to pressure from animals and humans unusual for Réunion forests and has also for some time been threatened with being cleared and planted with commercial trees. The main threats to the Cuckoo-shrike's ecology are:

1. introduced deer, and deer-hunting activities;
2. poaching;
3. clearance and reafforestation;
4. tourism.

Deer and the hunt

Unlike Mauritius, deer in Réunion are confined to a keepered and hunted herd at the Plaine des Chicots (the Cuckoo-shrike's only refuge) and two other places (Moutou 1981). Deer had been introduced to the island in the eighteenth century, but were eliminated by overhunting (Chapter 1). A few from later introductions still survived at the Plaine des Chicots before the present hunt was set up in 1954 with new stock introduced from Mauritius (Cazal 1974).

The devastating effect of deer and other introduced herbivores on endemic vegetation with no native herbivores, and hence no adaptations against grazing or browsing, has been well documented for New Zealand (Howard 1965, 1967, Challies 1974 and references therein). The situation in Réunion is comparable: the upland forests had no herbivores (tortoises were predominantly lowland animals) and the steep volcanic topography, easily eroded soils and high rainfall are similar to New Zealand.

There can be no doubt that the deer represent a threat to the long-term survival of the forest in its present form at the Plaine des Chicots. In New Zealand initial devastation of the understorey and elimination of highly palatable plants leads to two possible results: (a) a forest of very different floristic composition in equilibrium with a reduced population of deer, the physiognomy of the forest being permanently changed to a more open one with little understorey, or, (b) a progressive degradation of the forest to grassland or bare eroded slopes. The factors determining the result in any particular area are complex, and not necessarily predictable (Poole *et al.* 1959). At the Plaine des Chicots the early stages of vegetation alteration are clearly visible (Cheke 1975f, 1976), and if allowed to continue might well either destroy the native forest or alter its composition enough to render it unsuitable for the Cuckoo-shrike. As the bird is restricted to this small area one must assume that there is some special feature of it that it requires for its existence – hence any long-term trend out of keeping with the natural ecology of the area must be seen as threatening its survival. Although the deer are controlling the invasive exotic bramble and ginger, they are also compromising the long-term survival of the forest. The *tamarin*/bamboo stage so well represented at the Plaine des Chicots will very soon lose the bamboo altogether, and regeneration of mixed evergreen trees is very poor (Cheke 1975*f* and pers. obs. 1978).

In the Mascarenes it was until recently almost a heresy to suggest that deer might have a significant effect on forest regeneration. The failure of the Mauritian forests to maintain themselves has long been attributed entirely to invasive plants, monkeys and pigs (see e.g. Wiehé 1969, Owadally 1973), and the deer, though abundant for over 300 years, were disregarded. Deer hunters insist that not only do deer do no damage, but that ancillary activities (gamekeeping etc.) preserve the forest, although foresters now accept that deer are a problem (Owadally 1980, Jones & Owadally 1982). Deer hunting (and recently also ranching (Rivière 1982)) is big business, and there is considerable vested interest in overlooking the animals' disadvantages; at the same time one must recognise the real and possibly synergistic effects of the other factors mentioned. It is this traditional Mauritian view of deer, rather than that prevailing in Europe or New Zealand, that has been adopted in Réunion, and needs to be challenged. As deer damage is not recognised to exist there are no policies in Réunion to contain it.

It might be possible to find a level of deer density compatible with the survival of the forest, and hence of the Cuckoo-shrike; such a stocking rate could be reliably determined only by enclosure and exclosure experiments. Present evidence suggests that the deer population at the Plaine des Chicots is far too large in relation to forest growth and dynamics, with a stocking rate (*c.* 30/100 ha: 350 head (Moutou 1980*b*, 1981 on 1200 ha (Map 8)) about 12 times the maximum deemed tolerable in Europe (Mitchell *et al.* 1977; Chapter 1).

During 1974 a deer-proof fence was built across part of the lower edge of the Plaine des Chicots hunting area, and also across the Piton Grêle end of the Plaine d'Affouches. This was to keep deer within the preserve, thus reducing losses by wandering and subsequent poaching. While keeping deer out of the Plaine d'Affouches is very desirable, retaining the animals above 1400 m in the Plaine des Chicots has increased pressure on the forest, as excess animals could formerly escape more easily and would normally be killed by poachers lower down.

In addition to the long-term implications for the forest of a grossly overstocked herd of deer, the activities of the hunters present a threat to the birds. The hunting season is at present the four months from July to the end of October/early November, during which a hunting weekend is held every fortnight. The shouts of beaters and barking of their dogs are extremely disruptive to the peace of the forest, and as hunting takes place during the early part of the breeding season it is likely that birds trying to find a quiet safe place for their nests will be frequently disturbed, thus delaying breeding. This could have a damaging effect on the productivity of as rare a bird as the Cuckoo-shrike. Furthermore the beaters hired for the hunts have a reputation as poachers; it was probably not a coincidence that it was during a hunt weekend that we found fresh lime-sticks in the forests.

Poaching

Poachers in Réunion generally pursue three kinds of quarry: Tenrecs, birds and palm-cabbage. At the Plaine des Chicots there are also deer poachers. In spite of TB's efforts poachers are frequent in the latter area, though undoubtedly less so than in the Plaine des Fougères. The main quarry appears to be hibernating tenrecs, and poachers of these are said to set lime-sticks in suitable spots (e.g. *Hypericum* clumps) before starting their day's work with their dogs, sniffing out and digging up the tenrecs. They return later to collect any birds caught. Such poachers are unlikely to catch Cuckoo-shrikes but might do so occasionally. Merle poachers, however, use whistles or a call-bird to attract their quarry to the lime-stick, and this method can also be used by the expert to catch Cuckoo-shrikes (TB).

Indiscriminate poaching must be a threat to any endangered species. In recent years it has been contained somewhat by TB's enlightened keepering. It may be that the numbers of the Cuckoo-shrike in the 1950s and early 1960s, before the area was keepered, were kept low by poaching pressure. Le Brûlé and the Plaine des Cafres are the most noted places for poachers on the island.

Clearance and re-afforestation

At the Plaines de Chicots and d'Affouches the ONF have, since the early 1950s, pursued a policy of clear-felling native forest and replacing it with *Cryptomeria japonica*. By early 1975 about a fifth of the Plaine d'Affouches and a large area below 1200 m at the west side of the lower Plaine des Chicots were under *Cryptomeria* (Map 7). Plans published in 1968 (Moulin & Miguet 1968, anon. 1974*b*, Bourgenot 1974) suggested that *Cryptomerias* and *tamarins* were to be planted over most of the two *plaines*. This would leave only about 2 km² of Cuckoo-shrike habitat, enough for at most 15 pairs, which is far too few to ensure the survival of the species. More recent information is that this plan has been revised and that the area is no longer scheduled for plantations (Miguet *in litt.*), but the threat has yet to be removed (anon. 1981*d*).

Tourism

During the year several thousand people climb the path through the Plaine des Chicots to the Roche Ecrite, a mountain which gives a magnificent view over the centre of the island. Some return by way of the Plaine d'Affouches and the Dos d'Ane, and a proportion spend the night at the rest-house. Although numerous, these walkers have little effect on the flora and fauna as they rarely wander from the path.

APPENDIX
Bird density and relative encounter-frequency in different forest and heath habitats

Data for various localities and forest types are presented in three tables. For each a list is given of bird species in order of frequency of encounter in that habitat, and also a figure for total bird abundance. This latter is based on subjective estimates made at the time, which are correlated in the text table below with approximate densities based on calculations made from encounters per kilometre walking and from actual censuses of some species at the Plaine des Chicots. The scale is logarithmic with intervals increasing approximately fourfold. Note that Gill (1971*a*) estimated a total population of 2.5–3 birds per hectare over the whole island for the two white-eye species

added together (710 000 in 2512 km²). In the data given here habitats in which the Grey White-eye *Zosterops borbonicus* was the only native species occurring have been omitted; these were all lowland areas of secondary exotic vegetation.

Bird abundance scale

Scale value	Subjective impression	Approx. density of birds in middle of value range (individuals per ha)
1	Birds virtually absent	0.1–0.2
2	Birds very sparse	0.5–1
3	Birds infrequent to fairly frequent	2–3
4	Birds plentiful	10
5	Birds abundant	40

Where a locality was visited several times the dates are given by month through an annual cycle starting in August (beginning of breeding season) even if this cuts across the actual sequence of visits: i.e. a visit in August 1974 appears earlier in the list than one in November 1973.

In the tables bird and vegetation names are abbreviated as follows; introduced species are starred (*):

At = *Acridotheres tristis**	Sp = *Streptopelia picturata**
Cn = *Coracina newtoni*	St = *Saxicola tectes*
Ea = *Estrilda astrild**	Tb = *Terpsiphone*
Fm = *Foudia*	*bourbonnensis*
*madagascariensis**	Zb = *Zosterops borbonicus*
Hb = *Hypsipetes borbonicus*	Zo = *Zosterops olivaceus*
Sc = *Serinus canicollis**	BdC = *Bois de couleurs*
Sm = *Serinus*	(mixed evergreen
*mozambicus**	forest)

Where bird species were of similar abundance they are bracketed together. The symbol –/– between species indicates a major discontinuity in abundance.

Table A1. *Bird density and order of occurrence in primary and secondary forests in Réunion*

		Place	Altitude (m)	Type of forest	Date	Birds in order of frequency	Total bird abund.	General remarks
1	A	Pl. des Chicots Upper zone	1600–1800	Mixed *tamarin* and BdC with *Philippia* on ridges	9–16.8.74	Zb}/Zo} –St–Hb–Tb–Cn–Sp–/–Fm	4	Only 1 Fm seen
					12.10.78	Zo–Zb–St–Tb–Hb–/–Sc–Cn–Sp–Fm	4	
					6–7.11.73	Zb}/Zo} –St–Tb–Hb–Fm	4	
					5–8.18– 20.11.74	Zb}/Zo}–St–Tb, Hb} –Fm–Cn–/–Sp	4	
					14.2.75	Zo}–St–Tb–Hb–Fm–Cn	4	No Zo above deer fence
					8.5.74	Zb–Hb }, St, Tb, Cn	4	Zo few; no Fm, Birds otherwise plentiful
	B	Deer fence	1400	BdC with *Philippia* on ridges	13.8.74	Zb}/Zo} –St–Hb–Tb–Cn	4	No Fm
					12.10.78	Zb}/Zo}–, St, Tb} –Hb–/–Cn	4	No Fm
					8.5.74	Zb–St, Tb} –Hb–/–Cn	4	No Fm
2		Pl. d'Affouches	1200–1400	BdC with some *tamarin* and *Philippia*	9–16.8.74	Zb}/Zo} –St–Hb–Tb–Cn–Sp	4	No Fm
					6–7.5.74	Zo–Zb–Hb}, St} –/–Cn; Tb} –Hb–Fm	3	No Fm
3		Grande Montagne *Ligne domaniale* from Pl. d'Affouches road	1000–1100	BdC with some *Philippia*	15.11.74	Zb}/Zo} –St–Hb–Tb	4	
3	A	Ilet à Guillaume	700	Tall BdC; much *Rubus*	4–6.5.74	Zo–/–Hb–Zb–St–Tb–/–Fm–At	5	Zo very abundant, 5 times Zb
	B			Low secondary forest, largely BdC	4–6.5.74	Zb–Tb–St} Zo}	3	No Hb or Fm
4	A	Pl. des Fougères West end	1400–1500	Mixed *tamarin* and BdC, invaded by *Rubus*. Little *Philippia*	19–21.8.74	Zb}/Zo}–Hb, St, Tb	3	No Cn or Fm. Markedly fewer birds than at Pl. des Chicots

Table A1 (continued)

	Place	Altitude (m)	Type of forest	Date	Birds in order of frequency	Total bird abund.	General remarks
B	Eastend	900–1000	Dwarf cloud forest with much *Philippia*	17.8.74	Zb–Hb–St–Tb	2	No Fm or Zo. Birds few
5 A	Bébour	1300–1400	BdC, somewhat dwarf, wet; rich in *Fuchsia magellancia*. Tamarins towards Bélouve	31.10–5.11.73	St Zb Hb Zo }–Fm }–Tb–/–Sc–At	4	Density similar to Pl. des Chicots, perhaps slightly greater
5 B	Bélouve / Trail towards Piton des Neiges	1600–1800	As above	3.11.73	Zb Zo}–Hb} St }–Tb	4	Poor compared with Bébour
5 C	As above	1800–2000	Dwarf cloud forest	3.11.73	Zb–St	2	Very few birds
5 D	Wooded cliff down to Hellbourg	1300–1500	BdC	4.11.73	Zo–Zb–Hb–Tb–Fm	4	Stonly near top. Birds plentiful
6	Takamaka	800	BdC	10.5.74	Zb Zo}–Hb}–Tb St	4	
7	Grand Etang	500–550	BdC on cliffs, *Eugenia jambos* around lake edge + native remnants	11.11.73 / 14.11.74	Zb–Hb–Tb–Zo–Fm–St / Zb–Hb} Tb}–Fm–Zo–St	4	
8 A	Pl. des Palmistes / Hauts de Ste.-Anne	800–1100	Dwarf cloud forest	18.8.74	Zb–Tb–Hb–St	2	No Fm or Zo. Birds very few
8 B	As above	750–800	Secondary dwarf cloud forest; swampy. Rich in *Psidium*. Some *Eucalyptus robusta*	17–18.8.74	Zb–Zo–Hb–Tb	3	*Eucalyptus* in flower, attracting Zo and Hb. Few birds generally. No St. Fm only by main road
8 C	Grand Montée	1600	Low BdC rich in *Fuchsia magellanica*	12.5.74	Zo–/–Zb–St	4	Largely Zo. No Hb, Tb or Fm
9 A	Volcan / Slopes below Bellecombe	1800–2100	*Tamarin* with some BdC	25.8.74 / 10.11.74	Zb–Zo–St / Zb–Zo–St	3 / 2	} Birds few. No Hb or Tb
9 B	S. edge of Riv. de L'Est gorge	1500–1650	Dwarf cloud forest	10.11.74	Zb–Zo	2	Birds extremely scarce
10 A	St. Philippe / Mare Longue	600	Tall BdC	8.11.73	Zb–/–St–Tb–Hb–Zo	4	Birds plentiful

Table A1 (*continued*)

	Place	Altitude (m)	Type of forest	Date	Birds in order of frequency	Total bird abund.	General remarks
B	As above	400–600	Clear-felled *BdC* + selective managed regeneration	8.11.73	Zb-/-St-Fm-Zo-Sp	4	Also Ea, Sm and At, not quantified
C	Basse Vallée	700–800	Edge of tall *BdC* forest, clear-felled adjacent	11.11.74	Zo-Zb-St-Tb-Fm-Hb	4	
11	Riv. des Remparts	100–450	*BdC*, invaded by exotics on cliffs of canyon	22–23.11.74	Zb-At-Hb-/ {Tb}-St	3	No Zo, nor Fm beyond Goyave (near valley mouth)
12	Headwaters of Riv. d'Abord nr. Piton Mahot	1600	*Tamarin*, rich understorey	3.11.74	Zb-Zo-St-Tb	3	No Hb or Fm. Average density
13	Pl. des Cafres Trail down to Grand Bassin	1000–1350	*BdC*	9.11.73	Zb}-/-St-Tb {Zo} Fm	4	No Hb. Birds plentiful
14 A	Cilaos Trail to Piton des Neiges	1400–1900	*BdC*	15–24.1.74	St-Zb-Zo-Tb-Hb	3	Birds somewhat sparse (MB obs.)
				28–29.4.74	Zb-St-/-Hb}-Tb {Zo}	3	
B	Above Bras Sec	1500	*BdC*	27.4.74	Zb-St-Zo-Hb-Tb	3	
C	Path to Col du Taïbit	1600–1900	*BdC*	15–24.1.74	Zb-Hb-Zo-Tb	4	Birds plentiful. Position of St not noted (MB obs.)
				1.5.74	Zb-Zo-Hb-/-St-Tb	4	
D	Behind town	1200–1400	Secondary woodland of *Quercus, Eucalyptus, Pinus* & *BdC* relics	28.4.74	Zb-Tb-St-Fm-/-Sc	2	Birds infrequent. No Zo
E	Path to La Chapelle	900–1200	Secondary scrub, largely exotics	2.5.74	Zb-St-Tb-Hb	3	Birds fairly frequent. Also Sc and At, unclassified. No Zo
15	Plaine des Makes Bois de Bon Accueil	1000–1200	Tall *BdC*	17.11.74	Zb}-Tb-Hb-/-St-/-Fm {Zo}	4	Birds plentiful
16	Bras Mal Coté	900–1000	*Eugenia jambos* scrub with *BdC* relics	26.8.74	Zb-St-Zo-Hb	3	No Tb

Table A2. *Bird density and order of frequency of occurrence in plantation forests in Réunion*

	Place	Altitude (m)	Type of plantation	Date	Birds in order of frequency	Total bird abundance	General remarks
1	Le Brûlé Mamode Camp	1000–1200	Pure *Cryptomeria*, c. 20 yrs old (8–10 m tall). Sparse herb layer	16.8.74	St–Zb	1	3 prs of St and 1 small party of Zb in 1.5 km transect
				9.5.74	St	1	1 pr St in 1.5 km transect
2	Pl. d'Affouches	1300–1350	Pure *Cryptomeria*, c. 15 yrs old (5–7 m tall)	6.5.74	Zb–St	2	Very few birds. Also 1 Hb in a lone *Ficus* amongst the firs
3	Petite Plaine (nr Pl. des Palmistes)	1200	Pure *Cryptomeria*, c. 25 yrs old (10–12 m tall). Ground bare	2.11.73	St–Fm–Zb} Zo}	2	All birds scarce. Zo and Zb clearly passing through to and from *BdC* on either side
4 A	Bélouve	1400–1600	Pure *tamarin*, c. 15 yrs old (6–7 m tall). Good herb layer	2–4.11.73	St–/–Tb–Zo–Zb–Fm	3	St fairly plentiful, rest scarce; Hb only in copses and ravines with relict *BdC*
B		1400–1600	Pure *tamarin*, c. 25 yrs old with understorey of various exotic shrubs	2–4.11.73	St–Zo–Zb–Fm–Tb	3	Zb, Zo and Fm commoner than in above, but not plentiful

Table A3. *Bird density and frequency of occurrence in heath vegetation*

	Place	Altitude (m)	Type of heath	Date	Birds in order of frequency	Total bird abundance	General remarks
1	Pl. des Chicots	1800–2000	*Philippia* with *Stoebe*, *Phylica*, *Senecio ambavilla* and *Hypericum*. 1–2 m tall	9–16.8.74	St}–Zo–Sc Zb}	3	Above 2000 m vegetation sparse and very exposed. Only a few St and Zb
2	Bélouve	2000–2100	Tall *Philippia* (2–3 m), almost pure; no *Hypericum*	3.11.73	St} Zb}	1	
3 Volcan A	Savane Cimetière	1700–1800	*Philippia*, some *Stoebe*. Up to 3 m tall	10.11.74	St–Sc–Zb–/–At	2	Nearby swampy area, Les Mares, was nearly birdless (1 on scale)
B	Rempart de Bellecombe	2300–2350	Very sparse *Philippia Stoebe*. Low shrubs	25.8.74	St–Sc	1	
C	Pl. des Remparts	2350–2400	*Philippia* with *Stoebe*, *Phylica*, *Senecio* and some *Hypericum*. 1–2 m tall	25.8.74	St–Zo} Zb}	2	
4 Pl. des Cafres A	Around Piton Mahé	1600–1700	*Philippia* much invaded by *Ulex europaeus*; rare *Hypericum*	12.5.74	Sc} St}–/–Zo Zb}	2	
B	Savanne Mare à Boue	1600	Thin pasture with *Ulex* thickets; some *Philippia*	3.11.74	Sc–At–St–Fm	1–2	
5 Cilaos A	Coteau Kerveguen	2500–2800	*Philippia* and *Senecio* to 1 m tall	15–24.1.74 29.4.74	St–Zb St–Zb–(Zb–St at 2500 m)	2 2	Zb to 2750 m (MB obs.) Around rest-house abundance was 3, with Zb common; none seen above 2600 m
B	Piton des Neiges	2800–3069	Sparse grass and rare ericoid bushes	15–24.1.74 29.4.74	St}	1	St not seen above 2900 m in Jan. (MB obs.), but up to summit in Apr.

Acknowledgements

My first thanks go to Harry Gruchet and Auguste de Villèle and their families, without whose unfailing help with transport and accommodation work on the island would have been very difficult given the very limited means available to the expedition; Gruchet also very kindly allowed me to use the collection, archives and library of the Muséum d'Histoire Naturelle in St.-Denis. I am also grateful to the staff of the Office National des Forêts, and most particularly to the then Directeur Régional J.-M. Miguet, for, *inter alia*, allowing members of the expedition to use forest rest-houses free of charge. Théophane Bègue, gamekeeper at the Plaine des Chicots, was an invaluable help in exploring that area, and gave freely of his knowledge and experience. I had useful discussions on the bird life of the island with Armand Barau and Emile Hugot, and the former also helped the expedition in a number of ways. Thérésean Cadet was most helpful in identifying plants and discussing the vegetation, and also commented on a draft of the introductory part of this chapter. Finally I would like to thank my wife, Ruth Ashcroft, my brother Robert Cheke, and Michel Clouet for their support in the field and for data collected, and also Nicolas Barré, Michael Brooke, Mark Goode, Tony Diamond, Bertrand Trolliet and especially Frank Gill for allowing me to use their observations.

Notes added in press

The manuscript of Barré's important contribution (1983) reached the author too late to be included in the discussion. His findings on the distribution, density and general ecology of the native birds generally confirm those given here. In addition he gives evidence of a seasonal migration/displacement in the Grey White-eye that is the reverse of that of the Olive species. His discussion of possible selective pressures on the different colour morphs of the Grey White-eye is of interest in relation to sympatric speciation on islands, and would be well worth following up in the field. The paper also includes some additional information on breeding seasons and clutch size, and general ecological observations on introduced birds.

Since the above was written there have been important developments in conservation in Réunion, with the gradual establishment of large reserves, principally for vegetation, based on the report by Bosser (1982). A reserve at the Plaines des Chicots and d'Affouches for the Cuckoo-shrike is also being planned (P. de Montaignac pers. comm. 1985), probably of a size recommended by my study (Cheke 1976) rather than the rather restricted one proposed by Bosser.

7

Notes on the nesting of Procellariiformes in Réunion

C. JOUANIN

Four species of petrel breed in Réunion, two in the genus *Puffinus* and two in *Pterodroma*. A third gadfly-petrel, *Pterodroma arminjoniana*, nests on Round Is., off Mauritius, and brings the total number of petrels nesting in the Mascarenes to five. Breeding storm-petrels (Hydrobatidae) are absent, and this is true for the Indian Ocean as a whole, where members of this family occur only on migration.

The two endemic gadfly-petrels, *Pterodroma baraui* and *P. aterrima*, were the subject of earlier papers (Jouanin 1963, Jouanin & Gill 1967, Jouanin 1970a) where I tried to bring together all information that was then known. It seems useful also to publish details of the two shearwaters, although they are common and widespread species, and to add some unpublished observations on the *Pterodroma* species made since my previous papers were published.

Wedge-tailed Shearwater *Puffinus pacificus*

The Wedge-tailed Shearwater is widely distributed throughout the tropical and subtropical parts of the Indo-Pacific region. The various described races are doubtfully tenable (Jouanin & Mougin 1979), although there are slight statistical differences in measurements (see Murphy 1951), and colour phases occur in more or less stable ratios in different areas; only the dark phase is known from the Western Indian Ocean.

Créoles who live up in the mountains in Réunion often report cries of nocturnal birds resembling babies crying. I have been given this information at Hellbourg and Dos d'Ane, and by A. S. Cheke (*in litt.*, hereafter 'ASC') at St.-Joseph. Jadin & Billiet (1979) heard cries of this type at Pavillon, Petit Serré and from the road under the sea cliffs between St.-Denis and La Possession. These calls could apply to the Wedge-tailed Shearwater. Jadin and Billiet were told by their informants that it nests in the same cliffs as the Audubon's Shearwater, but at higher elevations. Nevertheless the only nesting place on Réunion I have seen myself, or heard reliably reported, is on Petite Ile, a stack a few dozen metres offshore in the south of Réunion, between St.-Pierre and St.-Joseph.

Petite Ile is divided from the mainland by a sea-gap only 50 m or so across. Despite its narrowness the crossing is impossible, and I believe that this is why the small breeding colony of shearwaters has persisted, for once one is on the island the birds are easy to catch. The stack is private property, and is so difficult to land on that the owners installed cables, bucket and a winch to get over from the mainland; however accidents occurred, and the apparatus was dismantled.

On 4 December 1964, a very calm day, H. Gruchet and I were able to land on the west side of the islet. We examined seven nests; all had a sitting adult on an egg. Two eggs were taken; one was barely incubated, but the other contained an embryo filling about a quarter of the volume, and already showing a little down. One of the nests was underneath a *Scaevola frutescens* bush, the others in holes inside rock caves.

My remaining notes on the breeding of this species are consistent with a summer breeding season on Réunion: a male which I caught on 25 October 1964 at sea off St.-Gilles had enlarged (10 mm × 7 mm) white testes, but its incubation patch was still completely feathered.

At the other end of the breeding season a fledgling male in uniformly fresh juvenile plumage was caught alive on the seafront at St.-Denis on 6 May 1963. Another was caught alive in the streets of the town at the end of May 1964, still with down on its flanks, and a young female still very downy on the belly was caught at St.-Denis in early May 1967.

On 20 November 1963 Appert (1965) found Wedge-tailed Shearwaters incubating on islets off the Madagascar coast near Morombe. Vezo fishermen told him that the species appeared only in the nesting season, approximately October to March. On Round

Is., off Mauritius, the annual cycle is very well defined, with egg-laying from the end of October to the third week in November (Vinson 1976*a*). In Western Australia the earliest laying date recorded is 29 October; most eggs are laid in late November or early December and the young leave the nests in May (Serventy *et al.* 1971). In the Seychelles, eggs are laid between late August and mid-October (A. W. Diamond *in litt.*). In that equatorial environment the population lays a little earlier than others but seems to fit broadly into the general pattern of the species. North of the Equator a similar remark applies to the Christmas Is. (Pacific Ocean) birds that lay in May, rather than in June as at Johnston Atoll and in the Hawaiian Islands (Gallagher 1960).

It thus seems there is good synchronisation between the breeding seasons of Wedge-tailed Shearwaters in Madagascar, in the Mascarenes and in Australia. The species is a summer nester throughout its vast range; in Hawaii it lays in June (Murphy 1951).

Audubon's Shearwater *Puffinus lherminieri*

Audubon's Shearwater belongs to a group with numerous forms widespread in the tropical and sub-tropical zones of the three oceans. Eighteen geographical races, traditionally divided between two species, are at present recognised (Jouanin & Mougin 1979). The Mascarene race, *P. lherminieri bailloni*, is clearly distinct from the one which nests in the Seychelles (*sensu lato*) and the Maldives (Jouanin 1970*b*). This race is by far the most common petrel on Réunion. It has also been known the longest: the Paris museum includes specimens of all ages, dating back nearly a century and a half. It is also the *fouquet** best known to the islanders, the only one whose nest they can find with certainty. Each time my colleagues and I have asked inhabitants of Réunion's interior for *fouquets* they have brought us Audubon's Shearwaters. It is difficult to explain how the birds have been able to remain abundant in Réunion when nests are so easy to find and the chicks are considered good eating.

Audubon's Shearwaters are found at a wide range of altitudes in the island. A. Barau has examined

---

* In Indian Ocean Créole based on French vocabulary the term *fouquet* has long been used broadly for all petrels (*Puffinus* and *Pterodroma*). Its etymology is uncertain though it may be a diminutive of *fou* = booby *Sula* spp. (Cheke 1982*b*).

nest burrows at 100 m altitude in the Rivière des Pluies, at Ste.-Marie, and another at 500 m on high ground behind St.-André. He was brought a specimen which had been unearthed at Bébour, 1500 m.

There is no doubt that it is to this species to which must be referred the call I have transcribed 'errr . . . errr' repeated several times, or 'rrr . . . oo . . . oo', a sort of mechanical trill followed by a more fluty note. ASC transcribed it 'who o who r' (with a rolled 'r'). I have often heard this call in the early part of the night: at Pavillon, outside the tunnel of the Pitons de Gueule Rouge, in front of the cliffs which dominate the Trois Bras, in the Ravine des Colimaçons, between St.-Leu and the Pointe des Châteaux, at Grand Bassin, in the Bras de Sainte Suzanne. ASC has heard it at Roche Plate in the Rivière des Remparts, Jadin & Billiet (1979) at Pavillon, Petit Serré and from the Route en Corniche between St.-Denis and La Possession.

These calls are always heard after nightfall. Listening at Grand Bassin to sounds from the rampart which dominates the gorge to the west and rises to 1300 m directly above the village (Grand Bassin is at 600 m), the *fouquets* called from 19.20 to about 20.45 hours. They started again before dawn from about 03.40 to 04.40 hours, just the moment when the night began to lighten. The calls are well known to the inhabitants of the village.

At least at the times when I have been able to make sea excursions off Réunion (mornings in October and November), Audubon's Shearwater has been with Brown Noddies *Anous stolidus*, the commonest of the birds which fish for cephalopods in groups 6–8 km off the coast at St.-Paul. On 29 November 1964 the sea was as smooth as a mirror, and I saw no birds, but on 25 October and 15 November I saw many Audubon's Shearwaters and Brown Noddies, and one each of Wedge-tailed Shearwater and Barau's Petrel. In the stomachs of three Audubon's Shearwaters caught on 25 October there were cephalopod tentacles 5 cm long and about 1.5 cm thick. I have described elsewhere (Jouanin & Gill 1967) how these flocks dominated by Audubon's Shearwaters guide the fishermen of St.-Gilles and St.-Paul who use drag-nets off the coast.

On 24 November 1974 ASC made a similar excursion off St.-Gilles to a distance of 9–11 km from the coast. On a calm sea he saw many Audubon's Shearwaters and Brown Noddies, and one Wedge-tailed Shearwater. Audubon's Shearwaters and Brown Noddies were also the most common species from 16.00 to 18.30 hours on 5 December 1964 a few hundred metres off the reef between St.-Leu and the Pointe des Aigrettes. Towards dusk on 2 November 1974 ASC saw a large number of Audubon's Shearwaters offshore at the Pointe du Gouffre (Etang Salé).

Not only is Audubon's Shearwater the commonest petrel in Réunion, but the above observations at sea also suggest that they forage nearer the coast than does Barau's Petrel. They also return to the nest later than Barau's Petrel, always after dark. Neither I nor ASC, nor Jadin & Billiet (1979), despite many hours spent at dusk at the seaside watching seabirds, have ever seen Audubon's Shearwaters making for land. They also return to sea in darkness. In contrast with Barau's Petrel, their activities on land are strictly nocturnal.

Breeding apparently can take place at any time of year. Fresh eggs have been dug up on 29 October, half-incubated ones on 9 November. Three chicks, of which one was already showing flight plumage, have been found on 3 December; also a 2-week-old nestling in a burrow on 31 March; while on 25 March a large young, with only a tuft of down on its belly, has been found grounded on its way from its nest to the sea. Other fledging young, free from down but still recognisable as young by various characters,* have been captured on 25 August, 5 and 30 November, and 5 December. Assuming incubation and rearing last about 4 months, the above information gives laying dates in January, April, July, August, September, October and November.

Breeding seems therefore to be spread throughout the year. However, my Créole informants, the

---

* As the moult of petrels is progressive and spread over a long period, the first flight plumage is the only one in which all feathers show the same amount of wear, or none at all at the moment of fledging. A young petrel thus has, for the only time in its life, uniformly fresh plumage. Several other characters can be used to confirm that the petrels cited here were young birds: (*a*) the outer primaries are more pointed than in subsequent plumages; (*b*) the greater coverts have a small ill-defined subterminal spot, rather than a terminal fringe; (*c*) beaks on skins of young birds kept for several years show a shrinkage between the nostrils and the forehead feathers with a little rumpled fold; (*d*) their primaries have not reached full length (wings *c.* 10 mm shorter than adults); the presence of tubes at the base of the rachis is confirmatory.

same people who brought me the birds, always affirmed the existence of a 'season' in which young could be found. People collecting for Jadin & Billiet (1979) told them that birds can be found in nests between 15 November and 15 January, but as reported above, a 2-week-old chick has been found on 31 March. In the Réunion population as a whole, it is possible that there is a peak of breeding during the southern summer, but the data so far are not sufficient to be sure of this. In any case, if there is such a peak, it is very spread out: the breeding cycles corresponding to the nestlings of 3 December and that of 31 March are more than 4 months apart.

Audubon's Shearwater is probably sedentary, i.e. it does not undertake long migrations in the non-breeding season but remains in the vicinity of Réunion.

I have given the usual mensural data for Réunion material of Audubon's Shearwater elsewhere (Jouanin 1970*b*). I will add here weights and wingspans: weight, 165–259 g (mean 226 g, S.D. 28.2, *n* = 12); wingspan, 63–68.5 cm (mean 66 cm, S.D. 19.2, *n* = 10).

Barau's Petrel *Pterodroma baraui*

Barau's Petrel also belongs to a group with tropical forms distributed in the three oceans. The characteristics of the eight known forms appear to justify their taxonomic treatment as several species. The form confined to Réunion, or at least to the Mascarenes (see below), was discovered only in 1963; this belated discovery illustrates the ability of petrels to remain unnoticed. This is all the more remarkable as *P. baraui* is not rare at Réunion. Once its habits are known it is easy to see; it is sufficient to sit on the beach at St. Gilles towards the end of the day to guarantee seeing some (Jouanin & Gill 1967).

Barau's Petrels apparently feed far out to sea, certainly well beyond the feeding range of Audubon's Shearwater. At around 16.30 hours they approach the land and fly along parallel to the coast, going back and forth just above the waves or rising briefly to a height of 10–20 m and dropping down again. The number of birds involved in these movements and visible with binoculars from the shore can exceed a hundred (in early November 1974 ASC saw over 200 at Pointe du Gouffre, Etang Salé, and several hundred at Rivière St.-Etienne; Jadin & Billiet (1979) saw over 1000 2 km west of St.-Pierre on 28 November 1977). From time to time one bird or a small group leaves the sea and makes for the interior of the island, usually flying in huge circles or spirals. ASC has seen birds behaving in this way at St.-Denis, Baie de la Possession, St.-Gilles, Pointe du Gouffre, mouth of the Rivière St.-Etienne and the Pointe de Langevin. Near St.-Pierre on 28 November 1977, Jadin & Billiet (1979) watched Barau's Petrels returning to land from 18.30 hours to darkness. To begin with the birds left the sea singly; as the light faded they flew in in small groups. Camping at the Plaine des Makes, at 970 m altitude, near the Ravine du Gol, ASC saw birds flying overhead towards the higher parts of the island: three at 1700 hours, then ten between 18.30 and 19.00, after which it was too dark to have seen any birds there might have been.

Between 15 and 23 January 1974 Brooke (1978) spent about 30 hours watching the behaviour of Barau's Petrels in the Cirque de Cilaos, from six observation points between 1200 and 2900 m altitude. He related the return of these birds to their nesting sites before nightfall to the pattern of air currents which rise up the precipices around the cirque and which are at their strongest when light is failing in early evening. These birds start off from sea level and need to reach nesting sites at about 2500 m altitude; using the air currents saves a considerable amount of flight energy (*ibid.*).

The presumed nesting sites of Barau's Petrels in Réunion are inaccessible on dangerous cliffs. Brooke (1978) confirmed our earlier conclusions (Jouanin & Gill 1967). Jadin & Billiet (1979) found a live bird on the ground under bushes along the path from Cilaos to the Caverne Dufour at about 2000 m, but nothing indicated the presence of a petrel colony nearby. To date a single nest only has been found and that in an unexpected place: Rodrigues (Cheke 1974a). This nest was in a cavity under a rock near the summit of the hill at Quatre Vents, only 320 m above sea level and very easily accessible. Cheke (1974a) suggested the nest represents a recent colonisation, which in turn suggests that the species is doing well at present; however the nest remains an isolated case, as no subsequent nests have been found, nor birds observed on the island since December 1974 (Chapter 1, ASC *in litt.*). In Mauritius ASC saw a Barau's Petrel at Tamarin Falls on 2 February 1974 making its way further inland; however this does not appear to represent a sure indication of nesting, as he watched unsuccessfully for the dusk passage of Barau's Petrels many other times at the same site. Temple (1976a) found a dead specimen on Round Is. in August 1974.

The breeding season is well defined. Barau's Petrels are absent from the interior of Réunion during the southern winter. We owe to F. B. Gill a precise date for their return inland. While he was living in Réunion from April to December 1967, the first petrels he saw were several flying over the Grand Matarum forest at Cilaos on 14 September. However it is not certain that the birds depart far from the nesting grounds in the non-breeding season, as Gill (*in litt.*) often saw them offshore (at Etang Salé) in July.

Basing our conclusions on the fledging dates of young birds (22–29 April), I suggested previously (Jouanin & Gill 1967) that *P. baraui* laid eggs at the end of November. Gill's observations (above) on the date of the birds' reappearance at the nest-sites at the end of the non-breeding interval fit perfectly into this timetable, leaving 9–10 weeks for courtship, preparation of the burrow and the pre-laying exodus.

The period of fledging has also been confirmed since. On 11 April 1968 a very fat bird with an unabsorbed bursa of Fabricius was caught on the shore at Bois Rouge in the north of Réunion (on the opposite side of the island from where the birds, locally called *taillevent*, have long been known to the inhabitants). Between 14 and 18 April 1974 several individuals were grounded in Cilaos in exactly similar circumstances to those we described previously (Jouanin & Gill 1967:17), two specimens examined by ASC still showed traces of down on their heads. A further fledgling with down on its head was captured at Le Port on 29 April 1981 (N. Barré pers. comm.).

Réunion Black Petrel *Pterodroma aterrima*
The affinities of this remarkable bird, described in 1857, have long remained a mystery; it seems to be related to the Tahiti Petrel *P. rostrata*, and the little-known Beck's Petrel *P. becki* (Jouanin 1970a, Jouanin & Mougin 1979). The recent finding of two specimens of this species, otherwise without authentic records since 1890 (Jouanin 1970a), have allayed the fears one might legitimately have had about its survival.

On 30 March 1970 an individual was caught on the ground in a farmyard at Entre-Deux (400 m above sea level) in the south-west of the island. On the day of capture, and the day before, the weather was overcast, with a very low cloud-base, and the village fogbound. It is known that petrels are disoriented under such weather conditions, are often attracted to lights, and fall to the ground where they are easily caught. On 29 and 30 March 1970, several *fouquets* were caught under these conditions, but all the other birds were apparently Audubon's Shearwaters. H. Gruchet and I questioned the farmers who had found the Black Petrel; they said that they knew two 'qualités' (species) of *fouquets*, but that the black one was much rarer than the other.

On 20 December 1973 a specimen was found freshly dead in Armand Barau's garden at Bois Rouge, at sea level in the north of the island.

These two specimens unfortunately bring us no information on the annual cycle of the species. The March bird was a male with white testes 4 mm × 3 mm in size, without subcutaneous fat, moulting on the breast, neck and head. The December bird was a female with an ovarian cluster 9 mm long (biggest oocytes 2.5 mm in diameter) and no subcutaneous fat, although there were two fairly solid strands of abdominal fat. Two very worn secondaries in the specimen contrast with the rest of the plumage. In neither specimen was the incubation patch unfeathered.

The measurements (mm) of the two specimens are as follows:

	Wing	Tail	Tarsus	Middle toe	Culmen
♂ 30.3.1970	251	97.5	38	49	28.5
♀ 20.12.1973	241	98	38	46.5	25

See Jouanin (1955: 160) for measurements of the other four known specimens. The male of 30 March 1970 also had a wingspan of 88 cm and weighed 232 g.

I have elsewhere (1970a) compared the Réunion Black Petrel with *P. rostrata* on Tahiti. *P. rostrata* makes its burrows in forested mountainous areas, nesting at very low density at the foot of cliffs and slopes under trees. Only in one or two areas do pairs nest within a few metres of each other. Thibault (1974) also indicates that the Tahitian bird is heard throughout the year, which suggests that it is more or less sedentary. The laying period is certainly very spread out, as nestlings have been found around 1 July (corresponding to eggs laid in early May) and a female ready to lay around 1 October, a range of 5 months.

The Réunion Black Petrel is apparently one of those tropical petrels of paradoxical habits which one might call 'forest petrels'. Réunion is still rich in forested areas which have been preserved from destruction by the astonishingly rugged topography of the mountains, which make access and penetration very difficult. In such a habitat the discovery of a nocturnal species nesting at best in very loose colonies can hardly be made except by chance. Such habits would explain both the infrequency of captures and the survival of the species. It seems very likely that *P. aterrima* nests under forest, very dispersed spatially, and with breeding spread out over a considerable time.

Acknowledgements

Once again I would like to express my gratitude to Messrs A. Barau and H. Gruchet, as much for their co-operation in the field and the help they have given me in Réunion itself, as for the observations and specimens that they have sent me over many years. Without their consistent and single-minded collaboration these pages would not have been written. I am most grateful to M. M. Malick, then director of the Météorologie Nationale in Réunion, for his efficient assistance in many circumstances. I would also like to express my keenest thanks to F. B. Gill who has given me permission to use the notes he took during his stay in Réunion in 1967; and A. S. Cheke, leader of the BOU's expedition to the Mascarenes, who has allowed me to publish all his observations on *P. baraui*, and who translated this chapter into English. Finally I would like to thank B. Jadin and F. Billiet who sent me their notes taken on visits to Réunion, as well as specimens they collected, and also N. Barré who added more recent data.

8

Observations on the surviving endemic birds of Rodrigues

A.S.CHEKE

Introduction

Rodrigues is the smallest and most ecologically devastated of the Mascarenes, retaining only three species of endemic land vertebrates from the wealth and diversity that existed there when the island was first discovered (North-Coombes 1980a, Chapter 1). The survivors are two birds, a warbler and a fody *Foudia*, considered here, and a flying-fox *Pteropus rodricensis* (Cheke & Dahl 1981). The ecological history of the island is reviewed in Chapter 1.

As the status of the Rodrigues Warbler *Acrocephalus* (*Bebrornis*) *rodericanus* was giving rise to "very grave anxiety" with three-star rating in the *Red Data Book* (Vincent 1966), I lost no time in visiting Rodrigues after establishing the BOU Expedition's base in Mauritius. Since the distribution of my original report (Cheke 1974a) on the status of the endemic vertebrates, Rodrigues has been the focus of more biological and conservation interest than at any time since the famous Transit of Venus expedition of 1874 (Hooker & Günther 1879). Flying-foxes (Durrell 1977a, b) and Fodies (Cheke 1979a) have been captured for captive breeding, and all the species have been the subject of some further study (Cheke 1979a, 1980, Mungroo 1979, Carroll 1982a, b; C. G. Jones unpubl.). Particular recommendations for conservation of the vertebrates are to be found in a variety of reports already produced (Cheke 1974a, 1978a, b, 1980, Carroll 1982b, Jones & Owadally 1982a). This chapter discusses the general biology of the two birds.

I visited Rodrigues four times in 1974: 16–20 January, 25 February–13 March, 12–18 July and 12–15 December. Dr A. W. Diamond (AWD) and A. S. Gardner (ASG), visiting 4–15 February and 18–24 September 1975 respectively, have very kindly allowed me to use their observations. Mrs. J. F. M. Horne was on the island in November 1974 making recordings of the birds' calls (Chapter 3). I have also used observations given to me by the following people whose initials are quoted in the text: Dr R. P. Alès (RPA), A. Forbes-Watson (AFW), C. G. Jones (CGJ), J. Montocchio (JM), J.-C. ('Roland') Raboude (RR) and Dr F. Staub (JJFS); Raboude's communications (*in litt.*) continued sporadically through to 1983.

From 17 to 29 September 1978 I made a further visit with John Hartley (JRMH) of the Jersey Wildlife Preservation Trust, and Yousoof Mungroo (YM), Mauritian conservation officer (Hartley 1979, Cheke 1979a); much of the data collected then is published here for the first time.

The island

Rodrigues (Map 1) is a small island of 42 square miles (109 km²), 11 miles by 5½ (17.7 km × 8.5 km), lying 356 miles (574 km) east of Mauritius. The centre is at about 19°42′ S, 63°25′ E. The terrain is hilly, mostly gently so, with a central ridge rising to 1300 ft (390 m); several of the valleys radiating from the ridge are steep-sided ravines with cliffs up to 350 ft (107 m) high. A wide, shallow lagoon protected by coral reefs and enclosing numerous islets surrounds the island apart from a short stretch of the east coast. The island's history has been written by North-Coombes (1971); most of the above details are taken from his book. McDougall *et al.* (1965) and Montaggioni (1970, 1973) have summarised what is known of the geological origins of Rodrigues, though faunal diversity suggests it is older than generally accepted (Moutou 1983a); most of the island is of Tertiary basalt, overlain with an extensive area of aeolian coral-sandstone ('dune-rock') at Plaine Corail, with very much smaller deposits elsewhere.

Climate

Rodrigues lies in the zone of the South-East Trades, which blow with varying intensity in most months of the year. Tropical cyclones are frequent in

Map 1. Rodrigues, showing the main island, satellite islets, and the extent of the reefs and lagoon. To avoid cluttering the map most place names have been given numbers and can be identified from the key. Only major roads are marked, and only those rivers that are mentioned in the text. Contours are in metres. The rectangle within dashed lines is enlarged as Map 2. (Adapted from Venkatasamy 1971.)

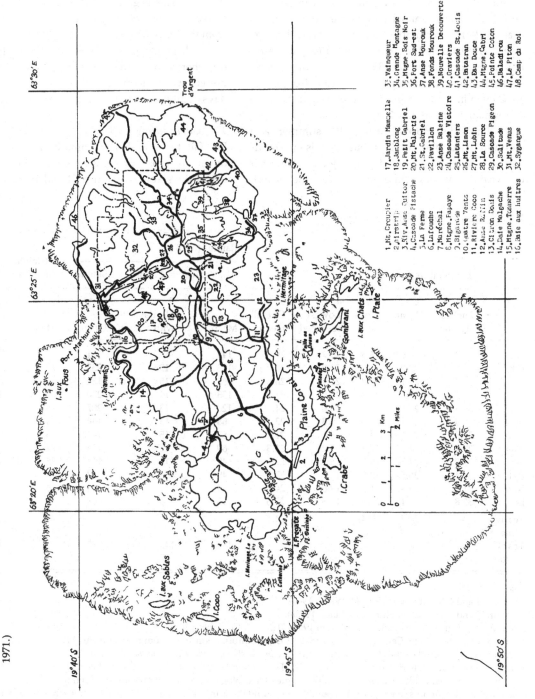

1. Mt. Croupier	17. Jardin Mamzelle	33. Vainqueur
2. Firstrip	18. Jambiong	34. Grande Montagne
3. Riv. Anse Quitor	19. Petit Gabriel	35. Mtgne. Bois Noir
4. Cascade Pistache	20. Mt. Malartic	36. Fort Sud-est
5. La Ferme	21. St. Gabriel	37. Anse Mourouk
6. Lafouche	22. Pavillon	38. Fonds Mourouk
7. Maréchal	23. Anse Baleine	39. Nouvelle Decouverte
8. Mtgne. Fanaye	24. Cascade Victoire	40. Graviers
9. Bigaigue	25. Lataniers	41. Cascade St. Louis
10. Quatre Vents	26. Mt. Limon	42. Batatran
11. Riviere Coco	27. Mt. Lubin	43. Eau Douce
12. Anse Raffin	28. La Source	44. Mtgne. Cabri
13. Citron Bois	29. Cascade Pigeon	45. Pointe Coton
14. Baie Malgache	30. Solitude	46. Baladirou
15. Mtgne. Tonnerre	31. Mt. Venus	47. Le Piton
16. Baie aux huitres	32. Sygangue	48. Camp du Roi

summer and are a very important source of rain, severe drought resulting if no cyclone passes near the island. Climate records are summarised in Tables 1 and 2. The mean annual temperature on the north coast is 24.1 °C. with about 5.5 deg C annual range. Wiehé (1949) quoted the daily range of temperatures: it is never very hot by day (33.7 °C max. during 1929–43) and seldom cold at night, very rarely dropping to 15 °C even in winter; night temperatures in the summer remain in the middle 20s.

Rain falls mostly during the summer and 'autumn' (December to May), but there is enormous variation from year to year depending on the number of cyclones passing close to or over the island.

Although there is no difference in rainfall between the north and south of the island (North-Coombes 1971), the north side is greener than the south, possibly due to reduced evapotranspiration on the side sheltered from the trade winds.

Mean relative humidity varies from 73% (October) to 81% (March), so is quite low even in the hotter, wetter summer. However, individual calm days occur in which it remains near 100% all day.

The weather during 1973–5
Rainfall for 1973–5 is summarised in Table 3. An unusually severe drought caused an almost total failure of the staple crops in 1974 and 1975. There were

Table 1. *Basic climatic data for Rodrigues*

| | Temperature (°C) | | | | | Rainfall (mm) | | | Relative humidity | Sunshine (hours) |
| | | | | | | Mt. Venus/Pte. Canon | | Solitude | | |
	1 Highest recorded	2 Mean max.	3 Mean min.	4 Lowest recorded	5 Overall mean	6 Mean	7 No. days with >1 mm	8 Mean	9 Mean	10 Daily mean
Jan.	32.2	28.6	23.5	20.7	26.3	144	14	171	79	8.7
Feb.	33.0	29.0	24.0	20.9	26.7	133	13	237	79	9.2
Mar.	33.7	29.1	23.9	21.9	26.6	162	15	182	81	8.4
Apr.	31.5	28.1	23.2	20.0	25.7	153	14	220	79	7.6
May	30.3	26.6	21.5	18.0	24.1	123	13	207	76	7.3
June	28.9	25.4	20.0	16.8	22.7	68	13	146	74	7.7
July	28.0	24.3	19.0	16.8	21.7	85	15	144	75	6.9
Aug.	27.2	24.0	18.3	15.2	21.3	69	15	122	74	7.1
Sept.	28.5	24.8	19.2	15.8	22.1	42	10	113	75	8.2
Oct.	30.4	26.5	20.0	17.2	22.8	49	9	73	73	8.6
Nov.	31.0	26.8	21.3	19.0	24.1	39	7	74	75	9.4
Dec.	32.8	28.1	22.5	18.5	25.6	113	9	68	78	7.7
Yearly mean	—	26.7	21.4	—	24.1				77	8.0
Yearly total						1180	147	1757		

Sources:
Columns 1 and 4: Mont Venus 1929–43 (Wiehé 1949).
Columns 2 and 3: Pointe Canon 1960–9 (Staub 1973*b*).
Columns 5, 6, 9 and 10: Mt. Venus/Pte. Canon 1951–70 (North-Coombes 1971).
Column 7: Mt. Venus/Pte. Canon 1951–70; from cyclostyled information sheet from the Mauritius Meteorological Department entitled *Climatological normals (Clino) and averages for climatic stations: Mauritius and the dependencies.*
Column 8: Solitude 1954–68 (North-Coombes 1971). There were 199 rainy days (Pte. Canon had 1145 mm and 218 rainy days over the same years). Note that in different places the Meteorological Department uses either 0.1 mm or 1 mm of rain to define a rainy day. The figures cited by North-Coombes (1971) and those in Table 3 are for 0.1 mm, whereas those above are for 1 mm (hence the different values for Pte. Canon).

Mt. Venus is at 140 ft (43 m) above sea level, Pte. Canon is near sea level; both are on the north coast near Port Mathurin. Solitude, an important bird site, is inland, on the north slope, at about 650 ft (198 m) above sea level.

four strong cyclones during 1972–73; rainfall in August 1973 was fairly normal, but thereafter the drought continued right through 1974 and 1975 and there were no cyclones in 1973–4 or 1974–5. Although the total rainfall was much less than usual, the number of days on which measurable rain fell was normal. Since the weather may be a proximate factor in initiating or maintaining breeding in birds, and has been shown to be so by Diamond (1980) for the Seychelles Warbler

Acrocephalus sechellensis, the closest relative of the Rodrigues Warbler, the times of breeding given in this paper may well be abnormal.

The forested areas

The rather open forest which once covered the island (Chapter 1, Wiehé 1949) has long since been destroyed, and at present most of the island is treeless grazing and arable land; a wide belt around the entire

Table 2. *Incidence and wind speeds^a of cyclones in Rodrigues 1959–82*

Date		Name	Mt. Venus		Maréchal	
			Max. gust	Sustained wind over 1 hour	Max. gust	Sustained wind over 1 hour
1959	12–14 Feb.		72	48		
	20–21 Apr.		115	—		
1962	1–3 Jan.	Chantal	95	40		
	3–6 Apr.	Maud	128	80		
	2–3 Dec.	Bertha	73	46		
1963	19 Feb.	Grace	127	79		
	9–11 Dec.	Amanda	100	55		
1965	7 Jan.	Freda	104	65		
1967	22–24 Dec.	Carmen	133	79		
1968	18–21 Jan.	Henriette	98	44	102	63
	28–31 Mar.	Monica	141	86	173	120
1970	27–28 Mar.	Louise	89	50	102	62
1971	22–23 Jan.	Ginette	93	47	103	67
	25–26 Feb.	Lise	92	55	89	56
1972	16–19 Feb.	Fabienne	148	93	158	123
	20–22 Dec.	Beatrice	110	59	104	63
1973	30–31 Jan.	Gertrude	105	49	(135)	70
	18–22 Feb.	Jessy	127	65	137	92
	28 Feb.–1 Mar.	Kitty	109	54	147	98
1979	8–9 Feb.^{b}	Celine II	134	85	n.a.	n.a.
1982	Jan.^{b}	Damia	113	65	n.a.	n.a.

Notes:
^{a} All wind speeds are in miles per hour. Only those storms with a sustained wind speed at Mt. Venus of over 40 mph are included. For discussions in relation to the effects on birds I have classified cyclones according to strength as follows:

Intensity scale	Sustained wind speed at Mt. Venus (mph)	Conventional description (Padya 1972)
1	40–54	Moderate depression
2	55–69	Severe depression
3	70–84	Intense cyclone
4	85–100	Intense cyclone

Note that Monica (1968) and Fabienne (1972) would reach 6 on this scale if measured at Maréchal, while Jessy and Kitty (both 1973) would both classify as 4, but rate only 2 and 1 respectively at Mt. Venus. Maréchal is on the south slope towards the west of the island, at about 600 ft (183 m) above sea level; it is not known whether it is the aspect or the altitude that makes it more vulnerable to cyclones.
^{b} Wind speed data from Pte. Canon not Mt. Venus.
Source: Data by courtesy of the Meteorological Department, Vacoas, Mauritius.

island from the coast inland is designated as 'cattle-walk' (North-Coombes 1971), within which trees cannot regenerate due to overgrazing. Some parts are planted with *Casuarina equisetifolia* or gums *Eucalyptus tereticornis*, but these woodlands are devoid of native birds. The birds are found only in areas under Rose-apple *Eugenia* (= *Syzygium*) *jambos*, *Tabebuia pallida*, Mango *Mangifera indica*, *Araucaria cunninghamii*, or a mixture of these and other native and introduced, especially evergreen, trees.

Cadet (1975) is the only published study of the vegetation cover since that of Wiehé (1949) but, like Wiehé's, his survey was rather cursory, so the details given here are my own.

The present distribution and composition of areas under evergreen or semi-deciduous forest are summarised in Map 2; pure stands of *Eucalyptus*, *Casuarina* and *Albizzia* spp., of no significance to the native birds, are excluded from detailed consideration. The history of the forests in relation to bird populations is discussed in the species accounts below (see also Cheke & Dahl 1981: 219), so the following descriptions refer to these areas as they were in 1974–8. Where relevant, Wiehé's (1949) terminology is used for the vegetation types; Cadet (1971, 1975) also provided useful information on some species. Strahm (1983) discussed the plight of the native flora.

Map 2. (*opposite*). The highlands of Rodrigues, with an indication of forest cover.
(Adapted from the 1 : 24 000 map (K.D./2ᵃ) of the Mauritius Survey Dept., and its source maps, the 1 : 5000 line maps from the aerial survey of 1951. The area covered by Map 2 is full of large cloud gaps on the aerial photographs, and detail has been inserted from the Admiralty Chart (No. 715). Where the two are irreconcilable, as is often the case, I have interpolated from my own knowledge of the topography. Many features are thus necessarily only roughly drawn and care should be taken in using the map in the parts indicated as doubtful. It has not been possible to amend this map in the light of the 1974 air survey (IGN 'Ile Rodrigue' at 1 : 10 000, 1978).)

Table 3. *Rainfall in Rodrigues during 1973–5*

	Pointe Canon						Solitude					
	1973		1974		1975		1973		1974		1975	
Month	mm	days[a]	mm	days	mm	days	mm	days	mm	days	mm	days
Jan.	343	19	45	20	68	13	381	20	58	23	87	19
Feb.	417	20	45	13	40	13	529	22	65	20	67	16
Mar.	171	15	46	15	38	11	189	14	60	15	65	15
Apr.	91	24	55	18	60	19	144	21	95	16	97	21
May	43	12	33	7	85	15	74	14	58	15	107	14
June	235	22	34	8	79	18	302	24	51	15	115	21
July	59	22	54	23	74	16	109	22	70	23	95	25
Aug.	127	19	58	24	50	24	168	23	73	24	70	27
Sept.	49	18	9	17	59	17	93	16	20	17	69	17
Oct.	16	11	15	13	12	16	23	7	30	14	27	16
Nov.	12	12	7	10	39	13	36	7	12	10	62	18
Dec.	106	19	32	4	31	12	127	18	55	8	29	16
Yearly total	1669	213	433	172	635	187	2175	208	647	200	890	225

[a] 'Days' is the number of days with more than 0.1 mm of rain recorded.

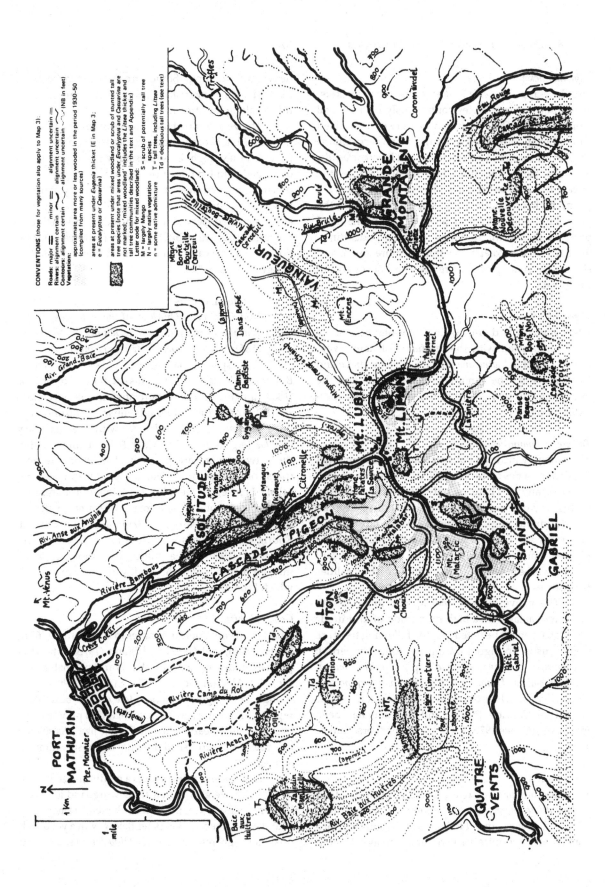

Eugenia thicket

The Rose-apple *Eugenia jambos* forms dense thickets, extending in some places continuously over several scores of acres. The stands may be pure or intermixed with other species, notably Mango, Strawberry Guava *Psidium cattleianum*, *Litsea glutinosa* and *Furcraea foetida*. In areas fairly free from human disturbance the thicket has a closed canopy varying from about 12 ft (4 m) high on south-facing slopes exposed to the prevailing winds, to over 30 ft (9 m) in sheltered parts of the northern slopes. Illegal woodcutting in many places keeps the thicket more open and shorter. In some places there are also houses within the forest.

Litsea thicket

Litsea glutinosa is dominant in a few places, usually much intermixed with the remnants of native vegetation, notably *Mathurina penduliflora*, *Olea lancea* and figs *Ficus* spp. The canopy is usually open and the trees often stunted (20–30 ft; 6–9 m), but taller individuals occur.

Tall tree communities

This category includes both plantation and spontaneous communities of trees over 30 ft (9 m) high and with a closed canopy. There are only relatively small areas of this kind of cover, and they include both evergreen and deciduous trees. In terms of area the most important trees are *Tabebuia pallida* and the Mango, both planted and subspontaneous. Other significant species will be noted in the descriptions of the localities themselves.

The native birds do not appear to be able to survive in areas with a high proportion of deciduous trees, so it should be noted that the following trees are deciduous in Rodrigues during any long dry period: *Adenanthera pavonina*, *Albizzia* spp., *Erythrina indica*, *Tabebuia pallida*, *Tectona grandis*, *Terminalia arjuna* and *Vitex cuneata*. Disturbance and degradation of this habitat in Cascade Pigeon and Solitude has been increasing (Cheke 1979*a*, CGJ, W. Strahm pers. comm. 1984).

Other vegetation

There are one or two areas planted in pure stands of *Callophyllum inophyllum*, sometimes very stunted. *Terminalia arjuna* and *Tabebuia pallida* both coppice readily, and a coppice scrub occurs in areas subject to

heavy woodcutting. Some stream beds have been planted with *Pongamia pinnata* which forms a dense thicket along the stream.

The vegetation of the patches of cover where the native birds still occur, and also of places on the island to which they might be successfully re-introduced, is described in full in the Appendix to this chapter. Each patch has its own characteristics of suitability or otherwise for the birds, and is given detailed coverage here as an aid to conservation on an island in which any potentially secure habitat is a rare resource; suggestions for conservation management were given elsewhere (Cheke 1978*b*).

Species accounts

Distribution and biology of the Rodrigues Warbler Acrocephalus rodericanus (A. Newton 1865); 'zoiseau longbec' (zwazo loñbek)

This species has generally been known as *Bebrornis rodericana*, in the genus established by Sharpe (1883) for this bird and its relative in the Seychelles *A.* (*B.*) *sechellensis*. Although museum taxonomists have tended to unite *Bebrornis* with the Malagasy *Nesillas* (e.g. Mayr 1971, Mayr & Vuilleumier 1984), these two genera are perfectly distinct in the field (pers. obs., Diamond 1980), especially in song and calls; their eggs are also completely different in pattern (pers. obs. of specimens in Cambridge). On the other hand, as Sharpe (1879) long ago noted, the Rodrigues Warbler is very close to *Acrocephalus*, and Diamond (1980), in a fuller discussion, formally merged *Bebrornis* with the reed warblers, a move with which I concur.

General habits

Apart from very brief summaries in E. Newton (1865*a*), Slater (1875), Gill (1967) and Staub (1973*a*), the only published descriptions of the Warblers' habits are those of Alès (*in* Gomy 1973*b*, Staub 1973*b*) which, though short, give a good idea of the bird. Both reports, and Staub (1973*a*), also included photographs by Alès which clearly show the bird's 'jizz'. As so little has been recorded I shall describe my observations in some detail.

Acrocephalus rodericanus is a medium-sized warbler (11–12 g weight) with a short wing, and long tail, legs and bill (see Chapter 9 for full measurements). It is dull grey-green above, paler below with a yellowish flush, strongly reminiscent of a *Hippolais* in shape, and in the way that at times the feathers of the head are raised to give a large-headed appearance; some birds, said by Alès (*in* Staub 1973*b*) to be males, raise the crown feathers only, giving the effect of a crest, but otherwise there are no detectable sex or age differences. The iris is rich bronze ('mordoré', Alès *in* Gomy 1973*b*); the bill is grey, as are the legs (not brown, *contra* Staub 1973*b*); the soles of the feet are yellowish. The legs and feet are very robust for a warbler of its size, though less so than in the Seychelles Warbler (pers. obs.). Fuller descriptions were given by A. Newton (1865*a*) and Staub (1973*b*).

In the field the birds were quiet and unobtrusive but inquisitive, single individuals or pairs approaching to within a metre or two of an observer if he or she made squeaking sounds by sucking the lips or the back of the hand. Investigating birds were usually silent, but sometimes gave scolding sounds.

Song was infrequent; during my visits singing was most in evidence in mid-January and February/March, less in July and September and virtually non-existent in December; AWD found the birds still very silent in February 1975. Even at their most vocal, individuals seldom sang more often than once every 5 minutes, and were frequently silent for a long time after two or three songs. Alès also remarked on the infrequency of song (*in* Gomy 1973*b*, but see also Chapter 3).

Although unobtrusive, the warblers were very active, continuously and very agilely gleaning the twigs and foliage for insects. Singing (or observer-watching) was only a brief interlude in feeding; birds sang from anywhere they happened to be with no favoured song-posts. Most of the time they kept within the canopy, occasionally flying up to 100 yards (92 m) over the tree-tops; I never saw one on the ground.

The long bill has given the species its Créole name, but the long tail is the bird's most characteristic feature in the field. The tail was frequently cocked up at 45–60° (photograph in Staub 1973*a*, *b*), a habit which reminded E. Newton (1865*a*) of an *Orthotomus* and me of an *Apalis*; much less frequently the tail was flicked from side to side.

Interactions were rare. I never saw any with another species, but once saw one warbler chase another out of its territory. Foraging pairs usually kept in visual contact, and a group of four seen at St.-Gabriel kept close together most of the time.

Distribution and numbers in 1974 and 1975

In January 1974 I was shown several pairs of warblers at St.-Gabriel, La Source and Cascade Pigeon, all in *Eugenia* thicket. Returning in late February I did a survey of all wooded areas and attempted to census the warbler population. I did not repeat the census on subsequent visits, but confined myself to checking certain sites. I also made use of information from RR, and AWD's notes (February 1975).

Finding warblers was not always easy; they are secretive and often silent. Frequency of calling and singing is variable from day to day and also seasonally, so sites required repeated visits to establish whether there were birds present and if so, how many. Some recent visitors to Rodrigues have had difficulty in finding the species at all (Vinson 1964*b*, Gill 1967), though Horne (Chapter 3) found them very vocal in November 1974. The species' secretive behaviour, and the lack of time available to some observers, have resulted in the past in fears that it was extinct.

The distribution of birds in February and March 1974 is shown on Map 3 and in Table 4; I have also included January observations where relevant. I could well have missed some birds in Cascade Pigeon, which is too diffuse an area for one person alone to cover very thoroughly, and possibly also on Mont Limon; elsewhere I am confident that very few could have escaped my searches. Subsequent observations by myself, AWD and RR in 1974 and 1975 suggest an initial increase in numbers and spread. By the end of 1974 a pair had colonised Sygangue, and there were probably two pairs at Solitude and one at Roseaux. During 1975 there appeared to be a considerable increase (unquantified, but supported by numerous observations of successful nesting) at Cascade Pigeon, and birds were seen at Vangar (RR). At Cascade Pigeon RR saw a colour-ringed bird which had been marked as a nestling at St.-Gabriel, where there was apparently a decline, possibly due to excessive human activity. In January 1977 RR estimated 30 pairs in Cascade Pigeon and eight in Solitude/Gros Mangue.

After 1975 the birds abandoned St.-Gabriel for some years (RR, Cheke 1979*a*), and were no longer to be found at Sygangue in September 1978, at which time I was also able to find only one bird at Mont Limon. The population (not censused) appeared to be around 20–25 pairs in 1978, concentrated in Solitude/Gros Mangue and La Source/Cascade Pigeon (Cheke 1979*a*), but was estimated to have fallen to eight or nine pairs after the passage of a severe cyclone in February 1979 (Mungroo 1979). All the remaining pairs were in La Source/Cascade Pigeon apart from one each in Solitude and St.-Gabriel. Carroll (1982*b*) suggested the population in late 1981 was still only around 10 pairs. CGJ counted 21 birds and estimated a total of 30–45 individuals in April 1983, with at least one bird back at Sygangue, and the first sighting on Grande

Montagne since the late 1960s. W. Strahm (pers. comm.) saw a bird on Montagne Cimetière in February 1984 and heard one on Mt. Malartic.

Recent history of the Warbler population
In the mid-nineteenth century, when the Warbler was first discovered by ornithologists, there was no suggestion that it was other than widespread and com-

Map 3. Distribution of the Rodrigues Warbler in 1974.
  Dot-centred circles indicate territories, the trio at St.-Gabriel being indicated by two dots. In 1978 the population was more evenly distributed through Cascade Pigeon and upper Solitude, but was no longer to be found at St.-Gabriel, Roseaux or Sygangue, and much reduced at Mount Limon (see text); Mont Malartic and Petit Gabriel were not checked. A pair was again seen at St.-Gabriel in 1979 (Mungroo 1979). Vegetation type is indicated within the area covered (see Map 2 for key). 'H' indicates habitations within the woodland. Wooded or scrubby areas are outlined by a dotted line. (Outline adapted from IGN 'Ile Rodrigue' at 1:10 000 (1978), from aerial photographs flown in 1974.)

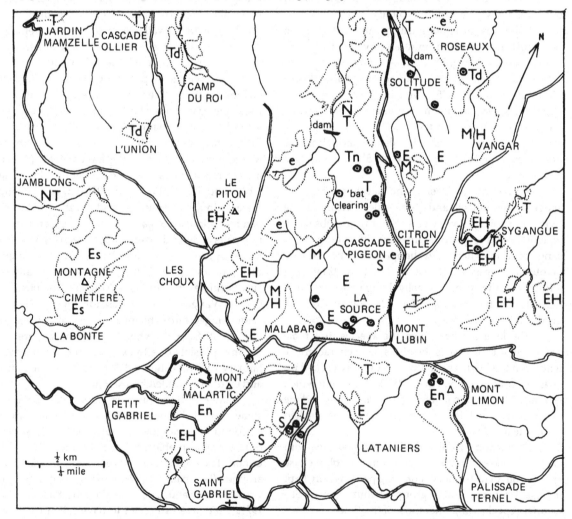

mon. Edward Newton (1865*a*) saw plenty, even as far from their present haunts as Plaine Corail, and Slater (1875) called it "very common". Between the Transit of Venus expedition (*ibid.*, Sharpe 1879) and 1930, when Jean Vinson visited the island (Vinson 1964*b*), there is no information on the Warbler. The decline in the intervening years was presumably related to the inexorable destruction and disturbance of the remaining habitat (Chapter 1); the position between 1930 and the present is examined below.

Vinson (1964*b*) described Warblers as "not rare" around St.-Gabriel. He collected two specimens which are now in the Paris museum; these are labelled 'Mont Malartic' and from this I infer that the bird was found throughout the forest (largely *Eugenia* thicket) then embracing La Source, Mont Lubin, Mont Malartic, Les Choux and St.-Gabriel. The late Dr P. O. Wiehé told me he had seen the birds in *Eugenia* off the road from St.-Gabriel to Mont Lubin in 1938 and 1941. By 1957, when C. M. Courtois collected a bird at La Source, the

Table 4. *Numbers of Rodrigues Warblers in early 1974*

Locality	Jan./Feb./Mar. census			Subsequent observations (1974–5)	Estimated active territories late 1975
	No. definite pairs	No. territories in which only one bird seen/ heard	Total active territories		
La Source/ Cascade Pigeon	5	7	12	No obvious change during 1974; marked increase in 1975 including young from St.-Gabriel	15–20
Gros Mangue/ Solitude	1	1	2	Single bird definitely solitary at first; paired by July, by which time another bird also present. At least 3 pairs in 1975	3+
Roseaux	—	—	—	Old nest found in Dec. 1974. Birds not seen	1
Vangar	—	—	—	Bird(s) seen in Aug. 1975	1
Sygangue	—	—	—	None seen till Oct. 1974, when a pair	1
Mount Limon	—	4	4	Very shy birds, often not detectable at all. At least 2 pairs present Feb. 1975	2+
St.-Gabriel	2 + '1'	—	3	One territory contained a group of 4 birds. Still 3 territories in July 1974, but? only one by Dec. 1974, Feb. 1975. No later observations. Much human disturbance	?1
Petit Gabriel	—	1	1	Not subsequently checked	?
Mt. Malartic/ Les Choux	—	1	1	Not subsequently checked	?
Totals	9+	14	23		24–29++

Agriculture Department had already begun their extensive clearances of upland afforested areas (Cheke 1974a, 1979a, Cheke & Dahl 1981). Courtois (pers. comm.) remembers no difficulty at this time in finding birds to shoot, but already by the next year Vinson (1964b) could find none at all, nor also in 1963. Both visits, as he admits, were short, and he did not have much time to search: yet knowing the birds from his earlier visit, and with the benefit of Courtois's recent observations, the fact that he could not find any suggests a big drop in numbers. There were no very severe cyclones between 1947 and 1955 (North-Coombes 1971), although close passes in 1953 and 1955 (Davy 1971) cannot have been negligible; one must conclude that the forest clearance policy and increased casual woodcutting then tolerated had had a most unfavourable effect on the Warbler's population. The numbers appear to have remained very low indeed until about 1970. Gill, who made a "thorough search" in October 1964, saw only one (which he collected) in Cascade Pigeon, though he heard another at Solitude (Gill 1967). Staub (1973b) saw Warblers only twice on four visits to Rodrigues in 1964, 1967, 1969 and 1970; JM told me that in December 1969 he, Staub and J. Guého searched Solitude and Cascade Pigeon for a week without finding any Warblers, which no doubt led Staub (1973b) to write that "cyclone Monica in 1968 nearly wiped out the species". December is probably a bad month to be looking: I found the birds particularly silent and hard to find in December 1974. Nevertheless Monica had the strongest gusts ever recorded in the Mascarenes (North-Coombes 1971) and undoubtedly had a nearly catastrophic effect on the Fody (see below), so may well have forced a perhaps slowly recovering population of Warblers down once again.

RPA was on the island for over a year from 1970 to 1972. He devoted the summer of 1971–2 to filming the warbler at the nest, obtaining some very remarkable pictures. He and RR found two pairs at Gros Mangue, and at least three at La Source; he searched St.-Gabriel (once only) and Cascade Pigeon without success, but did not check Sygangue or, more than cursorily, Solitude. He wrote (*in* Gomy 1973b) that "certainly no more than about 30 survive", which seems a very reasonable estimate.

RR also showed the birds at Gros Mangue to AFW and S. Keith in October 1971 (Forbes-Watson 1973). Shortly afterwards Rodrigues suffered another cyclone as violent as Monica; Fabienne struck in February 1972 and is still remembered with horror and awe. However, it appears that the present refugia of the Warblers are relatively cyclone-proof; the Gros Mangue birds were shown (by RR) to JM in December that year, and apparently since then the birds also colonised the relatively exposed site at St.-Gabriel, persisting for some years in spite of severe cyclones in the summer of 1972–3.

It is the habit of Rodriguans to attribute changes in their wildlife (and numerous other things!) to the effect of cyclones, and one has to be cautious in interpreting what one is told; but I think the two following pieces of evidence relating to the Warbler are worth repeating:

1. Grand Montagne is rich in apparently suitable habitat for the Warbler, but I could find none there on two thorough searches (January and March 1974). Mr. Tolvis, a forest worker who accompanied me on the second visit, and whose knowledge of nature and of the island was unusually thorough, assured me that until a few years earlier, "before the cyclones", *zoiseaux longbecs* were to be seen on Grande Montagne, and he was surprised not to find any there with me.

2. At Sygangue in March 1974 I spoke to an old woman whose house was in the middle of a *Eugenia* thicket. She told me that "before the big cyclones" there had been "lots" of *longbecs* around her house – and described the position of the nests ("often found by children") and the inquisitive nature of the birds. Unlike most Rodriguans, who have never seen or heard of the *longbec*, she clearly knew the bird very well. As noted above, Warblers were certainly absent from Sygangue during my first visits, and the first pair only appeared in October 1974.

From this brief survey of the Warbler's recent history I conclude that the fragmentation of the habitat that took place from 1955 to 1968, combined with increased human activity in the forests, reduced the population to perilously low levels; also that cyclones have played an important part in further curtailing the distribution and numbers, in spite of a high reproductive rate (see below). The increase in numbers during 1973–6 was due to the lack of cyclones after early 1973, and the next severe storm (1979: Mungroo 1979) produced a sharp setback. However, it is far from clear why the numbers stabilised meanwhile at such a low level.

Habitat: preferences and limitations
Most Warblers I found were in *Eugenia* thicket, usually

but not always, with up to 25% admixture of guava, Mango, *Litsea* and *Eucalyptus*. In the lower part of Cascade Pigeon the vegetation is more mixed, and birds at Solitude were often observed foraging in the mahogany plantation, as well as in other areas with little or no Rose-apple.

I had the impression, impossible to quantify in the present fragmented state of vegetation in Rodrigues, that they preferred dense vegetation with access to clearings or forest edge, and that they avoided the densest continuous areas of *Eugenia* thicket in the La Source/Cascade Pigeon valley.

It is difficult to pin-point any feature of the present habitat, except dense cover, which is essential to the birds – superficially one might still expect to find them in the more thickly vegetated parts of Anse Quitor valley, where the vegetation is dominated by native plants and where E. Newton (1865a) found the birds nesting in 1864 (his description of a "rivulet with steep sides" can refer only to Anse Quitor). Likewise they are not found in the dense *Pongamia* lining the lower parts of Cascade St.-Louis at Graviers. Other factors less obvious than food and evergreen vegetation are probably limiting the distribution of these birds. The limited areas of vegetation in these places may provide insufficient shelter during a cyclone; a severe storm defoliates *Pongamia*, though not *Elaeodendron*. However, cyclones also remove most of the leaves from *Eugenia*, which may in part explain the failure of the birds to exploit fully the available areas of this habitat. The density in 1974 was only about 0.2 pairs per hectare of occupied habitat, compared with up to 13 territories (often of three or four birds) per hectare of its closest relative *A. sechellensis* on Cousin Is. (G. & H. Bathe 1982). In cyclones birds would have difficulty avoiding exposure to wind and rain, and be faced with a greatly reduced food supply after the storm – assuming food abundance to be directly related to amount of foliage (see 'Feeding and food' below). Apart from many native trees, which provide reasonable cover only in Cascade St.-Louis and Anse Quitor, and Mango which is relatively resistant, all the tree species in Rodrigues are badly defoliated by cyclones (pers. obs. in Mauritius after cyclone Gervaise, February 1975). The fact that the present habitat provides far less protection than the now vanished native forest possibly explains why the birds are largely confined to the northern slopes which experience lower wind speeds

in cyclones (see Table 2), and are sheltered from the South-East Trades.

Given freedom from human interference the Warblers could, if the experience with *A. sechellensis* (*ibid.*) is any guide, increase to a population running well into four figures in the 200 or so hectares of *Eugenia* thicket and related vegetation. However, it seems that the frequency of severe cyclones would never allow this to occur, even if it were possible to keep people and disturbance out of the forests. The less seasonal climate in the Seychelles and the high productivity of a seabird island probably also contribute to a higher carrying capacity for *A. sechellensis* on Cousin Is.

Feeding and food

The Warblers fed primarily by gleaning from leaves but also frequently searched on twigs, less commonly along the upper side of larger branches. When feeding the birds were very active and agile, moving continuously.

Although I frequently watched Warblers feeding, the prey was usually too small to be identified. Of recognisable items, the birds at St.-Gabriel took large numbers of larvae of the common planthopper *Flatopsis nivea* (Homoptera: Fulguroidea) in January and March. This occurred on *Eugenia*, Mango, *Tabebuia* and probably other species, apparently throughout the year. ASG saw a Warbler take a caterpillar, and RR saw one brought to a nest. During a short watch at a nest, I saw a nestling at St. Gabriel in late February brought a white moth-like adult of *Flatopsis nivea* and another unidentified insect of the same size (*c.* 12 mm). A few days later the same young one, having left the nest, was being fed on *Flatopsis* larvae. One of the birds attending the nestling reached into and ate something in the nest (the young one was in the branches nearby at the time); after the fledgling had left, I found a second young, dead and full of fly larvae (Diptera); the adult may have been eating these.

When gleaning on twigs and leaves the Warblers often caught or tried to catch small insects that took flight on the bird's approach. These were pecked at rapidly as they took off, the bird remaining on its perch. The sharp 'clack' of the bill closing on these insects (or missing them) was easily heard from up to 15 ft (4.5 m) away, and is the sound thought by Staub (1973a) to be the beak tapping on wood and dry leaves.

Feeding behaviour was in some contrast to that of the larger (males 17 g, females 15 g) Seychelles

Warbler, which is slower and more deliberate in its movements, and can hang from twigs to pick food from below. In the Seychelles the Sunbird *Nectarinia dussumieri* (9–12 g), when feeding on insects, is more similar to the Rodrigues Warbler in its movements than is *A. sechellensis* (pers. obs. on Cousin Is.; weights from Diamond 1980).

Breeding

Breeding season and number of broods. E. Newton (1865*a*) saw a pair carrying food on 2 November 1864 and H. H. Slater (1875) gave December as the breeding time, and collected two nests and an egg (Sharpe 1879). Apart from these early records there were no observations on breeding until 1971 (Forbes-Watson 1973, Alès *in* Gomy 1973*b*, Staub 1973*b*), after which there were some records each season until 1975. These observations (Table 5) suggest that breeding normally begins late September or early October, and ends in March; however, the limits may be just those of the observers' visits. Gill's male, collected on 1 October 1964, had testes beginning to enlarge (4 mm × 2 mm). Two broods may be the norm; the same birds at St.-Gabriel nested successfully twice in the 1974–5 season, and others at Cascade Pigeon may have done so, as may one of the pairs at La Source In 1971–2. However, in view of the behaviour of the Seychelles Warbler, which is much better known (Diamond 1980), the data presented here should be treated with caution. On Cousin Is. the warbler re-nests within a given season only if the first nest has failed or the young have died soon after fledging; the fledgling (normally one) is fed by the parents for up to 3 months after leaving the nest, precluding a second brood. The information for Rodrigues is too scanty to be certain; the pair with an old young one seen on 26 January 1972 by RPA might equally well have been still supporting the same young seen in early November as the offspring of a second brood. The successive nesting at St.-Gabriel in 1974–5 may indicate only that the young from the first nest died shortly after fledging. Circumstantial evidence, however, suggests that the period of post-fledging support by the parents is not so prolonged as it is in the Seychelles, as birds were only seen once feeding young with fully grown tails (RPA's pair on 26 January 1972) and the juveniles lack the loud and persistent begging call so characteristic of *A. sechellensis* (Diamond 1980). The normal breeding period coincides exactly with the cyclone season; it is

most unlikely that a nest could survive such a storm, so the birds' ability to re-nest is no doubt *inter alia* an adaption to cyclone damage.

There was no sign of previous or actual breeding activity in January 1974, and it seems possible that for some reason nesting was greatly delayed in that season. The reason is unlikely to be connected with the drought, which continued throughout 1974 without preventing normal breeding in the 1974–75 season.

RR found a nest under construction in Cascade Pigeon on 18 June 1975. The Warbler may sometimes nest outside the usual months like the Rodrigues Fody, but the nests persist from one season to the next and an old nest beginning to fall apart could be mistaken for an uncompleted one. However RR has seen more nests, old and new, than anyone else, and is unlikely to have been mistaken.

Diamond (1980) has found for the Seychelles Warbler that breeding peaks are closely associated with peaks of rainfall, numbers of rainy days being more important than quantity of rain. The abundance of insects closely followed the curve of rainy days. I can see no connection between the incidence of rainy days and the nesting of the Rodrigues Warbler during 1973–75. As already mentioned, in spite of the drought, the number of days on which the rain fell during 1974 and 1975, and their distribution through the year, were normal. There are no data on insect abundance for Rodrigues. The only information on moult is Mungroo's observation (1979) that a number of territory-holding birds were tail-less or with part-grown tails in late April-early May 1979. The bird I trapped on 18 January 1979 lost its tail feathers during handling, but had a fully grown tail when seen again on 28 February.

Clutch and brood size. Slater (1875) gave the clutch size as four or five (no further comment in Sharpe 1879). There are two eggs in Cambridge (pers. obs.) with the same catalogue number, but no other information.

Of recently observed nests, only three have been found at the egg stage (Table 5); all had a clutch of three.

The egg has been described by Sharpe (1879), Oates *et al.* (1901–12, vol. 4), Alès (*in* Gomy 1973*b*) and Staub (1973*b*, with a photo). It is "dull white, spotted with grey and reddish-brown markings which become larger and more concentrated towards the blunt end" (Staub 1973*b*).

Table 5. *Summary of breeding records of the Rodrigues Warbler, 1971–5*

Year	Month: September	October	November	December	January	February	March	Observer
1971–2	Gros Mangue ✗	B	3B/_73e----253e	des } Not revisited		✗	✗	⎫ AFW
		...... B	3Pr+1fl.y					⎭
	La Source		8B---17B----		41y-61y. dead	26Pr+1o.fl.y		⎫ RPA
			8Pr + 1fl.y(........??.....). ..	133e_202e_291y-		26Pr+1j.fl.y		⎭
1972–3	Gros Mangue ✗	✗	......?Pr+1fl.y ✗		✗	✗		JM
1973–74		✗	✗	All sites: No sign of breeding; 16–20 ✗	St. Gabriel / Cascade Pigeon	.... 281y—4Grp+1j.fl.y*	.........8Pr+1j.fl.y ✗	⎫ ASC
1974–75	St.-Gabriel* ✗	......112y-182y*		*12B--------	232y----	2gone	✗	RR, ASC, AWD
	Cascade Pigeon					5Pr+j.fl.y		RR, AWD
	Mont Limon		......?y			{ 63e / 6B		AWD

Notes: A dotted line indicates the minimum time before a sighting that breeding activity must have been going on; dashed lines connect observations on the same nest. Asterisks (*) connect three successive nestings of the group at St.-Gabriel. Underlined numbers are dates within the month in question. A question mark (?) is used where the exact day of the month is unknown. Abbreviations used are as follows:

B = nest being built	des = nest deserted	e = eggs in nest (usually with number)
fl = fledged, flying	Grp = group	j = just, recently
o = older	Pr = pair	y = young, in or out of nest

Months for which there are no observations are indicated by ✗ in the top right-hand corner of the appropriate rectangle.

In the nest filmed by RPA, one egg vanished before hatching, and when he first found a hatchling there was no trace of the third egg. Each of the three or four family parties he saw had only one young accompanying the pair. He tentatively concluded (*in* Gomy 1973*b*) that the birds deliberately eliminated two of their eggs during incubation, and Staub (1973*b*), from the same data, suggested that the reduction (however performed) from three eggs to one chick was "presumably an adaptation to the present poorer potential of available insect food".

In December 1972 JM saw a pair with a single young; in March 1974 a single young one left the nest I watched at St.-Gabriel and I also saw a pair with one fledgling at Cascade Pigeon.

After the young bird had left the St.-Gabriel nest, I collected the structure and found in it a second nestling already decomposing which had died at about 7 or 8 days old. RPA found a nest at La Source on 8 November 1971 with a dead nestling 8 days old in it. He assumed at the time it was a failed nest, but in the light of my observation and the fact that he saw a pair with a fledged young nearby, I suggest it was probably a successful nest from which only one of two hatchlings had been reared.

The St.-Gabriel birds reared two young to fledging on each of their next two nestings (both broods colour-ringed by RR), though their contemporaries in Cascade Pigeon were feeding only one fledgling in February 1975 (AWD); AWD also saw another short-tailed bird, presumed to be a juvenile, in Cascade Pigeon, but it was unaccompanied and might have been an adult that had lost its tail.

Rodrigues Warblers can and do sometimes rear at least two young. However, one is clearly the most usual number, and it would be very interesting to see whether they do in fact destroy eggs or young they expect not to be able to rear. The Seychelles Warbler provides an interesting parallel; until 1975 it was always thought to lay one egg and raise one young, but in that year several broods of two were found, and the birds were able to rear both (Diamond 1980).

Incubation and fledging period. No nest has been observed sufficiently continuously to be certain of the incubation period; information is even more meagre about fledging, but an estimate can be made' from feather growth and analogy with similar birds.

Staub (1973*b*) quoted Alès as saying the incubation period was about 19 days; but having been through his notebooks with RPA himself, I find that it is not possible to support this figure. The nest for which he has most detail was not visited between 17 November (still building) and 13 December (three eggs); there were only two eggs on the 20th, and one young, unfeathered with closed eyes, on the 29th. Assuming the young was 3 days old on the 29th (as did Alès, *in* Staub 1973*b*), it hatched on the 26th or 27th, so the incubation period must at least be 10 or 11 days. Another nest watched by RPA was found deserted 19–20 days after the clutch was completed; the eggs were clear, although they had been incubated throughout (several visits by RPA). Since birds do not generally desert until they have sat well beyond the normal incubation period, I suggest that this must be well under 19 days. The incubation period of the slightly larger Seychelles Warbler is, however, 18 days (Diamond 1980).

RPA noted that both sexes incubate, and change over every 15 minutes or so. He also reported on the change-over display (*in* Staub 1973*b*): "the incoming bird calls gently; the other bird comes out and they both join in a gentle soft duet before the newcomer rejoins the nest".

The nestling period is greater than 11 days, but it is not clear by how much. The chick in RPA's nest died at about 11 days old, before it was ready to leave. The nest I found at St.-Gabriel on 28 February 1974 contained a nestling on the point of leaving; it was leaping about and chasing insects in the guava bush above the nest, occasionally returning to it. It could not yet fly, and I estimated the age as about 13 days assuming its feather-growth rate to be the same as European warblers of the same size (e.g. Reed Warbler *Acrocephalus scirpaceus*: Davies & Green 1976): the primaries were at growth stage 3–4 (Newton 1966) and the tail about $\frac{1}{2}$ in. (12.5 mm) long. Three days later it had left the nest, and the day after that (4 March 1974) I saw it, now flying, accompanied by its parents. A Seychelles Warbler chick takes 18 days to reach the stage reached by the Reed Warbler in 11 (AWD, pers. comm.), so the Rodrigues bird may also lag behind its European relatives. However, the Reed Warbler cannot fly properly until 18–20 days (a week or more after leaving the nest), whereas the St.-Gabriel nestling was doing so 2 or 3 days after fledging. I suggest the nestling period is about a fortnight.

Nesting success. Although the two nests he followed through were unsuccessful, RPA saw three or four family parties during the 1971–2 season. During 1974 and 1975 all four of the nests followed through reared young; however, only one of these had been found earlier than the nestling stage (St.-Gabriel, December 1974). Two other nests were found under construction or with eggs but were not followed through (Mont Limon, February 1975). At least two pairs accompanied by young were seen.

While some nests may have failed at the egg stage, the impression is that Rodrigues Warblers had a high nesting success in 1974–5, and this is supported by the continued expansion of the bird's range and numbers during 1974–7, described above.

There was no evidence of nest predation during 1971 to 1975.

Pair territories and group territories. All but one of the territories studied appeared to contain a pair of birds, the exception being one site at St.-Gabriel, where I saw four birds together in January 1974, and where at least three were attending the nest at the end of February; it was these birds which nested again successfully twice in the 1974–5 season. Unfortunately they became extremely hard to watch; they were only glimpsed in July 1974, and not seen at all in December 1974 or February 1975, when they must have been respectively building a nest and feeding two freshly fledged young. It was therefore impossible to establish whether or not the territory still held more than two birds. The birds' shyness may have been due to the intense human activity in and around their small patch of *Eugenia* thicket in the latter half of 1974; the other warblers there apparently abandoned the site entirely.

At its present very high density on Cousin Is., the Seychelles Warbler often holds group territories with up to eight birds in a territory (Sorensen 1982); it is not clear how far this was so before the enormous population incease that followed the establishment of the nature reserve on the island (Diamond 1976, 1980).

Nest structure and site. The nest has been described by Sharpe (1879), Alès (*in* Gomy 1973b) and Staub (1973b), who included a photograph, though not *in situ*. It is an ordinary small cup-shaped nest, rather deep inside, lined with a few feathers, and supported in the fork of two or more slender branches;

the external structure may be quite tall, giving the impression of an inverted cone, but often the nest is squat and more or less flat-bottomed.

Of ten nests for of which I have details, five were in Strawberry Guava, four in Rose-apple and one in *Tabebuia pallida*. Guava seems to be preferred, as all these nests were in *Eugenia* thicket, where there is far less guava than Rose-apple. The nests in *Eugenia* were in parts of La Source and Cascade Pigeon with a higher canopy and very little guava. The *Tabebuia* nest was at Roseaux, in a pure stand of young trees.

In the thickets containing much guava, the canopy height is low, only 8–12 ft (2.4–3.7 m), and the nests were near the top of the bushes, 1–2 ft (0.5 m) down. In taller *Eugenia* thicket, the nests were 4–15 ft up (1.2–4.6 m) under a 20–30 ft (6–9 m) canopy. The *Tabebuia* nest was 5 ft (1.5 m) up a 6 ft (1.8 m) sapling.

Nests in guava, near the top of short bushes, are much more exposed to the sun and elements than those in tall *Eugenia*. Guava is, however, more rigid than Rose-apple and has a more three-dimensional system of branching, so that a nest can be supported by three or four stems rather than the two usually available in *Eugenia*. A nest is therefore less likely to be upset in a strong wind; also guava is less defoliated in cyclones than Rose-apple, which might be an advantage in a mild storm.

Distribution and biology of the Rodrigues Fody *Foudia flavicans* A. Newton 1865; 'serin (zaune)' (sereñ zon)

Description of plumages and sexual dimorphism

The male in breeding plumage is bright yellow on head and breast, with red about the face, the back wings and tail being grey-brown tinged yellow; the belly is whitish, and the under tail coverts yellow. There is often a band of yellow on the rump. The beak is black and contrasts with the yellow plumage. The female is dull grey-brown, paler below, with a black beak; males in eclipse resemble females, though a few yellow feathers are usually retained. Fuller details of adult male and female plumages can be found in A. Newton (1865a), together with a coloured plate; Cheke (1979a) included a colour photograph of a male. Juveniles resemble females, but have a variable amount of yellow on the throat (not reliably related to the bird's sex) and pale horn-coloured bills, a character easily seen in the field. After a few months the bills of young birds turn black, although males may occupy territories before this happens (see Table 8, p. 392). Some males retain grey plumage for at least 7 months (to 2½ years in captivity: Jones 1980*b*); others apparently start acquiring yellow plumage when their bills are still horn-coloured; much more information is required from individually marked birds to clarify these observations. I have no evidence of grey-plumaged birds breeding, though such birds are often paired. There is a generally brief eclipse; in 1974 it was centred around May, but the timing may be variable in view of the lack of a well-defined breeding season (see below). The bill remains black throughout.

Males are larger than females, with little overlap in wing measurements, and a significant but lesser difference in weight (Chapter 9). Wing-length increases with age: 1–4 mm over 4 years, averaging 2.4 mm in six birds, both adult (3) and immature (3), ringed in 1974 and recaptured in 1978. Many birds in museum collections have apparently been sexed on plumage colour alone; Sharpe's description (1879) of the 'female' was made from young birds with pale bills and he wrongly assumed that all grey-plumaged birds with black bills were males. Full details of weights and measurements are given in Chapter 9.

Behaviour and general habits

The Rodrigues Fody is arboreal, lives alone, paired or in family parties, and is strictly territorial all the year round. Flocking of juveniles, reported by E. Newton (1865*a*), no longer occurs, and the significance of this will be discussed below in connection with the problem of coexistence with the Cardinal *F. madagascariensis*, introduced since Newton's visit. Gill (1967) and Staub (1973*b*) have summarised the behaviour of the Fody.

On my short visits to Rodrigues I was unable to follow the activities of a single pair of Fodies for any length of time, so the following observations on the amount of time spent in various activities are subjective. On all my visits some breeding activity was going on, and most of the male Fodies spent a good proportion of their time singing and displaying, interspersed with feeding. Females were less demonstrative; although they can produce most of the displays, songs and calls of the males, they do so less frequently and less intensively, so I discuss display largely with reference to males. Females tended to sing and display in conjunction with their mate, duetting in voice and action, and rarely did so alone. When

I played recordings of a male singing (with a female occasionally chipping in) to a female she would remain subdued though inquisitive until joined by her mate, when she would join in his displays, songs and calls. AWD, who visited the island when the Fodies were not breeding, found the birds far less demonstrative and vocal.

Most, though not all, male Fodies watched spent 50–70% of their time, even when feeding, with their wings partly open and quivering, frequently repeating staccato 'chip' calls. The birds were very inquisitive and investigated any intruder, human or avian, supplanting or chasing off other Fodies or Cardinals. Supplanting and chasing were silent. I saw no contact with Warblers or canaries *Serinus mozambicus*, the other two arboreal species on the island. I once saw a female Fody supplanted by a House Sparrow *Passer domesticus*, but the same bird also supplanted a male Sparrow, male Cardinal, and immature male Fody sitting on the same branch within a foot (30 cm) of each other; the Fody was 'chipping' vigorously, as usual, but this was not directed against the other birds. No communal or group activity of any kind was seen in the Rodrigues Fody.

In display, whether directed to an intruder or to a female, the male Fody used changes in body position relative to the perch, wing opening and shivering, and song. Body position varied from fully upright with wings just 'ajar' to upside-down with wings wide open, depending on the excitement of the bird. Birds seldom sang without the wings at least half open. A selection of these postures is shown in Fig. 1; photographs in Staub (1973*b*, 1976) and Cheke (1979*a*) also show typical display poses. The wings were often flapped slowly up and down when spread (Crook's (1961) 'Upright Wing Beating' display), directed as much to intruders as to females. Birds usually sang while doing this, but often continued flapping well after the end of the song. I have only one note on female behaviour in response to male sexual display: he hung upside-down with his wings open and shivering, and she approached very closely 'chipping' loudly and shivering her wings. Gill (1967) mentioned "an exaggerated moth-like flight" (= 'Glide Flight' of the Cardinal: Crook 1961), which I also saw but have no notes on. Crook described this, in the Cardinal, as a nest invitation flight; its function may be more generalised in the Rodrigues Fody where the female builds the nest (see below).

Territories were vigorously defended whether or not the birds had an active nest in them at the time; Crook (1961) suggested the Seychelles Fody does the same, but that Cardinals maintain a territory only while nesting, as do Aldabran Fodies *F. eminentissima aldabrana* (Frith 1976).

The Fodies spent most of their time in the crowns of trees in their territories, but came down quite freely, where there was an understorey, to feed in it or to investigate intruders. The only birds I ever saw on the ground were females collecting nest material. Males and females usually fed separately, but on occasions kept very close together over extended periods. Similar behaviour in the Mauritius Fody (Chapter 4) occurred just prior to nesting; a pair of Rodrigues Fodies seen feeding together at Solitude on 18 January 1974 were feeding a freshly fledged juvenile on 8 March, which suggests the same pattern.

A male was twice seen bathing in water trapped on a *Furcraea foetida* leaf after overnight rain (JRMH, September 1978).

Distribution and numbers 1974–83

On my first visit I was shown Fodies at Solitude and Cascade Pigeon but made no attempt to census them; I also found some at Sygangue. In March and July 1974 I counted all the Fodies I could find at all sites, following this up with a partial census at Solitude in December, repeated by AWD in February 1975. A further nearly complete census was made in September 1978 (Cheke 1979a) and repeated in April–May 1979 (Mungroo 1979), December 1981 (Carroll 1982a, b and *in litt.*), February 1982 and April 1983 (both CGJ).

In contrast to the Warbler, the Rodrigues Fody is noisy and conspicuous. It is highly territorial, and active and vocal in defending this territory; birds of either sex will almost always sing in response to a squeaking noise made by sucking through partly closed lips. The birds often make a repeated 'chip' call when feeding, which also serves to draw one's attention. Juveniles tend to be silent and inconspicuous, though inquisitive.

The Fody was easy to census, for, once stimulated to sing by the observer or a neighbouring bird, a

Fig. 1. Some display postures of the Rodrigues Fody (from field sketches and photographs).

male generally continued to do so, and often followed the observer to the edge of his territory, stimulating the next bird into song and display. Because of this convenient behaviour, I can be reasonably certain that my censuses were accurate, and also that if I failed to find birds in any locality they were in fact absent. This accuracy allowed me to plot a steady increase in numbers over the period 1974–8 and to note some expansion into new areas.

Occasionally a bird followed beyond its territory, and if the neighbour was elsewhere, this could have led to some territories being overlooked in the very high density at Solitude, where a territory may be only 30 m across. I believe I covered the areas often enough to eliminate any serious error from this source. In July censusing was easier as the males were all in different stages of change from eclipse to breeding plumage and were thus individually distinguishable. I also colour-ringed five birds then and a further 17 in December; these marked birds confirmed the validity of my previous method. Immature males occupied

territories long before they first acquired yellow plumage, and were often very silent at first – such birds may occasionally have been overlooked. Females and young males are indistinguishable in the field, and as both sexes sing, there was a risk of confusing singing females with territory-holding young males, erroneously increasing the apparent number of territories. In practice, a few minutes' watching always revealed the presence of a male in a territory with a singing female, whereas a young male would either be alone or accompanied by another grey bird. Males were often seen alone; in several cases I have reason to suppose that the females were incubating eggs at the time (See 'Breeding', below), but there did appear to be an overall surplus of males in the population in 1974–5.

Rodrigues Fodies were initially found only at Solitude, Cascade Pigeon and Sygangue, though by December there were birds also at Roseaux, Vangar, and above Gros Mangue, all close to Solitude. The census results are included in Table 6, and Map 4 gives the distribution over five censuses of the territories in

Table 6. *Distribution and numbers of Rodrigues Fody territories 1974–83[a]*

	1974			1975 Feb.	1978 Sept.	1979 Apr.–May	1981 Dec.	1982[b] Feb.–Mar.	1983 Apr.
	Mar.	July	Dec.						
Locality	(ASC)[c]	(ASC)	(ASC)	(AWD)	(ASC)	(Mungroo 1979)	(Carroll 1982b & *in litt.*)	(CGJ)	(CGJ)
Solitude[d]									
'Inner zone'	12	13	12	13	23	}41	}48	[29+]	15
'Outer zone' (including Gros Mangue)	8	13	[7+][e]	[4+]	[34+]				12
Vangar and Roseaux	—	—	2	n.v.	3	n.v.	—	6	7
Cascade Pigeon/La Source and associated valleys	5	7	[+]	[+]	26	8	24	[18+]	18
Sygangue	1	2	2	n.v.[f]	1	n.v.	—	2	2
No. of territories counted	26	35	[23]	?	87	49	72	[56]	57
No. of definite pairs	17	18	[17]	?	54	42	55	[41]	27
Estimated total no. of territories	26	35	40	?	100	49	72	76	60

[a] Figures given here supersede those in the equivalent table in Cheke 1979a.
[b] In 1982 a single territory in a new site, the copse between Mont Lubin and Citronelle, is included in the totals but not in the breakdown; likewise two grey birds at St. Gabriel in 1983.
[c] Observer.
[d] The 'inner zone' is shown in Map 5; the 'outer zone' is the rest of the area to the east of the main road above the upper Creve Coeur zigzag and as far as Gros Mangue. While the density in the 'inner zone' at Solitude was constant during 1974–5 it increased dramatically from 1975 to 1978. This shrinking of territory size was presumably in response to population pressure as the surrounding area ('outer zone') became increasingly saturated. The area of the 'inner zone' is about 5 ha.
[e] Square brackets enclose figures where only a partial census was carried out. A bracketed plus sign, [+], indicates birds were present but not censused.
[f] n.v., site not visited.

Map 4. Distribution of the Rodrigues Fody in 1974 and 1978.
The distribution in 1981–3 was similar to that in 1978 (Carroll 1982*b*, CGJ). (From Cheke 1979*a*.)

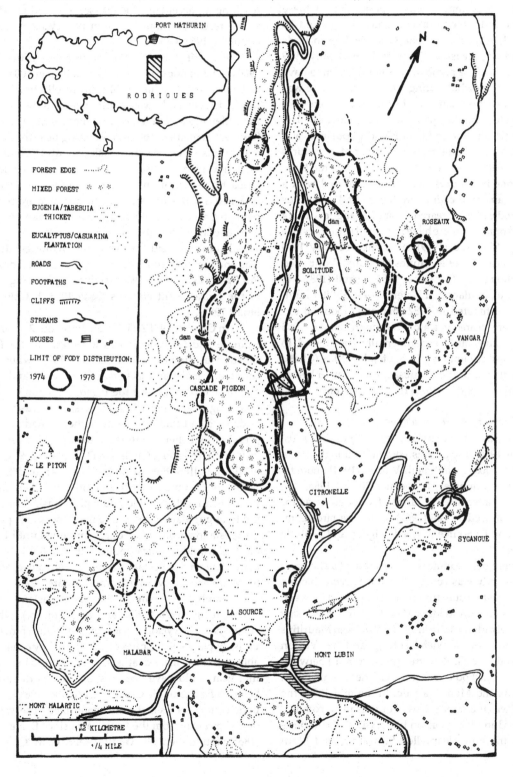

the 'inner zone' at Solitude. In summary in 1974 I found 26 territories in March, 35 in July and I estimated about 40 in December. Only a proportion of these contained undoubtedly mated pairs: 17, 18 and 25 (estimated) respectively. The Cascade Pigeon birds continued to increase until the cyclone of early 1979 (Cheke 1979a). The numbers at Solitude seem to have fluctuated somewhat during 1975–6 when there was occasional nest predation and some adults were shot (ASG and RR), but had increased substantially by late 1978, by which time the total island population was around 100 pairs, plus a good number of immatures (Cheke 1979a); four adult pairs and 16 unpaired birds (all but one immatures) were then captured with a view to captive breeding in Jersey and Mauritius (*ibid.*). Fodies disappeared from Sygangue during 1975, and numbers there have remained unstable since (Table 6). Mungroo (1979), after the February 1979 cyclone, recorded 49 territories (42 pairs, 7 unpaired males), though by December 1981 Carroll (1982a, b) estimated a population of 72 territories. Similar numbers were estimated in 1982, but only about 60 territories were located in a more thorough count in 1983 (Table 6). Five birds experimentally transferred to mixed forest at Jamblong in September 1978 (Cheke 1979a) could not be found subsequently, either before (RR) or after (Mungroo 1979) the 1979 cyclone.

Recent history of the Fody population

The Fody was "abundant" (Slater 1875) in the 1870s, and until 1964 at least was common enough not to arouse any anxiety, although Vinson (1964b) noted a diminution in numbers between 1930 and 1963, due directly to "intensive clearance of wooded areas for agricultural needs". Gill (1967) found the birds common around La Source, Cascade Pigeon and Solitude in late 1964.

Before the clearances of the late 1950s and early 1960s the Fody was common in the *Eugenia* thickets which clothed most of the spine of the island between Grand Montagne and Mont Malartic. P. O. Wiehé saw them in Cascade Victoire in 1938 (MS note in Mauritius Herbarium copy of Wiehé (1938)), and he, Vinson (1964b) and F. R. G. Rountree (pers. comm.) between them reported Fodies from St.-Gabriel, Mont Lubin Les Choux and Mont Malartic. In the mid 1950s the position was the same (C. M. Courtois pers. comm.), but by 1963 and 1964 (Vinson 1964b, Gill 1967) the birds had left the now clear-felled St.-Gabriel and Mont

Malartic areas, although they were newly to be found at Solitude, where the post-war plantations were just reaching an acceptable height (see Appendix). They were present on Mont Limon in 1963 (Vinson 1964b) but not in 1964 (Gill 1967).

It is difficult to be sure of numbers in the mid 1960s, but the population was probably over 100 pairs. Gill wrote (*in litt.*, 6 June 1974): "my recollection is that the Rodrigues Fody was distributed continuously up the valley [Cascade Pigeon] from Port Mathurin to La Source. I remember them as occurring near the lower end of this valley, certainly half way up to Solitude. They never were dense but there always seemed to be at least one pair around regardless of where I was." The area is about 100 ha, which would give around 50 pairs in Cascade Pigeon alone, even assuming only a quarter of the recent density at Solitude. Whatever the actual numbers, neither Gill nor Vinson thought that it was endangered, and it did not appear in the original *Red Data Book* (Vincent 1966): it is listed as 'endangered' in the more recent editions (King 1978–9, Collar & Stuart 1985).

In October 1971 AFW estimated about 25 pairs in Solitude and Cascade Pigeon together, but the first census was attempted in December 1972, when JM counted 10 males and six females at Solitude, and only one or two pairs at Cascade Pigeon. In spite of hard searching over a period of 12 days, that was all the birds he could find, but he nevertheless had the impression that there were more than on his previous visit (with JJFS and J. Guého) in December 1969, when in addition, JJFS shot one of the few pairs found (Staub 1973b). Some time between 1964 and 1969 there must have been a catastrophic population crash, bringing the total population down to a mere five or six pairs. Deforestation around La Source continued until 1968, but this would not have reduced the habitat by more than a few tens of hectares. A more likely cause, as for the Warbler, was cyclone Monica in March 1968.

Staub saw a pair on the offshore island of Ile Crabe in July 1970 (1973b and pers. comm.), which suggests that cyclones can disperse the birds to unlikely places; if cyclones can do that, they can probably also cause mass deaths. Cyclone Louise in March that year could have been responsible, as there had been no Fodies on Ile Crabe in December 1969 (Staub pers. comm.). This is the only confirmed observation of Fodies away from the 'usual' area since E. Newton (1865a) found them all over the island; C. M. Courtois (pers. comm.) found no birds at La Ferme, Camp du Roi, Baie aux Huitres and other 'likely' places during the mid 1950s, and the late P. O. Wiehé would surely have seen them during his botanical surveys in 1938 and 1941, but did not (pers. comm.).

As the reproductive rate of the Fody is very high (see 'Breeding' below) the recovery since 1969 may seem to have been slow. However, cyclone Fabienne in February 1972 probably wiped out any increase since Monica, and the four cyclones during the summer of 1972-3 (after JM's visit) may have delayed the effective start of the recovery until April 1973, only 9 months before my first visit. Cyclone Celine II in February 1979 apparently halved the population (Mungroo 1979). Recent information on the Fody population and the incidence of severe cyclones is summarised in Fig. 2.

Frequency and severity of cyclones at Rodrigues, and the risk to the future survival of the Fody

It appears that it requires a grade 4 cyclone (see Table 2) to have a serious effect on the Fodies, Monica and Fabienne being the most severe storms recorded at Rodrigues since an adequate anemometer was set up there in 1934. North-Coombes (1971) wrote that "Carmen, Henriette and Monique [*sic*] form the most severe cyclonic combination ever to strike Rodrigues in a period of just over 3 months." However, although infrequent, such combinations are not unique and occur at totally unpredictable intervals; North-Coombes cited the summers of 1863-4, 1875-6 and 1962-3 as being notable. That of 1875-6 had a particularly severe combination of four cyclones in 2 months, and may well have caused the extinction of the endemic parakeet *Psittacula exsul*; the last specimen

Fig. 2. The incidence of cyclones and the changes in Rodrigues Fody numbers since 1964 (redrawn and expanded from Cheke 1979a).

Notes:

1. Triangles represent the number of territories, crosses the number of definite pairs. Estimated numbers are in brackets.

2. During the period 1959-63 there were three grade 1 cyclones, one grade 2 and two grade 3. The last grade 4 storm before Monica was probably in 1876. Cyclone strengths are intensities recorded at Mont Venus (see Table 2).

Sources:

1964, Gill (1967 and *in litt.*); 1969 and 1972, JM; 1971, AFW; 1974 and 1978, this study and Cheke (1979a); 1979, Mungroo (1979); 1981, Carroll (1982b); 1982, CGJ (all pers. comm. unless otherwise indicated).

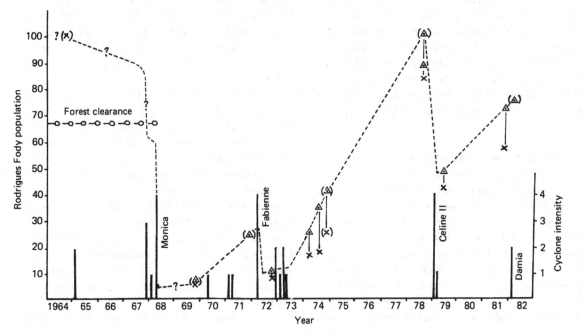

was killed on 14 August 1875 (A. & E. Newton 1876) and none have been seen since. The cyclone of 26 February 1876 was probably of grade 4, and it appears that there may have been no more of that intensity until Monica in 1968, although there were very severe ones also in 1896, 1903 and 1946 (North-Coombes 1971). Although Fabienne (February 1972) occurred in isolation, there was a combination of four cyclones in quick succession in the following summer, two of which, though grade 2 at Port Mathurin, were grade 4 at Maréchal (Table 2).

It appears that once a bird's population is sufficiently reduced and localised, a severe cyclone could wipe it out completely. In 1974, 75% of the Fody's population was in about 15 ha at Solitude, and a single localised gust might sweep away the entire colony, in the same way that all but two or three of the birds in Cascade Pigeon must have been destroyed in 1968. In February 1979 the population was again severely hit by a cyclone, Cascade Pigeon/La Source (69% drop) much more so than Solitude (27%) (Table 6). The near miss by a cyclone in January 1982 seems to have affected the birds relatively little (CGJ). It was because of the cyclone danger that I recommended in 1974 (Cheke 1974a, 1980); that a population of Rodrigues Fodies be established elsewhere, preferably in Mauritius. Were they more widespread in Rodrigues, some areas would be less hit than others by cyclones, and the species would survive as it did in the past; but as they are now so restricted, the risk of extinction is real and immediate unless a 'reserve' population can be established somewhere else. Captive birds caught in 1978 held in Jersey did not lay eggs until 1982 (Darby 1983a), successfully rearing two young (Darby 1983b) in 1983; birds held in Mauritius never bred (CGJ), and were transferred to Jersey in 1983 (Darby 1983b).

Habitat

It is difficult to define the habitat of the Rodrigues Fody, as different writers have said very different things. I shall describe where I found the birds and try to relate this to previous observations.

On my first visits, in early 1974, when the birds were fewest, they were all in mixed forest, tall by Mascarene standards, composed mostly of evergreen trees. The vegetation at Solitude is described in the Appendix; at Sygangue the birds were around the *Pterocarpus* trees and in the patch of *Terminalia arjuna*, and at Cascade Pigeon they were only in the area with

mixed tall trees (Appendix) except for one territory that was in Mango/*Eugenia* forest. By December birds were spreading into scrubbier areas (above Gros Mangue, and the edge of Vangar) or into tall but leafless trees (Roseaux). None were seen in pure *Eugenia* thicket until 1978.

When Edward Newton (1865a) first discovered the birds they were evidently widespread in all types of habitat, whether open country with scattered bushes, or woodland, which was probably native forest; *Eugenia* was a recent introduction (Chapter 1) and plantation trees except *Albizzia lebbek* and *Terminalia catappa* had yet to be introduced (Koenig 1914b). Vinson (1964b) found Fodies in the wooded uplands from 1930 to 1964, and Courtois (*in* Moreau 1960 and pers. comm.) specifically stated they were in the *Eugenia* thicket, though in the centre of the island this was certainly intermixed with several other species; Wiehé's birds at Cascade Victoire (see below) would then have been in native forest dominated by *Mathurina*, *Elaeodendron* and *Sideroxylon galeatum*. Gill (1967) found most of the pairs he observed closely in dense vegetation associated with *Araucaria* or streams or both, but he made his detailed observations at La Source and Solitude, which happen to be the only places with *Araucaria*, and both have streams. *Araucaria* is important for nesting (see below), but is far from essential. Gill also found one pair in "scrub vegetation at the edge of a maize field", as I did at Gros Mangue and Vangar. Gill (*in litt.*, 6 June 1974) wrote that the birds were not in dense *Eugenia* thicket (cf. Courtois!) but in "heavy forest" (i.e. taller trees), especially around clearings, disturbed areas and edges.

Apart from the area of tall *Eugenia* thicket at La Source and upper Cascade Pigeon, there is now no more than a hectare or so of tallish forest on Rodrigues with a continuous canopy, so all Fodies have access to clearing or edges and it is difficult to say whether these are necessary. I suspect they are not, though perhaps favoured when the numbers are low and the birds have a choice of not going into dense *Eugenia*, but variety of trees may be the crucial factor. Staub (1973b) stressed the occurrence of the endemic *Mathurina* in Cascade Pigeon where he found Fodies – there are few Fodies now where most of the *Mathurina* is, but there may have been *Mathurina* further up the valley where the birds now are.

Juveniles move extremely short distances from

their natal territory to establish a new one (see below), so one of the most important factors in habitat choice is where a bird was born. The Solitude colony expanded during 1974 but the fringe birds (Vangar, Roseaux, Gros Mangue) were still no more than 150 m from their nearest neighbours in the main block, and the smaller group at Cascade Pigeon consisted entirely of contiguous territories (Map 4). The separated populations in Solitude and Cascade Pigeon in 1974, presumably isolated as a result of climatic accident, had grown together by natural expansion by 1978, but the birds' absence from other areas may be due to a chronic unwillingness to disperse, which would leave an unfilled vacuum every time a local group is wiped out by a cyclone. Only Sygangue seems to be within regular colonisation range of the present population. Competition with the Cardinal (see below) is clearly an important factor, and would help confine the Fody to islands of woody vegetation once the forest had been fragmented.

Feeding and food

Gill (1967) and Staub (1973b) considered the Rodrigues Fody to be strictly insectivorous. Staub's discovery of the Fody's well-developed brush tongue suggested that nectar-feeding might also be important (Frith 1976), and my observations confirmed this. Feeding on *Casuarina* seed was recorded in 1983 (CGJ).

The Fody is very acrobatic and has the sturdiest legs for its size of any of its genus – a feature very noticeable in the hand but not so evident from museum specimens. It forages on leaves, twigs, branches and tree-trunks rather like a large tit (*Parus*), spending more time leaf-gleaning and less time 'nuthatching' than the similar Mauritius Fody *F. rubra* (Cheke 1980, 1983a, Chapter 4). It has the finest beak of any fody (Moreau 1960), and is also the lightest (Chapter 9).

Gleaning leaves and twigs was the most usual method of feeding, the birds regularly clinging (often upside-down) to the leaf-edges to cover the whole leaf area on large leaves (*Terminalia catappa, Tabebuia, Vitex* (pers. obs.); Mango (Staub 1973b)). Probing inflorescences for insects was also noted (Mango, old calyces of *Tabebuia*; Staub (1973b) mentioned Rose-apple). The birds can walk freely up and down vertical tree-trunks and sometimes fed on *Araucaria* and Mango in this way, probing in crevices. I once saw a

pair fly-catching, and also once only, a juvenile robbing a spider's web.

Birds were seen gleaning on leaves and twigs of most species of tree within their range, but never on *Eucalyptus* or, oddly, *Mathurina*.

I have little information on the insects eaten by the Fodies. In March I saw birds feeding themselves and fledglings on the adults and larvae of the common planthopper *Flatopsis nivea*; in July I again saw them taking the larvae. Also in July I saw a bird extracting leaf-rollers (Lepidoptera) from *Terminalia catappa* foliage. Gill (1967) and Staub (1973b) mentioned "insects and spiders" in the stomachs of the eight specimens collected.

Nectar-feeding was frequent where and when there were suitable flowers to feed on. There were only three such species (*Tabebuia*, Rose-apple and *Erythrina*), though very few of the last were within the Fody's range. In 1978 one bird was regularly seen feeding on flowers of a solitary *Hibiscus rosa-sinensis*. As nectar-feeding is so well developed, the original vegetation presumably provided a wide range of nectariferous species. Of those genera used by nectarivorous birds in Mauritius and Réunion (Chapters 4 and 6) the following are or were native to Rodrigues (Wiehé 1949, Friedmann & Guého 1977): *Aphloia, Dombeya* (2 spp.), *Eugenia* (2 spp.), *Hibiscus* (2 spp.) and *Psychotria. Morinda citrifolia*, so much used for nectar (and fruit) by the Seychelles Fody (Lloyd 1973, H. & G. Bathe 1982, Prŷs-Jones & Diamond 1984), seems not to have been recorded on Rodrigues.

The Fody uses sophisticated techniques for nectar-feeding. *Tabebuia* has a large soft blue-mauve corolla attached to a short rigid calyx, all on a longish pedicel; the birds worked on these either by holding the flower to the branch with a foot and piercing the corolla at its base with the bill, or by plucking the corolla out and inserting the beak into the calyx. The whole flower was often plucked and taken some distance before being fed on. This technique appears to be learnt; a juvenile I saw feeding on a *Tabebuia* flower tore the corolla clumsily apart to get at the nectar. The bird feeding on *Hibiscus* tore a hole in the calyx, then inserted the beak between the corolla-lobes to reach the nectar. Rose-apple has wide-open flowers on short stalks, and no complicated actions are needed to reach the nectar; it provides less nectar, though, and is less often fed on, than *Tabebuia*, and its flowers were often ignored by birds gleaning in the tree. By contrast

Tabebuia flowers were actively sought out, birds often sampling several in succession. Rose-apple was in flower from July to December in 1974 (though mostly in fruit by December), while *Tabebuia* was in full flower in January to March, again in December, and in February 1975. In July 1974 and September 1978 there were virtually no flowers, but the few that opened were rapidly found by the Fodies.

Fodies fed on *Casuarina* seeds by hanging onto the cones like a European Siskin *Carduelis spinus* on Alder *Alnus glutinosa* (CGJ). Until recently there had been little *Casuarina* available in their range, and the only observations were in April 1983, when some birds made extensive use of this food source (CGJ). Seychelles Fodies also feed on *Casuarina* seeds (H. & G. Bathe 1982, Prŷs-Jones & Diamond 1984).

Breeding

Timing. The little published information on breeding gives the impression of an early summer breeding season, but the data are biased as most observers were on the island at that time of year. The records are: nest-building in late September (Gill 1967), young in early November (E. Newton 1865a) and eggs in December (Sharpe 1879). Three similar unpublished observations were made prior to my own visits: a just-fledged young and a nest with eggs in late October 1971 (AFW) and a further nest being built in late December 1971 (RPA), and a nest with eggs in early December 1972 (JM).

In 1973–4 the situation seems to have been different. In January 1974 there was no evidence of recent breeding and there were no juveniles around. I saw a bird carrying nest material in mid-January and RR reported a nest with eggs later in the month. In early March I found three pairs at Solitude feeding newly fledged juveniles, and another building a nest, from which young flew in early April (RR). In July there were many extra young about, not being fed, which had probably also fledged in April. Many adults were in eclipse or had already regained yellow plumage – I saw one bird carrying food, another building, and found a nest at Sygangue where the young were just hatching. Between March and July there had also been some breeding at Cascade Pigeon, several juveniles being present in the July where there were none before. RR reported nests to me in September, October and November. There appeared to be no active breeding going on in December, but I did catch a gravid

female; there were plenty of young about, not being fed by adults, and mostly with dark bills. The pair with the gravid female in December (territory 4 on Map 5) was escorting a juvenile by February 1975 (AWD), as was another pair that had neither nest nor young in December.

This information is summarised in Table 7, which shows that during 1974 there was apparently no breeding activity only in May, although there was not much in August or December. From RR's letters it seems that in 1975 eclipse was at the same time, May/June, as in 1974, but there was little or no winter breeding (no nests found in July or August) although the birds were yellow again in July. Nesting started in late August (ASG saw flying young in mid-September) and continued through till at least December. There was an important peak in September: RR showed ASG six concurrently active nests in Cascade Pigeon, and others freshly destroyed by children at Solitude; there was no evidence of a similar peak in 1974. Nesting was just beginning in mid-September 1978, but the presence of numerous pale-billed juveniles suggested there had been, as in 1974, a spell of winter breeding; one pair were feeding a newly fledged young. Likewise in the following year, after the cyclone, birds were nesting in April–May (Mungroo 1979). In April 1983 CGJ found one active nest and one recently fledged brood. Given the inconsistency of the weather from one year to the next it is perhaps not surprising that breeding appears to be opportunistic and not clearly or regularly seasonal. At present it remains impossible to say what the 'normal' breeding season is for the Rodrigues Fody, or whether in fact it has one.

Crook (1961) suggested from limited evidence that Seychelles Fodies might breed in any month of the year: on Cousin Is. most of the population usually breeds synchronously in winter (Prŷs-Jones & Diamond 1984, Brooke 1985), but in some years breeding extends throughout the year (H. & G. Bathe 1982). The main season is prolonged, as it is in the Mauritius and Aldabra Fodies (Chapter 4, Frith 1976). Cardinals are much more seasonal (see below; Crook 1961).

Clutch size and brood size. Of the 14 clutches recorded by JM, RR, ASG and myself all but one were of three eggs; the exception was one of two, but this was then predated and may not have been complete. Staub (1973b) also gave the clutch as three, without any supporting details; Slater (1875) claimed "5–6". The eggs are unspotted pale blue.

Table 7. Nesting data for the Rodrigues Fody, 1973–6

Year	May	June	July	Aug.	Sept.	Oct.	Nov.	Dec.	Jan.	Feb.	Mar.	Apr.	
1973–4[a]		✗	✗	✗	✗	✗	✗	✗	20b ✗	…?2e–?pred	8Pr+1j.fl.y	?Pr+j.fl.y	} ASC } RR
										18Pr.ft.–––––––	8B_183e–253y–6Pr+3fl.y		ASC ASC, RR
											8♂+2j.fl.y		} ASC
										(no……juvs. seen)	12Pr+1j.fl.y		ASC
Plumage state:									Yellow	Yellow	Yellow		
1974–5[a]	✗	✗	12b …12♀cf 173ey–273y–3pred (many new juvs. since Mar.)			c103e–c20des ……113y–203y–fl …182y1e–fl …183e		13♀p ✗	(no active nests)	4Pr+1j.fl.y ✗ …4Pr+1j.fl.y	✗	✗	{ ASC, RR, AWD RR, ASC, AWD
Plumage state:	Eclipse		Eclipse to yellow					Yellow	Yellow	Yellow			
1975–6													
Solitude	✗	(no nests)	(no nests)				…?1y–fl ✗ …?mNe	✗	✗ (no nests)	✗	✗		{ RR, ASG
Cascade Pigeon					19B …19Pr+2j.fl.y …?2Npred …203e …203e …203e …203y …202N								} ASG RR, ASG
Plumage state:					Yellow								

Notes:

[a] All data for 1973–4 and 1974–5 are for Solitude.

A dotted line indicates the minimum time before a sighting that breeding activity must have been going on; dashed lines connect successive observations on the same nest.

Underlined numbers are dates within the month in question. A question mark (?) is used where the exact day of the month is unknown.

Abbreviations used are as follows:

B = nest being built	ey = eggs hatching	N = active nest, no other details
b = ♀ carrying nest material	fl = fledged, flying	pred = nest predated
cf = carrying food	ft = feeding together	p = pregnant
des = nest deserted	j = just, recently	Pr = pair
e = eggs in nest (usually with number found)	m = many	y = young

Months for which there are no observations are indicated by: ✗ in the top right-hand corner of the appropriate rectangle.

One, two or three young are reared; of six successful nests, four raised all three young and the others two and one young respectively. However I saw no pairs feeding more than two fledged juveniles, and it was more often only one (five cases out of eight), so there may be a high post-fledging mortality.

Incubation and fledging periods. I have no satisfactory information on incubation period. The fledging period, estimated from three nests with reasonable data, is about 14 days.

Post-fledging care and the status of juveniles in territories. The nature of my visits to Rodrigues, short with long intervals between, makes it very difficult to tie together my observations on the behaviour and dispersion of juveniles.

Immediately after fledging the young were fed by the parents, apparently for quite a short time. A very recently flown young in March, still with a marked whitish gape flange, was being largely ignored by its parents from whom it was soliciting food; I saw it being fed once and it was actively feeding itself as well.

In July and December none of the juveniles I saw were soliciting food or being fed, but were present in adult territories, mostly keeping quite silent, though as inquisitive as the adults. Unfortunately it was impossible to tell how long these young birds had been out of the nest, or whether the territory they were in was their natal one.

The presence of a juvenile in a territory did not prevent further nesting, and I observed two cases in July where pairs building or feeding hatched young still had a silent juvenile with them. RR colour-ringed two broods of three in October and November, but I could find none of these young in December; one of the pairs was alone (though had produced another young by February) and the other pair was accompanied by two unringed young birds.

Judging by such observations as I have, the young appear to do any of three things after becoming independent of their parents.

1. Remain in their parents' or others' territory, probably moving from one to another, waiting for a 'divorce' or a death. Either sex may do this. Mobility is suggested by AWD's failure in February 1975 to find any of the four juveniles I had colour-ringed in December 1974. They had probably moved to parts of Solitude he did not

check; at least two were alive as one was killed in August 1975, and another seen by ASG in September.

2. Leave the area of established territories and set up on their own; presumably only males do this. This can happen very soon (pale-billed birds) or later, after a period of following course 1 or 3.

3. Attempt to contest the ownership of the territory with an incumbent male. I observed this in two cases. One I followed through from July to February failed, the young bird (colour-ringed) eventually setting up a territory of its own on the edge of the 'inner zone' at Solitude; the incumbent male was still *in situ* in his old territory.

These behaviour patterns are discussed further below under 'Social structure of the population'.

The role of males and females in nesting. The female apparently normally builds the nest alone. Gill (1967), RPA and Mungroo (1979) saw only females bringing material to nests being built, and the only two birds I saw carrying nest material were females, as were two birds I watched adding to nearly complete nests; in captivity, males have been seen nest-building (Darby 1983). The female does most of the building in the Mauritius Fody (Newton 1959; Chapter 4) but the reverse is true for the Seychelles and Aldabran species (Crook 1961, Frith 1976); the sex of nest-builders in the Cardinal seems to vary with locality (Crook and Newton *locs. cit.*).

The female also incubates alone (Darby 1983: captive birds): I flushed only females from eggs, and JM, watching a nest in December 1972, saw only the female going in to incubate the eggs, which she did in spells of *c.* 15 minutes, interrupted by feeding sorties; the male never entered the nest. The Mauritius and Aldabra Fodies are similar (Chapter 5, Frith 1976).

I have no observations on which sex(es) feed the young in the nest, though the only bird I saw carrying food was a female. Both sexes, however, fed fledged young.

Nesting success. Of the 14 nests whose outcome is known, six produced fledged young, three (one with eggs, two with young) were deserted by the parents and five were destroyed by children. Of the predated nests, one was with eggs, one with large young, and two at an unknown stage. The nest with large young was at Sygangue, where there are several houses in and around the wood. At Solitude marauding children are reported by RR to destroy nests quite often, and he showed two such to ASG in September 1975 and CGJ saw another case in 1983. There are no children living there (except those of the forestry staff) though three are publicly used footpaths running through parts of the area, and Vangar is very close. At Cascade Pigeon there were also no residents only various passers-by, but the vegetation was mostly thicker and the birds may benefit from that. There was no evidence of predation other than by man.

Nest structure and site. The nest was described by Sharpe (1879), Gill (1967) and Staub (1973b). It is a typical *Foudia* nest, like a sparrow's (*Passer*) but tidier, rather roughly constructed, with a side entrance through a short porch of varying size. The body of the nest is generally made of flowering grass shoots (Staub 1973b, pers. obs.) with little lining but a few feathers.

The nest Gill saw being built (1967) was in an *Araucaria*. At Solitude, where these trees are plentiful, they seemed to be the favoured site (25 out of 33 nests, old and active: this study and Mungroo 1979) and indeed appeared almost always to be used if there was a tree in the territory. Elsewhere, and at Solitude in the absence of *Araucaria*, nests were found in Rose-apple (5), mahogany (3) *Ravenala* (3), guava (3), Mango (1) and *Eucalyptus* (1). Slater (1875) reported nests as "usually in young trees of the bois d'olive" [*Elaeodendron orientale*].

In *Araucarias* the nests were 6–15 ft (1.8–4.6 m) above the ground, usually at the end of a branch; in other species nests were lower (4–10 ft; 1.2–3 m), and the position variable. In *Ravenala* the nest was typically attached (often inadequately) to the lower end of a hanging dead leaf; in *Eugenia* it was generally at the end of a branch, but in guava sometimes in the middle of a bush. Several nests may be built in the same tree, though this was seen only in *Araucarias*; the new nest may be on top of the old one, or on the same branch, sometimes quite near the trunk if the old nest already occupied an extremity. Crook (1961) noted similar re-use of a favoured tree by the Seychelles Fody.

The prickly *Araucaria* is obviously more predator-proof than the other species, though it is not clear what predator the Fody is defending itself against. Rats (*Rattus norvegicus*; Cheke 1979b) are not excessively common now, but the more dangerous *R. rattus* were abundant in the past (North-Coombes 1971; Chapter 1) and have recently been recorded again (CGJ).

Social structure of the population

Map 5 shows the territories of Rodrigues Fodies in the 'inner zone' at Solitude, and Table 8 shows the make-up of those and other territories during my visits in 1974 and AWD's in 1975. Although the general configuration of territories was relatively constant, the boundaries were very plastic and there was considerable mobility of females and juveniles within the area.

Territory size, persistence and dispersion. Map 5 shows that territory size was variable within a certain range. The boundaries indicate the area defended, but cannot truly represent the actual situation, which varied from day to day. Also the birds occasionally crossed two or more of their neighbours' boundaries – for instance in December 1974 the pair in territory 10 were mistnetted in territory 5 together with that territory's holder; several such observations were made in 1978. Most of the time a particular bird or pair could be found within the area outlined.

Territories were held by males, with a high degree of site fidelity. None of the 11 territorial males colour-ringed in December 1974 had moved by February 1975, and at least five of them were still in the same places in September 1978. The loss of colour-rings in the interval prevented identification of two more survivors (retaining metal rings), both seen in territories where ringed birds had been present in February 1975. The only recorded change of site is the movement of one male about 150 m between February and September 1975 (Map 5).

Territory size ranged from under 1000 to about 4000 m^2 (Map 5); the 1974 average was just over 2000 m^2, representing a potential density of nearly five territories to the hectare. In 1974 patches of less suitable habitat and unoccupied areas reduced the actual density in even the best parts of Solitude to about 3 pairs/ha, though it rose to 4.6 in 1978, when many territories were well under 2000 m^2. The size of territories in Cascade Pigeon seemed to be of the same order, but I did not attempt to map them so thoroughly. One territory in the 'outer zone' at Solitude in December 1974 was as large as 0.75 ha.

Territory size apparently ceased to bear any particular relation to density below a certain level; fewer birds simply occupied less of the habitat, apparently in an aggregated way. This was particularly noticeable in Cascade Pigeon in 1974, where almost all the birds were together in one group in spite of the large area apparently available (Map 4). Where there

Table 8. *Composition of Rodrigues Fody territories in 1974 and 1975*

Column:	1	2	3	4	5	6	7	8	9	10	11	12
	♂ alone		♂ + pale-billed ♀		♂ + juv. ♂ (1st ♂ yellow)	♂ + juvs. of indeterminate status	Pair alone	Pr. + pale-billed juvs.	Pr. + black billed juvs.	Pr. + both Pale & black-billed juvs.	Lone juv. (pale bill)	Totals of territories checked
	Grey[a]	Yellow	Grey[a]	Yellow								
Mar. 1974	—	3[b] 6[b]	—	—	—	—	5 13	2 4	—	—	—	10 23
July 1974	—	4[b,c] 12[b,c]	—	1 2	1 2	1 1	4 10	1 5		—	1 3	13 35
Dec. 1974	1 3	—[d]	0 2	—	1 1	—	6 11[e]	1 1	2 2	1 1	—	12 20
Feb. 1975	1 1	2 2	1 2	1 1	—		6 9	2 2	—		—	13 17

Notes:

The non-underlined numbers are for the 'inner zone' at Solitude; underlined numbers show the total for all territories checked.

[a] Implies immature; eclipse adults in July not included.

[b] Some of these could have had incubating ♀♀.

[c] One at Solitude and one at Cascade Pigeon were entirely grey and might have been black-billed immatures (see note *a*).

[d] An additional lone yellow ♂ held a very small central territory briefly, but soon vanished.

[e] One of these had a ♂ which was only half yellow.

were gaps between territories the space was simply left vacant, rather than being used by the birds on either side; this occurred in many of the peripheral territories at Solitude and Cascade Pigeon. At high densities the territories become smaller (Map 5: 1978 distribution).

The Fodies may just be 'sociable', or it may be that they always concentrate as densely as possible into the best available habitat, thus becoming aggregated. The very well developed interactive and contact behaviour, however, suggests that presence of close neighbours plays an important part in the bird's biology, and that the aggregation is not primarily determined by microhabitat preferences.

The recorded density of territories, though high, is well within the range of densities exhibited by the Cardinal in optimal habitat (see below) and much lower than the density of the Seychelles Fody on Cousin Is. where there were c. 1000 birds in 27 ha in 1974 (AWD) and in 1980–81 (H. & G. Bathe 1982). By contrast the Mauritius Fody and the Comores Fody *F. eminentissima* on Mohéli (Chapter 4) occur at very much lower densities.

Females and juveniles: status, movements and dispersal. As shown above there were usually a number of juveniles present within the territorial area of established adults, often until well after the young were independent of their parents. In some cases pale-billed birds were seen in company with males but apparently in the absence of an old female (columns 4, 5 and 6 of Table 8). Some of these young were males and behaved territorially within the territory of the established birds, sometimes confronting them; others were females who appeared to be potential mates for the male, who might himself be a young bird in grey plumage.

My sparse evidence suggests that the pair bond is not permanent and that some females change mates, moving from one territory to another. Of five pairs colour-ringed in December, two females had moved to other mates by February; one of the deserted males was by then accompanied by a juvenile (a different bird from the one in their territory in December), the other was alone. One of the females had replaced an unringed bird in an adjacent territory, while the other had moved to the 'outer zone', to a territory I had not checked in December. By contrast the only pair I had colour-ringed in July was still together in February.

One of the mobile females was an adult, the other was a younger bird in post-juvenile plumage. None of the five males had moved. The old female caught in 1978 was in a new place, although her 1974 mate was still alive in his original territory (Map 5).

Some indication of the time juvenile males spend before setting up territories came from two birds ringed in July, in territories 1 and 7. Both had pale bills and had not started their post-juvenile moult. By December, the bird from 1 had a black bill and was holding a territory in the 'outer zone', 120 m up the main stream from his (presumed) parents; by February he had moved to a new site, still only 100 m from the parental territory. On both these last occasions he was accompanied by a juvenile (? female). The other bird was missed in December, but in February AWD found him in a new territory, 2*, in the 'inner zone', about 80 m from where he was first caught; he also had a black bill and was accompanied by a juvenile (? female) which he was "frequently displacing though generally tolerating" (AWD). Both these young males were still in grey plumage in February. As mentioned above, the four new juveniles colour-ringed in December were not found in February; three of these were relatively older than the juveniles ringed in July, having black beaks and undergoing or having finished their post-juvenile moult.

The evidence of the two July juveniles suggests a marked disinclination of Fodies to disperse far from their birthplace, which supports the view that the clumping of territories is social.

These observations show that the presence of a juvenile in a territory does not indicate recent breeding in that territory; the bird may be the male's new mate or a would-be rival. This complication made it difficult to know exactly what was happening in a population that was monitored only once every few months. The population structure over the year of observations is given in Table 8, which shows the various trends I have discussed. Fortunately there were no juveniles in the population before February 1974, which gave a baseline for subsequent developments.

I infer that none of the young males fledged in February and March had yet developed black bills and taken up territories by July (Table 8, column 1), although some of the young females had settled down with males (column 4). By December several of the young males had territories (column 1; one confirmed by colour-rings); young females were by then indistinguishable from older birds.

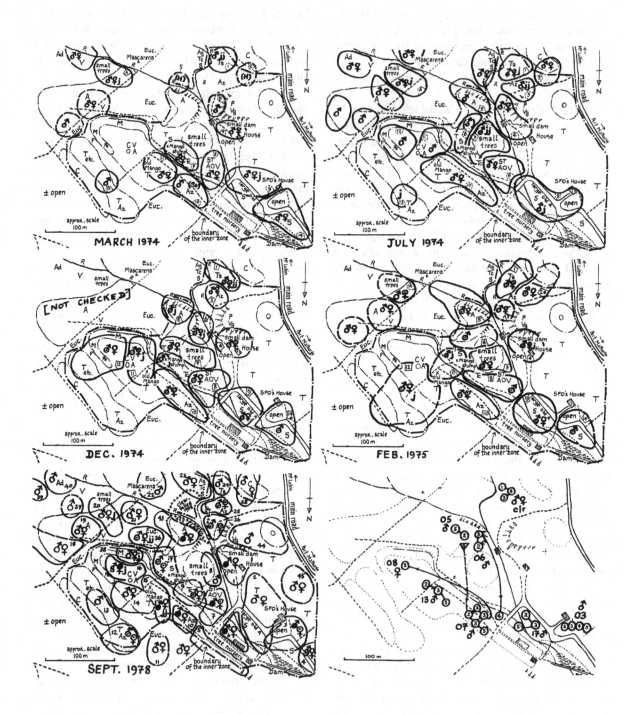

The old males without females early in the year had found them between July and December, so the only pale-billed females were with grey birds (column 3); meanwhile 'indecisive' juveniles (see choice 1, p. 390) were old enough to have gained black beaks but were still not established alone (column 9) and there was one pair with two generations of juveniles on its territory (column 10; the black-billed one was confirmed as juvenile by trapping; both of the territorial pair had been colour-ringed in July). By February 1975 all the black-billed juveniles that had been 'undecided' in December had gone (dispersed ?), though at least one was back, holding a territory, by September; another was killed in the 'inner zone' in August.

Table 9 shows the numbers of birds in relation to territory numbers in the 'inner zone' at Solitude. In 1974 the presence of juveniles was offset by the existence of unmated males until December, but the ratio was back to the previous level in February 1975. It is possible, however, that in March and July some apparently unmated males were concealing incubating

Table 9. *Number of Rodrigues Fodies per territory in the 'inner zone', Solitude*

Date		No. of territories	No. of birds	Birds per territory
1974	Mar.	10[b]	19	1.90
	July	13	24	1.85
	Dec.	12	28[a]	2.33
1975	Feb.	13	25	1.92
1978	Sept.	23	51	2.22

[a] Does not include the lone male who held a small territory briefly.
[b] There were two further territories where birds were only heard (Map 5).

Map 5. (*opposite*). Rodrigues Fody territories at Solitude, March 1974–September 1978. 'Inner zone' territories, and a few others that are within the range of the map, are shown.

Sex of birds seen in territories is indicated by conventional signs; j, juvenile. A dot within the circle of the male or female sign indicates a colour-ringed bird; juveniles are shown by two dots one above the other on the j. Filled circles of the conventional signs (1978) indicate 4-year old ringed birds; in the same map grey-plumaged males are indicated by ♂̣.

For simplicity territories are shown not overlapping; see text for comment. Numbered territories (1974–75) indicate presumed (number in dashed line) or certain (number in continuous line) continuation from one observation period to the next. An asterisk against the number indicates a re-occupation by new birds of a previously lapsed territory. The base map for the numbering is the July one, the March birds being traced backwards from it. Interrupted boundaries indicate some uncertainty as to the exact limits of the territory. Birds only heard are shown as '(H)'. The 1978 numbers are unrelated to the 1974–5 series.

Principal tree species at Solitude are indicated by code letters as follows:

A = *Araucaria*	F = *Furcraea*	S = *Swietenia*
Ad = *Adenanthera*	M = Mango (*Mangifera*)	T = *Tabebuia*
Ag = *Agathis*	O = *Elaeodendron*	Ta = *Terminalia arjuna*
Az = *Albizzia*	P = *Pandanus*	Tc = *Tectona*
C = *Calophyllum*	Ps = *Psidium*	V = *Vitex*
Euc = *Eucalyptus*	R = *Ravenala*	Vg = *Vangueria*

The last map shows positions where marked birds were trapped or sighted over the period 1974–8, with the following code: 1, July 1974; 2, December 1974, 3, February 1975 (AWD), 4, September 1975 (ASG); 5, September 1978. The birds are identified by the final two digits of the ring serial number (Museum Paris SA 2762**), except for the pair in territory 1 (1974–5) which bore only colour-rings. Birds 07 and 08 were paired in 1974; 07 was first caught 'trespassing' in 05's territory.

(Based on a freehand sketch by ASC fitted into the framework of a 'Plan of Crown Forests in Rodrigues' made by Davis in 1943 (copy in Forestry Services HQ, Curepipe, Mauritius; scale 6 inches to 1000 ft (1 : 2000)). Davis's map has virtually no detail apart from the position of the road and the line of the principal arm of the stream (Rivière Bambous). The detail I have added is adequate for finding one's way around Solitude, but is in no way an accurate survey of the site.)

females; on the other hand at least one of the new lone males in February had lost his mate to another male outside the area. Large numbers of young birds contributed to the high figure in September 1978, when again some females may have been missed while incubating.

Moult

I have a little information on moult from birds trapped in July and December, and field observations of eclipse plumage of males in July.

Adults. In mid-July 1974 most males were showing some degree of eclipse plumage; there was no sign of this at other times. Two adult males caught then, coming out of eclipse, showed no sign of wing- or tail-moult, suggesting that the complete moult takes place as the birds go into eclipse. This is the same sequence as in the Mauritius and Aldabra Fodies (Chapter 4, Frith 1976), but contrasts with the Seychelles Fody, where the birds breed straight after the full moult, which coincides with the eclipse plumage (Brooke 1985). Birds that had flying young in early March were back in full yellow plumage in July; they still had the juveniles with them, but two of the four pairs were already re-nesting. Pairs that produced young after mid-March were in full eclipse or showing only the first few yellow feathers (three males; in all cases I saw no female; a fourth pair disappeared in the interim). Unpaired males, and paired males without young, were at all stages of moult. Successful breeders had 4 months to complete the two moults, leaving a very short time in grey plumage, shorter even than the Mauritius Fody (Chapter 4). This may, however, be abnormal, for Slater (1875) and Guerin (1940–53), probably independently but citing no source, wrote that the first yellow feathers come through in early August, full breeding dress being acquired only in October (Guérin) or November (Slater). Staub (1973*b*) found a pair in what he called "winter plumage" in late July, but his description seems to refer to a dull-plumaged example in yellow dress. Captive birds have been known to remain 5–6 months in eclipse (Jones 1980*a*).

None of the three adults (two males, one female) caught in July, were moulting wings or tail. In December, however, six paired territorial birds were moulting their tails; three were males, and three females, two of which also displayed wing-moult, apparently arrested. All, except one of the females,

had the plumage characters of young birds. A further female moulting her tail was seen but not caught. Five non-moulting adult males and one female were also caught in December. No wing- or tail-moult was observed in September 1978.

Juveniles. Neither of the juveniles caught in July 1974 was moulting, nor any in September 1978, though in July I saw a pale-billed bird beginning to acquire yellow plumage. The two females among four juveniles caught in December 1974 were moulting their wings (one also the tail), while a young territorial male in grey plumage was moulting the tail only (cf. some 'adults' in December).

The sequence of moult after the young leave the nest is still not clear, and may be variable. Some males undoubtedly moult directly to yellow plumage while others have a post-juvenile grey plumage; some captive birds have retained grey plumage for 2½ years after fledging – young of *c*. July 1978 not becoming yellow until November 1980 (Jones 1980*b*). The primaries apparently are usually moulted during the post-juvenile body-moult, but the juvenile tail is often retained until later and then moulted alone. Juvenile primaries and tail feathers are distinctly narrower than those of later plumages; all the 'adults' moulting their tails in December had narrow outer feathers, except one female. There is also a textural difference between juvenile and later contour plumages, but it is less marked than the difference in flight feathers. Only one of the December juveniles was in pure juvenile body plumage, two were moulting out of it, and one was indeterminate; the territorial grey male had both adult primaries and contour plumage, and was moulting his narrow tail feathers.

Longevity, survival and predation

The marked population has not been sufficiently well monitored to measure survival accurately, but a few comments may be made. In September 1978 I found eight ringed birds at Solitude, out of 17 marked there in December 1974, of which two are known to have been killed during 1975 (RR *in litt.*). Of the eight, seven were males, out of 11 originally marked (one killed) – giving a natural survival rate of 70% over 4 years, or 91.5% per annum. On the other hand only one female out of six (one killed) was found – representing an annual survival of only 67.5%. If this difference is real it would account for the regular presence of considerable numbers of unmated males. Unfortunately ordinary plastic colour-rings mostly fall off this particular species (or are removed) after only a few months, so trapping is necessary for long-term observations, though none has been done since 1978. In April 1979 after cyclone Celine II, Mungroo (1979) saw only six ringed birds at Solitude (out of 35), only one of which was still colour-marked. CGJ found only one (male) in February 1982 after a second cyclone; he looked very hard for ringed birds, so mortality has apparently been very much higher since 1978, probably not solely due to the direct effect of cyclones.

Man is the only likely predator of adult Fodies in Rodrigues. In 1975 at least three birds (two of them ringed) were killed at Solitude (RR *in litt.*) – two by children with catapults, one by the retiring senior forest officer (for stuffing!). At least two nests, also at Solitude, were destroyed by children in that year. The birds at Sygangue disappeared during 1975. The three birds known to RR to have been killed may have represented only the tip of the iceberg. This level of human predation, if persisted in regularly, is clearly intolerable for so severely endangered a species as the Rodrigues Fody. If the birds are not safe at Solitude, the only well-patrolled forest area on the island, their chances of prospering elsewhere are not good. It would take only one determined boy with a catapult a week or less to eliminate the whole population. Fortunately predation of this kind has not been noted more recently. A new and possibly dangerous development first noticed in 1979 is the use of Fody nests by House Sparrows *Passer domesticus* (Mungroo 1979, RR *in litt.*). Further study will be needed to see whether Sparrows actually usurp the nests from breeding Fodies, or whether they merely occupy abandoned nests.

Relationship with the introduced Madagascar Fody or Cardinal Foudia madagascariensis

Gill (1967) noted that in contrast to other published observations on the Cardinal (Rand 1936, Crook 1961), the birds were apparently largely insectivorous at Solitude. My observations confirm Gill's and add a further element, nectar-feeding, to the Cardinal's repertoire.

At Solitude and Cascade Pigeon the territorial Cardinals were almost entirely insectivorous and nectarivorous. Apart from birds scavenging with House Sparrows *Passer domesticus* around the house, I never saw Cardinals feeding on the ground, nor on grass seeds, though AWD once did in Cascade Pigeon. I frequently saw the birds gleaning insects on the leaves and flowers of *Tabebuia* and other species, and have several observations of nectar-feeding on *Tabebuia* and Rose-apple flowers. On *Tabebuia* the Cardinal used the same methods as the Rodrigues Fody. I saw them feeding by holding the flower with the foot and tackling the corolla at the base with the bill, albeit clumsily biting rather than neatly piercing; I also once saw a bird take a whole flower off to another tree, but was unable to see what it did with it. The Cardinal also takes nectar from tree flowers in Mauritius (*Grevillea* sp., Chapter 4), and Réunion (*Eucalyptus robusta*, pers. obs.).

Away from the forested areas mentioned, the Cardinal apparently behaved more typically. Single birds or small groups were seen, often in company with waxbills *Estrilda astrild*, feeding on grass seeds in open areas. Both these species were, however, also each once seen nectar-feeding on *Lantana camara* flowers by plucking out the corolla, holding it crosswise in the bill, and manipulating it while squeezing the tube so that the flower eventually dropped out of the side of the bill and the bird retained a drop of nectar. Two or more flower heads were stripped in succession. The Cardinal also uses this method in Mauritius, Réunion and the Seychelles for *Lantana* (pers. obs., JFMH, Melville 1979), and in the Seychelles also for *Morinda citrifolia* (pers. obs., AWD *in* Prŷs-Jones & Diamond 1985); I never saw a Rodrigues Fody feeding on *Lantana* flowers, but there is very little *Lantana* within its range.

The Cardinal was abundant throughout most of Rodrigues, inland of the 'cattle-walk', not less so at Solitude and Cascade Pigeon than elsewhere. At Solitude, however, the native Fody outnumbered the Cardinal by about 2.5:1 in the 'inner zone' (1978; 1.8:1 in 'outer zone'); in Cascade Pigeon there were more Cardinals than Fodies, and elsewhere there were no Fodies at all. The overgrazed arid and bare areas in the 'cattle-walk' had very few Cardinals; I saw none along the east coast, in Cascade Victoire, Anse Raffin, Cascade Pistache, and on one of my two visits to Anse Quitor, but elsewhere the birds were very common in areas of mixed scrub, habitations and fields. Apparently they were even more common a few years ago: Gill (1967) recorded them as outnumbering Fodies by 10:1 even in the area where Fodies were densest (Solitude), and a local informant told me in 1974 that they were much more abundant before the 1972–3 cyclones. Certainly the density in 1974–8 nowhere rivalled that around La Saline in Réunion (5–8 territories/ha; Chapter 6), but it was similar to that recorded elsewhere on that island by Barré (1983; 3–5 territories/ha at low elevations) and by F. R. G. Rountree (2–4 territories/ha; unpubl. data 1949) in southern coastal Mauritius (Pointe d'Esny).

Nests or recently fledged young of the Cardinal were found in September (1978), October (1971: AFW), January and February (1974). In September 1978 all nests found had eggs but no young, suggesting that breeding had just started, although some birds in juvenile plumage were trapped. In 1974 breeding appeared to be finishing in late February, with a few birds feeding flying young, but others already moulting out of breeding plumage (one at least was in *fresh* red plumage, perhaps a young bird from earlier in the season). I saw no red-plumaged birds in July 1974, so there was obviously a longer eclipse period than in the native Fody. The same is true in Mauritius (Chapter 4).

It seems very probable, as Moreau (1960) suggested, that competition from the Cardinal has been directly responsible for the restriction of the Rodrigues Fody to a forest habitat. E. Newton (1865a) reported that not only were Fodies in pairs, but also that he saw "a flock of at least one hundred, all in the brown stage of plumage". This behaviour is completely absent today. I suggest that such flocks were exploiting grass seeds, and that competition from the better-adapted Cardinal has made this food source unavailable to the (presumed) juvenile (and ? non-breeding adult) native Fodies that once used it; in captivity Rodrigues Fodies take readily to grain (Jones 1980a). The remarkable plasticity of feeding behaviour in the Cardinal, reminiscent of the House Sparrow copying tits (*Parus* spp.) in English gardens (Summers-Smith 1963, pers. obs.), allows it to take advantage of new and different situations very readily. Gill (1967) suggested that the absence of a small local insectivore (e.g. a White-eye *Zosterops*) might have allowed the cardinal to expand its 'niche', but certainly the presence of the native Fody, an accomplished insectivore and nectar-feeder, has not prevented the Cardinal from adopting both these methods of feeding, while retaining its old ways as well.

In direct confrontations male Cardinals were subordinate to male Fodies and were displaced or chased off. Female Fodies were occasionally displaced

by male Cardinals, but most of the time the two species ignored each other. The Cardinal is a heavier bird (Chapter 9), which may give it some advantage. At Solitude the behaviour of Cardinals was very restrained, and they tended to be silent, refraining from their usual persistent high-pitched calling (Newton 1959, Moreau 1960), perhaps to avoid stimulating an aggressive response from the Fodies. In Mauritius, where interspecific competition appears to be less (Newton 1959, Chapter 4), the Cardinal is vocal in its normal way within Mauritius Fody territories.

Acknowledgements

This work could not have been carried out without the generous donations from all those organisations who contributed to the BOU Mascarene Island Expedition; they are listed in the Introduction to the book.

The 1978 trip was financed by Wildlife Preservation Trust International.

Wahab Owadally, Conservator of Forests, Mauritius, arranged for berths to be available for me on the M.V. *Mauritius*, and for me to stay with the Senior Forest Officers at Solitude. The SFO in 1974, the late Paul Lecordier, and his family were always extremely kind to me, and Mr Lecordier was most helpful in providing transport and guides to all parts of the island, as did his successor Hans Paupiah in 1978. Curt Reintsma also kindly provided accommodation for several days in February 1974. My companions in 1978, John Hartley and Yousoof Mungroo, contributed many useful observations.

I had fruitful ornithological discussions about Rodrigues with R. P. Alès, Claude Courtois, Tony Diamond, Alec Forbes-Watson, Carl Jones, Jean Montocchio, Frank Rountree, France Staub, Wendy Strahm and the late Octave Wiehé. Alès, Diamond, Forbes-Watson, Jones, Montocchio and Andrew Gardner very kindly allowed me to use in full their unpublished observations; Bryan Carroll and Carl Jones have also provided data from their recent visits. Claude Courtois identified *Flatopsis nivea* for me. I am particularly indebted to J. C. 'Roland' Raboude, who by continuous observation and correspondence enabled me to fill in what happened in the gaps between my visits, and carried the story on through subsequent years; he also found many of the nests that Alès, Diamond, Gardner and I subsequently examined. I would like to thank all the other forest staff who helped me in various ways, especially Messrs Ramakasin, Marie and Tolvis. Père R. Brown, Susan Donelly, Myke Jaynes, Mike Holland, Mary Smith and Curt Reintsma were all variously helpful, and Mr Nigel Heseltine, Resident Commissioner for the island in 1974, lent a sympathetic ear to the birds' plight. Finally I am grateful to the Mauritius Meteorological Department for data on cyclones and weather during the years of the study, to Drs G. E. Watson and G. F. Mees for details of bird specimens from Rodrigues in the Smithsonian Institution, Washington, and the Rijksmuseum, Leiden, respectively, and to Joseph Guého for checking over the section on the vegetation and for many identifications of botanical specimens I collected. Dr David Snow kindly commented on an early draft of this chapter.

APPENDIX

Description of the areas of forest cover on Rodrigues

Map 2 shows the localities mentioned and the extent of the vegetation.

The specific names of frequently mentioned plants are omitted on second and later mentions unless there is more than one species in the genus. The localities described are uninhabited unless otherwise stated.

Localities where endemic birds are still found
La Source/Cascade Pigeon

This is a well-vegetated valley running from Mont Lubin down to Port Mathurin. The region above the dam is of most interest ornithologically.

The upper parts of the valley are clothed in a tall dense blanket of Rose-apple *Eugenia jambos*, with a few large old *Araucaria cunninghamii*, and some *Terminalia arjuna* and Mango *Mangifera indica* around the spring at La Source. Further down Mangos become increasingly frequent, and in the middle section the stream bed is dominated by large Mangos and Travellers' Palms *Ravenala madagascariensis*; Rose-apple still dominates the slopes.

The east flank between Citronelle and the first hairpin on the main road is very mixed with tall trees

of *Pterocarpus indicus*, Camphor *Cinnamomum camphora*, Terminalia arjuna, T. catappa, Aleurites molucanna and others as well as Rose-apple and Mango. Near the dam the *Eugenia* on the upper slope gives way to a *Pandanus heterocarpus* thicket heavily invaded with *Furcraea foetida*. Down towards the stream bed the mixed forest continues with increasing admixture of planted *Vitex*, native *Mathurina penduliflora*, *Pyrostria trilocularis*, *Pandanus* and *Dracaena*, and spontaneous *Litsea glutinosa*.

The west slope ceases to have much wood on it after the stream from Les Choux enters the valley; around this point there are a lot of large old Mangos.

The flat area at the top of the east side between Montagne Patates and Citronelle is planted with *Eucalyptus tereticornis*, *Calophyllum inophyllum* and *Tabebuia pentaphylla*, mostly rather young. Some parts are invaded by *Furcraea* and Rose-apple.

Cascade Pigeon forms the main catchment supplying water to Port Mathurin; to prevent pollution human habitation is not allowed. Woodcutting is also officially prohibited.

Gros Mangue/Solitude; Vangar and Roseaux

At Gros Mangue *Eugenia* thicket covers the headwaters of the stream known as Rivière Bambous. The thicket is lower and in places more open than at La Source, and much of it outside the stream bed itself has been interplanted with *Eucalyptus*, about 6–8 years old in 1974. The stream bed itself is full of large mangos, with *Ravenala* coming in downhill towards Solitude.

Solitude itself is the headquarters in Rodrigues of the Mauritian Forestry Services and has been planted with a variety of trees. The oldest plots, around the houses and seedling nursery, were planted in the late 1940s and early 1950s (L. F. Edgerley, pers. comm.), and in 1974 the trees were 40–70 ft (12–21 m); there are a few older large trees, Araucarias and Mangos. The principal plantation species is *Tabebuia*, but mixed with it or in pure stands are Mahogany *Swietenia mahogani*, Araucaria, Teak *Tectona grandis*, Calophyllum, Albizzia spp., *Terminalia arjuna*, Vitex, and the native tree *Elaeodendron orientale*; there are Mangos and *Ravenala* along stream beds. On the slopes south-east of the central area there is quite a high proportion of the endemic tree *Mathurina*, with Mangos, Vitex, Rose-apple, *Adenanthera pavonina* and Strawberry Guava *Psidium cattleianum*, before *Eucalyptus* plantations are reached. Just behind the southernmost house on the west flank of the valley is a rocky outcrop covered in endemic *Pandanus heterocarpus*, an introduced Malagasy fruit tree *Vangueria madagascariensis*, and other shrubs. Below the dam, built in 1973, are plantations of *Tabebuia* and *Araucaria*, with *Terminalia catappa*, Mangos, guava and *Erythrina indica*. *Eucalyptus* plantations exist within the patchwork of other species (see Map 4) and, northwards, below the dam, the sides of the valley are mostly planted in this species.

To the east of Solitude is a fairly sheltered valley known as Vangar, which is thickly planted with Mangos, fringed on the eastern side with quite large *Vitex*. There is also in places an understorey of Rose-apple and guava. To the west the upper part of the area is linked to Solitude by a young *Tabebuia* plantation, further down by open ground and a patch of older trees (*Tabebuia* and *Albizzia*) fringing the south of the area known as Roseaux.

Vangar is studded with houses and has a substantial human population. Solitude is a River Reserve (there are only three houses); it is also the one place on the island completely free from illegal woodcutting, as there are forestry officials around all the time.

Mont Malartic/Les Choux

The long ridge of Mont Malartic has a very depauperate native flora along its summit, with much *Pandanus*. This is fringed by a dense, rather stunted growth of Rose-apple and *Litsea*. A spur of *Eugenia* thicket to the north connects with the more southerly of two streams feeding the Cascade Pigeon valley from the west. This stream bed is thickly vegetated with a wide band (200 yds (183 m) in places) of Mangos, *Ravenala*, Rose-apple and other trees. Further on, the other stream, coming from the west, joins it, and the water drops into Cascade Pigeon at a waterfall some 60 ft (18 m) high. The western stream and its flanks are vegetated with *Eugenia* thicket.

Mont Malartic is a Mountain Reserve, officially protected against woodcutting to prevent erosion; the wooded parts of Les Choux are part of the Cascade Pigeon catchment and protected as such, but there are still quite a number of families living in houses in the forest.

St.-Gabriel/Petit Gabriel

Between Mont Malartic and the road through St.-Gabriel there are patches of *Eugenia* scrub, *Pongamia*/Mango/guava thicket, stunted *Litsea*, *Terminalia arjuna* and *Tabebuia* scrub and a self-regenerated young growth of *Calophyllum*.

There is a concentration of large trees around St.-Gabriel church, many of them *Terminalia catappa*. The thicket at Petit Gabriel has houses in the upper part, but there are no others except around the church.

Mont Lubin/Lataniers

Behind Mont Lubin there is a small neglected plantation of *Araucaria*, *Aleurites*, *Tabebuia*, Mango and Rose-apple at the head of the stream leading through Lataniers to Cascade Victoire. Along the valley are patches of Mango and *Eugenia*, culminating in a dense thicket of *Pongamia*, *Erythrina* and guava before the stream goes into open arable land.

Mont Limon

Mont Limon is the highest hill on the island (1300 ft, 396 m). It is a Mountain Reserve, with dense vegetation extending along the ridge and down the slopes for some 650 yds (600 m). Along the south-west side the cover is largely dense, low Rose-apple and guava in the north becoming more mixed southwards, with *Litsea*, *Mathurina*, *Calophyllum* and *Mimusops bojeri*. The north-east flank has sparser vegetation with a higher proportion of native species.

Citronelle/Sygangue

There is a patch of tall *Araucarias* underplanted with *Vitex* and *Tabebuia* between Citronelle and Mont Lubin, to the east of the main road. Further down the stream from the *Araucarias* is an area of *Eugenia* thicket extending above and below a small catchment dam at Sygangue, by which there are some large old *Pterocarpus* and Mangos. Below the dam, on the east of the stream, is a patch of *Terminalia arjuna* and *Albizzia*. Below there is a waterfall, beyond which is a wooded patch with Mangos, *Terminalia catappa*, bamboo *Bambusa vulgaris* and a sprinkling of other species.

There are houses in or fringing all of the vegetation at Sygangue, and there is always a high level of human activity.

To the east of Sygangue there is a patch of *Eugenia* thicket, with houses in it.

Areas suitable for recolonisation by endemic birds

Outside the present distribution of endemic birds are the following wooded areas and copses which could be recolonised by one or the other species; their former distribution is given in the species discussions in the main text.

Grande Montagne

This area is well known to botanists for being the best place for surviving relics of the native upland vegetation (Friedmann & Guého 1977, Tirvengadum 1980, Strahm 1983). However apart from one area on the east side, the vegetation is very heavily invaded by exotics, notably *Litsea* and Mango on the summit plateau and Rose-apple on the slopes. The eastern flank is dominated by *Mathurina* and is one of the few places to give an idea of what the mountains of the island must once have looked like. *Olea lancea* is the commonest native species in the *Litsea* scrub on the plateau; there are also some pure stands of *Pandanus*. The south and west slopes are dominated by *Eugenia*. Grande Montagne is a Mountain Reserve.

Montagne Nouvelle Découverte

South of Grande Montagne there is a small area on the next hill-top of *Terminalia arjuna*, *Eugenia* and *Psidium*, and a stream to the south-east of it with a wood of *Albizzia*, *T. arjuna* and *Pongamia*, before the ground drops steeply into Cascade St.-Louis.

Cascade St.-Louis/Eau Rouge

This is another of the rare places where native vegetation is still found. At the north end, at the head of the valley, is much *Mathurina*, which rapidly gives way southwards to mixed evergreen forest, dense enough in places to have a closed canopy, dominated by *Elaeodendron*, *Pyrostria* and *Pandanus*. The forest is best developed down the east (sheltered) flank, and is much thinner on the west (exposed; see Cadet 1975). The valley is within the 'cattle walk' (North-Coombes 1971) and there is no chance for the trees to regenerate. Further south the forest thins out to isolated trees around the large dam which was under construction in 1974. Below this, to the sea at Graviers, the stream bed is clothed in *Pongamia* and the slopes have scattered Ebonies *Diospyros diversifolia*.

There are a few houses in the upper part of the valley, and more in Eau Rouge, the small valley immediately to the east which also has good patches of forest dominated by *Elaeodendron*.

Anse Mourouk

The heads of the streams leading to this bay, known as the Fonds Mourouk, are quite densely wooded with *Pandanus* and *Elaeodendron* with an admixture of other native trees.

Cascade Victoire/Montagne Bois Noir

This area was noted for having had the best endemic vegetation in the early 1940s (Wiehé 1949), but is now devastated; there are some quite dense patches of *Pandanus*.

Above the valley to the east of its head is Montagne Bois Noir, no longer covered, as its name would imply, in *Albizzia lebbek*. There is a patch of old *Mathurina* and Mangos, with other native trees, on the flank overlooking the valley and its eastern feeder stream.

Vainqueur/Rivière Brûlé

North-west of Grande Montagne is the large valley of Vainqueur, which is well wooded with Mangos, together with some *Terminalia catappa*. The area is well populated, with houses throughout amidst the trees, and plots of cultivated land as well.

Leading into Vainqueur from Grande Montagne is Rivière Brûlé, a stream along which there is a small wood of *Terminalia arjuna*.

Camp du Roi

At the head of the stream feeding the west of Port Mathurin's backwater is an area wooded with large *Terminalia arjuna* and *Albizzia*, some *T. catappa*, *Tamarindus indica* and *Litsea*.

Cascade Ollier

The other side of the ridge from Camp du Roi, to the west, is a small steep-sided valley that is well wooded, mostly with large *Terminalia catappa*, *Tamarindus*, and a native *Ficus*; there are also some *Litsea* and *Albizzia*.

L'Union

Here there is a plantation of *Albizzia* and a stream densely vegetated with *Pongamia*. There are several houses near the head of the stream.

Jamblong

Jamblong valley, feeding from the east the larger one leading to Baie aux Huitres, is wooded on its northern slope, largely with *Litsea*, *Albizzia*, native *Ficus* and *Mathurina*.

Jardin Mamzelle

The slopes of Baie aux Huitres valley above the Agriculture Department's experimental farm are lightly wooded with large old tamarinds and Mangos. Above these there is a young (8–10-year-old in 1974) plantation of *Tabebuia*. There are one or two houses in it.

La Ferme/Cascade Pistache

The area around La Ferme supports several *Eucalyptus* plantations; there are also some *Terminalia arjuna* and *Calophyllum* copses.

Cascade Pistache is a valley, headed by a waterfall, due west of La Ferme. On the cliff around the fall there are remnants of native vegetation, but the rest of the valley is thickly wooded with tall (50–60 ft; 15–18 m) *Terminalia arjuna* with some *Erythrina*.

Anse Quitor

The last vegetated area on Plaine Corail is Anse Quitor. The valley of Rivière Quitor, largely a collapsed limestone cavern, is well vegetated with a rich variety of native trees, *Elaeodendron* predominating. The whole area is heavily overgrazed by cattle and goats, and there is no regeneration of the native vegetation. *Lantana camara* has been invaded under the trees forming nearly impenetrable thickets in places. There are squatters' houses along the valley and the limestone is mined (sawn up into building blocks).

There are other copses on the island but so scattered and so small as not to merit description. Young plantations of *Tabebuia*, if given the chance to grow on, might be of significance to the birds in the future, especially if a variety of evergreen trees were interplanted with them. Finally there is a small copse of *Terminalia bentzoe*, and other scattered trees, on Ile Crabe.

Note added in proof

Captive breeding of the Rodrigues Fody at the Jersey Wildlife Preservation Trust is now well established (Darby, Jeggo & Redshaw, *Dodo* 21: 109–26, 1985). Some areas of native vegetation remnants (Grande Montagne, part of Mourouk) have now been fenced to keep out livestock under an EEC-aided agriculture and reafforestation programme (Thingsgaard, K. WWF/IUCN Project: Conservation of endemic plants, Rodrigues. Progress report January–April 1986).

9

Measurements and weights of the surviving endemic birds of the Mascarenes and their eggs

Part IV

A.S.CHEKE and C.G.JONES

MEASUREMENTS AND WEIGHTS

Details of weights and measurements are brought together in this chapter to facilitate comparisons and for easy reference. The data are given in Table 1, the text being in effect a commentary on the table; some species do not require further comment. The location of the museum specimens used in this study is given in Table 3. Conventions in text tables are the same as in Table 1.

Mauritius Kestrel Falco punctatus

In Table 1 weights of captive Kestrels are from birds in good body condition with empty crops weighed by CGJ prior to feeding. These weights are similar to the few records from wild birds. Temple's (1978c) weights (male: 178 g; 2 females: 221 g, 240 g) were of wild-caught adults weighed later in captivity (Temple *in litt.* to CGJ); these weights are well above the weight range we have, but two overweight young females have reached 252 and 253 g in captivity. Egg weights from wild birds are a clutch weighed after 8 days of incubation; details of calculated weights in relation to egg shell thinning are given in Chapter 5.

Ecological correlates of linear measurements and comparisons with other species are given in Chapter 5. The measurements given in Table 1 are as found, and include museum specimens with worn wings and/or tail, but in the majority of 50 skins the extent of wear ranges only from 1 to 4 mm, only 7 approaching the 20 mm reported by Temple (see Chapter 5). Only 23 of the 50 museum specimens were gonadally sexed; the others have not been used in the table.

Table 1. *Measurements and weights of surviving endemic Mascarene birds and their eggs*

Species	1 Wing length (mm)[a] (range; av. ± S.E. (*n*))	2 Tail length (mm)
Réunion Harrier *Circus m. maillardi* (Réunion)	♂ 342–360; 348.6 (8) ♀ 370–380; 375 (3)	218–230; 225.1 (8) 230–240; 235 (3)
Mauritius Kestrel *Falco punctatus* (Mauritius)	L ♂ 178, 179, 180 ♀ 186, 186, 190	132, 135, 136 136, 139, 140
	M ♂ 164–184; 172.1 ± 1.4 (14) ♀ 171–192; 181.3 ± 2.8 (9)	118–133; 127.5 ± 1.2 (14) 129–150; 134.1 ± 2.5 (9)
Pink Pigeon *Nesoenas mayeri* (Mauritius)	L ♂ 202–217; 209.4 ± 1.0 (14) ♀ 190–212; 203.0 ± 1.7 (15)	158–180; 165.4 ± 1.6 (14) 145–167; 157.6 ± 1.4 (16)
	M ♂ 209.8 ± 2.1 (7) ⎱ 198–217; ♀ 210.3 ± 1.2 (6) ⎰ 209.2 ± 0.9 (32)	157.8 ± 3.2 (8) ⎱ 144–176 160.5 ± 2.7 (6) ⎰ 159.5 ± 1.6 (34)
Echo Parakeet *Psittacula echo* (Mauritius)	♂ 171–195; 184.9 ± 1.8 (13) ♀ 169–190; 180.1 ± 1.5 (11)	171–215; 195.8 ± 3.2 (13) 150–196; 176.2 ± 4.3 (11)
Mascarene Swiftlet *Collocalia francica* (Mauritius)	L 108–117; 112.1 ± 0.21 (86)	—
	M ♂ 112–115; 113.1 (7) ♀ 108–112; 110.5 (4)	⎱ 50, 51, 53
(Réunion)	L 110	—
	M ♂ 106.5–113; 108.8 ± 0.72 (8) ♀ 106–112.5; 110.4 ± 0.78 (7)	⎱ 45–50
Mascarene Swallow *Phedina borbonica* (Mauritius)	L 115.5–126; 120.2 ± 0.99 (10)	—
	M 116–121; 118.8 ± 1.16 (5)	50, 53, 54, 55
(Réunion)	M 110–121; 115.7 ± 1.84 (6)	60–63/51
Mauritius Cuckoo-shrike *Coracina typica* (Mauritius)	M ♂ 103–109; 105.2 ± 0.46 (12) ♀ 105–110; 107.9 ± 0.6 (8)	72–84; 76.7 ± 0.95 (11) 73–90; 79.2 ± 2.96 (5)
Réunion Cuckoo-shrike *C. newtoni* (Réunion)	M ♂ 96–101; 98.9 ± 0.62 (9) ♀ 96–101; 97.3 ± 0.8 (6)	74–84; 78.7 ± 1.14 (9) 73–87; 77.8 ± 1.28 (11)
Mauritius Merle *Hypsipetes olivaceus* (Mauritius)	L 125.5, 127 (♀), 131, 132	—
	M ♂ 125–143; 133.6 ± 1.98 (8) ♀ 122.5, 128	107 ⎱ 99/100–115; 108.0 (6) —
Réunion Merle *H. borbonicus* (Réunion)	L 118	—
	M ♂ 111, 118 ⎱ 102–121; ♀ 102, 111, 116, 119 ⎰ 112.0 ± 1.76 (12)	⎱ 92–100/95.5

3 Tarsus (mm)	4 Bill (mm)[b]	5 Weight (g)[c]
}78–90	e: 31–34; 32.6 (8) e: 34–35; 34.3 (3)	— —
38, 39, 41 40, 41, 42	e: 14, 15 e: 14, 15	w: 123, 142, 142 c: 127–146 w: ad. 173, 204; juv. 144, 170 c: 152–184[e]
31–46; 37.79 ± 1.10 (14) 36–49; 41.67 ± 1.22 (9)	e: 13–17; 14.87 ± 0.36 (12) e: 14–17; 15.43 ± 0.37 (7)	— —
32.5–37; 35.4 ± 0.35 (14) 31–38; 34.3 ± 0.41 (17)	f: 21–25.5; 23.8 ± 0.57 (8) f: 22–25; 23.3 ± 0.26 (11)	240–410; 314.6 ± 7.5 (33) 213–369; 291.0 ± 7.1 (29)
35.43 ± 0.35 (7) ⎱(24)31–38 33.83 ± 0.87 (6) ⎰34.10 ± 0.33 (29)	f: 21.75 ± 0.40 (5) ⎱19.6–24.1; f: 21.75 ± 0.25 (4) ⎰21.91 ± 0.25 (25)	— —
18–23; 19.77 ± 0.40 (12) 18–21.5; 19.82 ± 0.30 (11)	e: 23–25; 24.17 ± 0.25 (13) e: 22–24; 22.84 ± 0.23 (11)	166.5 (immature) 162.7
—	—	7.8–13.5; 9.2–9.6*
}8.5–10	}e: 4–5	8.0–9.3; 8.8 ± 0.19 (7) 8.9, 8.9, 9.0, 9.5
8	s: 4	9
}7–8	}e: 2–3	7.9, 8.5, 8.5, 8.6 8.6, 8.9, 9.5, 11.4
—	—	20.9–25.8 (28); 23.9 ± 0.41 (19)
14	9	—
10–12/13	e: 7–8/9	—
24, 28 24, 26.5	s: 21.5–23.5; 22.4 ± 0.23 (9) s: 22.5–24.5; 23.6 ± 0.3 (6)	42.9 —
"24"[j] "25"[j]	s: 17.5–20; 19.0 ± 0.22 (9) s: 17–21; 19.2 ± 0.39 (9)	— —
—	—	70, 76 (♀), 79
}20, 24, 24, 26, 27, 27[k]	}e: 27–30; 28.7 (6)	— —
—	—	56
}"23"[j]/26.5	}e: 19–21 s: 25	51 57.2

(Cols. 6–8 continued overleaf)

Table 1 *continued*

Species	6 Clutch size	7 Egg measurements (mm) (ranges/avs.)	8 Egg weight (g)^d	Sources (details below table)
Réunion Harrier *Circus m. maillardi* (Réunion)	2–3	4.6–5.1 × 3.5–3.7 (12)	—	1, 2, 4, 6, 7: P 3: Q
Mauritius Kestrel *Falco punctatus* (Mauritius)	(2)–3 —	33.6–39.2 × 28.3–30.8^f 36.6 ± 0.35 × 29.6 ± 0.15 (15)	w: 14.8, 16.7, 17.2 c: 16.4, 16.4^g	1–8: t/s
	— —	34.7–38.8 × 29.2–31.6 36.9 ± 0.37 × 30.8 ± 0.18 (12)	—	1–4, 7: t/s
Pink Pigeon *Nesoenas mayeri* (Mauritius)	1–2	31.6–40.5 × 25.2–28.5 36.3 ± 0.23 × 26.7 ± 0.11 (50)	9.6–18.7 (450) 14.24 ± 0.05	1–8: t/s
	— —	— —	—	1–4: t/s
Echo Parakeet *Psittacula echo* (Mauritius)	2–3	32.2–33.7 × 25.3–26.8 33.08 ± 0.17 × 26.12 ± 0.18 (9)	[11.4]+	1–7: t/s 8: A
Mascarene Swiftlet *Collocalia francica* (Mauritius)	(1)–2	—	—	1: t/s (cf. B) 5, 6: t/s
	—	18.3–21.7 × 12.5–13.4 19.6 ± 0.51 × 12.9 ± 0.14 (6)	[1.53]+	1: B (–120, t/s) 2,7: t/s 3, 4: BB 5: B 8: A
(Réunion)	2	18.3–20 × 13–13.4 19.2 ± 0.34 × 13.1 ± 0.08 (5)	—	1, 3–5: V 6, 7: F, V
	—	—	—	1: t/s, B (–115, Q) 2–4: Q^h 5: B
Mascarene Swallow *Phedina borbonica* (Mauritius)	2	—	—	1, 5: t/s 6: E
	—	—	—	1: t/s 2: C, t/s 3, 4: C
(Réunion)	2–3^i	—	—	1: t/s 2–4: Q^h/C 6: F
Mauritius Cuckoo-shrike *Coracina typica* (Mauritius)	2	26.9–27.3 × 18.6–19 (2)	[4.6]+	1: t/s (cf. N) 2, 4: N 3: C, EE 5: t/s 6: ↑ 7: t/s 8: A
Réunion Cuckoo-shrike *C. newtoni* (Réunion)	2^i	—	—	1: t/s (cf. N) 2, 4: N 3: AA 6: t/s
Mauritius Merle *Hypsipetes olivaceus* (Mauritius)	—	—	—	1, 5: t/s
	2(?+)	29.7–32.9 × 20.2–22.5 31.2 ± 0.27 × 21.3 ± 0.18 (13)	[6.75]+	1: t/s (cf. K) 2: EE/L 3: C, EE 4: L 6: M 7: t/s 8: A
Réunion Merle *H. borbonicus* (Réunion)	2–3	—	—	1, 5: t/s 6: R, T
	2	—	—	1: t/s (cf. K) 2: Q/EE 3: DD/EE 4: Q, t/s 5: t/s 6: S

Table 1 *continued*

Species	1 Wing length (mm)[a] (range; av. ± S.E. (n))	2 Tail length (mm)
Réunion Stonechat *Saxicola tectes* (Réunion)	L ♂ 68, 69 ♀ 62, 67, 67	— —
	M ♂ 65, 67 ♀ [62.5–67; 65.0 ± 0.56 (8)][m]	45.5 38 }45–50
Rodrigues Warbler[n] *Acrocephalus rodericanus* (Rodrigues)	L 62 (?♂)	—
	M ♂ 58–66; 61.9 ± 0.85 (10) ♀ 60, 61 }58–66; 61.3 ± 0.55 (17)	62–75; 71.7 ± 1.14 (10) 73, 75
Mascarene Paradise Flycatcher *Terpsiphone bourbonnensis desolata* (Mauritius)	L ♂ 72–77; 73.9 ± 0.5 (11) ♀ 66–71; 68.8 ± 0.55 (9) M ♂ 70–76; 73.3 ± 0.67 (8) ♀ 69, 69, [71.5, 73][m]	— — 71–78; 74.2 (9) 68–73; 70.5 (4)
T. b. bourbonnensis (Réunion)	M ♂ 69, 70 ♀ [65–71; 68.2 ± 0.87 (6)][m]	75 [65–73; 69.5 (9)][m]
Mascarene Grey White-eye *Zosterops borbonicus mauritianus* (Mauritius)	L 50–59; 53.6/54.7* (88) M ♂ 52–57.5; 55.7 (9) ♀ 53.5–56; 54.9 (4)	— 37–43; 40.4 (9) 40–41; 40.5 (4)
Z. b. borbonicus (Réunion)	L ♂ 53–60.5; 55–58* ♀ 51–59; 53.5–57* }(494) M ♂ 52–60; 54–57.7* (40) ♀ 51–59; 53.6–56.1* (25)	}[p] (391) 39–45.5; 40.5–43.1* (38) 39–45; 39–41.6* (23)
Mauritius Olive White-eye *Zosterops chloronothos* (Mauritius)	L 52–54; 53.0 ± 0.44 (5) M ♂ 55, 55 ♀ 51, 52.5, 53, 54 }49–55; 52.5 ± 0.45 (14)	— }31.5–35; 32.6 (7)
Réunion Olive White-eye *Z. olivaceus* (Réunion)	L 57–63; 58.6 ± 0.33 (20) M ♂ 59 ♀ 57, 60 }55–60; 57.0 ± 0.35 (18)	— }36–43; 40.7 (19)
Mauritius Fody *Foudia rubra* (Mauritius)	L ♂ 67–73; 69.0 ± 0.52 (13) ♀ 63–65; 64 (3) M ♂ 66.5–74; 70.0 ± 0.53 (15) ♀ 63, 64 (2)	— — 37–43; 40.0 ± 0.89 (8) 37
Rodrigues Fody[n, r] *F. flavicans* (Rodrigues)	L ♂ 68–74; 70.6 ± 0.05 (29) ♀ 63–68; 65.6 ± 0.08 (19) M ♂ 68–73; 70.2 ± 0.06 (23) ♀ 62, 63, 65, 65	49, 49, 50, 51.5 45 43–52; 48.7 ± 0.48 (22) 40–49; 45.5 ± 0.76 (11)

(*Cols. 3–8 continued overleaf*)

Table 1 *continued*

Species	3 Tarsus (mm)	4 Bill (mm)[b]	5 Weight (g)[c]
Réunion Stonechat *Saxicola tectes* (Réunion)	— — $\left.\begin{matrix}22.3\\21.5\end{matrix}\right\}$21–22	— — $\left.\right\}$e: 8–9	11.2–13.8; 12.3 ± 0.20 (13)[*j*] 11.9–13.6 (14.8); 13.0 ± 0.27 (8) — —
Rodrigues Warbler[*n*] *Acrocephalus rodericanus* (Rodrigues)	25 (?♂) 22.5–25; 23.9 ± 0.36 (8) 22, 22.5	g: 20.5 (?♂) s: 18, 18, 18.4, 19 s: 18.5	11.4 (?♂) 12.2 —

Mascarene Paradise
  Flycatcher
*Terpsiphone bourbonnensis
  desolata*
(Mauritius)
	—
	
	$\left.\begin{matrix}15\\—\end{matrix}\right\}$17[*o*]
		—
		
		s: 16–17.5; 16.4 (9)
		s: 15–17; 16.4 (4)
			11.0–12.5; 11.7 ± 0.11 (12)
			10.4–12.1; 11.3 ± 0.16 (10)
			—
			—

T. b. bourbonnensis (Réunion)	$\left.\right\}$14–16	s: 15.5, 16 s: [15–16; 15.3 (9)][*m*]	10.1–11.4; 10.6 (5) 8.2–10.6; 9.7 (5)
Mascarene Grey White-eye *Zosterops borbonicus mauritianus* (Mauritius)	— 17–18.3; 17.6 (9) 17.5–17.9; 17.7 (4)	— e: 12–13.9; 13.0 (9) e: 13.4–13.6; 13.5 (4)	7.2–10.1 (10.5); 8.3–8.7* (95) 7.6–9.2; 8.3 (9) 7.8–8.4; 8.0 (4)
Z. b. borbonicus (Réunion)	$\left.\begin{matrix}17.3–21.2;\\18–19.5^{*}\end{matrix}\right.$ (448) 17.6–21.1; 18.7–20.4* (41) 17.1–22.5; 18.0–20.2* (26)	$\left.\begin{matrix}\text{e: }12.2–16.2;\\13.3–14.2^{*}\end{matrix}\right.$ (342) e: 12–15; 12.8–14.1* (40) e: 12.8–14; 13.0–13.9* (26)	$\left.\begin{matrix}6.6–9.9; 7–8.9^{*}\\6.1–10.6; 6.9–8.8^{*}\end{matrix}\right\}$(513) 6.8–10; 7.4–8.5* (34) 7.3–11.1; 7.9–9.4* (22)
Mauritius Olive White-eye *Zosterops chloronothos* (Mauritius)	— $\left.\right\}$15, 17	— $\left.\right\}$s: 15.5–17; 16.4 (7)	7.5–9.0 (11.5); 8.1 ± 0.30 (5) — —
Réunion Olive White-eye *Z. olivaceus* (Réunion)	— $\left.\right\}$17–19	— $\left.\right\}$s: 14–17; 15.6 (19)	7.9–11.2; 8.7–9.7* (22) 7.7–9.8; 8.6 ± 0.23 (9) 8.6–11.4; 9.7 ± 0.26 (11)
Mauritius Fody *Foudia rubra* (Mauritius)	— — 20–23; 21.2 ± 0.31 (10) 20, 21	— — e: 14–16.5; 15.1 ± 0.26 (8) e: 15.5	16.0–20.1; 17.8 ± 0.32 (13) 16.1–19.7; 17.3 ± 0.81 (4) — —
Rodrigues Fody[*n, r*] *F. flavicans* (Rodrigues)	21.5, 22, 22, 22.5 21 20–27; 22.5 ± 0.28 (26) 20.5–23.2; 21.7 ± 0.27 (11)	e: 14–14.5; 14.4 ± 0.10 (5) e: 14 e: 12–16.2; 14.3 ± 0.15 (29) e: 12.9–14; 13.3 ± 0.10 (12)	14.7–17; 15.9 ± 0.02 (26) 13.5–16 (19.6); 14.8 ± 0.06 (14) 15, 16.9, 16.9 14.9, (16.9)

6 Clutch size	7 Egg measurements (mm) (ranges/avs.)	8 Egg weight (g)^d	Sources (details below table)
2–3 (4)	—	—	1, 5: t/s 6: t/s, R, U
3	18–20 × 15	—	1: t/s 2, 3: EE/Q 4: Q 6: t/s 7: Q
3	"19 × 14" (= 18.3 × 14.4)	—	1,3–5: t/s 6: t/s, Z 7: Z
(?2)	17.8 × 12.4	[1.44]+	1–6: t/s 7: O, A 8: A
3	—	—	1, 5, 6: t/s
(?2)–3	18–20 × 15–15.5 (6+)	[2.3]	1: t/s 2, 4: N 3: C 6: C, E ↑ 7: O, C, A, Q 8: A
2–3^i	—	—	1: t/s (cf. N) 2, 4: N 3: Q 5: W 6: R, T, X
—	—	—	1, 5: t/s
2–(?3)	17.0 × 18.6	[1.18]	1–5: G (cf. H) 6: C ↑ 7: C 8: A
2–4	—	—	1, 3, 4: Y 5: t/s 6: U, R
—	—	—	1–5: G
—	—	—	1, 5: t/s
2–(?3)	16–16.5 × 12–12.5	[1.27]	t: t/s (cf. H, I) 2, 4: H 6: H, J ↑ 3, 7: C 8: A
2–3	—	—	1, 5: t/s 6: t/s, U, R
—	18–20 × 13–14 (?2)	^{q}	1: t/s 2, 4: H 3: Q 5: t/s 7: Q (*not* A)
—	—	—	1, 5: t/s
3	18.4–19.9 × 13.0–14.6; 19.2 ± 0.17 × 13.9 ± 0.16 (10)	—	1, 2, 3, 4, 7: t/s 6: C, D
(?2)–3	—	1.5, 1.5, 1.7	1–6, 8: t/s
—	17.1–19.4 × 13.0–14.1 18.3 ± 0.24 × 13.6 ± 0.12 (7)	—	1–5: t/s 7: t/s (cf. O)

(*Notes to Table 1 overleaf*)

Notes to Table 1:

*, See text in appropriate chapter; no simple mean available due to seasonal or geographical variation.

↑, See text in appropriate chapter.

[a] L, living birds; M, museum studies. This designation is omitted where data from both categories are combined.

[b] Birds' bills are measured differently by different workers; where it is known the measurement is qualified by 'e' (exposed culmen), 'f' (measured to feathers), 's' (measured to skull), or 'g' (measured to gape).

[c] Bracketed weights beyond the normal range usually refer to pregnant females. Weights cited as 't/s' (this study) include unpublished data collected by F. G. Gill.

[d] A '+' after an estimated weight (in square brackets) in column 9 indicates that the value was calculated by Schönwetter (1960–) from eggs smaller than the average given in column 7.

[e] w, wild, c, captive. Temple (*in litt.* to CGJ) has provided weights of two birds when freshly caught in 1973 (male: 142 g, female: 204 g) that supplement the captive weights quoted in Temple (1978c) and Cade (1982).

[f] Recent (1981–3) wild-laid eggs; the equivalent figures for eight laid in captivity are: 32.8–35.1 × 28.1–29.9 (34.25 ± 0.26 × 29.2 ± 0.25), significantly shorter ($P < 0.001$; t-test) and hence rounder.

[g] w, wild; c, captive; each is a full clutch.

[h] It is not clear that Berlioz (1946) measured only Réunion specimens.

[i] Information from field observations.

[j] An 'average' measurement without range or other supporting data is given in double inverted commas.

[k] A. Newton (1876) gave an average figure of 1.4 in (= 35.5 mm) which is not compatible with Hartlaub's (1877: 20–26 mm) or Cowper's (1984: 27, 27 mm) measurements.

[l] F. B. Gill collected three fully grown nestling Stonechats weighing 14.0, 14.3 and 14.5 g.

[m] Square-bracketed measurements for 'female' Flycatchers and Stonechats are from museum skins in grey/brown plumage labelled '♀' but not gonadally sexed and thus likely to include some males (see text in appropriate chapters).

[n] Staub's (1973b) measurements of the tail and culmen are not compatible with ours and are not used here; his Rodrigues Warbler egg has been re-measured from his photograph.

[o] Cowper (1984) gives ♂♂: 20.5, 21.5 mm; ?♀ 19 mm, suggesting a consistent difference in measuring technique.

[p] Details of Gill's tail measurements are not extractable from his paper (Gill 1973b).

[q] Schönwetter's (1960–) '*Zosterops o. olivacea*' egg must be from a *Z. chloronothos* as it is the same size as others of this form, but a good deal smaller than the measurements given by Berlioz (1946) for *Z. olivaceus*.

[r] Tail, tarsus and bill lengths for Rodrigues Fodies include birds assigned to sex on basis of wing length; see text.

Sources:

t/s	this study
A	Schönwetter (1960–)
B	Gill (1971b)
C	Hartlaub (1877)
D	Newton (1959)
E	Staub (1973a, 1976)
F	Jadin & Billiet (1979)
G	Storer & Gill (1966)
H	Moreau (1957a)
I	Gill (1970)
J	Carié (1904)
K	Benson (1970–1)
L	Benson (1960)
M	Guérin (1940–53)
N	Benson (1971)
O	Oates *et al.* (1901–12)
P	Clouet (1978)
Q	Berlioz (1946)
R	Barré (1983)
S	Schlegel & Pollen (1868)
T	Barré & Barau (1982)
U	Milon (1951)
V	Gruchet (1976)
W	Gill (1973b)
X	Focard (1861)
Y	Gill (1973b)
Z	Staub (1973b)
AA	Pollen (1866)
BB	Oberholser (1906)
CC	S. Parker (*in litt.*)
DD	A. Newton (1876)
EE	Cowper (1984)

For comparison, details of some other kestrels are given overleaf.

First-year European Kestrels are 8% lighter and 2–3% shorter in the wing than adults, but young Seychelles Kestrels do not differ from adults in size or weight. Siegfried & Frost's (1970) small sample of Madagascar Kestrels was of larger birds than Benson & Penny's long series (wings: 4 ♂♂ 183–201; 193 ± 3.9, 6 ♀♀ 198–220; 211.5 ± 3.0).

Réunion Harrier Circus m. maillardi

Ecological correlates of linear measurements are discussed in Chapter 6. Measurements of other forms of the Marsh Harrier complex (*Circus aeruginosus* superspecies) are given by Nieboer (1973).

Pink Pigeon Nesoenas mayeri

In birds measured live and sexed by behaviour in captivity, males are significantly larger than females in wing length, tail, tarsus and weight. This difference, except in the tarsus, is absent in museum specimens allegedly sexed gonadally, casting doubt on the accuracy of their gender assignment.

Young birds are smaller than adults up to the post-juvenile moult (see Chapter 5). Wing growth is shown in Fig. 1, and the pattern is similar for tail and

Fig. 1. Wing lengths of juvenile and adult Pink Pigeons.

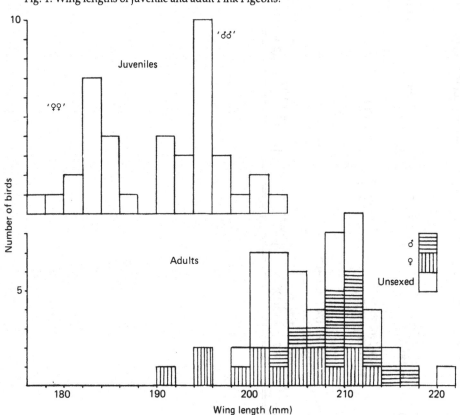

Weights and measurements of other kestrels

	Wing (mm)	Tail (mm)	Tarsus (mm)	Bill (exposed culmen, mm)	Weight (g)	Sources
F. timunculus timunculus (Europe, N. Africa)	♂♂ 246 ± 1.04 (37)	163 ± 0.92 (37)	39.6 ± 0.16 (36)	13.9 ± 0.10 (36)	202.2 ± 2.88 (34)	Cramp & Simmons (1980), Kirkwood (1981)
	♀♀ 256 ± 1.26 (45)	171 ± 1.04 (44)	39.6 ± 0.20 (37)	15.0 ± 0.13 (43)	214.8 ± 4.32 (42)	
F. t. rupicolus (S. Africa)	♂♂ 217 ± 248; 236 (?n)	142.5 ± 5.3 (?n)	35.8 ± 1.7 (?n)	14.1 ± 1.0 (?n)	145–247; 192 (99)	Brown & Amadon (1968), Biggs et al. (1979), Watson (1981)
	♀♀ 240–258; 247 (?n)					
F. neotoni neotoni (Madagascar)	♂♂ 171–196; 189.8 (33)	110–130 (?n)	32–40 (?n)	—	90–117; 105 ± 5.8 (4)	Brown & Amadon (1968), Siegfried & Frost (1970), Benson & Penny (1971)
	♀♀ 181–210; 197 (23)	115–131 (?n)		—	131–159; 145 ± 4.0 (6)	
F. n. aldabranus (Aldabra)	♂♂ 170–183; 174.6 ± 1.9 (7)	—	—	—	—	Benson & Penny (1971)
	♀♀ 174–197; 183.8 ± 2.9 (7)	—	—	—	—	
F. araea (Seychelles)	♂♂ 148.2 ± 1.7 (15)	102.1 ± 6.0 (17)	27.7 ± 3.3 (17)	11.5 ± 0.36 (46)	72.4 ± 4.35 (14)	Watson (1981)
	♀♀ 158.2 ± 1.83 (33)			12.4 ± 0.28 (77)	87.9 ± 4.73 (32)	

tarsus; the bill is full-grown on fledging. Details of juvenile measurements (all captive birds) are:

	Wing (mm)	Tail (mm)	Tarsus (mm)	Culmen (to feathers) (mm)
'♂♂' (RH histogram, Fig. 1)	191–203 195.3 ± 0.7 (24)	92–150	30–35	21–26.3
'♀♀' (LH, histogram, Fig. 1)	177–187 182.9 ± 0.6 (16)	123.1 (31)	32.9 ± 0.3 (33)	23.11 ± 0.48 (10)

The distribution of tail lengths is not normal, so no standard error has been calculated.

Adult birds in captivity vary an average of 9% (♂♂) and 13% (♀♀) in weight from month to month. One female weighed 41% more when pregnant, while the greatest male weight-change was 17% (taken as the ratio of higher weight over lower).

Captive birds show no seasonal fluctuations in weight, but wild birds (Chapter 5) probably show an annual cycle, being heaviest at the end of summer (March–June) when food is abundant and lightest at the end of winter (September–mid-November). Only one weight is available from a wild bird, an adult male with a full crop which was shot in the evening of 23 April 1973 and weighed 337 g (J.-M. Vinson *in litt.*) – a value comparable to captive birds in good condition. Wild and captive birds fledge at about 160–180 g and captive birds reach adult weight when 75–120 days old.

Echo Parakeet *Psittacula echo*

Comparative details for the sympatric introduced Ring-necked Parakeet *Psittacula krameri*, and the two extinct Indian Ocean forms, *P. exsul* (Rodrigues) and *P. wardi* (Seychelles), are as follows:

The Echo Parakeet, though differing little in linear measurements from the Ring-necked, is some 1.4 times heavier.

A. & E. Newton (1876) and Hartlaub (1877, Forshaw & Cooper 1978) gave slightly smaller figures for the long axis of the eggs.

Mascarene Swiftlet *Collocalia francica*

Birds in Mauritius were weighed each time they were caught (Fig. 2); no seasonal trend is discernible from the average weights. Weights of birds in the two colonies differed significantly only in mid-February 1974. The unusually low weights on 10 February at cave F followed a prolonged wet spell since 21 January. There was no significant change of weather before the catch on 13 February at N, when weights were normal, suggesting that birds from the two colonies were feeding in different areas. January and February were by far the wettest months of 1974 at Vacoas (M. Lee Man Yan *in litt.*).

Wing lengths of 86 birds averaged 112.09 ± 0.21 mm, the range (108–117) being greater than that usually quoted (112–114; Oberholser 1906, Medway

		Wing (mm)	Tail (mm)	Tarsus (mm)	Culmen (exp.) (mm)	Weight (g)	Sources
krameri	♂	170, 171, 175	(154), 235, 236	16, 18, 21	24, 25, 26	119	This study
	♀	160, 164, 169	200, 207, 211	16, 18, 18	22, 23, 26	120	
exsul	♂	199	207	22	25	—	
	♀	192	212	23	24	—	Forshaw & Cooper (1978)
wardi	♂	204, 208	184, 187	22	33, 34	—	
	♀	182–204; 194.5 (6)	200–261; 228.0 (6)	20–22; 21.0 (6)	29–34; 31.7 (6)	—	

Fig. 2. Weight patterns in Mauritian Swiftlets.

Notes:

1. All weights recorded during the study are included. Birds only caught once are given the smallest dots; those involved in multiple recaptures have larger dots, increasing in size where more than one such bird of the same weight was caught at the same time (up to 4). 'P' indicates a pregnant female.

2. The mean weight and 2 standard errors on either side are shown for each trapping occasion (cave F, inverted triangles; N, diamonds).

3. Recaptured birds are joined by continuous lines if caught again within 3 months, by dotted lines if the interval was longer.

4. Birds weighing over 10.5 g are omitted from the calculation of means, as even when included they lie outside 2 standard deviations. Most such were pregnant females, but one bird was always exceptionally heavy, and another very heavy bird (10.9 g) was caught in early April at a time when no birds in the two caves were nesting.

5. Times when egg-laying took place in the two caves are indicated by horizontal bars at the top of the graph (see Chapter 4).

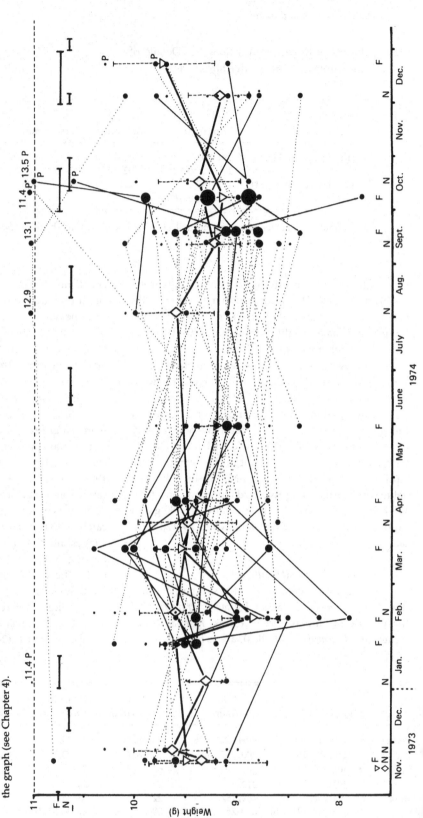

1966) but similar to that found by Gill (1971*b*) on a much smaller sample (108–116.5, *n* = 11). Using gonadally sexed specimens he found males (av. 113.1, *n* = 7) longer-winged than females (av. 110.5, *n* = 4) (*ibid.*), but our sample is clearly unimodal, and indeed the longest-winged bird caught was a female (pregnant on 16 October 1974):

Wing length (mm)	107	108	109	110	111	112
	–	4	2	7	22	19
Wing length (mm)	113	114	115	116	117	118
	15	5	8	3	1	–
						Total 86

Twenty-one birds were measured before and after moulting, showing a mean increase of 1.2 mm in wing length (112.33 ± 0.31 to 113.55 ± 0.38, $P < 0.002$, one-tailed sign test). This apparently represents the annual wear on the longest primary (second from outside) rather than an increase with age, although some physiological change from year to year is no doubt present also, as some birds increased 4 mm and others 'shrank' (up to 2 mm). Means of samples taken about 10, 5 and 3 months apart respectively, in two different moult cycles, gave an annual rate of shrinkage (= wear) of 1.25–1.6 mm.

Mascarene Swallow Phedina borbonica
Measurements of birds of the Malagasy race *P. b. madagascariensis* are as follows:

	Wing (mm)	Tail (mm)	Tarsus (mm)	Bill (mm)
♂♂	110 ⎫	46 ⎫	11 ⎫	8 ⎫
♀♀	– ⎭ 115, 117	– ⎭ 53, 54	– ⎭ 11, 12	– ⎭ 7, 7.5

	Weight (g)	Sources
♂♂	22.1 ± 0.3 (4) ⎫	Hartlaub (1877), Benson *et al.* (1976–7)
♀♀	20.3 ± 0.8 (5) ⎭	

Cuckoo-shrikes Coracina typica and C. newtoni
Réunion birds are smaller than Mauritian ones, and less sexually dimorphic in size. The sample is small, but Réunion males appear to be larger than females whereas the reverse is true in Mauritius.

Merles Hypsipetes olivaceus and H. borbonicus
For comparison, the measurements of other Indian Ocean merles are shown overleaf.

Réunion Stonechat Saxicola tectes
This species is markedly lighter than its relative *S. torquata sibilla* on Madagascar which has similar linear measurements but weighs 15–16.5 g (Benson *et al.* 1976–7). *S. t. ankaratrae*, also of Madagascar, is larger (*ibid.*).

Rodrigues Warbler Acrocephalus rodericanus
The Seychelles Brush Warbler *A. sechellensis* is a larger bird:

	Wing (mm)	Tail (mm)	Tarsus (mm)
♂♂	67.7 ± 0.5 (9)	51–60; 56.2 (13)	24.8 ± 0.5 (8)
♀♀	65.2 ± 0.3 (12)		23.0 ± 0.2 (10)
	Bill (mm)	Weight (g)	Sources
♂♂	14.6 ± 0.14 (8)	16.8 ± 0.12 (43)	Diamond (1980), Benson (1960)
♀♀	14.5 ± 0.2 (12)	15.0 ± 0.2 (25)	

Mascarene Paradise Flycatcher Terpsiphone bourbonnensis desolata and T. b. bourbonnensis
Réunion birds are lighter than Mauritian ones, as expected from their shorter wings (Salomonsen 1933). Salomonsen (1933) split wing and bill measurements by sex, but in view of the number of grey-headed males seen in Réunion (Chapter 6) we are inclined to doubt the sexing of museum specimens, especially as the Paris museum has seven Réunion '♀♀' to only three black-headed ♂♂ (Salomonsen; ASC saw only five '♀♀' and two ♂♂ there).

The Malagasy *T. mutata* appears to be the closest relative to this species, the Mohéli form *T. m. voelzkowiana* being nearest in size and appearance; two trapped in February 1975 weighed 13 g (♂) and 12.1 g (♀) (ASC), and the wing length is 72–76 mm for males (av. 74.3 ± 0.6, *n* = 6) and 68–73 mm for females (av. 70.7 ± 0.8, *n* = 7) (Benson 1960 and ASC data). The other races are slightly larger (Benson 1960, 1971) and heavier (♂♂ 11.5–15 g, 12.8(9) and ♀♀ 11–15.5 g, 12.8(8) in Madagascar: Benson *et al.* 1976–7).

Weights and measurements of other Indian Ocean merles

		Wing (mm)	Tail (mm)	Tarsus (mm)	Bill (to skull) (mm)	Weight (g)	Sources
H. madagascariensis madagascariensis (Madagascar)	♂♂	109.8 (5) } 104–118	} 86–95; 89.6 (13)	—	} 22–26; 23.2 (13)	44.2 ± 1.2 (6)	Benson (1960), Benson et al. (1976–7)
	♀♀	107.2 (4)		—		44.6 ± 2.6 (5)	
parvirostris (Comores)		101–119; c. 110 (62)	81–99; c. 90 (62)	—	21–26; c. 24 (62)	—	Benson (1960)
rostratus (Aldabra)	♂♂	105–111; 108.4 (7)	93–102; 96.6 (7)	—	23–25; 23.9 (7)	48.4 ± 0.9 (5)	Benson & Penny (1971)
	♀♀	102–107; 104.8 (6)	87–98; 91.8 (6)	—	22–24; 23.1 (6)	43.5 ± 0.8 (4)	
H. crassirostris crassirostris (Seychelles)	♂♂	139.3 ± 0.9 (16)	110.0 ± 0.9 (17)	28.3 ± 0.3 (17)	31.4 ± 0.4 (15)	79.5 ± 1.1 (32)	Grieg-Smith (1979b)
	♀♀	131.5 ± 1.0 (12)	102.5 ± 0.4 (14)	27.1 ± 0.4 (15)	29.6 ± 0.3 (14)		
moheliensis (Mohéli, Comores)		109–127; 117.7 (5)	101, 110	—	29.5, 29.5	58, 61, 65, 68	Benson (1960), ASC & AWD (unpubl.)

Grey White-eyes *Zosterops borbonicus mauritianus* and *Z. b. borbonicus*

The ecological correlates of variation in weights and measurements are discussed in Chapters 4 and 6 where the data are presented more fully.

Olive White-eyes *Zosterops chloronothos* and *Z. olivaceus*

Réunion Olive White-eyes weigh 7.9–11.4 g, the heaviest weight recorded being for a pregnant female collected by F. B. Gill on 2 December 1967. Sixteen caught at the Plaine des Chicots in August 1974 averaged 9.7 g (range 8–11.2), while six in the same area in November averaged only 8.7 g (7.9–9.3). Average weights of Mascarene birds usually increased during the breeding season (e.g. Grey White-eye, Chapters 4 and 6), but, given the altitude, it is possible that Olive White-eyes caught in August were carrying fat against the very cold winter nights (Chapter 6). Mauritius Olive White-eyes are a little lighter, but also reach 11.5 g when pregnant; one female put on 2.1 g (from 7.5 to 9.6 g) in a fortnight.

The Mauritian species is smaller in all measurements except the bill, the most marked difference being in tail length. Wing-lengths of 38 living and museum specimens of Réunion Olive White-eyes are distributed as follows, suggesting a 1.5 mm shrinkage in prepared skins (all measured by ASC):

Fodies *Foudia rubra* and *F. flavicans*

The range of afternoon weights of Mauritius Fodies trapped in the field was 16–20.1 g, similar for both sexes, and averaging 17.7 g. Captive birds taking a lot of nectar mixture could put on as much as 3 g during daylight hours, but when taking a mixed diet would normally be 0.5–1.5 g heavier at dusk than at dawn. Morning weights of individuals usually varied over a few days through a range of about 1 g, evening weights up to 2 g. The extreme weights (g) of seven captive birds were (capture weight in brackets): ♂ ♂: 16.1–19.1 (18.4), 17.2–19.7 (17.7), 17.1–19.2 (17.8), 15.3–17.7 (16.0), 13.9–16.2 (16.1); ♀ ♀ 15.1–16.3 (16.5), 15.3–17.0 (17.0).

Female Rodrigues Fodies are lighter than males, only pregnant birds (17.1, 17.8, 18.0, 18.4 g) exceeding 16 g. The 17.8 g bird, after being held in captivity for a day, reached 19.6 g before laying a 1.7 g egg. Only daytime weights after 07.00 hours local time are included in Table 1. Birds held overnight lost 0.6–1.9 g (av. 1.27 ± 0.04; $n = 18$), and weights of birds caught before 07.00 hours suggested that this loss had yet to be regained.

Brown-plumaged Fodies behaving as females had shorter wing lengths than full-plumaged males in both species. Wings of Mauritius Fody females did not exceed 65 mm while no red-plumaged male had a wing of less than 66.5 mm. I attempted to confirm this

Wing length (mm):	55	56	57	58	59	60	61	62	63	Total	Average
Living birds	—	—	4	7	5	2	1	—	1	20	58.65 ± 0.33
Skins	2	7	3	2	3	1	—	—	—	18	57.00 ± 0.35

The difference is significant ($P < 0.002$, *t*-test). Gill (1970) gave a mean of 57.9 mm for 20 specimens, and Moreau (1957*a*) 57.7 mm (56–60, $n = 19$). There seems to be little difference in wing length between the sexes; the only three sexed skins overlapped (♂: 59; ♀ ♀: 57, 60) and there seemed to be no relation between wing length and presence of brood patches in birds trapped in November 1974.

Measurements of other western Indian Ocean white-eyes, none of which are close to the Mascarene species (Gill 1971*a*), can be found in Benson (1960), Benson & Penny (1971), Benson *et al.* (1976–7) and Grieg-Smith (1979*a*).

relationship on museum specimens, but of 17 skins in Cambridge, Tring (BM(NH)) and Paris only four are in brown plumage and only one of these gonadally sexed. Staub (1973*b*) reported on further specimens, including a sexed female. Details of these measurements are given in Table 2.

No male Rodrigues Fodies caught by ASC or in museums had wings of less than 68 mm, nor any females greater than 68 mm (Fig. 3). Many long-winged but grey-plumaged birds in museums are labelled ' ♀ ', but there is no evidence that any collectors except Gill, E. Newton and possibly Gulliver looked at gonads; none of the four gonadally sexed females

Table 2. *Wing length and sex in Mauritius Fodies*

Category	Wing length (mm)												Totals
	63	64	65	66	67	68	69	70	71	72	73	74	
Red-plumaged ♂♂													
Museum specimens[a]				1		2	5	2	1	2	1	1	15
Living birds					1	5	3		1	1	1		12
Brown-plumaged birds													
Museum specimens sexed ♀	1	1											2
Living birds[b]	1	1	1		1								4
Unsexed museum specimens						1	1			1			3

[a] Museum material from Staub (1973*b*: 2 ♂♂ 1♀) and specimens examined in Cambridge (6 ♂♂, 1 ♀, 1 unsexed), London (BM (NH): 2 ♂♂, 2 unsexed) and Paris (5 ♂♂). The male in the 66 mm column actually measured 66.5 mm. Moreau (1960) gave a range of only 69–71 mm for 6 males.
[b] The three brown birds with wings 63–65 mm were all paired with red ♂♂ and behaving as ♀♀. The 67 mm bird was holding a territory in which a ♀ and a juvenile were present.

Fig. 3. Wing lengths of Rodrigues Fodies. The vertical axis represents numbers of individuals in each of the wing length intervals (mm) represented on the horizontal axis. Birds above the line are those of known sex, those below unsexed. For each interval the left-hand column consists of live-caught birds, the right-hand one museum specimens. Columns are coded as follows:

	Live birds	*Museum specimens*
Males	Black	White
Females	Diagonal hatching	Dots
Unsexed	Horizontal hatching	White

Museum data are from specimens in the British Museum (Natural History), Paris, Leiden, National Museum of Natural History (Smithsonian), Cambridge, Oxford, Liverpool, and (*ex* Staub 1973*b*) the Mauritius Institute. Moreau gave a range of 67–71 mm, evidently based on a much smaller museum sample.

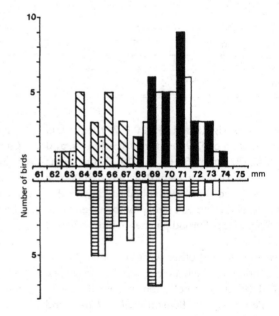

exceeds 65 mm in wing length. One grey-plumaged bird trapped in 1978 with wing length 65 mm was showing male plumage characters by 1981 (B 583 at Jersey Wildlife Preservation Trust; D. Jeggo pers. comm.). Another trapped at the same time with wing 70 mm was still in brown plumage when it died in June 1980, but does not appear to have been gonadally sexed post-mortem (Jones 1980*b*). These are the only two birds outside the normal wing length limits for the sexes.

Comparative data for other fodies are as follows:

	Wing (mm)	Tarsus (mm)	Bill (mm)	Weight (g)	Sources
F. madagascariensis					
(Mauritius)	♂♂ 69.1 ± 0.33 (26)	—	—	18.22 ± 0.20 (25)	⎫
(Rodrigues)	♂♂ 68.87 ± 0.63 (8)	—	—	17.79 ± 0.50 (8)	⎬ This study
(Réunion)	♂♂ 69.75 ± 0.48 (4)	—	—	18.65 ± 0.43 (4)	⎭
(Seychelles)	♂♂ 63.5–69; 68.2 (9)	16.2–19.2; 17.8 (?)	?: 11.4–14.5; 13.7 (?)	14.9–18.3; 16.8 (10)	Crook (1961)
(Madagascar)	♂♂ 65–69; 67.1 (20)	—	s: 14–15; 14.3 (20)	—	Benson *et al.* (1976–7)
(Mauritius)	♀♀ 63.3 ± 0.43 (14)	—	—	16.27 ± 0.36 (13)	⎫
(Rodrigues)	♀♀ 63.5 ± 0.87 (5)	—	—	15.82 ± 1.12 (5)	⎬ This study
(Réunion)	♀♀ 65.0 ± 0.84 (5)	—	—	15.58 ± 0.24 (5)	⎭
(Seychelles)	brn —	16.3–18.9; 17.5 (?)	?: 9.2–13.8; 13 (?)	13.5–16.5; 15.2 (9)	Crook (1961)
(Madagascar)	♀♀ 59–64; 62.1 (10)	—	s: 13–15; 14.3 (10)	—	Benson *et al.* (1976–7)
F. eminentissima	♂♂ 79.8 ± 0.12 (273)	22.7 ± 0.07 (98)	?: 19.0 ± 0.12 (98)	av. 24.5–26.3	⎫ Frith (1976)
(Aldabra)	♀♀ 73.7 ± 0.13 (175)	22.0 ± 0.08 (60)	?: 18.6 ± 0.15 (60)	av. 22.7–24.7	⎭
F. omissa	♂♂ 69–75; 72.6 (25)	—	s: 15.5–17; 16.5 (25)	16–21: 18.2 (5)	⎫ Benson *et al.* (1976–7)
(Madagascar)	♀♀ 65–70; 67.5 (15)	—	s: 15–17; 16.3 (15)	18–19.5; 18.7 (4)	⎭
F. seychellarum	♂♂ 74.5 ± 0.59 (15)	⎱16.5–20.2; 18.8 (17)	s: 16–17.5; 16.9 (22)	15.7–18.7; 17.5 (10)	Benson (1970–1),
(Seychelles)	brn 69.5 ± 0.67 (15)	⎰	s: 15–17; 16.4 (17)	15.5–19.1; 16.9 (17)	Crook (1961)

Neither Crook (1961) nor Frith (1976) specified how they measured bills, and Frith's weights for Aldabra Fodies are given as the range of averages, as he found a marked seasonality. Measurements of the Comorian forms of *F. eminentissima* can be found in Benson (1960).

There is minimal overlap in wing length between the sexes in all species, though in two cases this is obscured by the difficulty of distinguishing young males from females (Cardinals in the Seychelles and Seychelles Fodies). Only two birds out of 38 male and 24 female live-caught Cardinals in the Mascarenes were in the size range of the other sex – a male of 65 mm and a female of 68 mm (normal range: ♂♂: 67–72, ♀♀ 61–66, 67–72). Only 10.2% of Frith's (1976) large sample were in the overlap zone. The seasonal trends in weight for Aldabra Fodies (*ibid.*) differed somewhat between the sexes; this kind of data is not available for other species.

Mascarene Cardinals are consistently larger than samples from Madagascar and the Seychelles, suggesting slight differentiation since reaching the islands.

Acknowledgements

Warm thanks are due to all the museum staff (listed with Table 3) who sent us information and allowed us access to their collections; also to successive Directors of the Mauritius Institute (C. Michel, D. D. Tirvengadum & R. Gajeelee) for use of its collection and facilities.

Table 3. *Summary table of collections of Mascarene native land-birds in museums*

	UK					France	Netherlands	Switzerland	Austria
	BM (NH) Tring	UMZC Cambridge	MCML Liverpool	RSM Edinburgh	RCS[a] London	MNHN Paris	RMNH Leiden	MHNG Geneva	NMW Vienna
Réunion									
Butorides striatus	—	—	—	—	—	1	—	—	—
Circus maillardi	4[c]	2	—	—	—	12	5	—	—
Gallinula chloropus	—	—	—	—	—	2	2[d]	—	—
Mascarinus mascarinus†	—	—	—	—	—	1	—	—	1
Collocalia francica	2	1	—	—	—	13[e]	6[f]	—	1
Phedina b. borbonica	3	—	—	—	—	3[e]	2	—	—
Coracina newtoni[i]	6	4	—	—	—	5, 2a	8*[j]	—	—
Hypsipetes borbonicus	1	4	1	—	—	9	4	—	1
Saxicola tectes	5	5/6e	2	—	—	8/+e	3	—	—
Terpsiphone b. bourbonnensis	—	1	—	—	—	7+[k]	8	—	—
Zosterops b. borbonicus	6	6	(1[l])	—	—	39[e]/+e	8	—	—
Z. olivaceus	5	6	(1[l])	1[m]	—	7[e]/+e	5	—	—
Fregilupus varius†	1	—, 1s	—	—	—	2*	1	—	—
Mauritius									
Butorides striatus	3	7	—	—	—	+/+e	—	1	—
Falco punctatus	12, 2a/1e	9	1	1	—, 6a	3, 1a	4*	2	2
Gallinula chloropus	3	1 (*[o])	—	—	—	?/+e	—	4	1
Raphus cucullatus†	—	—	—	—	—	—	—	—	—
Nesoenas mayeri	6, 2a/9e[q]	4	1	—	—, 91a	5	2	1	—
Alectroenas nitidissima†	—	—	—	1	—	1*	—	—	—
Psittacula echo	4, 1a	4/2e	—	1[r]	—	6	4	1	—
Collocalia francica	2, 9a[h]	3/8e	1	—	—	2[e]/+e	—	5	—
Phedina b. borbonica	2	3	—	—	—	—	—	1	—
Coracina typica	4	11/2e	2	1	—	7	4	2	—
Hypsipetes olivaceus	6	22/13e	2	—	—	17/+e	5	2	3
Terpsiphone b. desolata	6/1e	7/19e	3	1	—	3+/+e	2[t]	2	1
Zosterops b. mauritianus	9	8/21e	(3[l])	1	—	2[e]	1	4	—
Z. chloronothos	4	7/4e	(2[l])	—[m]	—	4[e]	1	2	3
Foudia rubra	4, 1a, 1s	8/10e	—	1	—	4[e]	1	2	2
Rodrigues									
Butorides striatus	—	—	—	—	—	—	—	—	—
Psittacula exsul†	—	2*	—	—	—	—	—	—	—
Acrocephalus rodericanus	7/1e	3*/2e	—	—	—	2	—	—	—
Foudia flavicans	12/2e	17*/8e	1	—	—, 2a	3	4	—	—
Origin uncertain									
Necropsar leguati†	—	—	1*	—	—	—	—	—	—

Notes:

1. **Symbols.** This table includes details of all Mascarene bird specimens derived from living birds that we have been able to locate, although there are undoubtedly others scattered in smaller collections, especially in Europe. Osteological material from subfossil remains is excluded, but details can be found in Cowles (Chapter 2). Plain numbers indicate skins (mounted or not), spirit specimens are indicated by 'a' (= alcohol), skeletons by 's' and eggs by 'e'. Eggs are separated from bird specimens by a diagonal stroke.

Other symbols:

+   Specimens present, number uncertain

?   Not known whether any specimens present

—   No specimens

†   Species extinct

*   Indicates type material

2. **Full names of the museums** are as follows, with details of the sources of information:

UK

BM (NH) British Museum (Natural History), Tring: Sharpe *et al.* (1874–98), Oates *et al.* (1901–12), I. C. J. Galbraith & M. P. Walters (*in litt.*), this study (t/s)

UMZC University Museum of Zoology, Cambridge: Benson (1970–1, 1971 & *in litt.*), N. J. Collar (pers. comm.), t/s

able 3 *continued*

Germany DKUM Bremen	USA USNM Washington	AMNH New York	MCZH Harvard	ANSP Philadelphia	UMMZ Ann Arbor	Mauritius MIPL[b] Port Louis	Réunion MHNR St.-Denis	Other locations
—	—	—	—	—	—	—	—	
—	—	4	1	—	—	—/1e	7	ZMUO (1)
—	—	3	—	—	—	—	3	
—	—	—	—	—	—	—	—	
—	4, 7a, 6s[g,h]	—	—	—	4, 2a, 2s[g,h]	—	1	
—	—, 3s[g]	—	(1[s])	—	—, 1s[g]	—	2	
—	—	1	—	—	—	—	1	MNHU (1)
—	1, 1a, 1s[g]	3	(1[s])	—	—	1	2	
—	9	9	—	—	10, 2s	—	2	
—	3, 2s[g]	8	—	—	3, 3s[g]	—	2	MNHU (1)
—	+, 1a, 1s[g]	11	—	4	744, 2s[g]	—	1	
—	+, 1a, 2s	10	1	—	26, 1a, 5s	—	2	
—	—	—, 1a	—, 1a	—	—	1	1	See Hachisuka (1953) and Chapter 1
—	1	—	—	—	—	1+/10e	—	SAM (1), lost
1	—	2	3, 2s	2	—	5, 1s/17e	—	ZMK (1), MZN (1), MZS (1) MHH (1), WFVS (/3e), SAuM (/1e), ZMUH (/1e)[n]
—	1	7	?	—	—	2, 1s/6e	—	MNHU (1)
—	—	—	—	—	—	—	—	UMO (head & foot)[p]
—	1	4	2	—	—	6s, 1s/8e	—	SMTD (1), MZN (1), ZMUH (1) WFVZ (/3e)
—	—	—	—	—	—	1	—	
1	1	5, 1a	1	—	—	5/6e	1	SAM (1), lost; TMP (/1e)
—	13	—	—	—	3, 1s	2/8e	—	
—	9	—	—	1	3	2	—	SAM (1), lost; MNHU (1)
2	1	2	1	—	—	3/2e	1	
1	4	10	(4[s])	—	1	6/4e	—	MNHU (2)
—	2	8	(3[s])	—	—	3/2e	—	SAM (/3e), MNHU (1)
—	2 (+?)	9	2	—	4, 1s	6, 1s/6e	—	MNHU (1)
2	?	3	1	—	—	4/3e	—	
1	4, 1a, 1s	1	1	2	—	6/1e	—	
—	3	—	—	—	—	—	—	
—	—	—	—	—	—	—	—	
—	1	—	—	—	—	4/1e	—	
2	7	—	—	—	—	3+	—	UMO (2)
—	—	—	—	—	—	—	—	

MCML Merseyside County Museums, Liverpool: anon. (1981*b*), Fisher (1981 & *in litt.*), Cowper (1984), t/s
RSM Royal Scottish Museum, Edinburgh: Hachisuka (1953), I. H. J. Lister & R. McGowan (*in litt.*), t/s
RCS Royal College of Surgeons of England, London: anon (1981*c*), J. E. Cooper (*in litt.*)
France
    MNHN Muséum Nationale d'Histoire Naturelle, Paris: Berlioz (1946), Jouanin (1962 & *in litt.*), R. Bour & C. Voisin (*in litt.*), t/s
    MZN Musée de Zoologie, Nancy: A. Philippot (*in litt.*)
    MZS Musée Zoologique, Strasbourg: G. Hildwein (*in litt.*)
Netherlands
    RMNH Rijksmuseum van Natuurlijke Histoire, Leiden: Hachisuka (1953), G. F. Mees (*in litt.*)
Switzerland
    MHNG Muséum d'Histoire Naturelle, Geneva: C. Weber (*in litt.*)
Austria
    NMW Naturhistorisches Museum Wien (Vienna): Sassi (1940), H. Schifter (*in litt.*)

Notes continued overleaf

Notes to Table 3 continued:

Germany

   DKUM Deutsches Kolonial und Übersee-Museum, Bremen: Hartlaub (1877)

   MHH Museum Heineanum, Halberstadt: N. König (*in litt.*)

   SMTD Staatliches Museum für Tierkunde, Dresden: S. Eck (*in litt.*)

   MNHU Museum für Naturkunde der Humboldt-Universität, Berlin: G. Mauersberger (*in litt.*)

Denmark

   ZMK Zoologisk Museum, Københaven (Copenhagen): J. Fjeldså (*in litt.*)

Finland

   ZMUH Zoological Museum of the University, Helsinki: O. Hilden & T. Stjernberg (*in litt.*)

Norway

   ZMUO Zoological Museum of the University, Oslo: Hartlaub (1877)

USA

   USNM United States National Museum, Smithsonian Institution, Washington, DC: G. E. Watson & B. B. Farmer (*in litt.*)

   AMNH American Museum of Natural History, New York: R. F. Pasquier, J. Farrand & M. LeCroy (*in litt.*), Berger (1957)

   MCZH Museum of Comparative Zoology, Harvard, Cambridge, Mass.: Berger (1957), Greenway (1967), R. A. Paynter (*in litt.*), t/s.

   ANSP Academy of Natural Science, Philadelphia: F. B. Gill & M. B. Robbins (*in litt.*)

   UMMZ Museum of Zoology, University of Michigan, Ann Arbor: J. Hinshaw (*in litt.*)

   WFVZ Western Foundation for Vertebrate Zoology, Los Angeles: (*in litt.*)

South Africa

   SAM South African Museum, Cape Town: Brooke (1976)

   TMP Transvaal Museum, Pretoria: A. Kemp & T. Salinger (*in litt.*)

Australia

   SAuM South Australian Museum, Adelaide: S. A. Parker (*in litt.*)

Mauritius

   MIPL Mauritius Institute (formerly the Desjardins Museum), Port Louis: Rountree *et al.* (1952), Staub (1973b), t/s

Réunion

   MHNR Muséum d'Histoire Naturelle, St.-Denis: Milon (1951), H. Gruchet (*in litt.*), t/s

Details of spirit and skeletal specimens have been added from Wood *et al.* (1982a, b). These useful inventories do not, however, include the collections of certain important museums, notably Cambridge, Paris, Vienna and Bremen, and unaccountably omit the spirit specimens of *Fregilupus* in New York and Harvard, though the former is included as a skeleton (see Berger 1957). Their authors do not recognise *Saxicola tectes* or *Hypsipetes olivaceus* as full species, so material could be concealed under *S. torquata* and *H. borbonicus*; species occurring in more than one geographical area are not separated so some doubt must remain about some forms (see notes *n* and *o* below).

   3. **Comments on the table:**

[a] All specimens are derived from the captive breeding programme and remain the property of the Mauritius Government. Two of the kestrels and 52 of the Pink Pigeons are nestlings.

[b] Some specimens listed by Rountree *et al.* (1952) appear to be no longer present.

[c] Also a hatchling, without data.

[d] Also a chick.

[e] There may also be stuffed specimens (in closed galleries) and spirit specimens.

[f] Also three nestlings.

[g] Alcohol and skeletal specimens of *Collocalia francica*, *Phedina borbonica*, '*Hypsipetes borbonica*', *Terpsiphone bourbonnensis* and *Zosterops borbonicus* (Scott *et al.* 1982a, b) are assumed to have been collected by F. B. Gill and to come from Réunion.

[h] It is not clear whether all these are *Collocalia francica sensu stricto* from the Mascarenes, or whether Asian forms formerly listed under that name are included.

[i] The majority collected by Pollen & van Dam, distributed as follows: Leiden (8), Paris (2), Cambridge (2) and Tring (1).

[j] Benson (1971) mentioned only 7, G. F. Mees (*in litt.*) found 8.

[k] Salomonsen (1933) found more than we did. There are also three of uncertain origin.

[l] *Zosterops* specimens formerly in Liverpool but lost during the 1939–45 war (Fisher 1982).

[m] Specimen labelled 'Mauritius' but appears to be a *Z. olivaceus* from Réunion (pers. obs. & R. McGowan *in litt.*).

[n] Also one in the Kreuger Collection in Germany (Schönwetter 1960–).

[o] Type of *G. c. pyrrhorrhoa* A. Newton (Benson 1970–1).

[p] An egg in the East London Museum, South Africa, claimed as a Dodo's by Courtenay-Latimer (1953, Vinson 1956b), must remain, in the absence of good evidence of its origin, *incertae sedis*.

[q] The two alcohol specimens of Pink Pigeons in the BM (NH) are nestlings from the captive breeding programme on loan from the Mauritius Government.

[r] See Chapter 5.

[s] No longer traceable.

[t] Labelled 'Afrique'.

References

COMPILED BY A.S.CHEKE

Journal titles are abbreviated according to British Standard BS 4148(1970, 1975), except that *Orn.* is used for *Ornithol-* and *Ornitol-* and *Roy.* for *Royal*. The initials BOU, ICBP and IUCN are used for the British Ornithologists' Union, the International Council for Bird Preservation and the International Union for Conservation of Nature and Natural Resources. The *Recueil trimestriel de documents et travaux inédits pour servir à l'histoire des Mascareignes françaises* is shortened simply to *Rec. Trim.*

French names of the forms 'de X', 'd'X' and 'de la X' are listed under 'X' and 'La X'.

d'Aglosse, P. (pseud.) 1891. La prétendue relation du P. Brown. *Rev. Hist. Litt. Ile Maurice* 5: 337–40.

Ahimaz, P. 1984. The Echo Parakeet *(Psittacula (eques) echo)*. Project Report. Report to Vogelpark Walsrode. Xeroxed.

Albany, J. 1974. *P'tit glossaire. Le piment des mots créoles.* Paris: (author).

Alexander, J. B., Duck, D. D. & Gardner, A. S. 1978. *Edinburgh University Expedition to Round Island 1978. Preliminary Report.* Edinburgh: (authors). Xeroxed, 9 pp. + maps.

Ali, S. & Abdulali, H. 1938. The birds of Bombay and Salsette: IV. *J. Bombay Nat. Hist. Soc.* 40: 148–73.

Ali, S. & Ripley, S. D. 1968–74. *Handbook of the birds of India & Pakistan.* 10 vols. Bombay: Oxford University Press. [Also 2nd edn (1978–), Delhi: OUP, in progress]

Allan, W. & Edgerley, L. F. 1950. *White paper on Crown Forest Land (land utilization) and Forestry.* Port Louis, Mauritius: Government Printer.

Allen, R. B., Payton, I. J. & Knowton, J. E. 1984. Effects of ungulates on structure and species composition in the Urewera forests as shown by exclosures. *N.Z.J.Ecol.* 7: 119–130.

Andersen, K. 1907. Some remarks on *Pteropus mascarinus* Mason. *Ann. Mag. Nat. Hist.* (7)20: 351–5.

Andersen, K. 1912. *Catalogue of the Chiroptera in the collection of the British Museum,* 2nd edn. Vol. 1: *Megachiroptera.* London: Trustees of the British Museum.

Andersen, K. 1913. A sub-fossil bat's skull from Rodriguez Island. *Rec. Ind. Mus.* 9(5): 337.

Andrews, C. W. 1897. On some fossil remains of Carinate birds from central Madagascar. *Ibis* (7)3: 343–59.

anon. *c.* 1862. *The National Encyclopaedia: a dictionary of universal knowledge by writers of eminence in literature, science and art.* 13 vols. London: Wm. Mackenzie.

anon.('Z'). 1873. Wild animals dispersed by human agency. *The Field* 41: 215, 241–2, 269.

anon. 1961. *Progress report 1955–60 by the Forest Department of Mauritius prepared for the British Commonwealth Forestry Conference 1962.* Colony of Mauritius: Government Printer.

anon. 1962. *Petit atlas de Bourbon.* Nantes: Changreau & Cie.

anon. 1967 ('1968'). *Progress report 1961–65 by the Forest Department of Mauritius prepared for the Commonwealth Forestry Conference 1968.* Curepipe, Mauritius: Forest Department.

anon. 1971*a.* La Réunion. *Marchés Trop.* 27: 3635–56.

anon. 1971*b.* *4-year plan for social and economic development.* 2 vols. Port Louis, Mauritius: Government Printer.

anon. 1973. Caméléon de l'Ile Bourbon. *Info-Nature Ile Réunion* 10: 66–72.

anon. 1974*a.* *Progress report 1966–72 by the Forestry Services of the Ministry of Agriculture and Natural Resources, Mauritius, prepared for the Tenth Commonwealth Forestry Conference, 1974.* Curepipe, Mauritius: Forestry Services.

anon. 1974*b.* *Fiche sur la restauration de la forêt réunionnaise.* St.-Denis, Réunion: Office Nationale des Forêts. Cyclostyled.

anon. 1976. *Family planning in five continents: December 1976.* London: International Planned Parenthood Federation.

anon. 1981*a.* *ICBP Newsletter* 3(1). [Special issue on conservation priorities]

anon. 1981*b.* *A list of bird species represented in the collections of the Merseyside County Museums.* Liverpool: Merseyside County Museums.

anon. 1981*c.* *Directory resources of biomedical and zoological specimens.* Washington, DC: Registry of Comparative Pathology, Armed Forces Institute of Pathology.

anon. 1981*d.* Reflexions pour la création des réserves. Document SREPEN. *Info-Nature Ile Réunion* 18: 83–7.

anon. 1982. Conservation hat-trick. *On the edge (Wildl. Preserv. Trust Newsl.)* 44: 1–2.

anon. 1984. The echo of the parakeet: report from Mauritius. *Flying Free* 2(1): 1–2.

Antelme, H. 1914. Hunting. In Macmillan (1914), q.v.: 169–73.

Antoine, E. 1973. La forêt tropicale à la Réunion. *Info-Nature Ile Réunion.* No. spécial hors série, 'La Forêt': 39–47.

Appert, O. 1965. Découverte de la nidification de *Puffinus pacificus* (Gmelin) près de la côte ouest de Madagascar. *Oiseau Rev. Fr. Orn.* 35: 135–9.

Après de Mannevillette, J.B.N.D. d' 1775. *Instructions sur la navigation des Indes Orientales et de la Chine, pour servir au Neptune Oriental.* Paris/Brest: Demonville/Malassis.

Arlidge, E. Z. & Wong You Cheong, Y. 1975. *Notes on the land reserves and agricultural suitability map of Mauritius.* Reduit, Mauritius: Mauritius Sugar Industry Research Institute, Occ. Pap. 29.

Armstrong, E. A. 1953. Territory and birds. A concept which originated from study of an extinct species. *Discovery* [July]: 223–4.

Arnold, E. N. 1979. Indian Ocean giant tortoises: their systematics and island adaptations. *Phil. Trans. Roy. Soc., Lond.* 286B: 127–45.

Arnold, E. N. 1980. Recently extinct reptile populations from Mauritius and Réunion, Indian Ocean. *J. Zool., Lond.* 191: 33–47.

Athalante, log of the 1722. Journal de l'*Athalante*. *Rev. Retrospective Ile Maurice* 5: 3–14 (1954).

Atkinson, G. 1922. *The extraordinary voyage in French literature*, vol. 2, *From 1700 to 1720.* Paris: Champion. [Reprinted 1969. New York: Burt Franklin]

Atkinson, I. A. E. 1973. Spread of the ship rat (*Rattus r. rattus* L.) in New Zealand. *J. Roy. Soc. N.Z.* 3: 457–72.

Atkinson, I. A. E. 1977. A reassessment of factors, particularly *Rattus rattus* L., that influenced the decline of endemic forest birds in the Hawaiian Islands. *Pacific Sci.* 31: 109–33.

Atkinson, I. A. E. 1978. Evidence of effects of rodents on the vertebrate wildlife of New Zealand islands. In Dingwall *et al.* (1978), q.v.: 7–31.

Atkinson, I. A. E. 1979. What's so special about Kapiti Island? *Forest & Bird* 212: 12–15.

Azéma, G. 1859. *Histoire de l'Ile Bourbon depuis 1643 jusqu'au 20 décembre 1848.* Paris: Henri Plon.

Backhouse, J. 1844. *A narrative of a visit to the Mauritius and South Africa.* London: Hamilton Adams & Co.

Baker, J. G. 1877. *Flora of Mauritius and the Seychelles. A description of the flowering plants and ferns of those islands.* London: Reeve.

Baker, J. K. & Reeser, D. W. 1972. Goat management problems in Hawaii Volcanoes National Park. *U.S. Dept. Inter. Natl. Park Serv. Nat. Res. Rep.* 2.

Baker, H. G. & Stebbings, G. L. (eds.) 1965. *The genetics of colonizing species.* New York: Academic Press.

Balfour, I. B. 1879a. The physical features of Rodriguez. *Phil. Trans. Roy. Soc., Lond.* 168: 289–92.

Balfour, I. B. 1879b. [The collections from Rodriguez] Botany. *Phil. Trans. Roy. Soc., Lond.* 168: 302–87.

Barau, [C.] A. 1980. L'histoire des oiseaux de la Réunion du dodo à nos jours. *Bull. Acad. Réunion* 24: 41–72.

Barbehenn, K. R. 1962. The house shrew on Guam. In T. I. Storer (ed.), *Pacific island rat ecology. Report of a study made on Ponape and adjacent islands 1955–1958*, pp. 247–56. Bernice P. Bishop Mus. Bull. 225.

Barkly, H. 1870. Notes on the fauna and flora of Round Island. *Trans. Roy. Soc. Arts Sci. Mauritius*, N.S. 4: 109–30.

Barnwell, P. J. 1948. *Visits and despatches 1598–1948.* Port Louis, Mauritius: Standard Printing Establishment.

Barnwell, P. J. 1955. Early place-names of Mauritius. *Rev. Retrospective Ile Maurice* 6: 167–80.

Barnwell, P. J. & Toussaint, A. 1949. *A short history of Mauritius.* London: Longmans, Green.

Barré, N. 1982. Helminthes des animaux domestiques et sauvages de la Réunion. II. Oiseaux, reptiles, batraciens, poissons. *Rev. Elev. Med. Vet. Pays Trop.* 35: 245–53.

Barré, N. 1983. Distribution et abondance des oiseaux terrestres de l'Ile de la Réunion (Ocean Indien). *Rev. Ecol. (Terre Vie)* 37: 37–85.

Barré, N. & Barau, [C.]A. 1982. *Oiseaux de la Réunion.* St.-Denis, Réunion: (authors).

Barré, N. & Moutou, F. 1979. Sur la présence de *Bubulcus ibis* et *Butorides striatus* (Ardéidés) à la Réunion. *Info-Nature Ile Réunion* 17: 21–3.

Barré, N. & Moutou, F. 1982. Helminthes des animaux domestiques et sauvages de la Réunion. I. Mammifères. *Rev. Elev. Med. Vet. Pays Trop.* 35: 43–55.

Barré, N. 1984 ('1983'). Oiseaux migrateurs observés à la Réunion. *Oiseau Rev. Fr. Orn.* 53: 323–33.

Bathe, G. M. & Bathe, H. V. 1982. Territory size and habitat requirement of the Seychelles Brush Warbler *Acrocephalus (Bebrornis) sechellensis. Cousin Is. Res. Station Tech. Rep.* 18. Cambridge: ICBP.

Bathe, H. [V.] & Bathe, G. [M.] 1982. Feeding studies of three endemic landbirds, *Acrocephalus sechellensis, Foudia seychellarum* and *Nectarinia dussumieri* on Cousin Island, Seychelles. *Cousin Is. Res. Station Tech. Rep.* 26. Cambridge: ICBP.

Baumer, M. 1981. Le couvert végétal à la Réunion. *Info-Nature, Ile Réunion* 18: 15–25.

Bell, B. D. 1978. The Big South Cape Islands rat irruption. In Dingwall *et al.* (1978) q.v.: 33–45.

Bendire, C. 1894. Descriptions of nests and eggs of some new birds collected on the island of Aldabra, north-west of Madagascar, by Dr W. L. Abbott. *Proc. U.S. Nat. Mus.* 17: 39–41.

Benson, C. W. 1960. The birds of the Comoro Islands: results of the British Ornithologists' Union Centenary Expedition 1958. *Ibis* 103b: 5–106.

Benson, C. W. 1967. The birds of Aldabra and their status. *Atoll Res. Bull.* 118: 63–111.

Benson, C. W. 1970–1. The Cambridge Collection from the Malagasy Region. *Bull. Br. Orn. Club* 90: 168–172; 91: 1–7.

Benson, C. W. 1971. Notes on *Terpsiphone* and *Coracina* spp. in the Malagasy region. *Bull. Br. Orn. Club* 91: 56–64.

Benson, C. W., Colebrook-Robjent, J. F. R. & Williams, A. 1976–7. A contribution to the ornithology of Madagascar. *Oiseau Rev. Fr. Orn.* 46: 103–34, 209–42, 368–86; 47: 41–64, 168–91.

Benson, C. W. & Penny, M. J. 1971. The land birds of Aldabra. *Phil. Trans. Roy. Soc., Lond.* 260 B: 417–527.

Berger, A. J. 1957. On the anatomy and relationships of *Fregilupus varius*, an extinct starling from the Mascarene Islands. *Bull. Am. Mus. Nat. Hist.* 113: 225–72.

Berger, A. J. 1981. *Hawaiian bird life*, 2nd edn. Honolulu: University of Hawaii Press.

Berlioz, J. 1946. *Oiseaux de la Réunion.* Faune de l'Empire Français, 4. Paris: Larose.

Bernardin, Père 1687. Mémoire du R. P. Bernardin sur l'Ile de Bourbon (1687) [ed. A. Lougnon]. *Rec. Trim.* 4: 57–70 (1939).

Bernardin de St. Pierre, J. H. 1773. *Voyage à l'Isle de France, à l'Isle de Bourbon, au Cap de Bonne-Espérance; &c. par un Officer du Roi.* Neuchâtel: Société Typographique. [Two parts with separate pagination bound in one]

Berry, R. J. 1977. *Inheritance and natural history.* London: Collins.

Bertuchi, A. J. 1923. *The island of Rodrigues.* London: John Murray.

Biggs, H. C., Kemp, A. C., Mendelsohn, H. P. & Mendelsohn, J. M. 1979. Weights of southern African raptors and owls. *Durban Mus. Novit.* 12: 72–81.

Billiard, A. 1829. *Voyage aux colonies Orientales ou lettres écrites des îles de France et de Bourbon pendant les années 1818, 1819 et 1820.* Paris: J. L. J. Brière.

Blanc, C. P. 1972. Les reptiles de Madagascar et des îles voisines. In R. Battistini & G. Richard-Vindard (eds.) *Biogeography and ecology of Madagascar,* pp. 501–611. The Hague: W. Junk.

[Blenkinsop, A.] 1851. *A transport voyage to the Mauritius and back, touching at the Cape of Good Hope and St. Helena.* London: John Murray.

Bloxam, Q. M. C. 1977 ('1976'). Maintenance and breeding of the Round Island skink *Leiolopisma telfairii* (Desjardins). *Jersey Wildl. Preserv. Trust Ann. Rep.* 13: 53–6.

Bloxam, Q. [M. C.] 1982. Gunners Quoin and Round Island. *On the edge (Wildl. Preserv. Trust Newsl.)* 44: 7.

Bloxam, Q. M. C. 1983 ('1982'). Feasibility of reintroduction of captive Round Island Skink to Gunners' Quoin. *Dodo, J. Jersey Wildl. Preserv. Trust* 19: 37–42.

Bloxam, Q. M. C. & Vokins, A. M. A. 1979 ('1978'). Breeding and maintenance of *Phelsuma guentheri* (Boulenger 1885) at the Jersey Zoological Park. *Dodo, J. Jersey Wildl. Preserv. Trust* 15: 82–91.

Boisrouvray, F. du 1975. Arc-en-Barrois: vers un équilibre forêt-gibier à très haut niveau. *Rev. For. Fr.* 26: 99–108.

Bonaparte. R. 1890. *Le premier établissement des Néerlandais à Maurice.* Paris: (author).

Bonnell, M. L. & Selander, K. 1974. Elephant seals: genetic variation and near extinction. *Science* 184: 908–9.

Bory de St Vincent, J. B. G. M. 1804. *Voyage dans les quatres principales îles des mers d'Afrique fait par ordre du gouvernement pendant les années neuf et dix de la République (1801 et 1802).* 3 vols. Paris: F. Buisson.

Bosser, J. 1982. *Projet de constitution de réserves biologiques dans le domaine forestier à la Réunion.* Rapport de Mission J. Bosser. Paris: ORSTOM.

Bosser, J., Cadet, T., Guého, J., Julien, H. R. & Marais, W. (eds.) 1978– . *Flore des Mascareignes: La Réunion, Maurice, Rodrigues.* Reduit, Mauritius: Sugar Industry Research Institute/Paris: ORSTOM/Kew: Royal Botanic Gardens.

Boucher, A. 1710. Mémoire d'Antoine Boucher sur l'Ile Bourbon en 1710 [ed. A. Lougnon]. *Rec. Trim.* 5: 279–355 (1941).

Bour, R. 1978. Les tortues des Mascareignes; description d'une espèce nouvelle d'après un document

(Mémoires de l'Académie) de 1737 dans lequel le crâne est figuré. *C.R. Acad. Sci., Paris* 287D: 491–3.

Bour, R. 1979. Première découverte de restes osseux de la Tortue terrestre de la Réunion, *Cylindraspis borbonica. C.R. Acad. Sci., Paris* 288D: 1223–6. [Also in *Info-Nature Ile Réunion* 17: 53–9 (1979), with an additional illustration]

Bour, R. 1980. Systématique des tortues terrestres des îles Mascareignes: genre *Cylindraspis* Fitzinger 1835 (Reptilia, Chelonii). *Bull. Mus. Natl. Hist. Nat., Paris* (4)2A: 895–904.

Bour, R. 1981. Histoire de la tortue terrestre de Bourbon. *Bull. Acad. Ile Réunion* 25: 98–147.

Bour, R. 1983 Trois populations endémiques du genre *Pelusios* (Reptilia, Chelonii, Pelomedusidae) aux îles Seychelles; relations avec les espèces Africaines et Malgaches. *Bull. Mus. Natl. Hist. Nat., Paris* (4)5A: 343–82.

Bour, R. 1984a. L'identité de *Testudo gigantea* Schweigger, 1812 (Reptilia, Chelonii). *Bull. Mus. Natl. Nat. Hist., Paris* (4)6A: 159–75.

Bour, R. 1984b. Données sur la repartition géographique des tortues terrestres et d'eau douce aux îles Maurice et Rodrigues. *Info-Nature Ile Réunion* 21: 7–38.

Bour, R. & Moutou, F. 1982. Reptiles et Amphibiens de l'ile de la Réunion. *Info-Nature Ile Réunion* 19: 119–56.

Bourgat, R. [M.] 1967. Introduction à l'étude écologique sur le Caméléon de l'Ile de la Réunion, *Chamaeleo pardalis* Cuv. *Vie & Milieu* 18C: 221–30.

Bourgat, R. M. 1970. Recherches écologiques et biologiques sur le *Chamaeleo pardalis* Cuvier 1829 de l'Ile de la Réunion et de Madagascar. *Bull. Soc. Zool. France* 95: 259–269.

Bourgat, R. M. 1972. Biogeographical interest of *Chameleo pardalis* Cuvier 1829 (Reptilia, Squamata, Chamaeleonidae) on Réunion Island. *Herpetologica* 28: 22–4.

Bourgenot, [ ] 1974. [Interview with M. Bourgenot, Ingénieur général, Directeur technique de l'ONF.] *Journal de la Réunion,* 16 August.

Bourn, D. M. 1976. The giant tortoise population of Aldabra (Cryptodira: Testudinidae). I. Preliminary results. *Zool. Afr.* 11: 275–84.

Bourn, D. M. & Coe, M. J. 1979. Features of tortoise mortality and decomposition on Aldabra. *Phil. Trans. Roy. Soc., Lond.* 286B: 189–93.

Bourne, W. R. P. 1966. The birds of the Chagos Group, Indian Ocean. *Atoll Res. Bull.* 149: 175–207.

Bourne, W. R. P. 1968. The birds of Rodriguez, Indian Ocean. *Ibis* 110: 338–44.

Bourne, W. R. P. 1976. On subfossil bones of Abbott's Booby *Sula abbotti* from the Mascarene Islands, with a note on the proportions and distribution of the Sulidae. *Ibis* 118: 119–23.

Bourne, W. R. P. 1981. Rats as avian predators: discussion. *Atoll Res. Bull.* 255: 69–71.

Bourne, W. R. P., Bogan, J. A., Bullock, D., Diamond A. W. & Feare, C. J. 1977. Abnormal terns, sick sea and shore birds, organochlorines and arboviruses in the Indian Ocean. *Mar. Poll. Bull.* 8: 154–8.

Bouton, L. 1869. Séance du vendredi 19 juin 1868. *Trans. Roy. Soc. Arts Sci. Mauritius*, N.S. 3: 109–11.

Bouton, L. 1871. [P. V. du] Séance du vendredi 15 juillet 1870. *Trans. Roy. Soc. Arts Sci. Mauritius*, N.S. 5: 42.

Bouton, L. 1875. Report of the Secretary to the Royal Society of Arts and Sciences, 3 May 1873 to 10 October 1874. *Trans. Roy. Soc. Arts Sci. Mauritius*, N.S. 8: i–xxiii+.

Bouton, L. 1878. [Letter to Edward Newton on the Huppe de Bourbon and the Pigeon Hollandais, dated 21 Feb. 1878.] MS in bound quarto volume entitled *Indian Ocean 3. Madagascar – Mascarene Islands (MSS)*, Newton Library, Cambridge University Zoology Dept.

Bouton, L. 1883. Rapport annuel du Secrétaire [for 1877–8; dated 6.2.1878]. *Trans. Roy. Soc. Arts Sci. Mauritius*, N.S. 11: 126–36.

Bower, G. 1903. *Mauritius. Report for 1901*. Colonial Reports – Annual, 379. London: HMSO.

Brasil, L. 1912. Un oiseau étient de la Réunion *Fregilupus varius* (Bodd.). *Bull. Soc. Linn. Normandie* (6)4: 16–29.

Brisson, M. J. 1756. *Le Règne animale divisé en IX classes*. Paris: J.-B. Bauche.

Brisson, M. J. 1760. *Ornithologie*. 6 vols. Paris: J.-B. Bauche.

Brooke, M. de L. 1978. Inland observations of Barau's Petrel *Pterodroma baraui* on Réunion. *Bull. Br. Orn. Club* 98: 90–5.

Brooke, M. de L. 1985. The annual cycle of the Toc-toc *Foudia sechellarum* on Cousin Island, Seychelles. *Ibis* 127: 7–15.

Brooke, R. K. 1970. Zoogeography of the swifts. *Ostrich Suppl.* 8: 47–54.

Brooke, R. K. 1972. Generic limits in old world Apodidae and Hirundinidae. *Bull. Br. Orn. Club* 92: 53–7.

Brooke, R. K. 1976. Layard's extralimital records in his *Birds of South Africa* and in the South African Museum. *Bull. Br. Orn. Club* 96: 75–80.

Brosset, A. 1963. Statut actuel de la faune aux îles Galapagos. *Notic. Galapagos* 1: 5–9.

Brouard, N. R. 1960. A brief account of the 1960 cyclones and their effect upon exotic plantations in Mauritius. *Emp. For. Rev.* 39: 411–16.

Brouard, N. R. 1963. *A history of woods and forests in Mauritius*. Port Louis, Mauritius: Government Printer.

Brouard, N. R. 1966. Les problèmes forestiers de Maurice. *Rev. Agric. Sucrière Ile Maurice* 45: 220–30.

Brouard, N. R. 1967. *Damage by tropical cyclones to forest plantations, with particular reference to Mauritius*. Port Louis, Mauritius: Government Printer.

Brouard, N. R. 1969. *Annual report of the Forest Department for the year 1967*. Mauritius, No. 14 of 1969. Port Louis, Mauritius: Government Printer.

Brouard, N. R. 1970. *Annual report of the Forestry Service for the year 1968*. Mauritius, No. 13 of 1970. Port Louis, Mauritius: Government Printer.

Brown, Rév. Père 1773. Lettre du révérend père Brown, missionaire de la Compagnie de Jesus, à Madame la marquise de Benamont. *Lettres Edifiantes* (q.v.), *Mémoires des Indes* 30: 321–51 [also 13: 302–19 (1781 edn); 7: 450–60 (1819 edn)].

Brown, L. [H.] & Amadon, D. 1968. *Eagles, hawks and falcons of the world*. 2 vols. London: Country Life.

Bruce-Chwatt, L. J. 1974. *Anopheles gambiae* complex in Mauritius. *Trans. Roy. Soc. Trop. Med. Hyg.* 68: 497–8.

Buffon, G.-L. Leclerc, Comte de 1770–83. *Histoire naturelle des oiseaux*. Paris: Imprimerie Royale. [Three original editions with the variable numbers of volumes published over more or less the same dates; one, in 10 vols., includes the *Planches enluminées* by Martinet; numerous later reprints. See: E. Genet-Varcin & J. Roger (1954) *Bibliographie de Buffon*. Paris. Also N. Mayaud, *Alauda* 9: 18–32 (1939)]

Buffon, G.-L. Leclerc, Comte de 1776. *Histoire naturelle générale et particulière, servant de suite à l'histoire des animaux quadrupèdes*. Supplément, Vol. 3. Paris: Imprimerie Royal.

Buffon, G. L. LeClerc, Comte de 1789. *Ibid*. Supplément, Vol. 7. Paris: Imprimerie Royal.

Bullock, D. 1977. Round Island – a tale of destruction. *Oryx* 14: 51–8.

Bullock, D. (ed.) 1982. *Round Island Expedition 1982, Preliminary Report*. St Andrews: (the Expedition).

Bullock, D. J., Arnold, E. N. & Bloxam, Q. 1985. A new endemic gecko from Mauritius. *J. Zool., Lond. (A)* 206: 591–9.

Bullock, D. & North, S. n.d. (1977). Report of the Edinburgh University Expedition to Round Island, Mauritius, July & August 1975. Edinburgh: (authors). Xeroxed. [A preliminary report, 14 pp., produced in 1976, was more widely circulated]

Bullock, D. & North, S. 1984. Round Island in 1982. *Oryx* 18: 36–41.

Bullock, D., North, S. & Grieg, S. 1983. *Round Island Expedition 1982. Final Report*. St. Andrews, Scotland: (authors).

Bulpin, T. V. 1958 (1969). *Islands in a forgotten sea*. Cape Town: Books of Africa.

Byrd, G. V. & Moriarty, D. 1980. Treated chicken eggs reduce predation on shearwater eggs. *Elepaio* 41(2): 13–15.

'J. C.' 1861. Le Merle de Bourbon. In Roussin (1860–7), q.v., 2: 78–80.

Cade, T. [J.] 1982. *The falcons of the world*. London: Collins.

Cadet, [L. J.] T. 1971. Flore de l'Ile Rodrigue: espèces spontanées introduites depuis Balfour (1874). *Mauritius Inst. Bull.* 7: 1–12.

Cadet, [L. J.] T. 1973. Histoire d'une forêt de 'bois de couleur' dans l'île de la Réunion. *Info-Nature Ile Réunion*. No. spécial hors série, 'La Forêt: 29–37.

Cadet, [L. J.] T. 1974. Etude sur la végétation des hautes altitudes de l'île de la Réunion (Ocean Indien). *Vegetatio* 29: 121–30.

Cadet, [L. J.] T. 1975. Contribution à l'étude de la végétation de l'Ile Rodrigue. *Cah. Centre Univ. Réunion* 6: 5–29.

Cadet, L. J. T. 1977 (1980). La végétation de l'Ile de la Réunion: étude phytoécologique et phytosociologique. Thesis, University of Marseille. Published 1980: St.-Denis, Réunion: Imprimerie Cazal.

Cadet, [L. J.] T. 1981. *Fleurs et plantes de la Réunion et de l'île Maurice*. Papeete, Tahiti: Editions du Pacifique.

Caldwell, J. 1875. Notes on the zoology of Rodrigues. *Proc. Zool. Soc., Lond.* [1875]: 644–7.

Campbell, D. J. 1978. The effects of rats on vegetation. In Dingwall *et al.* (1978), q.v.: 99–120.

Cap, P. A. 1861. *Philibert Commerson, naturaliste voyageur. Etude biographique suivi d'un appendice.* Paris: Victor Masson et fils.

Carié, P. 1904. Observations sur quelques oiseaux de l'Ile Maurice. *Ornis* 12: 121–8.

Carié, P. 1910. Note sur l'acclimatation du Bulbul (*Otocompsa jocosa* L.) à l'Ile Maurice. *Bull. Soc. Natl. Acclim., Paris* 57: 462–4.

Carié, P. 1916. L'acclimatation à l'Ile Maurice. *Bull. Soc. Natl. Acclim., Paris* 63: 10–18, 37–46, 72–9, 107–10, 152–9, 191–8, 245–50, 355–63, 401–4. [Also reprinted by the Society with new pagination]

Carié, P. 1921. Le Merle Cuisinier de l'Ile Maurice (*Lalage rufiventer* Sw.). *Rev. Hist. Nat. Appl., Oiseau* 2: 2–5.

Carié, P. 1930. Le *Leguatia gigantea* Schlegel (Rallidae) a-t-il existé? *Bull. Mus. Natl. Hist. Nat., Paris* (2)2: 204–13.

Carroll, J. B. 1982a. Rodrigues. *On the edge (Wildl. Preserv. Trust Newsl.)* 43: [3–4].

Carroll, J. B. 1982b ('1981'). The wild status and behaviour of the Rodrigues Fruit Bat *Pteropus rodricensis*. A report of the 1981 field study. *Dodo* 18: 20–9.

Carver, J. E. A. 1945. *Annual report of the Forest Department for the year 1944.* Colony of Mauritius: Government Printer.

Carver, J. E. A. 1948. *Annual report of the Forest Department for the year 1946.* Colony of Mauritius: Government Printer.

Cauche, F. 1651. *Relation du voyage que François Cauche de Rouen a fait à Madagascar, isles adjacentes et coste d'Afrique.* Receuilly par le sieur Morisot, avec notes en marge. Paris: Augustin Courbe. [An English translation appeared in Vol. 2 of John Stevens, ed. (1711), *A new collection of voyages and travels . . .*, and the entire French text was reprinted in Grandidier *et al.* (1903–20), q.v., Vol. 7]

Caulier, Père 1764. Fragments sur l'île Bourbon par le R. P. Caulier en 1764. [ed. A. Lougnon]. *Rec. Trim.* 3: 149–69 (1938).

Cazal, C. 1974. Une réserve de chasse. *Info-Nature Ile Réunion* 11: 56–8.

Céré, N. *c.* 1781. [Letter to P. Poivre, undated] published in *Rev. Hist. Litt. Ile Maurice* 4 (Archives Coloniales): 281–320+.

Challies, C. N. 1974. Trends in red deer (*Cervus elephas*) populations in Westland forests. *Proc. N.Z. Ecol. Soc.* 21: 45–50.

Chapin, J. P. 1948. Variation and hybridization among the Paradise Flycatchers of Africa. *Evolution* 2: 111–26.

Chapotin, C. 1812. *Topographie médicale de l'Ile de France.* Paris: Didot jeune.

Charpentier de Cossigny, Jean F. 1732–1755. Treize lettres de Cossigny à Réaumur [ed. A. Lougnon]. *Rec. Trim.* 4: 168–96, 205–82, 305–16 (1939–40).

Charpentier de Cossigny, Jean F. 1764. *Mémoire sur l'Ile de France.* In P. Crépin (1922), q.v.: 51–99.

Charpentier de Cossigny, Jos. F. An VII (1799). *Voyage à Canton, capitale de la province de ce nom, à la Chine: par*

Gorée, le Cap de Bonne Espérance et les Iles de France et de la Réunion. Paris: André.

Chazal, M. de & Baissac, J. de B. 1950. Etude sur la géologie de l'île Maurice. *Proc. Roy. Soc. Arts Sci. Mauritius* 1: 53–72.

Chazal, M. de & Baissac, J. de B. 1953. Façonnement de l'île Maurice par les facteurs géologiques. *Proc. Roy. Soc. Arts Sci. Mauritius* 1: 173–83.

Cheke, A. S. 1974a. *Report on Rodrigues.* London: BOU. Cyclostyled.

Cheke, A. S. 1974b (1975). *Proposition pour introduire à la Réunion des oiseaux rares de Maurice.* MS privately circulated. Published 1975 in *Info-Nature Ile Réunion* 12: 25–9.

Cheke, A. S. 1974c. *Interim report on the Mauritius Fody* Foudia rubra. London: BOU. Cyclostyled.

Cheke, A. S. 1975a. Official report on the introduction of the Mauritius Fody *Foudia rubra* to Réunion. Oxford: BOU Mascarene Islands Expedition. Cyclostyled.

Cheke, A. S. 1975b. Cyclone Gervaise – an eye-witness comments. *Birds Int.* 1: 13–14.

Cheke, A. S. 1975c. Suggestions on the management of *Cassia* coppice at Bras d'Eau, Mauritius in relation to the population of Coqs de Bois (*Terpsiphone bourbonnensis*). *BOU Mascarene Is. Exped., Conserv. Memo* 1. Cyclostyled.

Cheke, A. S. 1975d. Pourquoi le 'Tec-tec' est-il si variable? *Info-Nature Ile Réunion* 13: 38–9.

Cheke, A. S. 1975e. [Letter to the editor about deer at the Plaine des Chicots.] *Info-Nature Ile Réunion* 13: 128–9.

Cheke, A. S. 1975f. Un lézard malgache introduit à la Réunion. *Info-Nature* Ile Réunion 12: 94–6.

Cheke, A. S. 1976 (1977). Rapport sur la distribution et la conservation du Tuit-tuit, oiseau rarissime de la Réunion. *BOU Mascarene Is. Exped., Conserv. Memo* 2. Cyclostyled. Reprinted 1977 in *Info-Nature Ile Réunion* 15; 21–42.

Cheke, A. S. 1978a. A summary of A. S. Cheke's recommendations for the conservation of Mascarene vertebrates. *BOU Mascarene Is. Exped., Conserv. Memo* 3. Reprinted 1978 (French translation) in *Info-Nature Ile Réunion* 16: 69–83.

Cheke, A. S. 1978b. Habitat management for conservation in Rodrigues. *BOU Mascarene Is. Exped., Conserv. Memo* 4.

Cheke, A. S. 1979a. The Rodrigues Fody *Foudia flavicans*. A brief history of its decline, and a report on the 1978 expedition. *Dodo, J. Jersey Wildl. Preserv. Trust* 15: 12–19.

Cheke, A. S. 1979b. The threat to the endemic birds of Rodrigues (Indian Ocean) from the possible introduction of Ship Rats *Rattus rattus* from vessels coming alongside the proposed new wharf at Port Mathurin. *BOU Mascarene Is. Exped., Conserv. Memo* 5.

Cheke, A. S. 1980. Urgency and inertia in the conservation of endangered island species, illustrated by Rodrigues. In *Proceedings of the 4th Pan-African Ornithological Congress*, pp. 355–9.

Cheke, A. S. 1982a ('1981'). A note on *Phelsuma* Gray 1825 of the Agalega Islands, Indian Ocean. *Senckenbergiana Biol.* 62: 1–3.

Cheke, A. S. 1982b. *Les noms créoles des oiseaux dans les îles francophones de l'Ocean Indien.* Paris: Inst. Int. Ethnosci (collection 'L'Homme et son Milieu').

Cheke, A. S. 1983a. Status and ecology of the Mauritius Fody *Foudia rubra*, an endangered species. *Natl. Geog. Soc. Res. Rep.* 15: 43–56.

Cheke, A. S. 1983b. The identity of Buffon's *Grand Traquet*, and other nomenclatural problems in eighteenth century descriptions of endemic Mascarene birds. *Bull. Br. Orn. Club* 103: 95–100.

Cheke, A. S. 1983c. A note on the *Album of a hundred birds* by Kono Bairei, a nineteenth century Japanese artist, with new light on the 'Avis Indica' of Collaert. *Arch. Nat. Hist.* 11(2): 291–7.

Cheke, A. S. 1984. Lizards of the Seychelles. In Stoddart (1984), q.v.: 331–60 (ch. 19).

Cheke, A. S. & Dahl, J. F. 1981. The status of bats on western Indian Ocean islands, with special reference to *Pteropus*. *Mammalia* 45: 205–38.

Cheke A. S., Gardner, T. [A. M.], Jones, C. G., Owadally, A. W. & Staub, [J. J.] F. 1984. Did the Dodo do it? *Animal Kingdom* 87(1): 4–6.

Cheke, A. S. & Lawley, J. C. 1984 ('1983'). Biological history of Agalega with special reference to birds and other land vertebrates. *Atoll Res. Bull.* 273: 65–107.

Clancey, P. A., Lawson, W. J. & Stuart Irwin, M. P. 1969. The Mascarene Martin *Phedina borbonica* (Gmelin) in Moçambique: a new species to the South African list. *Ostrich* 40: 5–8.

Clark, G. 1859. A ramble round Mauritius with some excursions in the interior of that island; to which is added a familiar description of its fauna and some subjects of its flora. In Palmer & Bradshaw (compilers) *The Mauritius register: historical, official & commercial, corrected to the 30th June 1859.*, pp. i–cxxxii. Port Louis, Mauritius: L. Channell.

Clark, G. 1866. Account of the late discovery of Dodos' remains in the island of Mauritius. *Ibis* (2)2: 141–6.

Clouet, M. 1976. La Papangue ou Busard de Maillard. *Info-Nature Ile Réunion* 14: 39–44.

Clouet, M. 1978. Le Busard de Maillard (*Circus aeruginosus maillardi*) de l'île de la Réunion. *Oiseau Rev. Fr. Orn.* 48: 95–106.

Coblentz, B. E. 1978. The effects of feral goats (*Capra hircus*) on island ecosystems. *Biol. Conserv.* 13: 279–86.

Coe, M. J., Bourn, D. & Swingland, I. R. 1979. The biomass, production and carrying capacity of giant tortoises on Aldabra. *Phil. Trans. Roy. Soc., Lond.* 286B: 163–76.

Collar, N. J. & Stuart, S. N. 1985. *Threatened birds of Africa and related islands. The ICBP/IUCN Red Data Book*, part 1. Cambridge: ICBP & IUCN.

Coode, M. J. E. 1979. 68: Burseracées. In Bosser et at. (1978–), q.v.

Cooper, J. E. 1978. *Veterinary aspects of captive birds of prey.* Saul, Gloucestershire: Standfast Press.

Cooper, J. E. 1979a. An oviduct adenocarcinoma in a Mauritius Kestrel. *Avian Pathol.* 7: 651–8.

Cooper, J. E. 1979b. Veterinary care of wild birds. *Anim. Regulation Stud.* 2: 21–9.

Cooper, J. E., Jones, C. G. & Owadally, A. W. 1981. Morbidity and mortality in the Mauritius Kestrel (*Falco punctatus*). In J. E. Cooper & A. G. Greenwood (eds.) *Recent advances in the study of raptor diseases*, pp. 31–5. UK: Chiron Publications.

Coquerel, C. 1864. Catalogue des animaux qui se rencontrent à la Réunion: Oiseaux. *Bull. Soc. Acclim. Hist. Nat. Ile Réunion* 2(1): 7–27.

Coquerel, C. 1865a. Des animaux perdus qui habitaient les îles Mascareignes. In Roussin (1860–7), q.v., 3: 74–86.

Coquerel, C. 1865b. Les fringilliens. In Roussin (1860–7), q.v., 3: 169–73.

Coquerel, C. 1867a. L'hirondelle des blés et la salangane. In Roussin (1860–7), q.v., 4: 17–24.

Coquerel, C. 1867b. Le serin et le moutardier. In Roussin (1860–7), q.v., 4: 169–72.

[Corby, T.] 1845. [Letter to J. A. Lloyd, HM Surveyor General in Mauritius, dated 30.10.1845, on the subject of Rodrigues.] MS in Public Record Office, London: CO 167, vol. 262. [The MS is unattributed, but the ship's log, HMS *Conway*, confirms the author's identity. See also North-Coombes (1971)]

Cormack, R. M. 1968. The statistics of capture–recapture methods. *Oceanogr. Mar. Biol. Ann. Rev.* 6: 455–506.

Cossigny: see Charpentier de Cossigny.

Courrier de Bourbon, log of the 1721. Extraits du Journal de Navigation du *Courrier de Bourbon* (1720–1722). *Rev. Retrospective Ile Maurice* (1952) 3: 149–66.

Courtenay-Latimer, M. 1953. A Dodo egg. *S. Afr. J. Sci.* 49: 208–10.

Cowles, G. S. in press. Extinct avifauna of the Mascarene Islands, the fossil record revised. *Bull. Brit. Mus. Nat. Hist.* (*Zool.*).

Cowper, S. G. 1984. Birds from the Mascarene Islands in the collections of the Merseyside County Museums. *Mauritius Inst. Bull.* 10: 7–14.

Cramp, S. & Simmons, K. E. L. (eds.) 1977, 1980. *The birds of the Western Palearctic.* Vol. 1: *Ostriches–Ducks.* Vol. 2: *Hawks–Bustards.* Oxford: Oxford University Press.

Crémont, H. de 1768. Relation du voyage au volcan de l'Isle de Bourbon les 27 et 28 octobre 1768 par Monsieur de Crémont, commissaire ordonnateur en la dite isle [ed. A. Lougnon]. *Rec. Trim.* 3: 15–18+ (1937).

Crépin 1887. Sur la Caille de Madagascar. *Bull. Soc. Natl. Acclim.*, Paris (4)4: 240–1.

Crépin, P. 1922. Charpentier de Cossigny, Fonctionnaire Colonial, d'après ses écrits et ceux de quelques-uns de ses contemporains. Thesis, Université de Paris, Faculté des Lettres.

Crépin, P. 1931. Paul Carié 1876–1930. *Bull. Soc. Natl. Acclim. France* 78: 302–4.

Crook, J. H. 1961. The Fodies (Ploceinae) of the Seychelles Islands. *Ibis* 103a: 517–48.

Cuvier, G. 1829. *Le règne animal . . .* 5 vols. Paris: Deterville & Crochard.

Dalrymple, A. c.1755. L'Ile de France vers 1755 [ed. A. Lougnon]. *Rec. Trim.* 1: 299–322, 359–70, 398–421, 458–72 (1934).

Darby, P. 1983a. Watching the Fody. *On the edge (Wildl. Preserv. Trust Newsl.)* 45: 4.

Darby, P. 1983b. A first for the fodies. *On the edge (Wildl. Preserv. Trust Newsl.)* 46: 8.

Daruty de Grandpré, A. 1878. [On *Margaroperdix striata*.] *Trans. Roy. Soc. Arts Sci. Mauritius*, N.S. 10: 55.

Daruty de Grandpré, A. 1883a. Rapport annuel du secrétaire, 5 fèvrier 1879. *Trans. Soc. Roy. Arts Sci. Mauritius*, N.S. 12: 134–47.

Daruty de Grandpré, A. 1883b. Rapport annuel du secrétaire, 3 fèvrier 1881. *Trans. Soc. Roy. Arts Sci. Mauritius*, N.S. 13: 72–86.

Das, A. K. 1973. The Dodo and the Mughals. *History Today* 23: 60–3.

Davies, N. B. & Green, R. E. 1976. The development and ecological significance of feeding techniques in the Reed Warbler *Acrocephalus scirpaceus*. *Anim. Behav.* 24: 213–29.

Davy, E. G. *c.*1971. *Tracks and intensities of tropical cyclones in the south-west Indian Ocean, 1939–1970.* Vacoas, Mauritius: Mauritius Meteorological Services.

Day, D. 1981. *The doomsday book of animals.* London: Ebury Press.

Decary, R. 1962. Sur des introductions imprudentes d'animaux aux Mascareignes et à Madagascar. *Bull. Mus. Natl. Hist. Nat., Paris* (2) 34(5): 404–7.

Defos du Rau, J. 1960. *L'Ile de la Réunion.* Bordeaux: Institut de Géographie.

Deignan, H. G. 1942. Nomenclature of certain Pycnonotidae. *Auk* 59: 313–15.

Delacour, J. 1933. Le Foudi rouge et son élevage. *Oiseau Rev. Fr. Orn.* N.S. 3: 404–8.

Desjardins, J. 1829. Description physique de l'Ile d'Ambre. MS published in *Trans. Roy. Soc. Arts Sci. Mauritius*, N.S. 6: 35–41 (1872) as 'L'Ile d'Ambre il y a 42 ans'.

Desjardins, J. 1830. Premier rapport annuel sur les travaux de la Société d'Histoire Naturelle de l'Ile Maurice. MS published in *Trans. Roy. Soc. Arts Sci. Mauritius*, N.S. 6: 179–98 (1872) and in Ly-Tio-Fane (1972), q.v.: 1–23.

Desjardins, J. 1831a. Sur trois espèces de lézard du genre Scinque qui habitent l'île Maurice (Ile de France). *Ann. Sci. Nat.* 22: 292–9.

Desjardins, J. 1831b. Deuxième Rapp. Ann. Trav. Soc. Hist. Nat. Ile Maurice. MS published in Ly-Tio-Fane (1972), q.v.: 25–51.

Desjardins, J. 1832. Troisième Rapp. Ann. Trav. Soc. Hist. Nat. Ile Maurice. MS published in Ly-Tio-Fane (1972), q.v.: 53–112.

Desjardins, J. 1834. Cinquième Rapp. Ann. Trav. Soc. Hist. Nat. Ile Maurice. MS Published in Ly-Tio-Fane (1972), q.v.: 139–66.

Desjardins, J. 1835. *Sixième Rapp. Ann. Trav. Soc. Hist. Nat. Ile Maurice.* [Mention of *Papangue* p. 28]

Desjardins, J. 1837. *Huitième Rapp. Ann. Trav. Soc. Hist. Nat. Ile Maurice.*

Desnoyers, J. L. 1837. Aperçu d'une topographie médicale de l'Ile Maurice. *Rapp. Ann. Trav. Soc. Hist. Nat. Ile Maurice* 8: 21–7.

Devillers, P. 1976– . Projet de nomenclature française des oiseaux du monde. *Gerfaut* 66: 153–68, 349–421; 67: 171–200, 337–65, 469–89; 68: 129–36, 233–40, 703–20; 70: 121–46. [Incomplete]

Diamond, A. W. 1976. Cousin Island nature reserve. Management Plan 1975–79. London: ICBP. Cyclostyled.

Diamond, A. W. 1980. Seasonality, population structure and breeding ecology of the Seychelles Brush Warbler *Acrocephalus sechellensis*. In *Proceedings of the Fourth Pan-African Ornithological Congress*, pp. 253–66.

Diamond, A. W. in press (1985). The conservation of land birds on islands in the tropical Indian Ocean. In P. J. Moors (ed) *Management of islands for bird conservation*, pp. 85–100. Cambridge: ICBP Tech. Pub. 3.

Diamond, A. W. & Feare, C. J. 1980. Past and present biogeography of central Seychelles birds. In *Proceedings of the Fourth Pan-African Ornithological Congress*, pp. 89–98.

Diamond, J. M. & Veitch, C. R. 1981. Extinctions and introductions in the New Zealand avifauna: cause and effect? *Science* 211: 499–501.

Dingwall, P. R., Atkinson, I. A. E. & Hay, C. (eds.) 1978. *The ecology and control of rodents in New Zealand nature reserves.* Wellington: Dept. of Lands and Survey, Information Series 4.

Dobson, G. E. 1879. [The collections from Rodriguez] Mammalia. *Phil. Trans. Roy. Soc., Lond.* 168: 457–8.

Douglas, M. J. W. 1982. Biology and management of rusa deer on Mauritius. *FAO/UN Tigerpaper* 9(3): 1–10.

Doyen, L. *c.*1870. *Notice sur l'ornithologie de l'Ile Maurice.* MS. Doyen Papers. Mauritius Institute, Port Louis.

Dubois 1674. *Les voyages faits par le sieur D.B. aux isles Dauphine ou Madagascar et Bourbon ou Mascarenne, ès années 1669, 70, 71 et 72.* Paris: Claude Barbin. 234pp. [Also English trans., see Oliver (1897). The section on Réunion is reprinted in full in Lougnon (1970), and the list of birds by, *inter alia*, Berlioz (1946), Barau (1980) and Barré & Barau (1982), q.v.]

Ducros, Père 1725 (1728). Lettre du Père Ducros, missionnaire de la Compagnie des Indes, à M. l'abbé Raguet, directeur de la Compagnie des Indes. *Lettres édifiantes* (q.v.), *Mémoires des Indes* 18: 1–26 [also 13: 320–40 (1781 edn), 7: 461–73 (1819 edn)].

Dumeril, A. M. C. & Bibron, G. 1834–54. *Erpétologie générale ou histoire naturelle complète des reptiles.* 9 vols. + atlas. Paris: Librairie Roret.

Duncan, J. 1857. Observations respecting the island of Rodrigues. In Mann (ed.) *Report on Rodrigues as a quarantine station*, pp. 11–14. Port Louis, Mauritius: Government Printer.

Dupon, J. F. 1969. *Recueil de documents pour servir à l'histoire de Rodrigues.* Port Louis: Mauritius Archives Publications. No. 10.

Dupon, J. F. 1978. Les facteurs humains de l'évolution de l'environnement dans les îles densement peuplées de l'Ocean Indien. In A. Lauret (ed) (1978), q.v.: 36–75.

Dupont, E. 1860. Moineaux et Corbeaux. *Trans. Roy. Soc. Arts Sci. Mauritius*, N.S. 1: 326–8.

Durrell, G. 1977a ('1976'). The Mauritian expedition. *Jersey Wildl. Preserv. Trust Ann. Rep.* 13: 7–11.

Durrell, G. 1977b. *Golden bats and pink pigeons.* London: Collins.

Durrell, G. 1978 ('1977'). The Honorary Director's report. *Dodo, J. Jersey Wildl. Preserv. Trust* 14: 3–8.

Durrell, G. 1984. The first release. *On the edge (Wildl. Preserv. Trust Newsl.)* 47: [3].

Durrell, G. & Durrell, L. M. 1980. Breeding Mascarene wildlife in captivity. *Int. Zool. Yearbk.* 20: 112–19.

Easteal, S. O. 1981. The history of introductions of *Bufo marinus*, a natural experiment in evolution. *Biol. J. Linn. Soc.* 16: 93–113.

Edgerley, L. F. 1957. *Annual report of the Forest Department for the year 1955.* Colony of Mauritius, No. 19 of 1957. Port Louis Mauritius: Government Printer.

Edgerley, L. F. 1961. *Annual Report of the Forest Department for the year 1960.* Colony of Mauritius, No. 24 of 1961. Cyclostyled.

Edgerley, L. F. 1962. The conifers in Mauritius. *Proc. Roy. Soc. Arts Sci. Mauritius* 2: 190–202.

Edwards, A. 1872. [Quelques renseignements au sujet d'une tortue . . .] *Trans. Roy. Soc. Arts Sci. Mauritius*, N.S. 6: 12–13.

Eisenberg, J. F. & Gould, E. 1970. The tenrecs: a study in mammalian behaviour and evolution. *Smithsonian Contrib. Zool.* 27.

Eisenberg, J. F. & Seidensticker, J. 1976. Ungulates in southern Asia: a consideration of biomass estimates for selected habitats. *Biol. Conserv.* 10: 293–308.

Elton, C. S. 1958. *The ecology of invasions by animals and plants.* London: Methuen.

Ely, C. A. & Clapp, R. B. 1973. The natural history of Laysan Island, northwest Hawaiian Islands. *Atoll Res. Bull.* 171.

d'Emmerez de Charmoy, A. 1941. Le Boul-boul (*Otocompsa jocosa peguensis* W. L. Sclater). *Rev. Agric. Ile Maurice* 20: 154–8.

d'Emmerez de Charmoy, D. 1914. Fauna. In Macmillan (1914), q.v.: 94–101.

d'Emmerez de Charmoy, D. 1928. Mauritius: the surviving fauna. *J. Soc. Preserv. Fauna Emp.* N.S. 8: 108–9.

Etienne, J. 1972. Lutte contre les mouches des fruits à la Réunion. *Info-Nature Ile Réunion* 8: 79–87.

Evans, P. G. H. 1979. Status and conservation of the Seychelles Black Parrot. *Biol. Conserv.* 16: 233–40.

Feare, C. J. 1978. The decline of Booby (Sulidae) populations in the western Indian Ocean. *Biol. Conserv.* 14: 295–305.

Feare, C. J., Temple, S. A. & Procter, J. 1974. The status, distribution and diet of the Seychelles Kestrel, *Falco araea. Ibis* 116: 548–51.

Fenton, M. B. 1975. Acuity of echolocation in *Collocalia hirundinacea* (Aves: Apodidae), with comments on the distribution of echolocating swiftlets and molossid bats. *Biotropica* 7: 1–7.

Feuilley 1705. Mission à l'Ile Bourbon du Sieur Feuilley en 1704. [ed. A. Lougnon]. *Rec. Trim.* 4: 3–56, 101–67 (1939). [The part pertaining to Réunion birds is reprinted in Barré & Barau (1982), q.v.]

Fisher, C. T. 1981. Specimens of endangered or rare birds in the Merseyside County Museums, Liverpool. *Bull. Br. Orn. Club* 101: 276–85.

Flinders, M. 1814. *A voyage to Terra Australis; undertaken for the purpose of completing the discovery of that vast country and prosecuted in the years 1801, 1802 and 1803 . . .* 3 vols. London: G. & W. Nicol.

Focard, [E.] V. 1861. L'Oiseau de la Vierge. In Roussin (1860–7), q.v.: 25–7.

Forbes, H. O. 1898. On an apparently new, and supposed to be now extinct, species of bird from the Mascarene Islands, provisionally referred to the genus *Necropsar. Bull. Liverpool Mus.* 1: 29–35.

Forbes-Watson, A. D. 1969. Notes on birds observed in the Comoros on behalf of the Smithsonian Institution. *Atoll Res. Bull.* 128: 23.

Forbes-Watson, A. [D.] 1973. *The surviving endemic birds of the Mascarenes.* Typescript MS. [Copy in Alexander Library, Edward Grey Institute, Oxford]

Forbes-Watson, A. D., Keith, G. S. & Turner, D. A. 1974. Mascarenes bird list. Nairobi: (privately circulated). Cyclostyled.

Forshaw, J. M. & Cooper, W. T. 1978. *Parrots of the world*, 2nd edn. Newton Abbott: David & Charles.

Foucher d'Obsonville 1783. *Essais philosophiques sur les moeurs de divers animaux étrangers, avec des observations relatives aux principes & usages de plusieurs peuples ou extraits des voyages de M*** en Asie.* Paris: Coutourier fils & Veuve Tilliard et fils.

Foucherolle, Soullet & David c.1714. Mémoire pour faire porter à l'isle de Bourbon des plants de café de Moka. Instruction pour la prise de possession de Cirne ou Maurice. *Rev. Rétrospective Ile Maurice* 5: 131–3 (1954).

Fox, N. C. 1977. The biology of the New Zealand Falcon (*Falco novaeseelandiae* Gmelin 1788). Thesis, Canterbury University, N.Z.

Fox, N. C., Fox, B. & Bailey, T. 1985. The predation ecology of the Mauritius Kestrel (*Falco punctatus*). Unpublished report to the Peregrine Fund, Cornell University, New York, and ICBP, Cambridge, England.

Frauenfeld, G. von 1868. *Neu aufgefundene Abbildung des Dronte und eines zweiten kurzflügeligen Vogels . . .* Vienna: C. Ueberreuter'sche Buchdrückerei. [Abridged versions also in *Ibis* (2)4: 480–2 and *J. Orn.* 16: 138–40, both 1868]

Fréri 1751. [A visit to the *Volcan* on Bourbon] [ed. A. Lougnon]. *Rec. Trim.* 3: 8–12 (1937).

Friedmann, F. 1981. 116: Sapotacées. In Bosser et al. (1978–), q.v.

Friedmann, F. & Guého, J. 1977. Guide des principales plantes indigènes de l'Ile Rodrigue. *Rev. Agric. Sucrière Ile Maurice* 56(1): 1–19.

Friedmann, H. 1956. New light on the Dodo and its illustrators. *Smithsonian Rep.* for 1955: 475–81.

Frith, C. B. 1976. A twelve-month field study of the Aldabran Fody *Foudia eminentissima aldabrana*. *Ibis* 118: 155–78.

Frith, C. B. 1979. Feeding ecology of land birds on West Island, Aldabra Atoll, Indian Ocean: a preliminary survey. *Phil. Trans. Roy. Soc., Lond.* 286B: 195–210.

Froberville, E. de 1848. Rodrigues, Galéga, Les Sechelles, Les Almirantes etc. In d'Avézac (ed.) *Iles de l'Afrique*, part III, *Iles Africaines de la Mer des Indes*, pp. 65–114. Paris: Didot Frères.

Froidevaux, H. 1899. Textes historiques inédits ou peu connus relatifs aux tortues de terre de l'île Bourbon. *Bull. Mus. Hist. Nat., Paris* 5: 214–18.

Gallagher, M. D. 1960. Bird notes from Christmas Island, Pacific Ocean. *Ibis* 102: 489–502.

Gandon, Abbé 1732. Les Mascareignes vues par l'abbé Gandon en 1732 [ed. A. Lougnon]. *Rec. Trim.* 5: 74–83 (1940).

Garnett, M. C. 1977. The effect of giant tortoises (*Geochelone gigantea*) on the vegetation and the density of nesting seabirds on Cousin Island, Seychelles. *Cousin Is. Res. Station Tech. Rep.* 24. Cambridge: ICBP.

Garnier du Fougeray 1722. Lettre du chevalier Garnier du Fougeray aux directeurs de la Compagnie des Indes. Groix, le 22 mars 1722. *Rev. Retrospective Ile Maurice* 6: 67–80 (1955).

Gennes de la Chancelière 1735. Observations sur les Iles de Rodrigue et de France en mars 1735 [ed. A. Lougnon]. *Rec. Trim.* 1: 201–38 (1933).

Geoffroy, L. 1790. Ile Plate [short section from Geoffroy's journal]. *Trans. Roy. Soc. Arts Sci. Mauritius*, N.S. 6:9 (1872).

Geoffroy Ste. Hilaire, E. 1806. Note sur quelques habitudes de la grande chauve-souris de l'île de France, connue sous le nom de roussette. *Ann. Mus. Hist. Nat., Paris* 7: 227–230.

Geoffroy Ste. Hilaire, E. & Cuvier, G. F. 1824–47. *Histoire naturelle des Mammifères.* 4 vols. Paris: Muséum d'Histoire Naturelle.

Geoffroy Ste. Hilaire, I. 1827. Mémoire sur quelques espèces nouvelles ou peu connues du genre Musaraigne. *Mém. Mus. Hist. Nat., Paris* 15: 117–44.

Gérard, G. & Picot, H. 1975. Protection d'un environnement insulaire contre des maladies. *Info-Nature Ile Réunion* 12: 15–24.

Gibbons, J. 1984. Iguanas of the South Pacific. *Oryx* 18: 82–91.

Gill, F. B. 1964. [Notes on Réunion birds] Typescript MS [copy in Alexander Library, Edward Grey Institute, Oxford].

Gill, F. B. 1967. Birds of Rodriguez Island (Indian Ocean). *Ibis* 109: 383–90.

Gill, F. B. 1970. The taxonomy of the Mascarene Olive White-eye *Zosterops olivacea* (L.). *Bull. Br. Orn. Club* 90: 81–2.

Gill, F. B. 1971a. Ecology and evolution of the sympatric Mascarene white-eyes *Zosterops borbonica* and *Zosterops olivacea*. *Auk* 88: 35–60.

Gill, F. B. 1971b. Endemic landbirds of Mauritius Island, Indian Ocean. Typescript MS [copy in Alexander Library, Edward Grey Institute, Oxford].

Gill, F. B. 1972. Causerie de Franck [*sic*] B. Gill [transcript of untitled talk given in Réunion in 1967]. *Info-Nature Ile Réunion* 6: 8–10.

Gill, F. B. 1973a. Passé, présent et avenir des oiseaux des forêts de l'Ile de la Réunion. *Info-Nature Ile Réunion*, No. spécial hors série, 'La Forêt': 62–4.

Gill, F. B. 1973b. Intra-island variation in the Mascarene White-eye *Zosterops borbonica*. *Orn. Monogr.* 12.

Gill, F. B., Jouanin, C. & Storer, R. W. 1970. Notes on the seabirds of Round Island, Mauritius. *Auk* 87: 514–21.

Gleadow, F. 1904. *Report on the forests of Mauritius with a preliminary working plan.* Port Louis, Mauritius: Government Printer. [There was a 'revised' (actually censored) edition in 1906.]

Gomy, Y. 1973a. L'insecte et la forêt. *Info-Nature Ile Réunion*, No. spécial hors série, 'La Forêt': 65–76.

Gomy, Y. 1973b. Voyage en île d'amertume. *Info-Nature Ile Réunion* 9: 72–99.

Goodwin, D. 1955. Notes on European wild pigeons. *Avicult. Mag.* 61: 54–85.

Goodwin, D. 1959. Taxonomy of the genus *Columba*. *Bull. Br. Mus. (Nat.Hist.) Zool.* 6(1).

Goodwin, D. 1967. *Pigeons & doves of the world.* London: Trustees of The British Museum (Nat. Hist.).

Goodwin, D. 1983. *Pigeons and doves of the world*, 3rd edn. New York: Cornell/London: British Museum (Natural History).

Gould, M. S. & Swingland, I. R. 1980. The tortoise and the goat: interactions on Aldabra Island. *Biol. Conserv.* 17: 267–79.

Gould, S. J. 1980. *The panda's thumb. More reflections in natural history.* New York & London: W. W. Norton & Co. [Also Harmondsworth: Penguin Books, 1983]

Grandidier, A. *et al.* 1903–20. *Collection des ouvrages anciens concernant Madagascar.* 9 vols. Paris: Comité de Madagascar.

Grant, C. 1801. *The History of Mauritius or the Isle of France and the neighbouring islands from their first discovery to the present time, composed principally from the papers and memoirs of Baron Grant, . . ., by his son Charles Grant, Viscount de Vaux.* London: W. Bulmer & Co.

Grant, P. R. 1966. Ecological compatibility of bird species on islands. *Am. Nat.* 100: 451–62.

Greathead, D. J. 1971. A review of biological control in the Ethiopian Region. *Tech. Comm. Commonw. Inst. Biol. Control* 5.

Greenway, J. C. 1967. *Extinct and vanishing birds of the world*, 2nd edn. New York: Dover.

Greenwood, P. J., Harvey, P. H. & Perrins, C. M. 1978. Inbreeding and dispersal in the Great Tit. *Nature* 271: 52–4.

Grieg-Smith, P. W. 1979a. Observations of nesting and group behaviour of Seychelles White-eyes *Zosterops modesta*. *Ibis* 121: 344–8.

Grieg-Smith, P. W. 1979b. Notes on the biology of the Seychelles Bulbul. *Ostrich* 50: 45–58.

Griffin, D. R. & Suthers, R. A. 1970. Sensitivity of echolocation in cave swiftlets. *Biol. Bull.* 139: 495–501.

Griffin, J. 1978. *Ecology of the feral pig on the island of Hawaii.* State of Hawaii Dept. of Land & Natural Resources, Division of Fish & Game.

Gruchet, H. 1973. Note sur la conservation de la nature intéressant plus spécialement la Réunion. *Info-Nature Ile Réunion* 10: 29–49.

Gruchet, H. 1975. Oiseaux nuisibles à la Réunion. *Info-Nature Ile Réunion* 13: 91–3.

Gruchet, H. 1976. Etude d'une petite colonie de Salangane, *Collocalia francica*, au 'Trou Z'Armand' dans la Ravine St-Gilles. *Info-Nature Ile Réunion* 14: 45–7.

Gruchet, H. 1977. Les causes du déboisement de la région ouest aride de la Réunion. *Info-Nature Ile Réunion* 15: 67–70.

Guého, J. & Staub, [J. J.]F. 1979. 'Le Mondrain' nature reserve. *Proc. Roy. Soc. Arts Sci. Mauritius* 5: 75–102.

Guérin, R. 1940–53. *Faune ornithologique ancienne et actuelle des iles Mascareignes, Seychelles, Comores et des iles avoisinantes.* 3 vols. Port Louis, Mauritius: General Printing & Stationery Co.

Guibé, J. 1949. Révision des Boidés de Madagascar. *Mem. Inst. Sci. Madagascar* A 3(1): 95–105.

Guibé, J. 1958. Les serpents de Madagascar. *Mem. Inst. Sci. Madagascar* A.12: 189–260.

Guignes, C. L. J. de 1808. *Voyages à Peking, Manille et l'Ile de France faits dans l'intervalle des années 1784 à 1801.* 3 vols + atlas. Paris: Imprimerie Imperiale.

Günther, A. 1869 (1870). [Letter to L. Bouton, dated 3.11.1869, on various zoological specimens.] *Trans. Roy. Soc. Arts Sci. Mauritius,* N.S. 4: 139 (1870).

Günther, A. 1874 (1875). [Letter to L. Bouton, dated 5.3.1874, on various zoological specimens.] *Trans. Roy. Soc. Arts Sci. Mauritius,* N.S. 8: 87 (1875).

Günther, A. 1877. Notice of two large extinct lizards formerly inhabiting the Mascarene Islands. *Zool. J. Linn. Soc.* 13: 321–7. [Reprinted in: *Trans. Roy. Soc. Arts Sci. Mauritius,* N.S. 11: 110–15 (1883)]

Günther, A. 1879. [The collections from Rodriguez] Reptiles. *Phil. Trans. Roy. Soc., Lond.* 168: 470.

Günther, A. & Newton, E. 1879. The extinct birds of Rodriguez. *Phil. Trans. Roy. Soc., Lond.* 168: 423–37.

Hachisuka, M. 1937. Revisional note on the Didine birds of Réunion. *Proc. Biol. Soc., Washington* 50: 69–72.

Hachisuka, M. 1953. *The Dodo and kindred birds, or the extinct birds of the Mascarene Islands.* London: H. F. & G. Witherby.

Hall, B. P. & Moreau, R. E. 1970. *An atlas of speciation in African Passerine birds.* London: British Museum.

Halliday, T. R. 1978. *Vanishing birds.* London: Sidgwick & Jackson.

Hamann, O. 1979. Regeneration of vegetation on Santa Fe and Pinta Islands, Galapagos, after the eradication of goats. *Biol. Conserv.* 15: 215–36.

Hancock, J. & Elliott, H. [F. I.] 1978. *The herons of the world.* London: London Editions.

Hardancourt 1712. Etat présent de l'Isle de Bourbon [annotated by A. Lougnon]. *Rec. Trim.* 1: 71–8 (1932).

Harrison, C. J. O. & Walker, C. A. 1978. Pleistocene bird remains from Aldabra Atoll, Indian Ocean. *J. Nat. Hist.* 12: 7–14.

Hart, log of the 1629. Log of the *Hart* (June–July 1629). *Rev. Retrospective Ile Maurice* 1: 325.

Hartlaub, G. 1877. *Die Vögel Madagascars und der benachbarten Inselgruppen.* Halle: H. W. Schmidt.

Hartley, J. R. M. 1978 ('1979'). The Mauritius Pink Pigeon *Columba mayeri. Dodo, J. Jersey Wildl. Preserv. Trust* 14: 23–6.

Hartley, J. [R. M.] 1979. Fodies from Rodrigues. *Wildl. Preserv. Trust, Jersey, Newsl.* 34: [3–4].

Hartley, J. [R. M.] 1984. Our mission in Mauritius. *On the edge (Wildl. Preserv. Trust Newsl.)* 47: [4–6].

Hébert, G. 1708. Rapport de G. Hébert sur l'Ile Bourbon en 1708 avec les apostilles de la Compagnie des Indes [ed. A. Lougnon]. *Rec. Trim.* 5: 34–73 (1940).

d'Héguerty, P. A. 1754. Discours prononcé devant le roi de Pologne Stanislas le 26 mars 1751 sur l'île Bourbon. *Mém. Soc. Roy. Sci. Belles Lettres Nancy* 1: 73–91.

Heim de Balsac, H. & Heim de Balsac, M-H. 1956. *Suncus murinus* (L.) à la Réunion et en Nouvelle Guinée. Considérations sur le commensalisme et la vie domiciliaire des Soricidées. *Nat. Malgache* 8: 143–7.

Heinzel, H., Fitter, R. S. R. & Parslow, J. F. 1972. *The birds of Britain and Europe, with North Africa and the Middle East.* London: Collins.

Hemsley, W. B. 1899. *Calvaria major,* 'Tambalocoque'. *Hooker's Icon. Plant.* 2512.

Henri 1865. Note sur l'acclimatation du moineau à l'île de la Réunion. *Mém. Soc. Imp. Sci. Nat. Cherbourg* 11: 252–6.

Herbert, T. 1634, 1638, etc. *A relation of some yeares' travaile, begunne Anno 1626, into Afrique and the greater Asia, especially the territories of the Persian Monarchie, and some parts of the Orientall Indies and Iles adiacent.* London: W. Stansby & J. Bloome. [Much of the text was expanded and revised in later editions, especially that of 1638]

Higgin, E. 1849. Remarks on the contrary, products, and appearance of the island of Rodriguez, with opinions as to its future colonisation. *J. Roy. Geog. Soc.* 19: 17–20.

High, J. 1976. *Natural history of the Seychelles.* Port Victoria, Seychelles: [Government Printer].

Hill, A. W. 1941. The genus *Calvaria,* with an account of the stony endocarp and germination of the seed, and description of a new species. *Ann. Bot.,* N.S. 5: 587–606.

Hill, J. L. 1974. *Peromyscus:* effects of early pairing on reproduction. *Science* 186: 1042–4.

Hnatiuk, R. J., Woodell, S. R. J. & Bourn, D. M. 1976. Giant tortoise and vegetation interactions on Aldabra. II. Coastal. *Biol. Conserv.* 9: 305–16.

Hnatiuk, S. H. 1978. Plant dispersal by the Aldabran giant tortoise *Geochelone gigantea* (Schweigger). *Oecologia* 36: 345–50.

Hoart, C. T. 1825. Observations générales sur l'île Rodrigue. MS, Mauritius Archives LH45. [Published in Mann (ed.) *Report on Rodrigues as a quarantine station,* pp. 22–7. Port Louis, Mauritius, 1857]

Hoffman, J. C. 1680. *Oost-Indianische Voyage . . .* [also in French]. Cassel: (author). [Reprinted by M. Nijhoff, The Hague, 1931. Section on Mauritius reproduced in Grandidier *et al.* (1903–20), q.v.: vol. 3 pp. 368–80]

Hoffstetter, R. 1946a. Remarques sur la classification des Ophidiens et particulièrement des Boidae des Mascareignes (Bolyerinae subfam. nov.). Bull. Mus. Natl. Hist. Nat., Paris (2)18: 132–5.

Hoffstetter, R. 1946b. Sur les Gekkonidae fossiles. Bull. Mus. Natl. Hist. Nat., Paris (2)18: 195–203.

Hoffstetter, R. 1946c. Les Typhlopidae fossiles. Bull. Mus. Natl. Hist. Nat., Paris (2)18: 309–15.

Hoffstetter, R. 1949. Les reptiles subfossiles de l'île Maurice. I. Les Scincidae. Ann. Paleontol. 35: 45–72.

Holloway, C. W. 1968. The protection of man-made forests from wildlife. In World symposium on man-made forests and their industrial importance, pp. 687–715. Rome: FAO.

Holthuis, L. B., Muller, H. E. & Smeenk, C. 1971. Vogels op Nederlandse 17de eeuwse tegels naar gravures van Adriaen Collaert en iets over Leguatia gigantea. Bull. Mus. Boymans van Beuningen 21(1): 3–19.

Holyoak, D. T. 1971. Comments on the extinct parrot Lophopsittacus mauritianus. Ardea 59: 50–1.

Holyoak, D. T. 1973. An undescribed extinct parrot from Mauritius. Ibis 115: 417–18.

Honegger, R. E. 1975. Amphibia and Reptilia. Red Data Book 3. Morges, Switzerland: IUCN.

Honegger, R. [E.] 1981. List of amphibians and reptiles either known or thought to have become extinct since 1600. Biol. Conserv. 19: 141–58.

Hooker, J. D. & Günther, R. A. (eds.) 1879. An account of the petrological, botanical and zoological collections made in Kerguelen's Land and Rodriguez, during the Transit of Venus Expeditions, carried out by Her Majesty's Government during the years 1874–75. Phil. Trans. Roy. Soc., Lond. 168: 1–579.

Horn, B. K. P. 1978. Dodo apocrypha. Science News, Washington 113(2): 19.

Horne, J. 1887. Notes on the flora of Flat Island. Trans. Soc. Roy. Arts Sci. Mauritius, N.S. 19: 116–51.

Horne, J. F. M. 1975. Mascarene Islands endemic bird vocalizations recording project: report on La Réunion for the Office National des Forêts. London: BOU. Cyclostyled.

Horne, J. [F. M.] 1979. Recording for the BOU Mascarene Islands Project. Rec. Sound 74/75: 74–7.

Howard, W. E. 1964. Introduced browsing mammals and habitat stability in New Zealand. J. Wildl. Manage. 28: 421–9.

Howard, W. E. 1965 (1966). Control of introduced mammals in New Zealand. N.Z. Dept. of Scientific and Industrial Research, Information Series 45.

Howard, W. E. 1967. Ecological changes in New Zealand due to introduced mammals. In Proceedings and Papers of the IUCN 10th Technical Meeting, part 3, sect. 2, IUCN Publications, N.S. no. 9, pp. 219–40.

Hutson, A. M. 1975. Observations on the birds of Diego Garcia, Chagos Archipelago, with notes on other vertebrates. Atoll Res. Bull. 175.

Imber, M. J. 1975. Petrels and predators. Bull. ICBP 12: 260–3.

Imber, M. J. 1978. The effects of rats on the breeding success of petrels. In Dingwall et al. (1978), q.v.: 67–72.

International Commission for Zoological Nomenclature 1974. Amendments to the International Code of Zoological Nomenclature adopted since the XVI International Congress of Zoology, Washington 1963. Bull. Zool. Nomencl. 31: 77–89.

Ivanov (='Iwanow'), I. 1958. An Indian picture of the Dodo. J. Orn. 99: 438–40.

'J. C.' (1861): see under 'C'.

Jacob de Cordemoy, E. 1895. Flore de l'Ile de la Réunion. Paris: Klincksieck.

Jadin, B. & Billiet, F. 1980. Observations ornithologiques à la Réunion. Gerfaut 69: 339–52.

Jeggo, D. 1978a ('1977'). Preliminary notes on the Mauritius Pink Pigeon at the Jersey Zoological Park. Dodo, J. Jersey Wildl. Preserv. Trust 14: 26–30.

Jeggo, D. 1978b. The Mauritius Pink Pigeon at the Jersey Zoological Park. Avicult. Mag. 84: 112–14.

Jeggo, D. 1979. Breeding the Mauritius Pink Pigeon at Jersey Zoological Park. Avicult. Mag. 85: 1–2.

Jerningham, H. E. H. 1892. [On the cyclone of April 29th, 1892] Blackwood's Mag. [Sept. 1892]. [Reprinted in Macmillan (1914), q.v.: 61–5]

Johnson, E. D. H. 1971. Observations on a resident population of Stonechats in Jersey. Br. Birds 64: 201–13, 267–79.

Johnson, N. K. & Peters, H. J. 1963. The systematic position of certain hawks in the genus Buteo. Auk 80: 417–46.

Jolly, G. M. 1965. Explicit estimates from capture–recapture data with both death and immigration – stochastic model. Biometrika 52: 225–47.

Joly, H. (ed.) 1977. Nomenclature de la faune et de la flore: latin, français, anglais. Afrique au sud du Sahara, Madagascar, Mascareignes. Paris: Hachette.

Jonchée de la Goleterie 1729. Mémoire envoyé à la Compagnie, le 19 avril 1729, par M. Jonchée de la Goleterie pour bien établir l'Ile de France, son gouvernement et un conseil supérieur. MS, copy in Doyen Collection, Mauritius Institute, Port Louis. [Extracts also published in Brouard (1963), q.v.]

Jones, C. G. 1979. Mauritius/Rodrigues conservation project report for January 22nd–June 20th 1979. Unpublished report to ICBP.

Jones, C. G. 1980a. The conservation of the endemic birds and bats of Mauritius and Rodrigues. A progress report and proposal for future activities. Washington: ICBP. Xeroxed.

Jones, C. G. 1980b. The conservation of the endemic birds and bats of Mauritius and Rodrigues. Progress report January 1980–November 1980. Mauritius: (author). Xeroxed.

Jones, C. G. 1980c. Parrot on the way to extinction. Oryx 15: 350–4.

Jones, C. G. 1980d. The Mauritius Kestrel. Hawk Trust Ann. Rep. 10: 18–29.

Jones C. G. 1981. Mauritius. Project 1082. Promotion of conservation [report for 1980]. World Wildl. Fund Yearbk. 1980–1: 206–12.

Jones, C. G. 1982. The conservation of the endemic birds and bats of Mauritius and Rodrigues. Annual Report [for] 1981, WWF Project 1082. Unpublished report to ICBP/WWF.

Jones, C. G. & Owadally, A. W. 1981. *The status, ecology and conservation of the Mauritius Kestrel*. Unpublished report submitted to ICBP.

Jones, C. G. & Owadally, A. W. 1982a. *Conservation priorities in Mauritius and Rodrigues*. Report submitted to ICBP. Xeroxed.

Jones, C. G. & Owadally, A. W. 1982b ('1981'). The world's rarest falcon: the Mauritius Kestrel. *Falconer* 7(5): 322–7.

Jones, C. G. & Owadally, A. W. 1985. The status, ecology and conservation of the Mauritius Kestrel. *In* I. Newton & R. D. Chancellor (eds.) *Conservation studies on raptors*, pp. 211–22. Cambridge: ICBP.

Jones, C. G., Steele, F. N. & Owadally, A. W. 1981. An account of the Mauritius Kestrel captive breeding project. *Avic. Mag.* 87: 191–207.

Jouanin, C. 1955. Une nouvelle espèce de Procellariidé. *Oiseau Rev. Fr. Orn.* 25: 155–61.

Jouanin, C. 1962. Inventaire des oiseaux éteints ou en voie d'extinction conservés au Muséum de Paris. *Terre-Vie* 109: 257–301.

Jouanin, C. 1964a ('1963'). Un pétrel nouveau de la Réunion, *Bulweria baraui*. *Bull. Mus. Natl. Hist. Nat., Paris* (2)35: 593–7.

Jouanin, C. 1964b. Notes sur l'avifaune de la Réunion. *Oiseau Rev. Fr. Orn.* 34: 83–4.

Jouanin, C. 1970a. Le Pétrel Noir de Bourbon, *Pterodroma aterrima* Bonaparte. *Oiseau Rev. Fr. Orn.* 40: 48–68.

Jouanin, C. 1970b. Note taxinomique sur les Petits Puffins, *Puffinus lherminieri*, de l'Océan Indien Occidental. *Oiseau Rev. Fr. Orn.* 40: 303–6.

Jouanin, C. & Gill, F. B. 1967. Recherche du Pétrel de Barau *Pterodroma baraui*. *Oiseau Rev. Fr. Orn.* 37: 1–19.

Jouanin, C. & Mougin, J.-L. 1979. Order Procellariiformes. In E. Mayr & G. W. Cottrell (eds.) *Check-list of Birds of the World*, 2nd edn, vol. 1, pp. 48–121. Cambridge, Mass.: Museum of Comparative Zoology.

Kaeppelin, P. 1908. *La Compagnie des Indes Orientales et François Martin*. Paris: Augustin Challamel.

Kemp, A. C. 1984. Preliminary description of the dynamics of a Greater Kestrel population. In J. M. Mendelsohn & C. W. Sapsford (eds.) *Proceedings of the second symposium on African predatory birds*, pp. 141–50. Durban: Natal Bird Club.

Kennedy, W. R. 1893. Notes on a visit to the islands of Rodriguez, Mauritius and Réunion. *J. Bombay Nat. Hist. Soc.* 7: 440–6.

Khan, S. A. (ed.) 1927. *John Marshall in India. Notes and observations in Bengal*. London: Oxford University Press.

King, H. C. 1945. *Pine plantations in Mauritius*. Working Plan. Port Louis, Mauritius: Government Printer.

King, H. C. 1946. *Interim report on indigenous species in Mauritius*. Colony of Mauritius, Forest Dept. Port Louis, Mauritius: Government Printer.

King, H. C. 1947. *Colony of Mauritius. Empire forests and the war*. Statement to British Empire Forestry Conference, 1947. Port Louis, Mauritius: Government Printer.

King, W. B. 1977. Notes on a conversation with David McKelvey, 31 August 1977. Typescript in ICBP files.

King, W. B. 1978–9 (1981). *Endangered birds of the world. The ICBP bird Red Data Book*. Washington: Smithsonian Press & ICBP.

King, W. B. 1980. Ecological basis of extinction in birds. *Acta Int. Orn. Cong.* 17: 905–11.

Kirkwood, J. K. 1981. Bioenergetics and growth in the Kestrel (*Falco tinnunculus*). Unpublished PhD thesis, University of Bristol.

Kluge, A. G. 1983. Cladistic relationships among Gekkonid lizards. *Copeia* [1983]: 465–75.

Koenig, P. 1895. *Suggestions for the management of the 'Grand Bassin' forests*. Mauritius [no place or publisher, but ? Port Louis, Mauritius: Government Printer].

Koenig, P. 1914a. Economic flora. In Macmillan (1914), q.v.: 102–9.

Koenig, P. 1914b. Report on forestry at Rodrigues. MS, Mauritius Archives SD166.

Koenig, P. 1924. *Annual report of the Forest Department for the year 1923*. Colony of Mauritius: Government Printer.

Koenig, P. 1926. *Annual report of the Forest Department for the year 1925*. Colony of Mauritius: Government Printer.

Koenig, P. 1932. Actes et Comptes-Rendus de la Société Royale des Arts et des Sciences de l'Ile Maurice. In *Centenaire de la Société Royale des Arts et des Sciences de l'Ile Maurice 1829–1929*, pp. 39–97. Port Louis, Mauritius: (Society).

La Caille, N. L., Abbé de 1763. *Journal Historique du Voyage fait au Cap de Bonne Espérance par feu M. l'Abbé de la Caille*. Paris: Guillyn.

Lafresnaye, F. 1850. Sur la nidification de quelques espèces d'oiseaux de la famille des Tisserins (Ploceinae). *Rev. Mag. Zool.* (2)2: 315–26.

Laissus, Y. 1978. Catalogue des manuscrits de Philibert Commerson (1727–1773) conservés à la Bibliothèque centrale du Muséum national d'Histoire Naturelle (Paris). *Rev. Hist. Sci., Paris* 31: 131–62.

La Motte, D. de 1754–7. *Voyages de Sieur DDLM contenant dix-neuf lettres écrites sur les lieux à un de ses amis en Europe dans les années 1754 à 1757*. MS (location unknown) of which the parts referring to the Mascarenes were published, ed. P. d'Aglosse (pseud.), in *Rev. Hist. Litt. Ile Maurice* 5: 205–9, 220–3, 231–5, 243–6, 257–60, 270–2, 284–8, 297–300, 303–7, 319–23 (1891).

Lamusse, M. J. R. 1958. The economic development of the Mauritius sugar industry. Unpublished thesis, University of Oxford [copy in Rhodes House Library].

Lane, C. 1946. *The laws of Mauritius in force on the 31st. day of July 1945*. 5 vols. London: Waterlow & Sons.

Lantz, A. 1887. Sur les mammifères et les oiseaux de l'île de la Réunion. *Bull. Soc. Nat. Acclim., Paris*. (4)4: 657–9.

La Nux, J.-B. de 1772. [Letter on *Roussettes* and *Rougettes*] Published apparently with only a few minor excisions, in Buffon (1776), q.v.: 253–62. [Reprinted in *Info-Nature Ile Réunion* 17: 35–41 (1979)]

La Roque, J. de 1716. *Voyage de l'Arabie Heureuse* . . . Paris: Cailleau. [La Merveille's account of Mauritius (pp. 171–85) is reprinted in Grandidier *et at.* (1903–20), q.v. 3: 320–6; his account of Réunion (pp. 185–211), *Ibid.* 3: 566–75 & Lougnon (1970), q.v.: 201–12.]

Lauret, A. ed. 1978*a*. *L'environnement dans les îles du sud-ouest de l'Océan Indien: Maurice–Réunion–Seychelles–Comores.* Travaux du Séminaire International organisé par l'ENDA et le Ministère de l'Agriculture et des Ressources Naturelles de l'Ile Maurice, Maurice 9–21 mai 1977. St.-Denis, Réunion: Fondation pour la recherche et le développement dans l'Océan Indien. *Documents et Recherches* 5.

Lauret, E. 1978*b*. Aspects géographiques, économiques et sociaux du programme général d'aménagement des hauts de la Réunion. In A. Lauret (ed.) (1978), q.v.: 77–90.

Laurie, A. 1983. Marine iguanas in Galapagos. *Oryx* 17: 18–25.

Lavergne, R. 1978. Les pestes végétales de l'Ile de la Réunion. *Info-Nature Ile Réunion* 16: 9–60.

Layard, E. L. 1867. *The birds of South Africa.* Capetown/London: J. C. Juta/Longman, Green & Co.

Leal, C. H. 1878. *Un voyage à la Réunion. Récits, souvenirs et anecdotes.* Port Louis, Mauritius: General Steam Printing Co.

Lebeau, A. & Lebrun, G. 1974. Elevage expérimental de la tortue marine *Chelonia mydas. Info-Nature Ile Réunion* 11: 7–27.

Lebel, Frère (attrib.) 1740. Lettre d'un frère de Saint-Lazare sur les paroisses de Bourbon en 1740 [MS, ed. A. Lougnon]. *Rec. Trim.* 3: 236–265+ (1938).

Legendre, M. 1929. La Huppe de la Réunion (*Fregilupus varius* Boddaert). *Rev. Hist. Nat. Appl., Oiseau* 10: 645–54, 729–43.

Le Gentil de la Barbinais, G. 1727. *Nouveau voyage autour du monde . . .* 3 vols. Paris: Flahault. [Also, apparently, Paris: Briasson, 1728 or 1729]

Le Gentil de la Galaisière, G. J. H. J.-B. 1779–81. *Voyage dans les mers de l'Inde fait par ordre du roi à l'occasion du passage de Vénus sur le disque du soleil, le 6 juin 1761 et le 3 du même mois 1769.* 2 vols. Paris: Imprimerie Royale.

Legras, A. 1861. Le Martin. In Roussin (1860–7), q.v., 2: 81–4.

Legras, [? C.] Dr 1863. De la Grande Caille de Madagascar et son acclimatation à l'île de la Réunion et en France. *Bull. Soc. Acclim. Hist. Nat.* 1: 75–6.

Leguat de la Fougère, F. 1708. *Voyage et avantures de François Leguat et de ses compagnons en deux isles désertes des Indes Orientales.* 2 vols. Amsterdam: J. L. de Lorme. [See also Oliver (1891)]

Lesouëf. 1975. *Proposal by Monsieur Lesouëf for a reserve in Rodrigues: restoration of the vegetation of Grande Montagne at Rodrigues.* Typescript report to IUCN Threatened Plant Committee [+ detailed comment by A. S. Cheke, *in litt.* to M. J. E. Coode, 14 Jan. 1976].

Lettres édifiantes et curieuses écrites des missions étrangères par quelques missionnaires de la Compagnie de Jésus. Many volumes. Various editions, the first spread over many dates, 1707–76: Paris: N. Leclerq. [Later definitive editions include one in 1781 (Paris: J. G. Merigot le Jeune) and another in 1819 (Lyon: J. Vernarel & E. Cabin)]

Levaillant 1801–5. *Histoire naturelle des perroquets.* 2 vols. Paris: Levrault.

Levi, W. M. 1945. *The pigeon,* 2nd. edn. Columbia, S. Carolina: R. L. Bryan Co.

Lewin, R. 1978. Gentle giants of the Galapagos. *New Scientist* 79: 334–6.

Liénard, F. 1843. Descriptions de reptiles, de poissons et de crustacés. *Rapp. Ann. Soc. Hist. Nat., Ile Maurice* 13 (for 1841–2): 55–7.

Lincoln, R. n.d. Philippe Eugène Jacob de Cordemoy (1835–1911). MS biographical note in the Archives de la Réunion.

Lionnet, [J. F.] G. 1972. *The Seychelles.* Newton Abbot: David & Charles.

List, R. J. 1951. *Smithsonian meteorological tables.* Washington: Smithsonian Institution.

Lloyd, D. E. B. 1973. Habitat utilisation by land birds of Cousin Island, and the Seychelles. [incomplete] *Cousin Is. Res. Station Tech. Rep.* 5. Cambridge: ICBP.

Lloyd, J. A. 1846. [Letter read to the society on 2.10.1845 on the subject of Round & Serpent Iss.] *P.V. Soc. Hist. Nat., Ile Maurice,* 6 Oct. 1842-28 Aug. 1845: 154–62.

Long, J. L. 1981. *Introduced birds of the world.* Sydney: A. H. & A. W. Reed.

Lougnon, A. (ed.) 1933–7. *Correspondence du conseil supérieur de Bourbon et de la Compagnie des Indes. 1724–1741.* St.-Denis, Réunion: (author).

Lougnon, A. 1953. *Documents concernant les îles de Bourbon et de France pendant la régie de la Compagnie des Indes, répertoire des pièces conservées dans divers dépôts d'archives à Paris.* Nérac, France: Archives de la Réunion.

Lougnon, A. 1956. *Classement et inventaire du fonds de la Compagnie des Indes (Série C°). 1665–1767.* Nérac, France: Archives de la Réunion.

Lougnon, A. 1957. *L'Ile Bourbon pendant la Régence.* Paris: Larose.

Lougnon, A. 1970. *Sous le signe de la tortue. Voyages anciens à l'Ile Bourbon (1611–1725),* 3rd edn. St.-Denis, Réunion: (author).

Lovejoy, T. E. 1978. Genetic aspects of dwindling populations. A review. In S. A. Temple (1978*a*). q.v.: 275–9.

Loveridge, A. 1947. Revision of the African lizards of the family Gekkonidae. *Bull. Mus. Comp. Zool., Harvard* 98: 1–469.

Loveridge, A. 1951. A new gecko of the genus *Gymnodactylus* from Serpent Island. *Proc. Biol. Soc., Washington* 64: 91–2.

Low, R. 1980. *Parrots. Their care and breeding.* Poole: Blandford Press.

Ly-Tio-Fane, M. 1958. *Mauritius and the spice trade. The odyssey of Pierre Poivre.* Mauritius Archives Publication 4. Port Louis, Mauritius: Esclapon.

Ly-Tio-Fane, M. 1968. Problèmes d'approvisionnement de l'Ile de France au temps de l'Intendant Poivre. *Proc. Roy. Soc. Arts. Sci. Mauritius* 3: 101–15.

Ly-Tio-Fane, M. (ed.) 1972. *Société d'Histoire Naturelle de l'Ile Maurice. Rapports annuels, I-V, 1830–1834.* Port Louis, Mauritius: Royal Society of Arts and Sciences of Mauritius.

Ly-Tio-Fane, M. 1978 ('1976'). *Pierre Sonnerat 1748–1814. An account of his life and work*. Mauritius: (author).

Lyon, log of the 1722. Log of the *Lyon*, 1722. *Rev. Retrospective Ile Maurice* 4: 133–4 (1953).

Macdonald, R. A. 1979. The biology of the Seychelles Cave Swiftlet *Aerodramus (francicus) elaphrus*. In P. A. Racey (ed.) *Aberdeen University expedition to the Seychelles 1977 Report*, pp. 92–112. Aberdeen: Dept of Zoology (Xeroxed).

McDougall, I. & Chamalaun, F. H. 1969. Isotopic dating and geomagnetic polarity studies on volcanic rocks from Mauritius, Indian Ocean. *Geol. Soc. A. Bull.* 80: 1419–42.

McDougall, I., Upton, B. G. J. & Wadsworth, W. J. 1965. A geological reconnaissance of Rodriguez Island, Indian Ocean. *Nature* 206: 26–7.

MacFarland, C. G., Villa, J. & Toro, B. 1974. The Galapagos giant tortoises *(Geochelone elephantopus)*, I: Status of the surviving populations. II: Conservation methods. *Biol. Conserv.* 6: 118–33, 198–212.

McKelvey, S. D. 1976. A preliminary study of the Pink Pigeon. *Mauritius Inst. Bull.* 8: 145–75.

McKelvey, S. D. 1977*a*. Observations on the Mauritian Pink Pigeon. *Birds Int.* 2: 36–8.

McKelvey, S. D. 1977*b*. The Meller's Duck on Mauritius. Its status in the wild and captive propagation. *Game Bird Breeders', Avicult. Zool. Conserv. Gaz.* 26(5): 11–13.

McKelvey, S. D. 1977*c*. The Mauritian Kestrel. *Hawk Trust Ann. Rep.* 8: 19–21.

McKelvey, S. D. 1978. The Mauritius Kestrel. *Wildlife 1978*: 46–51.

Macmillan, A. (ed.) 1914. *Mauritius illustrated*. London: W. H. & L. Collingridge.

Magon de Saint-Elier, F. 1839. *Tableaux historiques, politiques et pittoresques de l'Ile de France, aujourd'hui Maurice, depuis sa découverte jusqu'a nos jours*. Port Louis, Mauritius: (author).

Mahé de la Bourdonnais, B. F. 1740. *Mémoire des isles de France et de Bourbon* . . . [ed. A. Lougnon]. Published 1937, Paris: Gaston Daudé.

Maillard, L. 1862. *Notes sur l'Ile de la Réunion*. Paris: Dentu. [Also 2nd edn in 1863, 2 vols.]

Maisels, F. 1979. An attempt to evaluate predation on the Seychelles fruit bat *Pteropus seychellensis* by man. In P. A. Racey (ed.) *Aberdeen University Expedition to the Seychelles 1977. Report*, pp. 49–54. Aberdeen: Dept of Zoology (Xeroxed).

Mamet, J. R. 1979. Chronology of events in the control and eradication of malaria in Mauritius. *Rev. Agric. Sucrière Ile Maurice* 58: 107–46.

Mamet, J. R. & Webb-Gebert, F. 1980. Insects and other arthropoda of medical interest in Mauritius. *Mauritius Inst. Bull.* 9: 53–73.

Manders, N. 1911. An investigation into the validity of Müllerian and other forms of mimicry, with special reference to the islands of Bourbon, Mauritius and Ceylon. *Proc. Zool. Soc. Lond.* [1911]: 696–749.

Manly, B. F. J. & Parr, M. J. 1968. A new method of estimating

population size, survivorship and birth rate from capture–recapture data. *Trans. Soc. Br. Ent.* 18: 81–9.

Mann, J. R. 1860. Observations on the water supply of Mauritius. *Trans. Roy. Soc. Arts. Sci. Mauritius*, N.S. 2: 63–81.

Mannick, A. R. 1979. *Mauritius: the development of a plural society*. Nottingham: Spokesman Books.

Marragon, P. 1795. Mémoire sur l'Isle de Rodrigue. MS preserved in the Mauritius Archives TB 5/2. [Partially printed in Dupon (1969), q.v.]

Marragon, P. 1803. [Letter addressed to the 'Citoyens Administrateurs Généraux des Etablissements Français à l'Est du Cap de Bonne Espérance à l'Isle de France' dated '19 Thermidor an 10' (i.e. 6 August 1803)] MS preserved in the Mauritius Archives TB 5/1.

Mason, G. E. 1907. On an extinct undescribed fruit-bat of the genus *Pteropus* from the Mascarenes. *Ann. Mag. Nat. Hist.* (7)20: 220–2.

Mauduyt, P. J. E. 1784. Histoire naturelle des oiseaux. In *Encyclopédie Méthodique*, vol. 1, pp. 321–91 and vol. 2. Paris & Liège: Panckoucke & Plomteux.

Maynard-Smith, J. 1966. *The theory of evolution*. Harmondsworth: Penguin Books.

Mayr, E. 1971. New species of birds described from 1956 to 1965. *J. Orn.* 112: 302–16.

Mayr, E., Paynter, R. A., Jr & Traylor, M. A. 1968. Estrildidae. In Peters, J. L. & successors, *Check-list of birds of the world*, vol. 14, pp. 306–90. Cambridge, Mass.: Harvard University Press.

Mayr, E. & Vuilleumier, F. 1984 ('1983'). New species of birds described from 1966–1975. *J. Orn.* 124: 217–32.

Medway, Lord 1962*a*. The swiftlets *Collocalia* of Niah Cave, Sarawak. *Ibis* 104: 45–66, 228–45.

Medway, Lord 1962*b*. The relation between the reproductive cycle moult and changes in the sub-lingual salivary glands of the swiftlet *Collocalia maxima* Hume. *Proc. Zool. Soc. Lond.* 138: 305–15.

Medway, Lord 1966. Field characters as a guide to the specific relations of Swiftlets. *Proc. Linn. Soc. Lond.* 177: 151–72.

Medway, Lord 1969. Studies on the biology of the edible-nest swiftlets of South-east Asia. *Malay. Nat. J.* 22: 57–63.

Medway, Lord & Pye, D. 1977. Echolocation and the systematics of swiftlets. In B. Stonehouse & C. M. Perrins (eds.) *Evolutionary ecology*, pp. 225–38. London: Macmillan.

Mees, G. F. 1969. A systematic review of the Indo-Australian Zosteropidae: III. *Zool. Verhandl.* 102: 1–390.

Meinertzhagen, R. 1910–11. Diary [the 1910 & 1911 volumes cover his stay in Mauritius]. MS preserved in the library of Rhodes House, Oxford.

Meinertzhagen, R. 1912. On the birds of Mauritius. *Ibis* (9)6: 82–108.

Meldrum, C. 1868. *On the rainfall of Mauritius*. Mauritius: Observatory Dept. [Not seen; figure for forest acreage quoted by L. Bouton, *Ann. Rep.* [Roy. Soc. Arts. Sci. Mauritius] for 1868: xvi (1869, appended to *Trans. Roy. Soc. Arts. Sci. Mauritius* 3(2))]

Melville, D. S. 1979. Madagascar Fodies *Foudia madagascariensis* feeding on nectar of *Lantana camara*. *Ibis* 121: 361–2.

Melville, R. 1973. L'importance scientifique et le potentiel économique de la flore de la Réunion. *Info-Nature Ile Réunion*, No. spécial hors série, 'La Forêt': 56–9.

Mertens, R. 1934. Die Insel Reptilien ihre Ausbreitung, Variation und Artbildung. *Zoologica, Stuttgart* 32(84): 1–209.

Mertens, R. 1966. Die nichtmadagassischen Arten und Unterarten der Geckonengattung *Phelsuma*. *Senckenbergiana Biol.* 47: 85–110.

Merton, D. V., Morris, R. B. & Atkinson, I. A. E. 1984. Lek behaviour in a parrot: the Kakapo *Strigops habroptilus* of New Zealand. *Ibis* 126: 277–83.

Merton, L. F. H., Bourn, D. M. & Hnatiuk, R. J. 1976. Giant tortoise and vegetation interactions on Aldabra. I. Inland. *Biol. Conserv.* 9: 293–304.

Michel, C. 1972. [Annexe II] Zoologie. In Ly-Tio-Fane (1972), q.v.: 174–82.

Michel, C. 1981. *Notes on the birds of Mauritius*. Reduit, Mauritius: Mauritius Institute of Education.

Miguet, J.-M. 1957. Mise en valeur et régénération de la forêt de tamarins des hauts en zone d'altitude à la Forêt de Bélouve à la Réunion. *Rev. For. Fr.* 8: 285–310.

Miguet, J.-M. 1973. Forêt et équilibre biologique. *Info-Nature Ile Réunion*, No. Spécial hors série, 'La Forêt': 48–55.

Miguet, J.-M. 1980. Régénération et reconstitution de forêts naturelles à l'île de la Réunion. *Rev. Ecol. (Terre Vie)* 34: 3–22.

Milbert, J. G. 1812. *Voyage pittoresque à l'Ile de France, au Cap de Bonne Espérance et à l'Ile de Ténériffe*. 2 vols. + atlas. Paris: A. Nepveu.

Milies, H. C. 1868. Over eene nieuw ontdekte afbeelding van den Dodo (*Didus ineptus* L.). *Natuurk. Verh. K. Akad. Wet., Amsterdam* 11: 1–20.

Mills, J. A. & Williams, G. R. 1978. The status of endangered New Zealand birds. In M. J. Tyler (ed.) *The status of endangered Australasian wildlife*, pp. 147–68. Adelaide: Royal Zoological Society of South Australia.

Milne-Edwards, A. 1867a. Mémoire sur une espèce éteinte du genre *Fulica*. *Ann. Sci. Nat. (Zool.)* (5)8: 195–220.

Milne-Edwards, A. 1867b. Mémoire sur un Psittacien fossile de l'Ile Rodrigue. *Ann. Sci. Nat. (Zool.)* (5)8: 145–56.

Milne-Edwards, A. 1868 (1869). Researches into the zoological affinities of the bird recently described by Herr von Frauenfeld under the name of *Aphanapteryx imperialis*. *Ibis* N.S. 5: 256–275. [Trans. from French original in *Ann. Sci. Nat. (Zool)* (5)10: 325–46 (1868)]

Milne-Edwards, A. 1873 ('1874'). Recherches sur la faune ancienne des Iles Mascareignes. *Ann. Sci. Nat. (Zool.)* (5)19: Art. 3. [And in A. Milne-Edwards, 1873. *Recherches sur la faune ornithologique éteinte des Iles Mascareignes et de Madagascar*, pp. 117–47. Paris: Masson]

Milne-Edwards, A. & Grandidier, A. 1879. Oiseaux. In A. & G. Grandidier (eds.) *Histoire physique, naturelle et politique de Madagascar*, vol. 12 (text) & vols. 13–15 (plates). Paris: Imprimerie Nationale.

Milne-Edwards, A. & Oustalet, E. 1893. Notice sur quelques espèces d'oiseaux actuellement éteintes qui se trouvent représentées dans les collections du Muséum d'Histoire Naturelle. In *Centenaire de la fondation du Muséum d'Histoire Naturelle*, pp. 190–252. Paris: Muséum d'Histoire Naturelle.

Milon, P. 1951. Notes sur l'avifaune actuelle de l'île de la Réunion. *Terre Vie* 98: 129–78.

Milon, P., Petter, J.-J. & Randrianasolo, G. 1973. *Faune de Madagascar. XXXV. Oiseaux*. Tananarive: ORSTOM.

Ministry of Economic Planning & Development. 1980. *Two-year plan for economic and social development 1980–1982*. Port Louis, Mauritius: Government Printer.

Mitchell, B., Staines, B. W. & Welch, D. 1977. *Ecology of red deer: A research review relevant to their management in Scotland*. Cambridge: Institute of Terrestrial Ecology.

Miyadi, D. 1967. Differences in social behaviour among Japanese Macaque troops. In G. Fischer (ed.) *First Congress of the International Primatology Society*, pp. 228–31.

Montaggioni, J. 1970. Essai de reconstitution paléogéographique de l'Ile Rodrigue (Archipel des Mascareignes, Océan Indien). *C. R. Hebd. Acad. Sci., Paris* 271: 1741–4.

Montaggioni, J. 1973. Histoire géologique de l'Ile Rodrigue. *Info-Nature Ile Réunion* 9: 52–9.

Montaggioni, J. 1975. Histoire géologique des récifs coralliens de l'archipel des Mascareignes. *Cah. Centre Univ. Ile Réunion* 6: 97–110.

Moors, P. J. 1983. Predation by mustelids and rodents on the eggs and chicks of native and introduced birds in Kowhai Bush, New Zealand. *Ibis* 125: 137–54.

Moreau, R. E. 1957a. Variation in western Zosteropidae. *Bull. Br. Mus. (Nat. Hist.), Zool.* 4(7): 311–433.

Moreau, R. E. 1957b. The names of the Mascarene Olive White-eyes. *Bull. Br. Orn. Club* 77: 7–8.

Moreau, R. E. 1960. The ploceine weavers of the Indian Ocean Islands. *J. Orn.* 101: 29–49.

Moreau, R. E. 1966. *The bird faunas of Africa and its islands*. London & New York: Academic Press.

Moreau, R. E. 1967. Family Zosteropidae, African and Indian Ocean taxa. In Peters J. L. & successors (1931–), q.v. 12: 327–37.

Moreau, R. E., Perrins, M. & Hughes, J. H. 1969. Tongues of the Zosteropidae (white-eyes). *Ardea* 57: 29–47.

Morel 1778. Sur les oiseaux monstrueux nommés *Dronte, Dodo, Cygne capuchoné, solitaire, & Oiseau de Nazare, &* sur la petite isle de Sable, à 50 lieues environ de Madagascar. *Obs. Phys. Hist. Nat. & Arts* 12: 154–7.

Morel, L. 1861a. L'oiseau vert. In Roussin (1860–7), q.v. 2: 10–12.

Morel, L. 1861b. Le sénégali. In Roussin (1860–7), q.v. 2: 8–120.

Morony, J. J., Bock, W. J. & Farrand, J. 1975. *Reference list of the birds of the world*. New York: American Museum of Natural History.

Mortensen, T. 1934. On François Leguat and his 'Voyage et Avantures', with remarks on the dugong of Rodriguez and on *Leguatia gigantea* Schlegel. *Ardea* 23: 67–77.

Moulin, P. & Miguet, J.-M. 1968. L'Office National des Forêts au service de l'homme dans un département d'outre-mer. Unpaginated reprint from *La Revue Française*, suppl. to no. 213.

Mountfort, G. R. 1970. The need for research concerning endangered species. *Ibis* 112: 445–7.

Moutou, F. 1979. Les mammifères sauvages de l'île de la Réunion. *Info-Nature Ile Réunion* 17: 25–34.

Moutou, F. 1980. *Enquête sur la faune murine dans le département de la Réunion*. Réunion: DDASS.

Moutou, F. 1981. Les mammifères sauvages de l'île de la Réunion, notes complémentaires. *Info-Nature Ile Réunion* 18: 29–42.

Moutou, F. 1982a. Quelques aspects de la forêt et de la faune autochtones de l'Ile de la Réunion. *C. R. Soc. Biogeog.* 58(1): 3–20.

Moutou, F. 1982b. Note sur les chiroptères de l'île de la Réunion (Ocean Indien). *Mammalia* 46: 35–51.

Moutou, F. 1983a. Les peuplements des vertebrés terrestres des îles Mascareignes. *Rev. Ecol. (Terre Vie)* 37: 21–35.

Moutou, F. 1983b. Introduction dans les îles: l'exemple de l'île de la Réunion. *Info-Nature, Ile Réunion* 20: 39–48.

Moutou, F. 1983c. Identification des reptiles réunionnais + Carte de répartition de 3 espèces de reptiles réunionnais. *Info-Nature Ile Réunion* 20: 53–62, 63–4.

Moutou, F. 1984. Wildlife on Réunion. *Oryx* 18: 160–2.

Mueller, H. C. 1971. Displays and vocalizations of the Sparrow Hawk. *Wilson Bull.* 83: 249–54.

Müller, H. J. 1965. [Investigations regarding the evaluation of the economically acceptable density of game stocking in forests according to the damage caused by game to the site.] *Arch. Forstw.* 14: 533–61.

Mundy, P. 1608–67. *The Travels of Peter Mundy in Europe and Asia*. [Published in 5 vol., 1905–36, ed. R. C. Temple, London: Hakluyt Society]

Mungroo, Y. 1979. *Report of post-Celine II survey of the endemic Passeriformes, and bats and sea birds of Rodrigues*. Curepipe, Mauritius: Forest Department. Cyclostyled.

Murie, J. 1874. On the skeleton and lineage of *Fregilupus varius*. *Proc. Zool. Soc. Lond.* [1874]: 474–88.

Murphy, R. C. 1951. The populations of the Wedge-tailed Shearwater (*Puffinus pacificus*). *Am. Mus. Novit.* [1951]: 1512.

Murton, R. K. & Isaacson, A. J. 1962. The functional basis of some behaviour in the Wood-pigeon *Columba palumbus*. *Ibis* 104: 503–21.

Nelson, J. B. 1971. The biology of Abbott's Booby *Sula abbotti*. *Ibis* 113: 429–67.

Nelson, J. B. 1974. The distribution of Abbott's Booby *Sula abbotti*. *Ibis* 116: 368–9.

Nelson, J. B. 1978. *The Sulidae. Gannets and boobies*. Oxford: Oxford University Press.

Nelson, R. W. 1977. *Behavioral ecology of coastal Peregrines* (Falco peregrinus pealei). PhD thesis, University of Alberta, Calgary, Canada.

Newlands, W. A. 1975. *Mauritius Conservation Project. Monthly Reports*. February–October 1975. Cyclostyled.

[Newton, A.] 1861a. [Editorial note on specimens sent from Mauritius by Edward Newton] *Ibis* 3: 115–16.

Newton, A. 1861b. Description of a new species of Water-hen (*Gallinula*) from the island of Mauritius. *Proc. Zool. Soc. Lond.* [1861]: 18–19.

Newton, A. 1865a. On two new birds from the island of Rodriguez. *Proc. Zool. Soc. Lond.* [1865]: 46–8.

Newton, A. 1865b. [Editorial note] *Ibis*, N.S. 1: 530.

[Newton, A.] 1868b. Recent ornithological publications. 2. Dutch; 3. German. [Publications reviewed and abstracted by the editor, A. Newton] *Ibis* N.S. 4: 476–82.

Newton, A. 1872. On an undescribed bird from the island of Rodriguez. *Ibis* (3)2: 30–4.

Newton, A. 1875a. [On some unpublished sketches of the Dodo and other extinct birds of Mauritius] *Proc. Zool. Soc. Lond.* [1875]: 349–50. [Reprinted in *Trans. Roy. Soc. Arts Sci. Mauritius*, N.S. 11: 107–9 (1883)]

Newton, A. 1875b. Birds. *Enc. Br. (Ed. 9)* 3: 699–778. [Birds recently extirpated: 732–5]

Newton, A. 1875c. A note on *Palaeornis exsul*. *Ibis* (3)5: 343–4.

Newton, A. 1875d. Additional evidence as to the original fauna of Rodrigues. *Proc. Zool. Soc. Lond.* [1875]: 39–43.

Newton, A. 1876. On the species of *Hypsipetes* inhabiting Madagascar and the neighbouring islands. *Rowley's Orn. Miscellany* 2: 41–52.

Newton, A. 1907a. Leguat's giant bird. *Ornis* 14 (*Proc. 4th Int. Orn. Cong., Lond.*): 70–1.

Newton, A. 1907b. Books, letters and papers exhibited in the Philosophical Library by permission of the Committee, with notes by Professor Newton, 20th June 1905. *Ornis* 14 (*Proc. 4th Int. Orn. Congr., Lond.*): 72–84.

Newton, A. & Newton, E. 1870 ('1869'). On the osteology of the Solitaire or Didine bird of the island of Rodrigues, *Pezophaps solitaria* (Gmel.). *Phil. Trans. Roy. Soc., Lond.* 159: 327–62. [Abstract in *Proc. Roy. Soc., Lond.* 16: 428–33 (1868)]

Newton, A. & Newton, E. 1876. On the Psittaci of the Mascarene Islands. *Ibis* (3)6: 281–9.

Newton, A. & Gadow, H. 1896. *A dictionary of birds*. London: A. & C. Black.

Newton, E. 1861a. Ornithological notes from Mauritius. I. A visit to Round Island. *Ibis* 3: 180–2.

Newton, E. 1861b. Ornithological notes from Mauritius. II. A ten days sojourn at Savanne. *Ibis* 3: 270–7.

Newton, E. 1863. Notes of a second visit to Madagascar. *Ibis* 5: 333–50, 452–61.

Newton, E. 1865a. Notes of a visit to the island of Rodriguez. *Ibis*, N.S. 1: 146–53.

Newton, E. 1865b. [Address to the Society] In L. Bouton, Annual Report [of the Secretary]. *Trans. Roy. Soc. Arts Sci. Mauritius*, N.S. 2: 225.

Newton, E. 1871. [Letter on "Perdrix rouges"] *Trans Roy. Soc. Arts Sci. Mauritius*, N.S. 5: 51.

Newton, E. 1875 [Address to the Society] In L. Bouton, Report of the Secretary to the Royal Society of Arts and Sciences, 3 May 1873 to 10 October 1874, pp. xxii–xxiii. *Trans Roy. Soc. Arts Sci. Mauritius*, N.S. 8: i–xxiii+.

Newton, E. 1878a (1883). Discours de l'Hon. E. Newton [delivered to the Royal Society of Arts and Sciences, 6 February 1878]. *Trans. Roy. Soc. Arts Sci. Mauritius*, N.S. 11: 137–41.

Newton, E. 1878b (1883). [Letter to the Hon V. Naz, dated

26.2.1878] *Trans. Soc. Roy. Arts Sci. Mauritius*, N.S. 12: 70–3.

Newton, E. 1888. Address by the President, Sir Edward Newton K.C.M.G., F.L.S., C.M.Z.S., to the members of the Norfolk and Norwich Naturalists Society. *Trans. Norfolk & Norwich Nat. Soc.* 4: 537–54.

Newton, E. & Clark, J. W. 1879. On the osteology of the Solitaire (*Pezophaps solitaria*, Gmel.). *Phil. Trans. Roy. Soc., Lond.* 168: 438–51.

Newton, E. & Gadow, H. 1893. On additional bones of the Dodo and other extinct birds of Mauritius obtained by Mr. Théodore Sauzier. *Trans. Zool. Soc. Lond.* 13: 281–302.

Newton, I. 1966. The moult of the Bullfinch. *Ibis* 108: 41–67.

Newton, I. 1979. *Population ecology of raptors*. Berkhamsted: T. & A. D. Poyser.

Newton, R. 1956. Bird islands of Mauritius. *Ibis* 98: 296–302.

Newton, R. 1958a. Ornithological notes on Mauritius and the Cargados Carajos Archipelago. *Proc. Roy. Soc. Arts Sci. Mauritius* 2: 39–71.

Newton, R. 1958b. Bird Preservation in Mauritius. *Bull. ICBP* 7: 182–5.

Newton, R. 1959. Notes on the two species of *Foudia* in Mauritius. *Ibis* 101: 240–3.

Ng, F. S. P. 1983. Ecological principles of tropical rain forest conservation. In Sutton, S. L., Whitmore, T. C. & Chadwick, A. C. (eds.), 1983, *Tropical rain forest: ecology and management*. Br. Ecol. Soc. Special Publ. Ser. 2. Oxford: Blackwell Scientific Publications.

Nichelson, W. 1780. Mr Nichelson's account of a passage from Madras, to and from the island Diego Rayes, or Rodrigues, in the south-west monsoon. CCXVII. A description of the island Rodrigues. In S. Dunn (ed) *A new directory for the East Indies*, 5th edn, pp. 272–6. London: Henry Gregory.

Nieber, E. 1973. Geographical and ecological differentiation in the genus *Circus*. Thesis, Free University, Amsterdam.

Nieuhoff, J. 1682 (1752). John Nieuhoff's voyages. In Churchill's *Voyages and Travels*, 2nd edn, vol. 2, pp. 1–305. London: Thomas Osborne.

Noël, C. 1974. *Report of the National Food Production Committee*. Port Louis, Mauritius. Cyclostyled.

Norman, F. I. 1975. The murine rodents *Rattus rattus, exulans* and *norvegicus* as avian predators. *Atoll. Res. Bull.* 182.

North-Coombes, A. 1971. *The Island of Rodrigues*. Port Louis, Mauritius: (author).

North-Coombes, A. 1980a ('1979'). *The vindication of François Leguat*. Port Louis, Mauritius: Service Bureau.

North-Coombes, A. 1980b. *Le découverte des Mascareignes par les Arabes et les Portugais. Rétrospective et mise au point*. Port Louis, Mauritius: Service Bureau.

North-Coombes, A. 1983. François Leguat, le géant and the flamingo in the Mascarene Islands. *Proc. Roy. Soc. Arts Sci. Mauritius* 4(3): 1–30.

Nyon, de & Hauville, de 1722–3. [Correspondence between de Nyon in Grand Port and de Hauville in Port Louis, Mauritius] MSS in the Archives Nationales, Paris: Fonds des Colonies, Archives Anciennes, Col C4,1, ff. 13–46.

Oates, E. W., Reid, S. G. & Ogilvie-Grant, W. R. 1901–12. *Catalogue of the collection of birds' eggs in the British Museum*. 5 vols. London: Trustees [of the British Museum].

Oberholser, H. C. 1906. A monograph of the genus *Collocalia*. *Proc. Acad. Nat. Sci. Philadelphia* 53: 177–212.

O'Connor, R. 1976. Weight and body composition in nestling Blue Tits *Parus caeruleus*. *Ibis* 118: 108–12.

Olearius, A. 1670. De Beschryving der Reizen van Volkert Evertsz naar Oostindien. In J. H. Glazemaker (trans. & ed.) *Verhaal van due voorname Reizen naar Oostindien . . .*, pp. 91–130. Amsterdam: Jan Rieuwertsz & Pieter Arentsz.

Oliver, S. P., (ed.) 1891. *The voyage of François Leguat of Bresse to Rodriguez, Mauritius, Java, and the Cape of Good Hope, transcribed from the first English edition, edited and annotated by Capt. Pasfield Oliver*. 2 vols. London: Hakluyt Society. [Reprinted c.1974, New York: Burt Franklyn]

Oliver, S. P. (trans. & ed.) 1897. *The voyages made by the Sieur D.B. to the islands Dauphine or Madagascar and Bourbon or Mascarenne in the years 1669, 70, 71 and 72*. London: David Nutt.

Oliver, S. P. 1909. *The life of Philibert Commerson, D. M., Naturaliste du Roi. An old-world story of French travel and science in the days of Linnaeus* (ed. G. F. Scott Elliott). London: John Murray.

Oliver, W. D. 1896. *Crags and craters. Rambles in the island of Réunion*. London: Longman, Green & Co.

Olson, S. L. 1975. An evaluation of the supposed Anhinga of Mauritius. *Auk* 92: 374–6.

Olson, S. L. 1977. A synopsis of the fossil Rallidae. In Ripley (1977), q.v.: 339–73.

Oudemans, A. C. 1917. Dodo-Studien, naar aanleidingvan de vondst van een gerelsteen met Dodo-beeld van 1561 te Vere. *Verh. K. Akad. Wet. Amsterdam* 2nd. sec. 19(4).

Oustalet, E. 1897 ('1896'). Notice sur la faune ornithologique ancienne et moderne des Iles Mascareignes et en particulier de l'Ile Maurice. *Ann. Sci. Nat. Zool.* (8)3: 1–128.

Owadally, A. W. 1971a. *Annual report of the Forestry Service for the year 1969*. Mauritius, No. 13 of 1971. Port Louis, Mauritius: Government Printer.

Owadally, A. W. 1971b. *Annual report of the Forestry Service for the year 1970*. Mauritius, No. 22 of 1971. Port Louis, Mauritius: Government Printer.

Owadally, A. W. 1972. *Annual report of the Forestry Service for the year 1971*. Port Louis, Mauritius: Government Printer.

Owadally, A. W. 1973. Les forêts naturelles de l'Ile Maurice. *Info-Nature Ile Réunion*. No. spécial hors série, 'La Forêt': 88–94.

Owadally, A. W. 1974. *Annual report of Forestry Service for the year 1972*. Mauritius, No. 21 of 1974. Port Louis, Mauritius: Government Printer.

Owadally, A. W. 1975. *Annual report of the Forestry Service for* the year 1973. Mauritius, No. 8 of 1975. Port Louis, Mauritius: Government Printer.

Owadally, A. W. 1976a. *A guide to the Royal Botanic Gardens, Pamplemousses.* Port Louis, Mauritius: Government Printer.

Owadally, A. W. 1976b. *Annual report of the Forestry Service for the year 1974.* Mauritius, No. 6 of 1976. Port Louis, Mauritius: Government Printer.

Owadally, A. W. 1977. The Dodo and the Tambalacoque tree. Typescript MS. [Copy in Alexander Library, Edward Grey Institute, Oxford; greatly shortened version published as Owadally (1979), q.v.]

Owadally, A. W. 1979. The Dodo and the Tambalacoque tree. *Science* 203: 1363–4. [With a reply by S. A. Temple, p. 1364]

Owadally, A. W. 1980. Some forest pests and diseases in Mauritius. *Rev. Agric. Sucrière Ile Maurice* 59: 76–94.

Owadally, A. W. 1981a. Mauritius. In E. J. Kormondy & J. F. McCormick (eds.) *Handbook of contemporary developments in world ecology*, pp. 457–67. Westport, Conn. & London: Greenwood Press.

Owadally, A. W. 1981b. *Annual report of the Forestry Service for the year 1980.* Mauritius, No. 27 of 1981. Port Louis, Mauritius: Government Printer.

Owadally, A. W. 1984. *Annual report of the Forestry Service for the year 1981.* Mauritius, No. 3 of 1984. Port Louis, Mauritius: Government Printer.

Owadally, A. W. & Bützler, W. 1972. *The deer in Mauritius.* Port Louis, Mauritius: (authors).

Owen, R. 1866. Evidence of a species, perhaps extinct, of a large parrot (*Psittacus mauritianus*, Owen), contemporary with the Dodo in the island of Mauritius. *Ibis* (2)2: 168–71.

Padya, B. M. (ed). 1972. *Climate of Mauritius.* Vacoas, Mauritius: Mauritius Meteorological Services.

Padya, B. M. 1976. *Cyclones of the Mauritius region.* Port Louis: Mauritius Printing Co. & Government Printer.

Padya, B. M. 1984. *The climate of Mauritius*, 2nd edn. Port Louis, Mauritius: Government Printer. [This is really a completely new and different book rather than a '2nd edition']

Pasquier, R. 1980. *Report and management plan on ICBP's project for the conservation of forest birds of Mauritius.* Washington: ICBP. Xeroxed.

Pasquier, R. F. (ed.) (n.d.(1982)). *Conservation of New World parrots. Proceedings of the ICBP Working Group meeting, St. Lucia 1980.* Washington, DC: Smithsonian Institution Press, for ICBP.

Pasquier, R. [F.] & Jones, C. G. 1982. The lost and lonely birds of Mauritius. *Nat. Hist., New York* 91(3): 39–43.

Paulian, R. 1961. *La zoogéographie de Madagascar et des îles voisines.* Faune de Madagascar 13. Tananarive: Institut de Recherche Scientifique.

Peirce, M. A. 1979. Some additional observations on haematozoa of birds in the Mascarene Islands. *Bull. Br. Orn. Club* 99: 68–71.

Peirce, M. A., Cheke, A. S. & Cheke, R. A. 1977. A survey of blood parasites of birds in the Mascarene Islands,

Indian Ocean, with descriptions of two new species and taxonomic discussion. *Ibis* 119: 451–61.

Penny, M. 1974. *The birds of Seychelles and the outlying islands.* London: Collins.

Perrault, C. 1676. *Suite des mémoires pour servir à l'histoire naturelle des animaux*, pp. 92–205. Paris.

Perry, R. 1964. Santa Cruz tortoise reserve. *Notic. Galapagos* 4: 19–20.

Peters, J. L. & successors 1931– . *Checklist of birds of the world.* 15 vols. Cambridge, Mass.: Harvard University Press & Museum of Comparative Zoology.

Peterson, R. T., Mountfort, G. R. & Hollom, P. A. D. 1974. *A field guide to the birds of Britain and Europe*, 3rd edn. [in collab. with I. J. Ferguson-Lees & D. I. M. Wallace]. London: Collins.

Petit, F. P. du 1741. Description anatomique des yeux de la grenouille et de la tortue. *Mem. Acad. Roy. Sci., Paris* ['1737']: 142–69.

Petit-Radel, P. 1801. *De amoribus Pancharitis et Zoroae.* Paris: Didot. [The part of the introduction concerning the author's visit to Réunion, translated into French by F. Cazamian, was published under the title *Un Voyage a l'île Bourbon en 1794*, in Roussin (1860–7), q.v.]

Pijl, L. van der 1972. *Principles of dispersal in higher plants*, 2nd edn. Berlin etc.: Springer-Verlag.

Pike, N. 1870a. A visit to Round Island. *Trans. Roy. Soc. Arts Sci. Mauritius*, N.S. 4: 11–22.

Pike, N. 1870b. Notes on the fauna of Round Island with special reference to the prepared case sent to his excellency Sir Henry Barkly, K.C.B. *Trans. Roy. Soc. Arts Sci. Mauritius*, N.S. 4: 131–5.

Pike, N. 1873. *Subtropical rambles in the land of the Aphanapteryx. Personal experiences, adventures & wanderings in and around the island of Mauritius.* London: Sampson Low, Marston, Low & Serle.

Pingré, G. 1760–2. Relation de mon voyage de Paris à l'île Rodrigue. MS 1803, Bibliothèque Ste. Geneviève, Paris. [Day to day diary]

Pingré, G, c. 1763. *Voyage à l'île Rodrigue.* MS 1804, Bibliothèque Ste. Geneviève, Paris. [Edited and rearranged version of the diary, MS 1803 (1760–2). MS 1804 must date from 1763 at least as Pingré cites La Caille (1763), q.v.]

Pitot, A. 1899. *L'Ile de France. Equisses historiques (1715–1810).* Port Louis, Mauritius: Imprimerie Pezzani.

Pitot, A. 1905. *T'Eylandt Mauritius. Esquisses historiques (1598–1710). Précédées d'une notice sur la découverte des Mascareignes, et suivies d'une monographie du Dodo, des Solitaires de Rodrigue et de Bourbon et de l'oiseau Bleu.* Port Louis, Mauritius: Coignet frères & Cie.

Pitot, A. 1914. Extinct birds of the Mascarene Islands. In Macmillan (1914), q.v.: 82–93.

Piveteau, J. 1945. Etude sur l'*Aphanapteryx*, oiseau éteint de l'Ile Maurice. *Ann. Paleontol.* 31: 31–7.

Pollen, F. P. L. 1865a. Note sur l'*Oxynotus ferrugineus. Bull. Soc. Acclim. Hist. Nat. Ile Réunion* 3: 5–14.

Pollen, F. P. L. 1865b. Le tui-tuit. In Roussin (1860–7), q.v. 3: 193–4.

Pollen, F. P. L. 1866. On the genus *Oxynotus* of Mauritius and Réunion. *Ibis*, N.S. 2: 275–80.

Pollen, F. P. L. 1868. *Relation de Voyage*. Vol. 1 of *Recherches sur la faune de Madagascar et de ses dependences, d'après les decouvertes de François P. L. Pollen et D. C. van Dam*. Leiden: J. K. Steenhoff.

Poole, A. L. *et al.* 1959 (1972). Browsing animals in protective vegetation. *N.Z. For. Serv. Publicity Item* 43. [Five articles originally published in *N.Z. Sci. Rev.* 17(2), Apr. 1959]

Pope Hennessy, J. (ed.) 1886. *Letters from Mauritius in the Eighteenth Century by Grant, Baron de Vaux, including an account of Labourdonnais' capture of Madras with an introduction by Sir John Pope Hennessy*. Port Louis, Mauritius: 'Printed for private circulation'.

Pope Hennessy, J. 1889. Speech of his excellency the governor delivered at the annual meeting of the Royal Society of Arts and Sciences of Mauritius held at Reduit on the 28th January 1889. *C. R. Seanc. Ann. Soc. Roy. Arts' Sci. Maurice* [28.1.1889]: i–viii.

Pourquier, J. 1963. *Catalogue des parasites et parasitoses du bétail à la Réunion*. St.-Denis, Réunion: Min. Agriculture, Service veterinaire. Cyclostyled.

Prater, S. H. 1965. *A book of Indian animals*, 2nd edn. Bombay: Bombay Natural History Society.

President, log of the 1681. Log of the *President*, June–Sept. 1681. *Rev. Rétrospective Ile Maurice* 2: 139–44, 201–3 (1951).

Prior, J. 1820. *Voyage in the Indian Seas in Nisus frigate, to the Cape of Good Hope, Isles of Bourbon, France and Seychelles; to Madras; and Isles of Java, St. Paul and Amsterdam, during the years 1810 and 1811*. Vol. 1 of *New voyages and travels*. London: Sir Richard Phillips & Co.

Procter, J. 1972. The nest and the identity of the Seychelles Swiftlet *Collocalia*. *Ibis* 114: 272–3.

Procter, J. & Salm, R. 1975. *Conservation in Mauritius 1974*. Morges, Switzerland: IUCN. Cyclostyled.

Prŷs-Jones, R. P. & Diamond, A. W. 1984. Ecology of land birds in the granitic and coralline islands of the Seychelles, with particular reference to Cousin Island and Aldabra Atoll. In Stoddart (1984), q.v.: 530–58 (ch. 27).

Querhoënt, Vicomte de 1773. [MS copies, in Buffon's hand, of notes made in 1773 of birds on the Ile de France (Mauritius), reserved for use in a supplementary volume on birds of the *Histoire naturelle* which was never published.] MS 369 in the library of the Muséum d'Histoire Naturelle, Paris.

Quoy, [ ] & Gaimard, P. 1824. Zoologie. Special vol. of L. C. D. de Freycinet (1824–44) *Voyage autour du monde* . . . 9 vols + atlas (in 4 vols.). Paris: Pillet ainé.

Racey, P. A. 1979. Two bats in the Seychelles. *Oryx* 15: 148–52.

Ralls, K., Brugger, K. & Ballow, J. 1979. Inbreeding and juvenile mortality in small populations of ungulates. *Science* 206: 1101–3.

Ramdin, T. 1969. *Mauritius, a geographical survey*. London: University Tutorial Press.

Rand, A. L. 1936. The distribution and habitats of Madagascar birds. *Bull. Ann. Mus. Nat. Hist.* 72: 142–499.

Rand, A. L. & Deignan, H. C. 1960. Pycnonotidae. In Peters *et al.* (1931–), q.v., 9: 221–300.

Rebeyrol, Y. 1966. Les problèmes forestiers à la Réunion. *Rev. Bois. Appl.* 21: 51–3.

Régnaud, C. 1878. Tec-tec. [MS letter to Sir Edward Newton dated 14 Feb. 1878] In the Newton Library, Cambridge University Zoology Dept. Published 1984 *Info-Nature, Ile Réunion* 21: 79–81.

Reydellet, D. 1978. *Bourbon et ses Gouverneurs*. St.-Denis, Réunion: Cazal.

Ricaud, C. 1975. Pesticide use and hazards in Mauritian agriculture. *Rev. Agric. Sucrière Ile Maurice* 54: 143–8.

Ripley, S. D. 1969. Comment on the Little Green Heron of the Chagos Archipelago. *Ibis* 111: 101–2.

Ripley, S. D. 1977. *Rails of the world. A monograph of the family Rallidae*. Boston, Mass.: David R. Godine.

Rising Sun, log of the 1702. Log of the *Rising Sun*, June–Sept. 1702. *Rev. Retrospective Ile Maurice* 3: 199–202 (1952).

Rivals, P. 1951. Effets des cyclones sur les arbres à la Réunion. *Trav. Lab. For. Toulouse*, tome 5, sec. 3e, vol. 1, art. 2, and *Rev. Agric. Ile Réunion* 1951 (Mar.–Apr.): 49–58.

Rivals, P. 1952. Etudes sur la végétation naturelle de l'île de la Réunion. Toulouse: Douladoure. Also *Trav. Lab. For. Toulouse*, tome 5, sec. 3e, vol. 1, art. 2.

Rivals, P. 1968. La Réunion. In I. & O. Hedberg (eds.) *Conservation of vegetation in Africa south of the Sahara:*, pp. 272–5. Acta Phytogeogr. Suec. 54.

Rivière, L. (ed.) 1982a. Rodrigues almanach 1982. Curepipe, Mauritius: Maurice Almanach Ltd.

Rivière, L. (ed.) 1982b. Maurice 1983. *L'almanach d'information générale sur l'Ile Maurice*. Port Louis: Maurice Almanach Ltd.

Robert, R. [1980]. *Géographie physique de l'île de la Réunion*. St.-Denis, Réunion: (author).

Roch, S. & Newton, E. 1862–3. Notes on birds observed in Madagascar. *Ibis* 4: 265–75; 5: 165–77.

Ronsil, R. 1948–9. *Bibliographie ornithologique française*. 2 vols. *Encyclopédie ornithologique*, vols. 8 and 9. Paris: P. Lechevalier.

Rothschild, W. 1907a. On extinct and vanishing birds. *Ornis* 14 (*Proc. 4th Int. Orn. Congr., Lond.*): 191–217.

Rothschild, W. 1907b. *Extinct birds*. London: Hutchinson.

Rouillard, J. 1866–9. *A collection of laws of Mauritius and its dependencies*. 9 vols. Port Louis, Mauritius: L. Channell.

Rountree, F. R.G. 1951. Some aspects of bird life in Mauritius. *Proc. Roy. Soc. Arts Sci. Mauritius* 1: 83–96.

Rountree, F. R. G., Guérin, R., Pelte, S. & Vinson, J. 1952. Catalogue of the birds of Mauritius. *Bull. Mauritius Inst.* 3: 155–217.

Roussin, A. (ed.) 1860–7. *Album de la Réunion*. 4 vols. St.-Denis, Réunion: (author). [The articles are quite differently distributed among the volumes in the second edition, Réunion & Paris 1879–83]

Rowlands, B. W. 1982. Tropic birds and other seabirds in Mauritius. *Bokmakierie* 34: 9–12.

Roy, B. D. 1969. The development of the tea industry in Mauritius. *Rev. Agric. Sucrière Ile Maurice* 48: 156–64.

Roy, J. N. 1960. *Mauritius in transition*. Curepipe, Mauritius: (author).

Sale, G. N. 1935a. *Exotics in Mauritius* [paper given at the British Empire Forestry Conference, South Africa, 1935]. Port Louis, Mauritius: Government Printer.

Sale, G. N. 1935b. *British Empire Forestry Conference, South Africa 1935. Quinquennial statement on forestry in Mauritius*. Port Louis, Mauritius: Government Printer.

Salomonsen, F. 1933. Les gobe-mouches de Paradis de la région Malgache avec description d'une nouvelle espèce de l'île Maurice. *Oiseau Rev. Fr. Orn.*, N.S. 3: 603–14.

Salomonsen, F. 1934. Notes on some Lemurian birds. *Proc. Zool. Soc. Lond.* [1934]: 219–24.

Sassi, M. 1940. Die wertvollsten Stücke der Wiener Vogel-sammlung. *Ann. Naturhist. Mus. Wien* 50: 395–409.

Saunders, D. A. 1982. The breeding behavior and biology of the short-billed form of the White-tailed Black Cocka-too *Calyptorhynchus funereus*. *Ibis* 124: 422–55.

Sauzier, T. 1893. *Les tortues de terre gigantesques des Mas-careignes et de certaines autres îles de la Mer des Indes*. Paris: G. Masson.

Scarborough, log of the 1703. Log of the *Scarborough*, Oct. 1703. *Rev. Retrospective Ile Maurice* 3: 271 (1952).

Scherer, A. 1965. *Histoire de la Réunion*. Que sais-je? 1164. Paris: Presses Univ. de France. [New edn 1980, No. 1846]

Schlegel, H. 1854. Ook en woordje over den Dodo (*Didus ineptus*) en zijine. *Verwandten Versl. Meded. K. Acad. Wet. Amsterdam* 2(1): 232–56.

Schlegel, H. 1858 (1866). On some extinct gigantic birds of the Mascarene Islands. *Ibis*, N.S. 2: 146–68. [A translation from the Dutch original in *Versl. Med. Kon. Ned. Akad. Wet., Afd. Natuurkunde* 7: 116–37 (1858)]

Schlegel, H. & Pollen, F. P. L. 1868. *Mammifères et Oiseaux*. Vol. 2 of *Recherches sur la Faune de Madagascar et de ses dépendances, d'après les découvertes de François P. L. Pollen et D. C. van Dam*. Leiden: J. K. Steenhoff.

Schönwetter, M. 1960– . *Handbuch der Oologie*, ed. W. Meise. 3 vols. (issued in parts, vol. 3 incomplete). Berlin: Adademie Verlag.

Sclater, W. L. 1915. The 'Mauritius Hen' of Peter Mundy. *Ibis* (10)3: 316–19.

Sclater, W. L. 1924–30. *Systema avium Aethiopicarum*. 2 vols. London: Taylor & Francis.

Scott, R. 1961. *Limuria. The lesser dependencies of Mauritius*. Oxford: Oxford University Press.

Scowcroft, P. G. & Wood, H. B. 1976. Reproduction of *Acacia koa* after fire. *Pacific Sci.* 30: 177–86.

Self, H. M. 1841. Report on the state of Rodrigues, 1841. MS in Public Record Office, London; C.O. 167, v. 231, art. 78.

Sélys-Longchamps, E. de 1848. Résumé concernant les oiseaux brévipennes mentionnés dans l'ouvrage de M. Strickland sur le Dodo. *Rev. Zool.* [1848]: 293.

Servat, J. (signatory) 1974. *Règlement permanent sur la police de la chasse dans le département de la Réunion*. Secretaire d'Etat à l'Environnement, Paris. 8 April 1974. Cyclo-styled.

Serventy, D. L., Serventy, V. & Warham, J. 1971. *The handbook of Australian seabirds*. Sydney: A. H. & A. W. Reed.

Seth-Smith, D. 1903. *Parakeets. A handbook to the imported species*. London: R. H. Porter. [Reprinted 1979 as *Small Parrots (Parakeets)*. New York: TFH Publications.]

Sharpe, R. B. 1879. Birds [of Rodrigues]. *Phil. Trans. R. Soc., Lond.* 168: 459–69.

Sharpe, R. B. 1883. *Catalogue of the birds in the British Museum*, vol. 7. London: Trustees of the British Museum.

Sharpe, R. B. *et al.* 1874–1898. *Catalogue of the birds in the British Museum*. 27 vols. London: Trustees of the British Museum.

Sheppard, P. M. 1975. *Natural selection and heredity*, 4th edn. London: Hutchinson.

Siegfried, W. R. & Frost, P. G. H. 1970. Notes on the Madagascar Kestrel. *Ibis* 112: 400–2.

Siegfried, W. R. & Temple, S. A. 1975. Mauritius, Paradise lost? *African Wildl.* 29: 34–8.

Siltman, K., Abplanalp, H. & Fraser, R. A. 1966. Inbreeding depression in Japanese Quail. *Genetics* 54: 371–9.

Simpson, E. S. W. 1951. The geology and mineral resources of Mauritius. *Colonial Geol. Min. Res.* 1: 217–38. [Also reprinted 1951, London: HMSO]

Skutch, A. F. 1976. *Parent birds and their young*. Austin & London: University of Texas Press.

Slater, H. H. n.d. (c. 1875). Notes on the birds of Rodriguez. MS in bound quarto volume entitled *Indian Ocean 3. Madagascar – Mascarene Islands (MSS)*, Newton Library, Cambridge University Zoology Dept.

Slater, H. H. 1879a. Reports of the proceedings of the naturalists. 2. Report of Henry H. Slater, Esq., B.A. *Phil. Trans. Roy. Soc., Lond.* 168: 294–5.

Slater, H. H. 1879b. Observations on the bone caves of Rodrigues. *Phil. Trans. Roy. Soc., Lond.* 168: 420–2.

Smith, G. A. 1975. Systematics of parrots. *Ibis* 117: 18–68.

Smith, G. A. (1975–8). Notes on some species of parrot in captivity, *Avicult. Mag.* 81: 200–11; 82: 22–32, 73–83, 143–50; 83: 21–7, 160–6; 84: 200–5 [*Psittacula* spp. in 82: 73–81 (1978)]

Smith, M. A. 1935. *The Fauna of British India including Ceylon and Burma. Reptilia and Amphibia, vol. 2 (Sauria)*. London: Taylor & Francis. [Reprinted 1974, New Delhi: Today and Tomorrow's Printers & Publishers]

Sonnerat, P. 1782. *Voyage aux Indes Orientales et à la Chine . . .* 2 vols. Paris: Froulé.

Sorensen, A. E. 1982. The spatial distribution and foraging behaviour of the Seychelles Brush Warbler *Acrocephalus (Bebrornis) sechellensis*. Cousin Is. Res. Station Tech. Rep. 21. Cambridge: ICBP.

Starmühlner, F. 1976. Contribution to the knowledge of the fauna of running waters of Mauritius. *Mauritius Inst. Bull.* 8: 105–28.

Starmühlner, F. 1979. Results of the Austrian Hydrobiological Mission. 1974, to the Seychelles- Comores- and Mascarene Archipelagoes. Part I: Preliminary report . . . *Ann. Naturhist. Mus. Wien* 82: 621–742.

Staub, [J. J.] F. 1971. Actual situation of the Mauritius endemic birds. *Bull. ICBP* 11: 226–7.

Staub, [J. J.] F. 1973a. *Oiseaux de l'Ile Maurice et de Rodrigue.* Port Louis, Mauritius: Mauritius Printing Co.

Staub, [J. J.] F. 1973b. Birds of Rodriguez Island. *Proc. Roy. Soc. Arts Sci. Mauritius* 4: 17–59.

Staub, [J. J.] F. 1976. *Birds of the Mascarenes and Saint Brandon.* Port Louis, Mauritius: Organisation Normale des Entreprises.

Staub, [J. J.] F. 1977a. Dodos et Tambalacoques. *Le Cerneen, Mauritius,* 15 [Oct. 1977]: 1 & 3.

Staub, [J. J.] F. 1980 ('1979'). Mauritius national section. Report 1977–8. *Bull. ICBP* 13: 188–90.

Staub, [J. J.] F. & Guého, J. 1968. The Cargados Carajos Shoals or St. Brandon: resources, avifauna and vegetation. *Proc. Roy. Soc. Arts Sci. Mauritius* 3: 7–46.

Steele, F. N. 1978. Conservation of birds on Mauritius, bimonthly reports February–November 1978. Unpublished typescript reports to ICBP.

Steele, F. N. 1979. Conservation of birds on Mauritius, final report for 1978. Unpublished report to ICBP.

Steinbacher, J. 1977. Vogelleben auf Inseln im Indischen Ozean, 2. Mauritius und Réunion. *Gefiederte Welt* 101: 174–6.

Stoddart, D. R. 1972. Pinnipeds or sirenians at western Indian Ocean islands? *J. Zool., Lond.* 167: 207–17.

Stoddart, D. R. (ed.) 1984. *Biogeography and ecology of the Seychelles Islands.* The Hague: W. Junk.

Stoddart, D. R. & Peake, J. F. 1979. Historical records of Indian Ocean giant tortoise populations. *Phil. Trans. Roy. Soc., Lond.* 286B: 147–61.

Stoddart, D. R. & Walsh, R. P. D. 1979. Long term climatic change in the western Indian Ocean. *Phil. Trans. Roy. Soc., Lond.* 286B: 11–23.

Stoddart, D. R. & Wright, C. A. 1967. Geography and ecology of Aldabra Atoll. *Atoll Res. Bull.* 118: 11–52.

Storer, R. W. 1970. Independent evolution of the Dodo and the Solitaire. *Auk* 87: 369–70.

Storer, R. W. & Gill, F. B. 1966. A revision of the Mascarene White-eye *Zosterops borbonica* (Aves). *Occ. Pap. Mus. Zool. Univ. Michigan* 648.

Strahm, W. 1983. Rodrigues: can its flora be saved? *Oryx* 17: 122–5.

Stresemann, E. 1952. On the birds collected by Pierre Poivre in Canton, Manila, India and Madagascar (1751–1756). *Ibis* 94: 499–523.

Stresemann, E. 1958. Wie hat die Dronte (*Raphus cucullatus* L.) ausgesehen? *J. Orn.* 99: 441–59.

Strickland, H. E. 1844. On the evidence of the former existence of Struthious birds distinct from the Dodo in the islands near Mauritius. *Proc. Zool. Soc. Lond.* [1844] (12): 77–9.

Strickland, H. E. 1848. History and external characters of the Dodo, Solitaire and other extinct brevipennate birds of Mauritius, Rodriguez and Bourbon. Part 1 of Strickland & Melville (1848), q.v.: 3–65.

Strickland, H. E. 1862. On some bones allied to the Dodo, in the collection of the Zoological Society of London. *Trans. Zool. Soc. Lond.* 4: 187–96.

Strickland, H. E. & Melville, A. G. 1848. *The Dodo and its kindred.* London: Reeve, Benham & Reeve.

Summers-Smith, J. D. 1963. *The house sparrow.* London: Collins.

Sussmann, R. W. & Tattersall, I. 1980. A preliminary study of the Crab-eating Macaque (*Macaca fascicularis*) in Mauritius. *Mauritius Inst. Bull.* 9: 31–52.

Swingland, I. R. & Coe, M. J. 1978. The natural regulation of giant tortoise populations on Aldabra Atoll. Reproduction. *J. Zool., Lond.* 186: 285–309.

Tafforet (attrib.). c. 1726. Relation de l'isle Rodrigue. Anonymous MS in the Archives Nationales in Paris. [Printed whole or in part as follows, annotated by the authors indicated: Newton, A. 1875 (*Proc. Zool. Soc., Lond.* [1875]: 39–43), Milne-Edwards, A. 1875 (*Ann. Sci. Nat. Zool.* 6e ser. art. 4: 1–20), Dupon, J. F. 1969 (*Mauritius Archiv Pub.* 10: 19–23) and 1973 (*Proc. Roy. Soc. Arts Sci. Mauritius.* 4: 1–16). The last is the only unaltered version]

Tattersall, I. M. 1977. Lemurs of the Comoro Islands. *Oryx* 13: 445–8.

Tatton, J. 1625. *A journall of a Voyage made by the Pearle to the East India wherein went as Captain Master Samuel Castleton of London, and Captaine George Bathurst as Lieutenant: written by John Tatton, Master.* In S. Purchas (ed.) (1625) *Hakluytus Posthumus, or Purchas his Pilgrimes containing a history of the World in sea voyages and lande travels by Englishmen and others,* vol. 3, pp. 343–54. Reprinted 1905–7. 20 vols. Glasgow: James Maclehose & Sons.

Telfair, C. 1833 (1832). [Letter dated 8 Nov. 1832 to the Zoological Society of London, about Rodrigues] *Proc. Zool. Soc. Lond.* [1833]: 31–2.

Temple, S. A. 1973–4. Faune et Flore. [Series of articles appearing weekly in the Mauritian daily newspaper *L'Express* from late 1973 to mid-1974]

Temple, S. A. 1974a. Wildlife in Mauritius today. *Oryx* 12: 584–90.

Temple, S. A. 1974b. Last chance to save Round Island. *Wildlife* 16: 370–4.

Temple, S. A. 1974c. Saving the Mauritius Kestrel from extinction. *Mauritius Soc. Prevent. Cruelty Anim. Newsl.* 3(1): 7.

Temple, S. A. 1975. The native faunas of Mauritius. I. The land birds. In Procter & Salm (1975), q.v., appendix 6.

Temple, S. A. 1976a. Observations of seabirds and shorebirds on Mauritius. *Ostrich* 47: 117–25.

Temple, S. A. 1976b. *Conservation of endemic birds and other wildlife on Mauritius (a progress report and proposal for future activities).* Cornell University, Ithaca: (author). Xeroxed.

Temple, S. A. 1976c. Status of endangered birds on Western Indian Ocean islands. Pre-print, 4th Pan-Afr. Cong., Seychelles. [Not published in *Proceedings*]

Temple, S. A. 1977a. The status and conservation of endemic Kestrels on Indian Ocean islands. In R. D. Chancellor (ed.) *World conference on birds of prey, Vienna 1–3 October 1975, Report on proceedings,* pp. 74–81. London: ICBP.

Temple, S. A. 1977b. Plant–animal mutualism: co-evolution with Dodo leads to near extinction of plant. *Science* 197: 885–6.

Temple, S. A. (ed.) 1978a. *Endangered birds. Management techniques for preserving threatened species.* Madison/London: University of Wisconsin Press/Croom Helm.

Temple, S. A. 1978b. Manipulating behavioral patterns of endangered birds: a potential management technique. In Temple (1978a), q.v.: 435–43.

Temple, S. A. 1978c. The life histories of the indigenous landbirds of Mauritius [Kestrel, Pink Pigeon, Parakeet]. MS, copy in Alexander Library Edward Grey Institute, Oxford.

Temple, S. A. 1979. [Reply to Owadally (1979), q.v.] *Science* 203: 1364.

Temple, S. A. 1981. Applied island biogeography and the conservation of endangered island birds in the Indian Ocean. *Biol. Conserv.* 20: 147–61.

Temple, S. A. 1983. The Dodo haunts a forest. *Animal Kingdom* 86(1): 20–5.

Temple, S. A. 1984. Author Stanley Temple replies [to Cheke *et al.* (1984), q.v.]. *Animal Kingdom* 87(1): 7 & 51.

Temple, S. A., Staub, J. J. F. & Antoine, R. 1974 (1975). *Some background information and recommendations on the preservation of the native flora and fauna of Mauritius.* MS published as Appendix 7 of Procter & Salm (1975), q.v.

Thibault, J.-C. 1974. Le peuplement avien des îles de la Société (Polynésie). Thesis presented for the Diplôme de l'École Pratique des Hautes Etudes, upheld 17 May 1974.

Thomas, P. L. U. 1828. *Essai de statistique de l'Ile Bourbon, considéré dans sa topographie, sa population, son agriculture, son commerce etc . . . suivi d'un projet de colonisation de l'intérieur de cette île.* 2 vols. Paris: Bachelier.

Thompson, R. 1880. *Report on the forests of Mauritius, their present condition and future management.* Port Louis, Mauritius: Mercantile Record Co.

Thomson, A. L. (ed.) 1964. *A new dictionary of birds.* London: Nelson.

Thornback, J. 1978. *Red Data Book 1. Mammalia.* Morges, Switzerland: IUCN.

Tilbrook, E. M. 1968. *Annual report of the Forest Department for the year 1966.* Colony of Mauritius, No. 4 of 1968. Port Louis, Mauritius, Government Printer.

Tirvengadum, D. D. 1980. On the possible extinction of *Randia heterophylla*, a Rubiaceae of great taxonomic interest from Rodrigues Island. *Mauritius Inst. Bull.* 9: 1–19.

Toussaint, A. 1956. *Répertoire des Archives de l'Ile de France pendant la régie de la Compagnie des Indes (1715–1768).* Mauritius Arch. Publ. 4.

Toussaint, A. 1965. *L'administration française de l'Ile Maurice et ses Archives (1721–1810).* Mauritius Arch. Publ. 8.

Toussaint, A. 1966 (1973). *Port Louis, a tropical city* [trans. W. E. F. Ward from the French edn. of 1966]. London: George Allen & Unwin.

Toussaint, A. 1972. *Histoire des Iles Mascareignes.* Paris: Berger-Levrault.

Toussaint, A. & Adolphe, H. 1956. *Bibliography of Mauritius (1502–1954), covering the printed period, manuscripts,* archivalia and cartographic material. Port Louis, Mauritius: Esclapon Ltd.

Tricot, B. 1978. Volcanisme à la Réunion. *Info-Nature Ile Réunion* 16: 85–97.

Tuck, G. S. & Heinzel, H. 1978. *A field guide to the seabirds of Britain and the world.* London: Collins.

Turner, B. D. 1979 ('1977'). Psocids as prey for Mascarene Swiftlets. *Ent. Mon. Mag.* 113: 210.

Twigg, G. 1975. *The brown rat.* Newton Abbott: David & Charles.

d'Unienville, R. 1982. Histoire politique de l'île de France (1791–1794). *Mauritius Archives Publ.* 14.

Upton, B. G. J. & Wadsworth, W. J. 1966. The basalts of Réunion island, Indian Ocean. *Bull. Volcanol.* 29: 7–24.

Vaillant, L. 1898. Dessins inédits de chéloniens tirés des manuscrits de Commerson. *Bull. Mus. Hist. Nat., Paris* 4: 133–9.

Vaillant, L. 1899. Nouveaux documents historiques sur les tortues terrestres des Mascareignes et des Seychelles. *Bull. Mus. Hist. Nat., Paris* 5: 19–23.

Vaillant, L. & Grandidier, G. 1910. *Histoire naturelle des Reptiles, part 1, Crocodiles et Tortues.* In A. & G. Grandidier (eds.) *Histoire, physique, naturelle et politique de Madagascar* 17. Paris: Imprimerie Nationale.

Valadon, Y. 1980. *Report on tropical cyclone Claudette, 22nd–23rd December, 1979.* Vacoas, Mauritius: Meteorological Service. Cyclostyled. [Reprinted as appendix 2 of Jones (1980a), q.v.]

Vaughan, R. E. 1958. Wenceslaus Bojer 1795–1856. *Proc. Roy. Soc. Arts Sci. Mauritius* 2: 73–98.

Vaughan, R. E. 1968. Mauritius and Rodrigues. In I. & O. Hedberg (eds.) *Conservation of vegetation in Africa South of the Sahara,* pp. 265–72. *Acta Phytogeogr. Suec.* 54.

Vaughan, R. E. 1984. [Untitled letter.] *Animal Kingdom* 87(1): 6–7.

Vaughan, R. E. & Wiehé, P.O. 1937. Studies on the vegetation of Mauritius. I. A preliminary survey of the plant communities. *J. Ecol.* 25: 289–343.

Vaughan, R. E. & Wiehé, P. O. 1939. Note on 'The Plant Communities of Mauritius'. *J. Ecol.* 27: 281.

Vaughan, R. E. & Wiehé, P. O. 1941. Studies on the vegetation of Mauritius. III. The structure and development of the upland climax forest. *J. Ecol.* 19: 127–60.

Venkatasamy, D. 1971, 1980. *Atlas for Mauritius.* London: Macmillan Educational Ltd. [The 2nd edn, 1980, has inferior maps]

Verreaux, J. 1863. [Description of the *papangue Circus maillardi*.] In 2nd edn (1863) of Maillard (1862), q.v.: 160.

Viader, R. 1939. In memoriam Georges Antelme. *Mauritius Inst. Bull.* 1(5): 1–3.

Vincent, J. 1966. *Red Data Book 2. Aves.* Morges, Switzerland: IUCN.

Vinson, A. 1861. *De l'apparition d'oiseaux étrangers aux Iles de la Réunion et Maurice.* [Apparently published in Réunion in 1861, *fide* Coquerel (1864, q.v.), but no copy seems to have survived. A MS text submitted to the Académie des Sciences in Paris was noticed in 1862 (*C. R. Hebd.*

Séanc. Acad. Sci., Paris 54: 275) and is held in the Académie's archives. A photocopy is deposited in the Edward Grey Institute, Alexander Library, Oxford.]

Vinson, A. 1863. [List of birds brought to Réunion] *Bull. Soc. Acclim. Nat. Hist. Ile Réunion* 1: 197.

Vinson, A. 1867. Le Martin (*Acridotheres tristis* Vieillot), son utilité pour les pays exposés à l'invasion des sauterelles. *Bull. Soc. Imp. Zool. Acclim., Paris* (2)4: 181-9.

Vinson, A. 1868. De l'acclimatation à l'île de la Réunion. *Bull. Soc. Imp. Zool. Acclim., Paris* (2)5: 579-90, 625-38. [Also published in *Bull. Soc. Sci. Arts Ile Réunion* [1868]: 35-65]

Vinson, A. 1870 (1871). [Letter dated 1870 to L. Bouton on "*Agama versicolor*"] *Trans. Roy. Soc. Arts Sci. Mauritius*, N.S. 5: 31-4 (1871).

Vinson, A. 1876 (1878). [Letter dated 27 July 1876 to the Royal Society of Arts and Sciences on "perdrix pintadées"] *Trans. Roy. Soc. Arts Sci. Mauritius*, N.S. 10:41 (1878).

Vinson, A. 1877. Faune détruite: Les Aepiornidés et les Huppes de l'île Bourbon. *Bull. Hebd. Assoc. Sci., France* 20: 327-31.

Vinson, A. 1887. Etude sur les Colombes des Mascareignes et les espèces importées. *Bull. Soc. Nat. Acclim., Paris* (4)4: 640-51.

Vinson, J. 1950. L'Ile Ronde et l'Ile aux Serpents. *Proc. Roy. Soc. Arts Sci. Mauritius* 1: 32-52.

Vinson, J. 1953*a*. The fauna of Mauritius. I. The vertebrates. *Mauritius Police Gaz.* 1: 37-41.

Vinson, J. 1953*b*. Some recent data on the fauna of Round and Serpent Islands. *Proc. Roy. Soc. Arts Sci. Mauritius* 1: 253-7.

Vinson, J. 1956*a*. The problem of bird protection in the island of Mauritius. *Proc. Roy. Soc. Arts Sci. Mauritius* 1: 387-92.

Vinson, J. 1956*b*. Notes d'histoire naturelle: I. L'oeuf du Dodo. *Proc. Roy. Soc. Arts Sci. Mauritius* 1: 313-15.

Vinson, J. 1963. The extinction of endemic birds in the island of Mauritius, with a possible way of saving some of the remaining species. *Bull. ICBP* 9: 99-101.

Vinson, J. 1964*a*. Sur la disparition progressive de la flore et de la faune de l'Ile Ronde. *Proc. Roy. Soc. Arts Sci. Mauritius* 2: 247-61.

Vinson, J. 1964*b*. Quelques remarques sur l'Ile Rodrigue et sur sa faune terrestre. *Proc. Roy. Soc. Arts Sci. Mauritius* 2: 263-77.

Vinson, J. & Vinson, J.-M. 1969. The saurian fauna of the Mascarene islands. *Mauritius Inst. Bull.* 6: 203-320.

Vinson, J.-M. 1973. A new skink of the genus *Gongylomorphus* from Macabé Forest (Mauritius). *Rev. Agric. Sucrière Ile Maurice* 52: 39-40.

Vinson, J.-M. 1974. Round Island: conservation problems. *Maur. Soc. Prevent. Cruelty Anim. Newsl.* 3(1): 2-3.

Vinson, J.-M. 1975. Notes on the reptiles of Round Island. *Bull. Mauritius Inst.* 8: 49-67.

Vinson, J.-M. 1976*a*. Notes sur les procellariens de l'Ile Ronde. *Oiseau Rev. Fr. Orn.* 46: 1-24.

Vinson, J.-M. 1976*b*. The saurian fauna of the Mascarene Islands. II. The distribution of *Phelsuma* species in

Visdelou-Guimbeau, G de 1948. *La découverte des Iles Mascareignes*. Port Louis, Mauritius: General Printing & Stationery Co.

Vitry, H. 1883. [Quelques remarques sur la tortue indigène . . .] *Trans. Soc. Roy. Arts Sci. Mauritius*, N.S. 13: 8.

Voous, K. H. 1973. List of recent Holarctic bird species. Non-passerines. *Ibis* 115: 612-38.

Wagstaffe, R. 1978. *Type specimen of birds in the Merseyside County Museums*. Liverpool: Merseyside County Museums.

Wallace, A. R. 1865. On the pigeons of the Malay Archipelago. *Ibis* N.S.1: 365-400.

Walter, A. 1914*a*. Climate. In Macmillan (1914), q.v.: 185-92.

Walter, A. 1914*b*. The sugar industry. In Macmillan (1914), q.v.: 208-31.

Warner, R. E. 1968. The role of introduced disease in the extinction of the endemic Hawaiian avifauna. *Condor* 70: 101-20.

Watling, R. J. 1975. Observations on the ecological separation of two introduced congeneric mynahs (*Acridotheres*) in Fiji. *Notornis* 22: 37-53.

Watson, A. *et al.* n.d. (*c.* 1970). *Grouse management*. Fordingbridge, Hants.: Game Conservancy.

Watson, G. F., Zusi, R. L. & Storer, R. W. 1963. *Preliminary field guide to the birds of the Indian Ocean. For use during the International Indian Ocean Expedition*. Washington: Smithsonian Institution.

Watson, J. 1979. Clutch sizes of Seychelles' endemic land birds. *Bull. Br. Orn. Club* 99: 102-5.

Watson, J. 1981. Population ecology, food and conservation of the Seychelles Kestrel (*Falco araea*) on Mahé. Ph.D. thesis, Aberdeen University.

Watson, J. 1984. Land birds: endangered species on the granitic Seychelles. In Stoddart (1984), q.v.,: 513-27 (Ch. 26).

Weir, D. N. & Picozzi, N. 1975. Aspects of social behaviour in the Buzzard. *Br. Birds* 68: 125-41.

Wermuth, H. & Mertens, R. 1977. Liste der rezenten Amphibien und Reptilien. Testudines, Crocodylia, Rhynchocephalia. *Tierreich* 100.

Whitaker, A. H. 1973. Lizard populations on islands with and without Polynesian rats, *Rattus exulans* (Peale). *Proc. N.Z. Ecol. Soc.* 20: 121-30.

Whitaker, A. H. 1978. The effect of rodents on reptiles and amphibians. In Dingwall *et al.* (1978), q.v.: 75-88.

White, C. M. N. 1951. Systematic notes on African birds. II. The affinities of the races of *Butorides striatus* in the eastern Indian Ocean. *Ibis* 93: 460-1.

White C. M. N. 1963. *A revised check list of African flycatchers, tits, tree-creepers, sunbirds, white-eyes, honey-eaters, buntings, finches, weavers and waxbills*. Lusaka: Government Printer.

White, C. M. N. 1965. *A revised check list of African non-passerine birds*. Lusaka: Dept. of Game & Fisheries.

Wiehé, P. O. 1938. *Report on a visit to Rodrigues*. Typescript MS; copy in the Mauritius Herbarium, Réduit, Mauritius.

Wiehé, P. O. 1946. L'herbe condé et la lutte contre les mauvaises herbes. *Rev. Agric. Ile Maurice* 25: 51-61.

Wiehé, P. O. 1949. The vegetation of Rodrigues Island. *Mauritius Inst. Bull.* 2: 279–305.

Wiehé, P. O. 1969. Visite de la forêt Macabé. *Rev. Agric. Sucrière Ile Maurice* 48: 223–6.

Williams, J. R. 1953. Field rats on sugar estates and methods for their control. *Rev. Agric. Ile Maurice* 27: 56–66.

Witherby, H. F., Jourdain, F. C. R., Ticehurst, N. F. & Tucker, B. W. 1938–41. *The handbook of British birds.* 5 vols. London: H. F. & G. Witherby.

Wood, D. S., Zusi, R. L. & Jenkinson, M. A. 1982a. *World inventory of avian spirit specimens 1982.* Norman, Oklahoma: Am. Orn. Union & Oklahoma Biol. Survey.

Wood, D. S., Zusi, R. L. & Jenkinson, M. A. 1982b. *World inventory of avian skeletal specimens 1982.* Norman, Oklahoma: Am. Orn. Union & Oklahoma Biol. Survey.

Wrege, P. H. & Cade, T. J. 1977. Courtship behaviour of large falcons. *Raptor Res.* 11: 1–27.

Yapp, W. B. 1962. *Birds and woods.* Oxford: Oxford University Press.

INDEX

The index contains all animal and plant names; both scientific and vernacular names (English and local) are included as used in the text. References to place names within the Mascarene Islands have been indexed only for Part I which contains general material. Accounts of individual species are indexed in **bold** and in general only these species are cross-referenced from the common to the scientific names. Maps, tables and appendices are not indexed.